Birkhäuser

Operator Theory: Advances and Applications

Volume 250

Founded in 1979 by Israel Gohberg

Wolfgang Arendt • Ralph Chill • Yuri Tomilov
Editors

Operator Semigroups Meet Complex Analysis, Harmonic Analysis and Mathematical Physics

Birkhäuser

Editors
Wolfgang Arendt
Institute of Applied Analysis
Ulm University
Ulm, Germany

Ralph Chill
Institute of Analysis
TU Dresden
Dresden, Germany

Yuri Tomilov
Institute of Mathematics
Polish Academy of Sciences
Warsaw, Poland

ISSN 0255-0156 ISSN 2296-4878 (electronic)
Operator Theory: Advances and Applications
ISBN 978-3-319-18493-7 ISBN 978-3-319-18494-4 (eBook)
DOI 10.1007/978-3-319-18494-4

Library of Congress Control Number: 2015958022

Mathematics Subject Classification (2010): 30, 35, 42, 46, 47

Springer Cham Heidelberg New York Dordrecht London

Printed on acid-free paper

Springer International Publishing AG Switzerland is part of Springer Science+Business Media (www.birkhauser-science.com)

Contents

Professor Charles Batty together with the editors of this volume, Jurata 2010 (from left to right: R. Chill, C. Batty, W. Arendt, Y. Tomilov)

Preface

The last fifteen years opened a new era for semigroup theory with the emphasis on applications of abstract results, often unexpected and often far away from traditional ones. The aim of the conference held in Herrnhut in June 2013 was to bring together prominent experts around modern semigroup theory, harmonic analysis, complex analysis and mathematical physics, and to show a lively interplay between all of those areas and even beyond them. In addition, the meeting honoured the sixtieth anniversary of Prof C.J.K. Batty, whose scientific achievements are an impressive illustration of the conference goal.

The present conference proceedings provide an opportunity to see the power of abstract methods and techniques dealing successfully with a number of applications stemming from classical analysis and mathematical physics. The sample of diverse topics treated by the proceedings include partial differential equations, martingale and Hilbert transforms, Banach and von Neumann algebras, Schrödinger operators, maximal regularity and Fourier multipliers, interpolation, operator-theoretical problems (concerning generation, perturbation and dilation, for example), and various qualitative and quantitative Tauberian theorems with an accent on transfinite induction and magics of Cantor.

The organizers express their sincere gratitude to Volkswagenstiftung for their generous support of the Herrnhut conference and to Thomas Hempfling of Birkhäuser for the enjoyable cooperation.

Ulm, Dresden and Warsaw, December 2014
Wolfgang Arendt, Ralph Chill, Yuri Tomilov

Operator Theory:
Advances and Applications, Vol. 250, 1–16

Polynomial Internal and External Stability of Well-posed Linear Systems

El Mustapha Ait Benhassi, Said Boulite, Lahcen Maniar and Roland Schnaubelt

Abstract. We introduce polynomial stabilizability and detectability of well-posed systems in the sense that a feedback produces a polynomially stable C_0-semigroup. Using these concepts, the polynomial stability of the given C_0-semigroup governing the state equation can be characterized via polynomial bounds on the transfer function. We further give sufficient conditions for polynomial stabilizability and detectability in terms of decompositions into a polynomial stable and an observable part. Our approach relies on a recent characterization of polynomially stable C_0-semigroups on a Hilbert space by resolvent estimates.

Mathematics Subject Classification (2010). Primary: 93D25. Secondary: 47A55, 47D06, 93C25, 93D15.

Keywords. Internal and external stability, polynomial stability, transfer function, stabilizability, detectability, well-posed systems.

1. Introduction

Weakly damped or weakly coupled linear wave type equations often have polynomially decaying classical solutions without being exponentially stable, see, e.g., [1], [2], [4], [5], [8], [15], [16], [17], [18], [23], and the references therein. In these contributions various methods have been used, partly based on resolvent estimates. Recently this spectral theory has been completed for the case of bounded semigroups $T(\cdot)$ in a Hilbert space with generator A. Here one can now characterize the 'polynomial stability' $\|T(t)(I-A)^{-1}\| \le ct^{-1/\alpha}$, $t \ge 1$, of $T(\cdot)$ by the polynomial bound $\|R(i\tau, A)\| \le c|\tau|^{\alpha}$, $|\tau| \ge 1$, on the resolvent of A. These results are due to Borichev and Tomilov in [7] and to Batty and Duyckaerts in [6], see also [5], [15] and [17] for earlier contributions. We describe this theory in the next section. In a

This work is part of a cooperation project supported by DFG (Germany) and CNRST (Morocco).

polynomial stable system the spectrum of the generator may approach the imaginary axis as $\operatorname{Im}\lambda \to \pm\infty$. This already indicates that this concept is more subtle than exponential stability. For instance, so far robustness results for polynomial stability are restricted to small regularizing perturbations, see [19].

At least for bounded semigroups in a Hilbert space one has now a solid background which can be used in other areas such as control theory. In the context of observability this was already done in [11] (based on [5] at that time). In this paper we start an investigation of polynomial stabilizability and detectability.

Stabilizability is one of the basic concepts and topics of linear systems theory. Let the state system be governed by a generator A on the state Hilbert space X, and let Y and U be the observation and the control Hilbert spaces, respectively. For a moment, we simply consider bounded control and observation operators and feedbacks. For a bounded control operator $B : U \longrightarrow X$ we obtain the system

$$x'(t) = Ax(t) + Bu(t), \quad t \geq 0, \qquad x(0) = x_0, \tag{1.1}$$

with the control $u \in L^2_{\mathrm{loc}}(\mathbb{R}_+, U)$, the initial state $x_0 \in X$ and the state $x(t) \in X$ at time $t \geq 0$. This system is exponentially stabilizable if one can find a (bounded) feedback $F : X \longrightarrow U$ such that the C_0-semigroup $T_{BF}(\cdot)$ solving the closed-loop system

$$x'(t) = Ax(t) + BFx(t), \quad t \geq 0, \qquad x(0) = x_0, \tag{1.2}$$

is exponentially stable. Observe that $A + BF$ generates $T_{BF}(\cdot)$.

For the dual concept of exponential detectability, one starts with a generator A and a bounded observation operator $C : X \longrightarrow Y$. The output of this system is $y = CT(\cdot)x_0$. One then looks for a (bounded) feedback $H : Y \longrightarrow X$ such that the C_0-semigroup $T_{HC}(\cdot)$ generated by $A + HC$ becomes exponentially stable.

In our paper we allow for unbounded observation operators C defined on $D(A)$ and control operators B mapping into the larger space $X_{-1} = D(A^*)^*$, where the domains are equipped with the respective graph norm. Here one has to assume that the output map $x_0 \mapsto y$ and the input map $u \mapsto x(t)$ are continuous. Such systems are called *admissible*, see the next section for a precise definition and further information. The monograph [24] investigates these notions in detail. In this framework one can in particular treat boundary control and observation of partial differential equations.

In order to use the full system (A, B, C), one also has to assume the boundedness of the input-output map $u \mapsto y$. This leads to the concept of a well-posed system, which was introduced by G. Weiss and others, see Section 2, the recent survey [25], and, e.g., [22], [27], [28]. In well-posed systems, the Laplace transform of the input-output map gives the transfer function of the system, which plays an important role in the present paper. For well-posed systems, it becomes more difficult to determine the generators of the feedback systems, cf. [28]. However, in our arguments we can avoid to use a precise description of these operators. For well-posed systems exponential stabilizability and detectability was discussed in many papers, see, e.g., [9], [12], [13], [20], [21], [29], and the references therein.

In this paper we will weaken the exponential stability of the feedback system in the above concepts to polynomial stability. Here the feedback systems are described by equations for the resolvents of the generators of given and the feedback semigroup which are coupled via a perturbation term involving the feedback, see Definitions 3.1 and 3.1. In the study of the resulting concepts of polynomial stabilizability and detectability we pursue two main questions, also treated in the above papers.

We show that a system possesses these properties if it can be decomposed into a polynomial stable and an observable part, see Theorem 4.6 and 4.7. In the exponential case, such results are often called pole-assignment if the stable part has a finite-dimensional complement. Actually one can derive exponential stabilizability from much weaker concepts (optimizability or the finite cost condition), see [9] or [29]. So far it is not clear whether such implications hold for the natural analogues of these concepts to the polynomial setting. Moreover, it is known that optimizability can be characterized by decompositions as above if the resolvent set of the generator contains a strip around $i\mathbb{R}$, see [12] or [21]. In the polynomial setting one here has to fight against the fact that the spectrum may approach the imaginary axis at infinity. So far we only have partial results in this context, not treated below.

The main part of our results is devoted to the relationship between polynomial stability of the given semigroup and polynomial estimates on the transfer function of the system. It is known that A generates an exponentially stable semigroup if (and only if) the system (A, B, C) is exponential stabilizable and detectable and its transfer function is bounded on the right half-plane, see [20] and also [29] for an extension to the concepts of optimizability and estimatibility. (Note that the 'only if' implication is easily shown with 0 feedbacks.) The boundedness of the transfer function is called *external stability*. In Theorem 4.3 we extend these results to our setting, thus requiring polynomial stabilizability and detectability and that the transfer function grows at most polynomially as $|\operatorname{Im}\lambda| \to \infty$. (The latter condition may be called *polynomial external stability.*) If the involved semigroups are bounded, we then obtain polynomial stability of the order one expects, i.e., the sum of the orders in the assumption. The proofs are based on various estimates and manipulations of formulas connecting resolvents, the transfer functions and their variants. We further use the results polynomial stability from [6] and [7] mentioned above.

If the given semigroup is not known to be bounded, then the available theory on polynomial stability does not give the above-indicated convergence order. However, in applications one can often check the boundedness of a semigroup by the dissipativity of its generator, possibly for an equivalent norm. Similarly one can characterize well-posed systems with energy dissipation (so-called scattering passive systems), see, e.g., [22]. Besides the given semigroup, here also the transfer function is contractive which leads to an improvement of our main result for scattering passive systems, see Corollary 4.4. In general, not much is known on the preservation of boundedness under perturbations. In Theorem 5 of the recent paper [19] one finds a result which requires smallness of the perturbations as maps into spaces between $D(A)$ and X. In Proposition 4.5 we show the boundedness in the framework of the present paper. Our approach is based on a characterization of bounded semigroups in terms of L^2-norms of the resolvents of A and A^* due to [10], see Proposition 2.4.

In the next section we discuss the background on polynomial stability and well-posed systems. In Section 3 we introduce polynomial stabilizability and detectability and establish several basic estimates. The last section contains our main results on external polynomial stability and on sufficient criteria for polynomial stabilizability and detectability.

2. Polynomial stability and well-posed systems

We first discuss polynomially stable semigroups. Throughout $T(\cdot)$ denotes a C_0-semigroup on a Banach space X with generator A. There are numbers $\varpi \in \mathbb{R}$ and $M \geq 1$ such that $\|T(t)\| \leq Me^{\varpi t}$ for all $t \geq 0$. The infimum of these numbers ϖ is denoted by $\omega_0(A)$. The semigroup is called *bounded* if $\|T(t)\| \leq M$ for all $t \geq 0$.

We fix some $\omega > \omega_0(A)$. It is well known that then the fractional powers $(\omega - A)^\beta$ exist for $\beta \in \mathbb{R}$. They are bounded operators for $\beta \leq 0$ and closed ones for $\beta > 0$. The domain X_β of $(\omega - A)^\beta$ for $\beta > 0$ is endowed with the norm given by $\|x\|_\beta = \|(\omega - A)^\beta x\|$. The fractional powers satisfy the power law and coincide with usual powers for $\beta \in \mathbb{Z}$. In particular, $(\omega - A)^{-\beta}$ is the inverse of $(\omega - A)^\beta$ for all $\beta \in \mathbb{R}$. We next recall a definition from [5].

Definition 2.1. A C_0-semigroup $T(\cdot)$ is called *polynomially stable* (of order $\alpha > 0$) if there is a constant $\alpha > 0$ such that

$$\|T(t)(\omega - A)^{-\alpha}\| \leq ct^{-1} \qquad \text{for all } \ t \geq 1.$$

(Here and below, we write $c > 0$ for a generic constant.) Note that a larger order α means a weaker convergence property. Due to Proposition 3.1 of [5], a bounded C_0-semigroup $T(\cdot)$ is polynomially stable of order $\alpha > 0$ if and only if

$$\|T(t)(\omega - A)^{-\alpha\gamma}\| \leq c(\gamma)\, t^{-\gamma}\,, \qquad t \geq 1, \tag{2.1}$$

for all/some $\gamma > 0$. (There is also a partial extension to general C_0-semigroups.)

Combined with (2.1), Proposition 3 of [6] yields the following necessary condition for polynomial stability of *bounded* C_0-semigroups. Here we set

$$\mathbb{C}_\pm = \{\lambda \in \mathbb{C} \,\big|\, \operatorname{Re}\lambda \gtrless 0\} \qquad \text{and} \qquad \mathbb{C}_r = r + \mathbb{C}_+ \quad \text{for } \ r \in \mathbb{R}.$$

Proposition 2.2. *Let $T(\cdot)$ be a bounded C_0-semigroup which is polynomially stable of order $\alpha > 0$. Then the spectrum $\sigma(A)$ of A belongs to $\mathbb{C}_-$ and its resolvent is bounded by*

$$\|R(\lambda, A)\| \leq c\,(1 + |\lambda|)^\alpha \qquad \textit{for all } \ \lambda \in \overline{\mathbb{C}_+}. \tag{2.2}$$

Due to Lemma 3.2 in [14], the estimate (2.2) is true if and only if

$$\|R(\lambda, A)(\omega - A)^{-\alpha}\| \le c \qquad \text{for all } \lambda \in \overline{\mathbb{C}_+}. \tag{2.3}$$

If one drops the boundedness assumption, the above result still holds with an epsilon loss in the exponent in the right-hand side of (2.2) by Proposition 3.3 of [5] and (2.3). We further note that condition (2.2) implies the inclusion

$$\{\lambda \in \sigma(A) \,\big|\, \operatorname{Re}\lambda \ge -\delta\} \subset \{\lambda \in \mathbb{C}_- \,\big|\, |\operatorname{Im}\lambda| \ge c(-\operatorname{Re}\lambda)^{-1/\alpha}\}$$

for some $c, \delta > 0$, see Proposition 3.7 of [5].

The next result from [7] provides the important converse of the above proposition for bounded semigroups on a Hilbert space, see Theorem 2.4 of [7].

Theorem 2.3. *Let $T(\cdot)$ be a bounded C_0-semigroup on a Hilbert space X such that $\sigma(A) \subset \mathbb{C}_-$ and (2.2) holds for all $\lambda \in i\mathbb{R}$. Then $T(\cdot)$ is polynomially stable of order $\alpha > 0$.*

For general Banach spaces X, in Theorem 5 in [6] this result was shown up to a logarithmic factor in the estimate in semigroup, see also [5], [15] and [17]. The paper [7] gives an example where this logarithmic correction actually occurs. Without assuming its boundedness, the semigroup is still polynomially stable if a holomorphic extension of $R(\lambda, A)(\omega - A)^{-\alpha}$ satisfies (2.3), but here one only obtains the stability order $2\alpha + 1 + \epsilon$ for any $\epsilon > 0$, see Proposition 3.4 of [5].

The proof of Theorem 2.3 is based on the following characterization of the boundedness of C_0-semigroups on Hilbert spaces, see Theorem 2 in [10] and also Lemma 2.1 in [7].

Proposition 2.4. *Let A generate the C_0-semigroup $T(\cdot)$ on the Hilbert space X. The semigroup is bounded if and only if $\mathbb{C}_+ \subset \rho(A)$ and*

$$\sup_{r>0} r \int_{\mathbb{R}} \left(\|R(r + i\tau, A)x\|^2 + \|R(r + i\tau, A^*)x\|^2\right) d\tau \le c\,\|x\|^2$$

for each $x \in X$.

We now turn our attention to the concept of well-posed systems. From now on, X, U and Y are always Hilbert spaces, A generates the C_0-semigroup $T(\cdot)$ on X and $\omega > \omega_0(A)$. Let X_{-1} be the completion of X with respect to the norm given by $\|x\|_{-1} = \|R(\omega, A)x\|$. We sometimes write X_{-1}^A instead of X_{-1} to stress that this *extrapolation space* depends on A. The operator A has a unique extension $A_{-1} \in \mathcal{B}(X, X_{-1})$ which generates a C_0-semigroup given by the continuous extension $T_{-1}(t) \in \mathcal{B}(X_{-1})$ of $T(t)$, $t \ge 0$. We often omit the subscript -1 here. One can define such a space for each linear operator with non-empty resolvent set. Recall that we have set $X_1 = D(A)$.

A bounded linear (observation) operator $B : U \longrightarrow X_{-1}$ is called *admissible* for A (or the system $(A, B, -)$ is called *admissible*) if the integral

$$\Phi_t u := \int_0^t T(t - s)Bu(s)\, ds$$

belongs to X for all $u \in L^2(0,t;U)$ and some $t > 0$. (The integral is initially defined in X_{-1}.) By Proposition 4.2.2 in [24], this property then holds for all $t \geq 0$ and $\Phi_t \in \mathcal{B}(L^2(0,t;U), X)$. Moreover, these operators are exponentially bounded, see Proposition 4.4.5 in [24].

A bounded linear (control) operator $C : X_1 \longrightarrow Y$ is called *admissible* for A (or the system $(A, -, C)$ is called *admissible*) if the map

$$\Psi_t x := CT(\cdot)x, \qquad x \in X_1,$$

has a bounded extension in $\mathcal{B}(X, L^2(0,t;Y))$ for some $t > 0$. Propositions 4.2.3 and 4.3.3 in [24] show that this fact then holds for all $t > 0$ and that the extensions are exponentially bounded. We still denote the extension by Ψ_t. One can extend an admissible observation operator C to the map C_Λ given by

$$C_\Lambda x = \lim_{\lambda \to \infty} C\lambda R(\lambda, A)x$$

with domain $D(C_\Lambda) = \{x \in X \mid \text{this limit exists in } Y\}$. For each $x \in X$ we have $T(s)x \in D(C_\Lambda)$ for a.e. $s \geq 0$ and $\Psi_t x = C_\Lambda T(\cdot)x$ a.e. on $[0,t]$ for all $t > 0$ by, e.g., (5.6) and Proposition 5.3 in [28].

Theorem 4.4.3 of [24] shows that an operator $B \in \mathcal{B}(U, X_{-1})$ is admissible for A if and only if its adjoint $B^* \in \mathcal{B}(D(A^*), U)$ is admissible for A^*. Here we recall that X_{-1} is the dual space of $D(A^*)$, if considered as a Banach space, see, e.g., Proposition 2.10.2 in [24].

Let system (A, B, C) be a system with a generator A and admissible control and observation operators B and C. One says that (A, B, C) is *well posed* if there are bounded linear operators $\mathbb{F}_t : L^2(0,t;U) \longrightarrow L^2(0,t;Y)$ such that

$$\mathbb{F}_{\tau+t}u = \begin{cases} \mathbb{F}_\tau u_1 & \text{on } [0,\tau], \\ \mathbb{F}_t u_2 + \Psi_t \Phi_\tau u_1 & \text{on } [\tau, \tau+t] \end{cases}$$

for all $t, \tau \geq 0$ and $u \in L^2(0,\tau+t;U)$, where $u = u_1$ on $(0,\tau)$ and $u = u_2$ on $(\tau, \tau+t)$, see [27]. Also these (input-output) operators are exponentially bounded by Proposition 4.1 of [27].

One can introduce versions of the maps Ψ_t and $\mathbb{F}_t$ on the time interval $\mathbb{R}_+$ using L^2_{loc} spaces. We denote these extensions by Ψ and $\mathbb{F}$ respectively. For $x_0 \in X$ and $u \in L^2_{loc}(\mathbb{R}_+, U)$ the output of the well-posed system (A, B, C) is then given by $y = \Psi x_0 + \mathbb{F}u$. In [27] it was shown that the Laplace transform $\hat{y}$ of y satisfies

$$\hat{y}(\lambda) = C(\lambda - A)^{-1}x_0 + G(\lambda)\hat{u}(\lambda)$$

for all $\lambda \in \mathbb{C}_\omega$, where $G : \mathbb{C}_\omega \to \mathcal{B}(U, Y)$ is a bounded analytic function. It satisfies $G'(\lambda) = -CR(\lambda, A)^2 B$ and it is thus determined by A, B and C up to an additive constant. (See, e.g., Theorem 2.7 in [22].) We call G the *transfer function* of (A, B, C).

Set $Z = D(A) + R(\omega, A_{-1})BU$ and endow it with the norm $\|z\|_Z$ given by the infimum of all $\|x\|_1 + \|R(\omega, A_{-1})Bv\|$ with $z = x + R(\omega, A_{-1})Bv$, $x \in D(A)$ and

$v \in U$. Theorem 3.4 and Corollary 3.5 of [22] then yield an extension $\overline{C} \in \mathcal{L}(Z,U)$ of C such that the transfer function is represented as

$$G(\lambda) = \overline{C}R(\lambda, A_{-1})B + D, \qquad \lambda \in \mathbb{C}_\omega, \tag{2.4}$$

for a *feedthrough* operator $D \in \mathcal{L}(U,Y)$. Hence, the operators $\overline{C}R(\lambda, A_{-1})B$ are uniformly bounded on $\overline{\mathbb{C}}_\omega$.

This representation of G is not unique in general since $D(A)$ need not to be dense in Z. Under the additional assumption of *regularity*, one can replace here $\overline{C}$ by C_Λ (possibly for a different D), see Theorem 5.8 in [27] and also Theorem 4.6 in [22] for refinements. We will not use regularity below.

3. Polynomial stabilizability and detectability

In this section we introduce our new concepts and establish their basic properties. We start with the main definitions.

Definition 3.1. The admissible system $(A, B, -)$ is *polynomially stabilizable* (*of order* $\alpha > 0$) if there exists a generator A_{BF} of a polynomially stable C_0-semigroup $T_{BF}(\cdot)$ on X (of order $\alpha > 0$) and an admissible observation operator $F \in \mathcal{L}(D(A_{BF}), U)$ of A_{BF} such that

$$R(\lambda, A_{BF}) = R(\lambda, A) + R(\lambda, A)BFR(\lambda, A_{BF}) \tag{3.1}$$

for all $\operatorname{Re}\lambda > \max\{\omega_0(A), \omega_0(A_{BF})\}$.

Definition 3.2. The admissible system $(A, -, C)$ is *polynomially detectable* (*of order* $\alpha > 0$) if there exists a generator A_{HC} of a polynomially stable C_0-semigroup $T_{HC}(\cdot)$ (of order $\alpha > 0$) and an admissible control operator $H \in \mathcal{L}(Y, X_{-1}^{A_{HC}})$ of A_{HC} such that

$$R(\lambda, A_{HC}) = R(\lambda, A) + R(\lambda, (A_{HC})_{-1})HCR(\lambda, A) \tag{3.2}$$

for all $\operatorname{Re}\lambda > \max\{\omega_0(A), \omega_0(A_{HC})\}$.

Here F, resp. H, plays the role of a feedback. These definitions are inspired by the Definition 3.2 in [12] for the exponentially stable case. For this case, in, e.g., [29] concepts of exponential stabilizability or detectability were used which are (at least formally) a bit stronger than those in [12], cf. Remark 3.3(b). In our context, one could also include the boundedness of the feedback semigroup $T_{BF}(\cdot)$ or $T_{HC}(\cdot)$ in the above definitions since the theory of polynomial stability works much better in the bounded case, as seen in the previous section. Instead, we make additional boundedness assumptions in some of our results. In applications one can check the boundedness or $T_{BF}(\cdot)$ or $T_{HC}(\cdot)$ by showing that the generators A_{BF} or A_{HC} are dissipative, respectively, where one may use their representation given in the next remark.

Remark 3.3. (a) Let $(A,B,-)$, $(A_{BF},-,F)$, $(A,-,C)$ and $(A_{HC},H,-)$ be admissible. Proposition 4.11 in [13] (with $\beta=\gamma=1$ and $b=c=0$) then shows that the equations (3.1) and (3.2) are equivalent to

$$T_{BF}(t)x = T(t)x + \int_0^t T(t-s)BF_\Lambda T_{BF}(s)x\,ds = T(t)x + \Phi_t F_\Lambda T_{BF}(\cdot)x, \quad (3.3)$$

$$T_{HC}(t)x = T(t)x + \int_0^t T_{HC}(t-s)HC_\Lambda T(s)x\,ds \quad (3.4)$$

for all $t\geq 0$ and $x\in X$, respectively.

(b) Applying $\lambda - A_{-1}$ to (3.1), we see that A_{BF} is a restriction of the part $(A_{-1}+BF)|X$ of $A_{-1}+BF$ in X. Similarly, multiplication of (3.2) by $\lambda - A_{HC,-1}$ leads to $A\subset (A_{HC,-1}-HC)|X$. See Proposition 6.6 in [28]. We note that in [29] exponential stabilizability and detectability was defined in such a way that $A_{BF}=(A_{-1}+BF_\Lambda)|X$ and $A_{HC}=(A_{-1}+CH_\Lambda)|X$.

(c) The system $(A,B,-)$ is polynomially stabilizable of order $\alpha>0$ (with feedback F) if and only if $(A^*,-,B^*)$ is polynomially detectable of order $\alpha>0$ (with feedback $H=F^*$). Moreover, the semigroups of the feedback systems are dual to each other.

(d) Let L be a closed operator with $\emptyset\neq\Lambda\subset\rho(L)$ and $\Omega\supset\Lambda$ be connected. If $R(\cdot,L)$ has a holomorphic extension R_λ to Ω, then $\Omega\subset\rho(L)$ and $R_\lambda = R(\lambda,L)$ for every $\lambda\in\Omega$. (See Proposition B5 in [3].)

In a sequence of lemmas we relate the growth properties of several operators arising in (3.1) or (3.2). We use the spectral bound $s(L)=\sup\{\operatorname{Re}\lambda \mid \lambda\in\sigma(L)\}\in[-\infty,\infty]$ for a closed operator L, where $\sup\emptyset=-\infty$

Lemma 3.4. *Let $C\in\mathcal{B}(X_1,Y)$ and $B\in\mathcal{B}(U,X_{-1})$ be admissible observation and control operators for A, respectively and let*

$$\|R(r+i\tau,A)\|\leq c\,|\tau|^\alpha \quad (3.5)$$

for some $r>s(A)$ and $\alpha>0$ and all $|\tau|\geq 1$. We then obtain the estimates

$$\|CR(r+i\tau,A)\|\leq c\,|\tau|^\alpha \qquad \text{and} \qquad \|R(r+i\tau,A)B\|\leq c\,|\tau|^\alpha$$

for all $|\tau|\geq 1$. Moreover, if (A,B,C) is also well posed, we have

$$\|\overline{C}R(r+i\tau,A)B\|\leq c\,|\tau|^\alpha$$

for all $|\tau|\geq 1$. Here the constants are uniform for r in bounded intervals.

Proof. Let $\lambda=r+i\tau$ and $\mu=\omega+i\tau$ for $\tau\in\mathbb{R}$ and some $\omega>\max\{0,\omega_0(A)\}$. The resolvent equation yields

$$CR(\lambda,A)=CR(\mu,A)+(\omega-r)CR(\mu,A)R(\lambda,A). \quad (3.6)$$

Let $x\in D(A)$. Since the resolvent is the Laplace transform of $T(\cdot)$, from the admissibility of C and exponential bound of $T(\cdot)$ we deduce

$$\|CR(\mu,A)x\|^2 \leq \Big[\int_0^\infty e^{-\frac{\omega}{2}t}e^{-\frac{\omega}{2}t}\,\|CT(t)x\|\,dt\Big]^2 \leq c\int_0^\infty e^{-\omega t}\,\|CT(t)x\|^2\,dt \quad (3.7)$$

$$\leq c\sum_{n=0}^{\infty} e^{-\omega n}\,\|CT(\cdot)T(n)x\|^2_{L^2(0,1;Y)} \leq c\sum_{n=0}^{\infty} e^{-\omega n}\|T(n)x\|^2 \leq c\,\|x\|^2.$$

By density, the formulas (3.5), (3.6) and (3.7) imply

$$\|CR(\lambda,A)\| \leq c + c\,|\tau|^\alpha \leq c\,|\tau|^\alpha$$

for $|\tau| \geq 1$. The second asserted inequality then follows by duality because B^* is an admissible observation operator for A^* and $\|R(\lambda,A)B\| = \|B^*R(\lambda,A^*)\|$. For the final claim, we start from the equation

$$\overline{C}R(\lambda,A)B = \overline{C}R(\mu,A)B + (\omega - r)CR(\mu,A)R(\lambda,A)B$$

for $\lambda = r + i\tau$, $\mu = \omega + i\tau$, $\tau \in \mathbb{R}$ and some $\omega > \max\{0, \omega_0(A)\}$. As noted in the previous section, $\overline{C}R(\mu,A)B : U \to Y$ is uniformly bounded. The third assertion now is a consequence of the two previous ones. □

In the next lemma we deduce resolvent estimates for A from those for A_{BF}.

Lemma 3.5. *Let $B \in \mathcal{L}(U, X_{-1})$ be an admissible control operator for A. Assume that there exist a generator A_{BF} of a C_0-semigroup $T_{BF}(\cdot)$ on X and an admissible observation operator $F \in \mathcal{L}(D(A_{BF}), U)$ of A_{BF} such that* (3.1) *holds. Assume that*

$$\|R(\lambda, A_{BF})\| \leq c\,(1 + |\lambda|^\alpha)$$

for $r < \operatorname{Re}\lambda \leq r + \delta$ and some $r \geq s(A_{BF})$, $\delta > 0$, $\alpha \geq 0$. Suppose that $R(\lambda,A)B$ has a holomorphic extension R^B_λ to $\mathbb{C}_r$ satisfying

$$\|R^B_\lambda\| \leq c\,(1 + |\lambda|^\beta)$$

for $r < \operatorname{Re}\lambda \leq r + \delta$ and some $\beta \geq 0$. Then $R(\cdot, A)$ can be extended to a neighborhood of $\overline{\mathbb{C}_r}$, and we obtain

$$\|R(\lambda,A)\| \leq c\,(1 + |\lambda|^{\alpha+\beta}) \tag{3.8}$$

for $r \leq \operatorname{Re}\lambda \leq r + \delta$. Moreover, (3.1) *holds on $\overline{\mathbb{C}_r}$. If $r = 0$, then $T(\cdot)$ is polynomially stable with order $2(\alpha + \beta) + 1 + \eta$ for any $\eta > 0$.*

Proof. By the assumption, (3.1) and Remark 3.3, the resolvent $R(\cdot, A)$ has the extension

$$R(\lambda,A) = R(\lambda, A_{BF}) - R^B_\lambda F R(\lambda, A_{BF})$$

to $\lambda \in \mathbb{C}_r$. Lemma 3.4 and the assumption then imply that

$$\|R(\lambda,A)\| \leq c\,(1 + |\lambda|^{\alpha+\beta})$$

for $r < \operatorname{Re}\lambda \leq r + \delta$. A standard power series argument allows us to extend this inequality to $\lambda \in \overline{\mathbb{C}_r}$ and to deduce that a neighborhood of $\overline{\mathbb{C}_r}$ belongs to $\rho(A)$. The uniqueness of the holomorphic extension now yields that $R^B_\lambda = R(\lambda,A)B$ on $\overline{\mathbb{C}_r}$ and that (3.1) holds on $\overline{\mathbb{C}_r}$. The last assertion then follows from estimate (3.8) and Propositions 3.4 and 3.6 in [5]. □

The next result is proved in the same manner as the above lemma.

Lemma 3.6. *Let the operators A, C and H satisfy the assumptions of Definition 3.2 except for the polynomial stability of $T_{HC}(\cdot)$. Assume that*

$$\|R(\lambda, A_{HC})\| \le c\,(1 + |\lambda|^{\alpha})$$

for $r < \operatorname{Re}\lambda \le r + \delta$ and some $r \ge s(A_{HC})$, $\delta > 0$ and $\alpha \ge 0$. Let $CR(\lambda, A)$ have a holomorphic extension R^C_λ to $\mathbb{C}_r$. Suppose that

$$\|R^C_\lambda\| \le c\,(1 + |\lambda|^{\beta})$$

for $r < \operatorname{Re}\lambda \le r + \delta$ and some $\beta > 0$. Then $\rho(A)$ contains a neighborhood of $\overline{\mathbb{C}_r}$, the equality (3.2) holds on $\overline{\mathbb{C}_r}$, and we obtain

$$\|R(\lambda, A)\| \le c\,(1 + |\lambda|^{\alpha+\beta})$$

for $r \le \operatorname{Re}\lambda \le r + \delta$. If $r = 0$, then $T(\cdot)$ is polynomially stable with order $2(\alpha + \beta) + 1 + \eta$ for any $\eta > 0$.

To apply Proposition 2.4, we will need a variant of the above estimates.

Lemma 3.7. *Let A generate a bounded C_0-semigroup and C be an admissible observation operator for A. Then*

$$\sup_{r>0} r \int_{\mathbb{R}} \|CR(r + i\tau, A)x\|^2 \, d\tau \le c\,\|x\|^2$$

for all $r > 0$ and $x \in X$.

Proof. Take $r > 0$ and $x \in D(A)$. Since $A - r$ generates the exponentially stable semigroup $(e^{-rt}T(t))_{t\ge0}$, Plancherel's theorem and the assumption yield

$$\begin{aligned}
\|CR(r + i\cdot, A)x\|^2_{L^2(\mathbb{R}_+,Y)} &= \|Ce^{-r\cdot}T(\cdot)x\|^2_{L^2(\mathbb{R}_+,Y)} \\
&= \sum_{n\ge0} \int_0^1 e^{-2rn} e^{-2rs} \|CT(s)T(n)x\|^2 \, ds. \\
&\le c \sum_{n\ge0} e^{-2rn} \|T(n)x\|^2 \le \frac{c\,\|x\|^2}{1 - e^{-2r}} \le \frac{c}{r}\,\|x\|^2.
\end{aligned}$$

The assertion follows by density. □

4. Main results

We show that external polynomial stability in the frequency domain, i.e., a polynomial estimate on the transfer function, imply polynomial stability of the state system. We begin with a result involving only the control operator B.

Proposition 4.1. *Let $(A, B, -)$ be admissible and polynomially stabilizable of order $\alpha > 0$. Assume that $R(\lambda, A)B$ has a holomorphic extension to $\mathbb{C}_+$ which is bounded by $c\,(1 + |\lambda|^{\beta})$ for $0 < \operatorname{Re}\lambda \le \delta$ and some $\beta \ge 0$, $\delta > 0$. The following assertions hold.*

a) *The resolvent* $R(\cdot, A)$ *can be extended to a neighborhood of* $\overline{\mathbb{C}_+}$ *and*

$$\|R(\lambda, A)\| \le c_\varepsilon \, (1 + |\lambda|^{\alpha+\beta+\varepsilon}) \tag{4.1}$$

for $0 \le \operatorname{Re}\lambda \le \delta$ *and every* $\varepsilon > 0$. *If* $T_{BF}(\cdot)$ *is bounded, we can choose* $\varepsilon = 0$.

b) *The semigroup* $T(\cdot)$ *is polynomially stable. If* $T(\cdot)$ *is also bounded, then it is polynomially stable of order* $\alpha + \beta + \varepsilon$. *If in addition* $T_{BF}(\cdot)$ *is bounded, we can take* $\varepsilon = 0$.

Proof. a) Propositions 3.3 and 3.6 in [5] imply that $\sigma(A_{BF}) \subset \mathbb{C}_-$ and

$$\|R(\lambda, A_{BF})\| \le c_\varepsilon (1 + |\lambda|^{\alpha+\varepsilon})$$

holds for $\operatorname{Re}\lambda \ge 0$ and every $\varepsilon > 0$. Using Lemma 3.5, we infer $\sigma(A) \subset \mathbb{C}_-$ and (4.1). If $T_{BF}(\cdot)$ is bounded, we can use Proposition 2.2 instead of the results from [5] and obtain the above estimates with $\varepsilon = 0$.

b) Proposition 3.4 of [5] and (4.1) imply the polynomial stability of $T(\cdot)$. If also $T(\cdot)$ is bounded, it is polynomially stable of order $\alpha+\beta+\varepsilon$ due to Theorem 2.3 and (4.1). □

By duality, the above proposition implies the next one for the observation system $(A, -, C)$.

Proposition 4.2. *Let* $(A, -, C)$ *be admissible and polynomially detectable of order* $\alpha > 0$. *Assume that* $CR(\cdot, A)$ *has a holomorphic extension to* $\mathbb{C}_+$ *which is bounded by* $c\,(1 + |\lambda|^{\beta})$ *for* $0 < \operatorname{Re}\lambda \le \delta$ *and some* $\beta \ge 0$. *The following assertions hold.*

a) *The resolvent* $R(\cdot, A)$ *can be extended to a neighborhood of* $\overline{\mathbb{C}_+}$ *and estimate* (4.1) *holds for every* $\varepsilon > 0$. *If* $T_{HC}(\cdot)$ *is bounded, we can take* $\varepsilon = 0$.

b) *The semigroup* $T(\cdot)$ *is polynomially stable. If* $T(\cdot)$ *is also bounded, then it is polynomially stable of order* $\alpha + \beta + \varepsilon$. *If in addition* $T_{HC}(\cdot)$ *is bounded, we can take* $\varepsilon = 0$.

We now can state our main result which uses the full system (A, B, C) and the transfer function G.

Theorem 4.3. *Let* (A, B, C) *be a well-posed system which is polynomially stabilizable of order* $\alpha > 0$ *and polynomially detectable of order* $\beta > 0$. *Assume that* G *has a holomorphic extension to* $\mathbb{C}_+$ *which is bounded by* $c\,(1 + |\lambda|^{\gamma})$ *for* $0 < \operatorname{Re}\lambda \le \delta$ *and some* $\gamma \ge 0$ *and* $\delta > 0$. *The following assertions hold.*

a) *The extension* $\overline{C}$ *of* C *is an admissible observation operator for* A_{BF}, $\sigma(A) \subset \mathbb{C}_-$, *and*

$$\|R(\lambda, A)\| \le c_\varepsilon (1 + |\lambda|^{\alpha+\beta+\gamma+\varepsilon})$$

for $0 < \operatorname{Re}\lambda \le \delta$ *and all* $\varepsilon > 0$. *If* $T_{BF}(\cdot)$ *is bounded, we can take* $\varepsilon = 0$.

b) *The semigroup* $T(\cdot)$ *is polynomially stable. If* $T(\cdot)$ *is bounded, then it is polynomially stable of order* $\alpha + \beta + \gamma + \varepsilon$. *If in addition* $T_{BF}(\cdot)$ *is bounded, we can take* $\varepsilon = 0$.

Proof. a) Due to (3.1) and (2.4), we have $D(A_{BF}) \subset Z$ and

$$\overline{C}R(\lambda, A_{BF}) = CR(\lambda, A) + \overline{C}R(\lambda, A)BFR(\lambda, A_{BF}),$$
$$\overline{C}R(\lambda, A_{BF}) = CR(\lambda, A) + G(\lambda)FR(\lambda, A_{BF}) - DFR(\lambda, A_{BF}) \qquad (4.2)$$

for $\mathrm{Re}\,\lambda > \max\{\omega_0(A), \omega_0(A_{BF})\}$. Taking the inverse Laplace transform of this equation, we define

$$\Psi_{BF}x := \mathcal{L}^{-1}(\overline{C}R(\cdot, A_{BF})x) = \Psi x + \mathbb{F}FT_{BF}(\cdot)x - DFT_{BF}(\cdot)x \qquad (4.3)$$

for $x \in D(A_{BF})$. By assumption, $\Psi_{BF} : X \longrightarrow L^2_{\mathrm{loc}}(\mathbb{R}_+, Y)$ is continuous. For $\tau \geq 0$ and $x \in D(A_{BF})$, the properties of a well-posed system and (3.3) yield

$$\begin{aligned}\Psi_{BF}x(\cdot+\tau) &= \Psi T(\tau)x + \mathbb{F}FT_{BF}(\cdot)T_{BF}(\tau)x + \Psi\Phi_\tau FT_{BF}(\cdot)x \\ &\qquad - DFT_{BF}(\cdot)T_{BF}(\tau)x \\ &= \Psi T_{BF}(\tau)x + \mathbb{F}FT_{BF}(\cdot)T_{BF}(\tau)x - DFT_{BF}(\cdot)T_{BF}(\tau)x \\ &= \Psi_{BF}T_{BF}(\tau)x.\end{aligned}$$

As a result, (Ψ_{BF}, T_{BF}) is an observation system in the sense of [26] or Section 4.3 in [24]. The proof of Theorem 3.3 of [26] and (4.3) thus show that $\Psi_{BF}x = \widetilde{C}T_{BF}(\cdot)x$ for $x \in D(A_{BF})$ and the admissible control operator $\widetilde{C} \in \mathcal{L}(D(A_{BF}), Y)$ for A_{BF} given by

$$\widetilde{C}x = \widehat{\Psi_{BF}}(\lambda)(\lambda - A_{BF})x = \overline{C}R(\lambda, A_{BF})(\lambda - A_{BF})x = \overline{C}x \quad \text{for } x \in D(A_{BF});$$

i.e., $\Psi_{BF}x = \overline{C}T_{BF}(\cdot)x$ for $x \in D(A_{BF})$. Proposition 3.4 of [5] and Lemma 3.4 then yield

$$\|\overline{C}R(\lambda, A_{BF})\| \leq c\,(1 + |\lambda|^{\alpha+\varepsilon}) \qquad \text{and} \qquad \|FR(\lambda, A_{BF})\| \leq c\,(1 + |\lambda|^{\alpha+\varepsilon})$$

for $\mathrm{Re}\,\lambda \geq 0$ and any $\varepsilon > 0$. If $T_{BF}(\cdot)$ is bounded, we can use Proposition 2.2 instead of the results in [5] and derive these estimates with $\varepsilon = 0$. By means of (4.2) and the bound on G, we now extend $CR(\cdot, A)$ (using the same symbol) to $\mathbb{C}_+$ and obtain

$$\|CR(\lambda, A)\| \leq c\,(1 + |\lambda|^{\alpha+\gamma+\varepsilon})$$

for $0 < \mathrm{Re}\,\lambda \leq \delta$. Proposition 4.2 then gives

$$\|R(\lambda, A)\| \leq c_\varepsilon(1 + |\lambda|^{\alpha+\beta+\gamma+\varepsilon})$$

for $0 < \mathrm{Re}\,\lambda \leq \delta$ and all $\varepsilon > 0$, where we can take $\varepsilon = 0$ if $T_{BF}(\cdot)$ is bounded.

b) Proposition 3.4 of [5] and part a) imply the polynomial stability of $T(\cdot)$. If $T(\cdot)$ is bounded, it is polynomially stable of order $\alpha+\beta+\gamma+\varepsilon$ due to Theorem 2.3 and part a), where we can take $\varepsilon = 0$ if $T_{BF}(\cdot)$ is bounded. □

In the above results one obtains the expected stability order of $T(\cdot)$ only if this semigroup is bounded. This property automatically holds in the important case of a *scattering passive* system (A, B, C); i.e., if we have

$$\|y\|^2_{L^2(0,t;Y)} + \|x(t)\|^2 \leq \|u\|^2_{L^2(0,t;U)} + \|x_0\|^2$$

for all $u \in L^2(0,t;U)$, $x_0 \in X$ and $t \geq 0$, where $x(t) = T(t)x_0 + \Phi_t u$ is the state and $y = \Psi x_0 + \mathbb{F}u$ is the output of (A, B, C). This class of systems has been characterized and studied in, e.g., [22]. In this case $T(t)$ and $G(\lambda)$ are contractions for $t \geq 0$ and $\lambda \in \mathbb{C}_+$ by Proposition 7.2 and Theorem 7.4 of [22].

Corollary 4.4. *Let (A, B, C) be a scattering passive system which is polynomially stabilizable of order $\alpha > 0$ and polynomially detectable of order $\beta > 0$. Then $T(\cdot)$ is polynomially stable of order $\alpha + \beta + \varepsilon$ for each $\varepsilon > 0$. We can take $\varepsilon = 0$ if $T_{BF}(\cdot)$ is bounded.*

Proposition 2.4 yields another sufficient condition for the boundedness of $T(\cdot)$ in the framework of the first two propositions of this section.

Proposition 4.5. *Assume that the assumptions of both Propositions 4.1 and 4.2 hold for some $\alpha > 0$ and for $\beta = 0$. Let $T_{BF}(\cdot)$ and $T_{HC}(\cdot)$ be bounded. Then $T(\cdot)$ is bounded, and hence polynomially stable of order $\alpha > 0$.*

Proof. Definitions 3.1 and 3.2 yield

$$R(r+i\tau, A)x = R(r+i\tau, A_{BF})x - R(r+i\tau, A)BFR(r+i\tau, A_{BF})x, \tag{4.4}$$

$$R(r+i\tau, A^*)x = R(r+i\tau, A^*_{HC})x - R(r+i\tau, A^*)C^*H^*R(r+i\tau, A^*_{HC})x \tag{4.5}$$

for all $r > \max\{\omega_0(A), 0\}$, $\tau \in \mathbb{R}$ and $x \in X$. We can extend these equations to $r > 0$ using the bounded extensions of $R(\lambda, A)B$ and $R(\lambda, A^*)C^* = (CR(\lambda, A))^*$ which are provided by our assumption. Since $T_{BF}(\cdot)$ and $T_{HC}(\cdot)$ are bounded, Lemma 3.7 implies that the terms on the right-hand sides belong to $L^2(\mathbb{R}, X)$ as functions in τ, with norms bounded by $cr^{-1/2}\|x\|$. Employing Proposition 2.4, we then deduce the boundedness of $T(\cdot)$ from (4.4) and (4.5). The final assertion now follows from Proposition 4.1. □

We finally present sufficient conditions for polynomial stabilizability and for polynomial detectability by means of a decomposition into a polynomial stable and an observable part. An admissible system $(A, B, -)$ is called *null controllable in finite time* if for each initial value $x_0 \in X$ there is a time $\tau > 0$ and a control $u \in L^2(0,\tau;U)$ such that $x(\tau) = T(\tau)x_0 + \Phi_\tau u = 0$. We further note that one can extend an operator S to X_{-1} if it commutes with $T(t)$ for all $t \geq 0$ since then $SR(\omega, A) = R(\omega, A)S$.

Theorem 4.6. *Let $(A, B, -)$ be admissible and let $P^2 = P \in \mathcal{B}(X)$ satisfy $T(t)P = PT(t)$ for all $t \geq 0$. Set $X_s = PX$, $X_u = (I-P)X$, $T_s(t) = T(t)P$, $A_u = (I-P)A$ and $B_u = (I-P)B$. Assume that*

(i) *the C_0-semigroup $T_s(\cdot)$ is polynomially stable of order $\alpha > 0$ on X_s and*
(ii) *the system $(A_u, B_u, -)$ is null controllable in finite time on X_u.*

Then the system $(A, B, -)$ is polynomially stabililizable of order $\alpha > 0$.

Proof. First observe that $T_u(\cdot)$ is the C_0-semigroup on X_u generated by A_u and that B_u is admissible for A_u. Due to (ii), for each $x_0 \in X_u$ there is a time $\tau > 0$ and a control $u \in L^2(0,\tau;U)$ such that $x_u(\tau) = T_u(\tau)x_0 + (I-P)\Phi_\tau u = 0$. Extending

x_u and u by 0 to (τ, ∞), we see that the system $(A_u, B_u, -)$ is optimizable in the sense of Definition 3.1 in [29]. Propositions 3.3 and 3.4 of [29] (or Theorem 2.2 of [9]) then give an operator F_u which satisfies the conditions of Definition 3.1 where $T_{B_uF_u}(\cdot)$ is even exponentially stable, i.e., $\omega_0(A_{B_uF_u}) < 0$. We thus have

$$R(\lambda, A_{B_uF_u}) = R(\lambda, A_u) + R(\lambda, A_u)B_uF_uR(\lambda, A_{B_uF_u}) \tag{4.6}$$

for all $\operatorname{Re}\lambda > \max(\omega_0(A), \omega_0(A_{B_uF_u}))$. We now set

$$F = \begin{pmatrix} 0 \\ F_u \end{pmatrix} \quad \text{and} \quad A_{BF} := \begin{pmatrix} A_s & 0 \\ 0 & A_{B_uF_u} \end{pmatrix}.$$

It is then straightforward to check that these operators fulfill the conditions of Definition 3.1. □

The next result follows by duality from Theorem 4.6.

Theorem 4.7. *Let $(A, -, C)$ be admissible and let $P^2 = P \in \mathcal{B}(X)$ satisfy $T(t)P = PT(t)$ for all $t \geq 0$. Set $X_s = PX$, $X_u = (I-P)X$, $T_s(t) = T(t)P$, $A_u = (I-P)A$ and $C_u = C(I-P)$. Assume that*

(i) *the C_0-semigroup $T_s(\cdot)$ is polynomially stable of order $\alpha > 0$ on X_s and*
(ii) *the system $(A_u^*, C_u^*, -)$ is null controllable in finite time on X_u.*

Then the system $(A, -, C)$ is polynomially detectable of order $\alpha > 0$.

Remark 4.8. The results of Theorem 4.6 and 4.7 also hold if we replace the condition (ii) by (ii)$'$: The system $(A_u, B_u, -)$ (resp., $(A_u^*, C_u^*, -)$) is polynomially stabilizable of order α.

References

[1] F. Alabau, P. Cannarsa and V. Komornik, *Indirect internal stabilization of weakly coupled evolution equations.* J. Evol. Equ. **2** (2002), 127–150.

[2] K. Ammari and M. Tucsnak, *Stabilization of second-order evolution equations by a class of unbounded feedbacks.* ESAIM Control Optim. Calc. Var. **6** (2001), 361–386.

[3] W. Arendt, C.J.K. Batty, M. Hieber and F. Neubrander, *Vector-Valued Laplace Transforms and Cauchy Problems.* Birkhäuser Verlag, Basel, 2001.

[4] G. Avalos and R. Triggiani, *Rational decay rates for a PDE heat-structure interaction: a frequency domain approach.* Evol. Equ. Control Theory **2** (2013), 233–253.

[5] A. Bátkai, K.-J. Engel, J. Prüss and R. Schnaubelt, *Polynomial stability of operator semigroups.* Math. Nachr. **279** (2006), 1425–1440.

[6] C.J.K. Batty and T. Duyckaerts, *Non-uniform stability for bounded semi-groups on Banach spaces.* J. Evol. Eq. **8** (2008), 765–780.

[7] A. Borichev and Y. Tomilov, *Optimal polynomial decay of functions and operator semigroups.* Math. Ann. **347** (2010), 455–478.

[8] N. Burq and M. Hitrik, *Energy decay for damped wave equations on partially rectangular domains.* Math. Res. Lett. **14** (2007), 35–47.

[9] F. Flandoli, I. Lasiecka and R. Triggiani, *Algebraic Riccati equations with nonsmoothing observation arising in hyperbolic and Euler–Bernoulli boundary control problems.* Ann. Mat. Pura Appl. (4) **153** (1988), 307–382.

[10] A.M. Gomilko, *On conditions for the generating operator of a uniformly bounded C_0-semigroup of operators.* Funct. Anal. Appl. **33** (1999), 294–296.

[11] B. Jacob and R. Schnaubelt, *Observability of polynomially stable systems.* Systems Control Lett. **56** (2007), 277–284.

[12] B. Jacob and H. Zwart, *Equivalent conditions for stabilizability of infinite-dimensional systems with admissible control operators.* SIAM J. Control Optim. **37** (1999), 1419–1455.

[13] Y. Latushkin, T. Randolph and R. Schnaubelt, *Regularization and frequency-domain stability of well-posed systems.* Math. Control Signals Systems **17** (2005), 128–151.

[14] Y. Latushkin and R. Shvidkoy, *Hyperbolicity of semigroups and Fourier multipliers.* In: A.A. Borichev and N.K. Nikolski (eds.), 'Systems, Approximation, Singular Integral Operators, and Related Topics' (Bordeaux, 2000), Oper. Theory Adv. Appl. **129**, Birkhäuser Verlag, Basel, 2001, pp. 341–363.

[15] G. Lebeau, *Équation des ondes amorties.* In: A. Boutet de Monvel and V. Marchenko (eds.), 'Algebraic and Geometric Methods in Mathematical Physics' (Kaciveli, 1993), Math. Phys. Stud. **19** Kluwer Acad. Publ., Dordrecht, 1996, pp. 73–101.

[16] G. Lebeau and E. Zuazua, *Decay rates for the three-dimensional linear system of thermoelasticity.* Arch. Rational Mech. Anal. **148** (1999), 179–231.

[17] Z. Liu and B. Rao, *Characterization of polynomial decay rate for the solution of linear evolution equation.* Z. Angew. Math. Phys. **56** (2005), 630–644.

[18] S. Nicaise, *Stabilization and asymptotic behavior of dispersive medium models.* Systems Control Lett. **61** (2012), 638–648.

[19] L. Paunonen, *Robustness of strongly and polynomially stable semigroups.* J. Funct. Anal. **263** (2012), 2555–2583.

[20] R. Rebarber, *Conditions for the equivalence of internal and external stability for distributed parameter systems.* IEEE Trans. Automat. Control **38** (1993), 994–998.

[21] R. Rebarber and H.J. Zwart, *Open-loop stabilization of infinite-dimensional systems.* Math. Control Signals Systems **11** (1998), 129–160.

[22] O. Staffans and G. Weiss, *Transfer functions of regular linear systems II. The system operator and the Lax–Phillips semigroup.* Trans. Amer. Math. Soc. **354** (2002), 3229–3262.

[23] L. Tébou, *Well-posedness and energy decay estimates for the damped wave equation with L^r localizing coefficient.* Comm. Partial Differential Equations **23** (1998), 1839–1855.

[24] M. Tucsnak and G. Weiss, *Observation and Control for Operator Semigroups.* Birkhäuser, Basel, 2009.

[25] M. Tucsnak and G. Weiss, *Well-posed systems – the LTI case and beyond.* Automatica J. IFAC **50** (2014), 1757–1779.

[26] G. Weiss, *Admissible observation operators for linear semigroups.* Israel J. Math. **65** (1989), 17–43.

[27] G. Weiss, *Transfer functions of regular linear systems. Part I: Characterization of regularity*, Trans. Amer. Math. Soc. **342** (1994), 827–854.

[28] G. Weiss, *Regular linear systems with feedback.* Math. Control Signals Systems **7** (1994), 23–57.

[29] G. Weiss and R. Rebarber, *Optimizability and estimatability for infinite-diemnsional systems.* SIAM J. Control Optim. **39** (2000), 1204–1232.

El Mustapha Ait Benhassi and Lahcen Maniar
Cadi Ayyad University
Faculty of Sciences Semlalia
B.P. 2390
Marrakesh, Morocco
e-mail: m.benhassi@ucam.ac.ma
maniar@ucam.ac.ma

Said Boulite
Hassan II University
Faculty of Sciences Ain Chock
B.P. 5366 Maarif
20100 Casablanca, Morocco
e-mail: s.boulite@fsac.ac.ma

Roland Schnaubelt
Department of Mathematics
Karlsruhe Institute of Technology
D-76128 Karlsruhe, Germany
e-mail: schnaubelt@kit.edu

Operator Theory:
Advances and Applications, Vol. 250, 17–29

Minimal Primal Ideals in the Multiplier Algebra of a $C_0(X)$-algebra

R.J. Archbold and D.W.B. Somerset

Abstract. Let A be a stable, σ-unital, continuous $C_0(X)$-algebra with surjective base map $\phi : \mathrm{Prim}(A) \to X$, where $\mathrm{Prim}(A)$ is the primitive ideal space of the C^*-algebra A. Suppose that $\phi^{-1}(x)$ is contained in a limit set in $\mathrm{Prim}(A)$ for each $x \in X$ (so that A is *quasi-standard*). Let $C_{\mathbf{R}}(X)$ be the ring of continuous real-valued functions on X. It is shown that there is a homeomorphism between the space of minimal prime ideals of $C_{\mathbf{R}}(X)$ and the space $\mathrm{MinPrimal}(M(A))$ of minimal closed primal ideals of the multiplier algebra $M(A)$. If A is separable then $\mathrm{MinPrimal}(M(A))$ is compact and extremally disconnected but if $X = \beta\mathbf{N} \setminus \mathbf{N}$ then $\mathrm{MinPrimal}(M(A))$ is nowhere locally compact.

Mathematics Subject Classification (2010). Primary 46L05, 46L08, 46L45; Secondary 46E25, 46J10, 54C35.

Keywords. C^*-algebra, $C_0(X)$-algebra, multiplier algebra, minimal prime ideal, minimal primal ideal, primitive ideal space, quasi-standard.

1. Introduction

Let A be a C^*-algebra with multiplier algebra $M(A)$ [10] and with primitive ideal space $\mathrm{Prim}(A)$. The ideal structure of $M(A)$ has been widely studied, and is typically much more complicated than that of A, see for example [1], [13], [21], [25],[27]. One approach, which the authors used in an earlier paper [7], is to endow A with a $C_0(X)$-structure (this can always be done, sometimes in many different ways). Let A be a σ-unital $C_0(X)$-algebra (defined below) with base map $\phi : \mathrm{Prim}(A) \to X$, and let X_ϕ denote the image of $\mathrm{Prim}(A)$ under ϕ. The authors showed that there is a map from the lattice of z-ideals of $C_{\mathbf{R}}(X_\phi)$ into the lattice of closed ideals of $M(A)$, and that this map is injective if A is stable [7, Theorem 3.2]. If X_ϕ is infinite then z-ideals generally exist in great profusion – for example, $C_{\mathbf{R}}(\mathbf{R})$ has uncountable chains of prime z-ideals associated with each point of $\mathbf{R}$ [22], [26] –

so this yields a vast multiplicity of closed ideals in $M(A)$ and indicates something of the complexity of $\mathrm{Prim}(M(A))$ [7, Theorem 5.3].

The most studied z-ideals are the minimal prime ideals and in this note we consider the image of the space of minimal prime ideals of $C_{\mathbf{R}}(X_\phi)$ under the injective map. We show in Theorem 3.4 that if A is stable, σ-unital, and quasi-standard (defined below) then the image of the space of minimal prime ideals is precisely $\mathrm{MinPrimal}(M(A))$, the space of minimal closed primal ideals of $M(A)$ (see below). It follows that $\mathrm{MinPrimal}(M(A))$ is totally disconnected and countably compact (Corollary 4.1). If A is also separable – for example if A equals $C[0,1] \otimes K(H)$ (where $K(H)$ is the algebra of compact operators on a separable infinite-dimensional Hilbert space) – then $\mathrm{MinPrimal}(M(A))$ is compact and extremally disconnected (Corollary 4.3).

All ideals in this paper will be two-sided, but not necessarily closed unless stated to be so. An ideal J in a C^*-algebra A is *primal* if whenever $I_1, \ldots, I_n$ is a finite collection of ideals of A with the product $I_1 \ldots I_n = \{0\}$ then $I_i \subseteq J$ for at least one $i \in \{1, \ldots, n\}$. An equivalent definition, when J is closed, is that the hull of J should be contained in a limit set in $\mathrm{Prim}(A)$ [3, Proposition 3.2]. Every primitive ideal is prime and hence primal. Each closed primal ideal of a C^*-algebra A contains one or more minimal closed primal ideals [2, p. 525]. The space of minimal closed primal ideals with the τ_w-topology (defined in Section 3) is denoted $\mathrm{MinPrimal}(A)$. This Hausdorff space is often identifiable in situations where the primitive ideal space is non-Hausdorff and highly complicated. Indeed, the multiplier algebras considered in this paper are a case in point.

A C^*-algebra A is said to be *quasi-standard* if the relation $\sim$ of inseparability by disjoint open sets is an open equivalence relation on $\mathrm{Prim}(A)$ [5]. This condition is a wide generalisation of the special case where $\mathrm{Prim}(A)$ is Hausdorff. Examples include, in the unital case, von Neumann and AW*-algebras, local multiplier algebras of C^*-algebras [29], and the group C^*-algebras of amenable discrete groups [17]; and in the non-unital case, many other group C*-algebras, see [4]. A basic non-unital example, however, is simply $A = C_0(X) \otimes K(H)$, where X is a locally compact Hausdorff space, and even in this case the ideal structure of $M(A)$ is not well understood, see [20], [7]. The connection between quasi-standard C^*-algebras and $C_0(X)$-algebras is explained in Lemma 2.1 and the remarks preceding it.

The structure of the paper is that in Section 2, we set up some machinery; in Section 3, we prove the main homeomorphism result; and in Section 4, we give some applications.

2. Preliminaries

First we collect the information that we need on $C_0(X)$-algebras. Recall that a C^*-algebra A is a $C_0(X)$-algebra if there is a continuous map ϕ, called the base map, from $\mathrm{Prim}(A)$, the primitive ideal space of A with the hull-kernel topology, to the locally compact Hausdorff space X [31, Proposition C.5]. Then X_ϕ, the image of ϕ in X, is completely regular; and if A is σ-unital, X_ϕ is σ-compact and hence normal [7, Section 1]. If ϕ is an open map then X_ϕ is locally compact.

For $x \in X_\phi$, set $J_x = \bigcap\{P \in \mathrm{Prim}(A) : \phi(P) = x\}$, and for $x \in X \setminus X_\phi$, set $J_x = A$. For $a \in A$, the function $x \to \|a + J_x\|$ $(x \in X)$ is upper semi-continuous [31, Proposition C.10]. The $C_0(X)$-algebra A is said to be *continuous* if, for all $a \in A$, the norm function $x \to \|a + J_x\|$ $(x \in X)$ is continuous. By Lee's theorem [31, Proposition C.10 and Theorem C.26], this happens if and only if the base map ϕ is open.

An important special case (through which all other cases factor) is when ϕ is the complete regularization map ϕ_A for $\mathrm{Prim}(A)$ [14, Theorem 3.9]. In this case, the ideals J_x $(x \in X_\phi)$ are called the *Glimm* ideals of A, and the set of Glimm ideals with the complete regularization topology is called $\mathrm{Glimm}(A)$. Each minimal closed primal ideal of A contains a unique Glimm ideal [5, Lemma 2.2]. If A is quasi-standard then the complete regularization map ϕ_A is open [5, Theorem 3.3], so $\mathrm{Glimm}(A)$ is locally compact and A is a continuous $C_0(X)$-algebra with $X = X_{\phi_A} = \mathrm{Glimm}(A)$. Furthermore, if A is quasi-standard then each Glimm ideal of A is actually primal and indeed the topological spaces $\mathrm{Glimm}(A)$ and $\mathrm{MinPrimal}(A)$ coincide [5, Theorem 3.3]. It then follows from [3, Proposition 3.2] that $\phi_A^{-1}(x)$ is a maximal limit set in $\mathrm{Prim}(A)$ for all $x \in X$. The following result is closely related to [5, Theorem 3.4].

Lemma 2.1. *For a C^*-algebra A, the following are equivalent:*

(i) *A is quasi-standard;*

(ii) *A is a continuous $C_0(X)$-algebra over a locally compact Hausdorff space X with base map ϕ such that $\phi^{-1}(x)$ is contained in a limit set in* $\mathrm{Prim}(A)$ *for all $x \in X_\phi$.*

When these equivalent conditions hold, there is a homeomorphism ψ : $\mathrm{Glimm}(A) \to$ *X_ϕ such that $\phi = \psi \circ \phi_A$, where ϕ_A is the complete regularization map for A. Moreover, for all $x \in X_\phi$, $\phi^{-1}(x)$ is a maximal limit set in* $\mathrm{Prim}(A)$ *and J_x is a minimal closed primal ideal of A.*

Proof. We have seen that (i) implies (ii). Conversely, suppose that (ii) holds. Since X is a locally compact Hausdorff space, for $P, Q \in \mathrm{Prim}(A)$, $P \sim Q$ if and only if $\phi(P) = \phi(Q)$. It follows that $\sim$ is an equivalence relation. Let Y be a non-empty open subset of $\mathrm{Prim}(A)$. Then $Y' := \phi^{-1}(\phi(Y))$ is the $\sim$-saturation of Y, and Y' is open since ϕ is open. Hence $\sim$ is an open equivalence relation so (i) holds.

When (ii) holds, we have that ϕ is continuous and open with image X_ϕ, and that it factors as $\phi = \psi \circ \phi_A$, where $\psi : \mathrm{Glimm}(A) \to X_\phi$ is continuous [14, Theorem 3.9]. Then ψ is surjective, and the limit set hypothesis easily shows that ψ is injective. Since ϕ is open and ϕ_A is continuous, ψ is open. Thus ψ is a homeomorphism.

Finally, let $x \in X_\phi$ and let Ω be a net in $\mathrm{Prim}(A)$ whose limit set L contains $\phi^{-1}(x)$. Since ϕ_A is constant on L, $\phi(L) = \{x\}$. Thus $L = \phi^{-1}(x)$ and $\phi^{-1}(x)$ is a maximal limit set. It follows from [3, Proposition 3.2] that J_x is a minimal closed primal ideal of A. □

Now let J be a proper, closed ideal of a C^*-algebra A. The quotient map $q_J : A \to A/J$ has a canonical extension $\tilde{q_J} : M(A) \to M(A/J)$ such that $\tilde{q_J}(b)q_J(a) = q_J(ba)$ and $q_J(a)\tilde{q_J}(b) = q_J(ab)$ $(a \in A, b \in M(A))$. We define a proper, closed ideal $\tilde{J}$ of $M(A)$ by

$$\tilde{J} = \ker \tilde{q_J} = \{b \in M(A) : ba, ab \in J \text{ for all } a \in A\}.$$

Various properties of $\tilde{J}$ were established in [6, Proposition 1.1]. For example, $\tilde{J}$ is the strict closure of J in $M(A)$ and $\tilde{J} \cap A = J$.

The following proposition was proved in [6, Proposition 1.2].

Proposition 2.2. *Let A be a $C_0(X)$-algebra with base map ϕ. Then ϕ has a unique extension to a continuous map $\overline{\phi}$: $\mathrm{Prim}(M(A)) \to \beta X$ such that $\overline{\phi}(\tilde{P}) = \phi(P)$ for all $P \in \mathrm{Prim}(A)$. Hence $M(A)$ is a $C(\beta X)$-algebra with base map $\overline{\phi}$ and* $\mathrm{Im}(\overline{\phi}) = \mathrm{cl}_{\beta X}(X_\phi)$.

Now let A be a $C_0(X)$-algebra with base map ϕ and let $\overline{\phi} : \mathrm{Prim}(M(A)) \to \beta X$ be as in Proposition 2.2. For $x \in \beta X$, we define

$$H_x = \bigcap \{Q \in \mathrm{Prim}(M(A)) : \overline{\phi}(Q) = x\},$$

a closed two-sided ideal of $M(A)$. Thus H_x is defined in relation to $(M(A), \beta X, \overline{\phi})$ in the same way that J_x (for $x \in X$) is defined in relation to (A, X, ϕ). It follows that for each $b \in M(A)$, the function $x \to \|b + H_x\|$ $(x \in \beta X)$ is upper semi-continuous. If ϕ is the complete regularization map for $\mathrm{Prim}(A)$ and $X = \beta\mathrm{Glimm}(A)$ then $\mathrm{Glimm}(M(A)) = \{H_x : x \in X\}$; see the comment after [9, Proposition 4.4].

The next proposition is contained in [7, Proposition 2.3].

Proposition 2.3. *Let A be a $C_0(X)$-algebra with base map ϕ, and set $X_\phi = \mathrm{Im}(\phi)$.*

(i) *For all $x \in X$, $J_x \subseteq H_x \subseteq \tilde{J}_x$ and $J_x = H_x \cap A$.*
(ii) *For all $b \in M(A)$,*

$$\|b\| = \sup\{\|b + \tilde{J}_x\| : x \in X_\phi\} = \sup\{\|b + H_x\| : x \in X_\phi\}.$$

In the case when $A = C_0(X) \otimes K(H) \cong C_0(X, K(H))$ and $\phi : \mathrm{Prim}(A) \to X$ is the homeomorphism such that $\phi^{-1}(x) = \{f \in C_0(X) : f(x) = 0\} \otimes K(H)$ $(x \in X)$, the multiplier algebra $M(A)$ is isomorphic to the C^*-algebra of bounded, strong*-continuous functions from X to $B(H)$ (the algebra of bounded linear operators on the Hilbert space H) [1, Corollary 3.5]. Then for $x \in X$,

$$\tilde{J}_x = \{f \in M(A) : f(x) = 0\}.$$

On the other hand, by Proposition 2.2 $M(A)$ is a $C(\beta X)$-algebra, and for $x \in \beta X$,

$$H_x = \{f \in M(A) : \lim_{y \to x} \|f(y)\| = 0\}.$$

We shall recall in Theorem 2.4 below that when A is a σ-unital $C_0(X)$-algebra with base map ϕ there is an order-preserving map from the lattice of z-ideals of $C_{\mathbf{R}}(X_\phi)$ into the lattice of closed ideals of $M(A)$. To describe this map, we give a brief account of the theory of z-ideals.

Let X be a completely regular topological space and let $C_{\mathbf{R}}(X)$ denote the ring of continuous real-valued functions on X. For $f \in C_{\mathbf{R}}(X)$, let

$$Z(f) = \{x \in X : f(x) = 0\},$$

the zero set of f. We note for later that every zero set clearly arises as the zero set of a bounded continuous function. A non-empty family $\mathcal{F}$ of zero sets of X is called a z-filter if: (i) $\mathcal{F}$ is closed under finite intersections; (ii) $\emptyset \notin \mathcal{F}$; (iii) each zero set which contains a member of $\mathcal{F}$ belongs to $\mathcal{F}$. Each ideal $I \subseteq C_{\mathbf{R}}(X)$ yields a z-filter $Z(I) = \{Z(f) : f \in I\}$. An ideal I is called a z-ideal if $Z(f) \in Z(I)$ implies $f \in I$; and if $\mathcal{F}$ is a z-filter on X then the ideal $I(\mathcal{F})$ defined by

$$I(\mathcal{F}) = \{f \in C_{\mathbf{R}}(X) : Z(f) \in \mathcal{F}\}$$

is a z-ideal. There is a bijective correspondence between the set of z-ideals of $C_{\mathbf{R}}(X)$ and the set of z-filters on X, given by $I = I(Z(I)) \leftrightarrow Z(I)$.

Now let A be a σ-unital $C_0(X)$-algebra with base map ϕ, and let $u \in A$ be a strictly positive element. For $a \in A$, set $Z(a) = \{x \in X_\phi : a \in J_x\}$. Unless norm functions of elements of A are continuous on X_ϕ, $Z(a)$ will not necessarily be a zero set of X_ϕ. However, since $Z(u) = \emptyset$ and A is closed under multiplication by $C^b(X_\phi)$, every zero set $Z(f)$ of X_ϕ arises as $Z(a)$ for the element $a = f \cdot u \in A$ ($f \in C^b_{\mathbf{R}}(X_\phi)$). For $b \in M(A)$, set $Z(b) = \{x \in X_\phi : b \in \tilde{J}_x\}$. Note that if $b \in A$ then this definition is consistent with the previous one because $\tilde{J}_x \cap A = J_x$ ($x \in X_\phi$). It is also useful to note that for $b \in M(A)$ and $x \in X_\phi$, $b \in \tilde{J}_x$ if and only if $bu \in \tilde{J}_x$ if and only if $bu \in J_x$. Hence $Z(b) = Z(bu)$.

For a z-filter $\mathcal{F}$ on X_ϕ define $L^{\mathrm{alg}}_{\mathcal{F}} = \{b \in M(A) : \exists Z \in \mathcal{F},\ Z(b) \supseteq Z\}$, and let $L_{\mathcal{F}}$ be the norm-closure of $L^{\mathrm{alg}}_{\mathcal{F}}$ in $M(A)$. Let $b \in L^{\mathrm{alg}}_{\mathcal{F}}$. Then for $a \in M(A)$, $Z(ab) \supseteq Z(b)$ and $Z(ba) \supseteq Z(b)$, while for $a \in L^{\mathrm{alg}}_{\mathcal{F}}$, $Z(a+b) \supseteq Z(a) \cap Z(b)$. Hence $L^{\mathrm{alg}}_{\mathcal{F}}$ is an ideal of $M(A)$, so $L_{\mathcal{F}}$ is a closed ideal of $M(A)$.

Theorem 2.4. ([7, Theorem 3.2]) *Let A be a σ-unital $C_0(X)$-algebra with base map ϕ. Suppose that A/J_x is non-unital for all $x \in X_\phi$. Let I and J be z-ideals of $C_{\mathbf{R}}(X_\phi)$ and suppose that there exists a zero set Z of X_ϕ such that $Z \in Z[I]$ but $Z \notin Z[J]$. Then $L_{Z[I]} \nsubseteq L_{Z[J]}$. Hence the assignment $I \to L_{Z[I]}$ defines an order-preserving injective map L from the lattice of z-ideals of $C_{\mathbf{R}}(X_\phi)$ into the lattice of closed ideals of $M(A)$.*

To identify what happens to some of the most important z-ideals of $C_{\mathbf{R}}(X_\phi)$ under this map, we use the following notation. For $x \in X$, let M_x be the maximal ideal given by $M_x = \{f \in C_{\mathbf{R}}(X) : f(x) = 0\}$, and let

$$O_x = \{f \in C_{\mathbf{R}}(X) : x \in \operatorname{int}(Z(f))\}$$

where $\mathrm{int}(Z(f))$ denotes the interior of $Z(f)$. Then M_x and O_x are z-ideals, and O_x is the smallest ideal of $C_{\mathbf{R}}(X)$ which is not contained in any maximal ideal other than M_x. The definitions just given can be extended as follows. For $p \in \beta X$, let $M^p = \{f \in C_{\mathbf{R}}(X) : p \in \mathrm{cl}_{\beta X} Z(f)\}$ and define O^p to be the set of all $f \in C_{\mathbf{R}}(X)$ for which $\mathrm{cl}_{\beta X} Z(f)$ is a neighbourhood of p in βX. Then for $x \in X$, $M^x = M_x$ and $O^x = O_x$. The embedding map takes M_x to $\tilde{J}_x$ and O^p to H_p (and hence O_x to H_x).

Proposition 2.5 **([7, Theorem 4.3]).** *Let A be a σ-unital $C_0(X)$-algebra with base map ϕ.*

(i) *For $x \in X_\phi$, $L_{Z[M_x]} = \tilde{J}_x$.*
(ii) *For $p \in \mathrm{cl}_{\beta X} X_\phi$, $L_{Z[O^p]} = H_p$.*

Proposition 2.5 shows that the embedding map of Theorem 2.4 is mainly shedding light on the lattice of closed ideals of $M(A)$ between $\tilde{J}_x$ and H_x; see [7, Section 4] for further discussion. Before presenting a simple example to illustrate Theorem 2.4 and Proposition 2.5, we need further terminology.

A z-filter $\mathcal{F}$ on a completely regular space X is said to be *prime* if $Z_1 \cup Z_2 \in \mathcal{F}$ implies that either $Z_1 \in \mathcal{F}$ or $Z_2 \in \mathcal{F}$, for zero sets Z_1 and Z_2. Let $PF(X)$ denote the set of prime z-filters, and let $PZ(X)$ be the set of prime z-ideals (recall that an ideal $P \subseteq C_{\mathbf{R}}(X)$ is prime if $fg \in P$ implies $f \in P$ or $g \in P$). The bijective correspondence between z-ideals and z-filters restricts to a bijective correspondence $j : PZ(X) \to PF(X)$ given by $j(P) = \{Z(f) : f \in P\}$ (see [14, Chapter 2]). If $P \in PZ(X)$ and $P \subseteq M_x$ for some $x \in X$ then $O_x \subseteq P$ [14, 4I], and hence $H_x \subseteq L_{Z[P]} \subseteq \tilde{J}_x$ by Proposition 2.5. Every z-ideal of $C_{\mathbf{R}}(X)$ is an intersection of prime z-ideals and the minimal prime ideals of $C_{\mathbf{R}}(X)$ are z-ideals [14, 2.8, 14.7]. The prime ideals containing a given prime ideal form a chain [14, 14.8].

Example. Let $X = \mathbf{N} \cup \{\omega\}$ be the one-point compactification of $\mathbf{N}$ and set $A = C(X) \otimes K(H)$. Then $M_x = O_x$ for $x \in \mathbf{N}$, but $M_\omega \neq O_\omega$. The assignment

$$\mathcal{F} \to P_{\mathcal{F}} = \{f \in C_{\mathbf{R}}(X) : Z(f) \setminus \{\omega\} \in \mathcal{F}\}$$

gives a bijection between the family of free ultrafilters on $\mathbf{N}$ (every ultrafilter on $\mathbf{N}$ is trivially a z-ultrafilter) and the family of non-maximal prime z-ideals contained in M_ω. Each $P_{\mathcal{F}}$ is a minimal prime z-ideal [14, 14G] and we shall see in Section 4 that its image $L_{\mathcal{F}}$ under the mapping of Theorem 2.4 is a minimal closed primal ideal of $M(A)$. The ideal $H_\omega = L_{Z(O_\omega)}$ is a Glimm ideal but is not primal.

3. The homeomorphism onto MinPrimal$(M(A))$

In this section we specialize to the case when A is a σ-unital quasi-standard C^*-algebra. We will be assuming that A is canonically represented as a $C_0(X)$-algebra with the base map ϕ as the complete regularization map for $\mathrm{Prim}(A)$ and with $X = X_\phi = \mathrm{Glimm}(A)$. For the main result we will also need to assume that A/J_x is non-unital for $x \in X$ (note that this is automatically satisfied if A is stable).

The reasons for restricting to quasi-standard C^*-algebras are twofold. The first is the fact, already mentioned, that when A is quasi-standard, MinPrimal(A) and Glimm(A) coincide as sets (and indeed as topological spaces). This has the implication that, for $x \in X = \text{Glimm}(A)$, the ideal $\tilde{J}_x$ is primal in $M(A)$ [6, Lemma 4.5]; and hence there must be minimal closed primal ideals of $M(A)$ lying between $\tilde{J}_x$ and the Glimm ideal H_x of $M(A)$. But secondly, if A is quasi-standard then norm functions of elements of A are continuous on Glimm(A), so for $a \in A$, $Z(a)$ is a zero set of Glimm(A). Furthermore if A is also σ-unital and $u \in A$ is a strictly positive element then, as we have already mentioned, for $b \in M(A)$ $Z(b) = Z(bu)$, so $Z(b)$ is also a zero set of Glimm(A). Thus the elaborate machinery of zero sets works smoothly for this class of algebras.

For a ring R let Min(R) be the space of minimal (algebraic) primal ideals of R with the *lower topology* generated by sub-basic sets of the form

$$\{P \in \text{Min}(R) : a \notin P\}$$

as a varies through elements of R. If R is a commutative ring then an argument of Krull shows that every minimal primal ideal of R is prime, and Min(R) is the usual space of minimal prime ideals of R with the hull-kernel topology, see [28] and the references given there. If P is a minimal prime ideal of $C_{\mathbf{R}}(X)$ then P is a z-ideal, as we have mentioned, so an obvious step is to identify the image of Min($C_{\mathbf{R}}(X)$) under the embedding map L of Theorem 2.4. We shall show that the embedding map L carries Min($C_{\mathbf{R}}(X)$) homeomorphically onto MinPrimal($M(A)$) with the τ_w-topology (where the τ_w-topology is defined on MinPrimal(A) by taking sets of the form $\{P \in \text{MinPrimal}(A) : a \notin P\}$ $(a \in A)$ as sub-basic; see [2, p. 525] where an equivalent definition is given).

It is convenient to proceed in two stages. In Theorem 3.2 we show that the assignment $P \mapsto L^{\text{alg}}_{Z[P]}$ defines a homeomorphism Θ from Min($C_{\mathbf{R}}(X)$) onto Min($M(A)$). For this theorem we do not require the quotients A/J_x $(x \in X)$ to be non-unital. Then in Theorem 3.4 we show that, if these quotients are non-unital, the assignment $L^{\text{alg}}_{\mathcal{F}} \mapsto L_{\mathcal{F}}$ defines a homeomorphism Φ from Min($M(A)$) onto MinPrimal($M(A)$). The method of proof of Theorem 3.2 is similar to that of [28, Theorem 3.2] except that we are here working with filters of zero sets rather than with ideals of cozero sets. For further work on the space of minimal (algebraic) primal ideals of a C^*-algebra, see [29] and [30].

For a C^*-algebra B and $a \in B$, let I_a be the closed ideal of B generated by a. The following lemma is a special case of [28, Theorem 2.3], which itself is a special case of a more general result due to Keimel [18]. Recall that ideals are not necessarily closed unless stated to be so.

Lemma 3.1. *Let B be a C^*-algebra and let P be a primal ideal of B. Then P is a minimal primal ideal if and only if for all $a \in P$ there exist $b_1, \ldots, b_n \in B \setminus P$ such that $I_a I_{b_1} \ldots I_{b_n} = \{0\}$.*

Let $I_a^\perp$ be the largest ideal of B such that $I_a I_a^\perp = \{0\}$. Then Lemma 3.1 implies that if P is a minimal primal ideal of B and $a \in P$ then $I_a^{\perp\perp} \subseteq P$.

Theorem 3.2. *Let A be a σ-unital quasi-standard C^*-algebra and set $X = \mathrm{Glimm}(A)$. Then the assignment $P \mapsto L^{\mathrm{alg}}_{Z[P]}$ defines a homeomorphism Θ from* $\mathrm{Min}(C_{\mathbf{R}}(X))$ *onto* $\mathrm{Min}(M(A))$.

Proof. First we show that if $\mathcal{F} = Z[P]$ for $P \in \mathrm{Min}(C_{\mathbf{R}}(X))$ then $L^{\mathrm{alg}}_{\mathcal{F}}$ is a minimal primal ideal of $M(A)$. Let $b_i \in M(A) \setminus L^{\mathrm{alg}}_{\mathcal{F}}$ $(1 \le i \le n)$. Then $Z(b_i) \notin \mathcal{F}$ for each i, so $Z(b_1) \cup \cdots \cup Z(b_n) \notin \mathcal{F}$ since $\mathcal{F}$ is a prime z-filter. Hence $Z(b_1) \cup \cdots \cup Z(b_n) \neq X$, so there exists $x \in X$ such that $b_i \notin \tilde{J}_x$ $(1 \le i \le n)$. Since $\tilde{J}_x$ is primal, $b_1 M(A) \dots M(A) b_n \neq \{0\}$. Hence $L^{\mathrm{alg}}_{\mathcal{F}}$ is primal. Now let $b \in L^{\mathrm{alg}}_{\mathcal{F}}$ with $b \neq 0$. Then $Z(b) \in \mathcal{F}$, so by [19, Lemma 3.1] there exists $f \in C_{\mathbf{R}}(X)$ such that $Z(f) \cup Z(b) = X$ and $Z(f) \notin \mathcal{F}$. Let $c \in A$ with $Z(c) = Z(f)$. Then $Z(c) \notin \mathcal{F}$ so $c \notin L^{\mathrm{alg}}_{\mathcal{F}}$, and $Z(c) \cup Z(b) = X$, so $bM(A)c = \{0\}$ by Proposition 2.3(ii). This shows that $L^{\mathrm{alg}}_{\mathcal{F}}$ is a minimal primal ideal of $M(A)$ and hence that Θ maps into $\mathrm{Min}(M(A))$.

Now let P and Q be distinct elements of $\mathrm{Min}(C_{\mathbf{R}}(X))$. Then $Z[P] \neq Z[Q]$, and since for each zero set Z there exists $c \in A$ with $Z(c) = Z$, it follows that $L^{\mathrm{alg}}_{Z[P]} \neq L^{\mathrm{alg}}_{Z[Q]}$. This shows that Θ is injective.

Now suppose that $Q \in \mathrm{Min}(M(A))$ and let $\mathcal{G} = \{Z(b) : b \in Q\}$. We show that $\mathcal{G}$ is a minimal prime z-filter on X. First note that if $b \in Q$ then $I_b^{\perp}$ is non-zero by Lemma 3.1, and indeed $I_b^{\perp} = \{a \in M(A) : Z(a) \cup Z(b) = X\}$ by the primality of the ideals $\tilde{J}_x$ $(x \in X)$. Hence $Z(b)$ is non-empty, so $\emptyset \notin \mathcal{G}$. For $b, c \in Q$, $Z(b) \cap Z(c) = Z(bb^* + cc^*) \in \mathcal{G}$. If $b \in Q$ and $c \in M(A)$ with $Z(c) \supseteq Z(b)$ then $Z(a) \cup Z(c) = X$ for all $a \in I_b^{\perp}$, so $c \in I_b^{\perp\perp} \subseteq Q$, as observed after Lemma 3.1. Hence $Z(c) \in \mathcal{G}$. This shows that $\mathcal{G}$ is a proper z-filter, and also that $Q = L^{\mathrm{alg}}_{\mathcal{G}}$. To show that $\mathcal{G}$ is a prime z-filter, let Z_1 and Z_2 be zero sets of X such that $Z_1 \cup Z_2 = X$. Let $b, c \in A$ such that $Z_1 = Z(b)$ and $Z_2 = Z(c)$. Then $bM(A)c = \{0\}$, so at least one of b and c (b say) belongs to Q since Q is primal. Hence $Z_1 \in \mathcal{G}$. This shows that $\mathcal{G}$ is prime [14, 2E].

To see that $\mathcal{G}$ is minimal prime, let $Z \in \mathcal{G}$ and let $b \in Q$ such that $Z(b) = Z$. Then by Lemma 3.1 there exist $c_1, \dots, c_n \in M(A) \setminus Q$ such that $I_b I_{c_1} \dots I_{c_n} = \{0\}$. Hence $Z(c_i) \notin \mathcal{G}$ $(1 \le i \le n)$, by an argument in the previous paragraph, and $Z(b) \cup Z(c_1) \cup \cdots \cup Z(c_n) = X$ by the primality of the ideals $\tilde{J}_x$ $(x \in X)$. Set $Y = Z(c_1) \cup \cdots \cup Z(c_n)$. Then Y is a zero set in X, being a finite union of zero sets, and $Y \notin \mathcal{G}$ since $\mathcal{G}$ is prime. Since $Z \cup Y = X$ it follows that no z-filter strictly smaller than $\mathcal{G}$ can be prime. Hence $\mathcal{G}$ is a minimal prime z-filter, and $Q = L^{\mathrm{alg}}_{\mathcal{G}}$ belongs to the range of Θ. Thus Θ is a bijection.

Finally, for $f \in C_{\mathbf{R}}(X)$ we can find $a \in A$ such that $Z(a) = Z(f)$; and conversely, given $a \in M(A)$, since A is σ-unital and quasi-standard we can find $f \in C_{\mathbf{R}}(X)$ such that $Z(a) = Z(f)$. Hence in either case

$$\begin{aligned}\Theta(\{P \in \mathrm{Min}(C_{\mathbf{R}}(X)) : f \notin P\}) &= \Theta(\{P \in \mathrm{Min}(C_{\mathbf{R}}(X)) : Z(f) \notin Z[P]\}) \\ &= \{L^{\mathrm{alg}}_{Z[P]} \in \mathrm{Min}(M(A)) : Z(a) \notin Z[P]\} \\ &= \{L^{\mathrm{alg}}_{Z[P]} \in \mathrm{Min}(M(A)) : a \notin L^{\mathrm{alg}}_{Z[P]}\}.\end{aligned}$$

Since the hull-kernel topology on $\mathrm{Min}(C_{\mathbf{R}}(X))$ can be defined either using ideals or using elements, it follows that Θ is a homeomorphism. □

A comparison of the proof of Theorem 3.2 with that of [28, Theorem 3.2] shows that when A is a σ-unital quasi-standard C^*-algebra, the assignment $Q \mapsto Q \cap A$ gives a homeomorphism from $\mathrm{Min}(M(A))$ onto $\mathrm{Min}(A)$.

For the next theorem, we need the following family of functions which is useful for relating $L_{\mathcal{F}}$ and $L_{\mathcal{F}}^{\mathrm{alg}}$. For $0 < \epsilon < 1/2$, define the continuous piecewise linear function $f_\epsilon : [0, \infty) \to [0, \infty)$ by: (i) $f_\epsilon(x) = 0$ $(0 \le x \le \epsilon)$; (ii) $f_\epsilon(x) = 2(x - \epsilon)$ $(\epsilon \le x \le 2\epsilon)$; (iii) $f_\epsilon(x) = x$ $(2\epsilon \le x)$. Note that for $b \in M(A)^+$, if $b \in L_{\mathcal{F}}$ then $f_\epsilon(b)$ belongs to the Pedersen ideal of $L_{\mathcal{F}}$ for all ϵ [24, 5.6.1], and hence $f_\epsilon(b) \in L_{\mathcal{F}}^{\mathrm{alg}}$. On the other hand, $\|b - f_\epsilon(b)\| \le \epsilon$. Thus we have the following lemma.

Lemma 3.3. *Let A be $C_0(X)$-algebra with base map ϕ and let $\mathcal{F}$ be a z-filter on X_ϕ. Let $b \in M(A)^+$. Then with the notation above, $b \in L_{\mathcal{F}}$ if and only if $f_\epsilon(b) \in L_{\mathcal{F}}^{\mathrm{alg}}$ for all $\epsilon \in (0, 1/2)$.*

Theorem 3.4. *Let A be a σ-unital, quasi-standard C^*-algebra with A/G non-unital for all $G \in \mathrm{Glimm}(A)$ and set $X = \mathrm{Glimm}(A)$. Then the assignment $P \mapsto L_{Z[P]}$ defines a homeomorphism from $\mathrm{Min}(C_{\mathbf{R}}(X))$ onto $\mathrm{MinPrimal}(M(A))$.*

Proof. By Theorem 3.2, it is enough to show that the assignment

$$L_{Z[P]}^{\mathrm{alg}} \mapsto L_{Z[P]} \qquad (P \in \mathrm{Min}(C_{\mathbf{R}}(X)))$$

defines a homeomorphism Φ from $\mathrm{Min}(M(A))$ onto $\mathrm{MinPrimal}(M(A))$. If R is a minimal closed primal ideal of $M(A)$ then R contains some $L_{Z[P]}^{\mathrm{alg}} \in \mathrm{Min}(M(A))$, and hence $R = L_{Z[P]}$. Thus the range of Φ certainly contains $\mathrm{MinPrimal}(M(A))$. Furthermore, Theorem 2.4 implies that Φ is injective and also that if $P, Q \in \mathrm{Min}(C_{\mathbf{R}}(X))$ with $P \ne Q$ then $L_{Z[P]} \not\subseteq L_{Z[Q]}$. Suppose that $Q \in \mathrm{Min}(C_{\mathbf{R}}(X))$. Then $L_{Z[Q]}^{\mathrm{alg}} \in \mathrm{Min}(M(A))$ so $L_{Z[Q]}$ is a closed primal ideal of $M(A)$. Hence $L_{Z[Q]}$ contains a minimal closed primal ideal of $M(A)$, which we have just seen is of the form $L_{Z[P]}$ for $P \in \mathrm{Min}(C_{\mathbf{R}}(X))$. Thus $P = Q$, so the range of Φ equals $\mathrm{MinPrimal}(M(A))$. Hence Φ is a bijection.

Now let $a \in M(A)^+$ and let $Z = Z(a)$, a zero set in X. Then by [7, Corollary 3.1] there exists $c \in M(A)^+$ such that $\|c + \tilde{J}_x\| = 1$ for $x \in X \setminus Z$ and $c \in \tilde{J}_x$ for $x \in Z$. Hence $Z(f_\epsilon(c)) = Z$ for all $\epsilon \in (0, 1/2)$. Thus

$$\begin{aligned}\Phi(\{L_{\mathcal{F}}^{\mathrm{alg}} \in \mathrm{Min}(M(A)) : a \notin L_{\mathcal{F}}^{\mathrm{alg}}\}) &= \Phi(\{L_{\mathcal{F}}^{\mathrm{alg}} \in \mathrm{Min}(M(A)) : Z \notin \mathcal{F}\}) \\ &= \{L_{\mathcal{F}} \in \mathrm{MinPrimal}(M(A)) : c \notin L_{\mathcal{F}}\},\end{aligned}$$

by Lemma 3.3. On the other hand, again by Lemma 3.3,

$$\begin{aligned}&\Phi^{-1}(\{L_{\mathcal{F}} \in \mathrm{MinPrimal}(M(A)) : a \notin L_{\mathcal{F}}\}) \\ &\quad = \bigcup_{\epsilon \in (0,1/2)} \{L_{\mathcal{F}}^{\mathrm{alg}} \in \mathrm{Min}(M(A)) : f_\epsilon(a) \notin L_{\mathcal{F}}^{\mathrm{alg}}\}).\end{aligned}$$

Thus it follows that Φ is a homeomorphism. □

Corollary 3.5. *Let A be a σ-unital, continuous $C_0(X)$-algebra with base map ϕ such that A/J_x is non-unital and $\phi^{-1}(x)$ is contained in a limit set in* $\mathrm{Prim}(A)$ *for all $x \in X_\phi$. Then* $\mathrm{Min}(C_{\mathbf{R}}(X_\phi))$ *is homeomorphic to* $\mathrm{MinPrimal}(M(A))$.

Proof. By Lemma 2.1, A is quasi-standard and there is a homeomorphic map $\psi : \mathrm{Glimm}(A) \to X_\phi$. For $G \in \mathrm{Glimm}(A)$, there exists $x \in X_\phi$ such that $\psi^{-1}(x) = G$. Hence $J_x = G$, so A/G is non-unital. The result now follows from Theorem 3.4. □

4. Applications

The space of minimal prime ideals of $C_{\mathbf{R}}(X)$ has been studied in numerous papers, e.g., [19], [15], [12], [11], [16], so Theorem 3.4 has various immediate corollaries. We present a sample of these. Recall that a topological space Y is *countably compact* if every countable open cover of Y has a finite subcover. If Y is a T_1-space then Y is countably compact if and only if every infinite subset of Y has a limit point in Y [23, p. 181].

Corollary 4.1. *Let A be a σ-unital, quasi-standard C^*-algebra with A/G non-unital for all $G \in$* $\mathrm{Glimm}(A)$.

(i) *The Hausdorff space* $\mathrm{MinPrimal}(M(A))$ *is totally disconnected and countably compact.*
(ii) *If* $\mathrm{MinPrimal}(M(A))$ *is locally compact then it is basically disconnected.*

Proof. (i) The space of minimal closed primal ideals of a C^*-algebra is always Hausdorff in the τ_w-topology [2, Corollary 4.3]. The total disconnectedness and countable compactness follow from Theorem 3.4 and from [15, Corollary 2.4] and [15, Theorem 4.9] respectively.

(ii) This follows from Theorem 3.4 and [15, Theorem 4.7]. □

In the context of Corollary 4.1, recall that a necessary and sufficient condition for $M(A)$ to be quasi-standard is that $\mathrm{Glimm}(M(A))$ and $\mathrm{MinPrimal}(M(A))$ should coincide both as sets and as topological spaces [5, Theorem 3.3]. Since $M(A)$ is unital, $\mathrm{Glimm}(M(A))$ is compact, so $\mathrm{MinPrimal}(M(A))$ would also have to be compact. By Corollary 4.1(ii), this implies that $\mathrm{MinPrimal}(M(A))$, and hence $\mathrm{Glimm}(M(A))$, would have to be basically disconnected; and this in turn implies that $\mathrm{Glimm}(A)$ would have to be basically disconnected [14, 6M.1]. Thus we recover the necessity of $\mathrm{Glimm}(A)$ being basically disconnected if $M(A)$ is quasi-standard. In point of fact, it was shown in [6, Corollary 4.9] that if A is a σ-unital quasi-standard C*-algebra with centre equal to $\{0\}$ then $M(A)$ is quasi-standard if and only if $\mathrm{Glimm}(A)$ is basically disconnected.

Corollary 4.2. *Let A be a σ-unital, quasi-standard C^*-algebra and suppose that A/G is non-unital for all $G \in$* $\mathrm{Glimm}(A)$. *Then the following are equivalent:*

(i) $\mathrm{MinPrimal}(M(A))$ *is compact;*

(ii) Glimm(A) *is cozero-complemented; that is, for every cozero set* U *in* Glimm(A) *there exists a cozero set* V *in* Glimm(A) *such that* $U \cap V = \emptyset$ *and* $U \cup V$ *is dense in* Glimm(A).

Proof. This follows by Theorem 3.4 and the characterization in [15, Corollary 5.5]. □

For example, if Glimm(A) is basically disconnected or is homeomorphic to an ordinal space then Glimm(A) is cozero complemented [16, Examples 1.6], so the space MinPrimal$(M(A))$ is compact. On the other hand, if Glimm(A) is the Alexandroff double of a compact metric space without isolated points then Glimm(A) is compact and first countable but not cozero complemented [16, Examples 1.7]. Hence MinPrimal$(M(A))$ is not compact.

If A is separable, much more can be said. Recall that a *regular closed* set is one that is the closure of its interior. If A is separable then Glimm(A) is perfectly normal [8, Lemma 3.9] (i.e., every closed subset of Glimm(A) is a zero set) so A certainly satisfies condition (ii) of the next corollary.

Corollary 4.3. *Let* A *be a* σ*-unital, quasi-standard* C^**-algebra. Suppose that* A/G *is non-unital for* $G \in$ Glimm(A). *Then the following are equivalent:*

(i) MinPrimal$(M(A))$ *is compact and extremally disconnected;*

(ii) *Every regular closed set in* Glimm(A) *is the closure of a cozero set.*

In particular, if A *is separable then* A *satisfies these equivalent conditions.*

Proof. This follows by Theorem 3.4 and the characterization in [15, Theorems 4.4 and 5.6]. □

More generally, recall that a topological space X has the *countable chain condition* if every family of non-empty pairwise disjoint open subsets of X is countable. It is easily seen that a completely regular topological space with the countable chain condition has property (ii) of Corollary 4.3. If a C*-algebra A has a faithful representation on a separable Hilbert space, then Glimm(A) satisfies the countable chain condition [30, p. 85].

We conclude with one further application of Theorem 3.4.

Corollary 4.4. *Set* $A = C(\beta\mathbf{N} \setminus \mathbf{N}) \otimes K(H)$. *Then* MinPrimal$(M(A))$ *is nowhere locally compact. If Martin's Axiom holds then* MinPrimal$(M(A))$ *is not an F-space.*

Proof. Both statements follow from Theorem 3.4, the first by [15, Example 5.9], and the second by [12, Corollary 4]. □

References

[1] C.A. Akemann, G.K. Pedersen and J. Tomiyama, *Multipliers of C^*-algebras.* J. Funct. Anal. **13** (1973), 277–301.

[2] R.J. Archbold, *Topologies for primal ideals.* J. London Math. Soc. (2) **36** (1987), 524–542.

[3] R.J. Archbold and C.J.K. Batty, *On factorial states of operator algebras, III.* J. Operator Theory **15** (1986), 53–81.

[4] R.J. Archbold, E. Kaniuth and D.W.B. Somerset, *Norms of inner derivations for multipliers of C^*-algebras and group C^*-algebras.* J. Functional Analysis **262** (2012), 2050–2073.

[5] R.J. Archbold and D.W.B. Somerset, *Quasi-standard C^*-algebras.* Math. Proc. Camb. Phil. Soc **107** (1990), 349–360.

[6] R.J. Archbold and D.W.B. Somerset, *Multiplier algebras of $C_0(X)$-algebras.* Münster J. Math. **4** (2011), 73–100.

[7] R.J. Archbold and D.W.B. Somerset, *Ideals in the multiplier and corona algebras of a $C_0(X)$-algebra.* J. London Math. Soc. (2) **85** (2012), 365–381.

[8] R.J. Archbold and D.W.B. Somerset, *Spectral synthesis in the multiplier algebra of a $C_0(X)$-algebra.* Quart. J. Math. Oxford **65** (2014), 1–24.

[9] R.J. Archbold and D.W.B. Somerset, *Separation properties in the primitive ideal space of a multiplier algebra.* Israel J. Math. **200** (2014), 389–418.

[10] R.C. Busby, *Double centralizers and extensions of C^*-algebras.* Trans. Amer. Math. Soc. **132** (1968), 79–99.

[11] A. Dow, *The space of minimal prime ideals of $C(\beta\mathbf{N} \setminus \mathbf{N})$ is probably not basically disconnected*, pp. 81–86, General Topology and Applications, (Lecture Notes in Pure and Applied Math., 123), Dekker, NY, 1990.

[12] A. Dow, M. Henriksen, R. Kopperman and J. Vermeer, *The space of minimal prime ideals of $C(X)$ need not be basically disconnected.* Proc. Amer. Math. Soc. **104** (1988), 317–320.

[13] G.A. Elliott, *Derivations of matroid C^*-algebras, II.* Ann. Math.(2) **100** (1974), 407–422.

[14] L. Gillman and M. Jerison, *Rings of Continuous Functions.* Van Nostrand, New Jersey, 1960.

[15] M. Henriksen and M. Jerison, *The space of minimal prime ideals of a commutative ring.* Trans. Amer. Math. Soc. **115** (1965), 110–130.

[16] M. Henriksen and R.G. Woods, *Cozero complemented space; when the space of minimal prime ideals of a $C(X)$ is compact.* Topology and Its Applications **141** (2004), 147–170.

[17] E. Kaniuth, G. Schlichting and K. Taylor, *Minimal primal and Glimm ideal spaces of group C^*-algebras.* J. Funct. Anal. **130** (1995), 45–76.

[18] K. Keimel, *A unified theory of minimal prime ideals.* Acta Math. Acad. Sci. Hungaricae **23** (1972), 51–69.

[19] J. Kist, *Minimal prime ideals in commutative semigroups.* Proc. London. Math. Soc. (3) **13** (1963), 31–50.

[20] D. Kucerovsky and P.W. Ng, *Nonregular ideals in the multiplier algebra of a stable C^*-algebra.* Houston J. Math. **33** (2007), 1117–1130.

[21] H.X. Lin, *Ideals of multiplier algebras of simple AF-algebras.* Proc. Amer. Math. Soc. **104** (1988), 239–244.

[22] M. Mandelker, *Prime z-ideal structure of $C(\mathbf{R})$.* Fund. Math. **63** (1968), 145–166.

[23] J.R. Munkres, *Topology.* 2nd Edition, Prentice-Hall, New Jersey, 1999.

[24] G.K. Pedersen, *C^*-algebras and their Automorphism Groups.* Academic Press, London, 1979.

[25] F. Perera, *Ideal structure of multiplier algebras of simple C^*-algebras with real rank zero.* Can. J. Math. 53 (2001), 592–630.

[26] H.L. Pham, *Uncountable families of prime z-ideals in $C_0(\mathbf{R})$.* Bull. London Math. Soc. **41** (2009), 354–366.

[27] M. Rørdam, *Ideals in the multiplier algebra of a stable C^*-algebra.* J. Operator Th. **25** (1991), 283–298.

[28] D.W.B. Somerset, *Minimal primal ideals in Banach algebras.* Math. Proc. Camb. Phil. Soc. **115** (1994), 39–52.

[29] D.W.B. Somerset, *The local multiplier algebra of a C^*-algebra.* Quart. J. Math. Oxford (2) **47** (1996), 123–132.

[30] D.W.B. Somerset, *Minimal primal ideals in rings and Banach algebras.* J. Pure and Applied Algebra **144** (1999), 67–89.

[31] D.P. Williams, *Crossed Products of C^*-algebras.* American Mathematical Society, Rhode Island, 2007.

R.J. Archbold and D.W.B. Somerset
Institute of Mathematics
University of Aberdeen
Kings College
Aberdeen AB24 3UE
Scotland, UK
e-mail: `r.archbold@abdn.ac.uk`
`dwbsomerset@gmail.com`

Operator Theory:
Advances and Applications, Vol. 250, 31–48

Countable Spectrum, Transfinite Induction and Stability

Wolfgang Arendt

Dedicated to Charles Batty on the occasion of his sixtieth birthday

Abstract. We reconsider the contour argument and proof by transfinite induction of the ABLV-Theorem given in [AB88]. But here we use the method to prove a Tauberian Theorem for Laplace transforms which has the ABVL-Theorem about stability of a semigroup as corollary and also gives quantitative estimates.

It is interesting that considering countable spectrum leads to the same problems Cantor encountered when he tried to prove a uniqueness result for trigonometric series. It led him to invent ordinal numbers and transfinite induction. We explain these connections in the article.

Mathematics Subject Classification (2010). 47D06, 44A10.

Keywords. Semigroups, asymptotic behavior, Laplace transform, Tauberian theorem, countable spectrum, transfinite induction, uniqueness theorem for trigonometric series.

1. Introduction

Frequently, it is worthwhile to revisit a mathematical result with the benefit of several years' hindsight. Things may appear in a different light, different methods might be known. The result I am talking about here is the stability theorem I proved together with Charles Batty 25 years ago, which says the following. Let $(T(t))_{t\geq 0}$ be a bounded C_0-semigroup with generator A. If $\sigma(A) \cap i\mathbb{R}$ is countable and $\sigma_p(A') \cap i\mathbb{R} = \emptyset$ (where $\sigma_p(A')$ denotes the point spectrum of the adjoint), then the semigroup is *stable*; i.e., $\lim_{t\to\infty} T(t)x = 0$ for all $x \in X$.

It was not necessary to wait 25 years for new methods to appear. In fact, exactly at the same time, this stability result was obtained independently by Ljubich and Vu [LV88] at the University of Kharkov in the Soviet Union by completely different methods. For this reason the result is frequently called the ABLV-Theorem. Ljubich and Vu use a quotient method, and later Fourier methods were developed by Esterle, Strouse and Zouakia [ESZ92] (spectral synthesis) and it was Chill

[Chi98] who shed new light onto the old methods of Ingham – 80 years after their first appearance. These different approaches all have their advantages and merits. The proof by Ljubich and Vu is the most functional analytical in nature. Its disadvantage is that it merely works in the context of semigroups and not more generally for Laplace transforms. The advantage of Chill's approach is that it is valid for Laplace transforms (see [Chi98], [ABHN11, Theorem 4.9.7]). We refer to the survey by Chill and Tomilov [CT07] for more information and also to Section 5.5 of [ABHN11].

Still, we want to revisit our proof from 1986, which used two ingredients, a contour argument and transfinite induction. Compared with the other methods, there are two advantages: our proof is completely elementary and it also gives quantitative results (which have grown in importance recently, see Batty [Bat90], Batty and Duykaerts [BD08], Borichev and Tomilov [BT10], Batty, Chill and Tomilov [BCT], [BBT14] as well as Section 4.4 in [ABHN11]).

Concerning elegance and esthetics, the opinions of colleagues are not unanimous. Most people believe that our method is quite technical and even we did not use it in our book [ABHN11] to prove the ABLV-Theorem. Still, we believe that the transfinite induction argument we used in 1986 is quite striking and even elegant. Once the inductive statement is formulated in the right way, its proof is automatic. Our aim in this article is to make this transparent by formulating the technical part in an abstract and easy way (Lemma 3.6). But we also arrange the arguments differently and obtain a new interesting result, namely a (quantitative) Tauberian theorem for Laplace transforms (Theorem 3.1) where an exceptional countable set occurs in the hypothesis. It is this result which we prove by transfinite induction in the present article (in contrast to our original proof [AB88] where the argument by transfinite induction was done on the level of the semigroup). The powerful Mittag-Leffler Theorem, a topological argument in the spirit of Baire's theorem, allows one to pass from the Tauberian theorem to the ABLV-Theorem, see Section 4. Our Tauberian theorem gives also an improvement of a Tauberian theorem for power series by Allan, O'Farrell and Randsford [AOR87] which was motivated by the Katznelson–Tzafriri theorem.

Concerning the contour estimates, they demonstrate the power of Cauchy's Theorem and are most elegant when the spectrum on the imaginary axis is empty. As an appetizer we consider this case in Section 2 emphasizing the quantitative character. In Section 3 we prove the general Tauberian theorem elaborating the use of transfinite induction. It is interesting that Cantor encountered similar problems as we did in the context of countable spectrum when he tried to prove a uniqueness result for trigonometric series where a closed, countable exceptional set has to be mastered. It was this problem which led him to develop set theory, ordinal numbers and transfinite induction. In Section 5 we take the opportunity to present the solution of Cantor's problem by transfinite induction, a striking resemblance to our proof, a resemblance of which we were not aware in 1986.

Cantor must have been aware of the argument, but he never published the end of the proof.

2. Empty spectrum

This section is an introduction to the subject where we consider the simplest case of a complex Tauberian theorem, the Newman-trick for contour integrals and the special case of the ABLV-Theorem where the spectrum on the imaginary axis is empty. The results are contained in [AB88], [AP92] (see also [ABHN11]). Here however, we put them together in a way which makes transparent the quantitative nature of the results and which demonstrates the power of the contour argument in a simple case. The more refined techniques are then presented in Section 3.

We consider a function $f \in L^\infty(\mathbb{R}_+, X)$ where X is a complex Banach space, $\mathbb{R}_+ = [0, \infty)$. By

$$\hat{f}(\lambda) := \int_0^\infty e^{-\lambda t} f(t)\,\mathrm{d}t \qquad (\operatorname{Re}\lambda > 0)$$

we denote the Laplace transform of f. It is a holomorphic function defined on the right half-plane $\mathbb{C}_+$.

If $F(t) := \int_0^t f(s)\,\mathrm{d}s$ converges to F_∞ as $t \to \infty$, then $\lim_{\lambda \to 0} \hat{f}(\lambda) = F_\infty$. This Abelian theorem is easy to see. The converse is false in general: If $\lim_{\lambda \to \infty} \hat{f}(\lambda) = F_\infty$ exists, then $\int_0^t f(s)\,\mathrm{d}s$ need not converge as $t \to \infty$. But if a theorem says that it does under some additional hypothesis then we call it a *Tauberian theorem* and the additional hypothesis a *Tauberian condition*. An interesting Tauberian theorem is the following.

Theorem 2.1 (Newman–Korevaar–Zagier). *Assume that $\hat{f}$ has a holomorphic extension to an open set containing $\overline{\mathbb{C}_+}$. Then*

$$\lim_{t \to \infty} \int_0^t f(s)\,\mathrm{d}s$$

exists.

It follows that $\lim_{t \to \infty} \int_0^t f(s)\,\mathrm{d}s = \hat{f}(0)$ by the remark above. Here the Tauberian condition is that $\hat{f}$ can be extended to a holomorphic function on an open set containing $\overline{\mathbb{C}_+}$. A theorem of this type had already been proved by Ingham [Ing35] in the thirties (see also Korevaar's book [Kor04, p. 135]). But Newman [New80] found an elegant contour argument (which he applied to Dirichlet series), that was used by Korevaar [Kor82] and Zagier [Zag97] for Laplace transforms to give beautiful proofs of the prime number theorem. Here is an estimate, which implies Theorem 2.1 and which shows the simplicity of the argument as well as its quantitative aspect.

We let $\|f\|_\infty := \sup_{t \geq 0} \|f(t)\|$.

Proposition 2.2. *Let $R > 0$. Assume that $\hat{f}$ has a holomorphic extension to a neighborhood of $\mathbb{C}_+ \cup i[-R, R]$. Then*

$$\limsup_{t \to \infty} \left\| \int_0^t f(s) - \hat{f}(0) \right\| \leq \frac{\|f\|_\infty}{R}.$$

Proof. Let $g = \hat{f}$ and for $t > 0$ let

$$g_t(z) = \int_0^t e^{-zs} f(s)\, \mathrm{d}s.$$

Thus g_t is an entire function. Let U be an open, simply connected set containing $i[-R, R] \cup \mathbb{C}_+$. Denote by γ a path going from iR to $-iR$ lying entirely in $U \cap \{z \in \mathbb{C} : \operatorname{Re} z < 0\}$ besides the endpoints.

We apply Cauchy's Theorem to this contour. The introduction of an additional fudge factor under the following integral is the ingenious trick due to Newman.

$$\begin{aligned}\int_0^t f(s)\, \mathrm{d}s - \hat{f}(0) &= g_t(0) - g(0)\\ &= \frac{1}{2\pi i} \int_{\substack{|z|=R\\ \operatorname{Re} z>0}} (g_t(z) - g(z)) e^{tz} \left(1 + \frac{z^2}{R^2}\right) \frac{\mathrm{d}z}{z}\\ &\quad + \frac{1}{2\pi i} \int_\gamma (g_t(z) - g(z)) e^{tz} \left(1 + \frac{z^2}{R^2}\right) \frac{\mathrm{d}z}{z}\\ &=: I_1(t) + I_2(t).\end{aligned}$$

It follows from the Dominated Convergence Theorem that $\lim_{t\to\infty} I_2(t) = 0$.

In order to estimate $I_1(t)$ let $z = Re^{i\theta}$, $|\theta| < \frac{\pi}{2}$, be on the right-hand semicircle. Then on the one hand

$$\begin{aligned}\|(g_t(z) - g(z))e^{tz}\| &= \left\| \int_t^\infty e^{-zs} f(s)\, \mathrm{d}s\, e^{tz} \right\|\\ &\le \|f\|_\infty \int_t^\infty e^{-sR\cos\theta}\, \mathrm{d}s\, e^{tR\cos\theta}\\ &\le \frac{\|f\|_\infty}{R\cos\theta}\end{aligned}$$

and on the other

$$\begin{aligned}|1 + \frac{z^2}{R^2}| &= |1 + e^{i2\theta}| = |e^{-i\theta} + e^{i\theta}|\\ &= 2\cos\theta.\end{aligned}$$

Thus

$$\|I_1(t)\| \le \frac{1}{2\pi} \pi \frac{\|f\|_\infty}{R\cos\theta} 2\cos\theta = \frac{\|f\|_\infty}{R}$$

and the proposition is proved. □

In 1986 when we worked in Oxford on stability of semigroups we knew a version of Theorem 2.1 from an unpublished manuscript by Zagier (cf. [Zag97]). It was easy to apply it to semigroups:

Let $(T(t))_{t\ge0}$ be a C_0-semigroup with generator A. Assume that $\|T(t)\| \le M$ for all $t \ge 0$. For $x \in X$ let $f(t) = T(t)x$. Then $\hat{f}(\lambda) = R(\lambda, A)x$. Now assume that

$\sigma(A) \cap i\mathbb{R} = \emptyset$. Then $\hat{f}(0) = -A^{-1}x$ and the Newman–Korevaar–Zagier Theorem 2.1 implies that

$$\int_0^t f(s)\,\mathrm{d}s = \int_0^t T(s)AA^{-1}x\,\mathrm{d}s$$
$$= T(t)A^{-1}x - A^{-1}x$$

converges to $-A^{-1}x$ as $t \to \infty$.

Hence $T(t)A^{-1}x \to 0$ as $t \to \infty$ for all $x \in X$. Since $\operatorname{rg} A^{-1} = D(A)$ is dense in X and $\|T(t)\| \leq M$ it follows that $\lim_{t\to\infty} T(t)x = 0$ for all $x \in X$; i.e., the semigroup is *stable*. We have proved the following.

Theorem 2.3. *Assume that $(T(t))_{t\geq 0}$ is a bounded C_0-semigroup with generator A. If $\sigma(A) \cap i\mathbb{R} = \emptyset$, then $\lim_{t\to\infty} T(t)x = 0$ for all $x \in X$.*

It is natural to ask what happens if $\sigma(A) \cap i\mathbb{R} \neq \emptyset$. If $i\eta \in \sigma_p(A')$, the point spectrum of the adjoint A' of A, then $T(t)'x' = e^{i\eta t}x'$ for all $t \geq 0$ and some $x' \in X' \setminus \{0\}$. Let $x \in X$ be such that $\langle x', x\rangle = 1$. Then $\langle T(t)x, x'\rangle = e^{i\eta t}$ for all $t \geq 0$ and so the semigroup is definitely not stable. Thus

$$\sigma_p(A') \cap i\mathbb{R} = \emptyset \tag{2.1}$$

is a necessary condition for stability.

By the Hahn–Banach Theorem, condition (2.1) is equivalent to

$$\operatorname{rg}(i\eta - A) \text{ being dense in } X \tag{2.2}$$

where rg stands for the range of the operator. What is special for $x \in \operatorname{rg}(i\eta - A)$? Let $x = (i\eta - A)y$ where $y \in D(A)$, $f(t) = T(t)x$ as before. Then

$$\int_0^t f(s)e^{-i\eta s}\,\mathrm{d}s = y - e^{-i\eta t}T(t)y.$$

Thus

$$\sup_{t\geq 0}\left\|\int_0^t f(s)e^{-i\eta s}\,\mathrm{d}s\right\| < \infty. \tag{2.3}$$

This condition turns out to be useful for proving a Tauberian theorem by the contour method if $i\eta$ is a singular point.

Before discussing this in the next section we point out a generalization of the Tauberian Theorem 2.1. It is not necessary to assume that a holomorphic extension exists, a continuous extension suffices.

Theorem 2.4. *Let $f \in L^\infty(\mathbb{R}_+, X)$, $R > 0$, $F_\infty \in X$. Assume that $\frac{1}{\lambda}(\hat{f}(\lambda) - F_\infty)$ has a continuous extension to $\mathbb{C}_+ \cup i[-R, R]$. Then*

$$\limsup_{t\to\infty}\left\|\int_0^t f(s)\,\mathrm{d}s - F_\infty\right\| \leq \frac{2\|f\|_\infty}{R}. \tag{2.4}$$

This is obtained by a modification of the contour argument above (cf. [AP92, Lemma 5.2], where a more complicated situation is considered). We give the proof of Theorem 2.4 in order to be complete. It is interesting that now, instead of the Dominated Convergence Theorem, we use the Riemann–Lebesgue Theorem for Fourier coefficients. The price is a factor 2 appearing in the estimate (2.4) in contrast to the better estimate given in Proposition 2.2.

Proof. First case: $F_\infty = 0$.

Let $g = \hat{f}$. Thus $\frac{g(z)}{z}$ has a continuous extension to $\mathbb{C}_+ \cup i[-R, R]$. By (a slight extension of) Cauchy's Theorem one has

$$\int_\gamma \frac{g(z)}{z}(1 + \frac{z^2}{R^2})e^{tz}\,\mathrm{d}z + \int_{\substack{|z|=R \\ \operatorname{Re} z>0}} \frac{g(z)}{z}(1 + \frac{z^2}{R^2})e^{tz}\,\mathrm{d}z = 0 \tag{2.5}$$

where γ is the straight line from iR to $-iR$.

For $t > 0$ consider the entire function

$$g_t(z) = \int_0^t e^{-sz} f(s)\,\mathrm{d}s.$$

Thus by (2.5),

$$\begin{aligned}
\int_0^t f(s)\,\mathrm{d}s &= \frac{1}{2\pi i}\int_{|z|=R} g_t(z)\left(1 + \frac{z^2}{R^2}\right)e^{tz}\frac{\mathrm{d}z}{z} \\
&= \frac{1}{2\pi i}\int_{\substack{|z|=R \\ \operatorname{Re} z>0}} (g_t(z) - g(z))\left(1 + \frac{z^2}{R^2}\right)e^{tz}\frac{\mathrm{d}z}{z} \\
&\quad - \frac{1}{2\pi i}\int_\gamma g(z)e^{tz}\left(1 + \frac{z^2}{R^2}\right)\frac{\mathrm{d}z}{z} + \frac{1}{2\pi i}\int_{\substack{|z|=R \\ \operatorname{Re} z<0}} g_t(z)\left(1 + \frac{z^2}{R^2}\right)e^{tz}\frac{\mathrm{d}z}{z} \\
&=: I_1(t) + I_2(t) + I_3(t).
\end{aligned}$$

By the Riemann–Lebesgue Theorem, $\lim_{t\to\infty} I_2(t) = 0$.

One has $\|I_1(t)\| \le \frac{1}{R}\|f\|_\infty$ for all $t \ge 0$ as in Proposition 2.2.

The integral $I_3(t)$ can be estimated in a similar way,

$$\limsup_{t\to\infty} \|I_3(t)\| \le \frac{1}{R}\|f\|_\infty.$$

Thus $\limsup_{t\to\infty} \|\int_0^t f(s)\,\mathrm{d}s\| \le \frac{2}{R}\|f\|_\infty$.

Second case: $F_\infty \in X$ *is arbitrary.*

Let $\varphi\colon [0,\infty) \to \mathbb{R}$ be continuous with compact support satisfying $\int_0^1 \varphi(s)\mathrm{d}s = 1$. Let $f_1(t) := f(t) - \varphi(t)F_\infty$. Then $\hat{f}_1(\lambda) = \hat{f}(\lambda) - \hat{\varphi}(\lambda)F_\infty$, $\hat{\varphi}(0) = 1$.

Thus

$$\frac{\hat{f}_1(\lambda)}{\lambda} = \frac{\hat{f}(\lambda) - F_\infty}{\lambda} - \frac{\hat{\varphi}(\lambda) - \hat{\varphi}(0)}{\lambda}F_\infty$$

has a continuous extension to $\mathbb{C}_+ \cup i[-R, R]$.

By the first case

$$\limsup_{t\to\infty}\left\|\int_0^t f(s)\,\mathrm{d}s - F_\infty\right\| \le \frac{2}{R}\|f\|_\infty. \qquad \square$$

Applying the preceding results to $f(\cdot+s)$ instead of f one even obtains the estimate

$$\limsup_{t\to\infty}\left\|\int_0^t f(s)\,\mathrm{d}s - F_\infty\right\| \le \frac{1}{R}\limsup_{t\to\infty}\|f(t)\|, \tag{2.6}$$

in Proposition 2.2 and the estimate

$$\limsup_{t\to\infty}\left\|\int_0^t f(s)\,\mathrm{d}s - F_\infty\right\| \le \frac{2}{R}\limsup_{t\to\infty}\|f(t)\|, \tag{2.7}$$

instead of (2.4), cf. [AP92, Remark 9.2].

We finish this section by going back to the origins of Tauberian theory. Given a bounded sequence $(a_n)_{n\in\mathbb{N}_0}$ consider the power series $p(z)=\sum_{n=0}^\infty a_n z^n$ which is defined for $|z|<1$. If $\sum_{n=0}^\infty a_n =: b_\infty$ exists then Abel showed in 1826 that

$$\lim_{x\nearrow 1} p(x) = b_\infty. \tag{2.8}$$

The converse is not true in general. Additional assumptions are needed. It was Tauber who proved in 1897 that the series converges if in addition to (2.8) one assumes that

$$\lim_{n\to\infty} n\|a_n\| = 0, \tag{2.9}$$

thus proving the first "Tauberian theorem".

Littlewood showed in 1911 that the "Tauberian condition" (2.9) can be relaxed to $\sup_{n\in\mathbb{N}} n\|a_n\| < \infty$.

Another Tauberian theorem is due to Riesz. It is actually a consequence of the estimate (2.6).

Theorem 2.5 (Riesz). *Let $a_n \in X$, $n\in\mathbb{N}_0$, such that $\lim_{n\to\infty}\|a_n\|=0$. Assume that the power series*

$$p(z)=\sum_{n=0}^\infty a_n z^n \qquad (|z|<1)$$

has a holomorphic extension to an open neighborhood of 1. *Then*

$$\sum_{n=0}^\infty a_n = p(1).$$

Proof. Let $f(t)=a_n$ if $t\in[n,n+1)$. Then $f\in L^\infty(\mathbb{R}_+,X)$ and

$$\hat{f}(\lambda)=\frac{1-e^{-\lambda}}{\lambda}p(e^{-\lambda}) \qquad (\operatorname{Re}\lambda>0).$$

Thus $\hat{f}$ has a holomorphic extension to a disc of radius $2R$ centered at 0 for some $R > 0$ and $\hat{f} = p(1)$. Thus (2.6) implies that

$$\lim_{n\to\infty} \left\| \sum_{k=0}^{n} a_k - p(1) \right\| \le \frac{1}{R} \limsup_{n\to\infty} \|a_n\| = 0. \qquad \square$$

This proof is taken from [AP92, Remark 3.4].

3. A complex Tauberian theorem

Let $f \in L^\infty(\mathbb{R}_+, X)$. The Laplace transform $\hat{f}$ of f is a holomorphic function from the open right-hand half-plane $\mathbb{C}_+$ into X.

If $F(t) := \int_0^t f(s)\,ds$ converges to F_∞ as $t \to \infty$, then $\lim_{\lambda\searrow 0} \hat{f}(\lambda) = F_\infty$ by an easy Abelian theorem. As in Section 2, we want to prove the converse. But here we will relax the assumptions considerably. As in Theorem 2.4 we will estimate

$$\limsup_{t\to\infty} \|F(t) - F_\infty\|.$$

The Tauberian condition is expressed in terms of the boundary behavior of $\hat{f}(\lambda)$ as $\lambda \to i\eta$.

Theorem 3.1. *Let $R > 0$, $F_\infty \in X$. Let $E \subset (-R, 0) \cup (0, R)$ be closed and countable. Assume that*

(a) $\frac{\hat{f}(\lambda)-F_\infty}{\lambda}$ *has a continuous extension to $\mathbb{C}_+ \cup i([-R, R] \setminus E)$ and that*

(b) $\sup_{t\ge 0} \| \int_0^t e^{-i\eta s} f(s)\,ds\| < \infty$ *for all $\eta \in E$.*

Then

$$\limsup_{t\to\infty} \left\| \int_0^t f(s)\,ds - F_\infty \right\| \le \frac{2\|f\|_\infty}{R}. \tag{3.1}$$

Our point is that the bound in (b) may depend on $\eta \in E$. In the case where it is independent, (3.1) can be proved purely by a contour argument (see [AP92, Theorem 3.1], and [AB88, Theorem 4.1] for a slightly more special case). Since we do not assume a uniform bound in (b) our proof needs an argument of transfinite induction. It is similar to the transfinite induction argument given for the proof of the ABLV-Theorem in [AB88] and, we think, an interesting mathematical argument in its own right. Here it is.

As in the proof of Proposition 2.4 we may assume that $F_\infty = 0$ which we do now. We assume the hypotheses of Theorem 3.1.

For the proof we denote by J_n the set of all $(\eta_1, \ldots, \eta_n, \epsilon_1, \ldots, \epsilon_n)$ with $\eta_j \in E$, $\epsilon_j > 0$ such that the intervals $(\eta_j - \epsilon_j, \eta_j + \epsilon_j)$ are pairwise disjoint and $0 \notin \bigcup_{j=1}^n [\eta_j - \epsilon_j, \eta_j + \epsilon_j] \subset (-R, R)$. Given a set $K \subset (-R, 0) \cup (0, R)$ we say that $(\eta_1, \ldots, \eta_n, \epsilon_1, \ldots, \epsilon_n)$ *covers* K, if $K \subset \bigcup_{j=1}^n (\eta_j - \epsilon_j, \eta_j + \epsilon_j)$. With the help of these notations the basic estimate can be formulated as follows.

Lemma 3.2 (basic estimate). *There exist functions $a_n, b_n \colon J_n \to (0, \infty)$ satisfying for all $n, p \in \mathbb{N}$*

(a) $a_n(\eta_1, \ldots, \eta_n, \epsilon_1, \ldots, \epsilon_n) \to 1$ *as* $(\epsilon_1, \ldots, \epsilon_n) \to 0$ *in* $\mathbb{R}^n$
(b) $a_{n+p}(\eta_1, \ldots, \eta_{n+p}, \epsilon_1, \ldots, \epsilon_{n+p}) \to a_n(\eta_1, \ldots, \eta_n, \epsilon_1, \ldots, \epsilon_n)$
as $(\epsilon_{n+1}, \ldots, \epsilon_{n+p}) \to 0$ *in* $\mathbb{R}^p$
(c) $b_n(\eta_1, \ldots, \eta_n, \epsilon_1, \ldots, \epsilon_n) \to 0$ *as* $(\epsilon_1, \ldots, \epsilon_n) \to 0$ *in* $\mathbb{R}^n$
(d) $b_{n+p}(\eta_1, \ldots, \eta_{n+p}, \epsilon_1, \ldots, \epsilon_{n+p}) \to b_n(\eta_1, \ldots, \eta_n, \epsilon_1, \ldots, \epsilon_n)$
as $(\epsilon_{n+1}, \ldots, \epsilon_{n+p}) \to 0$ *in* $\mathbb{R}^p$

such that the following holds:

If E *is covered by* $(\eta_1, \ldots, \eta_n, \epsilon_1, \ldots, \epsilon_n) \in J_n$ *then*

$$\limsup_{t\to\infty} \left\| \int_0^t f(s)\,\mathrm{d}s \right\| \\ \le \frac{2\|f\|_\infty}{R} a_n(\eta_1, \ldots, \eta_n, \epsilon_1, \ldots, \epsilon_n) + b_n(\eta_1, \ldots, \eta_n, \epsilon_1, \ldots, \epsilon_n).$$

These estimates are obtained by changing the contour in the proof of Theorem 2.4 on the straight line $i[-R, R]$ by introducing semicircles of radius ϵ_j, $j = 1, \ldots, n$. For the proof we refer to [AP92, Lemma 5.2] (which is a modification of [AB88, Lemma 3.1]).

Remark. The reader might better understand the proof of [AP92, Lemma 5.2] by replacing "and 0 =" on line 11, 12 of p. 430 by a "−" and lifting the term to the end of line 10. Also the signs "+" on lines 15 and 17 should be replaced by a "−".

Proof of Theorem 3.1. Let $E_0 := E \cap [-R, R]$. Thus E_0 is compact and countable. Given an ordinal α we define E_α inductively by

$$E_\alpha = \begin{cases} \text{the set of all cluster points} \\ \text{of } E_\alpha, \text{ if } \alpha \text{ is a successor ordinal;} \\ \bigcap\limits_{\beta<\alpha} E_\beta, \text{ if } \alpha \text{ is a limit ordinal.} \end{cases}$$

We will prove that the following statement $S(\alpha)$ holds for all ordinals α:

$S(\alpha):$ if $E_\alpha = \emptyset$, then

$$\limsup_{t\to\infty} \left\| \int_0^t f(s)\,\mathrm{d}s \right\| \le \frac{2}{R}\|f\|_\infty \tag{3.2}$$

and if E_α is covered by $(\eta_1, \ldots, \eta_n, \epsilon_1, \ldots, \epsilon_n) \in J_n$ then

$$\limsup_{t\to\infty} \left\| \int_0^t f(s)\,\mathrm{d}s \right\| \\ \le \frac{2\|f\|_\infty}{R} a_n(\eta_1, \ldots, \eta_n, \epsilon_1, \ldots, \epsilon_n) + b_n(\eta_1, \ldots, \eta_n, \epsilon_1, \ldots, \epsilon_n). \tag{3.3}$$

Once this statement is proved the proof of the theorem is completed as follows:

Since E_α is compact and countable it possesses an isolated point whenever E_α is non empty. Thus $E_{\alpha+1} \subsetneq E_\alpha$ whenever $E_\alpha \neq \emptyset$. This implies that $E_{\alpha_0} = \emptyset$ for some α_0 (see Proposition 5.2). Hence statement $S(\alpha_0)$ gives the result.

Now we prove that $S(\alpha)$ holds for all ordinals α.

$\alpha = 0$: If $E_0 = \emptyset$, this is Theorem 2.4. If $E_0 \neq \emptyset$, then this follows immediately from the basic estimate Lemma 3.2.

$\alpha > 0$: Assume that $S(\beta)$ holds for all $\beta < \alpha$. We show that $S(\alpha)$ holds.

First case: α is a limit ordinal.

Then $E_\alpha = \bigcap_{\beta<\alpha} E_\beta$.

If $E_\alpha = \emptyset$, then there exists $\beta < \alpha$ such that $E_\beta = \emptyset$. The inductive hypothesis implies that (3.2) holds.

If E_α is covered by $(\eta_1, \dots, \eta_n, \epsilon_1, \dots, \epsilon_n) \in J_n$, then there exists $\beta < \alpha$ such that E_β is covered by $(\eta_1, \dots, \eta_n, \epsilon_1, \dots, \epsilon_n)$. Thus (3.3) follows by the inductive hypothesis.

Second case: α is a successor ordinal.

If $E_\alpha = \emptyset$, then $E_{\alpha-1}$ is finite, say $E_{\alpha-1} = \{\eta_1, \dots, \eta_n\}$. Choose $\epsilon_j > 0$ so small that $(\eta_1, \dots, \eta_n, \epsilon_1, \dots, \epsilon_n) \in J_n$. Then it follows by the inductive hypothesis that

$$\begin{aligned}&\limsup_{t\to\infty} \left\| \int_0^t f(s)\,\mathrm{d}s \right\| \\ &\quad \le \frac{2\|f\|_\infty}{R} a_n(\eta_1, \dots, \eta_n, \epsilon_1, \dots, \epsilon_n) + b_n(\eta_1, \dots, \eta_n, \epsilon_1, \dots, \epsilon_n).\end{aligned}$$

Letting $(\epsilon_1, \dots, \epsilon_n) \to 0$ in $\mathbb{R}^n$ yields (3.2).

If E_α is covered by $(\eta_1, \dots, \eta_n, \epsilon_1, \dots, \epsilon_n) \in J_n$, then

$$E_{\alpha-1} \setminus \bigcup_{j=1}^{n} (\eta_j - \epsilon_j, \eta_j + \epsilon_j)$$

is finite, consisting of, say, $\{\eta_{n+1}, \dots, \eta_{n+p}\}$. Choose $\epsilon_j > 0$, $j = n+1, \dots, n+p$, so small that $(\eta_1, \dots, \eta_{n+p}, \epsilon_1, \dots, \epsilon_{n+p}) \in J_{n+p}$. Then by the inductive hypothesis

$$\begin{aligned}&\limsup_{t\to\infty} \left\| \int_0^t f(s)\,\mathrm{d}s \right\| \\ &\le \frac{2\|f\|_\infty}{R} a_{n+p}(\eta_1, \dots, \eta_{n+p}, \epsilon_1, \dots, \epsilon_{n+p}) + b_{n+p}(\eta_1, \dots, \eta_{n+p}, \epsilon_1, \dots, \epsilon_{n+p}).\end{aligned}$$

Sending $(\epsilon_{n+1}, \dots, \epsilon_{n+p})$ to 0 in $\mathbb{R}^p$ gives the desired estimate (3.3).

Thus $S(\alpha)$ is proved. □

Remark 3.3. Applying Theorem 3.1 to the function $f(\cdot + s)$ instead of f one obtains the estimate

$$\limsup_{t\to\infty} \left\| \int_0^t f(s)\,\mathrm{d}s - F_\infty \right\| \le \frac{2}{R} \limsup_{t\to\infty} \|f(t)\| \tag{3.4}$$

which improves (3.1), cf. [AP92, Remark 3.2].

The following is an immediate consequence of Theorem 3.1.

Corollary 3.4. *Let $E \subset \mathbb{R}$ be closed and countable such that $0 \notin E$. Let $F_\infty \in X$. Assume that*

(a) $\frac{\hat{f}(\lambda)-F_\infty}{\lambda}$ *has a continuous extension to* $\overline{\mathbb{C}_+} \setminus iE$ *and that*

(b) $\sup_{t\geq 0} \| \int_0^t e^{-i\eta s} f(s)\, \mathrm{d}s\| < \infty$ *for all* $\eta \in E$.

Then $\lim_{t\to\infty} \int_0^t f(s)\, \mathrm{d}s = F_\infty$.

Remark. In the case where (a) is replaced by the stronger hypothesis

(a′) $\hat{f}$ has a holomorphic extension to an open set containing $\overline{\mathbb{C}_+} \setminus iE$,

Corollary 3.4 is proved by Batty, van Nerven and Räbiger [BvNR98, Theorem 4.3], where a slightly weaker hypothesis than (b) is considered (cf. [BvNR98, Remark 2]. The methods are very different though.

We may transform Corollary 3.4 into a Tauberian theorem of a different type where convergence of $f(t)$ as $t \to \infty$ is the conclusion.

Corollary 3.5. *Let $f \in L^\infty(0,\infty;X)$ and $f_\infty \in X$. Assume that*

$$\hat{f}(\lambda) - \frac{f_\infty}{\lambda} \qquad (\operatorname{Re}\lambda > 0)$$

has a continuous extension to $\overline{\mathbb{C}_+}\setminus iE$ where $E \subset \mathbb{R}$ is closed, countable and $0 \notin E$. Assume that

$$\sup_{t\geq 0} \left\| \int_0^t e^{-i\eta s} f(s)\, \mathrm{d}s \right\| < \infty \text{ for all } \eta \in E. \tag{3.5}$$

Then $\lim_{t\to\infty} \frac{1}{\delta} \int_t^{\delta+t} f(s)\, \mathrm{d}s = f_\infty$ *for all* $\delta > 0$.

If f is uniformly continuous on $[\tau,\infty)$ for some $\tau > 0$, then

$$\lim_{t\to\infty} f(t) = f_\infty.$$

This follows from Corollary 3.4 as [AP92, Theorem 3.5] follows from [AP92, Theorem 3.1]. We refer to Chill [Chi98], [ABHN11, Theorem 4.9.7] for a different approach via Fourier Analysis to such Tauberian theorems.

Finally, we apply Corollary 3.5 to power series. By $\mathbb{D} := \{z \in \mathbb{C} : \ |z| < 1\}$ we denote the unit disc and by $\Gamma := \{z \in \mathbb{C} : \ |z| = 1\}$ the unit circle.

Corollary 3.6. *Let $a_n \in X$, $\sup_{n\in\mathbb{N}_0} \|a_n\| < \infty$, $p(z) = \sum_{n=0}^\infty a_n z^n$ for $z \in \mathbb{D}$. Let F be a closed, countable subset of Γ such that*

(a) *p has a continuous extension to $\overline{\mathbb{D}} \setminus F$ and*

(b) $\sup_{N\in\mathbb{N}} \| \sum_{n=0}^N a_n z^n\| < \infty$ *for all* $z \in \Gamma$.

Then $\lim_{n\to\infty} a_n = 0$

It follows from Riesz' Theorem 2.5 that

$$\sum_{n=0}^\infty a_n z^n = p(z)$$

for all $z \in \overline{\mathbb{D}} \setminus F$.

Proof. Replacing a_n by $w^{-n}a_n$ where $w \in \Gamma \setminus F$ we may assume that $1 \notin F$.

Let $f(t) = a_n$ if $t \in [n, n+1)$. Then $f \in L^\infty(\mathbb{R}_+, X)$ and

$$\hat{f}(\lambda) = \frac{1-e^{-\lambda}}{\lambda} p(e^{-\lambda}) \qquad (\operatorname{Re}\lambda > 0)$$

has a continuous extension to $\overline{C_+} \setminus iE$ where $E := \{\eta \in \mathbb{R} : \; e^{-i\eta} \in F\}$.

Moreover, for $t \in [n, n+1)$ we have

$$\int_0^t e^{-i\eta s} f(s)\, \mathrm{d}s = \sum_{m=0}^{n} a_m e^{-i\eta m} \frac{1-e^{-i\eta}}{i\eta} + a_n e^{-i\eta n} \frac{1-e^{-i\eta(t-n)}}{i\eta}.$$

Thus (3.4) is satisfied. It follows from Corollary 3.5 that

$$a_n = \int_n^{n+1} f(s)\, \mathrm{d}s \to 0 \text{ as } n \to \infty. \qquad \square$$

Corollary 3.6 improves Theorem 4 of Allan–O'Farrell and Ransford [AOR87] where a uniform bound is assumed in (b). This is possible by the transfinite induction argument we gave.

4. The ABLV-Theorem

It is easy to deduce the ABLV-Theorem from the Tauberian Theorem 3.1, if one has at hand a powerful topological tool, the Mittag-Leffler Theorem, see below.

Theorem 4.1 (ABLV-Theorem). *Let $(T(t))_{t\geq 0}$ be a C_0-semigroup on a Banach space X and A its generator. Assume*

(a) $\|T(t)\| \leq M \qquad (t \geq 0)$
(b) $\sigma(A) \cap i\mathbb{R}$ *is countable*
(c) $\sigma_p(A') \cap i\mathbb{R} = \emptyset$.

Then $\lim_{t\to\infty} T(t)x = 0$ for all $x \in X$.

Proof. Replacing A by $A - i\eta$ if necessary, we may assume that $0 \notin \sigma(A)$. By hypothesis the space $\operatorname{rg}(A - i\eta)$ is dense in X. The Mittag-Leffler Theorem implies that even

$$Y := \bigcap_{\eta \in E} \operatorname{rg}(A - i\eta) \tag{4.1}$$

is dense in X (see Proposition 4.2 below). Let $x \in Y$ and consider $f(t) = T(t)x$. Then $\hat{f}(\lambda) = R(\lambda, A)x$. Hence $\hat{f}(0) = -A^{-1}x$. Thus f satisfies the hypothesis (a) of Corollary 3.4 with $F_\infty = -A^{-1}x$, $iE = \sigma(A) \cap i\mathbb{R}$.

We prove that condition (b) of Corollary 3.4 is satisfied. Let $\eta \in E$. Since $x \in Y$, there exists $z \in D(A)$ such that $(A - i\eta)z = x$. Hence

$$\int_0^t e^{-i\eta s} T(s)x\, \mathrm{d}s = e^{-i\eta t} T(t)z - z.$$

It follows that $\| \int_0^t e^{-i\eta s} T(s)x \, ds\| \leq (M+1)\|z\|$ for all $t \geq 0$. Now we can deduce from Corollary 3.4 that

$$T(t)A^{-1}x = \int_0^t AT(s)A^{-1}x \, ds + A^{-1}x = \int_0^t f(s)x \, ds - \hat{f}(0)$$

converges to 0 as $t \to \infty$.

Since Y is dense in X it follows that $\lim_{t\to\infty} T(t)A^{-1}x = 0$ for all $x \in X$. Since $\operatorname{rg} A^{-1} = D(A)$ is dense in X, we finally deduce that $\lim_{t\to\infty} T(t)x = 0$ for all $x \in X$. □

We remark that conditions (a) and (c) of Theorem 4.1 are necessary as we have already noted in Section 2. Condition (b) is not necessary, but optimal by [AB88, Example 2.5 (a)]. Up to date, there seems to be no complete characterization of stability. However, Tomilov [Tom01] obtained an interesting result which (in view of the Mittag-Leffler Theorem) generalizes the ABLV-Theorem. A bounded C_0-semigroup with generator A is stable whenever the space

$$\bigcap_{i\eta \in i\mathbb{R}\cap\sigma(A)} \operatorname{rg}(i\eta - A)$$

is dense in X. But also this condition is not necessary, see [CT03].

Next we state the Mittag-Leffler Theorem and prove the density of the space Y in (4.1). A very good reference for the Mittag-Leffler Theorem is the article by Esterle [Est84]. In Esterle–Strouse–Zouakia [ESZ92] the Mittag-Leffler Theorem is used in a similar way as we do here.

Let M_n be a complete metric space and $\Theta_n \colon M_{n+1} \to M_n$ a continuous map with dense image ($n \in \mathbb{N}$). A sequence $(y_n)_{n\in\mathbb{N}}$ with $y_n \in M_n$ is called *projective* if $\Theta_n y_{n+1} = y_n$ for all $n \in \mathbb{N}$. In that case we call y_1 the *final point* of the projective sequence.

Let $F := \{y_1 \in M_1 : \text{ there exists a sequence } (y_n)_{n\in\mathbb{N}} \text{ such that } \Theta_n y_{n+1} = y_n \text{ for all } n \in \mathbb{N}\}$ be the set of all final points of a projective sequence.

Theorem 4.2 (Mittag-Leffler). *The set F is dense in M_1.*

As application we prove the density of Y.

Proposition 4.3. *Let A be an operator on X such that $\rho(A) \neq \emptyset$. Let $\lambda_n \in \mathbb{C}$ such that* $\operatorname{rg}(A - \lambda_n)$ *is dense in X.*

Then

$$Y := \bigcap_{n\in\mathbb{N}} \operatorname{rg}(A - \lambda_n)$$

is dense in X.

Proof. We may assume that $0 \in \rho(A)$ replacing A by $A - \mu$ and λ_n by $\lambda_n - \mu$ otherwise. Then $D(A^n)$ is a Banach space for the norm $\|x\|_n := \|A^n x\|$.

If for $\lambda \in \mathbb{C}$, $(A - \lambda)D(A)$ is dense in X, then $(A - \lambda)D(A^{n+1})$ is dense in $D(A^n)$ with respect to $\|\cdot\|_n$. In fact, let $x \in D(A^n)$. Then there exists $y_k \in D(A)$

such that $(\lambda - A)y_k \to A^n x =: y$ in X as $k \to \infty$. Thus $A^{-n}y_k \in D(A^n)$ and $(\lambda - A)A^{-n}y_k \to x$ in $D(A^n)$ as $k \to \infty$.

Let $\Theta_n \colon D(A^{n+1}) \to D(A^n)$ be defined by $\Theta_n x = (A - \lambda_n)x$. Then Θ_n is continuous with dense image. Thus, by the Mittag-Leffler Theorem the set F of all final points is dense in X. Since $F \subset Y$, the claim follows. □

The discrete stability theorem [AB88, Theorem 5.1] is a direct consequence of Corollary 3.5. Recall that $\Gamma = \{z \in \mathbb{C} : |z| = 1\}$.

Theorem 4.4. *Let $T \in \mathcal{L}(X)$ such that $M := \sup_{n\in\mathbb{N}} \|T^n\| < \infty$. Assume that*
(a) *$\sigma(T) \cap \Gamma$ is countable and*
(b) *$\sigma_p(T') \cap \Gamma = \emptyset$.*

Then $\lim_{n\to\infty} T^n x = 0$ for all $x \in X$.

Proof. By hypothesis $\operatorname{rg}(\lambda - T)$ is dense for all $\lambda \in \sigma(T) \cap \Gamma$. It follows from the Mittag-Leffler Theorem that

$$Y := \bigcap_{\lambda\in\sigma(T)\cap\Gamma} \operatorname{rg}(\lambda - T)$$

is dense in X.

Let $p(z) := \sum_{n=0}^{\infty} z^n T^n = (I - zT)^{-1}$, $|z| < 1$, $F := \{\overline{z} : z \in \sigma(T) \cap \Gamma\}$. Then p has a continuous extension to $\overline{\mathbb{D}} \setminus F$. Let $y \in Y$. Then for $z \in F$ there exists $x \in X$ such that $y = (I - zT)x$. Thus

$$\left\| \sum_{n=0}^{N} z^n T^n y \right\| = \|x - z^{N+1}T^{N+1}x\| \le (M+1)\|x\|.$$

It follows from Corollary 3.6 that $\lim_{n\to\infty} T^n y = 0$. Since Y is dense in X, the proof is finished. □

5. Cantor's work on trigonometric series

In this section we describe Cantor's work on the uniqueness property for trigonometric series.

Definition 5.1. A subset of $E \subset [0, 2\pi]$ is called a *set of uniqueness* if the following holds: If $c_k \in \mathbb{C}$, $k \in \mathbb{Z}$, and

$$\lim_{N\to\infty} \sum_{k=-N}^{N} c_k e^{ikt} = 0$$

for all $t \in [0, 2\pi] \setminus E$, then $c_k = 0$ for all $k \in \mathbb{Z}$.

In 1870 Cantor showed that finite sets are sets of uniqueness. Two years later he extended his result to a certain class of countable closed sets. More precisely, in his article [Can72] he considered a closed countable subset E of $[0, 2\pi]$ and defined

$$E' := \{t \in E : t \text{ is a cluster point of } E\}.$$

Today we sometimes call E' the *derivative* of E *in the sense of Cantor*. Successively, one may define $E^{(1)} := E'$, $E^{(n)} = (E^{(n-1)})'$. Now Cantor was able to show that a countable closed set E is a set of uniqueness whenever $E^{(n)} = \emptyset$ for some $n \in \mathbb{N}$.

It was tempting to conjecture that all countable, closed sets are sets of uniqueness. Cantor's method of proof had conceptual limits, though. And in fact, Cantor turned away from the problem to develop set theory, ordinal numbers and transfinite induction. Having transfinite induction as a tool one can indeed show that closed, countable sets are sets of uniqueness.

We want to explain this in more detail in order to point out the similarity to the proof of the ABLV-Theorem in [AB88] (and also our proof of the Tauberian theorem in Section 3).

Let $E_0 \subset \mathbb{R}$ be a compact, countable set. Given an ordinal $\alpha > 0$ we define E_α inductively as follows:

$$E_\alpha := \begin{cases} (E_{\alpha-1})' & \text{if } \alpha \text{ is a successor ordinal} \\ \bigcap_{\beta<\alpha} E_\beta & \text{if } \alpha \text{ is a limit ordinal.} \end{cases}$$

Then E_α is compact, countable and $E_{\alpha_2} \subset E_{\alpha_1}$ if $\alpha_1 \leq \alpha_2$.

Denote by ω_1 the first uncountable ordinal.

Proposition 5.2. *There exists $\alpha_0 < \omega_1$ such that $E_{\alpha_0} = \emptyset$.*

Proof. Let $E_0 = \{q_n : n \in \mathbb{N}\}$ with $q_n \neq q_m$ for $n \neq m$. Assume that $E_\alpha \neq \emptyset$ for all $\alpha < \omega_1$.

It follows from Baire's Theorem that E_α has isolated points for all $\alpha < \omega_1$. Thus $E_\alpha \setminus E_{\alpha+1} \neq \emptyset$ for all $\alpha < \omega_1$.

Define $f\colon [0, \omega_1) \to \mathbb{N}$ by

$$f(\alpha) = \min\{n \in \mathbb{N} : \ q_n \in E_\alpha \setminus E_{\alpha+1}\}.$$

Then f is injective. In fact, assume that $\alpha < \beta$ and $f(\alpha) = f(\beta)$. Then $\alpha + 1 \leq \beta$. Thus $E_\beta \subset E_{\alpha+1}$. Then $q_{f(\beta)} = q_{f(\alpha)} \notin E_{\alpha+1}$. Hence $q_{f(\beta)} \notin E_\beta$, a contradiction.

Thus f is injective. Since $[0, \omega_1)$ is uncountable, this is not possible. $\square$

Remark. To each $\alpha < \omega_1$ there exists a compact, countable set $E_0 \subset \mathbb{R}$ such that $E_\alpha \neq \emptyset$ and $E_{\alpha+1} = \emptyset$.

Now we discuss the uniqueness problem. We consider a trigonometric series

$$\sum_{n=-\infty}^{\infty} c_n e^{int} \tag{5.1}$$

where $c_n \in \mathbb{C}$, $n \in \mathbb{Z}$. We assume that $E \subset [0, 2\pi]$ is closed and countable and that the series converges to 0 (i.e., $\lim_{N\to\infty} \sum_{i=-N}^{N} c_n e^{int} = 0$) for all $t \in [0, 2\pi] \setminus E$. We want to prove that $c_n = 0$ for all $n \in \mathbb{Z}$.

The assumption implies that $\lim_{|n|\to\infty} c_n = 0$. It was an idea of Riemann to consider the formal second anti-derivative

$$F(t) := c_0 \frac{t^2}{2} - \sum_{\substack{n\neq 0\\ n\in\mathbb{Z}}} \frac{c_n}{n^2} e^{int}$$

which defines a continuous function $F\colon \mathbb{R} \to \mathbb{C}$. The following two lemmas are due to Riemann, Schwarz and Cantor. For the proof and further details we refer to the beautiful lecture notes of Kechris [Kec92] which inspired our presentation.

Lemma 5.3. *If the series* (5.1) *converges to* 0 *on an interval* $(a,b) \subset \mathbb{R}$, *then* F *is affine on* $[a,b]$.

See [Kec92, 2.2 and 3.3].

Lemma 5.4. *Let* $t_1 < t_2 < t_3$ *be real numbers. If* F *is affine on* (t_1,t_2) *and on* (t_2,t_3), *then* F *is affine on* $[t_1,t_3]$.

This follows from [Kec92, 2.6]. The next lemma is easy to prove.

Lemma 5.5. *If* F *is affine on* $[-2\pi, 2\pi]$, *then* $c_n = 0$ *for all* $n \in \mathbb{Z}$.

Proof. We have

$$c_0 \frac{t^2}{2} - \sum_{n\neq 0} \frac{c_n}{n^2} e^{int} = at + b$$

on $[-2\pi, 2\pi]$. Evaluating at $t = \pm\pi$ and subtracting yields $a = 0$. Evaluating at $t = 0$ and $t = 2\pi$ yields $c_0 = 0$. Thus

$$\sum_{\substack{n\neq 0\\ n\in\mathbb{Z}}} \frac{c_n}{n^2} e^{int} = b.$$

Since this series converges uniformly, it follows that $\frac{c_n}{n^2} = \frac{1}{2\pi}\int_0^{2\pi} be^{-int}\,dt = 0$ for all $n \neq 0$. □

Now we can prove that closed, countable sets are sets of uniqueness.

Theorem 5.6. *Assume that* (5.1) *converges to* 0 *outside a closed, countable set* $E \subset [0, 2\pi]$. *Then* $c_n = 0$ *for all* $n \in \mathbb{Z}$.

Proof. We may assume that the series converges to 0 at 0 (translations of sets of uniqueness are sets of uniqueness). Since the series is 2π-periodic we find a compact, countable subset E_0 of $(-2\pi, 2\pi)$ such that the series converges to 0 on $[-2\pi, 2\pi] \setminus E_0$.

For an ordinal α define the set E_α as before and consider the statement

$$S(\alpha): \text{ if } [a,b] \subset [-2\pi, 2\pi] \text{ such that } E_\alpha \cap [a,b] = \emptyset \text{ then } F \text{ is affine on } [a,b].$$

We will prove this statement by transfinite induction.

$\alpha = 0$: This follows from Lemma 5.3.

Let $\alpha > 0$ be an ordinal such that $S(\beta)$ holds for all $\beta < \alpha$. We have to prove $S(\alpha)$.

First case: α is a limit ordinal.

Then $E_\alpha = \bigcap_{\beta<\alpha} E_\beta$. Let $E_\alpha \cap [a,b] = \emptyset$. Then there exists $\beta < \alpha$ such that $E_\beta \cap [a,b] = \emptyset$. Thus the claim follows from the inductive hypothesis.

Second case: α has a predecessor.

Let $E_\alpha \cap [a,b] = \emptyset$. Since E_α is the set of all limit points of $E_{\alpha-1}$, it follows that $E_{\alpha-1} \cap [a,b]$ is finite consisting of, say, $\{t_1, \ldots, t_n\}$ with $a \le t_1 < t_2 < \cdots < t_n < b$. Then $E_{\alpha-1} \cap (t_{i-1}, t_i) = \emptyset$. It follows from the inductive hypothesis that F is affine on each strict subinterval of (t_{i-1}, t_i). Since F is continuous, F is affine on $[t_{i-1}, t_i]$. Now Lemma 5.4 implies that F is affine on $[a,b]$.

Thus $S(\alpha)$ is true for all ordinals α. There exists an ordinal α_0 such that $E(\alpha_0) = \emptyset$. Hence F is affine on $[-2\pi, 2\pi]$. It follows from Lemma 5.5 that $c_n = 0$ for all $n \in \mathbb{Z}$. □

It is strange that after his development of set theory and ordinals Cantor never came back to his original problem. It was Lebesgue [Leb03] who gave a proof of Theorem 5.6. Today it is known that all countable sets are sets of uniqueness. Moreover a measurable set of positive Lebesgue measure is not a set of uniqueness. But there exist closed, non-empty sets without isolated points which are sets of uniqueness. We refer to [Kec92] for much more information on this subject.

References

[AB88] W. Arendt and C.J.K. Batty, *Tauberian theorems and stability of one-parameter semigroups*, Trans. Amer. Math. Soc. **306** (1988), no. 2, 837–852.

[ABHN11] W. Arendt, C.J.K. Batty, M. Hieber, and F. Neubrander, *Vector-valued Laplace Transforms and Cauchy Problems*, Springer, 2011.

[AOR87] G.R. Allan, A.G. O'Farrell, and T.J. Ransford, *A Tauberian theorem arising in operator theory*, Bull. London Math. Soc. **19** (1987), no. 6, 537–545.

[AP92] W. Arendt and J. Prüss, *Vector-valued Tauberian theorems and asymptotic behavior of linear Volterra equations*, SIAM J. Math. Anal. **23** (1992), no. 2, 412–448.

[Bat90] C.J.K. Batty, *Tauberian theorems for the Laplace–Stieltjes transform*, Trans. Amer. Math. Soc. **322** (1990), no. 2, 783–804.

[BBT14] C.J.K. Batty, A. Borichev, and Yu. Tomilov, *L^p-tauberian theorems and L^p-rates for energy decay*, arXiv:1403.6084 (2014).

[BCT] C.J.K. Batty, R. Chill, and Yu. Tomilov, *Fine scales of decay of operator semigroups*, J. Eur. Math. Soc., to appear.

[BD08] C.J.K. Batty and T. Duyckaerts, *Non-uniform stability for bounded semi-groups on Banach spaces*, J. Evol. Equ. **8** (2008), no. 4, 765–780.

[BT10] A. Borichev and Yu. Tomilov, *Optimal polynomial decay of functions and operator semigroups*, Math. Ann. **347** (2010), no. 2, 455–478.

[BvNR98] C.J.K. Batty, J. van Neerven, and F. Räbiger, *Tauberian theorems and stability of solutions of the Cauchy problem*, Trans. Amer. Math. Soc. **350** (1998), no. 5, 2087–2103.

[Can72] G. Cantor, *Ueber die Ausdehnung eines Satzes aus der Theorie der trigonometrischen Reihen*, Math. Ann. **5** (1872), no. 1, 123–132.

[Chi98] R. Chill, *Tauberian theorems for vector-valued Fourier and Laplace transforms*, Studia Math. **128** (1998), no. 1, 55–69.

[CT03] R. Chill and Yu. Tomilov, *Stability of C_0-semigroups and geometry of Banach spaces*, Math. Proc. Cambridge Philos. Soc. **135** (2003), no. 3, 493–511.

[CT07] ______, *Stability of operator semigroups: ideas and results*, Perspectives in operator theory, Banach Center Publ., vol. 75, Polish Acad. Sci., Warsaw, 2007, pp. 71–109.

[Est84] J. Esterle, *Mittag-Leffler methods in the theory of Banach algebras and a new approach to Michael's problem*, Proceedings of the conference on Banach algebras and several complex variables (New Haven, Conn., 1983) (Providence, RI), Contemp. Math., vol. 32, Amer. Math. Soc., 1984, pp. 107–129.

[ESZ92] J. Esterle, E. Strouse, and F. Zouakia, *Stabilité asymptotique de certains semigroupes d'opérateurs et idéaux primaires de $L^1(\mathbf{R}^+)$*, J. Operator Theory **28** (1992), no. 2, 203–227.

[Ing35] A.E. Ingham, *On Wiener's Method in Tauberian Theorems*, Proc. London Math. Soc. **S2-38** (1935), no. 1, 458.

[Kec92] A.S. Kechris, *Set theory and uniqueness for trigonometric series.* Lecture Notes, CalTech (1992), `http://www.math.caltech.edu/~kechris/papers/uniqueness.pdf`.

[Kor82] J. Korevaar, *On Newman's quick way to the prime number theorem*, Math. Intelligencer **4** (1982), no. 3, 108–115.

[Kor04] ______, *Tauberian Theory. a Century of Developments*, Springer, 2004.

[Leb03] H. Lebesgue, *Sur les séries trigonométriques*, Ann. Sci. École Norm. Sup. (3) **20** (1903), 453–485.

[LV88] Yu.I. Lyubich and Phóng Vũ, *Asymptotic stability of linear differential equations in Banach spaces*, Studia Math. **88** (1988), no. 1, 37–42.

[New80] D.J. Newman, *Simple analytic proof of the prime number theorem*, Amer. Math. Monthly **87** (1980), no. 9, 693–696.

[Tom01] Yu. Tomilov, *A resolvent approach to stability of operator semigroups*, J. Operator Theory **46** (2001), no. 1, 63–98.

[Zag97] D. Zagier, *Newman's short proof of the prime number theorem*, Amer. Math. Monthly **104** (1997), no. 8, 705–708.

Wolfgang Arendt
Institute of Applied Analysis
University of Ulm
D-89069 Ulm, Germany
e-mail: `wolfgang.arendt@uni-ulm.de`

Operator Theory:
Advances and Applications, Vol. 250, 49–66

Maximal Regularity in Interpolation Spaces for Second-order Cauchy Problems

Charles J.K. Batty, Ralph Chill and Sachi Srivastava

Abstract. We study maximal regularity in interpolation spaces for the sum of three closed linear operators on a Banach space, and we apply the abstract results to obtain Besov and Hölder maximal regularity for complete second-order Cauchy problems under natural parabolicity assumptions. We discuss applications to partial differential equations.

Mathematics Subject Classification (2010). Primary 34G10, 47D09, 35B65; Secondary 35L10, 35K10, 35K90.

Keywords. Maximal regularity, interpolation spaces, second order, Cauchy problems.

1. Introduction

We study maximal regularity results in certain time interpolation spaces for the second-order Cauchy problem

$$\begin{aligned} \ddot{u} + B\dot{u} + Au &= f \quad \text{in } [0,T], \\ u(0) = u_0, \ \dot{u}(0) &= u_1. \end{aligned} \tag{1.1}$$

Here A and B are closed linear operators defined on a complex Banach space X with domains D_A and D_B respectively. By a maximal regularity result we mean a result which asserts that for every f in a certain function space $\mathbb{E} \subseteq L^1(0,T;X)$ and homogeneous initial values $u_0 = u_1 = 0$ the problem (1.1) admits a unique strong solution u satisfying $\ddot{u}$, $B\dot{u}$, $Au \in \mathbb{E}$. In particular, the three terms on the left-hand side of (1.1) have the same regularity as the given right-hand side.

The notion of L^p maximal regularity (that is, $\mathbb{E} = L^p(0,T;X)$) and, closely connected with it, of maximal regularity in rearrangement invariant Banach function spaces for this abstract second-order Cauchy problem has been studied in Chill & Srivastava [13, 14] and Chill & Król [12]. See also Arendt et al. [1], Batty, Chill & Srivastava [4], Cannarsa, Da Prato & Zolésio [10], Dautray & J.-L. Lions [18, Chapter XVIII, Section 5], Favini [19] and Yakoubov [36] for generalisations to non-autonomous problems, Bu & Fang [8, 9], Keyantuo & Lizama [26] for second-order problems with periodic boundary conditions, Fernández, Lizama & Poblete [22] for third-order problems, Zacher [37] for Volterra equations and Bu [6, 7], Keyantuo & Lizama [27], Lizama & Poblete [28] for fractional-order problems with periodic boundary conditions.

The notion of maximal regularity for the second-order problem generalises in a natural way the notion of maximal regularity for the first-order problem

$$\begin{aligned} \dot{u} + Au &= f \quad \text{in } [0,T], \\ u(0) &= 0, \end{aligned} \tag{1.2}$$

which in turn goes back to the notion of maximal regularity of the sum of two closed linear operators on a Banach space by Da Prato & Grisvard [16]; see also the monograph by Lunardi [30, Theorem 3.18]. In particular, Da Prato and Grisvard showed that if A, D are two sectorial operators with domains D_A and D_D and sectoriality angles φ_A and φ_D, respectively, and if $\varphi_A + \varphi_D < \pi$, then for every x in a real interpolation space between X and D_D (or D_A) there is a unique solution y of the operator equation $Ay + Dy = x$ lying in the space $\mathsf{D}_A \cap \mathsf{D}_D$ with Ay and Dy belonging to the same interpolation space. This result was then applied to the first-order Cauchy problem (1.2) by taking D to be the differentiation operator on, for example, $L^p(0,T;X)$ or $C([0,T];X)$. The real interpolation spaces then include the Besov spaces and the Hölder spaces, respectively, that is, one obtains Besov or Hölder maximal regularity, the latter also being called optimal regularity in the literature.

Analogously, maximal regularity of the sum of three operators corresponds with the definition of maximal regularity of the second-order problem (1.1) as mentioned above. In this paper we follow the idea of [16] to prove a maximal regularity result on interpolation spaces for the second-order abstract problem (1.1) for $u_0 = u_1 = 0$. Here no assumptions are needed on the space X, and the operators A and B are not required to satisfy assumptions about functional calculus or R-boundedness. Moreover, we extend the result in such a way that in principle we can also treat initial value problems of the general form (1.1), although here the identification of the associated trace spaces remains an open problem. The conclusions provide Besov or Hölder maximal regularity of second-order Cauchy problems; see also Favini et al. [11, 20, 21], Mezeghrani [31] for similar results for elliptic problems with inhomogeneous Dirichlet boundary conditions or Bu [5, 6], Bu & Fang [9], Keyantuo & Lizama [26, 25], Poblete [33, 34] for second-order problems with periodic boundary conditions or on the line.

The paper is organised as follows. Section 2 is of a preliminary nature where we recall relevant definitions and facts. In Section 3 an abstract result concerning maximal regularity of certain sums of three closed operators is proven. As an application we obtain our main result on the maximal regularity for the second-order Cauchy problem in Section 4. Section 5 is devoted to the initial value problem, while examples of applications can be found in Section 6.

2. Preliminaries

Let X be a complex Banach space. Whenever (D, D_D) is a closed linear operator on X, $\theta \in (0,1)$, $1 \leq p \leq \infty$, we denote by

$$\mathsf{D}_D(\theta,p) := (X, \mathsf{D}_D)_{\theta,p} \quad \text{and}$$
$$\mathsf{D}_D(\theta) := (X, \mathsf{D}_D)_{\theta}$$

the real interpolation spaces between X and D_D (the latter space is equipped with the graph norm), as defined by the K-method or the trace method. Recall that $\mathsf{D}_D(\theta)$ is a closed subspace of $\mathsf{D}_D(\theta,\infty)$.

The operator D is called *sectorial of angle* $\varphi \in (0,\pi)$ if $\sigma(D) \subseteq \overline{\Sigma_\varphi}$, where

$$\Sigma_\varphi = \{\lambda \in \mathbb{C} : \lambda \neq 0,\ |\arg \lambda| < \varphi\}$$

is the open sector of opening angle φ, and for every $\varphi' \in (\varphi, \pi)$ one has

$$\sup_{\lambda \notin \overline{\Sigma_{\varphi'}}} \|\lambda R(\lambda, D)\| < \infty.$$

For a sectorial operator D, for $\theta \in (0,1)$ and $1 \leq p \leq \infty$ we have the equalities

$$\mathsf{D}_D(\theta,p) = \left\{x \in X : \|t^\theta D(t+D)^{-1}x\| \in L^p\left(0,\infty;\frac{\mathrm{d}t}{t}\right)\right\} \quad \text{and}$$
$$\mathsf{D}_D(\theta) = \left\{x \in \mathsf{D}_D(\theta,\infty) : \lim_{t\to\infty} t^\theta D(t+D)^{-1}x = 0\right\},$$

and

$$\|x\|_{\theta,p} := \|x\| + \|t^\theta D(t+D)^{-1}x\|_{L^p(0,\infty;\frac{\mathrm{d}t}{t})}$$

is an equivalent norm on the interpolation space $\mathsf{D}_D(\theta,p)$ [29, Propositions 2.2.2 and 2.2.6]. Note that if $\varphi' \in (\varphi,\pi)$, then $-e^{\pm i\varphi'}D$ is sectorial and its domain is D_D with an equivalent graph norm. Hence

$$\mathsf{D}_D(\theta,p) = \left\{x \in X : \|t^\theta DR(te^{\pm i\varphi'},D)^{-1}x\| \in L^p\left(0,\infty;\frac{\mathrm{d}t}{t}\right)\right\}. \tag{2.1}$$

Example 2.1. Let X be a Banach space, and fix $1 \leq p < \infty$. If $D_{\max}$ is the differentiation operator on $L^p(0,T;X)$ with maximal domain, that is,

$$\mathsf{D}_{D_{\max}} := W^{1,p}(0,T;X),$$
$$D_{\max}u := \dot{u},$$

then we have

$$\mathsf{D}_{D_{\max}}(\theta,q) = B^\theta_{pq}(0,T;X) \qquad (\theta \in (0,1),\ 1 \leq q < \infty),$$

and, in particular,

$$\mathsf{D}_{D_{\max}}(\theta,p) = W^{\theta,p}(0,T;X) \qquad (\theta \in (0,1));$$

compare with [30, Exercise 6, p. 18] or [35, Theorem, p. 204]. Here, B^θ_{pq} and $W^{\theta,p}$ are the Besov spaces and fractional-order Sobolev spaces defined respectively by

$$B^\theta_{pq}(0,T;X) := \left\{ u \in L^p(0,T;X) : \int_0^T \left(\int_0^T \frac{\|u(t)-u(s)\|^p}{|t-s|^{\theta p + p/q}} \, ds \right)^{q/p} dt < \infty \right\},$$

$$W^{\theta,p}(0,T;X) := \left\{ u \in L^p(0,T;X) : \int_0^T \int_0^T \frac{\|u(t)-u(s)\|^p}{|t-s|^{\theta p+1}} \, ds \, dt < \infty \right\}.$$

If D is the restriction of the differentiation operator $D_{\max}$ on $L^p(0,T;X)$ to the domain

$$\mathsf{D}_D = \mathring{W}^{1,p}(0,T;X) := \{u \in W^{1,p}(0,T;X) : u(0) = 0\},$$

then

$$\mathsf{D}_D(\theta,q) = \begin{cases} \mathring{B}^\theta_{pq}(0,T;X) & \text{if } \theta > \frac{1}{p}, \\ B^\theta_{pq}(0,T;X) & \text{if } \theta < \frac{1}{p}; \end{cases}$$

compare with [35, Theorem, p. 210], where actually two-sided homogeneous boundary conditions were considered. Here $\mathring{B}^\theta_{pq}$ is the space of all functions $u \in B^\theta_{pq}$ with trace $u(0) = 0$; note that the trace is well defined whenever $\theta > \frac{1}{p}$, since then the Besov space B^s_{pq} is embedded into the space of continuous functions. On the other hand,

$$\mathsf{D}_{D_{\max}}(\theta,q) = \mathsf{D}_D(\theta,q) \text{ whenever } \theta < \frac{1}{p}.$$

Recall that the operator D is sectorial of angle $\frac{\pi}{2}$, while $D_{\max}$ is not sectorial (in fact, $\sigma(D_{\max}) = \mathbb{C}$).

Example 2.2. If $D_{\max}$ is the differentiation operator on $C([0,T];X)$ with maximal domain, that is,

$$\mathsf{D}_{D_{\max}} := C^1([0,T];X),$$
$$D_{\max} u := \dot{u},$$

then we have

$$\mathsf{D}_{D_{\max}}(\theta,\infty) = C^\theta([0,T];X) \text{ and}$$
$$\mathsf{D}_{D_{\max}}(\theta) = h^\theta([0,T];X);$$

compare with [30, Example 1.9 and Exercise 5, p. 18]. Here, C^θ and h^θ are the Hölder and little Hölder spaces defined respectively by

$$C^\theta([0,T];X) := \left\{ u \in C([0,T];X) : \sup_{\substack{t,s \in [0,T] \\ s \neq t}} \frac{\|u(t)-u(s)\|}{|t-s|^\theta} < \infty \right\} \text{ and}$$

$$h^\theta([0,T];X) := \left\{ u \in C^\theta([0,T];X) : \lim_{|t-s| \to 0} \frac{\|u(t)-u(s)\|}{|t-s|^\theta} = 0 \right\}.$$

If D is the restriction of the differentiation operator $D_{\max}$ to the domain

$$\mathsf{D}_D := \mathring{C}^1([0,1];X) := \{u \in C^1([0,T];X) : u(0) = 0\}$$

(so that D is no longer densely defined), then

$$\begin{aligned} \mathsf{D}_D(\theta,\infty) &= \mathring{C}^\theta([0,T];X) \\ &:= \{u \in C^\theta([0,T];X) : u(0) = 0\} \text{ and} \\ \mathsf{D}_D(\theta) &= \mathring{h}^\theta([0,T];X) \\ &:= \{u \in h^\theta([0,T];X) : u(0) = 0\}. \end{aligned}$$

Also in this example, D is sectorial of angle $\frac{\pi}{2}$ while $\sigma(D_{\max}) = \mathbb{C}$.

3. An abstract theorem

Our main abstract result is a maximal regularity result for the sum of three closed, linear, commuting operators. We say that an operator A *commutes* with an invertible operator D if $AD \subseteq DA$, or equivalently if $D^{-1}A \subseteq AD^{-1}$. Here the compositions such as AD have their natural domains.

Theorem 3.1. *Let A, B and D be three closed, linear operators on a Banach space X with both A, B commuting with D. Assume that*

(a) *the operator D is invertible and sectorial of angle $\varphi_1 \in (0,\pi)$,*
(b) *there exists $\varphi_2 \in (\varphi_1,\pi)$, such that $H(\lambda) := (\lambda^2 + \lambda B + A)^{-1}$ exists in $\mathcal{L}(X)$ for every $\lambda \in \Sigma_{\varphi_2}$,*
(c) *H is holomorphic from Σ_{φ_2} to $\mathcal{L}(X)$, and*
(d) *the functions*

$$\begin{aligned} &\lambda \mapsto \lambda^2 H(\lambda), \\ &\lambda \mapsto \lambda B H(\lambda), \text{ and} \\ &\lambda \mapsto A H(\lambda) \end{aligned}$$

are uniformly bounded in Σ_{φ_2} with values in $\mathcal{L}(X)$.

Then, for every $\theta \in (0,1)$ and every $1 \le p \le \infty$ the operator $L_{\theta,p}$ on $\mathsf{D}_D(\theta,p)$ given by

$$\begin{aligned} \mathsf{D}_{L_{\theta,p}} &:= \{x \in \mathsf{D}_{D^2} \cap \mathsf{D}_{BD} \cap \mathsf{D}_A : D^2x,\ BDx,\ Ax \in \mathsf{D}_D(\theta,p)\}, \\ L_{\theta,p}x &:= D^2x + BDx + Ax, \end{aligned}$$

is closed and boundedly invertible. More precisely, if we define

$$Sx := \frac{1}{2\pi i}\int_\Gamma R(\lambda,D)H(\lambda)x \,\mathrm{d}\lambda, \quad x \in X, \tag{3.1}$$

where Γ is a path connecting $e^{i\varphi'}\infty$ with $e^{-i\varphi'}\infty$ for some $\varphi' \in (\varphi_1,\varphi_2)$ and surrounding $\sigma(D)$, then $S \in \mathcal{L}(X)$, S is a left-inverse of $D^2 + BD + A$ in X, and *for every* $x \in \mathsf{D}_D(\theta, p)$ *one has*

$$Sx \in \mathsf{D}_{D^2} \cap \mathsf{D}_{BD} \cap \mathsf{D}_A,$$
$$D^2Sx,\ BDSx,\ ASx \in \mathsf{D}_D(\theta, p) \text{ and}$$
$$L_{\theta,p}Sx = (D^2 + BD + A)Sx = x,$$

that is, S *restricted to* $\mathsf{D}_D(\theta, p)$ *is the bounded inverse of* $L_{\theta,p}$*. A similar result holds for* $\mathsf{D}_D(\theta)$ *instead of* $\mathsf{D}_D(\theta, p)$*.*

Proof. It follows from the assumptions, more precisely from the estimates on $R(\lambda, D)$ and H, that the integral in (3.1) converges absolutely and that S is thus a bounded operator on X. Note that since $0 \notin \sigma(D)$, the path Γ may be chosen so that $0 \notin \Gamma$ and lying in the sector Σ_{φ_2}. Since A and B commute with D, the bounded operators $H(\lambda)$ and $R(\lambda, D)$ commute with each other.

Let us first prove that S is a left-inverse of $D^2 + BD + A$. By definition of S, for every $x \in \mathsf{D}_{D^2} \cap \mathsf{D}_{BD} \cap \mathsf{D}_A$ we can calculate, using Lebesgue's dominated convergence theorem,

$$\begin{aligned}
&S(D^2 + BD + A)x \\
&= \frac{1}{2\pi i}\int_\Gamma R(\lambda, D)H(\lambda)(D^2 + BD + A)x \,\mathrm{d}\lambda \\
&= \lim_{s\to+\infty} \frac{1}{2\pi i}\int_\Gamma \frac{s}{s+\lambda} R(\lambda, D)H(\lambda)(D^2 - \lambda^2 + B(D-\lambda) + \lambda^2 + \lambda B + A)x \,\mathrm{d}\lambda \\
&= -\lim_{s\to+\infty} \frac{1}{2\pi i}\int_\Gamma \frac{s}{s+\lambda} H(\lambda)(D + \lambda + B)x \,\mathrm{d}\lambda + \lim_{s\to+\infty} \frac{1}{2\pi i}\int_\Gamma \frac{s}{s+\lambda} R(\lambda, D)x \,\mathrm{d}\lambda \\
&= x.
\end{aligned}$$

In the last step we have used the identities

$$\lim_{s\to+\infty} \frac{1}{2\pi i}\int_\Gamma \frac{s}{s+\lambda} R(\lambda, D)x \,\mathrm{d}\lambda = x$$

and

$$\frac{1}{2\pi i}\int_\Gamma \frac{s}{s+\lambda} H(\lambda)(D + \lambda + B)x \,\mathrm{d}\lambda = 0 \text{ for every } s > 0,$$

which follow from a simple application of Cauchy's residue theorem, remembering that $x \in \mathsf{D}_{D^2} \cap \mathsf{D}_{BD}$ and the estimate on the function H in assumption (d); compare also with [29, Proposition 2.1.4 (i)].

Now, fix $\theta \in (0, 1)$ and $1 \le p \le \infty$. We show that S is a right inverse of $L_{\theta,p}$ in $\mathsf{D}_D(\theta, p)$. By definition of S, for every $t > 0$ and every $x \in X$ we have, by the resolvent identity,

$$\begin{aligned}
(t + D)^{-1}Sx &= \frac{1}{2\pi i}\int_\Gamma \frac{1}{t+\lambda}(t + D)^{-1}H(\lambda)x \,\mathrm{d}\lambda \\
&\quad + \frac{1}{2\pi i}\int_\Gamma \frac{1}{t+\lambda} R(\lambda, D)H(\lambda)x \,\mathrm{d}\lambda
\end{aligned}$$

$$= \frac{1}{2\pi i}\int_\Gamma \frac{1}{t+\lambda} R(\lambda, D)H(\lambda)x \, d\lambda$$

$$= \frac{1}{2\pi i}\int_\Gamma \frac{1}{t+\lambda} H(\lambda)R(\lambda, D)x \, d\lambda. \tag{3.2}$$

For every $t > 0$ and every $x \in \mathsf{D}_D(\theta, p)$, we have

$$g_1(r) := \frac{(re^{\pm i\varphi'})^{1-\theta}}{t + re^{\pm i\varphi'}} \in L^q\left(0, \infty; \frac{dr}{r}\right), \qquad 1 \le q \le \infty,$$

$$g_2(r) := \|r^\theta R(re^{\pm i\varphi'}, D)x\| \in L^p\left(0, \infty; \frac{dr}{r}\right),$$

by (2.1), and $\|AH(re^{\pm i\varphi'})\|$ is bounded by assumption (d). By Hölder's inequality, the integral

$$\frac{1}{2\pi i}\int_\Gamma \frac{1}{t+\lambda} AH(\lambda)\, DR(\lambda, D)x \, d\lambda = \frac{1}{2\pi i}\int_\Gamma \frac{\lambda^{1-\theta}}{t+\lambda} AH(\lambda)\, \lambda^\theta DR(\lambda, D)x \, \frac{d\lambda}{\lambda}$$

converges absolutely. Since A is closed, we conclude from this and (3.2) that, for every $x \in \mathsf{D}_D(\theta, p)$, $y := D(t+D)^{-1}Sx \in \mathsf{D}_A$. Since A commutes with D, $Sx = tD^{-1}y + y \in \mathsf{D}_A$ and

$$D(t+D)^{-1}ASx = AD(t+D)^{-1}Sx = \frac{1}{2\pi i}\int_\Gamma \frac{1}{t+\lambda} AH(\lambda)DR(\lambda, D)x \, d\lambda.$$

Let

$$g_3(r) = \frac{r^\theta}{|\cos\varphi'| + r} \in L^1\left(o, \infty; \frac{dr}{r}\right).$$

By assumption (d), we can estimate

$$\|t^\theta D(t+D)^{-1}ASx\| \le \frac{C}{\pi}\int_0^\infty g_3(t/r)g_2(r)\, \frac{dr}{r},$$

It follows from Young's inequality (applied to the multiplicative group $(0, \infty)$ with the Haar measure $\frac{1}{t}dt$) that

$$\|t^\theta D(t+D)^{-1}ASx\| \in L^p\left(0, \infty; \frac{dt}{t}\right)$$

as well. This proves that $ASx \in D_D(\theta, p)$.

Similarly, we deduce that for every $x \in \mathsf{D}_D(\theta, p)$ one has $Sx \in \mathsf{D}_{BD}$ and

$$\begin{aligned} D(t+D)^{-1}BDSx &= \lim_{s\to+\infty} \frac{1}{2\pi i}\int_\Gamma \frac{s}{s+\lambda}\frac{1}{t+\lambda} BDH(\lambda)\, DR(\lambda, D)x \, d\lambda \\ &= \lim_{s\to+\infty} \frac{1}{2\pi i}\int_\Gamma \frac{s}{s+\lambda}\frac{1}{t+\lambda} \lambda BH(\lambda)\, DR(\lambda, D)x \, d\lambda \\ &\quad - \lim_{s\to+\infty} \frac{1}{2\pi i}\int_\Gamma \frac{s}{s+\lambda}\frac{1}{t+\lambda} BDH(\lambda)x \, d\lambda \\ &= \int_\Gamma \frac{1}{t+\lambda} \lambda BH(\lambda)\, DR(\lambda, D)x \, d\lambda. \end{aligned}$$

This computation is also justified by (3.2) (for the first equality), by Cauchy's theorem and since the integral on the right-hand side converges absolutely for every $x \in \mathsf{D}_D(\theta,p)$, using again assumption (d). One can proceed similarly as above and one obtains $BDSx \in \mathsf{D}_D(\theta,p)$. Similar arguments prove that $Sx \in \mathsf{D}_{D^2}$ and $D^2Sx \in \mathsf{D}_D(\theta,p)$. Indeed, by Cauchy's theorem,

$$\begin{aligned}
D(t+D)^{-1}D^2Sx &= \lim_{s\to+\infty} \frac{1}{2\pi i}\int_\Gamma \frac{s}{s+\lambda}\frac{1}{t+\lambda} D^2H(\lambda)DR(\lambda,D)x \,\mathrm{d}\lambda \\
&= \lim_{s\to+\infty} \frac{1}{2\pi i}\int_\Gamma \frac{s}{s+\lambda}\frac{1}{t+\lambda} D^2H(\lambda)x \,\mathrm{d}\lambda \\
&\quad + \lim_{s\to+\infty} \frac{1}{2\pi i}\int_\Gamma \frac{s}{s+\lambda}\frac{1}{t+\lambda} \lambda DH(\lambda)x \,\mathrm{d}\lambda \\
&\quad - \lim_{s\to+\infty} \frac{1}{2\pi i}\int_\Gamma \frac{s}{s+\lambda}\frac{1}{t+\lambda} \lambda^2 DH(\lambda)R(\lambda,D)x \,\mathrm{d}\lambda \\
&= \frac{1}{2\pi i}\int_\Gamma \frac{1}{t+\lambda}\lambda^2 H(\lambda)\, DR(\lambda,D)x \,\mathrm{d}\lambda.
\end{aligned}$$

Assumption (d) allows us to proceed as in the previous case.

From what we have proved above, it follows that the operator S leaves $\mathsf{D}_D(\theta,p)$ invariant. By the closed graph theorem, the restriction of S to $\mathsf{D}_D(\theta,p)$ is bounded. Moreover, the above equalities show that $SL_{\theta,p}x = x$ for all $x \in \mathsf{D}_{L_{\theta,p}}$ and $L_{\theta,p}Sy = y$ for all $y \in \mathsf{D}_D(\theta,p)$. Thus $L_{\theta,p} : \mathsf{D}_{L_{\theta,p}} \to \mathsf{D}_D(\theta,p)$ is boundedly invertible (and necessarily closed) with $L_{\theta,p}^{-1} = S|_{\mathsf{D}_D(\theta,p)}$. □

4. Maximal regularity of the second-order Cauchy problem in interpolation spaces

In this section we consider the second-order Cauchy problem with homogeneous initial data:

$$\begin{aligned}
\ddot u + B\dot u + Au &= f \quad \text{in } [0,T], \\
u(0) = \dot u(0) &= 0.
\end{aligned} \tag{4.1}$$

Theorem 4.1. *Let A and B be two closed linear operators on a Banach space X. Assume that*

(b) *there exists $\varphi \in (\frac{\pi}{2},\pi)$, such that $H(\lambda) := (\lambda^2+\lambda B+A)^{-1}$ exists in $\mathcal{L}(X)$ for every $\lambda \in \Sigma_\varphi$,*
(c) *H is holomorphic from Σ_φ to $\mathcal{L}(X)$, and*
(d) *the functions*

$$\begin{aligned}
&\lambda \mapsto \lambda^2\, H(\lambda), \\
&\lambda \mapsto \lambda\, B\, H(\lambda), \ \ \textit{and} \\
&\lambda \mapsto A\, H(\lambda)
\end{aligned}$$

are uniformly bounded in Σ_φ with values in $\mathcal{L}(X)$.

Then the problem (4.1) *has Besov and Hölder regularity in the following sense: for every* $f \in B^\theta_{pq}(0,T;X), \theta < \frac{1}{p}$ *(resp.* $f \in \mathring{B}^\theta_{pq}(0,T;X)$ *if* $\theta > \frac{1}{p}$, $f \in \mathring{C}^\theta([0,T];X)$, $f \in \mathring{h}^\theta([0,T];X)$) *the problem* (4.1) *admits a unique solution* u *satisfying*

$$u,\ \dot{u},\ \ddot{u},\ B\dot{u},\ Au \in B^\theta_{pq}(0,T;X)$$
$$(\textit{resp.}\ \in \mathring{B}^\theta_{pq}(0,T;X),\ \mathring{C}^\theta([0,T];X),\ \mathring{h}^\theta([0,T];X)).$$

Proof. Let the operators $\bar{A}$, $\bar{B}$ and D defined on the space $L^p(0,T;X)$ (resp. $C([0,T];X)$) be given as follows:

$$(\bar{A}u)(t) := Au(t),$$
$$(\bar{B}u)(t) := Bu(t),$$
$$(Du)(t) := \dot{u}(t).$$

Here the multiplication operators $\bar{A}$ and $\bar{B}$ have their natural domains, that is,

$$\mathsf{D}_{\bar{A}} = L^p(0,T;\mathsf{D}_A)$$

and similarly for $\mathsf{D}_{\bar{B}}$, and

$$\mathsf{D}_D := \mathring{W}^{1,p}(0,T;X) \qquad (\text{resp. } \mathsf{D}_D := \mathring{C}^1([0,T];X)\ );$$

compare with Examples 2.1 and 2.2. Recall that D is sectorial of angle $\frac{\pi}{2}$. Applying Theorem 3.1 to the above operators and noting that (compare with Examples 2.1 and 2.2)

$$\begin{aligned} \mathsf{D}_D(\theta,p) &= B^\theta_{pq}(0,T;X) \quad \text{if } \theta < \tfrac{1}{p},\\ (\text{resp. } \mathsf{D}_D(\theta,p) &= \mathring{B}^\theta_{pq}(0,T;X) \quad \text{if } \theta > \tfrac{1}{p},\\ \mathsf{D}_D(\theta,p) &= \mathring{C}^\theta([0,T];X),\\ \mathsf{D}_D(\theta) &= \mathring{h}^\theta([0,T];X)\ ), \end{aligned}$$

we obtain the required maximal regularity. □

The fractional power A^ε in the following corollary is defined by using any standard functional calculus for sectorial operators; see, for example, [23].

Corollary 4.2. *Consider the abstract second-order Cauchy problem*

$$\begin{aligned} \ddot{u}(t) + \alpha A^\varepsilon \dot{u}(t) + Au(t) &= f(t), \quad t \in [0,T],\\ u(0) = \dot{u}(0) &= 0, \end{aligned} \tag{4.2}$$

where A *is a sectorial operator of angle* $\varphi \in (0,\pi)$ *on a Banach space* X, *and* $\varepsilon \in \{\frac{1}{2}, 1\}$, $\alpha > 0$. *Assume that one of the following conditions holds:*

(a) $\varepsilon = \frac{1}{2}$, $\alpha \geq 2$, *and* $\varphi \in (0,\pi)$.
(b) $\varepsilon = \frac{1}{2}$, $\alpha \in (0,2)$, *and* $\varphi \in (0, \pi - 2\arctan\frac{\sqrt{4-\alpha^2}}{\alpha})$.
(c) $\varepsilon = 1$, $\alpha > 0$ *and* $\varphi \in (0,\frac{\pi}{2})$.

Then the problem (4.2) *has Besov and Hölder maximal regularity in the sense of Theorem* 4.1 *above.*

Proof. This follows directly from Theorem 4.1. The assumptions (b), (c) and (d) from Theorem 4.1 are easy to verify in the case $\varepsilon = 1$. If $\varepsilon = \frac{1}{2}$, then one factorizes $\lambda^2 + \alpha A^{\frac{1}{2}} + A = (c_1 + A^{\frac{1}{2}})(c_2 + A^{\frac{1}{2}})$ similarly as in the proof of [13, Lemma 4.1], and uses the fact that $A^{\frac{1}{2}}$ is sectorial of angle $\frac{\varphi}{2}$. □

There is an important difference between Corollary 4.2 and [13, Theorem 4.1], which is cited in the proof above and which asserts L^p-maximal regularity of the problem (4.2). Compared to [13, Theorem 4.1], Corollary 4.2 contains no further assumptions on the Banach space X and the operator A. In particular, X need not be a UMD space and A need not have a bounded RH^∞-functional calculus. The above result applies in general Banach spaces and for general sectorial operators. However, the conclusion of Corollary 4.2 is not L^p-maximal regularity, but rather Besov and Hölder maximal regularity.

5. The initial value problem

In this section we solve the abstract second-order Cauchy problem with initial values in certain trace spaces. Before turning to the Cauchy problem, however, we formulate an abstract theorem in the spirit of Theorem 3.1.

Theorem 5.1. *Take the assumptions of Theorem 3.1, fix $\theta \in (0,1)$, $1 \leq p \leq \infty$, and let $L_{\theta,p}$ be the operator on $\mathsf{D}_D(\theta,p)$ as defined in Theorem 3.1. Let $\hat{D}$ be a closed extension of D, and let $\hat{L}_{\theta,p}$ be the operator on $\mathsf{D}_{\hat{D}}(\theta,p)$ given by*

$$\begin{aligned} \mathsf{D}_{\hat{L}_{\theta,p}} &:= \{x \in \mathsf{D}_{\hat{D}^2} \cap \mathsf{D}_{B\hat{D}} \cap \mathsf{D}_A : \hat{D}^2 x,\, B\hat{D}x,\, Ax \in \mathsf{D}_{\hat{D}}(\theta,p)\}, \\ \hat{L}_{\theta,p}x &:= \hat{D}^2 x + B\hat{D}x + Ax. \end{aligned}$$

Then, for every $f \in \mathsf{D}_{\hat{D}}(\theta,p)$ and every $x_0 \in \mathsf{D}_{\hat{L}_{\theta,p}}$ satisfying the compatibility condition $f - \hat{L}_{\theta,p}x_0 \in \mathsf{D}_D(\theta,p)$ there exists a unique solution $x \in \mathsf{D}_{\hat{L}_{\theta,p}}$ of the problem

$$\begin{aligned} \hat{D}^2 x + B\hat{D}x + Ax &= f, \\ x - x_0 &\in \mathsf{D}_{L_{\theta,p}}. \end{aligned} \tag{5.1}$$

Proof. Uniqueness follows from the injectivity of the operator $L_{\theta,p}$ (Theorem 3.1): in fact, if x_1, $x_2 \in \mathsf{D}_{\hat{L}_{\theta,p}}$ are two solutions of (5.1), then $\hat{L}_{\theta,p}(x_1 - x_2) = 0$. On the other hand, $x_1 - x_2 = (x_1 - x_0) - (x_2 - x_0) \in \mathsf{D}_{L_{\theta,p}}$, so that $\hat{L}_{\theta,p}(x_1 - x_2) = L_{\theta,p}(x_1 - x_2)$ since $\hat{L}_{\theta,p}$ is an extension of $L_{\theta,p}$. Now the injectivity of $L_{\theta,p}$ yields $x_1 = x_2$.

Existence: Let $g := \hat{D}^2 x_0 + B\hat{D}x_0 + Ax_0 = \hat{L}_{\theta,p}x_0 \in \mathsf{D}_{\hat{D}}(\theta,p)$. By assumption, $f - g \in \mathsf{D}_D(\theta,p)$. By Theorem 3.1, there exists a unique $x_1 \in \mathsf{D}_{L_{\theta,p}}$ such that

$$L_{\theta,p}x_1 = D^2 x_1 + BDx_1 + Ax_1 = f - g \in \mathsf{D}_D(\theta,p).$$

Then $x := x_0 + x_1$ is a desired solution of the problem (5.1). □

The assumptions of Theorem 5.1 are quite general, and in view of an application to the second-order Cauchy problem, we shall successively impose further assumptions. Let $\hat{D}$ be a closed extension of D such that the operator $\hat{D}^2$ with the natural domain is closed, too. Then the domain $\mathsf{D}_{\hat{L}_{\theta,p}}$ equipped with the norm

$$\|x\|_{\mathsf{D}_{\hat{L}_{\theta,p}}} := \|x\|_{\mathsf{D}_{\hat{D}}(\theta,p)} + \|\hat{D}^2 x\|_{\mathsf{D}_{\hat{D}}(\theta,p)} + \|B\hat{D}x\|_{\mathsf{D}_{\hat{D}}(\theta,p)} + \|Ax\|_{\mathsf{D}_{\hat{D}}(\theta,p)},$$

becomes a Banach space. The domain $\mathsf{D}_{L_{\theta,p}}$ is a closed subspace for the induced norm

$$\|x\|_{\mathsf{D}_{L_{\theta,p}}} := \|x\|_{\mathsf{D}_D(\theta,p)} + \|D^2 x\|_{\mathsf{D}_D(\theta,p)} + \|BDx\|_{\mathsf{D}_D(\theta,p)} + \|Ax\|_{\mathsf{D}_D(\theta,p)}$$

or for the equivalent graph norm

$$\|x\|_{\mathsf{D}_{L_{\theta,p}}} := \|x\|_{\mathsf{D}_D(\theta,p)} + \|L_{\theta,p}x\|_{\mathsf{D}_D(\theta,p)}.$$

Consider now the bounded operators

$$\begin{aligned} S_{\theta,p} : \mathsf{D}_{\hat{L}_{\theta,p}} &\to \mathsf{D}_{\hat{D}}(\theta,p) \times \mathsf{D}_{\hat{L}_{\theta,p}}/\mathsf{D}_{L_{\theta,p}} \\ u &\mapsto (\hat{L}_{\theta,p}u, [u]) \end{aligned}$$

and

$$\begin{aligned} T_{\theta,p} : \mathsf{D}_{\hat{D}}(\theta,p) \times \mathsf{D}_{\hat{L}_{\theta,p}}/\mathsf{D}_{L_{\theta,p}} &\to \mathsf{D}_{\hat{D}}(\theta,p)/\mathsf{D}_D(\theta,p), \\ (f, [u_0]) &\mapsto \left[f - \hat{L}_{\theta,p}u_0\right], \end{aligned}$$

where $u \mapsto [u]$ denotes various quotient maps. Note that $T_{\theta,p}$ is well defined in the sense that the definition does not depend on the choice of the representative u_0. With these definitions one easily sees that the kernel $\ker T_{\theta,p}$ is exactly the (closed) space of all pairs $(f, [u_0])$ satisfying the compatibility condition from Theorem 5.1 (which does not depend on the representative u_0), and that the compatibility condition is necessary for the existence of a solution of (5.1) since $S_{\theta,p}$ maps into $\ker T_{\theta,p}$. Theorem 5.1 implies that $S_{\theta,p}$ is an isomorphism onto $\ker T_{\theta,p}$.

The drawback of this abstract situation is, however, that in general we have no general description of either the kernel of $T_{\theta,p}$, or the quotient space $\mathsf{D}_{\hat{L}_{\theta,p}}/\mathsf{D}_{L_{\theta,p}}$.

Example 5.2 (The second-order Cauchy problem). Let A and B be two closed, linear operators on a Banach space X, and let $1 \le p < \infty$, $1 \le q \le \infty$ and $\theta \in (0, \frac{1}{p})$. On the space $L^p(0,1;X)$, let the differentiation operators D and $\hat{D} := D_{max}$ be given as in Example 2.1. Then $\hat{D}^2$ is closed as one easily verifies. Unlike in the proof of Theorem 4.1, we denote the multiplication operators on $L^p(0,T;X)$ again by A and B, respectively.

Recall from Example 2.1 that

$$\mathsf{D}_{\hat{D}}(\theta,p) = \mathsf{D}_D(\theta,p) = B^\theta_{pq}(0,T;X).$$

Hence,

$$\mathsf{D}_{\hat{L}_{\theta,q}} = \{u \in W^{2,p}(0,T;X) : \ddot{u},\ B\dot{u},\ Au \in B^\theta_{pq}(0,T;X)\}$$

and

$$\mathsf{D}_{L_{\theta,q}} = \{u \in \mathsf{D}_{\hat{L}_{\theta,q}} : u(0) = \dot{u}(0) = 0\}.$$

Note that in this situation, the quotient $\mathsf{D}_{\hat{L}_{\theta,q}}/\mathsf{D}_{L_{\theta,q}}$ can be naturally identified with the *trace space*

$$(X, \mathsf{D}_B, \mathsf{D}_A)_{B^\theta_{pq}} := \{(u_0, u_1) \in X \times X : \exists u \in \mathsf{D}_{\hat{L}_{\theta,q}} \text{ s.t. } u(0) = u_0,\ \dot{u}(0) = u_1\},$$

and the quotient map is then the natural *trace operator* $u \mapsto (u(0), \dot{u}(0))$. For the trace space we use a notation which is similar to the notation of classical real interpolation spaces between a *pair* of Banach spaces. This is appropriate because the classical real interpolation spaces can be identified with trace spaces involving weighted L^p spaces. However, we point out that here we "interpolate" between three Banach spaces and that the trace space is a subspace of the product space $X \times X$.

Corollary 5.3. *Let A and B be two closed, linear operators on a Banach space X satisfying the hypotheses* (b), (c) *and* (d) *of Theorem* 4.1. *Let $1 \le p < \infty$, $1 \le q \le \infty$ and $\theta \in (0, \frac{1}{p})$. Then for every $f \in B^\theta_{pq}(0,T;X)$ and every $(u_0, u_1) \in (X, \mathsf{D}_B, \mathsf{D}_A)_{B^\theta_{pq}}$ the second-order Cauchy problem*

$$\begin{aligned} \ddot{u} + B\dot{u} + Au &= f \quad \text{in } [0,T], \\ u(0) = u_0, &\quad \dot{u}(0) = u_1, \end{aligned} \tag{5.2}$$

admits a unique solution $u \in B^\theta_{pq}(0,T;X)$ satisfying

$$\dot{u},\ \ddot{u},\ B\dot{u},\ Au \in B^\theta_{pq}(0,T;X).$$

Proof. Note that for the particular choice of p, q and θ we have $T_{\theta,q} = 0$, and hence the compatibility condition from Theorem 5.1 is empty. In other words, by Theorem 5.1, the operator

$$\begin{aligned} S_{\theta,q} : \mathsf{D}_{\hat{L}_{\theta,q}} &\to B^\theta_{pq}(0,T;X) \times (X, \mathsf{D}_B, \mathsf{D}_A)_{\theta,q} \\ u &\mapsto (\hat{L}_{\theta,q}u, u(0), \dot{u}(0)) \end{aligned}$$

is invertible, and this implies the claim. □

We point out that in the particular case $p = 1$ there is no restriction on the value of $\theta \in (0,1)$.

The identification of the trace space $(X, \mathsf{D}_B, \mathsf{D}_A)_{B^\theta_{pq}}$, even for particular choices of X, B and A, is left as an open problem.

6. Examples

Example 6.1 (Strong damping I). Let $\Omega \subseteq \mathbb{R}^n$ be an open set, and let $\alpha > 0$. We consider the following initial-boundary value problem:

$$\begin{aligned} u_{tt} - \alpha\Delta u_t - \Delta u &= f && \text{in } (0,T)\times\Omega, \\ u &= 0 && \text{in } (0,T)\times\partial\Omega, \\ u(0,x) &= 0 && \text{in } \Omega, \\ u_t(0,x) &= 0 && \text{in } \Omega. \end{aligned} \tag{6.1}$$

For $1 \le r \le \infty$, we consider the space

$$X_r := \begin{cases} L^r(\Omega) & \text{if } 1 \le r < \infty, \\ C_0(\Omega) & \text{if } r = \infty. \end{cases}$$

On $X_2 = L^2(\Omega)$ we consider the negative Dirichlet–Laplace operator B_2 given by

$$\begin{aligned} \mathsf{D}_{B_2} &:= \{u \in H_0^1(\Omega) : \exists f \in L^2(\Omega)\, \forall v \in H_0^1(\Omega) : \int_\Omega \nabla u \overline{\nabla v} = \int_\Omega f \bar{v}\}, \\ B_2 u &:= f. \end{aligned}$$

It is known that B_2 is selfadjoint, nonnegative, and thus sectorial of angle $\varphi = 0$. Hence, the operator $-B_2$ generates an analytic C_0-semigroup which is known to have Gaussian upper bounds [2], [17], [32]. Thus, if $1 \le r < \infty$, the operator B_2, restricted to $X_r \cap L^2(\Omega)$, extends consistently to a sectorial operator B_r on X_r of angle $\varphi = 0$ [24, Theorem 2.3]. For domains with uniform C^2-boundary, and if $1 \le r < \infty$, one may also refer to [29, Theorem 3.1.3], where one finds also the characterization of the domain

$$\mathsf{D}_{B_r} = W^{2,r}(\Omega) \cap W_0^{1,r}(\Omega) \text{ if } 1 < r < \infty.$$

However, we are particularly interested in the end points $r = 1$ and $r = \infty$.

If $1 \le r < \infty$, and if we put $A := B := B_r$ and $\varepsilon = 1$, then we see that this example is a special case of Corollary 4.2. We thus obtain the following result.

Corollary 6.2. *Fix $\theta \in (0,1)$ $1 \le p$, $q \le \infty$, and $1 \le r < \infty$. Then for every $f \in B^\theta_{pq}(0,T;L^r(\Omega))$ the problem (6.1) admits a unique strong solution*

$$u \in B^{\theta+1}_{pq}(0,T;\mathsf{D}_{B_r}) \cap B^{2+\theta}_{pq}(0,T;L^r(\Omega)).$$

On the space $X_\infty = C_0(\Omega)$ we take the following realization of the negative Dirichlet–Laplace operator:

$$\begin{aligned} \mathsf{D}_{B_\infty} &:= \{u \in C_0(\Omega) : \Delta u \in C_0(\Omega)\}, \\ B_\infty u &:= -\Delta u. \end{aligned}$$

It has been shown in [3, Theorem 1.1] that $-B_\infty$ is the generator of an analytic semigroup if and only if Ω is Wiener regular, that is, if and only at each point

$x \in \partial\Omega$ there exists a barrier [3, Definition 3.1]. A bounded open set Ω is Wiener regular if and only if the Dirichlet problem

$$\begin{aligned} -\Delta u &= 0 \quad \text{in } \Omega, \\ u &= g \quad \text{in } \partial\Omega, \end{aligned}$$

admits for each $g \in C(\partial\Omega)$ a unique solution $u \in C(\bar{\Omega})$. Note that in $\mathbb{R}^2$ every bounded, simply connected domain is Wiener regular [15, Corollary 4.18, p. 276].

If Ω is Wiener regular, then the operator B_∞ is sectorial of angle $< \frac{\pi}{2}$. Again, if we put $A := B := B_\infty$ and $\varepsilon = 1$, then we see that Corollary 4.2 applies to problem (6.1) and we obtain the following corollary.

Corollary 6.3. *Assume that Ω is open and Wiener regular, and fix $\theta \in (0,1)$. Then for every $f \in C^\theta([0,T]; C_0(\Omega))$ the problem* (6.1) *admits a unique strong solution*

$$u \in C^{1,\theta}([0,T]; \mathsf{D}_{B_\infty}) \cap C^{2,\theta}([0,T]; C_0(\Omega)).$$

Remark 6.4. Note again that the above maximal regularity results apply in particular in the spaces $L^1(\Omega)$ (Corollary 6.2) and $C_0(\Omega)$ (Corollary 6.3) which are not UMD spaces. Moreover, in Corollary 6.2, the time regularity allows us to consider also the space $B^\theta_{1,q}$ and in particular $B^\theta_{1,1}$.

Example 6.5 (Strong damping II). Let $\Omega \subseteq \mathbb{R}^n$ be an open set. We consider now the following initial-boundary value problem:

$$\begin{aligned} &u_{tt} - \mathcal{A}(x,D)u_t - \mathcal{A}(x,D)u = f && \text{in } (0,T) \times \Omega, \\ &u = 0 && \text{in } (0,T) \times \partial\Omega, \\ &u(0,x) = 0 && \text{in } \Omega, \\ &u_t(0,x) = 0(x) && \text{in } \Omega. \end{aligned} \tag{6.2}$$

Here $\mathcal{A}(x,D)$ is formally given by

$$\mathcal{A}(x,D)u = \sum_{i,j=1}^n D_i(a_{ij}D_ju) + \sum_{i=1}^n (D_i(b_iu) + c_iD_iu) + du$$

with real coefficients a_{ij}, b_i, c_i, $d \in L^\infty(\Omega)$ satisfying the ellipticity condition

$$\sum_{i,j=1}^n a_{ij}(x)\xi_i\xi_j \geq \eta\,|\xi|^2$$

for some $\eta > 0$ and all $x \in \Omega$, $\xi \in \mathbb{R}^n$, and the dissipativity condition

$$\sum_{i=1}^n D_ib_i + d \leq 0 \text{ in } \mathcal{D}(\Omega)'.$$

Under these assumptions, we have an operator $\mathcal{A} : H^1(\Omega) \to H^{-1}(\Omega)$ given by

$$\langle \mathcal{A}u, v\rangle_{H^{-1},H^1_0} := \sum_{i,j=1}^n \int_\Omega a_{ij}D_ju\overline{D_iv} + \sum_{i=1}^n \int_\Omega (b_iu\overline{D_iv} - c_iD_iu\bar{v}) - \int_\Omega du\bar{v}.$$

We consider the same scale of spaces as in Example 6.1. We now define an operator B_2 on $X_2 = L^2(\Omega)$ by

$$\mathsf{D}_{B_2} := \{u \in H_0^1(\Omega) : \exists f \in L^2(\Omega)\, \forall v \in H_0^1(\Omega) : \langle \mathcal{A}u, v\rangle_{H^{-1},H_0^1} = \int_\Omega f\bar{v}\},$$
$$B_2 u := f.$$

The operator B_2 is associated with an elliptic form, it is sectorial of angle $\varphi < \frac{\pi}{2}$, and hence $-B_2$ generates an analytic C_0-semigroup. Again, this semigroup has Gaussian upper bounds [2], [17], [32], and if $1 \le r < \infty$, then the operator B_2, restricted to $X_r \cap L^2(\Omega)$, extends consistently to a sectorial operator B_r on X_r of the same angle φ [24, Theorem 2.3]. In particular, if $1 \le r < \infty$, and if we put $A := B := B_r$ and $\varepsilon = 1$, then we see that this example is also a special case of Corollary 4.2. We thus obtain the following result.

Corollary 6.6. *Fix $\theta \in (0,1)$ $1 \le p$, $q \le \infty$, and $1 \le r < \infty$. Then for every $f \in B^\theta_{pq}(0,T;L^r(\Omega))$ the problem (6.2) admits a unique strong solution*

$$u \in B^{\theta+1}_{pq}(0,T;\mathsf{D}_{B_r}) \cap B^{2+\theta}_{pq}(0,T;L^r(\Omega)).$$

On the space $X_\infty = C_0(\Omega)$ we consider the following operator:

$$\mathsf{D}_{B_\infty} := \{u \in C_0(\Omega) \in H^1_{loc}(\Omega) : \mathcal{A}(x,D)u \in C_0(\Omega)\},$$
$$B_\infty u := -\mathcal{A}(x,D)u.$$

It has been shown in [3, Corollary 4.7] that if Ω is bounded and Wiener regular, then $-B_\infty$ is the generator of an analytic semigroup. Hence, if Ω is bounded and Wiener regular, then the operator B_∞ is sectorial of angle $< \frac{\pi}{2}$. Again, if we put $A := B := B_\infty$ and $\varepsilon = 1$, then we see that Corollary 4.2 applies to problem (6.2) and we obtain the following corollary.

Corollary 6.7. *Assume that Ω is open, bounded and Wiener regular, and fix $\theta \in (0,1)$. Then for every $f \in C^\theta([0,T];C_0(\Omega))$ the problem (6.2) admits a unique strong solution*

$$u \in C^{1,\theta}([0,T];\mathsf{D}_{B_\infty}) \cap C^{2,\theta}([0,T];C_0(\Omega)).$$

Example 6.8 (Intermediate damping). Let $\Omega \subseteq \mathbb{R}^n$ be an open set. We consider the following initial-boundary value problem:

$$\begin{aligned} &u_{tt} - \alpha\Delta u_t + \Delta^2 u = f && \text{in } (0,T)\times\Omega, \\ &u = \Delta u = 0 && \text{in } (0,T)\times\partial\Omega, \\ &u(0,x) = 0 && \text{in } \Omega, \\ &u_t(0,x) = 0 && \text{in } \Omega. \end{aligned} \tag{6.3}$$

This problem is in fact a special case of the problem (4.2) from Corollary 4.2 if we let $1 \le r \le \infty$, B_r be the negative Dirichlet–Laplace operator on X_r (see Example 6.1), and if we put $A = B_r^2$ and $\varepsilon = \frac{1}{2}$. Then A is still sectorial with angle $\varphi = 0$ if $1 \le r < \infty$ and $\varphi \in (0,\pi)$ if $r = \infty$. Moreover, $B_r = A^{\frac{1}{2}}$, and we obtain the following two corollaries.

Corollary 6.9. *Fix $\theta \in (0,1)$ $1 \le p$, $q \le \infty$, and $1 \le r < \infty$. Assume that $\alpha > 0$. Then for every $f \in B^{\theta}_{pq}(0,T;L^r(\Omega))$ the problem (6.3) admits a unique strong solution*

$$u \in B^{\theta}_{pq}(0,T;\mathsf{D}_{B_r^2}) \cap B^{1+\theta}_{pq}(0,T;\mathsf{D}_{B_r}) \cap B^{2+\theta}_{pq}(0,T;L^r(\Omega)).$$

Corollary 6.10. *Assume $\alpha \ge 2$, that Ω is open and Wiener regular, and fix $\theta \in (0,1)$. Then for every $f \in C^{\theta}([0,T];C_0(\Omega))$ the problem (6.3) admits a unique strong solution*

$$u \in C^{\theta}([0,T];\mathsf{D}_{B_\infty^2}) \cap C^{1,\theta}([0,T];\mathsf{D}_{B_\infty}) \cap C^{2,\theta}([0,T];C_0(\Omega)).$$

References

[1] W. Arendt, R. Chill, S. Fornaro, and C. Poupaud, *L^p-maximal regularity for non-autonomous evolution equations*, J. Differential Equations **237** (2007), 1–26.

[2] W. Arendt and A.F.M. ter Elst, *Gaussian estimates for second-order elliptic operators with boundary conditions*, J. Operator Theory **38** (1997), 87–130.

[3] W. Arendt and Ph. Bénilan, *Wiener regularity and heat semigroups on spaces of continuous functions*, Topics in nonlinear analysis, Progr. Nonlinear Differential Equations Appl., vol. 35, Birkhäuser, Basel, 1999, pp. 29–49.

[4] C.J.K. Batty, R. Chill, and S. Srivastava, *Maximal regularity for second-order non-autonomous Cauchy problems*, Studia Math. **189** (2008), no. 3, 205–223.

[5] Shang Quan Bu, *Maximal regularity of second-order delay equations in Banach spaces*, Acta Math. Sin. (Engl. Ser.) **25** (2009), no. 1, 21–28.

[6] Shang Quan Bu, *Well-posedness of equations with fractional derivative*, Acta Math. Sin. (Engl. Ser.) **26** (2010), no. 7, 1223–1232.

[7] Shang Quan Bu, *Well-posedness of fractional differential equations on vector-valued function spaces*, Integral Equations Operator Theory **71** (2011), no. 2, 259–274.

[8] Shang Quan Bu and Yi Fang, *Periodic solutions for second-order integro-differential equations with infinite delay in Banach spaces*, Studia Math. **184** (2008), no. 2, 103–119.

[9] Shang Quan Bu and Yi Fang, *Maximal regularity of second-order delay equations in Banach spaces*, Sci. China Math. **53** (2010), no. 1, 51–62.

[10] P. Cannarsa, G. Da Prato, and J.-P. Zolésio, *The damped wave equation in a moving domain*, J. Differential Equations **85** (1990), no. 1, 1–16.

[11] M. Cheggag, A. Favini, R. Labbas, St. Maingot, and A. Medeghri, *Abstract differential equations of elliptic type with general Robin boundary conditions in Hölder spaces*, Appl. Anal. **91** (2012), no. 8, 1453–1475.

[12] R. Chill and S. Król, *Extrapolation of L^p-maximal regularity for second-order Cauchy problems*, Perspectives in operator theory, Banach Center Publ., Polish Acad. Sci., Warsaw, 2015, to appear.

[13] R. Chill and S. Srivastava, *L^p-maximal regularity for second-order Cauchy problems*, Math. Z. **251** (2005), 751–781.

[14] R. Chill and S. Srivastava, *L^p-maximal regularity for second-order Cauchy problems is independent of p*, Boll. Unione Mat. Ital. (9) **1** (2008), no. 1, 147–157.

[15] J.B. Conway, *Functions of One Complex Variable*, second ed., Graduate Texts in Mathematics, vol. 11, Springer-Verlag, New York, 1978.

[16] G. Da Prato and P. Grisvard, *Sommes d'opérateurs linéaires et équations différentielles opérationnelles*, J. Math. Pures Appl. **54** (1975), 305–387.

[17] D. Daners, *Heat kernel estimates for operators with boundary conditions*, Math. Nachr. **217** (2000), 13–41.

[18] R. Dautray and J.-L. Lions, *Analyse mathématique et calcul numérique pour les sciences et les techniques. Vol. VIII*, INSTN: Collection Enseignement, Masson, Paris, 1987.

[19] A. Favini, *Parabolicity of second-order differential equations in Hilbert space*, Semigroup Forum **42** (1991), no. 3, 303–312.

[20] A. Favini, R. Labbas, St. Maingot, and M. Meisner, *Boundary value problem for elliptic differential equations in non-commutative cases*, Discrete Contin. Dyn. Syst. **33** (2013), no. 11-12, 4967–4990.

[21] A. Favini, R. Labbas, St. Maingot, H. Tanabe, and A. Yagi, *Necessary and sufficient conditions for maximal regularity in the study of elliptic differential equations in Hölder spaces*, Discrete Contin. Dyn. Syst. **22** (2008), no. 4, 973–987.

[22] C. Fernández, C. Lizama, and V. Poblete, *Maximal regularity for flexible structural systems in Lebesgue spaces*, Math. Probl. Eng. (2010), Art. ID 196956, 15.

[23] M. Haase, *The Functional Calculus for Sectorial Operators*, Operator Theory: Advances and Applications, vol. 169, Birkhäuser Verlag, Basel, 2006.

[24] M. Hieber, *Gaussian estimates and holomorphy of semigroups on L^p spaces*, J. London Math. Soc. (2) **54** (1996), no. 1, 148–160.

[25] V. Keyantuo and C. Lizama, *Hölder continuous solutions for integro-differential equations and maximal regularity*, J. Differential Equations **230** (2006), no. 2, 634–660.

[26] V. Keyantuo and C. Lizama, *Periodic solutions of second-order differential equations in Banach spaces*, Math. Z. **253** (2006), no. 3, 489–514.

[27] V. Keyantuo and C. Lizama, *A characterization of periodic solutions for time-fractional differential equations in UMD spaces and applications*, Math. Nachr. **284** (2011), no. 4, 494–506.

[28] C. Lizama and V. Poblete, *Maximal regularity for perturbed integral equations on periodic Lebesgue spaces*, J. Math. Anal. Appl. **348** (2008), no. 2, 775–786.

[29] A. Lunardi, *Analytic Semigroups and Optimal Regularity in Parabolic Problems*, Progress in Nonlinear Differential Equations and Their Applications, vol. 16, Birkhäuser, Basel, 1995.

[30] A. Lunardi, *Interpolation theory*, second ed., Appunti. Scuola Normale Superiore di Pisa (Nuova Serie). [Lecture Notes. Scuola Normale Superiore di Pisa (New Series)], Edizioni della Normale, Pisa, 2009.

[31] F.Z. Mezeghrani, *Necessary and sufficient conditions for the solvability and maximal regularity of abstract differential equations of mixed type in Hölder spaces*, Osaka J. Math. **50** (2013), no. 3, 725–747.

[32] E.M. Ouhabaz, *Analysis of Heat Equations on Domains*, London Mathematical Society Monographs, vol. 30, Princeton University Press, Princeton, 2004.

[33] V. Poblete, *Solutions of second-order integro-differential equations on periodic Besov spaces*, Proc. Edinb. Math. Soc. (2) **50** (2007), no. 2, 477–492.

[34] V. Poblete, *Maximal regularity of second-order equations with delay*, J. Differential Equations **246** (2009), no. 1, 261–276.

[35] H. Triebel, *Theory of Function Spaces*, Birkhäuser, Basel, 1983.

[36] Y. Yakubov, *Maximal L_p-regularity for second-order non-autonomous evolution equations in UMD Banach spaces and application*, Int. J. Evol. Equ. **3** (2009), no. 3, 379–393.

[37] R. Zacher, *Maximal regularity of type L_p for abstract parabolic Volterra equations*, J. Evol. Equ. **5** (2005), no. 1, 79–103.

Charles J.K. Batty
St. John's College
University of Oxford
Oxford OX1 3JP, United Kingdom
e-mail: charles.batty@sjc.ox.ac.uk

Ralph Chill
Technische Universität Dresden
Institut für Analysis
D-01062 Dresden, Germany
e-mail: ralph.chill@tu-dresden.de

Sachi Srivastava
Department of Mathematics
University of Delhi South Campus
Benito Juarez Road, New Delhi 110021, India
e-mail: sachi_srivastava@yahoo.com

Operator Theory:
Advances and Applications, Vol. 250, 67–85

Stability of Quantum Dynamical Semigroups

B.V. Rajarama Bhat and Sachi Srivastava

Dedicated to Prof. Charles Batty on his 60th birthday

Abstract. A one parameter semigroup of maps is said to be stable if it eventually decays to zero. Generally different topologies for convergence to zero give rise to different notions of stability. Stability is also connected with absence of fixed points. We examine these concepts in the context of quantum dynamical semigroups and dilation theory.

Mathematics Subject Classification (2010). 46L57; 47D03 .

Keywords. Quantum dynamical semigroups, stability.

1. Introduction

Stability and asymptotics of semigroups of bounded maps have been extensively studied in classical settings. A comprehensive survey may be found, for example, in [3]. This paper aims to look at the notion of stability for quantum dynamical semigroups on $B(\mathcal{H})$, the Banach space of bounded linear operators on a Hilbert Space $\mathcal{H}$. By a quantum dynamical semigroup (Q.D.S.) on $B(\mathcal{H})$ we shall mean a one parameter semigroup $\mathcal{T} = (\mathcal{T}_t)_{t\geq 0}$ of completely positive, contractive, normal maps from $B(\mathcal{H})$ to itself such that for each $X \in B(\mathcal{H})$ the map $t \mapsto \mathcal{T}_t(X)$ is continuous in the weak operator topology. The large time behaviour of quantum Markov semigroups, in particular their recurrence and transience, have been investigated in depth by Fagnola, Rebolledo, and Umanita [9, 13] by introducing the concept of potential associated with such semigroups. The Q.D.S. considered in this paper, however, are sub-Markovian and usually uniformly continuous and the aim is to connect the presence or absence of stability with the properties of the (coefficients of) the bounded generator. Our approach therefore is quite different from the above-mentioned articles. We link stability with the existence of fixed points of the Q.D.S. and then study the behaviour of these under minimal dilations of the Q.D.S. We refer to [6] for characterisations and liftings of fixed points of a single completely positive map on a von Neumann algebra. Recently, a characterisation of liftings of fixed points of quantum dynamical semigroups under dilations has also been obtained in [12]. Sections 2 and 3 discuss some necessary and sufficient conditions for the different notions of stability for quantum dynamical semigroups and the interplay with fixed points whereas behaviour of fixed points under dilations are dealt with in Section 4.

2. Stability

We begin by recalling some well-known facts about quantum dynamical semigroups and setting notation. Let $\mathcal{H}$ be a complex separable Hilbert space and $B(\mathcal{H})$ be the von Neumann algebra of all bounded operators on $\mathcal{H}$. By a quantum dynamical semigroup (Q.D.S.) on $B(\mathcal{H})$ we shall mean a one parameter semigroup $\mathcal{T} = (\mathcal{T}_t)_{t\geq 0}$ of completely positive, contractive, normal maps from $B(\mathcal{H})$ to itself such that for each $X \in B(\mathcal{H})$ the map $t \mapsto \mathcal{T}_t(X)$ is continuous in the weak operator topology.

If the Q.D.S. is uniformly continuous, the infinitesimal generator $\mathcal{L}$ of this semigroup is bounded and is given by

$$\mathcal{L}(X) \quad := \quad \lim_{t\to 0} \frac{\mathcal{T}_t(X) - X}{t}.$$

The limit above exists in the norm topology. Moreover, it is well known [4] that if the Q.D.S. is uniformly continuous then $\mathcal{L}$ is given by

$$\mathcal{L}(X) = KX + XK^* + \sum_j L_j^* X L_j, \; X \in B(\mathcal{H}), \tag{2.1}$$

where $K, L_j \in B(\mathcal{H})$ and the sum on the right-hand side above converges in strong operator topology. Note that such a decomposition is not unique. We will often write $\mathcal{L} = \mathcal{L}_1 + \mathcal{L}_0$ where $\mathcal{L}_1$ is the completely positive part of the generator, given by

$$\mathcal{L}_1(X) = \sum_j L_j^* X L_j, \tag{2.2}$$

while $\mathcal{L}_0$ is given by

$$\mathcal{L}_0(X) = KX + XK^*, \tag{2.3}$$

for all $X \in B(\mathcal{H})$. For the general theory of uniformly continuous completely positive semigroups we refer to [8] and [10].

In this note, we shall assume, unless otherwise stated, that the Q.D.S. $(\mathcal{T}_t)_{t\geq 0}$ is sub-Markovian, that is, $\mathcal{T}_t(I) \leq I$, for all $t \geq 0$. The generator for such a Q.D.S. necessarily satisfies $\mathcal{L}(I) \leq 0$, which also forces $K + K^* \leq 0$.

For any operator A, bounded or unbounded, acting on a Banach space, the *spectral bound* $s(A)$ of A is defined by

$$s(A) = \sup\{Re\lambda : \lambda \in \sigma(A)\}$$

where $\sigma(A)$ denotes the spectrum of A. The *exponential growth bound* $w_0(T)$ of a semigroup T is given by

$$w_0(T) = \inf\left\{w \in R : \text{ there exists } M_w > 0 \text{ with } \|T_t\| \leq M_w e^{wt}\right\}.$$

If A is the generator of a C_0 semigroup T, then $s(A) \leq w_0(T)$. For a uniformly continuous Q.D.S. $\mathcal{T}$, $s(\mathcal{L}) = w_0(\mathcal{T}) \leq 0$. The first equality is due to uniform continuity while the second inequality holds because $\mathcal{T}$ is contractive.

Borrowing from the theory of stability for classical semigroups, we may define analogous notions of stability for quantum dynamical semigroups. Here are some possibilities. We shall call a Q.D.S. $\mathcal{T}$

(i) Uniformly exponentially stable if there is an $M > 0$ and an $\epsilon > 0$ such that $\|\mathcal{T}_t\| \leq Me^{-\epsilon t}, t \geq 0$.
(ii) Uniformly stable if $\lim_{t\to\infty} \|\mathcal{T}_t\| = 0$.
(iii) Strongly stable if $\lim_{t\to\infty} \|\mathcal{T}_t(X)\| = 0$, for every $X \in \mathbf{B}(\mathcal{H})$.
(iv) Stable if $\lim_{t\to\infty} \mathcal{T}_t(I) = 0$ in the strong operator topology and
(v) Weakly stable if $\lim_{t\to\infty} \mathcal{T}_t(I) = 0$ in the weak operator topology.

It is obvious from the definitions above that (i) $\Rightarrow$ (ii) $\Rightarrow$ (iii) $\Rightarrow$ (iv) $\Rightarrow$ (v). But more is true. As in the case of C_0 semigroups on Banach spaces, [5, Proposition V.1.2], the class of uniformly exponentially stable Q.D.S. coincides with that of uniformly stable Q.D.S.. Further, since the operators $\mathcal{T}_t$ are completely positive, $\|\mathcal{T}_t\| = \|\mathcal{T}_t(I)\|$, so that strong stability of a Q.D.S. is equivalent to uniform stability. Thus the first three definitions are equivalent. Moreover, the positivity of $\mathcal{T}_t(I)$ implies that the stable and weakly stable Q.D.S. coincide. Therefore, it suffices to have the following definition.

Definition 2.1. A quantum dynamical semigroup is said to be *uniformly stable* if (i) holds and *stable* if (iv) holds.

To begin with, we note down some basic properties of these two notions.

Remark 2.2.

(i) From the definition, $(\mathcal{T}_t)_{t\geq 0}$ is uniformly stable if and only if $w_0(\mathcal{T}) < 0$.
(ii) Note that if $\mathcal{T}$ is a quantum dynamical semigroup, then since $\mathcal{T}_t$ is contractive, $(e^{-\alpha t}\mathcal{T}_t)_{t\geq 0}$ is uniformly stable for every $\alpha > 0$.
(iii) If $\mathcal{T}$ is stable then s-$\lim_{t\to\infty} \mathcal{T}_t(X) = 0$ for all $X \in B(\mathcal{H})$. Indeed, for $0 \leq X \leq I$, the positivity of $\mathcal{T}_t$ for each t implies $0 \leq \mathcal{T}_t(X) \leq \mathcal{T}_t(I)$. This forces $\mathcal{T}_t(X)$ to converge to 0 as t tends to infinity, in the strong operator topology. Since every operator in $B(\mathcal{H})$ is a finite linear combination of positive elements, the claim follows.
(iv) If $\mathcal{T}$ is a uniformly continuous Q.D.S. with generator $\mathcal{L}$, then $s(\mathcal{L}) = w_0(\mathcal{T})$ as remarked before. Therefore, if $\mathcal{L} = \mathcal{L}_1$ then the semigroup $\mathcal{T}$ cannot be uniformly stable. In fact in this case, $\mathcal{L}$ being positive, $s(\mathcal{L}) \geq 0$. Therefore, $w_0(\mathcal{T}) \geq 0$.
(v) On the other hand, if $\mathcal{L} = \mathcal{L}_0$ then the semigroup is uniformly stable if and only if the semigroup on $\mathcal{H}$ generated by the operator K is uniformly stable. Indeed, if $(P_t)_{t\geq 0}$ is the semigroup generated by K, that is, $P_t = e^{Kt}, t \geq 0$ and $\mathcal{T}_t^0(X) = e^{\mathcal{L}_0 t}(X) = e^{K^*t}Xe^{Kt}$, $X \in \mathbf{B}(\mathcal{H})$ is the quantum dynamical semigroup, then $\|\mathcal{T}_t^0\| = \|\mathcal{T}_t^0(I)\| = \|P_t\|^2$. So $w_0(\mathcal{T}) = 2w_0(P)$. Since $\mathcal{T}^0$ is uniformly stable if and only if $w_0(\mathcal{T}^0) < 0$, the claim holds.

The two notions of stability coincide, of course, if the underlying Hilbert space is finite dimensional. For the general case, as the following example shows, while uniform stability implies stability, the converse need not be true.

Example. Let $\mathcal{H} = L^2(-1,0)$ and A be the multiplication operator given by $(Af)(s) = q(s)f(s)$, where $q(s) = s, s \in (-1,0)$. Let $(P_t)_{t\geq 0}$ be the uniformly continuous semigroup generated by A. Then $(P_t f)(s) = e^{tq(s)}f(s)$, for $f \in \mathcal{H}, s \in (-1,0), t \geq 0$. Since $\sigma(A) = \overline{\text{range } q}, s(A) = 0 = w_0(P)$. Let $\mathcal{T}_t(X) = P_t X P_t^*, X \in B(\mathcal{H}), t \geq 0$. Then $\mathcal{T}_t$ is a quantum dynamical semigroup and since $w_0(\mathcal{T}) = w_0(P) = 0$, it is not uniformly stable in view of Remark 2.2,(i). However,

$$\lim_{t\to\infty} \mathcal{T}_t(I)f = \lim_{t\to\infty} \int_{-1}^{0} |e^{2ts}f(s)|^2 ds = 0,$$

for every $f \in \mathcal{H}$. Therefore, $\mathcal{T}$ is stable.

There is a close connection between the invertibility of the operator coefficient K that appears in the expression (2.1) for $\mathcal{L}$ and the stability of the semigroup generated by $\mathcal{L}$.

Theorem 2.3. *Suppose $\mathcal{H}$ is finite dimensional and $(\mathcal{T}_t)_{t\geq 0}$ is a uniformly continuous Q.D.S. on $B(\mathcal{H})$ with generator $\mathcal{L}$ given by* (2.1). *If the semigroup is stable, then $s(K) < 0$. In particular K has full rank.*

Proof. We write $\mathcal{L} = \mathcal{L}_0 + \mathcal{L}_1$ where $\mathcal{L}_0, \mathcal{L}_1$ are as in (2.2) and (2.3) denote by $\mathcal{T}^0$ the Q.D.S. generated by $\mathcal{L}_0$, so that $\mathcal{T}_t^0(X) = e^{Kt}Xe^{K^*t}, t \geq 0, X \in B(\mathcal{H})$. We may consider the Q.D.S. $\mathcal{T}$ as the semigroup obtained by perturbing the generator $\mathcal{L}_0$ by the completely positive operator $\mathcal{L}_1$. Then the following relation holds:

$$\mathcal{T}_t(X) = \mathcal{T}_t^0(X) + \int_0^t \mathcal{T}_{t-s}^0 \mathcal{L}_1 \mathcal{T}_s(X)\, ds,$$

for all $X \in B(\mathcal{H})$. Due to the positivity of $\mathcal{T}^0, \mathcal{L}_1, \mathcal{T}$ it follows that $\mathcal{T}_t \geq \mathcal{T}_t^0, t \geq 0$. Therefore, $\mathcal{T}_t(I) \geq \mathcal{T}_t^0(I), t \geq 0$. Since $\mathcal{T}$ is stable, $s - \lim_{t\to\infty} \mathcal{T}_t(I) = 0$, so that $s - \lim_{t\to\infty} \mathcal{T}_t^0(I) = 0$. As the underlying space is finite dimensional this implies that

$$\begin{aligned} \lim_{t\to\infty} \left\| e^{Kt} \right\|^2 &= \lim_{t\to\infty} \left\| e^{Kt}e^{K^*t} \right\| \\ &= \lim_{t\to\infty} \left\| \mathcal{T}_t^0(I) \right\| \\ &= 0. \end{aligned}$$

Thus the semigroup $\left(e^{Kt}\right)_{t\geq 0}$ is uniformly stable, so that $s(K) = w_0(K) < 0$. In particular, this means that $0 \notin \sigma(K)$ so that K is invertible. □

Remark 2.4.

(i) Theorem 2.3 is no longer true if the underlying Hilbert space is not finite dimensional. This is clear from Example 2 with K taken to be the multiplication operator A defined there. The Q.D.S. in this example is stable, the generator $\mathcal{L}$ is given by $\mathcal{L}(X) = KX + XK^*$ but $s(K) = 0$.

(ii) Recall that a matrix K for which $s(K) < 0$ is called stable. We also note here that it is possible that $K + K^*$ is not invertible even if $s(K) < 0$. In fact, if $\mathcal{H} = \mathbb{C}^2$ and

$$K = \begin{pmatrix} -1 & 2 \\ 0 & -1 \end{pmatrix},$$

then $s(K) = -1 < 0$ but the matrix

$$K + K^* = \begin{pmatrix} -2 & 2 \\ 2 & -2 \end{pmatrix}$$

is not invertible. Note that the quantum dynamical semigroup $\mathcal{T}^0$ acting on $B(\mathcal{H})$ generated by $\mathcal{L}_0$ where $\mathcal{L}_0(X) = KX + XK^*$ is uniformly stable.

The following example exhibits a Q.D.S. in a finite-dimensional setting such that $s(K) < 0$ yet the Q.D.S. is not stable. Thus the converse of Theorem 2.3 is not true in general.

Example. Let $\mathcal{H} = \mathbb{C}^2$ and

$$K = \begin{pmatrix} -1 & 2 \\ 0 & -1 \end{pmatrix}, \quad L = \begin{pmatrix} 1 & -1 \\ -1 & 1 \end{pmatrix}.$$

Then $s(K) = -1 < 0$. Let $\mathcal{L}$ be the bounded operator acting on $B(\mathcal{H})$ as

$$\mathcal{L}(X) = KX + XK^* + LXL^*, \; X \in B(\mathcal{H}).$$

Then the Q.D.S. $\mathcal{T}$ generated by $\mathcal{L}$ satisfies $\mathcal{L}(I) \leq 0$. In fact, routine calculations show that

$$\mathcal{L}\left(\begin{pmatrix} x & y \\ z & w \end{pmatrix}\right) = \begin{pmatrix} -x+y+z+w & -x-y+w+z \\ -x+y-z+w & x-y-z-w \end{pmatrix},$$

so that $\mathcal{L}(I) = 0$. This implies that $\mathcal{T}_t(I) = I$ for all $t \geq 0$. Thus $\mathcal{T}$ is not stable.

As we have already seen, one way of looking at $\mathcal{L}$ is to consider it as the generator of the semigroup obtained by perturbing the generator $\mathcal{L}_0$ by the bounded, completely positive map $\mathcal{L}_1$. A conditional converse of Theorem 2.3 can be obtained on invoking a well-known perturbation result from the theory of classical semigroups:

Theorem 2.5 (**[7, Theorem 3.1.1]**). *Let $\mathcal{X}$ be a Banach space and let θ be the infinitesimal generator of a C_0 semigroup $\mathcal{T}_t$ on $\mathcal{X}$, satisfying $\|\mathcal{T}_t\| \leq Me^{wt}$. If ϕ is a bounded linear operator on $\mathcal{X}$, then $\theta + \phi$ is the infinitesimal generator of a C_0 semigroup $\mathcal{S}$ on $\mathcal{X}$, satisfying $\| \mathcal{S}_t \| \leq Me^{(w+M\|\phi\|)t}$, for all $t \geq 0$.*

As a direct consequence of this result we have:

Theorem 2.6. *Suppose that $\mathcal{L}$ is a bounded operator on $B(\mathcal{H})$ with $\mathcal{L}(I) \leq 0$ and there exists $b < 0$ such that $\sum_j L_j^* L_j < -bI < -(K + K^*)$, where $K, L_j \in B(\mathcal{H})$ and $\mathcal{L}(X) = KX + XK^* + \sum_j L_j^* X L_j, \; X \in B(\mathcal{H})$. Then the Q.D.S. $\mathcal{T}$ generated by $\mathcal{L}$ is uniformly stable.*

Proof. As before, we write $\mathcal{L} = \mathcal{L}_0 + \mathcal{L}_1$ where $\mathcal{L}_1(X) = \sum_j L_j^* X L_j$, and $\mathcal{L}_0(X) = KX + XK^*$, for all $X \in B(\mathcal{H})$. Now K generates a uniformly continuous semigroup $\left(e^{tK}\right)_{t\geq 0}$, being a bounded operator. Since $K+K^* < bI$, $K - \frac{b}{2}$ is a bounded, dissipative operator. Therefore, by the Lumer–Phillips Theorem, [5, Theorem 3.15] $K - \frac{b}{2}$ generates a semigroup of contractions. Hence,

$$\|e^{tK}\| \leq e^{t\frac{b}{2}},\ t \geq 0. \tag{2.4}$$

Since $b < 0$, (2.4) implies that the semigroup generated by K is exponentially stable. Thus, the semigroup $\mathcal{T}^0$ generated by $\mathcal{L}_0$ is also uniformly stable (see Remark 2.2 (v)) and $\left\|\mathcal{T}_t^0\right\| \leq e^{tb}$. Applying Theorem 2.5 to the uniformly continuous semigroup $\mathcal{T}^0$ acting on $B(\mathcal{H})$, with generator $\mathcal{L}_0$ we have that $\mathcal{L}_0 + \mathcal{L}_1$ generates a uniformly continuous semigroup $\mathcal{T}$ satisfying

$$\|\mathcal{T}_t\| \leq e^{(b+\|\mathcal{L}_1\|)t},\ t \geq 0. \tag{2.5}$$

Since $\mathcal{L}_1$ is completely positive, the hypothesis implies that

$$\|\mathcal{L}_1\| = \|\mathcal{L}_1(I)\| = \left\|\sum\nolimits_j L_j^* L_j\right\| < -b.$$

Then (2.5) implies that $\mathcal{T}$ is uniformly stable. □

The following example illustrates Theorem 2.6.

Example. Let $\mathcal{H} = \mathbb{C}^2$ and $a \in \mathbb{C}$. Set

$$K = \begin{pmatrix} -1+i & 1 \\ 0 & -1+i \end{pmatrix},\quad L = \begin{pmatrix} a & 0 \\ 0 & a \end{pmatrix}.$$

It is easy to see that $K + K^* < -\frac{1}{2}I$. Then choosing a so that $|a|^2 < \frac{1}{2}$ ensures that $L^*L < \frac{1}{2}I < -(K+K^*)$. Thus the hypothesis of Theorem 2.6 is satisfied. Therefore, if $\mathcal{L}(X) = KX + XK^* + L^*XL$, $X \in B(\mathcal{H})$, then $\mathcal{L}$ must generate a stable Q.D.S. On the other hand, actual computation shows that

$$\mathcal{L}\left(\begin{pmatrix} x & y \\ z & w \end{pmatrix}\right) = \begin{pmatrix} (|a|^2-2)x + y + z & (|a|^2-2)y + w \\ (|a|^2-2)z + w & (|a|^2-2)w \end{pmatrix}.$$

This implies that $\mathcal{L}$ is represented by the 4×4 matrix

$$\begin{pmatrix} (|a|^2-2) & 1 & 1 & 0 \\ 0 & (|a|^2-2) & 0 & 1 \\ 0 & 0 & (|a|^2-2) & 1 \\ 0 & 0 & 0 & (|a|^2-2) \end{pmatrix}.$$

Therefore, $s(\mathcal{L}) < 0$, so that the semigroup generated by $\mathcal{L}$ is uniformly stable.

Corollary 2.7. *Suppose that $\mathcal{L} = \mathcal{L}_1 + \mathcal{L}_0$ generates a quantum dynamical semigroup $\mathcal{T}$ which is uniformly stable, so that $\|\mathcal{T}_t\| \leq Me^{-\epsilon t}$, for some $M, \epsilon > 0$. If $\|\mathcal{L}_1\| < \frac{\epsilon}{M}$ then $s(K) < 0$.*

Proof. Perturbing the generator $\mathcal{L}$ by $-\mathcal{L}_1$ and invoking Theorem 2.5, we get that $\mathcal{L}_0$ generates a uniformly continuous semigroup $\mathcal{T}^0$, satisfying,

$$\|\mathcal{T}_t\| \le Me^{(-\epsilon+M\|\mathcal{L}_1\|)t},\ t \ge 0.$$

Due to the condition on $\mathcal{L}_1$ this implies that the semigroup $\mathcal{T}$ is exponentially stable. Therefore, the semigroup generated by K is exponentially stable and in turn $s(K) < 0$. □

3. Fixed points and stability

In this section we shall discuss the role played by fixed points of a quantum dynamical semigroup. Recall that an operator $C \in B(\mathcal{H})$ is called a *fixed point* of the semigroup $\mathcal{T}$ defined on $B(\mathcal{H})$ if

$$\mathcal{T}_t(C) = C \text{ for all } t \ge 0.$$

Denote by $\mathcal{F}(\mathcal{T})$ the set of all fixed points of $\mathcal{T}$. If for some $X \in B(\mathcal{H})$, $\mathcal{T}_\infty(X) := s - \lim_{t\to\infty} \mathcal{T}_t(X)$ exists, then $\mathcal{T}_\infty(X)$ is a fixed point of $\mathcal{T}$. Moreover, every fixed point of $\mathcal{T}$ is of this form. Further, note that C is a fixed point of a uniformly continuous Q.D.S. semigroup $\mathcal{T}$, if and only if $\mathcal{L}(C) = 0$. In other words the set of fixed points of the semigroup is exactly the kernel of $\mathcal{L}$. Moreover, if there exists $C \in B(\mathcal{H})$, satisfying $\mathcal{L}(C) = 0, C \ne 0$, then the quantum dynamical semigroup generated by $\mathcal{L}$ cannot be stable: In fact, if $C \ne 0$ is a fixed point, then $\mathcal{T}_t(C) \not\to 0$ as $t \to \infty$. We have

Theorem 3.1. *Let $\mathcal{T}$ be a Q.D.S. on $B(\mathcal{H})$ with generator $\mathcal{L}$. The semigroup $\mathcal{T}$ is stable if and only if the only fixed point of the family $(\mathcal{T}_t)_{t\ge0}$ is the operator* 0.

Proof. First note that since $0 \le \mathcal{T}_{s+t}(I) \le \mathcal{T}_t(I) \le I$ for all $t, s \ge 0$, the family $(\mathcal{T}_t(I))_{t\ge0}$ must converge strongly in $B(\mathcal{H})$ as $t \to \infty$. Let $C := s - \lim_{t\to\infty} \mathcal{T}_t(I)$. The operators $\mathcal{T}_t$ are normal, the net $\mathcal{T}_t(I)$ decreases strongly to C, and the normality of $\mathcal{T}_s$ for any $s > 0$ implies that $\mathcal{T}_{t+s}(I)$ converges strongly to $\mathcal{T}_s(C)$. Thus C is a fixed point of $\mathcal{T}$.

Now suppose that the only fixed point of the family $(\mathcal{T}_t)_{t\ge0}$ is the operator 0. Then $C = 0$, so that the Q.D.S. is stable.

Conversely, suppose that $\mathcal{T}$ is stable, so that $s - \lim_{t\to\infty} \mathcal{T}_t(X) = 0$ for every $X \in B(\mathcal{H})$ (Remark 2.2 (iii)) and let Y be a fixed point of $B(\mathcal{H})$. Then $0 = s - \lim_{t\to\infty} \mathcal{T}_t(Y) = Y$. Thus the only fixed point of $\mathcal{T}$ is the zero operator. □

Remark 3.2. Suppose that K, L_i are selfadjoint operators in $B(\mathcal{H})$ and $\mathcal{L}$ given by $\mathcal{L}(X) = KX + XK^* + \sum_i L_i X L_i,\ X \in B(\mathcal{H})$ generates the quantum dynamical semigroup $\mathcal{T}$. If Ker $K \ne \{0\}$, then for every $x_0 \in \text{Ker } K, C := |x_0\rangle\langle x_0|$ is a fixed point for $\mathcal{T}$. Indeed, if $x_0 \in \text{Ker } K$, $\|x_0\| = 1$, then $\mathcal{L}(I) \le 0$ implies

$$\left\langle \left(2K + \sum\nolimits_i L_i^2\right) x_0, x_0 \right\rangle \le 0 \quad \text{ or } \quad \sum\nolimits_i \|L_i(x_0)\|^2 \le 0,$$

which ensures that $L_i(x_0) = 0$ for all i. Let $C = |x_0\rangle\langle x_0|$. Then

$$\begin{aligned}\mathcal{L}(C) &= K\,|x_0\rangle\langle x_0| + |x_0\rangle\langle x_0|\,K + \sum_i L_i\,|x_0\rangle\langle x_0|\,L_i \\ &= |K(x_0)\rangle\langle x_0| + |x_0\rangle\langle K(x_0)| + \sum_i |L_i(x_0)\rangle\langle L_i(x_0)| = 0.\end{aligned}$$

Thus such a semigroup cannot be stable.

For a completely positive map ϕ acting on $B(\mathcal{H})$, the set

$$\{X \in B(\mathcal{H}), X \geq 0 : \phi(X) = X\}$$

of positive fixed points of ϕ and the set $\{X \in B(\mathcal{H}), X \geq 0 : \phi(X) \leq X\}$ have been studied by Popescu [6] in detail. Analogously, we set, in addition to the already defined set $\mathcal{F}(\mathcal{T})$ of fixed points of $\mathcal{T}$, $\hat{\mathcal{F}} = \{X \in B(\mathcal{H}) : X = X^*, \mathcal{T}_t(X) \leq X, t \geq 0\}$, for the quantum dynamical semigroup $\mathcal{T}$ on $B(\mathcal{H})$. For some characterisations of the subspace $\mathcal{F}(\mathcal{T})$, and $\hat{\mathcal{F}}$ we refer to [13].

Note that positive elements of the subspace $\mathcal{F}(\mathcal{T})$ are also called *harmonic operators* with respect to the Q.D.S. $\mathcal{T}$ while positive operators in $\hat{\mathcal{F}}$ are called *super-harmonic*. A positive operator $X \in B(\mathcal{H})$ is said to be *sub-harmonic* if $\mathcal{T}_t(X) \geq X$ for all $t \geq 0$. Clearly a stable Q.D.S does not have any non trivial sub-harmonic operators. A Markovian quantum dynamical semigroup is said to be *irreducible* if it has no non-trivial sub-harmonic projections. Extending this concept to our case of sub-Markovian Q.D.S. we see that a stable Q.D.S. is *irreducible* (see [9]). Moreover, for E semigroups, that is, quantum dynamical semigroups consisting of automorphisms, the converse is also true, because in this case $\mathcal{T}^\infty(I) = s - \lim_{t\to\infty} \mathcal{T}_t(I)$ is a sub-harmonic orthogonal projection.

The proof of the following works exactly as in Theorem 3.1 [6]. For a different approach to a similar decomposition see [13, Theorem 2.13].

Theorem 3.3. *Let $\mathcal{T}$ be a quantum dynamical semigroup on $B(\mathcal{H})$ and let $A \in B(\mathcal{H})$ be a selfadjoint operator satisfying*

$$\mathcal{T}_t(A) \leq A \text{ for all } t \geq 0.$$

Then A admits a decomposition $A = B + C$ where $B, C \in B(\mathcal{H})$ and

(i) *$B = B^*$ and $\mathcal{T}_t(B) = B$ for all $t \geq 0$.*
(ii) *$C \geq 0$ and $\mathcal{T}_t(C) \downarrow 0$ in the strong operator topology as $t \to \infty$.*

If $\mathcal{T}$ has a bounded generator $\mathcal{L}$ then $\mathcal{L}(A) \leq 0$ if and only if $\mathcal{T}_t(A) \leq A$ for all $t \geq 0$.

Proof. Since $\{\mathcal{T}_t(A)\}_{t\geq 0}$ is decreasing and bounded, it follows that it converges strongly in $B(\mathcal{H})$. Let $B = s-\lim_{t\to\infty} \mathcal{T}_t(A)$. Then B is selfadjoint and $\mathcal{T}_t(B) = B$ for all $t \geq 0$. Set $C = A - B$. Then clearly $C \geq 0$ and for every $t \geq 0, \mathcal{T}_t(C) = \mathcal{T}_t(A) - B$. This implies that $\mathcal{T}_t(C) \leq C$ and $\lim_{t\to 0} \mathcal{T}_t(C) = 0$. □

Next, we have an ergodic type result, similar to Theorem 3.2 [6].

Corollary 3.4. *Let $\mathcal{T}$ be a quantum dynamical semigroup on $B(\mathcal{H})$ and let $A \in B(\mathcal{H})$ be a selfadjoint operator satisfying*

$$\mathcal{T}_t(A) \leq A \text{ for all } t \geq 0.$$

Then

$$w - \lim_{t\to\infty} \frac{1}{t}\int_0^t \mathcal{T}_t(A)\, dt = B,$$

where $A = B + C$ as obtained in the last theorem.

Proof. Using the decomposition of A obtained above, we have

$$\frac{1}{t}\int_0^t \mathcal{T}_t(A)\, dt = B + \frac{1}{t}\int_0^t \mathcal{T}_t(C)\, dt.$$

Therefore, it suffices to show that the integral on the right-hand side above converges strongly to 0. Since $\mathcal{T}_t(C)$ converges strongly to zero we have that $w - \lim_{t\to\infty} \frac{1}{t}\int_0^t \mathcal{T}_t(C)\, dt = 0$. □

Recall that a strongly continuous semigroup $A = (A_t)_{t\geq 0}$ of bounded operators on $\hat{H}$ is said to be a unit of the Q.D.S. $\mathcal{T}$ if there exists a $c > 0$ such that $\mathcal{T}$ dominates the elementary quantum dynamical semigroup $(e^{-ct}\alpha_t^A)_{t\geq 0}$ where $\alpha_t^A(X) = A_t X A_t^*$, $X \in B(\mathcal{H})$. The unit A is said to be normalised if c can be taken to be zero. Every normalised unit is contractive. The following results bring out the strong connection between super-harmonic operators in $B(\mathcal{H})$ with respect to the Q.D.S. and the invariant subspaces of the units of $\mathcal{T}$. The proofs are along the same lines as [6, Section 4].

Theorem 3.5. *Let $\mathcal{T}$ be a quantum dynamical semigroup on $B(\mathcal{H})$. If $C \geq 0$ satisfies $\mathcal{T}_t(C) \leq C$ then the subspace $Ker\, C$ is invariant under each $(A_t^*)_{t\geq 0}$, where $(A_t)_{t\geq 0}$ is any unit for $\mathcal{T}$. In particular, if $\mathcal{M}$ is a subspace of $\mathcal{H}$ and $\mathcal{T}_t(P_{\mathcal{M}}) \leq (P_{\mathcal{M}})$, then $\mathcal{M}$ is invariant under each A_t, $t \geq 0$.*

Proof. It is enough to establish the result for normalised units. Suppose $\mathcal{T}_t(C) \leq C$ and let $(A_t)_{t\geq 0}$ be a normalised unit for $\mathcal{T}$. For $h \in \operatorname{Ker} C$, and any $t \geq 0$,

$$\langle A_t C A_t^* h, h\rangle \leq \langle \mathcal{T}_t(C)h, h\rangle \leq \langle Ch, h\rangle = 0.$$

It follows therefore that $CA_t^*h = 0$. Thus, $A_t^*(\text{Ker } C) \subset \text{Ker } C$, for all $t \geq 0$. Equivalently, $A_t((Ker\, C)^{\perp}) \subset (\text{Ker } C)^{\perp}$. In particular, this holds for $C = P_{\mathcal{M}}$, so that $\mathcal{M}$ is invariant under each A_t. □

The proofs for the following two results work along the same lines as in [6, Corollary 4.2 and Theorem 4.3] and are not included here. We shall call an operator $C \in B(H)$ a *pure solution* of the operator inequality $\mathcal{T}_t(X) \leq X$, $t \geq 0$ if $\mathcal{T}_t(C) \leq C$, $t \geq 0$ and $s - \lim_{t\to\infty} \mathcal{T}_t(C) = 0$. This is consistent with the definition in [6] for the discrete case. Note that a pure solution is always positive.

Corollary 3.6. *If $X \in B(\mathcal{H})$ is a sub-harmonic operator with respect to the Q.D.S. $\mathcal{T}$ and $\|X\| = 1$, then the fixed point set of X is invariant under $A_t^*, t \geq 0$, where A is a unit of $\mathcal{T}$.*

Theorem 3.7. *Let C be a non-zero positive operator in $B(\mathcal{H})$ such that $\mathcal{T}_t(C) \leq C$. Every unit A of $\mathcal{T}$ has a non-trivial invariant subspace provided one of the following statements hold:*

(i) *C is not injective.*
(ii) *C is not pure with respect to $\mathcal{T}$ and there is a $h \in B(\mathcal{H}), h \neq 0$, such that $\lim_{t\to\infty} \mathcal{T}_t(C)h = 0$.*
(iii) *C is not a fixed point of $\mathcal{T}$ and there exists a $h \neq 0$ such that $\mathcal{T}_t(C)h = Ch$, for all $t \geq 0$.*

If $(\mathcal{T}_t)_{t\geq 0}$ is a quantum dynamical semigroup satisfying $\mathcal{T}_t(I) \leq I$ for all $t > 0$, then $(\mathcal{T}_t)_{t\geq 0}$ is a bounded, decreasing family of positive operators in $B(\mathcal{H})$. Therefore, $\lim_{t\to\infty} \mathcal{T}_t(I)$ exists in the strong operator topology in $B(\mathcal{H})$. We list some properties of this operator. The first two of these follow directly from the definition while the remaining can be deduced by adapting the proofs for the discrete case given in [6, Proposition 4.5].

Theorem 3.8. *Let $(\mathcal{T}_t)_{t\geq 0}$ be a quantum dynamical semigroup satisfying $\mathcal{T}_t(I) \leq I$, $t \geq 0$. Then*

$$\mathcal{T}^\infty(I) := s - \lim_{t\to\infty} \mathcal{T}_t(I)$$

exists and has the following properties:

(i) *$0 \leq \mathcal{T}^\infty(I) \leq I$.*
(ii) *$\mathcal{T}_t(\mathcal{T}^\infty(I)) = \mathcal{T}^\infty(I)$.*
(iii) *If $\mathcal{T}^\infty(I) \neq 0$ then $\|\mathcal{T}^\infty(I)\| = 1$.*
(iv) *If $\mathcal{T}^\infty(I)h \neq 0$, then $\mathcal{T}_t(I)h \neq 0$ for any $t \geq 0$.*
(v) Ker $\mathcal{T}^\infty(I) = \{h \in \mathcal{H} : \lim_{t\to\infty} \mathcal{T}_t(I)h = 0\}$.
(vi) Ker $(I - \mathcal{T}^\infty(I)) = \{h \in \mathcal{H} : \mathcal{T}_t(I)h = h,\ t \geq 0\}$.

The following is a continuous version of the Wold type decomposition theorem proved in [6, 4.7]. Again, the proof is similar to the discrete case, just using Theorem 3.5 instead of [6, Lemma 4.1] and we omit the details.

Theorem 3.9. *Let $(\mathcal{T}_t)_{t\geq 0}$ be a Q.D.S. with $\mathcal{T}_t(I) \leq I$ for all $t \geq 0$. Then $\mathcal{H}$ admits a decomposition of the form*

$$\mathcal{H} = \mathcal{M} \oplus \operatorname{Ker}(I - \mathcal{T}^\infty(I)) \oplus \operatorname{Ker} \mathcal{T}^\infty(I),$$

and the subspaces $\operatorname{Ker}(I - \mathcal{T}^\infty(I))$ and $\operatorname{Ker} \mathcal{T}^\infty(I)$ are invariant under $(A_t^*)_{t\geq 0}$, where $A = (A_t)_{t\geq 0}$ is any unit for $\mathcal{T}$. Further, $\mathcal{M} = \{0\}$ if and only if $\mathcal{T}^\infty(I)$ is an orthogonal projection.

Remark 3.10. If $\mathcal{T}$ is a semigroup of endomorphisms then $\mathcal{T}^\infty(I)$ is automatically an orthogonal projection. However, the following example shows that this may not be true for a general quantum dynamical semigroup.

Example. Let $\mathcal{L}$ be the operator on $B(\mathbb{C}^2)$ given by $\mathcal{L}(X) = KX + XK^* + LXL^*$ where

$$K = \begin{pmatrix} -1 & 0 \\ 0 & 0 \end{pmatrix}, L = \begin{pmatrix} c & d \\ 0 & 0 \end{pmatrix}$$

where $|c|^2 + |d|^2 - 2 < 0$, and $c, d \neq 0$. Then, since

$$\mathcal{L}\left(\left(\begin{array}{cc} x & y \\ z & w \end{array}\right)\right) = \left(\begin{array}{cc} (|c|^2 - 2)x + cdy + cdz + |d|^2 w & y \\ z & 0 \end{array}\right),$$

for $x, y, z, w \in \mathbb{C}$, it follows that

$$\operatorname{Ker} \mathcal{L} = \left\{ \left(\begin{array}{cc} \frac{|d|^2}{2-|c|^2} w & 0 \\ 0 & w \end{array}\right) : w \in \mathbb{C} \right\}.$$

Now $\mathcal{T}^{\infty}(I)$ is an element of $\operatorname{Ker} \mathcal{L}$. Therefore, it cannot be an orthogonal projection.

4. Fixed points and dilations

Just as every C_0 semigroup of contractions $(T_t)_{t\geq 0}$ on a Hilbert space $\mathcal{H}$ admits a minimal dilation consisting of semigroup of isometries [11, Section 11.18], every quantum dynamical semigroup has a minimal dilation (unique, up to unitary equivalence) consisting of E semigroups. We recall here that a Q.D.S. on $B(\mathcal{H})$ is called an E semigroup if it consists of $*$-endomorphisms of $B(\mathcal{H})$. Suppose $(\mathcal{T}_t)_{t\geq 0}$ is a quantum dynamical semigroup on $B(\mathcal{H})$. If $\hat{\mathcal{H}}$ is a Hilbert space containing $\mathcal{H}$ as a closed subspace and if $\theta = (\theta_t)_{t\geq 0}$ is an E semigroup on $\mathcal{B}(\hat{\mathcal{H}})$, such that

$$\mathcal{T}_t(X) = P\theta_t(X)P, t \geq 0, X \in B(\mathcal{H}) = P\mathcal{B}(\hat{\mathcal{H}})P \subset \mathcal{B}(\hat{\mathcal{H}}),$$

where P is the orthogonal projection of $\hat{\mathcal{H}}$ onto $\mathcal{H}$, then θ is called a *dilation* of $\mathcal{T}$. The dilation θ is said to be minimal if

$$\overline{\operatorname{span}}\{\theta_{r_1}(X_1)\ldots\theta_{r_n}(X_n)u : r_i \geq 0, X_i \in B(\mathcal{H}), u \in \mathcal{H}, 1 \leq i \leq n, n \geq 0\}$$

is all of $\hat{\mathcal{H}}$. Note that a semigroup $(\theta_t)_{t\geq 0}$ of $*$-endomorphisms is called an E_0 semigroup if it is unital, that is, $\theta_t(I) = I \;\forall\; t \geq 0$. Minimal dilation of a Q.D.S. is an E_0 semigroup if and only if the Q.D.S. is unital. We refer to [1] and [2] for details concerning minimal dilations of Q.D.S..

Theorem 4.1. *A quantum dynamical semigroup is stable if and only if its minimal dilation is stable.*

Proof. Let θ be the minimal E dilation of the stable quantum dynamical semigroup $(\mathcal{T}_t)_{t\geq 0}$. Now the unitisation of $(\mathcal{T}_t)_{t\geq 0}$ is the unital semigroup $(\tilde{\mathcal{T}}_t)_{t\geq 0}$ acting on $\mathbb{C} \oplus B(\mathcal{H})$ according to the rule

$$\tilde{\mathcal{T}}_t\left(\left(\begin{array}{cc} a & 0 \\ 0 & X \end{array}\right)\right) = \left(\begin{array}{cc} a & 0 \\ 0 & \mathcal{T}_t(X) + a(I - \mathcal{T}_t(I)) \end{array}\right),$$

where $a \in \mathbb{C}, X \in B(\mathcal{H})$. If $(\theta_t)_{t\geq 0}$ is the minimal dilation of $\mathcal{T}$ acting on $B(\hat{\mathcal{H}})$ then its unitisation, acting on $\mathbb{C} \oplus B(\hat{\mathcal{H}})$ is similarly given by $(\tilde{\theta}_t)_{t\geq 0}$ with

$$\tilde{\theta}_t\left(\left(\begin{array}{cc} a & 0 \\ 0 & Z \end{array}\right)\right) = \left(\begin{array}{cc} a & 0 \\ 0 & \theta_t(Z) + a(I - \theta_t(I)) \end{array}\right),$$

for $Z \in B(\hat{\mathcal{H}})$. Also,

$$\tilde{\theta}_t\left(\begin{pmatrix} 1 & 0 \\ 0 & 0 \end{pmatrix}\right) \uparrow \begin{pmatrix} 1 & 0 \\ 0 & I_{\hat{\mathcal{H}}} \end{pmatrix},$$

strongly as $t \to \infty$. Therefore, $s - \lim_{t\to\infty} \theta_t(I) = 0$, so that θ is stable.

Conversely, if θ is stable, then $\theta_t(I) \to 0$, strongly as $t \to \infty$. This forces $\mathcal{T}_t(I) \to 0$ strongly. So $\mathcal{T}$ is stable. □

Now we look at the behaviour of fixed points of a quantum dynamical semigroup. The following theorem obtains a characterization of fixed points.

Theorem 4.2. *Let $(\mathcal{T}_t)_{t\geq 0}$ be a uniformly continuous quantum dynamical semigroup on $B(\mathcal{H})$. A positive operator $C \in B(\mathcal{H})$ satisfies the equation $\mathcal{T}_t(X) = X$ for all $t \geq 0$, (respectively, the inequality $\mathcal{T}_t(X) \leq X$ for all $t \geq 0$) if and only if there exists a quantum dynamical semigroup $(\beta_t)_{t\geq 0}$ on $B(\mathcal{H})$, such that $\beta_t(I) = I$, (resp. $\beta_t(I) \leq I$) and*

$$\mathcal{T}_t(C^{\frac{1}{2}}XC^{\frac{1}{2}}) = C^{\frac{1}{2}}\beta_t(X)C^{\frac{1}{2}} \tag{4.1}$$

for all $t \geq 0$ and $X \in B(\mathcal{H})$. Moreover, C is a pure solution of $\mathcal{T}_t(X) \leq X$ if and only if there exists a Q.D.S. β which is stable and satisfies (4.1).

Proof. If (4.1) holds with $\beta_t(I) = I$ then it is easy to see that $\mathcal{T}_t(C) = C$. Now assume conversely that the positive operator $C \in B(\mathcal{H})$ satisfies $\mathcal{T}_t(C) = C$ for all $t \geq 0$. Let θ be the minimal E_0 dilation of $\mathcal{T}$ acting on $B(\hat{\mathcal{H}})$, where $\hat{\mathcal{H}}$ is a Hilbert space and $\mathcal{H} \subset \hat{\mathcal{H}}$, as discussed at the end of Section 2.

Let $\mathcal{K} := \overline{\text{Range}(C^{\frac{1}{2}})}$. For $t \geq 0$, define

$$W_t : \mathcal{K} \longrightarrow \hat{\mathcal{H}}, \text{ by setting} \tag{4.2}$$

$$W_t(C^{\frac{1}{2}}h) = \theta_t(C^{\frac{1}{2}})h, h \in H. \tag{4.3}$$

Then

$$\begin{aligned}
\langle W_t(C^{\frac{1}{2}}h), W_t(C^{\frac{1}{2}}h)\rangle &= \langle \theta_t(C^{\frac{1}{2}})h, \theta_t(C^{\frac{1}{2}})h\rangle \\
&= \langle \theta_t(C)h, h\rangle \\
&= \langle P\theta_t(C)Ph, h\rangle \\
&= \langle \mathcal{T}_t(C)h, h\rangle \\
&= \langle Ch, h\rangle.
\end{aligned}$$

Thus, $||W_t(C^{\frac{1}{2}}h)|| = ||C^{\frac{1}{2}}h||$, so that W_t is a well-defined isometry.

Define, for $t \geq 0, \gamma_t : B(\mathcal{K}) \longrightarrow B(\mathcal{K})$ by

$$\gamma_t(X) = PW_t^*\theta_t(X)W_tP|_{\mathcal{K}}, \tag{4.4}$$

for all $X \in B(\mathcal{K})$. Here $P := P_{\mathcal{K}}$ is the orthogonal projection of $\hat{\mathcal{H}}$ onto $\mathcal{K}$. We claim that $(\gamma_t)_{t\geq 0}$ is a unital quantum dynamical semigroup acting on $B(\mathcal{K})$ and satisfies $\mathcal{T}_t(C^{\frac{1}{2}}XC^{\frac{1}{2}}) = C^{\frac{1}{2}}\gamma_t(X)C^{\frac{1}{2}}$ for all $X \in B(\mathcal{K})$. Observe that, for $t \geq 0$ and $h, g \in \mathcal{H}$, and $X \in B(\mathcal{K})$,

$$\begin{aligned}
\langle C^{\frac{1}{2}}\gamma_t(X)C^{\frac{1}{2}}(h), g\rangle &= \langle C^{\frac{1}{2}}PW_t^*\theta_t(X)W_tPC^{\frac{1}{2}}(h), g\rangle \\
&= \langle (W_tPC^{\frac{1}{2}})^*\theta_t(X)\theta_t(C^{\frac{1}{2}})(h), g\rangle \\
&= \langle \theta_t(XC^{\frac{1}{2}})h, (W_tC^{\frac{1}{2}})g\rangle \\
&= \langle \theta_t(XC^{\frac{1}{2}})Ph, \theta_t(C^{\frac{1}{2}})g\rangle \\
&= \langle \theta_t(C^{\frac{1}{2}}XC^{\frac{1}{2}})Ph, Pg\rangle \\
&= \langle \mathcal{T}_t(C^{\frac{1}{2}}XC^{\frac{1}{2}})h, g\rangle.
\end{aligned}$$

Therefore

$$C^{\frac{1}{2}}\gamma_t(X)C^{\frac{1}{2}} = \mathcal{T}_t(C^{\frac{1}{2}}XC^{\frac{1}{2}}) \tag{4.5}$$

for $t \geq 0$ and $X \in B(\mathcal{K})$. Using the same argument as above we also obtain, for $h, g \in \mathcal{H}$,

$$\begin{aligned}
\langle \gamma_t(I)C^{\frac{1}{2}}h, C^{\frac{1}{2}}g\rangle &= \langle \mathcal{T}_t(C)h, g\rangle \\
&= \langle C\,h, g\rangle \\
&= \langle C^{\frac{1}{2}}\,h, C^{\frac{1}{2}}\,g\rangle.
\end{aligned}$$

Thus $\gamma_t(I) = I$ for all $t \geq 0$.

For each t, γ_t is clearly completely positive and $t \mapsto \gamma_t$ is continuous in the weak operator topology since θ is. Next we check the semigroup property for γ. Note first that, for $t, s \geq 0, h \in \mathcal{H}$,

$$\begin{aligned}
(\theta_s(W_tP)W_sP)(C^{\frac{1}{2}}h) &= \theta_s(W_tP)\theta_s(C^{\frac{1}{2}})(h) \\
&= \theta_s(W_tC^{\frac{1}{2}})h \\
&= \theta_s(\theta_t(C^{\frac{1}{2}}))h \\
&= \theta_{s+t}(C^{\frac{1}{2}})h \\
&= W_{s+t}P(C^{\frac{1}{2}}h).
\end{aligned}$$

It follows that

$$\theta_s(W_tP)W_sP = W_{t+s}P, \quad \text{for all } t \geq 0. \tag{4.6}$$

Therefore, for $X \in B(\mathcal{K})$,

$$\begin{aligned}
\gamma_s\gamma_t(X) &= \gamma_s\left(PW_t^*\theta_t(X)W_tP\right) \\
&= PW_s^*\theta_s\left(PW_t^*\theta_t(X)W_tP\right)W_sP \\
&= PW_s^*\theta_s(PW_t^*)\theta_{s+t}(X)\theta_s(W_tP)W_sP \\
&= PW_s^*\theta_s(PW_t^*)\theta_{s+t}(X)W_{s+t}P \\
&= PW_{s+t}^*\theta_{s+t}(X)W_{s+t}P \\
&= \gamma_{t+s}(X),
\end{aligned}$$

which establishes that γ is a semigroup. Thus γ is a quantum dynamical semigroup on $B(\mathcal{K})$. Now we will extend $(\gamma_t)_{t\geq 0}$ to a quantum dynamical semigroup on $B(\mathcal{H})$. Note that the uniform continuity of $\mathcal{T}$ implies the uniform continuity of θ which in turn makes the operator family W_t uniformly continuous and hence the Q.D.S. γ as well. Suppose the generator of the former is $\mathcal{L}$, given as usual by $\mathcal{L}(X) = KX+XK^*+\sum_i L_i^*XL_i$, $X \in B(\mathcal{H})$ while the generator of the latter semigroup is represented as $\mathcal{M}(X) = MX+XM^*+\sum_j N_j^*XN_j, X \in B(\mathcal{K})$. Here $K, L_i \in B(\mathcal{H})$ while $M, N_j \in B(\mathcal{K})$, and $j \in J$, say. From (4.5) it follows that

$$\begin{aligned} &KC^{\frac{1}{2}}XC^{\frac{1}{2}} + C^{\frac{1}{2}}XC^{\frac{1}{2}}K^* + \sum\nolimits_i L_i^*C^{\frac{1}{2}}XC^{\frac{1}{2}}L_i \\ &= C^{\frac{1}{2}}\left(MX + XM^* + \sum\nolimits_j N_j^*XN_j\right)C^{\frac{1}{2}} \end{aligned} \tag{4.7}$$

for all $X \in B(\mathcal{K})$. Choose operators $A, B_j \in B(\mathcal{K}^{\perp})$ such that $A+A^*+\sum_j B_j^*B_j = 0$. Set

$$Q := \begin{pmatrix} M & 0 \\ 0 & A \end{pmatrix}, R_j := \begin{pmatrix} N_j & 0 \\ 0 & B_j \end{pmatrix}.$$

Then $Q, R_j \in B(\mathcal{H})$ and $\tilde{L}$ given by $\tilde{L}(X) = QX+XQ^*+\sum_j R_jXR_j^*$ is a bounded operator on $B(\mathcal{H})$. Moreover, it generates a unital, quantum dynamical semigroup, say β. From (4.7) and the construction of Q, R_j it follows that $\mathcal{L}(C^{\frac{1}{2}}XC^{\frac{1}{2}}) = C^{\frac{1}{2}}\tilde{L}(X)C^{\frac{1}{2}}$ for all $X \in B(\mathcal{H})$. The proof for C satisfying the operator inequality $\mathcal{T}_t(C) \leq C$ works in an almost identical manner as above with equality replaced by an inequality in appropriate places. Alternatively, the argument used below to prove the last part of the result would also suffice.

To prove the last part, assume that $\mathcal{T}_t(C) \leq C$. Construct W_t as above and then extend it linearly to the whole of $\hat{\mathcal{H}}$ by taking it as zero on $\mathcal{K}^{\perp}$. Define $\beta_t : B(\mathcal{H}) \to B(\mathcal{H})$ by setting

$$\beta_t(X) = P_{\mathcal{H}}W_t^*\theta_t(X)W_tP_{\mathcal{H}},\ t \geq 0,\ X \in B(\mathcal{H}).$$

Then $(\beta_t)_{t\geq 0}$ is a Q.D.S satisfying (4.1) and $\beta_t(X)g = \beta_t(X)P_{\mathcal{K}}g$, for any $g \in \hat{\mathcal{H}}$. Here $P_{\mathcal{H}}$ and $P_{\mathcal{K}}$ respectively are the orthogonal projections of $\hat{\mathcal{H}}$ onto $\mathcal{H}$ and $\mathcal{K}$. Since for $h, g \in \mathcal{H}$,

$$\langle \mathcal{T}_t(C)h, g\rangle = \langle \beta_t(I)C^{\frac{1}{2}}h, C^{\frac{1}{2}}g\rangle$$

if β is a stable Q.D.S., C must be a pure solution of $\mathcal{T}_t(X) \leq X$. On the other hand, if C is a pure solution, then

$$\lim_{t\to\infty} \langle \beta_t(I)C^{\frac{1}{2}}h, C^{\frac{1}{2}}g\rangle = 0.$$

This implies that $s - \lim_{t\to\infty} \beta_t(I) = 0$. □

Remark 4.3. Note that in Theorem 4.2 above, the boundedness of the generator of $(\mathcal{T}_t)_{t\geq 0}$ is used just to extend the unital quantum dynamical semigroup $(\gamma_t)_{t\geq 0}$ acting on $B(\mathcal{K})$ to a unital quantum dynamical semigroup acting on all of $B(\mathcal{H})$. The existence of the unital semigroup $(\gamma_t)_{t\geq 0}$ satisfying the intertwining relation (4.5) is valid even without this assumption.

Next we explore the behaviour of fixed points of a quantum dynamical semigroup under dilations. The following lifting theorem holds.

Theorem 4.4 (Prunaru, [12]). *Let $(\mathcal{T}_t)_{t\geq 0}$ be a quantum dynamical semigroup acting on $B(\mathcal{H})$ and let $(\theta_t)_{t\geq 0}$ be its minimal dilation acting on $B(\hat{\mathcal{H}})$. A positive operator C is a fixed point of $\mathcal{T}$ iff $C = P_{\mathcal{H}}D|\mathcal{H}$, for some fixed point D of θ. Moreover, C is positive iff D is positive and $\|C\| = \|D\|$.*

Proof. This result for fixed points can be found in [12]. The construction of D is through Banach limits and it retains positivity and norm. □

At times lifting of fixed points to dilation can also be done through commutant lifting or intertwining lifting. We recall the Sz. Nagy-Foiaş commutant lifting theorem for contractions [11] and extend it to one parameter semigroups. We do not know as to whether this extension is already known or not. To this end we observe that the minimal isometric dilation of a contraction semigroup $(T_t)_{t\geq 0}$ defined on a Hilbert space $\mathcal{H}$ can also be obtained via the cogenerator of the semigroup. Let T be the cogenerator of the given semigroup and let U be its minimal unitary dilation on say $\mathcal{K}$. Then $\mathcal{K} = \bigvee_{n=-\infty}^{\infty} U^n\mathcal{H}$. It is shown in [11, Section III.9] that the semigroup $(U_t)_{t\geq 0}$ which has U as its cogenerator, is the minimal unitary dilation of $(T_t)_{t\geq 0}$. By considering U_+, the minimal isometric dilation of T on $\mathcal{K}_+$, given by $\mathcal{K}_+ = \bigvee_{n=0}^{\infty} U^n\mathcal{H}, U_+ = U\,|_{\mathcal{K}_+}$ we can similarly construct the minimal isometric dilation of the given semigroup. Indeed, since U is the minimal unitary dilation of T, and T being a cogenerator cannot have 1 as an eigenvalue, neither can U. But U is the minimal unitary dilation of U_+ as well, so that 1 cannot be an eigenvalue of U_+. Therefore, by Theorem III.8.1, [11], U_+ is the cogenerator of a semigroup $\left(U_t^+\right)_{t\geq 0}$ of isometries on $\mathcal{K}_+$. Using [11, Theorem III.2.3 (g) and Prop III. 9.2], it can be deduced that

$$\langle T_t h, \acute{h}\rangle = \langle U_t^+ h, \acute{h}\rangle$$

for all $h, \acute{h} \in \mathcal{H}$. Further

$$\mathcal{K}_+ = \bigvee_{n=0}^{\infty} U^n\mathcal{H} = \bigvee_{s\geq 0} U_s\mathcal{H} = \bigvee_{s\geq 0} U_s^+\mathcal{H}.$$

Thus, $\left(U_t^+\right)_{t\geq 0}$ is the minimal isometric dilation of $(T_t)_{t\geq 0}$. We also recall here that the cogenerator T and the semigroup$(T_t)_{t\geq 0}$ can be realized in terms of each other in the following manner:

$$T = \lim_{s\to 0+} \psi_s(T_s), \quad T_t = e_t(T) \text{ where}$$

$$\psi_s(\lambda) = \frac{1-s}{1+s} - \frac{2s}{1+s}\sum_{n=1}^{\infty} \frac{\lambda^n}{(1+s)^n}, \text{ and} \tag{4.8}$$

$$e_t(\lambda) = e^{\left(\frac{s(\lambda+1)}{\lambda-1}\right)}.$$

For details see [11, Chapter III].

Theorem 4.5. *Let $(R_t)_{t\geq 0}$ and $(S_t)_{t\geq 0}$ be two strongly continuous contraction semigroups acting on the Hilbert spaces $\mathcal{H}$ and $\mathcal{K}$, respectively. Suppose that $(V_t)_{t\geq 0}$ and $(W_t)_{t\geq 0}$ are the respective minimal isometric dilations acting on the Hilbert spaces $\hat{\mathcal{H}}$ and $\hat{\mathcal{K}}$. If a bounded operator $C : \mathcal{H} \to \mathcal{K}$ satisfies the equation*

$$CR_t = S_t C, \ t \geq 0 \tag{4.9}$$

then there exists an operator $\hat{C} : \hat{\mathcal{H}} \to \hat{\mathcal{K}}$ such that

$$\hat{C}V_t = W_t\hat{C}, \ t \geq 0, \tag{4.10}$$

and $\|\hat{C}\| = \|C\|, P_{\mathcal{H}}\hat{C}\,|_{\mathcal{H}} = C$.

Proof. Let R_0 and S_0 denote the respective cogenerators of the semigroups $(R_t)_{t\geq 0}$ and $(S_t)_{t\geq 0}$ and let V_0 and W_0 be their isometric dilations, acting on $\tilde{\mathcal{H}}$ and $\tilde{\mathcal{K}}$. Further, let (V_t^0) and (W_t^0) be the corresponding minimal isometric semigroup dilations. It is clear from (4.8) that $CR_0 = S_0C$. By the Sz. Nagy-Foiaş commutant lifting theorem [11, Theorem II.2.3], it follows that there exists a $\tilde{C} : \tilde{\mathcal{H}} \to \tilde{\mathcal{K}}$ such that $\tilde{C}V_0 = W_0\tilde{C}$, $\|\tilde{C}\| = \|C\|$, and $C = P_{\mathcal{H}}\tilde{C}\,|_{\mathcal{H}}$ Now,

$$V^0(t) = e_t(V_0) = s - \lim_{r\to 1-0} e_{t,r}(V_0)$$

where

$$e_{t,r}(\lambda) = e_t(r\lambda) = \sum_{k=0}^{\infty} c_{r,k}V_0^k$$

and $e_t \in H^\infty_{V_0}$, and $c_{k,r} \in \mathbb{C}$ with $\sum_0^\infty |c_{k,r}| < \infty$.

For the definition of $H^\infty_{V_0}$ we refer again to [11, Section III.2]. A similar expression holds for W_t^0. Thus for any $x \in \tilde{\mathcal{H}}$ we have

$$\begin{aligned}
\tilde{c}V_t^0 x = \tilde{C}\left[\lim_{r\to 1-0}\sum_{k=0}^{\infty} c_{k,r}V_0^k x\right] &= \lim_{r\to 1-0}\sum_{k=0}^{\infty} c_{k,r}\tilde{C}V_0^k x \\
&= \lim_{r\to 1-0}\sum_{k=0}^{\infty} c_{k,r}W_0^k\tilde{C}x = \left[\lim_{r\to 1-0}\sum_{k=0}^{\infty} c_{k,r}W_0^k\right](\tilde{C}x) \\
&= e_t(W_0)\tilde{C}x = W_t^0\tilde{C}x.
\end{aligned}$$

Thus $\tilde{C}V_t^0 = W_t^0\tilde{C}$ for all $t \geq 0$. Now V_t and V_t^0, being two minimal isometric dilations of R_t, are unitarily equivalent, that is, there exists a unitary operator $\phi_1 : \tilde{\mathcal{H}} \to \hat{\mathcal{H}}$ such that $V_t^0 = \phi_1^{-1}V_t\phi_1$ for all $t \geq 0$. Similarly, there exists a unitary operator $\phi_2 : \tilde{\mathcal{K}} \to \hat{\mathcal{K}}$ such that $W_t^0 = \phi_2^{-1}W_t\phi_2$ for all $t \geq 0$. Setting $\hat{C} = \phi_2\tilde{C}\phi_1^{-1}$, we get the required operator satisfying (4.10). □

The following result concerns implications of Theorem 4.2 for dilation theory. For any quantum dynamical semigroup η, we shall denote, for the sake of convenience, the operator $\eta_{r_1}(X_1)\eta_{r_2}(X_2)\ldots\eta_{r_n}(X_n)$ by $\eta(\underline{r}, \underline{X})$ for n-tuples $\underline{r} = (r_1, r_2, \ldots, r_n), r_i \geq 0$, and

$$\underline{X} = (X_1, X_2, \ldots, X_n), X_i \in B(\mathcal{H}).$$

Theorem 4.6. *Suppose $\mathcal{T}$ and β acting on $B(\mathcal{H})$ are Q.D.S. with β unital, and there exists an invertible operator $R \in B(\mathcal{H})$ satisfying*

$$\mathcal{T}_t(RXR^*) = R\beta_t(X)R^*,$$

for all $X \in B(\mathcal{H}), t \geq 0$. Let θ and η be the minimal dilations of $\mathcal{T}$ and β, acting on $B(\mathcal{K}_1)$ and $B(\mathcal{K}_2)$, respectively, where $\mathcal{H} \subset \mathcal{K}_1$ and $\mathcal{H} \subset \mathcal{K}_2$. Then there exists $\tilde{R} : \mathcal{K}_2 \to \mathcal{K}_1$ such that

$$\theta_t(\tilde{R}Y\tilde{R}^*) = \tilde{R}\eta_t(Y)\tilde{R}^*, \quad Y \in B(\mathcal{K}_2), t \geq 0,$$

with $R = \tilde{R}|_{\mathcal{H}}$, $\|\tilde{R}\| = \|R\|$.

Proof. Since θ and η are minimal dilations of $\mathcal{T}$ and β respectively, it follows that

$$\begin{aligned}
\mathcal{K}_1 &= \overline{\text{span}}\left\{\theta(\underline{r}, \underline{Y})u : (\underline{r}, \underline{Y}) \in N, u \in \mathcal{H}\right\}, \\
\mathcal{K}_2 &= \overline{\text{span}}\left\{\eta(\underline{r}, \underline{Y})u : (\underline{r}, \underline{Y}) \in N, u \in \mathcal{H}\right\}, \text{where} \\
N &= \left\{(\underline{a}, \underline{Y}) : a_1 \geq a_2 \geq \cdots a_n \geq 0, Y_i \in B(\mathcal{H}), n \in \mathbb{N} \cup \{0\}\right\}.
\end{aligned}$$

Define $S : \mathcal{K}_1 \to \mathcal{K}_2$ by setting

$$S(\theta(\underline{a}, \underline{B})v) = \eta_{a_1}(R^* B_1 R^{*^{-1}})\eta_{a_2}(R^* B_2 R^{*^{-1}}) \ldots \eta_{a_n}(R^* B_n R^{*^{-1}})R^* v \tag{4.11}$$

where $v \in B(\mathcal{H})$,

$$\begin{aligned}
\underline{B} &= (B_1, B_2, \ldots, B_n), B_i \in B(\mathcal{H}), \\
\underline{a} &= (a_1, a_2, \ldots, a_n), a_r \geq a_s \text{ if } r \geq s.
\end{aligned}$$

Then it is straightforward to check that for $u, v \in \mathcal{H}$,

$$\langle S(\theta(\underline{a}, \underline{A})u), S(\theta(\underline{a}, \underline{B})v)\rangle = \langle \theta_{a_1}(R^*R)\theta_{a_1}(A_1) \ldots \theta_{a_n}(A_n)u, \theta(\underline{a}, \underline{B})v\rangle,$$

and for $c, d \in \mathbb{C}$,

$$\begin{aligned}
&\langle S(c\theta(\underline{a}, \underline{A})u + d\theta(\underline{a}, \underline{B})v), S(c\theta(\underline{a}, \underline{A})u + d\theta(\underline{a}, \underline{B})v)\rangle \\
&\quad = \langle \theta_{a_1}(R^*R)(c\theta(\underline{a}, \underline{A})u + d\theta(\underline{a}, \underline{B})v), c\theta(\underline{a}, \underline{A})u + d\theta(\underline{a}, \underline{B})v\rangle.
\end{aligned}$$

In fact, the equation above holds for any finite linear combination of terms of the form $\theta(\underline{a}, \underline{Y})$, so that we have

$$\left\| S \sum_{i=1}^{k} c_i \theta(\underline{a}, \underline{A}^i)u^i \right\|^2 \leq \|R^*R\| \left\| \sum_{i=1}^{k} c_i \theta(\underline{a}, \underline{A}^i)u^i \right\|^2.$$

Therefore, S is a well-defined, bounded linear operator from $\mathcal{K}_1$ to $\mathcal{K}_2$ with $\|S\| \leq \|R\|$. Further,

$$S\theta(\underline{o}, \underline{I})v = \eta_0(R^* R^{*^{-1}})R^* v = R^* u$$

for all $u \in \mathcal{H}$. Thus $S|_{\mathcal{H}} = R^*$ and $\|S\| = \|R^*\|$. Moreover, routine calculations show that

$$\begin{aligned}
S^*(\eta(\underline{a}, \underline{B})u) = \theta_{a_1}(RB_1R^*)\theta_{a_2}(R^{*^{-1}} B_2 R^*)\theta_{a_3}(R^{*^{-1}} B_3 R^*) \cdots \\
\cdots \theta_{a_n}(R^{*^{-1}} B_n R^*) R^{*^{-1}} u.
\end{aligned} \tag{4.12}$$

Also, $S^*\eta(\underline{o},\underline{I})u = Ru$. We will show next that $\theta_t(S^*XS) = S^*\eta_t(X)S$ for all $X \in B(\mathcal{K}_2)$. It is enough to establish this formula for X of the form $X = \eta_{a_1}(X_1)\eta_{a_2}(X_2)\dots\eta_{a_n}(X_n)$.

For such an X, and $(\underline{s},\underline{B}) \in N$ we have, on using (4.12),

$$
\begin{aligned}
&(S^*XS)(\theta(\underline{s},\underline{B})v)\\
&= S^*X\eta_{s_1}(R^*B_1{R^*}^{-1})\eta_{s_2}(R^*B_2{R^*}^{-1})\dots\eta_{s_k}(R^*B_k{R^*}^{-1})R^*v\\
&= S^*\eta_{a_1}(X_1)\dots\eta_{a_n}(X_n)\eta_{s_1}(R^*B_1{R^*}^{-1})\eta_{s_2}(R^*B_2{R^*}^{-1})\cdots\eta_{s_k}(R^*B_k{R^*}^{-1})R^*v\\
&= \theta_{a_1}(RX_1R^*)\theta_{a_2}({R^*}^{-1}X_2R^*)\dots\theta_{a_n}({R^*}^{-1}X_nR^*)\theta_{s_1}(B_1)\theta_{s_2}(B_2)\cdots\theta_{s_k}(B_k)v.
\end{aligned}
$$

Therefore,

$$
\begin{aligned}
S^*XS &= \theta(\underline{a},\underline{\tilde{X}}) \text{ where}\\
\underline{\tilde{X}} &= (RX_1R^*, {R^*}^{-1}X_2R^*,\dots,{R^*}^{-1}X_nR^*).
\end{aligned} \tag{4.13}
$$

Therefore, for any $t \geq 0$, $\theta_t(\theta(\underline{a},\underline{\tilde{X}})) = \theta(\underline{a}+t,\underline{\tilde{X}})$. Thus,

$$
\begin{aligned}
&\theta_t(S^*XS)(\theta(\underline{s},\underline{B})u)\\
&= \theta_{a_1+t}(RX_1R^*)\theta_{a_2+t}({R^*}^{-1}X_2R^*)\cdots\theta_{a_n+t}({R^*}^{-1}X_nR^*)\\
&\quad\circ\theta_{s_1}({R^*}^{-1}R^*B_1{R^*}^{-1}R^*)\cdots\theta_{s_n}({R^*}^{-1}R^*B_n{R^*}^{-1}R^*){R^*}^{-1}R^*u\\
&= S^*\eta_{a_1+t}(X_1)\eta_{a_2+t}(X_2)\cdots\eta_{a_n+t}(X_n)\eta_{s_1}(R^*B_1{R^*}^{-1})\cdots\eta_{s_n}(R^*B_n{R^*}^{-1})R^*u\\
&= S^*\eta_{a_1+t}(X_1)\eta_{a_2+t}(X_2)\cdots\eta_{a_n+t}(X_n)S\theta(\underline{s},\underline{B})u\\
&= (S^*\eta_t(X)S)(\theta(\underline{s},\underline{B})u).
\end{aligned}
$$

Hence, $\theta_t(S^*XS) = S^*\eta_t(X)S$. Using similar arguments as above, it may be checked that S^* is actually invertible and

$$
{S^*}^{-1}\theta(\underline{a},\underline{B})v = \eta_{a_1}(R^{-1}B_1{R^*}^{-1})\eta_{a_2}(R^*B_2{R^*}^{-1})\cdots\eta_{a_n}(R^*B_n{R^*}^{-1})R^*v.
$$

Now $\tilde{R}$ may be chosen to be S^*. □

This theorem can be used for lifting positive, invertible fixed points of uniformly continuous quantum dynamical semigroups in a concrete way, as follows.

Corollary 4.7. *Let $(\mathcal{T}_t)_{t\geq 0}$ be a uniformly continuous quantum dynamical semigroup acting on $B(\mathcal{H})$ and let $(\theta_t)_{t\geq 0}$ be its minimal dilation acting on $B(\hat{\mathcal{H}})$. A positive, invertible operator $C \in B(\mathcal{H})$ is a solution of the equation $\mathcal{T}_t(X) = X$ for all $t \geq 0$, if and only if $C = P_{\mathcal{H}}D|\mathcal{H}$, where D is a positive, invertible solution of the equation $\theta_t(Y) = Y$, $t \geq 0$, such that $\|C\| = \|D\|$.*

Proof. From Theorem 4.2 it follows that the invertible positive operator C satisfies the equation $\mathcal{T}_t(X) = X$ for all $t \geq 0$, if and only if the Q.D.S. $\mathcal{T}$ is similar to another Q.D.S. β acting on $B(\mathcal{H})$ such that $\beta_t(I) = I$. By Theorem 4.6, there exists $\tilde{R} : \mathcal{K}_2 \to \hat{\mathcal{H}}$ such that

$$\theta_t(\tilde{R}Y\tilde{R}^*) = \tilde{R}\gamma_t(Y)\tilde{R}^*, Y \in B(\mathcal{K}_2), t \geq 0,$$

with $C^{\frac{1}{2}} = \tilde{R}|_{\mathcal{H}}, \|\tilde{R}\| = \|C^{\frac{1}{2}}\|$. Then $D = \tilde{R}\tilde{R}^*$ works. □

References

[1] B.V.R. Bhat, *An index theory for quantum dynamical semigroups*, Trans. of Amer. Math. Soc. **348** (1996), 561–583.
[2] B.V.R. Bhat, *Minimal dilations of quantum dynamical semigroups to semigroups of endomorphisms of C^* algebras*, J. Ramanujan Math. Soc. **14** (1999), 109–124.
[3] R. Chill and Y. Tomilov, *Stability of operator semigroups: ideas and results*, Perspectives in Operator Theory, Banach Center Publications, no. **75** (2007), 71–109.
[4] G. Lindblad, *On the generators of quantum dynamical semigroups*, Comm. Math. Phys. **48** (1976), 119–130.
[5] K.J. Engel and R. Nagel, *One-Parameter Semigroups for Linear Evolution Equations*, Springer, New York (2000).
[6] G. Popescu, *Similarity and ergodic theory of positive maps*, J. reine angew. Math. **561** (2003), 87–129.
[7] A. Pazy, *Semigroups of Linear Operators and Applications to Partial Differential Equations*, Springer Verlag, 1975.
[8] F. Fagnola, *Quantum Markov Semigroups and Quantum Flows*, Proyecciones **18** (1999), no. 3, 144 pp.
[9] F. Fagnola and R. Rebolledo, *Transience and recurrence of quantum Markov semigroups*, Probab. Theory Relat. Fields **126** (2003), 289–306.
[10] K.B. Sinha and D. Goswami, *Quantum Stochastic Processes and Non Commutative Geometry*, Cambridge Tracts in Mathematics, no. **169** (2007).
[11] Sz. Nagy and C. Foiaş, *Harmonic Analysis of Operators on Hilbert Space*, North-Holland, (1970).
[12] B. Prunaru, *Lifting fixed points of completely positive semigroups*, Integr. Equ. Oper. Theory 72 (2012), 219–222.
[13] V. Umanita, *Classification and decomposition of Quantum Markov Semigroups*, Probab. Theory Related Fields **134** (2006), no. 4, 603–623.

B.V. Rajarama Bhat
Indian Statistical Institute, Statistics and Mathematics Unit
R V College Post, 8th Mile Mysore Road
Bangalore 560059, India
e-mail: bhat@isibang.ac.in

Sachi Srivastava
Dept. of Mathematics, University of Delhi South Campus
Benito Juarez Road
New Delhi – 21, India
e-mail: sachi_srivastava@yahoo.com

Operator Theory:
Advances and Applications, Vol. 250, 87–105

Families of Operators Describing Diffusion Through Permeable Membranes

Adam Bobrowski

To Charles Batty with admiration

Abstract. We present generation and limit results for semigroups and cosine families for snapping out Brownian motion, a process modeling diffusion through permeable membranes.

Mathematics Subject Classification (2010). 47D06, 47D09, 35K05, 92B05, 60J35, 60J70.

Keywords. Method of images, cosine families, strongly-continuous semigroups, differential operators with integral conditions, transmission conditions, permeable membranes, snapping-out Brownian motion.

1. Introduction

Let U be the union of two compactified half-lines $U = [-\infty, 0-] \cup [0+, \infty]$ (with 0 split into two points $0-$ and $0+$, interpreted as representing positions to the immediate left and to the immediate right from a membrane situated at 0), and let $C(U)$ be the space of continuous functions on U, with the usual supremum norm. It will be convenient to identify $C(U)$ with the Cartesian product

$$\mathbb{X} := C[0, \infty] \times C[0, \infty]$$

via the formula

$$C(U) \ni f \mapsto (f_{-1}, f_1) \in C[0, \infty] \times C[0, \infty]$$

where $f_{-1}, f_1 \in C[0, \infty]$ are determined by $f_i(x) = f(ix), x > 0$, $i \in \mathfrak{I} := \{-1, 1\}$, and $C[0, \infty]$ is the space of continuous functions on $[0, \infty)$ having limits at infinity (with the supremum norm). Given four positive numbers σ_i, k_i, where $i \in \mathfrak{I}$ we

This research was supported in part by the Polish Government under Grant 6081/B/H03/2011/40.

define an operator A in $C(U)$ by

$$A\,(f_i)_{i\in\mathfrak{I}} = \left(\sigma_i^2 f_i''\right)_{i\in\mathfrak{I}} \tag{1.1}$$

with domain composed of $(f_i)_{i\in\mathfrak{I}} \in C^2[0,\infty] \times C^2[0,\infty]$ satisfying

$$f_i'(0) = k_i[f_i(0) - f_{-i}(0)], \qquad i \in \mathfrak{I}; \tag{1.2}$$

the $C^2[0,\infty] \subset C[0,\infty]$ denoting twice continuously members of $C[0,\infty]$ with second derivative in $C[0,\infty]$.

The related Cauchy problem in $C(U)$:

$$u'(t) = Au(t), \qquad t \geq 0, u(0) = f \in C(U) \tag{1.3}$$

models heat flow in two media (two half-lines), separated by a semi-permeable membrane located at $x = 0$, and transmission conditions (1.2) describe heat flow through the membrane.

These conditions may be plausibly interpreted: according to Newton's Law of Cooling, the temperature at $x = 0$ changes at a rate proportional to the difference of temperatures on either sides of the membrane, see [12, p. 9]. In this context, J. Crank uses the term *radiation boundary condition.* (Although, strictly speaking, these are not *boundary*, but *transmission* conditions, see [13–15].)

In the context of passing or diffusing through membranes, analogous transmission conditions were introduced by J.E. Tanner [31, Eq. (7)], who studied diffusion of particles through a sequence of permeable barriers (see also Powles et al. [29, Eq. (1.4)], for a continuation of the subject). In [1] (see, e.g., Eq. (4) there) similar conditions are used in describing absorption and desorption phenomena. We refer also to [19], where a compartment model with permeable walls (representing, e.g., cells, and axons in the white matter of the brain in particular) is analyzed, and to Equation [42] there.

In the context of neurotransmitters, conditions (1.2) were (re)-invented in [9] and [7], and interpreted in probabilistic terms (see [27] for a thorough stochastic analysis). To summarize the analysis presented in [9], we note that these conditions are akin to the elastic barrier condition: An elastic Brownian motion on $\mathbb{R}^+ := [0,\infty)$ (see, e.g., [23, 24]) is the process with generator $Gf = \frac{1}{2}f''$ defined on the domain composed of $f \in C^2[0,\infty]$ satisfying the *Robin boundary condition* (known also as *elastic barrier condition*):

$$f'(0) = kf(0).$$

In this process, the state-space is $\mathbb{R}^+$, and each particle performs a standard Brownian motion while away from the barrier $x = 0$. Upon touching the barrier a particle is reflected, but its time spent at the boundary (the so-called *local time*) is measured and after an exponential time with parameter k with respect to the local time, the particle is killed and no longer observed. Condition (1.2) expresses the fact that in the stochastic process described by the operator (1.1), each particle, instead of being killed, is transferred to the other side of the membrane $x = 0$. (Such a process, following Lejay [27], will be referred to as *snapping out Brownian motion.*) In particular, we see that k_i are permeability coefficients: the larger they are the shorter is the time for particles to diffuse through the membrane.

The paper is devoted to families of operators related to the snapping out Brownian motion. We show existence of semigroups and cosine families in the spaces of continuous and integrable functions. The generation theorem for cosine families involves Lord Kelvin's method of images modified so as to cover the case of transmission conditions. Moreover, we study the limit behavior of these families, as permeability coefficients k_1 and k_{-1} tend to ∞ or to 0.

2. Generation theorems for semigroups

2.1. A semigroup in $C(U)$

We start our considerations by showing that A defined in Introduction, generates a Feller semigroup $\left(\mathrm{e}^{tA}\right)_{t\geq 0}$ in $C(U)$, i.e., a semigroup leaving the non-negative cone of $C(U)$ invariant, and such that $\mathrm{e}^{tA}1 = 1$ where 1 is the constant function on U being equal to one everywhere. A well-known necessary and sufficient condition [4, 18, 30] for a densely defined operator to generate a Feller semigroup is that it satisfies the *positive maximum principle* (if the maximum of an $f \in D(A)$ is attained at $x \in U$, and $f(x) \geq 0$, then $Af(x) \leq 0$) and the *range condition* (see below). Our A is densely defined and satisfies the positive maximum principle. For if the maximum is attained at $x \notin \{0-, 0+\}$, then $Af(x)$ has the same sign as $f''(x) \leq 0$; in the other case, for example if $x = 0+$, we have $f'(0+) \leq 0$ while $f(0+) \geq f(0-)$. Therefore, by (1.2), $f'(0+) = f_1'(0+) = 0$ and the even extension of f_1 to the whole of $\mathbb{R}$ is twice continuously differentiable. Since its maximum is attained at $x = 0$, we have $f_1''(0) \leq 0$, as desired.

The range condition requires that, given $(g_i)_{i\in\mathfrak{I}} \in \mathbb{X}$ and $\lambda > 0$, we may find $(f_i)_{i\in\mathfrak{I}} \in D(A)$ such that $\lambda\,(f_i)_{i\in\mathfrak{I}} - A\,(f_i)_{i\in\mathfrak{I}} = (g_i)_{i\in\mathfrak{I}}$, i.e.,

$$\lambda f_i - \sigma_i^2 f_i'' = g_i, \qquad i \in \mathfrak{I}.$$

We will look for $(f_i)_{i\in\mathfrak{I}}$ of the form

$$\begin{aligned} f_i(x) &= C_i \mathrm{e}^{\frac{\sqrt{\lambda}}{\sigma_i}x} + D_i \mathrm{e}^{-\frac{\sqrt{\lambda}}{\sigma_i}x} - \frac{1}{\sigma_i\sqrt{\lambda}}\int_0^x \sinh\frac{\sqrt{\lambda}}{\sigma_i}(x-y)g_i(y)\,\mathrm{d}y \\ &= \frac{1}{2\sigma_i\sqrt{\lambda}}\int_0^\infty \mathrm{e}^{-\frac{\sqrt{\lambda}}{\sigma_i}|x-y|}g_i(y)\,\mathrm{d}y + D_i\mathrm{e}^{-\frac{\sqrt{\lambda}}{\sigma_i}x}, \qquad x \geq 0, \end{aligned} \tag{2.1}$$

where $C_i := \frac{1}{2\sigma_i\sqrt{\lambda}}\int_0^\infty \mathrm{e}^{-\frac{\sqrt{\lambda}}{\sigma_i}x} g_i(y)\,\mathrm{d}y$ and D_i are to be determined. Conditions (1.2) now impose $(\sigma_i\sqrt{\lambda} + k_i)D_i - k_iD_{-i} = (\sigma_i\sqrt{\lambda} - k_i)C_i + k_iC_{-i}, i \in \mathfrak{I}$. This is satisfied iff

$$D_i = \frac{\sqrt{\lambda} + k_{-i}\sigma_{-i} - k_i\sigma_i}{\sqrt{\lambda} + k_{-1}\sigma_{-1} + k_1\sigma_1}C_i + \frac{2k_i\sigma_i}{\sqrt{\lambda} + k_{-1}\sigma_{-1} + k_1\sigma_1}C_{-i}, \tag{2.2}$$

completing our task.

To summarize, we have the following theorem.

Theorem 2.1. *Let $k_i, \sigma_i, i \in \Im$ be positive numbers. The operator A defined by* (1.1) *and* (1.2) *is the generator of a Feller, conservative semigroup in $C(U)$.*

2.2. A semigroup in $L^1(\mathbb{R})$

The semigroup of the previous subsection describes (weighted) expected values of the snapping out Brownian motion. More specifically,

$$\mathrm{e}^{tA} f(x) = E_x f(w(t)), \qquad x \in U, f \in C(U),$$

where $w(t), t \geq 0$ is the said Brownian motion and E_x denotes expected value conditional on the Brownian motion starting at x. In this section, we want to study a semigroup in $L^1(\mathbb{R})$, that is in a sense dual to e^{tA} – this semigroup describes dynamics of the processes' *distributions* or, more precisely, of their densities. Hence, a natural concept here is that of a *Markov operator* which is a linear operator in $L^1(\mathbb{R})$, the space space of (equivalence classes) of Lebesgue integrable functions on $\mathbb{R}$. An operator P is said to be Markov iff it leaves the positive cone of $L^1(\mathbb{R})$ invariant (i.e., $P\phi \geq 0$ for $\phi \geq 0$) and preserves the integral there, i.e., $\int P\phi = \int \phi$, for $\phi \geq 0$. It is easy to see that Markov operators are contractions. For a densely defined operator A in $L^1(\mathbb{R})$ to generate a semigroup of Markov operators it is necessary and sufficient for its resolvent to be Markov, which means by definition that all $\lambda(\lambda - A)^{-1}, \lambda > 0$ are Markov. This result may be deduced from the Hille–Yosida theorem, see [26].

As in the previous section, it will be convenient to identify a member ϕ of $L^1(\mathbb{R})$ with the pair $(\phi_i)_{i \in \Im}$ of functions on $\mathbb{R}^+$ defined by $\phi_i(x) = \phi(ix), x \geq 0$. Here, as before, $\Im = \{-1, 1\}$. Certainly $\phi_i \in L^1(\mathbb{R}^+)$, i.e., we identify $L^1(\mathbb{R})$ with $L^1(\mathbb{R}^+) \times L^1(\mathbb{R}^+)$ (with norm $\| (\phi_i)_{i \in \Im} \| = \|\phi_{-1}\| + \|\phi_1\|$.)

With this identification in mind, and given positive constants $k_i, \sigma_i, i \in \Im$ of Introduction, we define the operator A^* in $L^1(\mathbb{R}^+)$ by

$$A^* (\phi_i)_{i \in \Im} = \left(\sigma_i^2 \phi_i''\right)_{i \in \Im} \tag{2.3}$$

with domain composed of $(\phi_i)_{i \in \Im} \in W^{2,1}(\mathbb{R}^+) \times W^{2,1}(\mathbb{R}^+)$ satisfying the *transmission conditions*:

$$\sigma_1^2 \phi_1'(0) = k_1 \sigma_1^2 \phi_1(0) - k_{-1} \sigma_{-1}^2 \phi_{-1}(0), \qquad \sigma_1^2 \phi_1'(0) + \sigma_{-1}^2 \phi_{-1}'(0) = 0. \tag{2.4}$$

Here, $W^{2,1}(\mathbb{R}^+)$ is the set of differentiable functions on $\mathbb{R}^+$ whose derivatives are absolutely continuous with second derivatives belonging to $L^1(\mathbb{R}^+)$. The operator A^* is dual to A introduced in (1.1) in the sense that,

$$\int_{\mathbb{R}} f A^* \phi = \int_{\mathbb{R}} \phi A f,$$

for all $\phi \in D(A^*)$ and $f \in D(A)$. The key factor in the necessary calculations (using integration by parts formula), is of course conditions (1.2) and (2.4). In other words, (2.4) is a dual version of (1.2). Again, these conditions describe the way the membrane allows the traffic from one half-axis to the other. Interestingly, as

Lemma 2.2 (later on) shows, in $L^1(\mathbb{R})$ the second relation in (2.4) has an additional interpretation: It is a balance condition saying that the mass inflow into one half-axis is equal to the mass outflow out of the other one, and guarantees that the resolvent of A^* preserves the integral.

For an alternative proof of this result, assume that A^* is already proved to be the generator of a Markov semigroup, and consider the functionals F_- and F_+ on $L^1(\mathbb{R}^+)$ given by

$$F_-\phi = \int_{-\infty}^0 \phi, \qquad F_+\phi = \int_0^\infty \phi.$$

Let $\phi \in D(A^*)$, where A^* is given by (2.3) and (2.4), be a density (i.e., $\phi \geq 0$ and $\int_\mathbb{R} \phi = 1$). Also, let $\alpha(t) := F_-(\mathrm{e}^{tA^*}\phi)$, be the proportion of probability mass in $\mathbb{R}^-$ at time $t \geq 0$. Since α satisfies

$$\frac{\mathrm{d}}{\mathrm{d}t}\alpha(t) = F_-(A^*\mathrm{e}^{tA^*}\phi) = \int_{-\infty}^0 \sigma_{-1}^2 \frac{\mathrm{d}^2}{\mathrm{d}x^2}\mathrm{e}^{tA^*}\phi(x)\,\mathrm{d}x = \sigma_{-1}^2\left(\frac{\mathrm{d}}{\mathrm{d}x}\mathrm{e}^{tA^*}\phi\right)(0-),$$

the quantity $\sigma_{-1}^2(\frac{\mathrm{d}}{\mathrm{d}x}\mathrm{e}^{tA^*}\phi)(0-)$ describes the intensity of mass inflow into (outflow out of) $\mathbb{R}^-$ at time t. Analyzing $\beta(t) = F_+(\mathrm{e}^{tA^*}\phi)$ in a similar way, we conclude that the second relation in (2.4) is a balance condition saying that the mass inflow into one half-axis is equal to the mass outflow out of the other one. This was our task.

Coming back to the main subject, we claim that A^* defined above generates a semigroup of Markov operators in $L^1(\mathbb{R})$. For the proof, we will need the following two lemmas.

Lemma 2.2. *For $\lambda > 0$ let the operator $R_\lambda\,(\psi_i)_{i\in\mathfrak{I}} = (\phi_i)_{i\in\mathfrak{I}}$ in $L^1(\mathbb{R})$ be given by*

$$\begin{aligned}\phi_i(x) &= C_i\mathrm{e}^{\frac{\sqrt{\lambda}}{\sigma_i}x} + D_i\mathrm{e}^{-\frac{\sqrt{\lambda}}{\sigma_i}x} - \frac{1}{\sigma_i\sqrt{\lambda}}\int_0^x \sinh\frac{\sqrt{\lambda}}{\sigma_i}(x-y)\psi_i(y)\,\mathrm{d}y\\ &= \frac{1}{2\sigma_i\sqrt{\lambda}}\int_0^\infty \mathrm{e}^{-\frac{\sqrt{\lambda}}{\sigma_i}|x-y|}\psi_i(y)\,\mathrm{d}y + D_i\mathrm{e}^{-\frac{\sqrt{\lambda}}{\sigma_i}x}, \qquad x \geq 0. \end{aligned} \tag{2.5}$$

Here, $C_i = C_i(\psi_i) = \frac{1}{2\sigma_i\sqrt{\lambda}}\int_0^\infty \mathrm{e}^{-\frac{\sqrt{\lambda}}{\sigma_i}y}\psi_i(y)\,\mathrm{d}y$, and D_i are some functionals on $L^1(\mathbb{R})$. The operators λR_λ preserve the integral iff D_i are chosen so that ϕ_i satisfy the second condition in (2.4).

Proof. By definition, λR_λ preserve the integral iff $\int_\mathbb{R}\lambda R_\lambda\phi = \int_\mathbb{R}\phi, \phi \in L^1(\mathbb{R}), \lambda > 0$. Integrating (2.5),

$$\int_{\mathbb{R}^+}\phi_i = \frac{\sigma_i}{\sqrt{\lambda}}D_i + \frac{1}{\lambda}\int_{\mathbb{R}^+}\psi_i - \frac{\sigma_i}{\sqrt{\lambda}}C_i.$$

Hence the integral is preserved iff

$$\sigma_1(D_1 - C_1) + \sigma_{-1}(D_{-1} - C_{-1}) = 0. \tag{2.6}$$

On the other hand, $\phi_i'(0) = \frac{\sqrt{\lambda}}{\sigma_i}(C_i - D_i)$ showing that the second condition in (2.4) is equivalent to (2.6). □

Lemma 2.3. *The R_λ's from the previous lemma leave the positive cone invariant iff*

$$C_i + D_i \geq 0, \qquad i \in \mathfrak{I}. \tag{2.7}$$

Proof. The above condition means that $C_i(\psi) + D_i(\psi) \geq 0$ provided $\psi \in L^1(\mathbb{R})$ is non-negative. Assuming (2.7) we have

$$\begin{aligned}\phi_i(x) &\geq \frac{1}{2\sigma_i\sqrt{\lambda}}\int_0^\infty \mathrm{e}^{-\frac{\sqrt{\lambda}}{\sigma_i}|x-y|}\psi_i(y)\,\mathrm{d}y - C_i\mathrm{e}^{-\frac{\sqrt{\lambda}}{\sigma_i}x} \\ &\geq \frac{1}{2\sigma_i\sqrt{\lambda}}\int_0^\infty \left[\mathrm{e}^{-\frac{\sqrt{\lambda}}{\sigma_i}|x-y|} - \mathrm{e}^{-\frac{\sqrt{\lambda}}{\sigma_i}(x+y)}\right]\psi_i(y)\,\mathrm{d}y \\ &\geq 0,\end{aligned}$$

as long as $\psi_i \geq 0$. Conversely, suppose $\phi_i \geq 0$ if $\psi_i \geq 0$. Then, $C_i + D_i = \phi_i(0) \geq 0$ since ϕ_i is continuous. □

To show that A^* generates a semigroup of Markov operators, consider the resolvent equation for A^*:

$$\lambda\,(\phi_i)_{i\in\mathfrak{I}} - A^*\,(\phi_i)_{i\in\mathfrak{I}} = (\psi_i)_{i\in\mathfrak{I}} \tag{2.8}$$

where $\psi_i \in L^1(\mathbb{R}^+)$ and $\lambda > 0$ are given. The solution is given by (2.5) where D_i are to be determined so that $(\phi_i)_{i\in\mathfrak{I}} \in D(A^*)$. In particular, by the second condition in (2.4), we must have (2.6). The other condition in (2.4) forces

$$\sigma_1\sqrt{\lambda}(C_1 - D_1) = k_1\sigma_1^2(C_1 + D_1) - k_{-1}\sigma_{-1}^2(C_{-1} + D_{-1}).$$

These two are satisfied iff

$$D_i = \frac{\sqrt{\lambda} + k_{-i}\sigma_{-i} - k_i\sigma_i}{\sqrt{\lambda} + k_{-1}\sigma_{-1} + k_1\sigma_1}C_i + \frac{2k_{-i}\sigma_{-i}^3}{\sigma_1\sigma_{-1}\sqrt{\lambda} + k_{-1}\sigma_{-1}^2\sigma_1 + k_1\sigma_1^2\sigma_{-1}}C_{-i}. \tag{2.9}$$

This proves that the resolvent equation has a solution. Moreover, since the coefficient of C_i above is no less than -1, we have $D_i + C_i \geq 0$. By Lemmas 2.2 and 2.3, the resolvent of A^* is Markov. Hence, we are done provided we show that A^* is densely defined, but this is straightforward.

To summarize, we have the following theorem.

Theorem 2.4. *Let $k_i, \sigma_i, i \in \mathfrak{I}$ be positive numbers. The operator A defined by* (2.3) *and* (2.4) *is the generator of a semigroup of Markov operators in $L^1(\mathbb{R})$.*

3. Limit behavior (large permeability coefficients)

In this section, we study the limit of snapping out Brownian motions, as the membrane's permeability increases.

3.1. A limit in $L^1(\mathbb{R})$

First, we consider the semigroups in the set up of $L^1(\mathbb{R})$: Let A_n^* be defined by (2.3) and (2.4) with k_i replaced by nk_i. The resolvents $(\lambda - A_n^*)^{-1}$ are again given by (2.5) provided (2.9) is also modified by replacing k_i by nk_i. Then, the corresponding sequence of $D_i = D_i(n), n \ge 1$ (see (2.9)) converges to

$$D_i = \frac{k_{-i}\sigma_{-i} - k_i\sigma_i}{k_{-1}\sigma_{-1} + k_1\sigma_1} C_i + \frac{2k_{-i}\sigma_{-i}^3}{k_{-1}\sigma_{-1}^2\sigma_1 + k_1\sigma_1^2\sigma_{-1}} C_{-i}. \tag{3.1}$$

Therefore, the resolvents of A_n converge also. We check directly that (2.5) with the above D_i is the resolvent of the densely defined operator A_∞^* given by (2.3), and related to the transmission conditions

$$k_1\sigma_1^2\phi_1(0) = k_{-1}\sigma_{-1}^2\phi_{-1}(0), \qquad \sigma_1^2\phi_1'(0) + \sigma_{-1}^2\phi_{-1}'(0) = 0. \tag{3.2}$$

Since the limit of Markov operators is Markov, the resolvent of A_∞^* is Markov, and we conclude that A_∞^* generates a semigroup of Markov operators. By the Trotter–Kato theorem [2, 4, 16, 21, 28], the semigroups generated by A_n^* converge to the semigroup generated by A_∞^*, almost uniformly in $t \in \mathbb{R}^+$, i.e., uniformly in compact subintervals of $\mathbb{R}^+$.

We will argue that the limit process is not entirely 'free' – there is still a kind of barrier at $x = 0$. To this end, consider a process on $\mathbb{R}$ in which points of $\mathbb{R}^+$ move to the right with speed σ_1 and points of $\mathbb{R}^-$ move to the right with speed σ_{-1}. If ϕ is an initial distribution of such points, then

$$T(t)\phi(x) = \begin{cases} \phi(x - \sigma_1 t), & x \ge \sigma_1 t, \\ \frac{\sigma_{-1}}{\sigma_1}\phi(\frac{\sigma_{-1}}{\sigma_1}x - \sigma_{-1}t), & 0 < x < \sigma_1 t, \\ \phi(x - \sigma_{-1}t), & x \le 0, \end{cases} \tag{3.3}$$

is their distribution after time $t \ge 0$ (see Figure 1). It is easy to check that this formula defines a semigroup of Markov operators in $L^1(\mathbb{R})$, and that the generator of this semigroup is $B_{\sigma_{-1},\sigma_1}\phi(x) = \sigma_{\operatorname{sgn} x}\phi'(x)$ with domain composed of functions $\phi \in W^{1,1}(\mathbb{R}^+) \cap W^{1,1}(\mathbb{R}^-)$ where $\mathbb{R}^- = (-\infty, 0]$, satisfying

$$\sigma_{-1}\phi(0-) = \sigma_{+1}\phi(0+). \tag{3.4}$$

Next, let $J \in \mathcal{L}(L^1(\mathbb{R}))$ be given by $J\phi(x) = \phi(-x)$. Clearly, J is a Markov operator with $J^{-1} = J$. Moreover, for $\phi \in D(B_{\sigma_{-1},\sigma_1})$, we have $J\phi \in D(B_{\sigma_1,\sigma_{-1}})$ and $JB_{\sigma_{-1},\sigma_1}J\phi = -B_{\sigma_1,\sigma_{-1}}\phi$. Therefore, $-B_{\sigma_{-1},\sigma_1}$ is similar (or: isomorphic) to

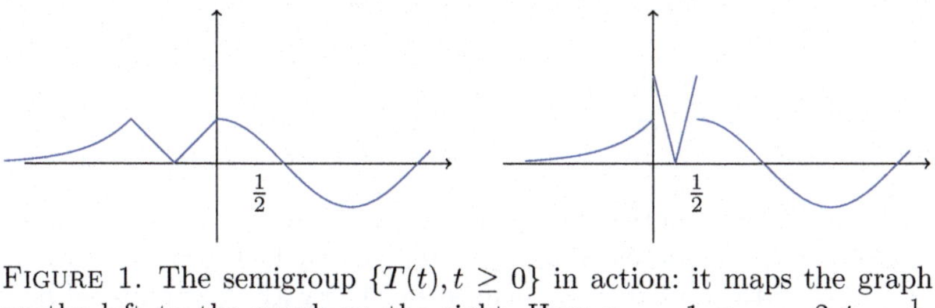

FIGURE 1. The semigroup $\{T(t), t \ge 0\}$ in action: it maps the graph on the left to the graph on the right. Here $\sigma_1 = 1, \sigma_{-1} = 2, t = \frac{1}{2}$. Points on the left half-axis move fast, and need to slow down on the right half-axis: hence, they are congested in the interval $[0, \frac{1}{2}]$.

$B_{\sigma_1,\sigma_{-1}}$ [4,16,17], and generates the semigroup of Markov operators $\{JT(t)J, t \geq 0\}$. By the generation theorem for groups (see, e.g., [16, p. 79]) B_{σ_{-1},σ_1} generates a group of Markov operators.

Combining this with Theorems 3.14.15 and 3.14.17 in [2], we see that $B^2_{\sigma_{-1},\sigma_1}$ generates the strongly continuous cosine family

$$C(t) = \frac{1}{2}\left(\mathrm{e}^{|t|B_{\sigma_{-1},\sigma_1}} + J\mathrm{e}^{|t|B_{\sigma_{-1},\sigma_1}}J\right), \quad t \in \mathbb{R},$$

and the related semigroup (both composed of Markov operators) defined by the Weierstrass formula.

The latter semigroup is a natural candidate for describing diffusion with different coefficients in the two half-axes and no barrier at $x = 0$. The domain of $B^2_{\sigma_{-1},\sigma_1}$ contains functions $\phi \in W^{2,1}(\mathbb{R}^+) \cap W^{2,1}(\mathbb{R}^-)$ with

$$\sigma_{-1}\phi(0-) = \sigma_1\phi(0+), \qquad \sigma^2_{-1}\phi'(0-) = \sigma^2_1\phi'(0+), \tag{3.5}$$

and we have $B^2_{\sigma_{-1},\sigma_1}\phi = \sigma^2_{\operatorname{sgn} x}\phi''$.

Finally, recalling the isomorphism of $L^1(\mathbb{R})$ and $L^1(\mathbb{R}^+)\times L^1(\mathbb{R}^+)$, we see that the isomorphic image in the latter space of the semigroup generated by $B^2_{\sigma_{-1},\sigma_1}$ is A^*_∞ provided $k_i\sigma_i = 1$ for $i \in \mathfrak{I}$. In other words, A^*_∞ describes the case of no barrier at $x = 0$ if the influences of diffusion and permeability coefficients cancel out. In other words, as we claimed, in general, conditions (3.2) do not describe the case of no barrier at $x = 0$, as there is an asymmetry in the way the particles filter in through the membrane from one half-axis to the other.

3.2. A limit in $C(U)$

Now, we would like to see what happens when permeability of the membrane becomes infinite, but this time we want to do analysis in $C(U)$. Replacing k_i with nk_i in (1.2) and defining the related operators A_n, we check that the corresponding sequence $D_i = D_i(n), n \geq 1$ (see (2.2)) converges to

$$D_i = \frac{k_{-i}\sigma_{-i} - k_i\sigma_i}{k_{-1}\sigma_{-1} + k_1\sigma_1}C_i + \frac{2k_i\sigma_i}{k_{-1}\sigma_{-1} + k_1\sigma_1}C_{-i}. \tag{3.6}$$

Therefore, the resolvents of A_n converge to the operator defined by (2.1) with D_i introduced above. However, here, we cannot apply the Trotter–Kato theorem in its classical form. The reason is that f_i defined by (2.1) and D_i given above satisfy:

$$\begin{aligned} f_1(0) = C_1 + D_1 &= \frac{2k_{-1}\sigma_{-1}}{k_{-1}\sigma_{-1} + k_1\sigma_1}C_1 + \frac{2k_1\sigma_1}{k_{-1}\sigma_{-1} + k_1\sigma_1}C_{-1} = C_{-1} + D_{-1} \\ &= f_{-1}(0). \end{aligned}$$

It follows that the range of the limit R_λ of $(\lambda - A_n)^{-1}$ is not dense in $\mathbb{X}$ and cannot be the resolvent of a densely defined operator. It may be shown though, that R_λ is the resolvent of the operator A_∞ defined by (1.1) with (1.2) replaced by

$$k_{-1}f_1'(0) + k_1f_{-1}'(0) = 0, \quad f_1(0) = f_{-1}(0). \tag{3.7}$$

Notably, A_∞ is not densely defined.

In such a case, we cannot claim almost uniform convergence of the semigroups on the whole of $\mathbb{X}$: rather the semigroups converge in this way merely on a subspace $\mathbb{X}_0$ defined as the closure of the range of the limit pseudoresolvent R_λ (see, e.g., [4, Section 8.4.3]).

As we will show now, $\mathbb{X}_0$ turns out to be the subspace of $(f_i)_{i\in\mathfrak{I}} \in \mathbb{X}$ satisfying $f_1(0) = f_{-1}(0)$, which may be identified with $C[-\infty,\infty]$, the space of continuous functions on $\mathbb{R}$ with limits at plus and minus infinity. To this end first note that, by (2.1) and (2.9), the coordinates of $R_\lambda\,(g_i)_{i\in\mathfrak{I}}$ are

$$f_i(x) = \frac{1}{2\sigma_i\sqrt{\lambda}} \int_{-\infty}^{\infty} \mathrm{e}^{-\frac{\sqrt{\lambda}}{\sigma_i}|x-y|} g_i^*(y)\,\mathrm{d}y, \qquad x \ge 0, \tag{3.8}$$

where

$$g_i^*(x) = \frac{k_{-i}\sigma_{-i} - k_i\sigma_i}{k_{-1}\sigma_{-1} + k_1\sigma_1} g_i(-x) + \frac{2k_i\sigma_i}{k_{-1}\sigma_{-1} + k_1\sigma_1} g_{-i}(-\frac{\sigma_{-i}}{\sigma_i}x), \qquad x < 0.$$

For $(g_i)_{i\in\mathfrak{I}} \in \mathbb{X}_0$, g_i^* is a continuous function, a member of $C[-\infty,\infty]$. Also, the operators $B_i = \sigma_i^2 \frac{\mathrm{d}^2}{\mathrm{d}x^2}$ with domain $C^2[-\infty,\infty]$ generate strongly continuous semigroups in $C[-\infty,\infty]$, and their resolvents are:

$$(\lambda - B_i)^{-1} g(x) = \frac{1}{2\sigma_i\sqrt{\lambda}} \int_{-\infty}^{\infty} \mathrm{e}^{-\frac{\sqrt{\lambda}}{\sigma_i}|x-y|} g(y)\,\mathrm{d}y, \qquad \lambda > 0, \quad x \in \mathbb{R}. \tag{3.9}$$

In particular $\lim_{\lambda\to\infty} \lambda\,(\lambda - B_i)^{-1} g = g$ in $C[-\infty,\infty]$. This implies that for $(g_i)_{i\in\mathfrak{I}} \in \mathbb{X}_0$, λf_i, where f_i is defined by (3.8), converges in $C[0,\infty]$ to g_i and shows that the range of R_λ is dense in $\mathbb{X}_0$, as claimed.

Hence, in $\mathbb{X}_0$ there is a semigroup being the limit of the semigroups generated by A_n. Its generator is the part A_p of A_∞ in $\mathbb{X}_0$ (see [4, Section 8.4.9]).

4. Limit behavior (small permeability coefficients)

If permeability coefficients are small, conditions (1.2) may be approximated by the Neumann boundary conditions

$$f_i'(0) = 0, i \in \mathfrak{I}, \tag{4.1}$$

describing perfectly non-permeable membrane. This intuition is confirmed by the following limit procedure.

For $n \ge 1$, let A_n be the operator defined by (1.1) and (1.2), with k_i replaced by $\frac{1}{n}k_i$. The resolvent of A_n is then given by (2.1) with D_i of (2.2) modified in the same way (i.e., with k_i replaced by $\frac{1}{n}k_i$). As $n \to \infty$, such D_i converges to C_i. Therefore, the corresponding f_i's of (2.1) converge to f_i's of (3.8) with $g_i^*(x) = g(-x), x \ge 0$. This means that the resolvents of A_n converge to the resolvent of A_0 defined by (1.1) with boundary conditions (4.1). A similar analysis shows that the same approximation works in $L^1(\mathbb{R})$.

To obtain a more interesting limit, we need to simultaneously let $k_i \to 0$ and $\sigma_i \to \infty$, while adding a reflecting barrier at $x = \pm 1$. More specifically, consider

$\widetilde{U} = [-1, 0-] \cup [0+, 1]$ and the space $C(\widetilde{U})$ identified with $C[0,1] \times C[0,1]$. Let $C^2[0,1]$ be the space of twice continuously differentiable functions on $[0,1]$, and let the operators A_n in $C[0,1] \times C[0,1]$ be defined by

$$A_n \left(f_i\right)_{i\in\mathfrak{I}} = \left(n\sigma_i^2 f_i''\right)_{i\in\mathfrak{I}} \tag{4.2}$$

for $f_i \in C^2[0,1], i \in \mathfrak{I}$ such that

$$f_i'(0) = \frac{1}{n}k_i f_i(0) - \frac{1}{n}k_i f_{-i}(0) \text{ and } f_i'(1) = 0. \tag{4.3}$$

These operators generate Feller semigroups in $C(\widetilde{U})$ and we have

$$\lim_{n\to\infty} \mathrm{e}^{tA_n} \left(f_i\right)_{i\in\mathfrak{I}} = \mathrm{e}^{tQ} P \left(f_i\right)_{i\in\mathfrak{I}} \tag{4.4}$$

almost uniformly in $t \in (0,\infty)$. Here, P defined by $P\left(f_i\right)_{i\in\mathfrak{I}} = \left(\int_0^1 f_i\right)_{i\in\mathfrak{I}}$ is a projection on the subspace $C_0(\widetilde{U})$ of functions that are constant on each of the intervals $[-1, 0-]$ and $[0+, 1]$; this subspace may of course be identified with $\mathbb{R}^2$ (with supremum norm). Moreover, Q is an operator in $C_0(\widetilde{U})$ identified with the matrix

$$\begin{pmatrix} -k_1\sigma_1 & k_1\sigma_1 \\ k_{-1}\sigma_{-1} & -k_{-1}\sigma_{-1} \end{pmatrix}.$$

Interpreted, (4.4) means that as diffusion coefficients increase while permeability coefficients decrease, points in each interval are lumped together to form two 'combined' points of the state-space of the limit process, and diffusions on two adjacent intervals are approximated by a two-state Markov chains with intensity matrix Q. Notably, the limit jump intensities are proportional to permeability and diffusion coefficients. This result is a simple case of a general theorem on fast diffusions on graphs – see [7]; an $L^1(\mathbb{R})$ analogue of this theorem may be found in [22].

5. A cosine family in $C(U)$

Our aim in this section is to prove that the operator A of Introduction with $\sigma_{-1} = \sigma_1$ generates a cosine family; without loss of generality we assume $\sigma_1 = \sigma_{-1} = 1$ (see Remark 5.5). To this end, we modify Lord Kelvin's method of images used previously in [5,6,10,11] in dealing with boundary conditions, to make it suitable in the case of transmission conditions, like (1.2).

The key step in the reasoning is finding extensions of members of $C(U)$, suitably coupled with transmission conditions (1.2). We start by introducing spaces where these extensions 'live'. Throughout this section $\omega > 0$ is a fixed parameter, $C[0,\infty]$, as before, is the space of continuous functions on $\mathbb{R}^+$ with limits at infinity (with usual supremum norm, denoted $\|\cdot\|$), and $C_\omega[0,\infty]$ is the space of functions on $\mathbb{R}^+$ such that $e_\omega f \in C[0,\infty]$, where

$$e_\omega(x) = \mathrm{e}^{-\omega x}, \quad x \geq 0.$$

The norm in $C_\omega[0,\infty]$ is

$$\|f\|_\omega := \|e_\omega f\|.$$

For a function $f:\mathbb{R}\to\mathbb{R}$, by Rf and Lf we denote its restrictions to $\mathbb{R}^+$ and $\mathbb{R}^-$, respectively. Moreover, for a function $f: D\to\mathbb{R}$ where $D\subset\mathbb{R}$, by Of (in Polish, 'reflection' starts with 'o', and the letter 'r' is already used for R='right'), we denote the function on $-D := \{x\in\mathbb{R}; -x\in D\}$ defined by

$$Of(x) = f(-x).$$

Let $C^L_\omega(\mathbb{R})$ be the space of $f:\mathbb{R}\to\mathbb{R}$ such that

$$Rf\in C[0,\infty] \quad \text{and} \quad OLf\in C_\omega[0,\infty];$$

the norm in this space is $\|f\|_L = \|Rf\|\vee\|OLf\|_\omega$. (This space will contain functions extended to the left.) Analogously, $C^R_\omega(\mathbb{R})$ is the space of functions with the following properties:

$$Rf\in C_\omega[0,\infty] \quad \text{and} \quad OLf\in C[0,\infty];$$

the norm here is $\|f\|_R = \|Rf\|_\omega\vee\|OLf\|$.

Let $\{C(t), t\in\mathbb{R}\}$ be the *basic cosine family* in $C^L_\omega(\mathbb{R})$ and $C^R_\omega(\mathbb{R})$ given formally by the same formula:

$$C(t)f(x) = \frac{1}{2}[f(x+t)+f(x-t)], \quad x\in\mathbb{R}. \tag{5.1}$$

We note that in both spaces the operator norm of $C(t)$ does not exceed $\mathrm{e}^{\omega t}$. Also, let

$$\mathcal{C}(t)(f,g) = (C(t)f, C(t)g)$$

be the Cartesian product cosine family in $C^R_\omega(\mathbb{R})\times C^L_\omega(\mathbb{R})$.

The main idea is to represent the searched-for cosine family $\{C_A(t), t\in\mathbb{R}\}$ in $\mathbb{X}$ in the form (see also (5.10))

$$C_A(t)(f_{-1},f_1) = (OLC(t)\widetilde{Of_{-1}}, RC(t)\widetilde{f_1}), \qquad (f_{-1},f_1)\in\mathbb{X}, \tag{5.2}$$

referred to as the *abstract Kelvin formula*, where $\widetilde{f_1}$ and $\widetilde{Of_{-1}}$ are suitable extensions of f_1 and Of_{-1} to members of $C^R_\omega(\mathbb{R})$ and $C^L_\omega(\mathbb{R})$, respectively. (See Figure 2.) We will argue that for $f=(f_{-1},f_1)\in D(A)$, these extensions are determined uniquely. To this end, we recall that a cosine family leaves the domain of its generator invariant. Hence, if (5.2) is to work, we must have

$$(OLC(t)\widetilde{Of_{-1}}, RC(t)\widetilde{f_1})\in D(A), \qquad t\ge 0. \tag{5.3}$$

We will formulate the existence and uniqueness result as a lemma.

Lemma 5.1. *For $f=(f_{-1},f_1)\in D(A)$, there is a unique pair $(\widetilde{Of_{-1}},\widetilde{f_1})\in C^R_\omega(\mathbb{R})\times C^L_\omega(\mathbb{R})$ such that (5.3) holds for $t\in\mathbb{R}$. Moreover, both extensions are twice continuously differentiable.*

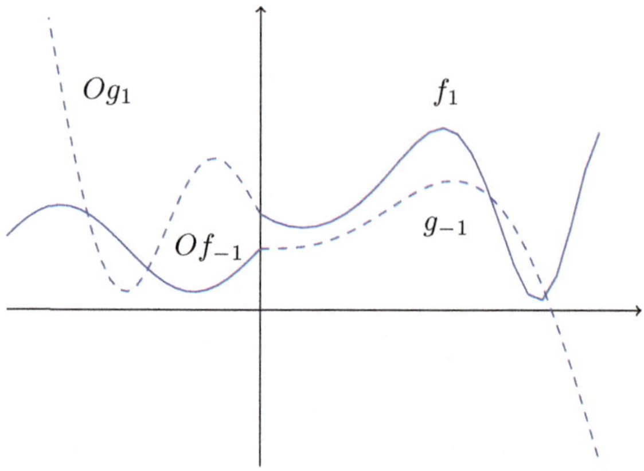

FIGURE 2. Extension of $f \in C(U)$ to a pair $(\widetilde{Of_{-1}}, \widetilde{f}_1) \in C^R_\omega(\mathbb{R}) \times C^L_\omega(\mathbb{R})$. Given $f_{-1}, f_1 \in C[0,\infty]$, we search for $g_{-1}, g_1 \in C_\omega[0,\infty]$. (In general, $g_{-1}, g_1 \notin C[0,\infty]$.)

Proof. We need to find

$$\begin{aligned} g_{-1}(x) &= \widetilde{Of_{-1}}(x), \\ g_1(x) &= \widetilde{f}_1(-x), \quad x \geq 0, \end{aligned} \tag{5.4}$$

and show that $g_{-1}, g_1 \in C_\omega[0,\infty]$ are determined uniquely. Of course, we need to have

$$g_1(0) = f_1(0) \text{ and } g_{-1}(0) = f_{-1}(0). \tag{5.5}$$

Moreover, since $C(t) = C(-t)$, condition (5.3) requires for all $t \geq 0$,

$$\begin{aligned} &\frac{\mathrm{d}}{\mathrm{d}x}\big[\widetilde{Of_{-1}}(-x+t) + \widetilde{Of_{-1}}(-x-t)\big]_{|x=0} \\ &\qquad = k_{-1}[\widetilde{Of_{-1}}(t) + \widetilde{Of_{-1}}(-t) - \widetilde{f}_1(t) - \widetilde{f}_1(-t)], \\ &\frac{\mathrm{d}}{\mathrm{d}x}\big[\widetilde{f}_1(x+t) + \widetilde{f}_1(x-t)\big]_{|x=0} \\ &\qquad = k_1[\widetilde{f}_1(t) + \widetilde{f}_1(-t) - \widetilde{Of_{-1}}(t) - \widetilde{Of_{-1}}(-t)], \end{aligned}$$

i.e.,

$$\begin{aligned} -g_{-1}'(t) + f_{-1}'(t) &= k_{-1}[g_{-1}(t) + f_{-1}(t) - f_1(t) - g_1(t)], \\ {f_1}'(t) - g_1'(t) &= k_1[f_1(t) + g_1(t) - g_{-1}(t) - f_{-1}(t)]. \end{aligned}$$

Moving unknowns to the left-hand side,

$$\begin{aligned} g_{-1}' + k_{-1}g_{-1} - k_{-1}g_1 &= f_{-1}' - k_{-1}f_{-1} + k_{-1}f_1, \\ g_1' + k_1g_1 - k_1g_{-1} &= f_1' - k_1f_1 + k_1f_{-1}. \end{aligned} \tag{5.6}$$

Pausing for the moment, we note that for any continuously differentiable h on $\mathbb{R}^+$, and a constant $k > 0$,

$$(h' + kh) * e_k = h - h(0)e_k,$$

where $*$ denotes convolution on $\mathbb{R}^+$. Hence, convolving both sides of equations (5.6) with $e_{k_{-1}}$ and e_{k_1}, respectively, and using conditions (5.5), we see that (5.6) implies

$$\begin{aligned} g_{-1} - k_{-1}e_{k_{-1}} * g_1 &= f_{-1} - 2k_{-1}e_{k_{-1}} * f_{-1} + k_{-1}e_{k_{-1}} * f_1, \\ g_1 - k_1 e_{k_1} * g_{-1} &= f_1 - 2k_1 e_{k_1} * f_1 + k_1 e_{k_1} * f_{-1}. \end{aligned} \tag{5.7}$$

Noting that $e_k * h = 0$ implies $h = 0$, we see that (5.7) is in fact equivalent to (5.6) coupled with (5.5).

Next, we equip $C_\omega[0,\infty] \times C_\omega[0,\infty]$ with the norm

$$\|(g_{-1}, g_1)\|_\omega = \|g_{-1}\|_\omega \vee \|g_1\|_\omega,$$

and consider a map T in this space, given by the formula:

$$T(g_{-1}, g_1) = (k_{-1}e_{k_{-1}} * g_1, k_1 e_{k_1} * g_{-1}).$$

We have (compare [3])

$$\begin{aligned} \|T(g_{-1}, g_1)\|_\omega &= \max_{i\in\Im} \sup_{x\geq 0} \left| e^{-\omega x} k_i \int_0^x e^{-k_i(x-y)} g_{-i}(y)\, dy \right| \\ &\leq \max_{i\in\Im} k_i \sup_{x\geq 0} \int_0^x e^{-(k_i+\omega)(x-y)} e^{-\omega y} |g_{-i}(y)|\, dy \\ &\leq \max_{i\in\Im} k_i \sup_{x\geq 0} \int_0^x e^{-(k_i+\omega)(x-y)} \|(g_{-1}, g_1)\|_\omega\, dy \\ &\leq \max_{i\in\Im} \frac{k_i}{\omega + k_i} \|(g_{-1}, g_1)\|_\omega \\ &= \frac{k}{\omega + k} \|(g_{-1}, g_1)\|_\omega, \end{aligned}$$

where $k = k_{-1} \vee k_1$. This proves that T is a bounded linear operator in $C_\omega[0,\infty] \times C_\omega[0,\infty]$ with norm at most $\frac{k}{\omega+k} < 1$. Hence, $I - T$ is invertible with bounded inverse, $\|(I-T)^{-1}\| \leq \frac{\omega+k}{\omega}$. Therefore, there is a unique solution of (5.7), given by

$$(g_{-1}, g_1) = (I - T)^{-1}(h_{-1}, h_1)$$

where h_{-1} and h_1 are right-hand sides of (5.7). (We note that $h_{-1}, h_1 \in C[0,\infty]$ and $\|h_{-1}\| \vee \|h_1\| \leq 4(\|f_{-1}\| \vee \|f_1\|$.) Hence,

$$\begin{aligned} \|(g_{-1}, g_1)\|_\omega &\leq \frac{\omega + k}{\omega} \|(h_{-1}, h_1)\|_\omega \leq \frac{\omega + k}{\omega} \|(h_{-1}, h_1)\| \\ &\leq 4\frac{\omega + k}{\omega} \|(f_{-1}, f_1)\|; \end{aligned} \tag{5.8}$$

we will need this information later.) By (5.6), it follows that g_{-1}, g_1 are twice continuously differentiable in $t \ge 0$ (with right-hand derivative at $t = 0$) provided so are f_{-1}, f_1, proving the first part of the lemma.

Turning to the second claim: we note that, by assumption $(f_{-1}, f_1) \in D(A)$, the right-hand sides of (5.6), evaluated at $t = 0$, are zero. Therefore, in view of (5.5),

$$\begin{aligned} g_{-1}'(0) &= k_{-1}[g_1(0) - g_{-1}(0)] = -f_{-1}'(0), \\ g_1'(0) &= k_1[g_{-1} - g_1(0)] = -f_1'(0), \end{aligned} \tag{5.9}$$

proving that the left-hand and right-hand derivatives of $\widetilde{Of_{-1}}$ and $\widetilde{f_{-1}}$ agree at $t = 0$, and so these functions are differentiable everywhere. Finally, differentiating the first equation in (5.6),

$$g_{-1}''(0) + k_{-1}[g_{-1}'(0) - g_1'(0)] = f_{-1}''(0) + k_{-1}[f_1'(0) - f_{-1}'(0)],$$

which, by (5.9) implies $g_{-1}''(0) = f_{-1}''(0)$, i.e., that $\widetilde{Of_{-1}}$ is twice differentiable at $t = 0$. A similar calculation of $g_1''(0)$ completes the proof. □

Definition 5.2. Let $f = (f_{-1}, f_1) \in \mathbb{X} = C[0,\infty] \times C[0,\infty]$ (not necessarily in $D(A)$); the pair $(\widetilde{Of_{-1}}, \widetilde{f_1}) \in C_\omega^R(\mathbb{R}) \times C_\omega^L(\mathbb{R})$, defined by (5.4) and (5.7), and denoted $\mathfrak{E}f$, will be termed the *transmission-mirror extension* of f, or the extension of f through transmission mirror. Of course, for $f \notin D(A)$, these extensions are not differentiable. We note that for $k_1 = k_{-1} = 0$, transmission mirror extensions reduce to classical even extensions.

Using (5.8) we see that $\mathfrak{E}$ is a bounded linear operator from $\mathbb{X}$ to $C_\omega^R(\mathbb{R}) \times C_\omega^L(\mathbb{R})$ with norm not exceeding $4\frac{\omega+k}{\omega}$. Introducing the map $\mathfrak{R} : C_\omega^R(\mathbb{R}) \times C_\omega^L(\mathbb{R}) \to \mathbb{X} = C[0,\infty] \times C[0,\infty]$ given by $\mathfrak{R}(f,g) = (OLf, Rg)$, we see that $\mathfrak{R}$ is bounded with norm equal to 1, and that the abstract Kelvin formula (5.2) may be written as

$$C_A(t) = \mathfrak{R}C(t)\mathfrak{E}, \tag{5.10}$$

so that $C_A(t)$ is a bounded operator with norm not exceeding $4\frac{\omega+k}{\omega}e^{\omega t}$.

Lemma 5.3. *Formula* (5.10) *defines a strongly continuous cosine family in* $\mathbb{X}$.

Proof. Fix $s \in \mathbb{R}$ and $f \in D(A)$. Lemma 5.1 says now that $\mathfrak{R}C(u)\mathfrak{E}f \in D(A)$ for all $u \in \mathbb{R}$. By d'Alembert's formula for $\{C(t), t \in \mathbb{R}\}$,

$$\mathfrak{R}C(t)C(s)\mathfrak{E}f = \frac{1}{2}\mathfrak{R}C(t+s)\mathfrak{E}f + \frac{1}{2}\mathfrak{R}C(t-s)\mathfrak{E}f \in D(A), \qquad t \in \mathbb{R}.$$

Since, by the same lemma, extensions through transmission mirror are unique for members of $D(A)$, and $\mathfrak{R}C(s)\mathfrak{E}f$ belongs to $D(A)$, we obtain

$$C(s)\mathfrak{E}f = \mathfrak{E}\mathfrak{R}C(s)\mathfrak{E}f.$$

It follows that

$$\begin{aligned}2C_A(t)C_A(s)f &= 2[\mathfrak{R}\mathcal{C}(t)\mathfrak{E}][\mathfrak{R}\mathcal{C}(s)\mathfrak{E}] = 2\mathfrak{R}\mathcal{C}(t)[\mathfrak{E}\mathfrak{R}\mathcal{C}(s)\mathfrak{E}] = 2\mathfrak{R}\mathcal{C}(t)\mathcal{C}(s)\mathfrak{E}f\\ &= \mathfrak{R}\mathcal{C}(t+s)\mathfrak{E}f + \mathfrak{R}\mathcal{C}(t+s)\mathfrak{E}f\\ &= C_A(t+s)f + C_A(t-s)f.\end{aligned}$$

Since $D(A)$ is dense and the operators involved are bounded, this proves d'Alembert's formula for $\{C_A(t), t \in \mathbb{R}\}$. Strong continuity of the cosine family follows by analogous property of the Cartesian product cosine family $\{\mathcal{C}(t), t \in \mathbb{R}\}$. □

We are in position to prove our main theorem in this section.

Theorem 5.4. *The operator*

$$A\,(f_i)_{i\in\mathfrak{I}} = (f_i'')_{i\in\mathfrak{I}} \tag{5.11}$$

with domain composed of $(f_i)_{i\in\mathfrak{I}} \in C^2[0,\infty] \times C^2[0,\infty]$ *satisfying* (1.2) *is the generator of the cosine family* (5.10).

Proof. For $(f_{-1}, f_1) \in D(A)$, both coordinates of $\mathfrak{E}(f_{-1}, f_1)$ are twice continuously differentiable functions on $\mathbb{R}$ (see Lemma 5.1). The Taylor formula implies now that

$$\lim_{t\to 0}\frac{2}{t^2}[C(t)\widetilde{f_1}(x) - \widetilde{f_1}(x)] = \widetilde{f_1}''(x), \qquad x \in \mathbb{R},$$

and the limit is uniform in $x \in [0,\infty)$ since $\widetilde{f_1}''$ is uniformly continuous in, say, $[-1,\infty)$, f_1 having a limit at ∞ by assumption. This shows that the second coordinate of $\frac{2}{t^2}[C_A(t)f - f]$ converges to f_1'' in $C[0,\infty]$. Similarly, the first coordinate converges to f_{-1}''. Hence, the generator, say G, of the cosine family (5.10) extends A. However, G cannot be a proper extension of A, since for $\lambda > \omega$, both $\lambda - A$ and $\lambda - G$ are onto and injective – for G this is clear since G is the generator, and for A this has been proved in Section 2.1. □

Remark 5.5. The case $\sigma_{-1} \neq \sigma_1$ should be treated by replacing the basic cosine family (5.1) by a cosine constructed, as in Section 3.1, from a 'dual' to (3.3). Since calculations are more extensive, they will be presented elsewhere. The main idea, though, remains the same.

6. A cosine family in $L^1(\mathbb{R})$

We proceed to showing that the operator A^* defined in (2.3) and (2.4) with $\sigma_1 = \sigma_{-1} = 1$, generates a cosine family in $L^1(\mathbb{R})$ identified with $L^1(\mathbb{R}^+) \times L^1(\mathbb{R}^+)$. The argument is quite analogous to that presented in Section 5; hence, we merely sketch it.

Instead of $C[0,\infty]$, we consider $L^1(\mathbb{R}^+)$. The role of $C_\omega[0,\infty]$ will be played by $L^1_\omega(\mathbb{R}^+)$ composed of (classes of) measurable functions ϕ with $e_\omega\phi \in L^1(\mathbb{R}^+)$; the norm here is $\|\phi\|_\omega = \|e_\omega\phi\|_{L^1(\mathbb{R}^+)} = \int_0^\infty \mathrm{e}^{-\omega t}|\phi(t)|\,\mathrm{d}t$.

Given $(\phi_i)_{i\in\mathfrak{I}} \in D(A^*)$ we search for smooth extensions $\widetilde{O\phi_{-1}}$ and $\widetilde{\phi_1}$ such that

$$(OLC(t)\widetilde{O\phi_{-1}}, RC(t)\widetilde{\phi_1}) \in D(A^*), \quad t \geq 0. \tag{6.1}$$

This leads to the following system of differential equations:

$$\begin{aligned} \psi_{-1}' + k_{-1}\psi_{-1} - k_1\psi_1 &= \phi_{-1}' - k_{-1}\phi_{-1} + k_1\phi_1, \\ \psi_1' + k_1\psi_1 - k_{-1}\psi_{-1} &= \phi_1' - k_1\phi_1 + k_{-1}\phi_{-1} \end{aligned} \tag{6.2}$$

for $\psi_{-1}(x) = \widetilde{O\phi_{-1}}(x)$ and $\psi_1(x) = \widetilde{\phi_1}(-x), x \geq 0$. As before, by $\psi_1(0) = \phi_1(0)$ and $\psi_{-1}(0) = \phi_{-1}(0)$, this system is equivalent to

$$\begin{aligned} \psi_{-1} - k_1 e_{k_{-1}} * \psi_1 &= \phi_{-1} - 2k_{-1}e_{k_{-1}} * \phi_{-1} + k_1 e_{k_{-1}} * \phi_1, \\ \psi_1 - k_{-1}e_{k_1} * \psi_{-1} &= \phi_1 - 2k_1 e_{k_1} * \phi_1 + k_{-1}e_{k_1} * \phi_{-1}. \end{aligned} \tag{6.3}$$

Noting that, for every $k > 0$ and $\phi \in L^1(\mathbb{R}^+)$,

$$\begin{aligned} \|e_k * \phi\|_\omega &\leq \int_0^\infty \mathrm{e}^{-\omega t} \int_0^t \mathrm{e}^{-k(t-s)} |\phi(s)| \,\mathrm{d}s\,\mathrm{d}t \\ &= \int_0^\infty \mathrm{e}^{ks} |\phi(s)| \int_s^\infty \mathrm{e}^{-(\omega+k)t} \,\mathrm{d}t\,\mathrm{d}s \\ &= \frac{1}{\omega+k} \int_0^\infty \mathrm{e}^{-\omega s} |\phi(s)| \,\mathrm{d}s = \frac{1}{\omega+k} \|\phi\|_\omega, \end{aligned}$$

we conclude that a solution to (6.3) exists in $L^1_\omega(\mathbb{R}^+)$ provided ω is large enough. More precisely, the map $(\psi_i)_{i\in\mathfrak{I}} \to (k_1 e_{k_{-1}} * \psi_1, k_{-1}e_{k_1} * \psi_{-1})$, as an operator in $L^1_\omega(\mathbb{R}^+) \times L^1_\omega(\mathbb{R}^+)$ (with norm $\| (\psi_i)_{i\in\mathfrak{I}} \| = \|\psi_{-1}\|_\omega + \|\psi_1\|_\omega$), has norm at most

$$K = \frac{k_1}{\omega + k_{-1}} \vee \frac{k_{-1}}{\omega + k_1}. \tag{6.4}$$

Hence, it suffices to take

$$\omega > |k_1 - k_{-1}| \tag{6.5}$$

to obtain the norm < 1. In other words, $\widetilde{O\phi_{-1}}$ is a member of the space $L^R_\omega(\mathbb{R})$, composed of $\varphi : \mathbb{R} \to \mathbb{R}$ such that

$$R\varphi \in L^1_\omega(\mathbb{R}^+) \quad \text{and} \quad OL\varphi \in L^1(\mathbb{R}^+);$$

(the norm in this space is $\|\varphi\|_L = \|R\varphi\|_\omega + \|OL\varphi\|_{L^1(\mathbb{R}^+)}$). Analogously, $\widetilde{\phi_1}$ belongs to $L^R_\omega(\mathbb{R})$, the space of functions φ with the following properties:

$$R\varphi \in L^1(\mathbb{R}^+) \quad \text{and} \quad OL\varphi \in L^1_\omega(\mathbb{R}^+);$$

the norm here is $\|\varphi\|_R = \|R\varphi\|_{L^1(\mathbb{R}^+)} \vee \|OL\varphi\|_\omega$.

We note that the pair formed by the right-hand sides in (6.3), as the member of $L^1(\mathbb{R}^+) \times L^1(\mathbb{R}^+)$ has norm at most $(3+K)\|\phi\|_{L^1(\mathbb{R})}$, where K was defined in (6.4). Therefore, the extension operator defined as in Definition 5.2, has norm at most $\frac{3+K}{1-K}$. The related restriction operator $\mathfrak{R}$ has norm one, and the formula

$$C_{A^*}(t) = \mathfrak{R}C(t)\mathfrak{E},$$

where $\mathcal{C}$ is the Cartesian product of the basic cosine families in $L^R_\omega(\mathbb{R})$ and $L^L_\omega(\mathbb{R})$, defines a cosine family with

$$\|C_{A^*}(t)\|_{\mathcal{L}(L^1(\mathbb{R}))} \le \frac{3+K}{1-K} \mathrm{e}^{\omega t}. \tag{6.6}$$

(As in the spaces of continuous functions of the previous section, the basic cosine families have norms not exceeding $\mathrm{e}^{\omega \cdot}$.) This is precisely the cosine family generated by A^*.

Remark 6.1. It is tempting to study convergence of the cosine families obtained in this section and the section preceding it, as permeability coefficients converge to infinity. For one thing, convergence of resolvents has been already established in Section 3. However, the Trotter–Kato analogue for cosine operators cannot be used here [20, 21, 25], since we do not have a stability condition at our disposal. For, if we take nk_i instead of k_i, and want to find ω independent of n, then (6.5) requires $k_1 = k_{-1}$, and even if this condition is fulfilled, still K in (6.6) converges to 1, and $\frac{3+K}{1-K}$ converges to ∞. A similar problem is faced when we try to study convergence of cosine families in $C(U)$ (see estimate (5.8)).

Fortunately, such problems do not occur in studying convergence as permeability coefficients tend to zero. Both in $C(U)$ and in $L^1(\mathbb{R})$, the cosine families involved converge to the cosine families related to the von Neumann conditions.

A potential cosine analogue of the limit (4.4) would be quite uninteresting, since cosine families by nature cannot converge outside of the regularity space defined as the closure of the limit pseudoresolvent (see [8]). In our case, this means that we cannot have convergence of cosine families outside of $C_0(\widetilde{U})$.

References

[1] S.S. Andrews, *Accurate particle-based simulation of adsorption, desorption and partial transmission*, Phys. Biol. **6** (2010), 046015.

[2] W. Arendt, C.J.K. Batty, M. Hieber, and F. Neubrander, *Vector-Valued Laplace Transforms and Cauchy Problems*, Birkhäuser, Basel, 2001.

[3] A. Bielecki, *Une remarque sur la méthode de Banach–Cacciopoli–Tikhonov*, Bull. Polish Acad. Sci. **4** (1956), 261–268.

[4] A. Bobrowski, *Functional Analysis for Probability and Stochastic Processes*, Cambridge University Press, Cambridge, 2005.

[5] ______, *Generation of cosine families via Lord Kelvin's method of images*, J. Evol. Equ. **10** (2010), no. 3, 663–675.

[6] ______, *Lord Kelvin's method of images in the semigroup theory*, Semigroup Forum **81** (2010), 435–445.

[7] ______, *From diffusions on graphs to Markov chains via asymptotic state lumping*, Ann. Henri Poincaré **13** (2012), 1501–1510.

[8] A. Bobrowski and W. Chojnacki, *Cosine families and semigroups really differ*, J. Evol. Equ. **13** (2013), no. 4, 897–916.

[9] A. Bobrowski and K. Morawska, *From a PDE model to an ODE model of dynamics of synaptic depression*, Discr. Cont. Dyn. Syst. B **17** (2012), no. 7, 2313–2327.

[10] A. Bobrowski and D. Mugnolo, *On moments-preserving cosine families and semigroups in* $C[0,1]$, J. Evol. Equ. **13** (2013), no. 4, 715–735.

[11] R. Chill, V. Keyantuo, and M. Warma, *Generation of cosine families on* $L^p(0,1)$ *by elliptic operators with Robin boundary conditions*, Functional Analysis and Evolution Equations. The Günter Lumer Volume, 2007, pp. 113–130. Amann, H. et al. (eds.)

[12] J. Crank, *The mathematics of diffusion*, Second Edition, Clarendon Press, Oxford, 1975.

[13] R. Dautray and J.-L. Lions, *Mathematical analysis and numerical methods for science and technology. Vol.* 2, Springer-Verlag, Berlin, 1988. Functional and variational methods, With the collaboration of Michel Artola, Marc Authier, Philippe Bénilan, Michel Cessenat, Jean Michel Combes, Hélène Lanchon, Bertrand Mercier, Claude Wild and Claude Zuily, translated from the French by Ian N. Sneddon.

[14] ______, *Mathematical analysis and numerical methods for science and technology. Vol.* 1, Springer-Verlag, Berlin, 1990. Physical origins and classical methods, with the collaboration of Philippe Bénilan, Michel Cessenat, André Gervat, Alain Kavenoky and Hélène Lanchon, translated from the French by Ian N. Sneddon, with a preface by Jean Teillac.

[15] ______, *Mathematical analysis and numerical methods for science and technology. Vol.* 3, Springer-Verlag, Berlin, 1990. Spectral theory and applications, With the collaboration of Michel Artola and Michel Cessenat, translated from the French by John C. Amson.

[16] K.-J. Engel and R. Nagel, *One-Parameter Semigroups for Linear Evolution Equations*, Springer, New York, 2000.

[17] ______, *A short course on operator semigroups*, Springer, New York, 2006.

[18] S.N. Ethier and T.G. Kurtz, *Markov processes. Characterization and convergence*, Wiley, New York, 1986.

[19] E. Fieremans, D.S. Novikov, J.H. Jensen, and J.A. Helpern, *Monte Carlo study of a two-compartment exchange model of diffusion*, NMR in Biomedicine **23** (2010), 711–724.

[20] J.A. Goldstein, *On the convergence and approximation of cosine functions*, Aequationes Math. **11** (1974), 201–205.

[21] ______, *Semigroups of Linear Operators and Applications*, Oxford University Press, New York, 1985.

[22] A. Gregosiewicz, *Asymptotic behaviour of diffusions on graphs*, Probability in Action, Banek, T., Kozlowski, E., eds., Lublin University of Technology, 2014, pp. 83–96.

[23] K. Itô and McKean, Jr. H.P., *Diffusion Processes and Their Sample Paths*, Springer, Berlin, 1996. Repr. of the 1974 ed.

[24] I. Karatzas and S.E. Shreve, *Brownian Motion and Stochastic Calculus*, Springer, New York, 1991.

[25] Y. Konishi, *Cosine functions of operators in locally convex spaces*, J. Fac. Sci. Univ. Tokyo Sect. IA Math. **18** (1971/1972), 443–463.

[26] A. Lasota and M.C. Mackey, *Chaos, fractals, and noise. Stochastic aspects of dynamics*, Springer, 1994.

[27] A. Lejay, *The snapping out Brownian motion*, 2013. hal–00781447.

[28] A. Pazy, *Semigroups of linear operators and applications to partial differential equations*, Springer, 1983.

[29] J.G. Powles, M.J.D. Mallett, G. Rickayzen, and W.A.B. Evans, *Exact analytic solutions for diffusion impeded by an infinite array of partially permeable barriers*, Proc. Roy. Soc. London Ser. A **436** (1992), no. 1897, 391–403.

[30] D. Revuz and M. Yor, *Continuous martingales and Brownian motion*, Springer, 1999. Third edition.

[31] J.E. Tanner, *Transient diffusion in a system partitioned by permeable barriers. Application to NMR measurements with a pulsed field gradient*, The Journal of Chemical Physics **69** (1978), no. 4, 1748–1754.

Adam Bobrowski
Lublin University of Technology
Nadbystrzycka 38A
PL-20-618 Lublin, Poland
e-mail: a.bobrowski@pollub.pl

Operator Theory:
Advances and Applications, Vol. 250, 107–118

Multiscale Unique Continuation Properties of Eigenfunctions

Denis Borisov, Ivica Nakić, Christian Rose, Martin Tautenhahn and Ivan Veselić

Abstract. Quantitative unique continuation principles for multiscale structures are an important ingredient in a number applications, e.g., random Schrödinger operators and control theory.

We review recent results and announce new ones regarding quantitative unique continuation principles for partial differential equations with an underlying multiscale structure. They concern Schrödinger and second-order elliptic operators. An important feature is that the estimates are scale free and with quantitative dependence on parameters. These unique continuation principles apply to functions satisfying certain 'rigidity' conditions, namely that they are solutions of the corresponding elliptic equations, or projections on spectral subspaces. Carleman estimates play an important role in the proofs of these results.

Mathematics Subject Classification (2010). 35J10, 35J15, 35B60, 35B45.

Keywords. Scale free unique continuation property, equidistribution property, observability estimate, uncertainty relation, Carleman estimate, Schrödinger operator, elliptic differential equation.

1. Introduction

Motivation: Retrival of global properties from local data

In several branches of mathematics, as well as in applications, one often encounters problems of the following type: Given a region in space $\Lambda \subset \mathbb{R}^d$, a subset $S \subset \Lambda$, and a function $f \colon \Lambda \to \mathbb{R}$, what can be said about certain properties of $f \colon \Lambda \to \mathbb{R}$ given certain properties of $f|_S \colon S \to \mathbb{R}$? In specific cases one may want to reconstruct f

D.B. was partially supported by Russian Science Foundation (project no. 14-11-00078) and the fellowship of Dynasty foundation for young mathematicians. I.N., Ch.R., M.T., and I.V. have been partially supported by the DAAD and the Croatian Ministry of Science, Education and Sports through the PPP-grant 'Scale-uniform controllability of partial differential equations'. M.T. and I.V. have been partially supported by the DFG.

as accurately as possible based on knowledge of $f|_S$, in others it may be sufficient to estimate some features of f.

It is clear that for this task additional global information on f is needed. Indeed, if f is one of the indicator functions χ_S or $\chi_{\Lambda\setminus S}$, an estimate based on $f|_S$ would yield wrong results. The first helpful property which comes to one's mind is some regularity or smoothness property of f. However, since there are C^∞-functions supported inside S (or inside $\Lambda \setminus S$) this is not quite the right condition. The required property of f is more adequately described as *rigidity*, as we will see in specific theorems formulated below.

In this paper we are mainly concerned with problems with a multiscale structure. For this reason it is natural to require that the set S is in some sense equidistributed within Λ. At this point we will not give a precise definition of such sets. It will become clear that such a set S should be relatively dense in $\mathbb{R}^d$ or Λ, and should have positive density. A particularly nice set S would be a periodic arrangement of balls, and we want to include small perturbations of such a configuration. Thus, equidistributed sets could be seen as a generalization of such a situation, cf. Fig. 1.

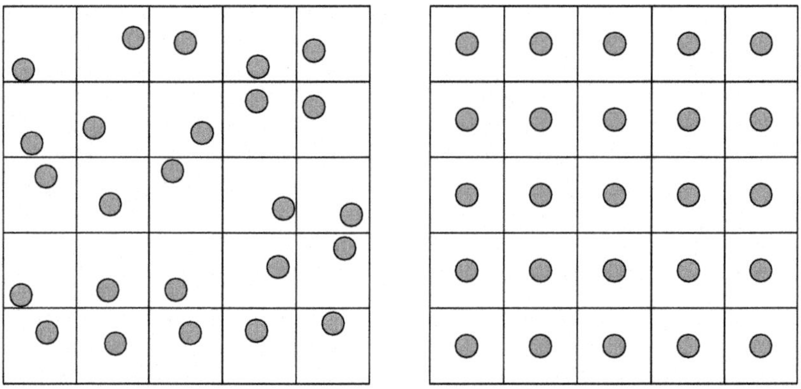

(a) non-periodic (b) periodic arrangement of balls

FIGURE 1. Examples of equidistributed sets S within region $\Lambda \subset \mathbb{R}^2$.

Example: Shannon sampling theorem

We recall a well-known theorem as an example or benchmark, see, e.g., [4]. This way we will see what is the best we can hope for in the task of reconstructing a function. Moreover, we will encounter one possible interpretation of what the term *rigidity* means, and see major differences between the reconstruction problem in dimension one and higher dimensions.

The Shannon sampling theorem states: Let $f \in C(\mathbb{R}) \cap L^2(\mathbb{R})$ be such that the Fourier transform

$$\hat{f}(p) = \frac{1}{\sqrt{2\pi}} \int_{\mathbb{R}} e^{-i\,x\,p}\, f(x)\, \mathrm{d}x$$

vanishes outside $[-\pi K, \pi K]$. Then the series

$$(S_K f)(x) = \sum_{j \in \mathbb{Z}} f\left(\frac{j}{K}\right) \frac{\sin \pi(K x - j)}{\pi(K x - j)} \tag{1.1}$$

converges absolutely and uniformly for $x \in \mathbb{R}$ and

$$S_K f = f \text{ on } \mathbb{R}.$$

Thus we can reconstruct the original function f from the sample values $f(j/K)$, which are multiplied with weights depending on the distance to the point $x \in \mathbb{R}$ and summed up. Here the rigidity condition is implemented by the requirement $\operatorname{supp} \hat{f} \subset [-\pi K, \pi K]$, which implies that f is entire. A remarkable feature of this exact result is that it is stable under perturbations: If the nodes j deviate slightly from the integers, or if the measurement data $f(\frac{j}{K})$ are inaccurate, the error $f - S_K f$ can still be controlled. If the support condition $\operatorname{supp} \hat{f} \subset [-\pi K, \pi K]$ is violated, the *aliasing error* is estimated as

$$\sup_{x \in \mathbb{R}} |f(x) - S_K f(x)| \leq \sqrt{\frac{2}{\pi}} \int_{|p| > \pi K} |\hat{f}(p)| \, dp. \tag{1.2}$$

This will give, for instance, good results for centered Gaussians with appropriate variance.

Statements (1.1) and (1.2) are strong with respect to the sampling set $S = \mathbb{Z}$, which is very thin. It has zero Lebesgue measure, in fact, it is discrete. Albeit, it is relatively dense in $\mathbb{R}$, so it has some of the properties we associated with an equidistributed set. Compared to Shannon's theorem, the results we present below appear much weaker. This is, among others, due to two features: we consider functions on multidimensional space, which, in addition, have low regularity, in fact are defined as equivalence classes in some L^2 or Sobolev space. In this situation evaluation of a function at a point may not have a proper meaning. This is one of the reasons why we have to consider samples S which are composed of small balls, rather than single points. A second aspect where dimensionality comes into play is the following: A polynomial of one variable of degree N vanishes identically if it has $N + 1$ zeros. A non-trivial polynomial in two variables may vanish on an uncountable set (albeit not on one of positive measure). This illustrates that reconstruction estimates for functions of several variables are more subtle than Shannon's theorem. Consequently, one has to settle for more modest goals than the full reconstruction of the function f. We want to derive an equidistribution property for functions satisfying some rigidity property. As will be detailed later this result is called – depending on the context and scientific environment – scale free unique continuation property, observability estimate, or uncertainty relation. A first result of this type is formulated in the next section.

2. Equidistribution property of Schrödinger eigenfunctions

The following result [15] was motivated by questions arising in the spectral theory of random Schrödinger operators. Later, it turned out that similar estimates are of relevance in the control theory of the heat equation.

We fix some notation. For $L > 0$ we denote by $\Lambda_L = (-L/2, L/2)^d$ a cube in $\mathbb{R}^d$. For $\delta > 0$ the open ball centered at $x \in \mathbb{R}$ with radius δ is denoted by $B(x, \delta)$. For a sequence of points $(x_j)_j$ indexed by $j \in \mathbb{Z}^d$ we denote the collection of balls $\cup_{j\in\mathbb{Z}^d} B(x_j, \delta)$ by S and its intersection with Λ_L by S_L. We will be dealing with certain self-adjoint operators on subsets of $\mathbb{R}^d$. Let Δ be the d-dimensional Laplacian, $V\colon \mathbb{R}^d \to \mathbb{R}$ a bounded measurable function, and $H_L = (-\Delta + V)_{\Lambda_L}$ a Schrödinger operator on the cube Λ_L with Dirichlet or periodic boundary conditions. The corresponding domains are denoted by $\mathcal{C}(\Delta_{\Lambda,0}) \subset W^{2,2}(\Lambda_L)$ and $\mathcal{C}(\Delta_{\Lambda,\mathrm{per}})$, respectively. Note that we denote a multiplication operator by the same symbol as the corresponding function.

Theorem 2.1 ([15]). *Let $\delta, K > 0$. Then there exists $C \in (0, \infty)$ such that for all $L \in 2\mathbb{N} + 1$, all measurable $V : \mathbb{R}^d \to [-K, K]$, all real-valued $\psi \in \mathcal{C}(\Delta_{\Lambda,0}) \cup \mathcal{C}(\Delta_{\Lambda,\mathrm{per}})$ with $(-\Delta + V)\psi = 0$ almost everywhere on Λ_L, and all sequences $(x_j)_{j\in\mathbb{Z}^d} \subset \mathbb{R}^d$, such that $\forall j \in \mathbb{Z}^d : B(x_j, \delta) \subset \Lambda_1 + j$ we have*

$$\int_{S_L} \psi^2 \geq C \int_{\Lambda_L} \psi^2. \tag{2.1}$$

To appreciate the result properly, the quantitative dependence of the constant C on model parameters is crucial. The very formulation of the theorem states that C is independent of position of the balls $B(x_j, \delta)$ within $\Lambda_1 + j$, and independent of the scale $L \in 2\mathbb{N} + 1$. The estimates given in Section 2 of [15] show moreover, that C depends on the potential V only through the norm $\|V\|_\infty$ (on an exponential scale), and it depends on the small radius $\delta > 0$ polynomially, i.e., $C \gtrsim \delta^N$, for some $N \in \mathbb{N}$ which depends on the dimension d and $\|V\|_\infty$. This shows that we are not able to control the integral $\int_{S_L} \psi^2$ by evaluating ψ at the midpoints $j \in \mathbb{Z}^d$ of the unit cubes. One sees with what rate the estimate diverges, as the balls become smaller and approximate a single point. The polynomial behavior $C \gtrsim \delta^N$ can be readily understood when looking at monomials $\psi_n(x) = x^n$ on the unit interval $(0, 1)$. There we have

$$\int_{(0,\delta)} \psi_n^2 = \frac{\delta^{2n+1}}{2n+1} = \delta^{2n+1} \int_{(0,1)} \psi_n^2.$$

We formulated the theorem only for the eigenvalue zero, but it is easily applied to other eigenfunctions as well since

$$H_L\psi = E\psi \Leftrightarrow (H_L - E)\psi = 0.$$

Consequently the constant $K = K_V$ has to be replaced with the possibly larger $K = K_{V-E}$.

There is a very natural question, which was spelled out in [15], namely does the same estimate (2.1) hold true for linear combinations $\psi \in \operatorname{Ran} \chi_{(-\infty,E]}(H_L)$ of eigenfunctions as well? The property in question can be equivalently stated as: Given $\delta > 0, K \geq 0, E \in \mathbb{R}$ there is a constant $C > 0$ such that for all measurable $V \colon \mathbb{R}^d \to [-K, K]$, all $L \in 2\mathbb{N}+1$, and all sequences $(x_j)_{j\in\mathbb{Z}^d} \subset \mathbb{R}^d$ with $B(x_j, \delta) \subset \Lambda_1 + j$ for all $j \in \mathbb{Z}^d$ we have

$$\chi_{(-\infty,E]}(H_L)\, W_L\, \chi_{(-\infty,E]}(H_L) \geq C\, \chi_{(-\infty,E]}(H_L), \tag{2.2}$$

where $W_L = \chi_{S_L}$ is the indicator function of S_L and $\chi_I(H_L)$ denotes the spectral projector of H_L onto the interval I. Here $C = C_{\delta,K,E}$ is determined by δ, K, E alone.

Note that all considered operators are lower bounded by $-K$ in the sense of quadratic forms. Thus the spectral projection on the energy interval $(-\infty, E]$ is the same as the spectral projection on the energy interval $[-K, E]$. The upper bound E in the energy parameter is crucial for preventing the corresponding eigenfunctions to oscillate too much.

One can pose a modified version of the question: Given $\delta > 0, K \geq 0, a < b \in \mathbb{R}$ is there is a constant $\tilde{C} > 0$ such that for all measurable $V \colon \mathbb{R}^d \to [-K, K]$, all $L \in 2\mathbb{N}+1$, and all sequences $(x_j)_{j\in\mathbb{Z}^d} \subset \mathbb{R}^d$ with $B(x_j, \delta) \subset \Lambda_1 + j$ for all $j \in \mathbb{Z}^d$ we have

$$\chi_{[a,b]}(H_L)\, W_L\, \chi_{[a,b]}(H_L) \geq \tilde{C}\, \chi_{[a,b]}(H_L). \tag{2.3}$$

Here $\tilde{C} = \tilde{C}_{\delta,K,a,b}$ depends (only) on δ, K, a, b. Note that inequality (2.2) implies (2.3) since

$$\begin{aligned} &\chi_{[a,b]}(H_L)\, W_L\, \chi_{[a,b]}(H_L) \\ &\quad = \chi_{[a,b]}(H_L)\, \chi_{(-\infty,b]}(H_L)\, W_L\, \chi_{(-\infty,b]}(H_L)\, \chi_{[a,b]}(H_L) \\ &\quad \geq C_{\delta,K,b}\, \chi_{[a,b]}(H_L)\, \chi_{(-\infty,b]}(H_L)\, \chi_{[a,b]}(H_L) \\ &\quad = C_{\delta,K,b} \chi_{[a,b]}(H_L). \end{aligned}$$

However, $C_{\delta,K,b}$ may be substantially smaller than $\tilde{C}_{\delta,K,a,b}$ due to the enlarged energy interval.

Klein obtained a positive answer to the question for sufficiently short intervals.

Theorem 2.2 ([8]). *Let $d \in \mathbb{N}$, $E \in \mathbb{R}$, $\delta \in (0, 1/2]$ and $V : \mathbb{R}^d \to \mathbb{R}$ be measurable and bounded. There is a constant $M_d > 0$ such that if we set*

$$\gamma = \frac{1}{2} \delta^{M_d \left(1 + (2\|V\|_\infty + E)^{2/3}\right)},$$

then for all energy intervals $I \subset (-\infty, E]$ with length bounded by 2γ, all $L \in 2\mathbb{N}+1$, $L \geq 72\sqrt{d}$ and all sequences $(x_j)_{j\in\mathbb{Z}^d} \subset \mathbb{R}^d$ with $B(x_j, \delta) \subset \Lambda_1 + j$ for all $j \in \mathbb{Z}^d$

$$\chi_I(H_L)\, W_L\, \chi_I(H_L) \geq \gamma^2 \chi_I(H_L). \tag{2.4}$$

Although this does not answer the above-posed question for arbitrary compact intervals, the result is sufficient for many questions in spectral theory of random Schrödinger operators. A generalization of Theorem 2.2 to intervals of arbitrary length is given in Section 4. This answers completely the question posed in [15].

Depending on the context and the area of mathematics the above-described estimates carry various names. If one speaks of an *equidistribution property of eigenfunctions*, one is interested in the comparison of the measure $|\psi(x)|^2 \mathrm{d}x$ with the uniform distribution on the cube Λ_L. The term *scale free unique continuation principle* is used in works concerning random Schrödinger operators. It refers to a quantitative version of the classical unique continuation principle, which is uniform on all large length scales. One can interpret Theorem 2.1 as an *uncertainty relation*: the condition $H_L\psi = E\psi$ corresponds to a restriction in momentum/Fourier-space and enforces a delocalization/flatness property in direct space. Similarly, the spectral projector $\chi_{(-\infty,E]}$ in Inequality (2.2) corresponds to a restriction in momentum space. Here we see a direct analogy to Shannon's theorem discussed above: If the Fourier transform of a function is sufficiently concentrated, the function itself cannot vary too much over short distances. Inequality (2.3) can also be interpreted as a *gain of positive definiteness*. It says that for a general self-adjoint operator $A \geq 0$, which may have a kernel, and an appropriately chosen spectral projector P of the Hamiltonian, the restriction $PAP \geq cP$ is strictly positive. In control theory results as we discuss them are sometimes called *observability* estimates. This term is more common for time-dependent partial differential equations, but sometimes used for stationary ones as well.

In the literature on random Schrödinger operators related results have been derived before in a number of papers. For more details we refer to Section 1 of [15].

3. Methods and background

A paradigmatic result for the *weak unique continuation principle* is the following. A solution of $\Delta f \equiv 0$ on $\mathbb{R}^d$ satisfying $f \equiv 0$ on $B(0,\delta)$ for arbitrary small, but positive δ, must vanish on all of $\mathbb{R}^d$. The restrictive conditions can be relaxed. First of all, the condition $f \equiv 0$ on $B(0,\delta)$ can be replaced by

$$\forall N \in \mathbb{N} \qquad \lim_{\delta \searrow 0} \delta^{-N} \int\limits_{B(0,\delta)} |f(x)| \mathrm{d}x = 0.$$

In this form the implication is called *strong unique continuation principle.* Moreover, the Laplacian Δ can be replaced by a rather general second-order elliptic operator. We will discuss related results in Sections 5 and 6. A powerful method to prove unique continuation statements, as well as quantitative versions thereof, are Carleman estimates. Originally, Carleman [5] derived them for functions of two variables. Later Müller [11] extended the estimates to higher dimensions. By now, there are hundreds of papers dealing with Carleman estimates. We will describe one explicit version in Section 5, which is an important tool for the quantitative unique continuation estimates discussed shortly for Schrödinger operators. In Section 6 we will present new results in this direction which deal with elliptic second-order operators with variable coefficients.

Quantitative unique continuation principle

In [1] Bourgain and Kenig derived the following pointwise quantitative unique continuation principle.

Theorem 3.1. *Assume* $(-\Delta+V)u = 0$ *on* $\mathbb{R}^d$ *and* $u(0) = 1$, $\|u\|_\infty \leq C$, $\|V\|_\infty \leq C$. *Let* $x_0 \in \mathbb{R}^d$, $|x_0| = R > 1$. *Then there exists a constant* $C' > 0$ *such that*

$$\max_{|x-x_0|\leq 1} |u(x)| > C' \exp\left(-C'(\log R)R^{4/3}\right).$$

In our context a version of this result with local L^2-averages is more appropriate. Various estimates of this type have been given in [7, 2, 15]. We quote here the version from the last mentioned paper.

Theorem 3.2. *Let* $K, R, \beta \in [0,\infty), \delta \in (0,1]$. *There exists a constant* $C_{\mathrm{qUC}} = C_{\mathrm{qUC}}(d, K_V, R, \delta, \beta) > 0$ *such that, for any* $G \subset \mathbb{R}^d$ *open, any* $\Theta \subset G$ *measurable, satisfying the geometric conditions*

$$\operatorname{diam}\Theta + \operatorname{dist}(0,\Theta) \leq 2R \leq 2\operatorname{dist}(0,\Theta), \quad \delta < 4R, \quad B(0,14R) \subset G,$$

and any measurable $V\colon G \to [-K, K]$ *and real-valued* $\psi \in W^{2,2}(G)$ *satisfying the differential inequality*

$$|\Delta\psi| \leq |V\psi| \quad \text{a.e. on } G \quad \text{as well as} \quad \int_G |\psi|^2 \leq \beta \int_\Theta |\psi|^2,$$

we have

$$\int_{B(0,\delta)} |\psi|^2 \geq C_{\mathrm{qUC}} \int_\Theta |\psi|^2.$$

4. Equidistribution property of linear combinations of eigenfunctions

In this section we present a result from a project of I. Nakić, M. Täufer, M. Tautenhahn and I. Veselić [13], namely which gives Inequality (2.1) also for linear combinations of eigenfunctions $\psi \in \operatorname{Ran}\chi_{(-\infty,E]}(H_L)$ for arbitrary $E \in \mathbb{R}$. As shown above, this implies Inequality (2.2) for arbitrary $E \in \mathbb{R}$ and hence Inequality (2.3) for $[a,b] \subset (-\infty, E]$. Indeed, our result gives a full answer to the open question in [15] whether Theorem 3.2 holds also for linear combinations of eigenfunctions, which was partially answered in [8], cf. Theorem 2.2.

Since we first show Inequality (2.2) for arbitrary $E \in \mathbb{R}$, the constant $\tilde{C}$ in Inequality (2.3) will not be optimal, since it does not depend on the lower bound a of the interval $[a,b] \subset (-\infty, E]$.

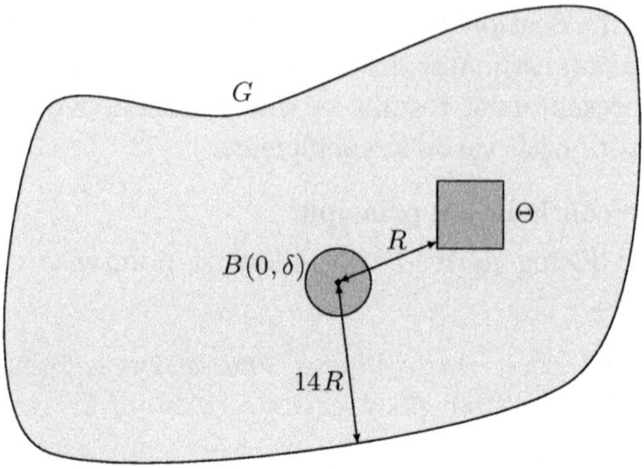

FIGURE 2. Assumptions in Theorem 3.2 on the geometric constellation of G, Θ, and $B(0,\delta)$

The following theorem was given in [13] and full proofs will be provided in [14].

Theorem 4.1 ([13]). *There is $N = N(d)$ such that for all $\delta \in (0,1/2)$, all measurable and bounded $V : \mathbb{R}^d \to \mathbb{R}$, all $L \in \mathbb{N}$, all $E \geq 0$ and all $\psi \in \operatorname{Ran}(\chi_{(-\infty,E]}(H_L))$ and all sequences $(x_j)_{j\in\mathbb{Z}^d} \subset \mathbb{R}^d$, such that for all $j \in \mathbb{Z}^d$ $B(x_j,\delta) \subset \Lambda_1 + j$, we have*

$$\int_{S_L} |\psi|^2 \geq C_{\mathrm{sfuc}} \int_{\Lambda_L} |\psi|^2 \tag{4.1}$$

where

$$C_{\mathrm{sfuc}} = C_{\mathrm{sfuc}}(d,\delta,E,\|V\|_\infty) := \delta^{N\left(1+\|V\|_\infty^{2/3}+\sqrt{E}\right)}.$$

Hence, as in Theorem 2.1, the constant is independent on the position of the balls $B(x_j,\delta)$, the scale L, and it depends on the potential V only through the norm $\|V\|_\infty$.

Here we give a sketch of the proof. We use two different Carleman inequalities in $\mathbb{R}^{d+1}$, one with a boundary term in $\mathbb{R}^d \times \{0\}$ and the other without boundary terms. From these Carleman estimates we deduce two interpolation inequalities for a solution of a Schrödinger equation in $\mathbb{R}^{d+1}$. In the final step we apply these interpolation inequalities to the function $F : \Lambda_L \times \mathbb{R} \to \mathbb{C}$ defined by

$$F(x) = \sum_{\substack{k\in\mathbb{N}\\ E_k\leq E}} \alpha_k \psi_k(x') s_k(x_{d+1}),$$

where $\alpha_k = \langle \psi_k, \psi\rangle$ with ψ_k denoting the eigenfunctions of H_L corresponding to the eigenvalues E_k, $\mathbb{R}^{d+1} \ni x = (x', x_{d+1})$, $x' \in \mathbb{R}^d$, $x_{d+1} \in \mathbb{R}$ and

$$s_k(x) = \begin{cases} \sinh(\sqrt{E_k}x)/\sqrt{E_k}, & E_k > 0,\\ x, & E_k = 0,\\ \sin(\sqrt{|E_k|}x)/\sqrt{|E_k|}, & E_k < 0. \end{cases}$$

This function F satisfies $\Delta F = VF$ on $\Lambda_L \times \mathbb{R}$ and $\partial_{d+1}F(x',0) = \psi(x')$ on Λ_L, and one can obtain upper and lower estimates for the H^1-norm of the function F in terms of the parameters K, E, d and $\sum_{E_k \le E} |\alpha_k|^2$.

5. Explicit Carleman estimates for elliptic operators

As mentioned above, Carleman estimates play a significant role in the results about unique continuation principles. In the case of quantitative unique continuation principles on multiscale structures, it is important to have a Carleman estimate with dependence on various parameters as precise as possible.

We consider the second-order elliptic partial differential operator

$$L = -\sum_{i,j=1}^{d} \partial_i \left(a^{ij} \partial_j\right),$$

acting on functions in $\mathbb{R}^d$. We introduce the following assumption on the coefficient functions a^{ij}.

Assumption (A). Let $r, \vartheta_1, \vartheta_2 > 0$. The operator L satisfies $A(r, \vartheta_1, \vartheta_2)$, if and only if $a^{ij} = a^{ji}$ for all $i, j \in \{1, \ldots, d\}$ and for almost all $x, y \in B(r)$ and all $\xi \in \mathbb{R}^d$ we have

$$\vartheta_1^{-1} |\xi|^2 \le \sum_{i,j=1}^{d} a^{ij}(x) \xi_i \xi_j \le \vartheta_1 |\xi|^2 \quad \text{and} \quad \sum_{i,j=1}^{d} |a^{ij}(x) - a^{ij}(y)| \le \vartheta_2 |x - y|.$$

Here $B(r) \subset \mathbb{R}^d$ denotes the open ball in $\mathbb{R}^d$ with radius r and center zero. Let the entries of the inverse of the matrix $(a^{ij}(x))_{i,j=1}^d$ be denoted by $a_{ij}(x)$.

We present the result for the ball $B(1)$, but by scaling arguments this result can be generalized to arbitrary large balls $B(R)$, now with a different weight function which depends also on R.

In the following theorem we formulate a Carleman estimate for elliptic partial differential operators with variable coefficients analogous to those given in [6] for parabolic operators. In the case of the pure Laplacian this has already been done in [1]. In particular, we establish that the estimate is valid on the whole domain (i.e. $\delta = 1$ holds in the notation of [6]) and give quantitative estimates for all the parameters. This is part of a recent work of I. Nakić, C. Rose and M. Tautenhahn [12].

For $\mu > 0$ let $\sigma : \mathbb{R}^d \to [0, \infty)$ and $\psi \colon [0, \infty) \to [0, \infty)$ be given by

$$\sigma(x) = \left(\sum_{i,j=1}^{d} a_{ij}(0) x_i x_j \right)^{1/2} \quad \text{and} \quad \psi(s) = s \cdot \exp\left[-\int_0^s \frac{1 - \mathrm{e}^{-\mu t}}{t} \mathrm{d}t \right].$$

We define the weight function $w\colon \mathbb{R}^d \to [0,\infty)$ by $w(x) = \psi(\sigma(x))$. Note that the weight function satisfies the bounds

$$\forall x \in B(1)\colon \quad \frac{|x|}{C_3\sqrt{\vartheta_1}} \le w(x) \le \sqrt{\vartheta_1}|x| \quad \text{with} \quad C_3 = \mathrm{e}\mu. \tag{5.1}$$

Theorem 5.1 ([6, 12]). *Let $\vartheta_1, \vartheta_2 > 0$ and Assumption $A(1,\vartheta_1,\vartheta_2)$ be satisfied. Then there exist constants $\mu, C_1, C_2 > 0$ depending only on ϑ_1, ϑ_2 and the dimension d such that for all $f \in C_0^\infty(B(0,1)\setminus\{0\})$ and all $\alpha > C_1$ we have*

$$\int \alpha w^{1-2\alpha}|\nabla f|^2 + \alpha^3 w^{-1-2\alpha} f^2 \le C_2 \int w^{2-2\alpha}(Lf)^2.$$

Explicit bounds on $\mu = \mu(\vartheta_1,\vartheta_2)$ are given in [12]. In particular,

$$\forall T > 0: \quad \mu_T = \sup\{\mu(\vartheta_1,\vartheta_2)\colon 0 < \vartheta_1, \vartheta_2 \le T\} < \infty. \tag{5.2}$$

With a regularization procedure (see, for example, [18, Theorem 1.6.1]) this result can be extended to the functions in $H_0^2(B(0,1))$ which are compactly supported away from the origin.

6. Quantitative unique continuation estimates for elliptic operators

In this section we announce a result from an ongoing work of D.I. Boris, M. Tautenhahn and I. Veselić [3]. It concerns a quantitative unique continuation principle for elliptic second-order partial differential operators with slowly varying coefficients.

As in the previous section we denote by L the second-order partial differential operator

$$Lu = -\sum_{i,j=1}^{d} \partial_i\left(a^{ij}\partial_j u\right),$$

acting on functions u on $\mathbb{R}^d$.

Theorem 6.1 ([3]). *Let $R, \vartheta_1, \vartheta_2 \in (0,\infty)$, $D_0 < 6R$, $K_V, \beta \in [0,\infty)$, $\delta \in (0,4R]$, let $C_3 = C_3(d,\vartheta_1,\vartheta_2)$ be the constant from Equation (5.1), and assume that*

$$A(12R + 2D_0, \vartheta_1, \vartheta_2) \quad \text{and} \quad \vartheta_1 C_3 < \frac{1}{4R}$$

are satisfied. Then there exists $C_{\mathrm{qUC}} = C_{\mathrm{qUC}}(d,\vartheta_1,\vartheta_2,R,D_0,K_V,\delta,\beta) > 0$, such that, for any $G \subset \mathbb{R}^d$ open, $x \in G$ and $\Theta \subset G$ measurable, satisfying

$$\operatorname{diam}\Theta + \operatorname{dist}(x,\Theta) \le 2R \le 2\operatorname{dist}(x,\Theta) \quad \text{and} \quad B(x, 12R + 2D_0) \subset G,$$

and any measurable $V : G \to [-K_V, K_V]$ and real-valued $\psi \in W^{2,2}(G)$ satisfying the differential inequality

$$|L\psi| \le |V\psi| \quad \text{a.e. on } G \quad \text{as well as} \quad \int_G |\psi|^2 \le \beta \int_\Theta |\psi|^2,$$

we have

$$\int_{B(x,\delta)} |\psi|^2 \ge C_{\mathrm{qUC}} \int_\Theta |\psi|^2.$$

Theorem 6.1 generalizes Theorem 2.1 to second-order elliptic operators with slowly varying coefficient functions. This is explicitly given by the assumption $\vartheta_1 C_3 < 1/(4R)$. Indeed, for fixed $R > 0$ the last inequality is satisfied for ϑ_1 sufficiently small, since (5.2) implies $\lim_{\vartheta_1 \to 0} \vartheta_1 \mu(\vartheta_1, \vartheta_2) = 0$. Furthermore, once one has a quantitative estimate on the dependence $(\vartheta_1, \vartheta_2) \mapsto \mu$, the assumption $4R\vartheta_1 C_3 < 1$ can be formulated as a condition involving ϑ_1, ϑ_2 and R only.

The proof of Theorem 6.1 is based on ideas developed in [15] for the pure Laplacian. The key tool for the proof is a Carleman estimate. For second-order elliptic operators there exist plenty of them in the literature, see, e.g., [9, 10, 16]. However, since we are interested in quantitative estimates, the Carleman estimate from Theorem 5.1 proved to be useful in this context.

Acknowledgment

I.V. would like to thank Thomas Duyckaerts and Matthieu Léautaud for discussions concerning the literature on Carleman estimates.

References

[1] J. Bourgain and C.E. Kenig. *On localization in the continuous Anderson-Bernoulli model in higher dimension.* Invent. Math., 161(2):389–426, 2005.

[2] J. Bourgain and A. Klein. *Bounds on the density of states for Schrödinger operators.* Invent. Math., 194(1):41–72, 2013.

[3] D.I. Borisov, M. Tautenhahn, and I. Veselić. *Equidistribution properties of eigenfunctions of divergence form operators.* In preparation.

[4] P.L. Butzer, W. Splettstoesser, and R.L. Stens. *The sampling theorem and linear prediction in signal analysis.* Jahresber. Deutsch. Math.-Verein. 90:1–70, 1988.

[5] T. Carleman. *Sur un problème d'unicité pour les systèmes d'équations aux dérivées partielles à deux variables indépendantes.* Ark. Mat. Astr. Fys., 26(17):1–9, 1939.

[6] L. Escauriaza and S. Vessella. *Optimal Three Cylinder Inequalities for Solutions to Parabolic Equations with Lipschitz Leading Coefficients*, volume 333 of Contemp. Math., pages 79–87. American Mathematical Society, 2003.

[7] F. Germinet and A. Klein. *A comprehensive proof of localization for continuous Anderson models with singular random potentials.* J. Eur. Math. Soc., 15(1):53–143, 2013.

[8] A. Klein. *Unique continuation principle for spectral projections of Schrödinger operators and optimal Wegner estimates for non-ergodic random Schrödinger operators.* Commun. Math. Phys., 323(3):1229–1246, 2013.

[9] H. Koch and D. Tataru. *Carleman estimates and unique continuation for second-order elliptic equations with nonsmooth coefficients.* Commun. Pur. Appl. Math., 54(3):339–360, 2001.

[10] J. Le Rousseau and G. Lebeau. *On Carleman estimates for elliptic and parabolic operators. Applications to unique continuation and control of parabolic equations.* ESAIM Contr. Optim. Ca., 18(3):712–747, 2012.

[11] C. Müller. *On the behavior of the solutions of the differential equation $\delta u = f(x,u)$ in the neighborhood of a point.* Commun. Pur. Appl. Math., 7(3):505–515, 1954.

[12] I. Nakić, C. Rose, and M. Tautenhahn. *A quantitative Carleman estimate for second order elliptic operators.* arXiv:1502.07575 [math.AP], 2015.
[13] I. Nakić, M. Täufer, M. Tautenhahn, and I. Veselić. *Scale-free uncertainty principles and Wegner estimates for random breather potentials.* arXiv:1410.5273 [math.AP], 2014.
[14] I. Nakić, M. Täufer, M. Tautenhahn, and I. Veselić. In preparation.
[15] C. Rojas-Molina and I. Veselić. *Scale-free unique continuation estimates and applications to random Schrödinger operators.* Commun. Math. Phys., 320(1):245–274, 2013.
[16] J. Le Rousseau. *Carleman estimates and some applications to control theory.* In *Control of Partial Differential Equations*, Lecture Notes in Mathematics, pages 207–243. 2012.
[17] S. Vessella. *Quantitative estimates of unique continuation for parabolic equations, determination of unknown time-varying boundaries and optimal stability estimates.* Inverse Probl., 24(2):023001, 2008.
[18] W.P. Ziemer. *Weakly differentiable functions.* Springer, New York, 1989.

Denis Borisov
Institute of Mathematics, USC RAS
Chernyshevskii st., 112
Ufa, 45008, Russia

and

Bashkir State Pedagogical University
October rev. st., 3a
Ufa, 45000, Russia
e-mail: borisovdi@yandex.ru

Ivica Nakić
University of Zagreb
Departement of Mathematics
10000 Zagreb, Croatia
e-mail: nakic@math.hr

Christian Rose, Martin Tautenhahn and Ivan Veselić
Chemnitz University of Technology
Faculty of Mathematics
D-09107 Chemnitz, Germany
e-mail: christian.rose@mathematik.tu-chemnitz.de
martin.tautenhahn@mathematik.tu-chemnitz.de
ivan.veselic@mathematik.tu-chemnitz.de

Operator Theory:
Advances and Applications, Vol. 250, 119–131

Dichotomy Results for Norm Estimates in Operator Semigroups

I. Chalendar, J. Esterle and J.R. Partington

Dedicated to Charles Batty on the occasion of his sixtieth birthday

Abstract. The results in this survey indicate that the quantitative behaviour of the semigroup at the origin provides additional qualitative information, such as uniform continuity or analyticity.

Mathematics Subject Classification (2010). Primary 47D03, 46J40, 46H30; Secondary 30A42, 47A60.

Keywords. Strongly continuous semigroup, functional calculus, Fourier–Borel transform, Laplace transform, analytic semigroup.

In line with an Oxford tradition,
Charles Batty was sent on a mission:
to instruct all the troops
in C_0 semigroups –
this indeed was a brilliant decision!

1. Introduction

We recall that a one-parameter family $(T(t))_{0<t<\infty}$ in a Banach algebra (often itself simply the algebra of bounded linear operators on a Banach space $\mathcal{X}$) is a semigroup if

$$T(s+t) = T(s)T(t) \qquad \text{for all } t, s > 0.$$

We shall be concerned here with semigroups that are strongly continuous on $\mathbb{R}_+ := (0, \infty)$, but not necessarily norm-continuous at the origin. As an example to bear in mind we mention the semigroup $T(t) : x \mapsto x^t$ in the algebra $C_0([0,1])$ of continuous functions on $[0,1]$ vanishing at the origin, which will be discussed later.

Later, we consider semigroups defined on a sector in the complex plane, in which case they will be assumed to be *analytic*: that is, complex-differentiable in the norm topology.

The results in this survey indicate that the quantitative behaviour of the semigroup at the origin provides additional qualitative information, such as uniform continuity or analyticity. Here are a few examples.

We recall the classical *zero–one law*, asserting that if the semigroup satisfies $\limsup_{t\to 0+} \|T(t) - I\| < 1$, then in fact $\|T(t) - I\| \to 0$ and hence the semigroup is uniformly continuous, and of the form e^{tA} for some bounded operator A. To see this, set $L = \limsup_{t\to 0+} \|T(t) - I\|$. Since

$$2(T(t) - I) = T(2t) - I - (T(t) - I)^2,$$

we have $2L \le L + L^2$, and thus $L = 0$ or $L \ge 1$. This proof is due to T. Coulhon.

Another result involving the asymptotic behaviour at 0 and providing a uniformly continuous semigroup is the following, proved in 1950 by Hille [12] (see also [13, Thm. 10.3.6]). This result is usually stated for $n = 1$, but Hille's argument works for any positive integer.

Theorem 1.1. *Let $(T(t))_{t>0}$ be a n-times continuously differentiable semigroup over the positive reals. If $\limsup_{t\to 0+} \|t^n T^{(n)}(t)\| < (\frac{n}{e})^n$, then the generator of the semigroup is bounded.*

In the direction of analyticity, a classical result of Beurling [3] is the following:

Theorem 1.2. *A C_0-semigroup $(T(t))_{0\le t<\infty}$ on a complex Banach space $\mathcal{X}$ is holomorphic if and only if there exists a polynomial p such that*

$$\limsup_{t\to 0+} \|p(T(t))\| < \sup\{|p(z)| : |z| \le 1\}. \tag{1.1}$$

Kato [15] and Neuberger [18] proved the sufficiency of (1.1) with $p(z) = z-1$, and $\sup |p(z)| = 2$, providing a zero–two law for analyticity. In general the converse is not true with $p(z) = z - 1$, although it holds if $\mathcal{X}$ is uniformly convex and the semigroup is contractive [19]. Some extensions of this result to arbitrary Banach spaces and for semigroups which are not necessarily contractive may be found in the very recent paper [11].

The more recent results considered in this survey concern estimates of the norm or spectral radius of quantities such as $T(t) - T((n+1)t)$ as t tends to 0. These are often formulated as dichotomy results, such as the zero–quarter law (the case $n = 1$ in the following theorem).

Theorem 1.3 ([10, 17]). *Let $n \ge 1$ be an integer, and let $(T(t))_{t>0}$ be a semigroup in a Banach algebra. If*

$$\limsup_{t\to 0+} \|T(t) - T((n+1)t)\| < \frac{n}{(n+1)^{1+1/n}},$$

then either $T(t) = 0$ for $t > 0$ or else the closed subalgebra generated by $(T(t))_{t>0}$ is unital, and the semigroup has a bounded generator A: that is, $T(t) = \exp(tA)$ for $t > 0$.

Another result that we mention here concerns the link between the norm and spectral radius ρ, and was motivated also by the Esterle–Katznelson–Tzafriri results on estimates for $\|T^n - T^{n+1}\|$, where T is a power-bounded operator (see [8, 16]). We have rewritten it in the notation of differentiable groups $(T(t)) = (\exp(tA))$, which may even be defined for $t \in \mathbb{C}$ if A is bounded; note that $T'(t) = AT(t)$.

Theorem 1.4 ([14]). *Let A be a bounded operator on a Banach space, and let $(T(t))$ be the group given by $T(t) = \exp(tA)$. Then each of the following conditions implies that $\rho(A) = \|A\|$.*

(i) $\sup_{t>0} t\|T'(t)\| \leq 1/e$;
(ii) $\sup_{t>0} \|T(t) - T((s+1)t)\| \leq s(s+1)^{-(1+1/s)}$ *for some* $s > 0$;
(iii) $\sup_{t>0} \|T((s+i)t) - T((s-i)t)\| \leq 2e^{-s\arctan(1/s)}/\sqrt{1+s^2}$ *for some* $s \geq 0$.

The third condition of Theorem 1.4 is linked to the Bonsall–Crabb proof of Sinclair's spectral radius formula for Hermitian elements of a Banach algebra, given in [4].

In Section 2 we review the existing literature on dichotomy laws for semigroups, first for semigroups on $\mathbb{R}_+$ and then for analytic semigroups defined on a sector in the complex plane; our main sources here are [1] and [5]. Then in Section 3 we present some very recent generalizations of these results, formulated in the language of functional calculus: this discussion is based on [6] and [7].

2. Dichotomy laws

2.1. Semigroups on $\mathbb{R}_+$

We begin with a result on quasinilpotent semigroups, that is, semigroups whose elements all have spectral radius 0. This is a commonly-occurring case, examples being found in the convolution algebra $L^1(0,1)$.

Theorem 2.1 ([9]). *Let $(T(t))_{t>0}$ be a C_0-semigroup of bounded quasinilpotent linear operators on a Banach space $\mathcal{X}$. Then there exists $\delta > 0$ such that*

$$\|T(t) - T(s)\| > \theta(s,t) \quad \textit{for} \quad 0 < t < s < \delta,$$

where

$$\theta(s,t) = (s-t)t^{t/(s-t)}s^{s/(t-s)}.$$

In particular, for all $\gamma > 0$, there exists $\delta > 0$ such that

$$\|T(t) - T((\gamma+1)t)\| > \frac{\gamma}{(\gamma+1)^{1+1/\gamma}}$$

for all $0 < t < \delta$.

This is a sharp result, in the sense that given a non-decreasing function $\epsilon : (0,1) \to (0,\infty)$ there exists a nontrivial quasinilpotent semigroup $(T_\epsilon(t))_{t>0}$ on a Hilbert space such that:

$$\|T_\epsilon(t) - T_\epsilon(s)\| \leq \theta(s,t) + (s-t)\epsilon(s)$$

(see [9]). Note that $\theta(s,t) = \max_{0\leq x\leq 1}(x^t - x^s)$.

It is a quantitative formulation of an intuitive fact: $T(t)$ cannot be uniformly too close to $T(s)$ for $s \neq t$, with s,t small when the generator is unbounded.

In the non-quasinilpotent case, it is possible to formulate similar results using the spectral radius. The following theorem is a strengthening of [10, Thm 2.3], which is expressed in terms of $\limsup_{t\to 0} \rho(T(t) - T(t(\gamma+1)))$. Note that $\operatorname{Rad} \mathcal{A}$ denotes the radical of the algebra $\mathcal{A}$, i.e., the set of elements with spectral radius zero.

Theorem 2.2 ([1]). *Let $(T(t))_{t>0}$ be a non-quasinilpotent semigroup in a Banach algebra, let $\mathcal{A}$ be the closed subalgebra generated by $(T(t))_{t>0}$, and let $\gamma > 0$ be a real number. If there exists $t_0 > 0$ such that*

$$\rho(T(t) - T(t(\gamma+1))) < \frac{\gamma}{(\gamma+1)^{1+\frac{1}{\gamma}}}$$

for $0 < t \leq t_0$, then $\mathcal{A}/\operatorname{Rad}(\mathcal{A})$ is unital, and there exist an idempotent J in $\mathcal{A}$, an element u of $J\mathcal{A}$ and a mapping $r : \mathbb{R}_+ \to \operatorname{Rad}(J\mathcal{A})$, with the following properties:

(i) *$\varphi(J) = 1$ for all $\varphi \in \widehat{\mathcal{A}}$;*
(ii) *$r(s+t) = r(s) + r(t)$ for all $s,t \in \mathbb{R}_+$;*
(iii) *$JT(t) = e^{tu+r(t)}$ for $t \in \mathbb{R}_+$, where $e^v = J + \sum_{k\geq 1} \frac{v^k}{k!}$ for $v \in J\mathcal{A}$;*
(iv) *$(T(t) - JT(t))_{t\in\mathbb{R}_+}$ is a quasinilpotent semigroup.*

If $\mathcal{A}$ is semi-simple (that is, $\operatorname{Rad}(\mathcal{A}) = \{0\}$), then the conclusion is much more straightforward.

Corollary 2.3 ([1]). *Let $(T(t))_{t>0}$ be a non-trivial semigroup in a commutative semi-simple Banach algebra, let $\mathcal{A}$ be the closed subalgebra generated by $(T(t))_{t\in\mathbb{R}_+}$ and let $\gamma > 0$. If there exists $t_0 > 0$ such that*

$$\rho(T(t) - T((\gamma+1)t)) < \frac{\gamma}{(\gamma+1)^{1+\frac{1}{\gamma}}}$$

for $0 < t \leq t_0$, then $\mathcal{A}$ is unital and there exists an element $u \in \mathcal{A}$ such that $T(t) = e^{tu}$ for $t \in \mathbb{R}_+$.

The following theorem needs no hypothesis on $\mathcal{A}$, but requires stronger estimates, based on the norm rather than the spectral radius.

Theorem 2.4 ([1]). *Let $(T(t))_{t>0}$ be a non-trivial semigroup in a Banach algebra, let $\mathcal{A}$ be the closed subalgebra generated by $(T(t))_{t>0}$ and let $n \geq 1$ be an integer.*

If there exists $t_0 > 0$ *such that*

$$\|(T(t) - T(t(n+1)))\| < \frac{n}{(n+1)^{1+\frac{1}{n}}}$$

for $0 < t \leq t_0$*, then* $\mathcal{A}$ *possesses a unit* J*,* $\lim_{t\to 0^+} T(t) = J$ *and there exists* $u \in \mathcal{A}$ *such that* $T(t) = e^{tu}$ *for all* $t > 0$*.*

If $(T(t)_{t>0}$ *is a quasinilpotent semigroup, then the condition*

$$\|T(t) - T((n+1)t)\| < \frac{n}{(n+1)^{1+1/n}} \quad \text{for } 0 < t \leq t_0$$

implies that $T(t) = 0$ *for all* $t > 0$*.*

The sharpness of the above result is shown by the following example [1], which involves a construction of appropriate sequences in the non-unital Banach algebra c_0.

Example. Let G be an additive measurable subgroup of $\mathbb{R}$ with $G \neq \mathbb{R}$. Then, given $(\gamma_n)_n$ in $\mathbb{R}^+$ such that $t\gamma_n \in G$ for all $t \in G$ with $t > 0$ and for all $n \in \mathbb{N}$, there exists a nontrivial semigroup $(S(t))_{t\in G, t>0}$ in c_0 such that

$$\|S(t) - S(t(\gamma_n + 1))\| < \frac{\gamma_n}{(\gamma_n + 1)^{1+1/\gamma_n}},$$

for all $t \in G$, $t > 0$.

2.2. Sectorial semigroups

In this subsection we discuss the behaviour of analytic semigroups defined on a sector

$$S_\alpha = \{z \in \mathbb{C} \setminus \{0\} : |\arg(z)| < \alpha\}.$$

with $0 < \alpha \leq \pi/2$. We begin with the case $\alpha = \pi/2$, so that $S_\alpha = \mathbb{C}_+$.

Theorem 2.5 ([1]). *Let* $(T(t))_{t\in\mathbb{C}_+}$ *be an analytic non-quasinilpotent semigroup in a Banach algebra. Let* $\mathcal{A}$ *be the closed subalgebra generated by* $(T(t))_{t\in\mathbb{C}_+}$ *and let* $\gamma > 0$*. If there exists* $t_0 > 0$ *such that*

$$\sup_{t\in\mathbb{C}_+, |t|\leq t_0} \rho(T(t) - T(\gamma+1)t)) < 2$$

then $\mathcal{A}/\operatorname{Rad}\mathcal{A}$ *is unital, and the generator of* $(\pi(T(t)))_{t>0}$ *is bounded, where* $\pi : \mathcal{A} \to \mathcal{A}/\operatorname{Rad}\mathcal{A}$ *denotes the canonical surjection.*

A semigroup $(T(t))$ defined on the positive reals or on a sector is said to be *exponentially bounded* if there exist $c_1 > 0$ and $c_2 \in \mathbb{R}$ such that $\|T(t)\| \leq c_1 e^{c_2|t|}$ for every t. Beurling [3], in his work described in the introduction, showed that there exists a universal constant k such that every exponentially bounded weakly measurable semigroup $(T(t))_{t>0}$ of bounded operators satisfying

$$\limsup_{t\to 0+} \|I - T(t)\| = \rho < 2$$

admits an exponentially bounded analytic extension to a sector S_α with $\alpha \geq k(2-\rho)^2$. From this one easily obtains the following result.

Theorem 2.6 ([1]). *Let $(T(t))_{t\in\mathbb{C}_+}$ be an analytic semigroup of bounded operators on a Banach space $\mathcal{X}$. If the generator of the semigroup is unbounded, then we have, for $-\frac{\pi}{2}<\alpha<\frac{\pi}{2}$,*

$$\limsup_{t\to 0+}\|I-T(t)\|\geq 2-\sqrt{\frac{\frac{\pi}{2}-|\alpha|}{k}},$$

where k is Beurling's universal constant.

We now consider similar results on smaller sectors than the half-plane, and in fact the result we prove will be stated in a far more general context.

Theorem 2.7 ([1]). *Let $0<\alpha<\pi/2$ and let f be an entire function with $f(0)=0$ and $f(\mathbb{R})\subseteq\mathbb{R}$, such that*

$$\sup_{\operatorname{Re} z>r}|f(z)|\to 0 \qquad \text{as} \quad r\to\infty, \tag{2.1}$$

and f is a linear combination of functions of the form $z^m\exp(-zw)$ for $m=0,1,2,\dots$ and $w>0$. Let $(T(t))_{t\in S_\alpha}=(\exp(tA))_{t\in S_\alpha}$ be an analytic non-quasi-nilpotent semigroup in a Banach algebra and let $\mathcal{A}$ be the subalgebra generated by $(T(t))_{t\in S_\alpha}$. If there exists $t_0>0$ such that

$$\sup_{t\in S_\alpha,|t|\leq t_0}\rho(f(-tA))<k(S_\alpha),$$

with $k(S_\alpha)=\sup_{t\in S_\alpha}|f(z)|$, then $\mathcal{A}/\operatorname{Rad}\mathcal{A}$ is unital and the generator of the semigroup $\pi(T(t))_{t\in S_\alpha}$ is bounded, where $\pi:\mathcal{A}\to\mathcal{A}/\operatorname{Rad}(\mathcal{A})$ denotes the canonical surjection.

Note that $f(-tA)$ is well defined in terms of $T(t)$ and its derivatives.

Suitable examples of $f(z)$ are linear combinations of functions $z^m\exp(-z)$, $m=1,2,3,\dots$, and $\exp(-z)-\exp(-(\gamma+1)z)$; also real linear combinations of the form $\sum_{k=1}^n a_k\exp(-b_kz)$ with $b_k>0$ and $\sum_{k=1}^n a_k=0$. This provides results analogous to those of [14, Thm. 4.12], where the behaviour of expressions such as $\|tA\exp(tA)\|$ and $\|\exp(tA)-\exp(stA)\|$ was considered for all $t>0$.

Remark 2.8. Another function considered in [14] is $f(z)=e^{-sz}\sin z$, where we now require $s>\tan\alpha$ for $f(-tA)$ to be well defined for $t\in S_\alpha$. This does not satisfy the condition (2.1), but we note that it holds for $z\in S_\alpha$, while for $z\notin S_\alpha$ there exists a constant $C>0$ with the following property: for each z with $\operatorname{Re} z>C$ there exists $\lambda\in(0,1)$ such that $|f(\lambda z)|\geq\sup_{z\in S_\alpha}|f(z)|$. Using this observation, it is not difficult to adapt the proof of Theorem 2.7 to this case.

The sharpness of the constants can be shown by considering examples in $C_0([0,1])$.

One particular case of the above is used in the estimates considered by Bendaoud, Esterle and Mokhtari [2, 10].

Corollary 2.9. *Let $\gamma > 0$ and $0 < \alpha < \pi/2$. Let $(T(t))_{t\in S_\alpha}$ be an analytic non-quasinilpotent semigroup in a Banach algebra and let $\mathcal{A}$ be the closed subalgebra generated by $(T(t))_{t\in S_\alpha}$. If there exists $t_0 > 0$ such that*

$$\sup_{t\in S_\alpha, |t|\le t_0} \rho(T(t) - T(t(\gamma+1))) < k(S_\alpha),$$

with $k(S_\alpha) = \sup_{t\in S_\alpha} |\exp(-t) - \exp(-(\gamma+1)t)|$, then $\mathcal{A}/\operatorname{Rad}\mathcal{A}$ is unital and the generator of $\pi(T(t))_{t\in S_\alpha}$ is bounded, where $\pi : \mathcal{A} \to \mathcal{A}/\operatorname{Rad}(\mathcal{A})$ denotes the canonical surjection.

Now set $f_n(z) = z^n e^{-z}$, and set $k_n(\alpha) = \max_{z\in S_\alpha} |f_n(z)|$. A straightforward computation shows that $k_n(\alpha) = \left(\frac{n}{e\cos(\alpha)}\right)^n$.

If A is the generator of an analytic semigroup $(T(t))_{t\in S_\alpha}$, then we have $f_n(-tA) = (-1)^n t^n T^{(n)}(t)$. So the following result, which may be deduced from Hille's work, described in Theorem 1.1, means that if

$$\sup_{t\in S_\alpha, 0<|t|<\delta} \|f_n(-tA)\| < k_n(\alpha)$$

for some $\delta > 0$, then the generator of the semigroup is bounded.

Theorem 2.10 ([1]). *Let $n \ge 1$ be an integer, let $\alpha \in (0, \pi/2)$ and let $(T(t))_{t\in S_\alpha}$ be an analytic semigroup. If*

$$\sup_{t\in S_\alpha, 0<|t|<\delta} \|t^n T^{(n)}(t)\| < \left(\frac{n}{e\cos(\alpha)}\right)^n$$

for some $\delta > 0$, then the closed algebra generated by the semigroup is unital, and the generator of the semigroup is bounded.

The remainder of this section is devoted to quasinilpotent semigroups. We let $D(0,r)$ denote $\{z \in \mathbb{C} : |z| < r\}$.

Remark 2.11. An analytic semigroup $(T(t))_{t\in S_\alpha}$ acting on a Banach space $\mathcal{X}$ and bounded near the origin can be extended to the closed sector $\overline{S_\alpha}$. Indeed, assume that there exists $r > 0$ such that

$$\sup_{t\in D(0,r)\cap S_\alpha} \|T(t)\| < +\infty.$$

Then $\lim_{\substack{t\to w\\ t\in S_\alpha}} T(t)x$ exists for every $x \in \mathcal{X}$ and every $w \in \partial S_\alpha$. Moreover if we set

$$T(w)x = \lim_{\substack{t\to w\\ t\in S_\alpha}} T(t)x,$$

then $(T(t))_{t\in\overline{S_\alpha}}$ is a semigroup of bounded operators which is continuous with respect to the strong operator topology. For we have $\lim_{\substack{t\to w\\ t\in S_\alpha}} T(t)T(t_0)x = T(t_0)x$ for every $t_0 > 0$ and every $x \in \mathcal{X}$. Now the result follows immediately from the fact that $\bigcup_{t>0} T(t)\mathcal{X}$ is dense in $\mathcal{X}$, given that

$$\sup_{z\in D(0,r)\cap S_\alpha} \|T(t)\| < +\infty.$$

The next lemma demonstrates that nontrivial quasinilpotent analytic semigroups cannot be bounded on the right half-plane $\mathbb{C}_+$. In fact, more is true.

Lemma 2.12 ([5]). *Let $(T(t))_{t\in\mathbb{C}_+}$ be a quasinilpotent analytic semigroup of bounded operators on a Banach space $\mathcal{X}$. Suppose that there exists $r>0$ such that*

$$\sup_{t\in D(0,r)\cap\mathbb{C}_+} \|T(t)\| < +\infty,$$

and define $T(iy)$ for $y\in\mathbb{R}$ using Remark 2.11. If

$$\int_{-\infty}^{\infty} \frac{\log^+ \|T(iy)\|}{1+y^2}\, dy < +\infty,$$

then $T(t)=0$ for $t\in\mathbb{C}_+$.

In the case when the semigroup is bounded near the origin, we may give appropriate estimates on the imaginary axis.

Theorem 2.13 ([5]). *Let $(T(t))_{t\in\mathbb{C}_+}$ be a nontrivial quasinilpotent analytic semigroup satisfying the conditions of Remark 2.11, and let $s>0$. Then*

$$\max(\rho(T(iy)-T(iy+is)), \rho(T(-iy)-T(-iy-is))) \geq 2,$$

for every $y>0$.

From this we may obtain estimates for semigroups satisfying a growth condition near the imaginary axis.

Corollary 2.14 ([5]). *Let $(T(t))_{t\in\mathbb{C}_+}$ be a quasinilpotent analytic semigroup such that*

$$\sup_{y\in\mathbb{R}} e^{-\mu|y|}\|T(\delta+iy)\| < +\infty$$

for some $\delta>0$ and some $\mu>0$, and let $\gamma>0$. Then

$$\sup_{t\in D(0,r)\cap\mathbb{C}_+} \|T(t)-T((1+\gamma)t)\| \geq 2,$$

for every $r>0$.

3. Lower estimates for functional calculus

In this section we summarise some very recent results from [6] and [7], which provide far-reaching generalizations of earlier work.

3.1. Semigroups on $\mathbb{R}_+$

3.1.1. The quasinilpotent case. Recall that if $(T(t))_{t>0}$ is a uniformly bounded strongly continuous semigroup with generator A, then

$$(A+\lambda I)^{-1} = -\int_0^\infty e^{\lambda t}T(t)\,dt, \tag{3.1}$$

for all $\lambda \in \mathbb{C}$ with $\operatorname{Re}\lambda < 0$. Here the integral is taken in the sense of Bochner with respect to the strong operator topology. If, in addition, $(T(t))_{t>0}$ is quasinilpotent, then we have (3.1) for all $\lambda \in \mathbb{C}$.

Similarly, if $\mu \in M_c(0,\infty)$ (the space of complex finite Borel measures on $(0,\infty)$) with Laplace transform

$$F(s) := \mathcal{L}\mu(s) = \int_0^\infty e^{-s\xi}\, d\mu(\xi) \qquad (s \in \mathbb{C}_+), \tag{3.2}$$

and $(T(t))_{t>0}$ is a strongly continuous semigroup of bounded operators on a Banach space $\mathcal{X}$, then we have a functional calculus for its generator A, defined by

$$F(-A) = \int_0^\infty T(\xi)\, d\mu(\xi),$$

in the sense of the strong operator topology; i.e.,

$$F(-A)x = \int_0^\infty T(\xi)x\, d\mu(\xi) \qquad (x \in \mathcal{X}),$$

which exists as a Bochner integral.

The following theorem applies to several examples studied in [1, 9, 10, 14]; these include $\mu = \delta_1 - \delta_2$, the difference of two Dirac measures, where $F(s) := \mathcal{L}\mu(s) = e^{-s} - e^{-2s}$ and $F(-sA) = T(t) - T(2t)$. More importantly, the theorem applies to many other examples, such as $d\mu(t) = (\chi_{[1,2]} - \chi_{[2,3]})(t)dt$ and $\mu = \delta_1 - 3\delta_2 + \delta_3 + \delta_4$, which are not accessible with the methods of [1, 9, 10, 14].

Theorem 3.1. *Let $\mu \in M_c(0,\infty)$ be a real measure such that $\int_0^\infty d\mu(t) = 0$, and let $(T(t))_{t>0}$ be a strongly continuous quasinilpotent semigroup of bounded operators on a Banach space $\mathcal{X}$. Set $F = \mathcal{L}\mu$. Then there exists $\eta > 0$ such that*

$$\|F(-sA)\| > \max_{x\geq 0} |F(x)| \qquad \text{for} \quad 0 < s \leq \eta.$$

If $\mu \in M_c(0,\infty)$ is a complex measure, then we write $\widetilde{F} = \mathcal{L}\overline{\mu}$, so that $\widetilde{F}(z) = \overline{F(\bar{z})}$. By considering the real measure $\nu := \mu * \overline{\mu}$, we obtain the following result.

Corollary 3.2. *Let $\mu \in M_c(0,\infty)$ be a complex measure such that $\int_0^\infty d\mu(t) = 0$, and let $(T(t))_{t>0}$ be a strongly continuous quasinilpotent semigroup of bounded operators on a Banach space $\mathcal{X}$. Set $F = \mathcal{L}\mu$. Then there exists $\eta > 0$ such that*

$$\|F(-sA)\widetilde{F}(-sA)\| > \max_{x\geq 0} |F(x)|^2 \qquad \text{for} \quad 0 < s \leq \eta.$$

3.1.2. The non-quasinilpotent case. Recall that a sequence $(P_n)_{n\geq 1}$ of idempotents in a Banach algebra $\mathcal{A}$ is said to be *exhaustive* if $P_n^2 = P_n P_{n+1} = P_n$ for all n and if for every $\chi \in \widehat{\mathcal{A}}$ there is a p such that $\chi(P_n) = 1$ for all $n \geq p$. Such sequences may often be found in non-unital algebras: for example, $P_n = e_1 + \cdots + e_n$ $(n = 1, 2, \ldots)$ in the Banach algebra c_0.

Theorem 3.3. *Let $(T(t))_{t>0}$ be a strongly continuous and eventually norm-continuous non-quasinilpotent semigroup on a Banach space $\mathcal{X}$, with generator A. Let $F = \mathcal{L}\mu$, where $\mu \in M_c(0,\infty)$ is a real measure such that $\int_0^\infty d\mu = 0$. If there exists $(u_k)_k \subset (0,\infty)$ with $u_k \to 0$ such that*

$$\rho(F(-u_k A)) < \sup_{x>0} |F(x)|,$$

then the algebra $\mathcal{A}$ generated by $(T(t))_{t>0}$ possesses an exhaustive sequence of idempotents $(P_n)_{n\geq 1}$ such that each semigroup $(P_n T(t))_{t>0}$ has a bounded generator.

If, further, $\|F(-u_k A)\| < \sup_{x>0} |F(x)|$, then $\bigcup_{n\geq 1} P_n \mathcal{A}$ is dense in $\mathcal{A}$.

3.2. Analytic semigroups

For $0 < \alpha < \pi/2$, let $H(S_\alpha)$ denote the Fréchet space of holomorphic functions on S_α, endowed with the topology of local uniform convergence; thus, if $(K_n)_{n\geq 1}$ is an increasing sequence of compact subsets of S_α with $\bigcup_{n\geq 1} K_n = S_\alpha$, we may specify the topology by the seminorms

$$\|F\|_n := \sup\{|f(z)| : z \in K_n\}.$$

We write $H(S_\alpha)'$ for its dual space; that is, the space of continuous linear functionals $\varphi : H(S_\alpha) \to \mathbb{C}$. This means that there is an index n and a constant $M > 0$ such that $|\langle f, \varphi \rangle| \leq M\|f\|_n$ for all $f \in H(S_\alpha)$.

We define the *Fourier–Borel transform* of φ by

$$\mathcal{FB}(\varphi)(z) = \langle e_{-z}, \varphi \rangle,$$

for $z \in \mathbb{C}$, where $e_{-z}(\xi) = e^{-z\xi}$ for $\xi \in S_\alpha$.

If $\varphi \in H(S_\alpha)'$, as above, then by the Hahn–Banach theorem, it can be extended to a functional on $C(K_n)$, which we still write as φ, and is thus induced by a Borel measure μ supported on K_n.

That is, we have

$$\langle f, \varphi \rangle = \int_{S_\alpha} f(\xi)\, d\mu(\xi),$$

where μ (which is not unique) is a compactly supported measure. For example, if $\langle f, \varphi \rangle = f'(1)$, then

$$\langle f, \varphi \rangle = \frac{1}{2\pi i} \int_C \frac{f(z)\, dz}{(z-1)^2},$$

where C is any sufficiently small circle surrounding the point 1. Note that

$$\mathcal{FB}(\varphi)(z) = \int_{S_\alpha} e^{-z\xi} d\mu(\xi).$$

Now let $T := (T(t))_{t\in S_\alpha}$ be an analytic semigroup on a Banach space $\mathcal{X}$, with generator A. Let $\varphi \in H(S_\alpha)'$ and let $F = \mathcal{FB}(\varphi)$.

We may thus define, formally to start with,

$$F(-A) = \langle T, \varphi \rangle = \int_{S_\alpha} T(\xi)\, d\mu(\xi),$$

which is well defined as a Bochner integral in $\mathcal{A}$. It is easy to verify that the definition is independent of the choice of μ representing φ.

Indeed, if $u \in S_{\alpha-\beta}$, where $\operatorname{supp}\mu \subset S_\beta$ and $0 < \beta < \alpha$, then we may also define

$$F(-uA) = \int_{S_\beta} T(u\xi)\, d\mu(\xi),$$

since $u\xi$ lies in S_α.

The following theorem extends [1, Thm. 3.6]. In the following, a symmetric measure is a measure such that $\mu(\overline{S}) = \overline{\mu(S)}$ for $S \subset S_\alpha$. A symmetric measure will have a Fourier–Borel transform f satisfying $f(z) = \widetilde{f}(z) := \overline{f(\overline{z})}$ for all $z \in \mathbb{C}$.

Theorem 3.4. *Let $0 < \beta < \alpha < \pi/2$. Let $\varphi \in H(S_\alpha)'$, induced by a symmetric measure $\mu \in M_c(S_\beta)$ such that $\int_{S_\beta} d\mu(z) = 0$, and let $f = \mathcal{FB}(\varphi)$. Let $(T(t))_{t\in S_\alpha} = (\exp(tA))_{t\in S_\alpha}$ be an analytic non-quasinilpotent semigroup in a Banach algebra and let $\mathcal{A}$ be the subalgebra generated by $(T(t))_{t\in S_\alpha}$. If there exists $t_0 > 0$ such that*

$$\sup_{t\in S_{\alpha-\beta}, |t|\le t_0} \rho(f(-tA)) < \sup_{z\in S_{\alpha-\beta}} |f(z)|,$$

then $\mathcal{A}/\operatorname{Rad}\mathcal{A}$ is unital and the generator of $\pi(T(t))_{t\in S_\alpha}$ is bounded, where $\pi : \mathcal{A} \to \mathcal{A}/\operatorname{Rad}(\mathcal{A})$ denotes the canonical surjection.

By considering the convolution of a functional $\varphi \in H(S_\alpha)'$, with Fourier–Borel transform f, and the functional $\widetilde{\varphi}$ with Fourier–Borel transform $\widetilde{f}$, we obtain the following result.

Corollary 3.5. *Let $0 < \beta < \alpha < \pi/2$. Let $\varphi \in H(S_\alpha)'$, induced by a measure $\mu \in M_c(S_\beta)$ such that $\int_{S_\beta} d\mu(z) = 0$, and let $f = \mathcal{FB}(\varphi)$. Let $(T(t))_{t\in S_\alpha} = (\exp(tA))_{t\in S_\alpha}$ be an analytic non-quasinilpotent semigroup in a Banach algebra and let $\mathcal{A}$ be the subalgebra generated by $(T(t))_{t\in S_\alpha}$. If there exists $t_0 > 0$ such that*

$$\sup_{t\in S_{\alpha-\beta}, |t|\le t_0} \rho(f(-tA)\widetilde{f}(-tA)) < \sup_{z\in S_{\alpha-\beta}} |f(z)||\widetilde{f}(z)|,$$

then $\mathcal{A}/\operatorname{Rad}\mathcal{A}$ is unital and the generator of $\pi(T(t))_{t\in S_\alpha}$ is bounded, where $\pi : \mathcal{A} \to \mathcal{A}/\operatorname{Rad}(\mathcal{A})$ denotes the canonical surjection.

It is possible to obtain a similar conclusion, based only on estimates on the positive real line.

Theorem 3.6. *Let $0 < \alpha < \pi/2$. Let $\varphi \in H(S_\alpha)'$, induced by a symmetric measure $\mu \in M_c(S_\alpha)$ such that $\int_{S_\alpha} d\mu(z) = 0$, and let $f = \mathcal{FB}(\varphi)$. Suppose that $f(\mathbb{R}_+) \subset \mathbb{R}$. Let $(T(t))_{t\in S_\alpha} = (\exp(tA))_{t\in S_\alpha}$ be an analytic non-quasinilpotent semigroup in a Banach algebra and let $\mathcal{A}$ be the subalgebra generated by $(T(t))_{t\in S_\alpha}$. If there exists $t_0 > 0$ such that*

$$\rho(f(-tA)) < \sup_{x>0} |f(x)|,$$

for all $0 < t \le t_0$, then $\mathcal{A}/\operatorname{Rad}\mathcal{A}$ is unital and the generator of $\pi(T(t))_{t\in S_\alpha}$ is bounded, where $\pi : \mathcal{A} \to \mathcal{A}/\operatorname{Rad}(\mathcal{A})$ denotes the canonical surjection.

The following example shows that the hypotheses of Theorem 3.6 are sharp.

Example. In the Banach algebra $\mathcal{A} = C_0([0,1])$ consider the semigroup $T(t) : x \mapsto x^t$. Clearly, $(T(t))$ is not norm-continuous at 0.

For $x \in (0,1]$ (which can be identified with the Gelfand space of $\mathcal{A}$) let $f = \mathcal{FB}(\mu)$ and

$$f(-tA)(x) = \int_{S_\alpha} x^{-t\xi}\, d\mu(\xi) = \int_{S_\alpha} e^{-t\xi \log x}\, d\mu(\xi),$$

where $\mu \in M_c(S_\alpha)$, supposing that $\int_{S_\alpha} d\mu(z) = 0$ and that $f(\mathbb{R}_+) \subset \mathbb{R}$.

Thus $f(-tA)(x) = f(-t\log x)$ and

$$\rho(f(-tA)) = \|f(-tA)\| = \sup_{x>0} |f(-t\log x)| = \sup_{r>0} |f(tr)|.$$

Clearly,

$$\sup_{t\in S_\alpha, |t|\le t_0} \rho(f(-tA)) = \sup_{t\in S_\alpha} |f(z)|$$

for all $t_0 > 0$.

Acknowledgment

This work was partially supported by the London Math. Society (Scheme 2).

References

[1] Z. Bendaoud, I. Chalendar, J. Esterle, and J.R. Partington. *Distances between elements of a semigroup and estimates for derivatives.* Acta Math. Sin. (Engl. Ser.), 26(12):2239–2254, 2010.

[2] Z. Bendaoud, J. Esterle, and A. Mokhtari. *Distances entre exponentielles et puissances d'éléments de certaines algèbres de Banach.* Arch. Math. (Basel), 89(3):243–253, 2007.

[3] A. Beurling. *On analytic extension of semigroups of operators.* J. Funct. Anal., 6:387–400, 1970.

[4] F.F. Bonsall and M.J. Crabb. *The spectral radius of a Hermitian element of a Banach algebra.* Bull. London Math. Soc., 2:178–180, 1970.

[5] I. Chalendar, J. Esterle, and J.R. Partington. *Boundary values of analytic semigroups and associated norm estimates.* In *Banach algebras* 2009, volume 91 of Banach Center Publ., pages 87–103. Polish Acad. Sci. Inst. Math., Warsaw, 2010.

[6] I. Chalendar, J. Esterle, and J.R. Partington. In preparation, 2014.

[7] I. Chalendar, J. Esterle, and J.R. Partington. *Lower estimates near the origin for functional calculus on operator semigroups*, J. Funct. Anal., to appear.

[8] J. Esterle. *Quasimultipliers, representations of H^∞, and the closed ideal problem for commutative Banach algebras.* In *Radical Banach algebras and automatic continuity* (*Long Beach, Calif.*, 1981), volume 975 of *Lecture Notes in Math.*, pages 66–162. Springer, Berlin, 1983.

[9] J. Esterle. *Distance near the origin between elements of a strongly continuous semigroup.* Ark. Mat., 43(2):365–382, 2005.

[10] J. Esterle and A. Mokhtari. *Distance entre éléments d'un semi-groupe dans une algèbre de Banach.* J. Funct. Anal., 195(1):167–189, 2002.

[11] S. Fackler. *Regularity of semigroups via the asymptotic behaviour at zero.* Semigroup Forum, 87(1):1–17, 2013.

[12] E. Hille. *On the differentiability of semi-group operators.* Acta Sci. Math. Szeged, 12:19–24, 1950.

[13] E. Hille and R.S. Phillips. *Functional analysis and semi-groups.* American Mathematical Society, Providence, R. I., 1974. Third printing of the revised edition of 1957, American Mathematical Society Colloquium Publications, vol. XXXI.

[14] N. Kalton, S. Montgomery-Smith, K. Oleszkiewicz, and Y. Tomilov. *Power-bounded operators and related norm estimates.* J. London Math. Soc. (2), 70(2):463–478, 2004.

[15] T. Kato. *A characterization of holomorphic semigroups.* Proc. Amer. Math. Soc., 25:495–498, 1970.

[16] Y. Katznelson and L. Tzafriri. *On power bounded operators.* J. Funct. Anal., 68(3):313–328, 1986.

[17] A. Mokhtari. *Distance entre éléments d'un semi-groupe continu dans une algèbre de Banach.* J. Operator Theory, 20(2):375–380, 1988.

[18] J.W. Neuberger. *Analyticity and quasi-analyticity for one-parameter semigroups.* Proc. Amer. Math. Soc., 25:488–494, 1970.

[19] A. Pazy. *Semigroups of linear operators and applications to partial differential equations*, volume 44 of *Applied Mathematical Sciences.* Springer-Verlag, New York, 1983.

I. Chalendar
Université de Lyon, UMR 5208
Institut Camille Jordan
43 bld. du 11 novembre 1918
F-69622 Villeurbanne Cedex, France
e-mail: `chalendar@math.univ-lyon1.fr`

J. Esterle
I.M.B., UMR 5251
Université de Bordeaux
351 cours de la Libération
F-33405 Talence Cedex, France
e-mail: `jean.esterle@math.u-bordeaux1.fr`

J.R. Partington
School of Mathematics
University of Leeds
Leeds LS2 9JT, U.K.
e-mail: `J.R.Partington@leeds.ac.uk`

Operator Theory:
Advances and Applications, Vol. 250, 133–146

Estimates on Non-uniform Stability for Bounded Semigroups

Thomas Duyckaerts

Dedicated to Charles J.K. Batty

Abstract. Let $S(t)$ be a bounded strongly continuous semigroup on a Banach space, with generator $-A$. Assume that the spectrum of A has empty intersection with the imaginary axis. In [6], Charles J.K. Batty and the author have given an estimate of the decay of the operator norm of $S(t)(1+A)^{-1}$, as t tends to infinity, in terms of asymptotic bounds of the resolvent of A on the imaginary axis. In this note, we give another proof of this result. The original proof relied on a trick appearing in an analytic proof of the prime number theorem by D. Newman, which we do not use here.

Mathematics Subject Classification (2010). Primary: 47D06; secondary 35B35, 35B40.

Keywords. Semigroups, stability.

1. Introduction

Let $(B, \|\cdot\|)$ be a Banach space and $S(t)$, $t \geq 0$ a strongly continuous semigroup on B, with generator $-A$. Recall that A is a closed, densely defined, operator on B. We assume that $S(t)$ is bounded, that is

$$\exists \widetilde{C} > 0, \quad \forall t \geq 0, \quad \|S(t)\| \leq \tilde{C}, \tag{1.1}$$

where $\|\cdot\|$ also denotes the operator norm on B.

In this note, we are interested in the relationship between the stability (that is the decay to 0 as $t \to +\infty$) of $S(t) = e^{-tA}$, and properties of the spectrum $\sigma(A)$ and of the resolvent of A. Recall that the boundedness (1.1) of $S(t)$ implies

$$\sigma(A) \subset \{\operatorname{Re} z \geq 0\}.$$

Arendt and Batty [1] and Lyubich and Vũ [14] have proved that if $\sigma(A) \cap i\mathbb{R}$ is countable and $\sigma(A^*) \cap i\mathbb{R}$ contains no eigenvalue (here A^* is the adjoint of A), then the semigroup is (pointwise) strongly stable, that is

$$\forall f \in B, \quad \lim_{t\to+\infty} \|S(t)f\| = 0. \tag{1.2}$$

In applications, in particular in the study of stability of linear partial differential equations, it is interesting to have stronger information on the decay of $S(t)$ to 0. When

$$\lim_{t\to+\infty} \|S(t)\| = 0,$$

the stability is said to be uniform, and the semigroup property implies that the decay of $\|S(t)\|$ to 0 is indeed exponential. We refer to [5], and to the books [2], [17] and [9] for surveys on pointwise and uniform stability, as well as other classical types of stability.

In this note, we are interested in another form of stability, stronger than the pointwise stability (1.2) but weaker than uniform stability. In [5] (this is also implicit in [1]), it was proved by C.J.K. Batty that

$$\sigma(A) \cap i\mathbb{R} = \emptyset \tag{1.3}$$

implies

$$\lim_{t\to+\infty} \|S(t)(1+A)^{-1}\| = 0 \tag{1.4}$$

(this was later called *semi-uniform stability* in [6]).

In applications, the rate of decay to 0 in (1.4) is of course important. This rate is related to the growth of the norm of the resolvent on the imaginary axis. Let

$$M(\xi) = \sup_{-\xi\leq\tau\leq\xi} \|(i\tau + A)^{-1}\|, \quad \xi \geq 0. \tag{1.5}$$

In [12], G. Lebeau has proved, when B is a Hilbert space, that an exponential bound of $M(\tau)$ for large τ implies

$$\forall t \gg 1, \quad \|S(t)(1+A)^{-1}\| \leq \frac{C}{\log(t)\log\log(t)}. \tag{1.6}$$

In [8], under the same assumption, the right-hand side of the bound (1.6) was improved to $\frac{C}{\log t}$. The proofs of [12] and [8] are based on a representation of the solution by an infinite integral of the resolvent, together with contour deformation arguments. An analogous result, with a closely related proof was given in [13] for polynomial stability in Hilbert spaces.

In [6], C. Batty and the author unified and generalized the preceding results on semi-uniform stabilities:

Theorem 1.1 (Batty, Duyckaerts). *Let $S(t) = e^{-tA}$ be a bounded semigroup, with generator $-A$, on the Banach space B. Then*

1. $\lim_{t\to\infty} \left\|S(t)(1+A)^{-1}\right\| = 0 \iff \sigma(A) \cap i\mathbb{R} = \emptyset.$

2. *Assume that the equivalent conditions in* (1) *hold. Let* M *be defined by* (1.5), *and*

$$M_{\log}(\eta) = M(\eta)\big[\log(1+M(\eta)) + \log(1+\eta)\big], \tag{1.7}$$

Let k *be an integer such that* $k \geq 1$. *Then there are constants* C_k, T_k, *such that:*

$$\forall t \geq T_k, \quad \left\| e^{-tA}(A+1)^{-k} \right\| \leq \frac{C_k}{\left(M_{\log}^{-1}\left(\frac{t}{C_k}\right)\right)^k}.$$

In the statement of the theorem, $M_{\log}^{-1}$ is the inverse function of $M_{\log}$, which maps $(T,+\infty)$ onto $(0,+\infty)$, where $T = M_{\log}(0)$.

In [6], the proof of point (2) is based on the method of [5] to prove the implication (1.3)$\Longrightarrow$(1.4). The main tool is a representation of $S(t)$ by an integral involving the resolvent of A on a closed path of the complex plane. It uses a trick due to Newman [16] and Korevaar [10], appearing in a proof of the prime number theorem. Carefully following the constants in the proof of [5], one obtains point (2) of the theorem.

The purpose of this note is to give another proof of Theorem 1.1 (2) that does not use the trick of Newman and Korevaar. Before this proof, we discuss the optimality of Theorem 1.1 (2) and more recent works on the subject. Let

$$m(t) = \sup_{s\geq t} \left\| e^{-sA}(1+A)^{-1} \right\|,\ t \geq 0,$$

so that Theorem 1.1 (2) means (in the case $k=1$ to fix ideas)

$$m(t) \leq \frac{C}{\left(M_{\log}^{-1}\left(\frac{t}{C}\right)\right)}. \tag{1.8}$$

Denote by m_r^{-1} a right inverse of the nonincreasing function m. It was observed in [6] that (if $\sigma(A) \cap i\mathbb{R} = \emptyset$), the following partial converse to (1.8) holds:

$$\forall \xi \geq \xi_0, \quad M(\xi) \leq 1 + C m_r^{-1}\left(\frac{1}{2(\xi+1)}\right), \tag{1.9}$$

There is a logarithmic gap between (1.8) and (1.9): in view of (1.9), the optimal possible decay result is the following improvement of (1.8):

$$m(t) \leq \frac{C}{\left(M^{-1}\left(\frac{t}{C}\right)\right)}. \tag{1.10}$$

In [6], it was conjectured that this logarithmic gap is necessary in general Banach space, but not if B is assumed to be a Hilbert space.

This was clarified by A. Borichev and Y. Tomilov [7], who constructed bounded semigroups, satisfying conditions (1) of Theorem 1.1, such that M has polynomial growth at infinity, and (1.8) is sharp (up to the constants). This proves the necessity of the log correction for general Banach spaces. Furthermore, in the same paper, the optimal bound (1.10) was proved in the case of a polynomial decay rate on Hilbert space. Namely, if B is a Hilbert space and $M(\xi) \leq C\xi^a$ for $\xi > 1$ (where $a > 0$) then (1.10) holds, i.e.,

$$\|S(t)(1+A)^{-1}\| \leq \frac{C}{t^{1/a}}.$$

This was already known (see [3]) for systems of commuting normal operators on Hilbert spaces.

In the recent paper [4], the bound (1.10) was proved for other decay rates on Hilbert spaces, including logarithmic perturbations of polynomial decay. Other types of decays (such as the decay to 0 of $\|(1+A)^{-1}AS(t)\|$ when $\sigma(A) \cap i\mathbb{R} = \{0\}$) are also considered. The proofs in [7] and [4] are based on operator theoretical arguments ultimately relying on the Plancherel theorem and are very different in nature from the arguments of [5], [12], [8], [6].

Let us mention that the method of [6] can also be applied when $\sigma(A) \cap i\mathbb{R}$ is finite, but not empty: see [15]. It is also possible, by similar methods, to estimate the decay rate of $P_1 S(t) P_2$, where P_1, P_2 are bounded operators on B, assuming bounds on $P_1(A-z)^{-1}P_2$ on the imaginary axis: see [8], [6, Section 4].

Theorem 1.1 (2) is proved in the next section. The proof is elementary and relies on the Fourier inversion formula and contour integration. As in the proof of G. Lebeau in the case of logarithmic decay (see [12]) the idea is to use a representation of $u(t) = S(t)f$ by an infinite contour integral. Following the proof of N. Burq in [8], we do not work directly on $u(t)$, but on a truncated function $v(t) = \varphi(t)u(t)$, where the smooth function φ is supported in $[1, +\infty)$, and equal to 1 for large t. The desired representation formula is then obtained through an immediate application of the Fourier inversion formula. Note that the truncation is important to avoid the additional log loss that appears in [12] and [13].

After some preliminaries (§2.1), we show, using that $i\mathbb{R} \cap \sigma(A)$ is empty, that the time Fourier transform $\hat{u}$ of u may be extended to an analytic function around the imaginary axis (see §2.2). In §2.3 we conclude the proof by estimating $u(t)$. The idea, that also goes back to [12], is to cut-off the integral $\int e^{it\tau}\hat{u}(\tau)d\tau$ into essentially two parts. For small τ (with respect to $M_{\log}^{-1}(t)$) we bound the integrand using the resolvent inequality $\|(A - i\tau)^{-1}\| \leq M(|\tau|)$, after a contour deformation in the half-plane $\operatorname{Im}\tau \geq 0$. For large τ, we only use the boundedness of the semigroup. The main novelty of the proof is in this part. In [12] and [8], B is assumed to be a Hilbert space and the bound follows from Plancherel Theorem. In the present note, thanks to a simple smoothing trick, we are able to use an elementary Fourier multipliers estimate which works in general Banach space.

Notation

The norm $\|\cdot\|$ denotes both the norm in the Banach space B and the operator norm from B to B.

We denote by $\hat{u}$ the Fourier–Laplace transform

$$\hat{u}(\tau) = \int_{-\infty}^{+\infty} e^{-it\tau}u(t)dt$$

of a function f with values in the Banach space B. We will only work with integrals that are absolutely convergent. Note that if u is a L^∞ function supported in $[0, +\infty)$, then $\hat{u}(z)$ is defined and analytic in the open half-space $\operatorname{Im} z < 0$, the Fourier transform of $e^{-\varepsilon t}u(t)$ is $\hat{u}(\tau - i\varepsilon)$, and the Fourier inversion formula reads

$$e^{-\varepsilon t}u(t) = \frac{1}{2\pi}\int_{-\infty}^{+\infty} e^{it\tau}\hat{u}(\tau - i\varepsilon)d\tau,$$

for any $\varepsilon > 0$ such that $\tau \mapsto \hat{u}(\tau - i\varepsilon)$ is L^1.

2. Proof of the main result

In this section we prove Theorem 1.1 (2).

2.1. Preliminaries

We first construct a C^1 bound of the resolvent on the imaginary axis.

Lemma 2.1. *Assume that $i\mathbb{R} \cap A = \emptyset$. Then there exists a non-decreasing function $Y \in C^1([0, +\infty))$ and a constant $t_0 > 0$ such that*

$$\forall \tau \in \mathbb{R}, \quad \|(i\tau + A)^{-1}\| \leq Y(|\tau|) \tag{2.1}$$

$$\forall t \geq t_0, \quad \frac{1}{Y_{\log}^{-1}(t)} \leq \frac{2}{M_{\log}^{-1}(t)}, \tag{2.2}$$

where the increasing function $Y_{\log}$ is defined as $M_{\log}$ by

$$Y_{\log}(\eta) = Y(\eta)\big[\log(1 + Y(\eta)) + \log(1 + \eta)\big]. \tag{2.3}$$

Proof. If M is C^1, we can just take $Y = M$. If not, using that M is continuous, thus locally integrable, we define

$$Y(\eta) = \int_0^\infty \psi(\eta - \xi)M(\xi)d\xi, \quad \eta \geq 0, \tag{2.4}$$

where

$$\psi \in C_0^\infty((-1, 0)), \quad \int_{-1}^0 \psi = 1, \quad \psi \geq 0.$$

Using that M is non-decreasing, we get that Y is non-decreasing and

$$M(\eta) \leq Y(\eta) \leq M(\eta + 1).$$

The left-hand side inequality and the definition (1.5) of M yield (2.1). The right-hand side inequality and the definitions (1.7) and (2.3) of $M_{\log}$ and $Y_{\log}$ imply

$$M_{\log}^{-1}(t) \leq Y_{\log}^{-1}(t) + 1,$$

which shows (2.2) using that $M_{\log}^{-1}(t)$ tends to $+\infty$ as t tends to ∞. $\square$

In all the proof we fix $k \in \mathbb{N}$, $k \geq 1$, and $u_0 \in D(A^k)$. The letter C denotes a positive constant, that may change from line to line and depending only on the function Y and the constant $\widetilde{C}$ in (1.1), but not on u_0 and k. If the constant also depends on k we will denote it by C_k.

Fix a non-decreasing function $\varphi \in C^\infty(\mathbb{R})$ such that $\varphi(t) = 0$ if $t \leq 1$ and $\varphi(t) = 1$ if $t \geq 2$ and consider

$$v(t) = \varphi(t) e^{-tA} u_0. \tag{2.5}$$

Note that

$$\forall t \leq 1,\ v(t) = 0, \quad \forall t \geq 2,\ v(t) = e^{-tA} u_0.$$

We will show (assuming $k \geq 2$) that there exist a time $T_k > 0$ such that

$$\forall t \geq T_k, \quad \|v(t)\| \leq \frac{C_k}{\left(Y_{\log}^{-1}\left(\frac{t}{C_k}\right)\right)^k} \|u_0\|_{D(A^k)}. \tag{2.6}$$

In view of (2.2), the desired conclusion will follow for $k \geq 2$. To deduce the case $k = 1$ from the case $k = 2$, one can write, using the moment inequality (see, e.g., [11, Theorem 15.14]):

$$\begin{aligned} \|(1+A)^{-1} e^{-tA} u_0\| &\leq C \|e^{-tA} u_0\|^{1/2} \|(1+A)^{-2} e^{-tA} u_0\|^{1/2} \\ &\leq C \|u_0\|^{1/2} \left(\frac{C_2 \|u_0\|}{\left(M_{\log}^{-1}(t/C_2)\right)^2} \right)^{1/2}, \end{aligned}$$

which yields the result for $k = 1$.

2.2. Analytic extension of the time-Fourier transform

For a function $y \in C^0(\mathbb{R}, [0, +\infty))$ we will denote

$$N_y = \{z \in \mathbb{C}, \quad -y(\operatorname{Re}(z)) < \operatorname{Im} z < y(\operatorname{Re}(z))\}. \tag{2.7}$$

Consider

$$\hat{v}(\tau) = \int_{-\infty}^{+\infty} e^{-it\tau} v(t) dt$$

the Fourier–Laplace transform of v. By (1.1), v is bounded and thus $\hat{v}$ is well defined, and analytic, in the open half-plane $\{\operatorname{Im} \tau < 0\}$.

Lemma 2.2. *Assume* (1.1) *and that* $\sigma(A) \cap i\mathbb{R} = \emptyset$. *Let*

$$y(\tau) := \frac{1}{2Y(|\operatorname{Re} \tau|)}, \tag{2.8}$$

where Y is given by Lemma 2.1. Then the function $\hat{v}(\tau)$ admits an analytic extension to N_y. Furthermore

$$\forall z \in N_y, \quad \|\hat{v}(z)\| \leq C_k \frac{Y(|\operatorname{Re} z|)}{(1+|z|)^k} \|u_0\|_{D(A^k)}. \tag{2.9}$$

Proof. The function v is solution of the following equation on $\mathbb{R}$:

$$\partial_t v + Av = f, \tag{2.10}$$

where $f(t) = \varphi'(t)e^{-tA}u_0$. We first note that

$$\hat{f}(\tau) = \int_{-\infty}^{+\infty} e^{-it\tau} f(t)dt$$

is the Fourier–Laplace transform of a continuous function with support in $[1,2]$, and hence that $\hat{f}$ is analytic in $\mathbb{C}$. Furthermore, the definition of $\hat{f}$ and (1.1) yield the bound

$$\|\hat{f}(z)\| \leq \widetilde{C}e^{2|\operatorname{Im}(z)|}\|\varphi'\|_\infty \|u_0\|.$$

Similarly, observing that

$$z^k \hat{f}(\tau) = (-i)^k \int_{-\infty}^{+\infty} e^{-itz}\partial_t^k f(t)dt,$$

and that (1.1) implies

$$\forall t \in \mathbb{R}, \quad \left\|\partial_t^k f(t)\right\| \leq C_k \|u_0\|_{D(A^k)}$$

we get the bound

$$\forall z \in \mathbb{C}, \quad (1+|z|^k)\hat{f}(z) \leq C_k e^{2|\operatorname{Im} z|}\|u_0\|_{D(A^k)}. \tag{2.11}$$

By (2.10), and the fact that v and f have support in $(0,+\infty)$ we have the formula $iz\hat{v}(z) + A\hat{v}(z) = \hat{f}(z)$ for $\operatorname{Im} z < 0$. As a consequence,

$$\hat{v}(z) = (iz + A)^{-1}\hat{f}(z), \quad \operatorname{Im} z < 0. \tag{2.12}$$

The imaginary axis is in the resolvent set of A, which shows that $(iz+A)^{-1}\hat{f}$ is analytic in a neighborhood of the real axis. Furthermore, the formula

$$\left(Id_B + \mu(i\tau + A)^{-1}\right)(i\tau + A) = (i\tau + \mu + A), \quad \tau, \mu \in \mathbb{R}$$

the bound (2.1), and the von Neumann expansion formula show that $(iz + A)^{-1}$ is analytic in N_y with the bound

$$\forall z \in N_y, \quad \left\|(iz + A)^{-1}\right\| \leq 2Y(|\operatorname{Re} z|). \tag{2.13}$$

By (2.12), $\hat{v}$ admits an analytic extension to N_y. By (2.11) and (2.13), we get the announced bound (2.9). □

2.3. Decay at infinity of functions with analytic Fourier transform and bounded derivatives

We now conclude the proof of Theorem 1.1 (2), by proving an abstract technical result relating the decay of a function of class C^k on $[0,+\infty)$, taking values in the Banach space B, to the behaviour of its Fourier transform around the real axis.

Proposition 2.3. *Consider two real-valued positive continuous functions P and y on $\mathbb{R}$. Assume furthermore that y is piecewise C^1. Let $k \geq 2$ and $v \in C^0(\mathbb{R}, B)$ such that*

$$\operatorname{supp} v \subset [0, +\infty), \quad v, \partial_t^k v \in L^\infty(\mathbb{R}, B). \tag{2.14}$$

Assume that the Fourier–Laplace transform $\hat{v}$ of v admits an analytic continuation to the set

$$\{z \in \mathbb{C}, \ \operatorname{Im} z < y(\operatorname{Re} z)\}$$

such that

$$\forall z, \quad 0 \leq \operatorname{Im} z < y(\operatorname{Re} z) \Longrightarrow \|\hat{v}(z)\| \leq P(\operatorname{Re} z). \tag{2.15}$$

Then

$$\forall t \geq 1, \quad \|v(t)\| \leq g_k(t), \tag{2.16}$$

where $g_k(t) := \inf_{\eta_0 > 0} G_k(t, \eta_0)$ *with*

$$\begin{aligned} G_k(t, \eta_0) &= \frac{1}{2\pi} \int_{-2\eta_0}^{2\eta_0} e^{-ty(\eta)} P(\eta) \sqrt{1 + |y'(\eta)|^2} d\eta \\ &+ \frac{1}{\pi t \eta_0} \int_{\eta_0 \leq |\eta| \leq 2\eta_0} \left(1 - e^{-ty(\eta)}\right) P(\eta) d\eta + \frac{C_k}{\eta_0^k} \left\|\partial_t^k v\right\|_{L^\infty(\mathbb{R}, B)}. \end{aligned} \tag{2.17}$$

Let us first prove Theorem 1.1 (2) assuming Proposition 2.3.

Proof of (2). By the arguments of §2.1, it is sufficient to show (2.6) for $u_0 \in D(A^k)$, $k \geq 2$.

By Lemma 2.2, v satisfies the assumptions of Proposition 2.3 with

$$P(\tau) = \frac{C_k Y(|\tau|)}{1 + |\tau|^k} \|u_0\|_{D(A^k)}, \quad y(\tau) = \frac{1}{2Y(|\tau|)}.$$

The function Y is C^1 and thus y is C^1 on $(0, +\infty)$ and $(-\infty, 0)$ and continuous in 0.

We fix $t \geq 2kY_{\log}(2)$. Let

$$\eta_0 = \frac{1}{2} Y_{\log}^{-1}\left(\frac{t}{2k}\right), \quad \text{i.e., } t = 2kY(2\eta_0)\big[\log(1 + Y(2\eta_0)) + \log(1 + 2\eta_0)\big].$$

Note that $\eta_0 \geq 1$. By (2.16),

$$\|v(t)\| \leq G_k(t, \eta_0).$$

It remains to bound each of the three terms in the definition (2.17) of $G_k(t, \eta_0)$

$$\begin{aligned} &\int_{-2\eta_0}^{2\eta_0} e^{-ty(\eta)} P(\eta) \sqrt{1 + |y'(\eta)|^2} d\eta \\ &\leq C_k \|u_0\|_{D(A^k)} \int_{-2\eta_0}^{2\eta_0} e^{-2ky(\eta)Y(2\eta_0)[\log(1+Y(2\eta_0)) + \log(1+2\eta_0)]} \\ &\qquad \times \frac{Y(|\eta|)}{1 + |\eta|^k} \left(1 + |y'(\eta)|\right) d\eta. \end{aligned}$$

As Y is increasing and $y(\eta) = \frac{1}{2Y(|\eta|)}$, we have $2y(\eta)Y(\eta_0) \geq 1$ if $-2\eta_0 \leq \eta \leq 2\eta_0$. Thus

$$\left| e^{-2ky(\eta)Y(2\eta_0)\left[\log(1+Y(2\eta_0))+\log(1+2\eta_0)\right]} \right| \leq \frac{1}{(1+Y(2\eta_0)^k)(1+\eta_0)^k}.$$

Hence

$$\begin{aligned} &\int_{-2\eta_0}^{2\eta_0} e^{-ty(\eta)} P(\eta) \sqrt{1+|y'(\eta)|^2} d\eta \\ &\quad \leq \frac{C_k}{(1+\eta_0)^k} \|u_0\|_{D(A^k)} \int_{-2\eta_0}^{+2\eta_0} \frac{1}{(1+|\eta|)^k} (1+|y'(\eta)|)\, d\eta \qquad (2.18) \\ &\quad \leq \frac{C_k}{(1+\eta_0)^k} \|u_0\|_{D(A^k)}. \end{aligned}$$

To get the second inequality we used that $k \geq 2$, so that $\int_{-\infty}^{+\infty} \frac{1}{(1+|\eta|)^k} d\eta$ converges, and that

$$\int_{-2\eta_0}^{+2\eta_0} |y'(\eta)| d\eta = \int_{-2\eta_0}^{0} y'(\eta) d\eta - \int_{0}^{2\eta_0} y'(\eta) d\eta = y(0) - y(-2\eta_0) + y(0) - y(2\eta_0),$$

which is bounded from above by $2y(0)$.

On the other hand, using that $Y(|\eta|) = \frac{1}{2y(\eta)}$

$$\frac{1}{t\eta_0} \int_{\eta_0}^{2\eta_0} \left(1 - e^{-ty(\eta)}\right) P(\eta) d\eta \leq C_k \|u_0\|_{D(A^k)} \frac{1}{\eta_0} \int_{\eta_0}^{2\eta_0} \frac{1-e^{-ty(\eta)}}{ty(\eta)} \times \frac{1}{1+\eta^k} d\eta.$$

The function $x \mapsto \frac{1-e^{-x}}{x}$ is bounded on $[0, +\infty)$, which yields

$$\frac{1}{t\eta_0} \int_{\eta_0}^{2\eta_0} \left(1 - e^{-ty(\eta)}\right) P(\eta) d\eta \leq \frac{C_k}{(1+\eta_0)^k} \|u_0\|_{D(A^k)}, \tag{2.19}$$

and similarly

$$\frac{1}{t\eta_0} \int_{-2\eta_0}^{-\eta_0} \left(1 - e^{-ty(\eta)}\right) P(\eta) d\eta \leq \frac{C_k}{(1+\eta_0)^k} \|u_0\|_{D(A^k)}. \tag{2.20}$$

By (1.1),

$$\frac{\|\partial_t^k v\|_{L^\infty(\mathbb{R},B)}}{\eta_0^k} \leq \frac{C_k}{\eta_0^k} \|u_0\|_{D(A^k)}. \tag{2.21}$$

Bounds (2.18), (2.19), (2.20) and (2.21) and the definition of η_0 yield (2.6). The proof of Theorem 1.1 (2) is complete. □

Proof of Proposition 2.3. We fix $t > 0$, $\eta_0 > 0$.

Since v is continuous and bounded, $e^{-\varepsilon t} v(t)$ is, for any $\varepsilon > 0$, a L^1 function of t, with Fourier transform

$$\hat{v}(\tau - i\varepsilon) = \int_{-\infty}^{+\infty} v(t) e^{-t\varepsilon - it\tau} dt.$$

Using that $v^{(k)}$ is bounded and $k \geq 2$, we deduce that $t \mapsto \partial_t^2 \left(e^{-\varepsilon t}v(t)\right)$ is in $L^1(\mathbb{R})$. As a consequence, $\tau \mapsto \hat{v}(\tau - i\varepsilon)$ is in $L^1(\mathbb{R})$. By the Fourier inversion formula,

$$e^{-\varepsilon t}v(t) = \frac{1}{2\pi}\int_{-\infty}^{+\infty} e^{it\tau}\hat{v}(\tau - i\varepsilon)d\tau.$$

Let $\eta \in [\eta_0, 2\eta_0]$. We define γ^η as the path, depending on η, oriented from left to right, and consisting in the union $\gamma_1^\eta \cup \cdots \cup \gamma_5^\eta$, where:

- $\gamma_1^\eta = \big\{\operatorname{Im} z = 0, \quad \operatorname{Re} z \leq -\eta\big\}$;
- $\gamma_2^\eta = \big\{\operatorname{Re} z = -\eta, \quad 0 \leq \operatorname{Im} z \leq y(-\eta)\big\}$;
- $\gamma_3^\eta = \big\{-\eta \leq \operatorname{Re} z \leq \eta, \quad \operatorname{Im} z = y(\operatorname{Re} z)\big\}$;
- $\gamma_4^\eta = \big\{\operatorname{Re} z = \eta, \quad 0 \leq \operatorname{Im} z \leq y(\eta)\big\}$;
- $\gamma_5^\eta = \big\{\operatorname{Im} z = 0, \quad \eta \leq \operatorname{Re} z\big\}$.

By a change of contour, using that $\hat{v}$ is analytic in $\{\operatorname{Im} z < y(\operatorname{Re} z)\}$, we get:

$$e^{-\varepsilon t}v(t) = \frac{1}{2\pi}\int_{\gamma^\eta} e^{it\tau}\hat{v}(\tau - i\varepsilon)d\tau = \sum_{j=1}^{5} v_j(t,\eta,\varepsilon).$$

where

$$v_j(t,\eta,\varepsilon) = \frac{1}{2\pi}\int_{\gamma_j^\eta} \hat{v}(\tau - i\varepsilon)e^{it\tau}d\tau.$$

We need to bound from above the norm of each of the terms v_j in B. We will use two different strategies.

To estimate v_2, v_3, v_4, we will use the bound (2.15) on $\hat{v}$.

To estimate v_1 and v_5 we would like to use the fact that $\partial_t^k v(t)$ is bounded and a Fourier multiplier estimate. However v_5 (respectively v_1) is the image of the function $s \mapsto e^{-\varepsilon s}v(s)$ by the Fourier multiplier whose symbol is the characteristic function of the set $\{z \geq \eta\}$ (respectively $\{z \leq -\eta\}$), which is not smooth. To obtain a smooth Fourier multiplier we will integrate with respect to the parameter η as follows.

Let

$$\chi \in C_0^\infty(1,2), \text{ such that } 0 \leq \chi \leq 2 \text{ and } \int_1^2 \chi(\eta)d\eta = 1.$$

We have:

$$e^{-\varepsilon t}v(t) = \int_{\eta_0}^{2\eta_0} \frac{1}{\eta_0}\chi\Big(\frac{\eta}{\eta_0}\Big)e^{-\varepsilon t}v(t)d\eta = \sum_{j=1}^{5} V_j(t,\eta_0,\varepsilon). \tag{2.22}$$

where

$$V_j(t,\eta_0,\varepsilon) = \int_{\eta_0}^{2\eta_0} \frac{1}{\eta_0}\chi\Big(\frac{\eta}{\eta_0}\Big)v_j(t,\eta,\varepsilon)d\eta.$$

We will prove the following bounds as $\varepsilon \to 0$, $\varepsilon > 0$

$$\left\|V_3(t,\eta_0,\varepsilon)\right\| \leq \frac{1}{2\pi}\int_{-2\eta_0}^{+2\eta_0} P(\tau)e^{-ty(\tau)}\sqrt{1+|y'(\tau)|^2}d\tau \tag{2.23}$$

$$\|V_2(t,\eta_0,\varepsilon)\| + \|V_4(t,\eta_0,\varepsilon)\| \leq \frac{1}{\pi t\eta_0}\int_{\eta_0\leq|t|\leq 2\eta_0}(1-e^{-ty(\tau)})P(\tau)d\tau + C(\eta_0)\varepsilon \tag{2.24}$$

$$\|V_1(t,\eta_0,\varepsilon)\| + \|V_5(t,\eta_0,\varepsilon)\| \leq \frac{C_k}{\eta_0^k}\left(\|\partial_t^k v\|_{L^\infty(\mathbb{R},B)} + C(v,k)\varepsilon\right), \tag{2.25}$$

where $C(\eta_0)$ depends only on η_0 and $C(v,k)$ on v and k.

Summing up (2.23), (2.24) and (2.25), and letting ε go to 0 we would get the bound $\|v(t)\| \leq G_k(t,\eta_0)$, where G_k is defined by (2.17), and thus, taking the infimum on all $\eta_0 > 0$, the announced result (2.16).

Proof of (2.23). Let $\eta \in [\eta_0, 2\eta_0]$. If $\varepsilon > 0$ is small (depending on η_0), and $\tau \in \gamma_3^\eta$, we have $0 \leq (\operatorname{Im}\tau) - \varepsilon < y(\operatorname{Re}\tau)$ and thus, by assumption (2.15), $\|\hat{v}(\tau - i\varepsilon)\| \leq P(\operatorname{Re}\tau)$. Hence

$$\begin{aligned}\|v_3(t,\eta,\varepsilon)\| &\leq \frac{1}{2\pi}\int_{-\eta}^{\eta} P(s)e^{-ty(s)}\sqrt{1+y'(s)^2}ds \\ &\leq \frac{1}{2\pi}\int_{-2\eta_0}^{2\eta_0} P(s)e^{-ty(s)}\sqrt{1+y'(s)^2}ds.\end{aligned}$$

Multiplying by $\frac{1}{\eta_0}\chi\left(\frac{\eta}{\eta_0}\right)$ and integrating on $(\eta_0, 2\eta_0)$ with respect to η we get (2.23).

Proof of (2.24). Let again $\eta \in [\eta_0, 2\eta_0]$. Then

$$v_4(t,\eta,\varepsilon) = \frac{1}{2\pi}\int_0^{\varepsilon} \hat{v}(\eta+is-i\varepsilon)e^{it(\eta+is)}\,ds + \frac{1}{2\pi}\int_{\varepsilon}^{y(\eta)} \hat{v}(\eta+is-i\varepsilon)e^{it(\eta+is)}\,ds.$$

By assumption (2.15),

$$\frac{1}{2\pi}\left\|\int_{\varepsilon}^{y(\eta)} \hat{v}(\eta+is-i\varepsilon)e^{it(\eta+is)}\,ds\right\| \leq \frac{1}{2\pi}\int_0^{y(\eta)} e^{-ts}P(\eta)ds = \frac{1-e^{-ty(\eta)}}{2\pi t}P(\eta).$$

Furthermore, since $\hat{v}$ is continuous on $\{\operatorname{Im} z \leq y(\operatorname{Re} z)\}$, $\|\hat{v}\|$ is bounded in the set

$$\{\eta_0 \leq \operatorname{Re} z \leq 2\eta_0, \quad -1 \leq \operatorname{Im} z \leq 0\},$$

and we deduce

$$\frac{1}{2\pi}\left\|\int_0^{\varepsilon} \hat{v}(\eta+is-i\varepsilon)e^{it(\eta+is)}\,ds\right\| \leq C(\eta_0)\varepsilon,$$

with a constant $C(\eta_0)$ depending only on η_0.

Multiplying by $\frac{1}{\eta_0}\chi\left(\frac{\eta}{\eta_0}\right)$ and integrating with respect to η, we obtain

$$\|V_4(t,\eta_0,\varepsilon)\| \leq \frac{1}{2\pi t\eta_0}\int_{\eta_0}^{2\eta_0}\left(1-e^{-ty(\eta)}\right)P(\eta)d\eta + C(\eta_0)\varepsilon.$$

Together with the analogous bound on V_2 we get (2.24).

Proof of (2.25). We have

$$V_5(t,\eta_0,\varepsilon) = \frac{1}{2\pi}\int_0^{+\infty}\frac{1}{\eta_0}\chi\left(\frac{\eta}{\eta_0}\right)\int_\eta^{+\infty} e^{it\tau}\hat{v}(\tau-i\varepsilon)d\tau d\eta$$

By Fubini's theorem:

$$\begin{aligned} V_5(t,\eta_0,\varepsilon) =& \frac{1}{2\pi}\int_0^{+\infty}\int_0^\tau \frac{1}{\eta_0}\chi(\frac{\eta}{\eta_0})d\eta\,\hat{v}(\tau-i\varepsilon)e^{it\tau}d\tau. \\ =& \frac{1}{2\pi}\int_0^{+\infty}\frac{1}{\tau^k}\zeta(\frac{\tau}{\eta_0})\underbrace{\tau^k\hat{v}(\tau-i\varepsilon)}_{W_k(\tau,\varepsilon)}e^{it\tau}d\tau. \end{aligned}$$

where:

$$\zeta(\xi) = \int_0^\xi \chi(\eta)d\eta$$

is smooth, non-decreasing, supported in $(1,\infty)$ and equals 1 for $\xi \geq 2$.

Thus $V_5(t,\eta_0,\varepsilon)$ is the inverse Fourier transform of $\frac{1}{\tau^k}\zeta(\frac{\tau}{\eta_0})W_k(\tau,\varepsilon)$. Moreover, $W_k(\tau,\varepsilon)$ is the Fourier Transform of $w_{k,\varepsilon} = \left(\frac{d}{idt}\right)^k(e^{-\varepsilon t}v)$ which satisfies

$$\forall t\geq 0,\quad \|w_{k,\varepsilon}(t)\| \leq \left\|\partial_t^k v\right\|_{L^\infty(\mathbb{R},B)} + C(v,k)\varepsilon. \tag{2.26}$$

where $C(v,k)$ depends only on k and $\|v\|_{L^\infty(\mathbb{R},B)},\ldots,\left\|\partial_t^{k-1}v\right\|_{L^\infty(\mathbb{R},B)}$.

Lemma 2.4. *Let ζ be a C^∞ function on $\mathbb{R}$, supported in $(0,+\infty)$ and bounded as well as all its derivatives. Let*

$$\zeta_{\eta_0}(\tau) = \frac{1}{\tau^k}\zeta\left(\frac{\tau}{\eta_0}\right)$$

Then

$$\forall k\geq 1,\ \widehat{\zeta}_{\eta_0}\in L^1,\quad \|\widehat{\zeta}_{\eta_0}\|_{L^1(\mathbb{R})} \leq \frac{C_k}{\eta_0^k}. \tag{2.27}$$

Proof. Observe that

$$\zeta_{\eta_0}(\tau) = \frac{1}{\eta_0^k}\Psi_k\left(\frac{\tau}{\eta_0}\right),\qquad \Psi_k := \frac{1}{\tau^k}\zeta(\tau).$$

Hence

$$\widehat{\zeta}_{\eta_0}(t) = \frac{1}{\eta_0^k}\,\eta_0\widehat{\Psi}_k(\eta_0 t).$$

It is sufficient to show that $\widehat{\Psi}_k$ is in L^1. As $k \geq 1$, Ψ_k is in H^s for all s. In particular:

$$\int |\widehat{\Psi}_k(t)|^2(1+t^2)^2 dt < \infty.$$

Now

$$\widehat{\Psi}_k = \underbrace{\widehat{\Psi}_k(1+t^2)}_{\in L^2}\underbrace{(1+t^2)^{-1}}_{\in L^2},$$

and hence $\widehat{\Psi}_k \in L^1$ which concludes the proof. □

Thus $V_5(t,\eta_0,\varepsilon)$, the Fourier inverse of $\zeta_{\eta_0}(\tau)W_k(\tau,\varepsilon)$, is the convolution of $w_{k,\varepsilon}$, which is in $L^\infty(\mathbb{R},B)$, and of the Fourier inverse of ζ_{η_0}, which is by the preceding Lemma in $L^1(\mathbb{R},\mathbb{C})$. By bounds (2.26) and (2.27), we obtain

$$\|V_5(t,\eta_0,\varepsilon)\| \leq \frac{C_k}{\eta_0^k}\left(\|\partial_t^k v\|_{L^\infty(\mathbb{R},B)} + C(v,k)\varepsilon\right).$$

Together with the analogous statement on V_1, we get the announced estimate (2.25). The proof of Proposition 2.3 is complete. □

Acknowledgment

The author would like to thank Nicolas Burq and Luc Miller for fruitful discussions on the subject.

References

[1] Arendt, W., and Batty, C.J.K. *Tauberian theorems and stability of one-parameter semigroups.* Trans. Amer. Math. Soc. 306, 2 (1988), 837–852.

[2] Arendt, W., Batty, C.J.K., Hieber, M., and Neubrander, F. *Vector-valued Laplace transforms and Cauchy problems*, vol. 96 of *Monographs in Mathematics.* Birkhäuser Verlag, Basel, 2001.

[3] Bátkai, A., Engel, K.-J., Prüss, J., and Schnaubelt, R. *Polynomial stability of operator semigroups.* Math. Nachr. 279, 13-14 (2006), 1425–1440.

[4] Batty, C.J., Chill, R., and Tomilov, Y. *Fine scales of decay of operator semigroups*, 2013. arXiv preprint 1305.5365, to appear in J. Eur. Math. Soc.

[5] Batty, C.J.K. *Asymptotic behaviour of semigroups of operators.* In *Functional analysis and operator theory (Warsaw,* 1992), vol. 30 of *Banach Center Publ.* Polish Acad. Sci., Warsaw, 1994, pp. 35–52.

[6] Batty, C.J.K., and Duyckaerts, T. *Non-uniform stability for bounded semigroups on Banach spaces.* J. Evol. Equ. 8, 4 (2008), 765–780.

[7] Borichev, A., and Tomilov, Y. *Optimal polynomial decay of functions and operator semigroups.* Math. Ann. 347, 2 (2010), 455–478.

[8] Burq, N. *Décroissance de l'énergie locale de l'équation des ondes pour le problème extérieur et absence de résonance au voisinage du réel.* Acta Math. 180, 1 (1998), 1–29.

[9] Eisner, T. *Stability of operators and operator semigroups*, vol. 209 of *Operator Theory: Advances and Applications.* Birkhäuser Verlag, Basel, 2010.

[10] KOREVAAR, J. *On Newman's quick way to the prime number theorem.* Math. Intelligencer 4, 3 (1982), 108–115.

[11] KUNSTMANN, P.C., AND WEIS, L. *Maximal L_p-regularity for parabolic equations, Fourier multiplier theorems and H^∞-functional calculus.* In *Functional analytic methods for evolution equations*, vol. 1855 of *Lecture Notes in Math.* Springer, Berlin, 2004, pp. 65–311.

[12] LEBEAU, G. *Équation des ondes amorties.* In *Algebraic and geometric methods in mathematical physics (Kaciveli,* 1993), vol. 19 of *Math. Phys. Stud.* Kluwer Acad. Publ., Dordrecht, 1996, pp. 73–109.

[13] LIU, Z., AND RAO, B. *Characterization of polynomial decay rate for the solution of linear evolution equation.* Z. Angew. Math. Phys. 56, 4 (2005), 630–644.

[14] LYUBICH, Y.I., AND VŨ QUỐC PHÓNG. *Asymptotic stability of linear differential equations in Banach spaces.* Studia Math. 88, 1 (1988), 37–42.

[15] MARTÍNEZ, M.M. *Decay estimates of functions through singular extensions of vector-valued Laplace transforms.* J. Math. Anal. Appl. 375, 1 (2011), 196–206.

[16] NEWMAN, D.J. *Simple analytic proof of the prime number theorem.* Amer. Math. Monthly 87, 9 (1980), 693–696.

[17] VAN NEERVEN, J. *The asymptotic behaviour of semigroups of linear operators*, vol. 88 of *Operator Theory: Advances and Applications.* Birkhäuser Verlag, Basel, 1996.

Thomas Duyckaerts
LAGA
Institut Galilée
Université Paris XIII
99, avenue Jean-Baptiste Clément
F-93430 – Villetaneuse, France
e-mail: duyckaer@math.univ-paris13.fr

Operator Theory:
Advances and Applications, Vol. 250, 147–154

Convergence of the Dirichlet-to-Neumann Operator on Varying Domains

A.F.M. ter Elst and E.M. Ouhabaz

Dedicated to Charles Batty on occasion of his 60th birthday

Abstract. We prove resolvent convergence for the Dirichlet-to-Neumann operator on domains which are uniformly starshaped with respect to a ball, when the domains converge appropriately.

Mathematics Subject Classification (2010). 35R30, 35J25, 35B20.

Keywords. Dirichlet-to-Neumann operator, resolvent convergence.

1. Introduction

The Dirichlet and Neumann Laplacian have been studied on varying domains by many authors. See [Dan], [BL] and references therein. The aim of this paper is to study the Dirichlet-to-Neumann operator if the domain changes.

If Ω is an open bounded set with Lipschitz boundary, then the Dirichlet-to-Neumann operator $\mathcal{N}_\Omega$ is an operator that lives on the boundary of Ω. It is defined as follows: if $\varphi, \psi \in L_2(\partial\Omega)$ then $\varphi \in D(\mathcal{N}_\Omega)$ and $\mathcal{N}_\Omega \varphi = \psi$ if and only if there exists a $u \in H^1(\Omega)$ such that $\mathrm{Tr}_\Omega u = \varphi$, $\Delta u = 0$ weakly on Ω and $\frac{\partial u}{\partial \nu} = \psi$. Since we wish to vary the domain, we transfer the operator $\mathcal{N}_\Omega$ to the boundary of a reference domain Ω_0 and then prove convergence on $L_2(\partial\Omega_0)$. For simplicity we assume that the domains are starshaped with respect to a common ball with centre at the origin. Then we can choose

$$\Omega_0 = \{x \in \mathbb{R}^d : |x| < 1\},$$

the unit ball, as a reference domain.

For all $n \in \mathbb{N} \cup \{\infty\}$ let $R_n \colon S^{d-1} \to (0, \infty)$ be a Lipschitz continuous function and define

$$\Omega_n = \{r\,\omega : \omega \in S^{d-1} \text{ and } r \in [0, R_n(\omega))\},$$

where $S^{d-1} = \Gamma_0 = \partial\Omega_0$ is the unit sphere. Throughout this paper we provide the boundaries $\partial\Omega_n$ and $\partial\Omega_0$ with the $(d-1)$-dimensional Hausdorff measure. Let $\mathcal{N}_n$ be the Dirichlet-to-Neumann operator on Ω_n. Define $V_n \colon L_2(\Gamma_0) \to L_2(\partial\Omega_n)$ by

$$(V_n\varphi)(x) = \varphi\left(\frac{1}{|x|}\,x\right).$$

Then V_n is invertible. Define the unbounded operator B_n in $L_2(\Gamma_0)$ by

$$B_n = V_n^{-1}\,\mathcal{N}_n\,V_n.$$

The main result of this paper is the following.

Theorem 1.1. *Suppose that* $\lim_{n\to\infty} R_n = R_\infty$ *in* $W^{1,\infty}(\Gamma_0)$. *Then*

$$\lim_{n\to\infty} (\lambda\,I + B_n)^{-1} = (\lambda\,I + B_\infty)^{-1}$$

in $\mathcal{L}(L_2(\Gamma_0))$ *for all* $\lambda > 0$.

In Section 2 we give background information on the Dirichlet-to-Neumann operator and in Section 3 we prove Theorem 1.1. For form methods we refer to Section 2 in [AE].

Note when the domains Ω_n vary, their boundaries $\partial\Omega_n$ vary as well. Therefore the operators $\mathcal{N}_n$ act on different spaces $L^2(\partial\Omega_n)$ endowed with the corresponding surface measures. It does not seem possible to obtain even strong convergence of the resolvents from any known result on convergence of the associated forms. The question whether in a general setting of a Lipschitz domain Ω one could approximate in the resolvent sense $\mathcal{N}$ by $\mathcal{N}_n$ with a sequence of Ω_n which converges in an appropriate sense is out of reach. This is an interesting question which would allow us for example to understand the heat kernel of $\mathcal{N}$ by using known results on the heat kernels of $\mathcal{N}_n$ when the Ω_n are C^∞, see [EO].

2. The Dirichlet-to-Neumann operator on starshaped domains

Let $\Omega \subset \mathbb{R}^d$ be open bounded with Lipschitz boundary. Next, define the form $\mathfrak{a}_\Omega \colon H^1(\Omega) \times H^1(\Omega) \to \mathbb{C}$ by

$$\mathfrak{a}_\Omega(u,v) = \int_\Omega \nabla u \cdot \overline{\nabla v}.$$

Let $\mathrm{Tr}_\Omega \colon H^1(\Omega) \to L_2(\Gamma)$ be the trace map, where $\Gamma = \partial\Omega$. The Dirichlet-to-Neumann operator on Ω is the operator $\mathcal{N}_\Omega$ associated with the pair $(\mathfrak{a}_\Omega, \mathrm{Tr}_\Omega)$ (see [AE] Section 4.4). So if $\varphi, \psi \in L_2(\Gamma)$, then $\varphi \in D(\mathcal{N}_\Omega)$ and $\psi = \mathcal{N}_\Omega\varphi$ if and only if there exists a $u \in D(\mathfrak{a}_\Omega)$ such that $\mathrm{Tr}_\Omega u = \varphi$ and $\mathfrak{a}_\Omega(u,v) = (\psi, \mathrm{Tr}_\Omega v)_{L_2(\Gamma)}$ for all $v \in D(\mathfrak{a}_\Omega)$.

We assume from now on that Ω is starshaped with respect to a ball with centre at the origin. We wish to use spherical coordinates. Define

$$\widetilde{\Omega} = \{(r,\omega) \in (0,\infty) \times S^{d-1} : r\,\omega \in \Omega\}$$

and if $u\colon \Omega \to \mathbb{C}$ is a function then define $\tilde{u}\colon \widetilde{\Omega} \to \mathbb{C}$ by $\tilde{u}(r,\omega) = u(r\,\omega)$. Further, if $\omega \in S^{d-1}$ define

$$\widetilde{\Omega}_\omega = \{r \in (0,\infty) : r\,\omega \in \Omega\}$$

and $\tilde{u}_\omega\colon \widetilde{\Omega}_\omega \to \mathbb{C}$ by $\tilde{u}_\omega(r) = u(r\,\omega)$. Similarly define $\widetilde{\Omega}_r$ and $\tilde{u}_r\colon \widetilde{\Omega}_r \to \mathbb{C}$ by $\tilde{u}_r(\omega) = u(r\,\omega)$. Then

$$\mathfrak{a}_\Omega(u,v) = \int_{S^{d-1}} \int_{\widetilde{\Omega}_\omega} \Big(\frac{\partial \tilde{u}}{\partial r}(r,\omega)\, \overline{\frac{\partial \tilde{v}}{\partial r}(r,\omega)} + \frac{1}{r^2}\, (\nabla \tilde{u}_r)(\omega) \cdot \overline{(\nabla \tilde{v}_r)(\omega)} \Big) r^{d-1}\, dr\, d\omega \quad (1)$$

for all $u,v \in H^1(\Omega) \cap C^\infty(\overline{\Omega})$, where ∇ is the gradient on S^{d-1}. This follows from Example 7 in Section 5.3 in [Hil] or (3.5) in [AH]. Define $R\colon S^{d-1} \to (0,\infty)$ by

$$R(\omega) = \sup \widetilde{\Omega}_\omega.$$

Then $R \in W^{1,\infty}(S^{d-1})$ since Ω is starshaped with respect to a ball with centre at the origin (see [Maz] Lemma 1.1.8). Define $\alpha\colon \Omega_0 \to \Omega$ by

$$\alpha(x) = \begin{cases} R(\frac{1}{|x|}\,x)x & \text{if } x \neq 0, \\ 0 & \text{if } x = 0. \end{cases}$$

Let $u \in H^1(\Omega) \cap C^\infty(\overline{\Omega})$ and write $v = u \circ \alpha$. Then $\tilde{u}(r,\omega) = \tilde{v}(\frac{r}{R(\omega)},\omega)$ and

$$(\nabla \tilde{u}_r)(\omega) = (\nabla \tilde{v}_{r/R(\omega)})(\omega) - (D_1\tilde{v})(\frac{r}{R(\omega)},\omega)\, \frac{r}{R(\omega)^2}\, (\nabla R)(\omega),$$

where D_1 is the partial derivative with respect to the first entry. Therefore (1), the chain rule and the substitution $r = R(\omega)\, r'$ give

$$\begin{aligned}
\mathfrak{a}_\Omega(u) &= \int_{S^{d-1}} \int_0^1 \Big(\frac{1}{R(\omega)^2} \Big| \frac{\partial \tilde{v}}{\partial r}(r,\omega) \Big|^2 \\
&\qquad + \frac{1}{R(\omega)^2}\, \frac{1}{r^2} \Big| (\nabla \tilde{v}_r)(\omega) - \frac{\partial \tilde{v}}{\partial r}(r,\omega)\, \frac{r}{R(\omega)}\, (\nabla R)(\omega) \Big|^2 \Big) R(\omega)^d\, r^{d-1}\, dr\, d\omega \\
&= \int_{S^{d-1}} R(\omega)^{d-2} \qquad\qquad (2) \\
&\qquad \cdot \int_0^1 \Big(\Big| \frac{\partial \tilde{v}}{\partial r}(r,\omega) \Big|^2 + \Big| \frac{1}{r}\, (\nabla \tilde{v}_r)(\omega) - \frac{\partial \tilde{v}}{\partial r}(r,\omega)\, \frac{(\nabla R)(\omega)}{R(\omega)} \Big|^2 \Big) r^{d-1}\, dr\, d\omega.
\end{aligned}$$

Since R is bounded above and below in $(0,\infty)$ and $|\nabla R|^2$ is essentially bounded on S^{d-1}, one has the bound

$$\begin{aligned}
\mathfrak{a}_\Omega(u) &\leq 2\|R\|_\infty^{d-2} \Big(1 + \Big\| \frac{|\nabla R|^2}{R^2} \Big\|_\infty \Big) \\
&\qquad \cdot \int_{S^{d-1}} \int_0^1 \Big(\Big| \frac{\partial \tilde{v}}{\partial r}(r,\omega) \Big|^2 + \Big| \frac{1}{r}\, (\nabla \tilde{v}_r)(\omega) \Big|^2 \Big) r^{d-1}\, dr\, d\omega \\
&= 2\|R\|_\infty^{d-2} \Big(1 + \Big\| \frac{|\nabla R|^2}{R^2} \Big\|_\infty \Big) \|\nabla v\|_{L_2(\Omega_0)}^2.
\end{aligned}$$

For a converse estimate we need a lemma.

Lemma 2.1. *Let $L > 0$. Then $|a|^2 + |x - a\,y|^2 \geq \frac{1}{4\max(L,1)}\,(|a|^2 + |x|^2)$ for all $a \in \mathbb{R}$ and $x, y \in \mathbb{R}^d$ with $|y|^2 \leq L$.*

Proof. Let $a \in \mathbb{R}$ and $x, y \in \mathbb{R}^d$ with $|y|^2 \leq L$. Then $|x| \leq |x - a\,y| + |a\,y|$ and hence $|x|^2 \leq 2|x - a\,y|^2 + 2|a|^2|y|^2 \leq 2\max(L,1)(|x - a\,y|^2 + |a|^2)$. Obviously, $|a|^2 \leq 2\max(L,1)(|x - a\,y|^2 + |a|^2)$ and the lemma follows. □

Let $L = 4\max\left(\||\nabla R|^2/R^2\|_\infty, 1\right)$. It follows from (2) and Lemma 2.1 that

$$\begin{aligned}
\mathfrak{a}_\Omega(u) &\geq \frac{1}{L}\int_{S^{d-1}} R(\omega)^{d-2}\int_0^1 \left(\Big|\frac{\partial \tilde v}{\partial r}(r,\omega)\Big|^2 + \Big|\frac{1}{r}\,(\nabla \tilde v_r)(\omega)\Big|^2\right) r^{d-1}\,dr\,d\omega \\
&\geq \frac{1}{L}\,\|R^{-1}\|_\infty^{-(d-2)}\int_{S^{d-1}}\int_0^1 \left(\Big|\frac{\partial \tilde v}{\partial r}(r,\omega)\Big|^2 + \Big|\frac{1}{r}\,(\nabla \tilde v_r)(\omega)\Big|^2\right) r^{d-1}\,dr\,d\omega \\
&= \frac{1}{L}\,\|R^{-1}\|_\infty^{-(d-2)}\,\|\nabla v\|^2_{L_2(\Omega_0)} \qquad (3)
\end{aligned}$$

for all $u \in H^1(\Omega) \cap C^\infty(\overline{\Omega})$, where $v = u \circ \alpha$. Since $H^1(\Omega) \cap C^\infty(\overline{\Omega})$ is dense in $H^1(\Omega)$ and $H^1(\Omega_0)$ is complete one deduces that $u \circ \alpha \in H^1(\Omega_0)$ for all $u \in H^1(\Omega)$. By a similar argument one obtains that $u \mapsto u \circ \alpha$ is continuous and invertible from $H^1(\Omega)$ onto $H^1(\Omega_0)$.

Note that $\Gamma = \partial\Omega = \{R(\omega)\omega : \omega \in S^{d-1}\}$. If $\varphi \in C(\Gamma)$, then

$$\int_\Gamma |\varphi|^2 = \int_{S^{d-1}} |\varphi(R(\omega)\omega)|^2 \sqrt{1 + \frac{|(\nabla R)(\omega)|^2}{R(\omega)^2}}\, R(\omega)^{d-1}\,d\omega.$$

The equality follows from [EG] Application 3.3.4D of Theorem 3.3.2. The square root of the determinant (g) in Application 3.3.4D can be calculated using the identity (5.3) in [DK1] to rewrite it as the norm of a cross product (cf. the integral formula on hypersurfaces on page 507 in [DK2]). Since the d-dimensional spherical coordinates give an orthogonal coordinate system, the result follows.

Define $\beta\colon \Gamma \to S^{d-1}$ by $\beta(z) = (1/|z|)\,z$. Let $\varphi \in C(S^{d-1})$. Then

$$\int_\Gamma |\varphi \circ \beta|^2 = \int_{S^{d-1}} |\varphi(\omega)|^2 \sqrt{R(\omega)^2 + |(\nabla R)(\omega)|^2}\, R(\omega)^{d-2}\,d\omega.$$

So by density the map $\varphi \mapsto \varphi \circ \beta$ extends to a continuous bijection from $L_2(S^{d-1})$ onto $L_2(\Gamma)$. Define $c\colon S^{d-1} \to (0,\infty)$ by

$$c = R^{d-2}\sqrt{R^2 + |\nabla R|^2}.$$

Then

$$\int_\Gamma (\varphi \circ \beta)\,\psi = \int_{S^{d-1}} c\,\varphi\,(\psi \circ \beta^{-1}) \qquad (4)$$

for all $\varphi \in L_2(S^{d-1})$ and $\psi \in L_2(\Gamma)$.

3. Convergence of the Dirichlet-to-Neumann operators

We adopt the notation and assumptions of Theorem 1.1. In addition we write $\Gamma_n = \partial\Omega_n$ for all $n \in \mathbb{N} \cup \{\infty\}$. Let $n \in \mathbb{N} \cup \{\infty\}$. Define $\alpha_n \colon \Omega_0 \to \Omega_n$ by

$$\alpha_n(x) = \begin{cases} R_n(\frac{1}{|x|}\,x)x & \text{if } x \neq 0, \\ 0 & \text{if } x = 0, \end{cases}$$

and $\Phi_n \colon H^1(\Omega_n) \to H^1(\Omega_0)$ by $\Phi_n u = u \circ \alpha_n$. On the reference domain we define the form $\mathfrak{b}_n \colon H^1(\Omega_0) \times H^1(\Omega_0) \to \mathbb{C}$ by

$$\mathfrak{b}_n(u,v) = \mathfrak{a}_{\Omega_n}(u \circ \alpha_n^{-1}, v \circ \alpha_n^{-1}).$$

Define $\beta_n \colon \Gamma_n \to \Gamma_0$ by $\beta_n(z) = (1/|z|)\,z$ and define $c_n \colon \Gamma_0 \to (0,\infty)$ by

$$c_n = R_n^{d-2}\sqrt{R_n^2 + |\nabla R_n|^2}.$$

Then $V_n\varphi = \varphi \circ \beta_n$ for all $\varphi \in L_2(\Gamma_0)$. We define the multiplication operator M_n in $L_2(\Gamma_0)$ by $M_n\varphi = \sqrt{c_n}\,\varphi$. Note that M_n is invertible. Finally we define $j_n \colon H^1(\Omega_0) \to L_2(\Gamma_0)$ by

$$j_n = M_n V_n^{-1} \operatorname{Tr}_{\Omega_n} \circ \Phi_n^{-1} = M_n \operatorname{Tr}_{\Omega_0}.$$

Lemma 3.1. *The form $\mathfrak{b}_n$ is j_n-elliptic.*

Proof. Set $L_n = 4\max\Big(\Big\|\frac{|\nabla R_n|^2}{R_n^2}\Big\|_\infty, 1\Big)$. It follows from (3) that

$$\mathfrak{b}_n(u) \geq \tfrac{1}{L_n}\,\|R_n^{-1}\|_\infty^{-(d-2)}\,\|\nabla u\|_{L_2(\Omega_0)}^2$$

for all $u \in H^1(\Omega_0)$. Moreover,

$$\|\operatorname{Tr}_{\Omega_0} u\|_{L_2(\Gamma_0)} \leq \|c_n^{-1/2}\|_\infty\,\|j_n(u)\|_{L_2(\Gamma_0)} \leq \|R_n^{-1}\|_\infty^{(d-1)/2}\,\|j_n(u)\|_{L_2(\Gamma_0)}$$

for all $u \in H^1(\Omega_0)$. It is a classical fact that there exists a $\mu_0 > 0$ such that

$$\mu_0\,\|u\|_{H^1(\Omega_0)}^2 \leq \|\nabla u\|_{L_2(\Omega_0)}^2 + \|\operatorname{Tr}_{\Omega_0} u\|_{L_2(\Gamma_0)}^2 \tag{5}$$

for all $u \in H^1(\Omega_0)$. Then

$$\mu_0\,\|u\|_{H^1(\Omega_0)}^2 \leq L_n\,\|R_n^{-1}\|_\infty^{d-2}\,\mathfrak{b}_n(u) + \|R_n^{-1}\|_\infty^{d-1}\,\|j_n(u)\|_{L_2(\Gamma_0)}^2 \tag{6}$$

which proves the lemma. □

Let $\widehat{B}_n$ be the operator associated with $(\mathfrak{b}_n, j_n)$.

Lemma 3.2. *We have $\widehat{B}_n = M_n\,B_n\,M_n^{-1}$.*

Proof. Let $\varphi \in D(\widehat{B}_n)$ and write $\psi = \widehat{B}_n\varphi$. Then there exists a $u \in H^1(\Omega_0)$ such that $j_n(u) = \varphi$ and $\mathfrak{b}_n(u,v) = (\psi, j_n(v))_{L_2(\Gamma_0)}$ for all $v \in H^1(\Omega_0)$. Therefore

$$\begin{aligned} \mathfrak{a}_{\Omega_n}(\Phi_n^{-1}u, v\circ\alpha_n^{-1}) &= \mathfrak{a}_{\Omega_n}(u\circ\alpha_n^{-1}, v\circ\alpha_n^{-1}) = \mathfrak{b}_n(u,v) \\ &= (\psi, j_n(v))_{L_2(\Gamma_0)} = (\psi, \sqrt{c_n}\,V_n^{-1}\operatorname{Tr}_{\Omega_n}(v\circ\alpha_n^{-1}))_{L_2(\Gamma_0)} \\ &= \left(V_n\left(\frac{1}{\sqrt{c_n}}\,\psi\right), \operatorname{Tr}_{\Omega_n}(v\circ\alpha_n^{-1})\right)_{L_2(\Gamma_n)} \end{aligned}$$

for all $v \in H^1(\Omega_0)$, where we used (4) in the last step. Hence

$$\mathfrak{a}_{\Omega_n}(\Phi_n^{-1}u, w) = (V_n\, M_n^{-1}\psi, \mathrm{Tr}_{\Omega_n} w)_{L_2(\Gamma_n)}$$

for all $w \in H^1(\Omega_n)$. So $\mathrm{Tr}_{\Omega_n}(\Phi_n^{-1}u) \in D(\mathcal{N}_n)$ and

$$\mathcal{N}_n\, \mathrm{Tr}_{\Omega_n}(\Phi_n^{-1}u) = V_n\, M_n^{-1}\psi = V_n\, M_n^{-1}\, \widehat{B}_n\varphi.$$

Note that $\varphi = j_n(u) = M_n\, V_n^{-1}\, \mathrm{Tr}_{\Omega_n}\, \Phi_n^{-1}u$. Therefore $V_n\, M_n^{-1}\varphi = \mathrm{Tr}_{\Omega_n}\, \Phi_n^{-1}u$ and

$$\mathcal{N}_n\, V_n\, M_n^{-1}\varphi = V_n\, M_n^{-1}\, \widehat{B}_n\varphi.$$

That is $\widehat{B}_n\varphi = M_n\, V_n^{-1}\, \mathcal{N}_n\, V_n\, M_n^{-1}\varphi$. So $\widehat{B}_n \subset (V_n\, M_n^{-1})^{-1}\, \mathcal{N}_n\, (V_n\, M_n^{-1})$. Since $(V_n\, M_n^{-1})$ is unitary and $\mathcal{N}_n$ is sectorial, also the operator $(V_n M_n^{-1})^{-1} \mathcal{N}_n (V_n M_n^{-1})$ is sectorial. But $\widehat{B}_n$ is m-sectorial. Hence $\widehat{B}_n = M_n V_n^{-1} \mathcal{N}_n V_n M_n^{-1} = M_n B_n M_n^{-1}$. □

Since $\lim_{n\to\infty} R_n = R_\infty$ in $W^{1,\infty}(\Gamma_0)$, there exist $\kappa_1 > 0$ and $\kappa_2 \geq 1$ such that $\|R_n^{-1}\|_\infty \leq \kappa_1$ and $4\left\|\frac{|\nabla R_n|^2}{R_n^2}\right\|_\infty \leq \kappa_2$ for all $n \in \mathbb{N} \cup \{\infty\}$. Let $\mu_0 > 0$ be as in (5).

Proposition 3.3. *If* $\lambda \in (\kappa_1^{d-1}, \infty)$, *then*

$$\lim_{n\to\infty} (\lambda I + \widehat{B}_n)^{-1} = (\lambda I + \widehat{B}_\infty)^{-1}$$

in $\mathcal{L}(L_2(\Gamma_0))$.

Proof. First note that $\lambda I + \widehat{B}_n$ is invertible by (6) since $\lambda > \kappa_1^{d-1}$. Secondly, let $\psi, \psi_1, \psi_2, \ldots \in L_2(\Gamma_0)$ and suppose that $\lim \psi_n = \psi$ weakly in $L_2(\Gamma_0)$. For all $n \in \mathbb{N}$ define $\varphi_n = (\lambda I + \widehat{B}_n)^{-1}\psi_n$. There exists a $u_n \in H^1(\Omega_0)$ such that $j_n(u_n) = \varphi_n$ and

$$\mathfrak{b}_n(u_n, v) + \lambda\, (j_n(u_n), j_n(v))_{L_2(\Gamma_0)} = (\psi_n, j_n(v))_{L_2(\Gamma_0)} \tag{7}$$

for all $v \in H^1(\Omega_0)$. Choose $v = u_n$. Then

$$\mathfrak{b}_n(u_n) + \lambda\, \|j_n(u_n)\|_{L_2(\Gamma_0)}^2 = (\psi_n, j_n(u_n))_{L_2(\Gamma_0)} \leq \|\psi_n\|_{L_2(\Gamma_0)}\, \|j_n(u_n)\|_{L_2(\Gamma_0)}.$$

So one obtains first the estimate $\|j_n(u_n)\|_{L_2(\Gamma_0)} \leq \lambda^{-1} \sup_{m\in\mathbb{N}} \|\psi_m\|_{L_2(\Gamma_0)}$ and then $\mathfrak{b}_n(u_n) \leq \lambda^{-1} \sup_{m\in\mathbb{N}} \|\psi_m\|_{L_2(\Gamma_0)}^2$. Hence

$$\|u_n\|_{H^1(\Omega_0)}^2 \leq \mu_0^{-1}\Big(\kappa_2\, \kappa_1^{d-2}\, \lambda^{-1} + \kappa_1^{d-1}\, \lambda^{-2}\Big) \sup_{m\in\mathbb{N}} \|\psi_m\|_{L_2(\Gamma_0)}^2$$

by (6). This is for all $n \in \mathbb{N}$. Therefore the sequence $(u_n)_{n\in\mathbb{N}}$ is bounded in $H^1(\Omega_0)$. Passing to a subsequence if necessary, there exists a $u \in H^1(\Omega_0)$ such that $\lim u_n = u$ weakly in $H^1(\Omega_0)$. Then $\lim u_n = u$ in $L_2(\Omega_0)$ and $\lim_{n\to\infty} \mathrm{Tr}_{\Omega_0} u_n = \mathrm{Tr}_{\Omega_0} u$ in $L_2(\Gamma_0)$, since the embedding of $H^1(\Omega_0)$ into $L_2(\Omega_0)$ and the trace map are compact. Since $(u_n)_{n\in\mathbb{N}}$ is also bounded in $D(\mathfrak{b}_\infty)$, we may assume that $(u_n)_{n\in\mathbb{N}}$ is weakly convergent in $D(\mathfrak{b}_\infty)$. Because $D(\mathfrak{b}_\infty)$ is continuously embedded in $L_2(\Omega_0)$ it follows that $\lim u_n = u$ weakly in $D(\mathfrak{b}_\infty)$.

Let $v \in H^1(\Omega_0)$. It follows from (7) that

$$\mathfrak{b}_n(u_n, v) + \lambda \int_{\Gamma_0} c_n \operatorname{Tr}_{\Omega_0} u_n \overline{\operatorname{Tr}_{\Omega_0} v} = \int_{\Gamma_0} \sqrt{c_n}\, \psi_n \overline{\operatorname{Tr}_{\Omega_0} v}$$

for all $n \in \mathbb{N}$. We wish to take the limit $n \to \infty$. Note that $|\mathfrak{b}_n(u_n, v) - \mathfrak{b}_\infty(u, v)| \leq |\mathfrak{b}_n(u_n, v) - \mathfrak{b}_\infty(u_n, v)| + |\mathfrak{b}_\infty(u_n, v) - \mathfrak{b}_\infty(u, v)|$. Clearly $\lim |\mathfrak{b}_\infty(u_n, v) - \mathfrak{b}_\infty(u, v)| = 0$. By (2) and polarisation one deduces that

$$\begin{aligned}
&\mathfrak{b}_n(u_n, v) - \mathfrak{b}_\infty(u_n, v) \\
&= \int_{S^{d-1}} (R_n^{d-2} - R_\infty^{d-2}) \\
&\quad \cdot \int_0^1 \Big(\frac{\partial \tilde{u}_n}{\partial r} \overline{\frac{\partial \tilde{v}}{\partial r}} + \Big(\frac{1}{r} \nabla \tilde{u}_{n,r} - \frac{\partial \tilde{u}_n}{\partial r} \frac{\nabla R_n}{R_n}\Big) \cdot \overline{\Big(\frac{1}{r} \nabla \tilde{v}_r - \frac{\partial \tilde{v}}{\partial r} \frac{\nabla R_n}{R_n}\Big)} \Big) r^{d-1}\, dr\, d\omega \\
&\quad - \int_{S^{d-1}} R_\infty^{d-2} \int_0^1 \Big(\frac{\partial \tilde{u}_n}{\partial r} \Big(\frac{1}{R_n} \nabla R_n - \frac{\nabla R_\infty}{R_\infty}\Big) \cdot \overline{\Big(\frac{1}{r} \nabla \tilde{v}_r - \frac{\partial \tilde{v}}{\partial r} \frac{\nabla R_n}{R_n}\Big)} \Big) r^{d-1}\, dr\, d\omega \\
&\quad - \int_{S^{d-1}} R_\infty^{d-2} \int_0^1 \overline{\frac{\partial \tilde{v}}{\partial r}} \Big(\frac{1}{r} \nabla \tilde{u}_{n,r} - \frac{\partial \tilde{u}_n}{\partial r} \frac{\nabla R_\infty}{R_\infty}\Big) \cdot \Big(\frac{1}{R_n} \nabla R_n - \frac{\nabla R_\infty}{R_\infty}\Big) r^{d-1}\, dr\, d\omega.
\end{aligned}$$

Since

$$\lim_{n\to\infty} \|R_n^{d-2} - R_\infty^{d-2}\|_\infty = \lim_{n\to\infty} \Big\| \frac{\nabla R_n}{R_n} - \frac{\nabla R_\infty}{R_\infty} \Big\|_\infty = \lim_{n\to\infty} \Big\| \frac{\nabla R_n}{R_n} - \frac{\nabla R_\infty}{R_\infty} \Big\|_\infty = 0$$

and

$$\sup_{n\in\mathbb{N}} \int_{S^{d-1}} \int_0^1 \Big(\Big|\frac{\partial \tilde{u}_n}{\partial r}\Big|^2 + \Big|\frac{1}{r} \nabla \tilde{u}_{n,r}\Big|^2 \Big) r^{d-1}\, dr\, d\omega = \sup_{n\in\mathbb{N}} \|\nabla u_n\|^2_{L_2(\Omega_0)} < \infty$$

it follows that $\lim[\mathfrak{b}_n(u_n, v) - \mathfrak{b}_\infty(u_n, v)] = 0$. So, $\lim \mathfrak{b}_n(u_n, v) = \mathfrak{b}_\infty(u, v)$. Obviously $\lim c_n = c_\infty$ in $L_\infty(\Gamma_0)$. Hence

$$\lim_{n\to\infty} \int_{\Gamma_0} c_n \operatorname{Tr}_{\Omega_0} u_n \overline{\operatorname{Tr}_{\Omega_0} v} = \int_{\Gamma_0} c_\infty \operatorname{Tr}_{\Omega_0} u \overline{\operatorname{Tr}_{\Omega_0} v} = (j_\infty(u), j_\infty(v))_{L_2(\Omega_0)}$$

and

$$\lim_{n\to\infty} \int_{\Gamma_0} \sqrt{c_n}\, \psi_n \overline{\operatorname{Tr}_{\Omega_0} v} = \int_{\Gamma_0} \sqrt{c_\infty}\, \psi \overline{\operatorname{Tr}_{\Omega_0} v} = (\psi, j_\infty(v))_{L_2(\Omega_0)}.$$

Combining the three limits one deduces that

$$\mathfrak{b}_\infty(u, v) + \lambda (j_\infty(u), j_\infty(v))_{L_2(\Omega_0)} = (\psi, j_\infty(v))_{L_2(\Omega_0)}.$$

This is for all $v \in H^1(\Omega_0)$. Therefore $\varphi := j_\infty(u) \in D(\widehat{B}_\infty)$ and $(\lambda I + \widehat{B}_\infty)\varphi = \psi$.

Finally, $\lim \operatorname{Tr}_{\Omega_0} u_n = \operatorname{Tr}_{\Omega_0} u$ in $L_2(\Gamma_0)$. So

$$\lim \varphi_n = \lim j_n(u_n) = \lim \sqrt{c_n} \operatorname{Tr}_{\Omega_0} u_n = \lim \sqrt{c_\infty} \operatorname{Tr}_{\Omega_0} u = j_\infty(u) = \varphi.$$

Since $(\lambda I + \widehat{B}_\infty)^{-1}$ is compact the proposition follows by Proposition B.1 in [Dan]. □

Proof of Theorem 1.1. Let $\lambda \in (\kappa_1^{d-1}, \infty)$. Then $\lambda I + \widehat{B}_n = M_n (\lambda I + B_n) M_n^{-1}$ for all $n \in \mathbb{N} \cup \{\infty\}$ by Lemma 3.2. So $(\lambda I + B_n)^{-1} = M_n^{-1} (\lambda I + \widehat{B}_n)^{-1} M_n$. Since $\lim M_n = M_\infty$ in $\mathcal{L}(L_2(\Gamma_0))$ the statement follows from Proposition 3.3. Finally, the convergence of $(\lambda I + B_n)^{-1}$ to $(\lambda I + B_\infty)^{-1}$ for all $\lambda > 0$ follows from Theorem IV.2.25 in [Kat]. □

Acknowledgement

We wish to thank the referee for his careful reading and his very good comments. Part of this work is supported by the Marsden Fund Council from Government funding, administered by the Royal Society of New Zealand.

References

[AE] Arendt, W. and Elst, A.F.M. ter, *Sectorial forms and degenerate differential operators.* J. Operator Theory **67** (2012), 33–72.

[AH] Atkinson, K. and Han, W., *Spherical harmonics and approximations on the unit sphere: an introduction.* Lecture Notes in Mathematics 2044. Springer, Heidelberg, 2012.

[BL] Burenkov, V.I. and Lamberti, P.D., *Spectral stability of general non-negative self-adjoint operators with applications to Neumann-type operators.* J. Differential Equations **233** (2007), 345–379.

[Dan] Daners, D., *Dirichlet problems on varying domains.* J. Differential Equations **188** (2003), 591–624.

[DK1] Duistermaat, J.J. and Kolk, J.A.C., *Multidimensional real analysis. I. Differentiation.* Cambridge Studies in Advanced Mathematics 86. Cambridge University Press, Cambridge, 2004.

[DK2] ——, *Multidimensional real analysis. II. Integration.* Cambridge Studies in Advanced Mathematics 87. Cambridge University Press, Cambridge, 2004.

[EO] Elst, A.F.M. ter and Ouhabaz, E.-M., *Analysis of the heat kernel of the Dirichlet-to-Neumann operator.* J. Funct. Anal. **267** (2014), 4066–4109.

[EG] Evans, L.C. and Gariepy, R.F., *Measure theory and fine properties of functions.* Studies in advanced mathematics. CRC Press, Boca Raton, 1992.

[Hil] Hildebrandt, S., *Analysis* 2. Springer-Verlag, Berlin, 2003.

[Kat] Kato, T., *Perturbation theory for linear operators.* Second edition, Grundlehren der mathematischen Wissenschaften 132. Springer-Verlag, Berlin etc., 1980.

[Maz] Maz'ja, V.G., *Sobolev spaces.* Springer Series in Soviet Mathematics. Springer-Verlag, Berlin etc., 1985.

A.F.M. ter Elst
Department of Mathematics
University of Auckland
Private bag 92019
Auckland 1142, New Zealand
e-mail:
`terelst@math.auckland.ac.nz`

E.M. Ouhabaz
Institut de Mathématiques de Bordeaux
Univ. Bordeaux, UMR 5251
351, Cours de la Libération
F-33405 Talence, France
e-mail:
`Elmaati.Ouhabaz@math.u-bordeaux1.fr`

Operator Theory:
Advances and Applications, Vol. 250, 155–170

A Banach Algebra Approach to the Weak Spectral Mapping Theorem for Locally Compact Abelian Groups

Jean Esterle and Eva Fašangová

Dedicated to Charles Batty on the occasion of his sixtieth birthday

Abstract. We give a general version of the weak spectral mapping theorem for non-quasianalytic representations of locally compact abelian groups which are weakly continuous in the sense of Arveson, based on a Banach algebra approach.

Mathematics Subject Classification (2010). Primary 47A16; Secondary 47D03, 46J40, 46H20.

Keywords. Spectral mapping theorem, group representation, Arveson spectrum, infinitesimal generator, regular Banach algebra, Bochner integral.

1. Introduction

Let $\mathbf{T} = (T(t))_{t\in\mathbb{R}}$ be a strongly continuous one-parameter group of bounded operators on a Banach space X, and let A be the infinitesimal generator of $\mathbf{T}$. Such a group is said to be non-quasianalytic if it satisfies the condition

$$\sum_{n\in\mathbb{Z}} \frac{\log^+ \|T(n)\|}{1+n^2} < +\infty.$$

This condition implies that the spectrum of $T(t)$ is contained in the unit circle. The weak spectral mapping theorem says that if the group is non-quasianalytic we have

$$\overline{e^{t\sigma(A)}} = \sigma(T(t)).$$

The weak spectral mapping theorem in this form was stated by Marschall in 1986, [25], Theorem 2.1-a, as a direct consequence of Theorem 1.3 of [24] concerning decomposable operators and local multipliers. In the more general context of Banach modules and with the notion of Beurling spectrum, the weak spectral mapping theorem for representations of a group was stated by Baskakov in 1979, [3], Lemma 3.

The theorem has been obtained independently and published by Lyubich and Vu in 1989 [35]. This paper is in fact a short note which shows how the theorem can be obtained as an easy consequence of results on "separability of the spectrum" of non-quasianalytic strongly bounded one-parameter groups proved by Lyubich and Matsaev in 1962 in the seminal paper [21], which was later generalized by Lyubich, Matsaev and Feldman [22], [23]. Finally, unaware of previous results, Nagel and Huang obtained again independently the weak spectral mapping theorem in [26]. Subsequently, Huang showed that the assumption of non-quasianalyticity is essential by giving an example in the quasianalytic case where the spectrum of the generator is empty [15], [16], which obviously prevents any form of spectral mapping theorem to hold.

Similar results can be obtained for more general groups of bounded operators on Banach spaces. Let G be a locally compact abelian group, let $\mathbf{T} : g \mapsto T(g) \in \mathcal{B}(X)$ be a representation of G on a Banach space X, and assume that the representation is *weakly continuous* in Arveson's sense, see Section 2. Set $\omega_{\mathbf{T}}(g) = \|T(g)\|$ for $g \in G$, denote by $\mathcal{M}_{\omega_{\mathbf{T}}}(G)$ the convolution algebra of all Borel measures μ on G such that $\int_G \omega_{\mathbf{T}}(g)d|\mu|(g) < +\infty$. Denote by $L^1_{\omega_{\mathbf{T}}}(G)$ the convolution algebra of all Haar measurable functions f on G such that $\int_G |f(g)|\omega_{\mathbf{T}}(g)dm(g) < +\infty$, where m denotes the Haar measure on G. Then $L^1_{\omega_{\mathbf{T}}}(G)$ is an ideal of $\mathcal{M}_{\omega_{\mathbf{T}}}(G)$, and for $\mu \in \mathcal{M}_{\omega_{\mathbf{T}}}(G), x \in X$, the formula

$$\phi_{\mathbf{T}}(\mu)x = \int_G T(g)xd\mu(g)$$

defines an algebra homomorphism $\phi_{\mathbf{T}} : \mathcal{M}_{\omega_{\mathbf{T}}}(G) \to \mathcal{B}(X)$.

Now assume that $\lim_{n\to+\infty} \|T(ng)\|^{\frac{1}{n}} = 1$ for every $g \in G$. Then $\omega_{\mathbf{T}}(g) \geq 1$ for $g \in G$, $\mathcal{M}_{\omega_{\mathbf{T}}}(G) \subset \mathcal{M}(G)$ and the Fourier transform $\mu \mapsto \widehat{\mu}$,

$$\widehat{\mu}(\chi) = \int_G \langle g, \chi\rangle d\mu(g) \quad (\chi \in \widehat{G}),$$

is well defined on $\mathcal{M}_{\omega_{\mathbf{T}}}(G)$. The *Arveson spectrum* of $\mathbf{T}$ is defined by the formula

$$\mathrm{spec}(\mathbf{T}) := \{\chi \in \widehat{G} : \hat{f}(\chi) = 0 \ \forall f \in \ker(\phi_{\mathbf{T}}) \cap L^1_{\omega_{\mathbf{T}}}(G)\}.$$

In Section 5 we describe a well-known result, given in [10] or in Proposition 3.18 of [9] in the case of bounded strongly continuous groups, which shows that if we identify $\widehat{\mathbb{R}}$ with $i\mathbb{R}$ then the spectrum $\sigma(A)$ of the generator of a one-parameter non-quasianalytic C_0-group $(T(t))_{t\in\mathbb{R}}$ equals the Arveson spectrum spec$(\mathbf{T})$, and the weak spectral mapping theorem means that the set $\{\chi(t) : \chi \in \mathrm{spec}(\mathbf{T})\}$ is dense in $\sigma(T(t))$ for every $t \in \mathbb{R}$.

The representation $\mathbf{T}$ is said to have the *weak spectral mapping property* if the set $\{\chi(g) : \chi \in \mathit{spec}(\mathbf{T})\}$ is dense in $\sigma(T(g))$ for every $g \in G$. In his celebrated paper on classification of type III factors [5], Connes shows that this is indeed the case for bounded representations (this result is stated in Lemma 2.3.8 of [5] for unitary representations of locally compact abelian groups on Hilbert spaces, but the same argument works for bounded strongly continuous representations on general Banach spaces).

In the same direction d'Antoni, Longo and Zsidó observed in 1981 in [6] that if $\mathbf{T} : g \mapsto T(g)$ is a weakly continuous bounded representation of a locally compact abelian group on a Banach space, then $\sigma(\phi_{\mathbf{T}}(f))) = \widehat{f}(\mathrm{spec}(\mathbf{T})) \cup \{0\}$ for every $f \in L^1_{\omega_{\mathbf{T}}}(G)$.

The weak spectral mapping theorem means that if δ_g denotes the Dirac measure associated to $g \in G$, then the set $\widehat{\delta}_g(\mathrm{spec}(\mathbf{T}))$ is dense in $\sigma(\phi_{\mathbf{T}}(\delta_g))$ for $g \in G$. Takahashi and Inoue showed in [33] that $\widehat{\mu}(\mathrm{spec}(\mathbf{T}))$ is dense in $\sigma(\phi_{\mathbf{T}}(\mu))$ if μ is contained in the largest regular subalgebra $\mathcal{M}_0(G)$ of $\mathcal{M}(G)$ provided $\mathbf{T}$ is a weakly continuous bounded representation of a compact abelian group G, and Seferoğlu extended this result to locally compact abelian groups in [30] (see also his previous paper [29] for bounded one-parameter groups).

In the present paper we show in Theorem 4.3 that, more generally, if $\mathbf{T} = (T(g))_{g\in G}$ is a representation of a locally compact abelian group G on $\mathcal{B}(X)$ which is weakly continuous in the sense of Arveson with respect to a dual pairing (X, X_*) (X_* is a subspace of the dual space X') and satisfies the non-quasianalyticity condition

$$\sum_{n=0}^{+\infty} \frac{\log \|T(ng)\|}{1+n^2} < +\infty \quad (g \in G),$$

then $\widehat{\mu}(\mathrm{spec}(\mathbf{T}))$ is dense in $\sigma(\phi_{\mathbf{T}}(\mu))$ for every measure $\mu \in \mathcal{M}_{\omega_{\mathbf{T}}}(G)$ which is contained in some closed regular subalgebra of $\mathcal{M}_{\omega_{\mathbf{T}}}(G)$, or, equivalently, which is contained in the largest closed regular subalgebra of $\mathcal{M}_{\omega_{\mathbf{T}}}(G)$. This result might be seen as well-known, but we could only find a reference for this in the case of one-parameter non-quasianalytic groups of bounded operators [31], in a slightly less general form: the result of [31] is stated for one-parameter $\mathcal{C}_0$-groups $(T(t))_{t\in\mathbb{R}}$ such that $\|T(t)\| \le \omega(t)$ for some non-quasianalytic weight w, while our result is valid for all weakly continuous group representations where the weight $\omega_{\mathbf{T}} : g \mapsto \|T(g)\|$ itself is non-quasianalytic.

This paper offers also a Banach algebra approach of these spectral mapping theorems, which is a development of the methods used by the second author in [11], [12], [13]. In Section 3 we give a general theorem concerning continuous unital homomorphisms $\phi : \mathcal{A} \to \mathcal{B}(X)$, where $\mathcal{A}$ is a commutative semisimple unital Banach algebra. Let $\mathcal{I}$ be a closed ideal of $\mathcal{A}$, and for $\chi \in \widehat{\mathcal{I}}$ (the Gelfand space) denote by $\tilde{\chi} \in \widehat{\mathcal{A}}$ the unique extension of χ to $\mathcal{A}$. Assume that $\mathcal{I}$ is a regular Banach algebra which satisfies spectral synthesis (see Section 3 for the definition), and that we have

$$\inf_{a\in\mathcal{I}} |\langle \phi(a)x - x, l\rangle| = 0 \quad (x \in X,\, l \in X_*) \tag{1.1}$$

for some dual pairing (X, X_*) (see Definition 2.1).

We have for the *hull* (see Section 3)

$$h_{\mathcal{I}}(\ker(\phi)\cap\mathcal{I}) = h_{\mathcal{I}}(\ker(\phi_{|\mathcal{I}})) = \{\chi \in \widehat{\mathcal{I}} : \chi(u) = 0\ \forall u \in \ker(\phi)\cap\mathcal{I}\} \subset \widehat{\mathcal{I}}.$$

We can summarize the spectral properties obtained in Theorem 3.2 in the following table (in this table $\mathcal{U}$ denotes a closed regular subalgebra of $\mathcal{A}$ such that $\mathcal{I} \subset \mathcal{U}$, $\widehat{\ }$ is the Gelfand transform, $\tilde{\ }$ denotes the extension as above and $\mathrm{spec}(\phi(a))$ is the spectrum of the operator $\phi(a) \in \mathcal{B}(X)$).

	$h_{\mathcal{I}}(\ker(\phi)\cap\mathcal{I})$ compact	$h_{\mathcal{I}}(\ker(\phi)\cap\mathcal{I})$ noncompact
$a \in \mathcal{A}$	$\mathrm{spec}(\phi(a)) = \widehat{a}\left(\tilde{h}_{\mathcal{I}}(\ker(\phi)\cap\mathcal{I})\right)$	$\mathrm{spec}(\phi(a)) \subset \overline{\widehat{a}\left(\tilde{h}_{\mathcal{I}}(\ker(\phi)\cap\mathcal{I})\right)} \ni 0$
$a \in \mathcal{U}$	$\mathrm{spec}(\phi(a)) = \widehat{a}\left(\tilde{h}_{\mathcal{I}}(\ker(\phi)\cap\mathcal{I})\right)$	$\mathrm{spec}(\phi(a)) = \overline{\widehat{a}\left(\tilde{h}_{\mathcal{I}}(\ker(\phi)\cap\mathcal{I})\right)} \ni 0$
$a \in \mathcal{I}$	$\mathrm{spec}(\phi(a)) = \widehat{a}\left(h_{\mathcal{I}}(\ker(\phi)\cap\mathcal{I})\right)$	$\mathrm{spec}(\phi(a)) = \widehat{a}\left(h_{\mathcal{I}}(\ker(\phi)\cap\mathcal{I})\right) \cup \{0\}$

Theorem 4.3 concerning non-quasianalytic weakly continuous representations of locally compact abelian groups is an application of Theorem 3.2 in a concrete situation: $\mathcal{A} := \mathcal{M}_{\omega_{\mathbf{T}}}(G)$, $\phi := \phi_{\mathbf{T}}$, $\mathcal{I} := L^1_{\omega_{\mathbf{T}}}(G)$, $\mathcal{U} :=$ a regular closed subalgebra of $\mathcal{M}_{\omega_T}(G)$. Note that in this context the Arveson spectrum $\mathrm{spec}(\mathbf{T})$ is actually $h_{L^1_{\omega_{\mathbf{T}}}(G)}(\ker(\phi_{\mathbf{T}}) \cap L^1_{\omega_{\mathbf{T}}}(G))$.

The authors wish to thank Y. Tomilov for very valuable discussions concerning the history of the weak spectral mapping theorem and for pointing to our attention the references [22], [23], [24], [25], and to I. Kryshtal for drawing our attention to the works of A.G. Baskakov.

2. Representations of locally compact abelian groups

Let $X = (X, \|.\|)$ be a Banach space. We denote by $\mathcal{B}(X)$ the Banach algebra of bounded linear operators $R : X \to X$ with composition, by $\mathcal{GL}(X)$ the group of invertible elements of $\mathcal{B}(X)$ and by $I = I_X$ the identity map on X. We also denote by $\|.\|$ the operator norm on $\mathcal{B}(X)$ associated to the given norm on X, and by $\rho(R)$ the spectral radius of $R \in \mathcal{B}(X)$. If Y is a subspace of the dual space X' of X we will denote by $\sigma(X, Y)$ the weak topology on X associated to Y.

We will use the following notion, introduced by Arveson in [2]. Notice that if we omit condition 2 we obtain the class of norming dual pairs used recently in [14], [20] to study Markov semigroups.

Definition 2.1. Let X be a Banach space, and let X_* be a subspace of the dual space X'. We will say that (X, X_*) is a dual pairing if the two following conditions are satisfied:

1. $\|x\| = \sup\{\langle x, l\rangle \ :\ l \in X_*, \|l\| \le 1\}$ for every $x \in X$.
2. The $\sigma(X, X_*)$-closed convex hull of every $\sigma(X, X_*)$-compact subset of X is $\sigma(X, X_*)$-compact.

For example, (X, X') is a dual pairing. Also if $X = Y'$ for some Banach space Y, and if we identify Y with a subspace of $X' = (Y')'$ in the obvious way, then (X, Y) is a dual pairing, see [2]. Notice that condition 1 means that if we set $\tilde{x}(l) = \langle x, l\rangle$ for $x \in X, l \in X_*$ then the map $x \mapsto \tilde{x}$ is an isometry from X into the dual space $(X_*)'$.

Let S be a locally compact space, and let (X, X_*) be a dual pairing. A map $u : S \to X$ is said to be *weakly continuous with respect to* (X, X_*) if the map $s \mapsto \langle u(s), l\rangle$ is continuous on S for every $l \in X_*$, and we will often just say that u is weakly continuous when no confusion may occur. In this situation it follows from the Banach–Steinhaus theorem and from condition 1 of Definition 2.1 that we have, for every compact subset K of S,

$$\sup_{s\in K} \|u(s)\| = \sup_{s\in K} \|\widetilde{u(s)}\| = \sup_{s\in K} \sup_{l\in X_*, \|l\|\leq 1} \langle u(s), l\rangle < +\infty. \tag{2.1}$$

Also since $\|u(s)\| = \sup_{l\in X_*,\|l\|\leq 1} |\langle u(s), l\rangle|$ for $s \in S$, the function $\omega_u : s \mapsto \|u(s)\|$ is lower semicontinuous on S, which allows to compute the upper integral

$$\int_S^* \|u(s)\| d|\mu|(s) := \sup_{\substack{f\in \mathcal{C}_c^+(S)\\ f\leq \omega_u}} \int_S f(s) d|\mu|(s) \in [0, +\infty]$$

for every regular measure μ on S, where $\mathcal{C}_c^+(S)$ denotes the space of all nonnegative compactly supported continuous functions on S.

The following proposition is an immediate generalization of Proposition 1.2 of [2].

Proposition 2.2. *Let S be a locally compact space, let (X, X_*) be a dual pairing, and let $u : S \to X$ be a weakly continuous map. Set $\omega_u(s) = \|u(s)\|$ for $s \in S$, and denote by $\mathcal{M}_{\omega_u}(S)$ the set of all regular measures μ on S such that $\|\mu\|_{\omega_u} := \int_S^* \|u(s)\| d|\mu|(s) < +\infty$. Then for every $\mu \in \mathcal{M}_{\omega_u}(S)$ there exists $x \in X$ satisfying*

$$\langle x, l\rangle = \int_S \langle u(s), l\rangle d\mu(s) \qquad (l \in X_*). \tag{2.2}$$

Proof. Since $\int_S |\langle u(s), l\rangle| d|\mu|(s) \leq \|l\| \|\mu\|_{\omega_u} < +\infty$ for every $l \in X_*$, the formula $f_\mu(l) := \int_S \langle u(s), l\rangle d\mu(s)$ for $l \in X_*$ defines an element $f_\mu \in (X_*)'$, and we have to show that $f_\mu = \tilde{x}$ for some $x \in X$. It follows from condition 1 of Definition 2.1 that we have

$$\|f_\mu\| = \sup_{l\in X_*, \|l\|\leq 1} \left| \int_S \langle u(s), l\rangle d\mu(s) \right| \leq \|\mu\|_{\omega_u}.$$

Denote by $\mathcal{M}_c(S)$ the space of all regular measures on S supported by some compact subset of S. It follows from (2.1) that $\mathcal{M}_c(S) \subset \mathcal{M}_{\omega_u}(S)$, and the fact that property (2.2) holds for every $\mu \in \mathcal{M}_c(S)$ follows directly from [2], Proposition 2.1. Set $\tilde{X} := \{\tilde{x} : x \in X\}$. It follows from condition 1 of Definition 2.1 that $\tilde{X}$ is closed in $(X_*)'$. Let $\mu \in \mathcal{M}_{\omega_u}(S)$. There exists a sequence $(\mu_n)_{n\geq 1}$ of elements of $\mathcal{M}_c(S)$ such that $\lim_{n\to+\infty} \|\mu - \mu_n\|_{\omega_u} = 0$. Hence $\lim_{n\to+\infty} \|f_\mu - f_{\mu_n}\| = 0$, and $f_\mu \in \tilde{X}$. □

When the conditions of Proposition 2.2 are satisfied, we will use the notation

$$x = \int_S u(s) d\mu(s), \tag{2.3}$$

where the integral is a Pettis integral computed with respect to the $\sigma(X, X_*)$ topology, which defines an element of X since (X, X_*) is a dual pairing.

Notice that since ω_u is lower semicontinuous, it follows from the theory of integration on locally compact spaces, see [4], Chapter 4, that $\mathcal{M}_{\omega_u}(S)$ is the space of regular measures μ such that ω_u is integrable with respect to the total variation $|\mu|$ of μ, and it follows from [4], Proposition 1, that we have for $\mu \in \mathcal{M}_{\omega_u}(S)$

$$\|\mu\|_{\omega_u} = \int_S \|u(s)\| d|\mu|(s) = \int_S^* \|u(s)\| d|\mu|(s) = \sup_{K \subset S, K \text{ compact}} \int_K^* \|u(s)\| d|\mu|(s).$$

Let G be a topological group. A representation of G on a Banach space X is a unital homomorphism $\mathbf{T} : G \to \mathcal{GL}(X)$, i.e., a map $g \mapsto T(g)$ satisfying $T(0_G) = I$ and $T(g_1 + g_2) = T(g_1)T(g_2)$ for $g_1, g_2 \in G$. We now introduce the notion of weakly continuous representation.

Definition 2.3. Let G be a locally compact abelian group and let X be a Banach space. A representation $\mathbf{T}$ of G on X is said to be weakly continuous with respect to a dual pairing (X, X_*) if the map $g \mapsto \langle T(g)x, l \rangle$ is continuous on G for every $x \in X$ and every $l \in X_*$.

We will often write $\mathbf{T} = (T(g))_{g \in G}$ when $\mathbf{T} : g \mapsto T(g)$ is a representation of G on X.

Let (X, X_*) be a dual pairing, and let $\mathbf{T} = (T(g))_{g \in G}$ be a weakly continuous representation. We have, for $g \in G$,

$$\|T(g)\| = \sup\{\langle T(g)x, l \rangle : x \in X, \|x\| \leq 1, l \in X_*, \|l\| \leq 1\}$$

and so the weight $\omega_{\mathbf{T}} : g \mapsto \|T(g)\|$ is lower semicontinuous on G. Let K be a compact subset of G. Since $\sup_{g \in K} \|T(g)x\| < +\infty$ for every $x \in X$, it follows again from the Banach–Steinhaus theorem that $\sup_{g \in K} \|T(g)\| < +\infty$. In this situation we can define the weighted space $\mathcal{M}_{\omega_{\mathbf{T}}}(G)$ consisting of all regular measures μ on G such that the upper integral $\int_G^* \|T(g)\| d|\mu|(g)$ is finite. Since $\omega_{\mathbf{T}}(g_1 + g_2) \leq \omega_{\mathbf{T}}(g_1)\omega_{\mathbf{T}}(g_2)$ for $g_1, g_2 \in G$, it follows from [4], Chapter 8, Proposition 2 that the convolution product $\mu * \nu$ is well defined and belongs to $\mathcal{M}_{\omega_{\mathbf{T}}}(G)$ for $\mu, \nu \in \mathcal{M}_{\omega_{\mathbf{T}}}(G)$ and that $(\mathcal{M}_{\omega_{\mathbf{T}}}(G), \|.\|_{\omega_{\mathbf{T}}})$ is a Banach algebra with respect to convolution which contains the convolution algebra $\mathcal{M}_c(G)$ of compactly supported regular measures on G as a dense subalgebra.

Denote by $L^1_{\omega_{\mathbf{T}}}(G)$ the convolution algebra of all Haar-measurable (classes of) functions f on G such that $f\omega_{\mathbf{T}}$ is integrable with respect to the Haar measure m on G, identified with the space of all measures $\mu \in \mathcal{M}_{\omega_{\mathbf{T}}}(G)$ which are absolutely continuous with respect to the Haar measure.

The following result is an easy extension of [2], Proposition 1.4.

Proposition 2.4. *Let G be a locally compact abelian group, let (X, X_*) be a dual pairing, and let $\mathbf{T} = (T(g))_{g \in G}$ be a weakly continuous representation of G on X. The formula*

$$\phi_{\mathbf{T}}(\mu)x = \int_G T(g)x d\mu(g) \tag{2.4}$$

defines for every $x \in X$ and every $\mu \in \mathcal{M}_{\omega_{\mathbf{T}}}(G)$ an element of X, $\phi_{\mathbf{T}}(\mu) \in \mathcal{B}(X)$ for every $\mu \in \mathcal{M}_{\omega_{\mathbf{T}}}(G)$, and $\phi_{\mathbf{T}} : \mu \mapsto \phi_{\mathbf{T}}(\mu)$ is a norm-decreasing unital algebra homomorphism from the convolution algebra $\mathcal{M}_{\omega_{\mathbf{T}}}(G)$ into $\mathcal{B}(X)$.

Moreover we have, for $x \in X, l \in X_*$,

$$\inf_{f \in L^1_{\omega_{\mathbf{T}}}(G)} |\langle \phi_{\mathbf{T}}(f)x - x, l\rangle| = 0.$$

Proof. Since $\|T(g)x\| \leq \|T(g)\|\|x\|$, the fact that formula (2.4) defines an element of X for $x \in X$ and $\mu \in \mathcal{M}_{\omega_{\mathbf{T}}}(G)$ follows directly from Proposition 2.2. We have

$$\begin{aligned}\|\phi_{\mathbf{T}}(\mu)x\| &= \sup_{l \in X_*, \|l\| \leq 1} \left| \int_G \langle T(g)x, l\rangle d\mu(g) \right| \\ &\leq \sup_{l \in X_*, \|l\| \leq 1} \int_G |\langle T(g)x, l\rangle| \, d|\mu|(g) \leq \|x\| \|\mu\|_{\omega_{\mathbf{T}}},\end{aligned}$$

and so $\phi_{\mathbf{T}}(\mu) \in \mathcal{B}(X)$ for $\mu \in \mathcal{M}_{\omega_{\mathbf{T}}}(G)$ and $\|\phi_{\mathbf{T}}(\mu)\| \leq \|\mu\|_{\omega_{\mathbf{T}}}$. As observed in [2], a routine application of Fubini's theorem shows that $\phi_{\mathbf{T}}(\mu * \nu) = \phi_{\mathbf{T}}(\mu)\phi_{\mathbf{T}}(\nu)$ for $\mu, \nu \in \mathcal{M}_c(G)$. Since $\mathcal{M}_c(G)$ is dense in $\mathcal{M}_{\omega_{\mathbf{T}}}(G)$, this shows that $\phi_{\mathbf{T}}$ is an algebra homomorphism.

The last assertion follows from the existence of bounded approximate identities in $L^1_{\omega_{\mathbf{T}}}(G)$, see, for example, [7], Theorem 3.3.23 or [28], Section 5.1.9, as indicated in [2], but we give the details for the sake of completeness. Let K be a compact neighbourhood of 0_G. For every open set $U \subset K$ containing 0_G set $f_U(g) = m(U)^{-1}$ if $g \in U$, $f_U(g) = 0$ otherwise, so that $\int_G f_U(g)dm(g) = 1$. We have, for $x \in X, l \in X_*$,

$$\begin{aligned}|\langle \phi_{\mathbf{T}}(f_U)x - x, l\rangle| &= \left| \int_U \langle T(g)x, l\rangle f_U(g)dm(g) - \langle x, l\rangle \int_U f_U(g)dm(g) \right| \\ &\leq \sup_{g \in U} |\langle T(g)x - T(0_G)x, l\rangle| \int_U f_U(g)dm(g) \\ &= \sup_{g \in U} |\langle T(g)x - T(0_G)x, l\rangle|,\end{aligned}$$

and so there exists a sequence $(U_n)_{n \geq 1}$ of open subsets of K containing 0_G such that $\lim_{n \to +\infty} \langle \phi_{\mathbf{T}}(f_{U_n})x - x, l\rangle = 0$, since the representation is weakly continuous. $\square$

3. A general spectral mapping theorem

Let $\mathcal{A}$ be a Banach algebra, and let $\rho(a) := \lim_{n\to+\infty} \|a^n\|^{\frac{1}{n}}$ be the spectral radius of $a \in \mathcal{A}$, so that $\rho(a) = \sup_{\chi\in\widehat{\mathcal{A}}\cup\{0\}} |\chi(a)|$ if $\mathcal{A}$ is commutative, where $\widehat{A}$ denotes the space of characters of A, endowed with the Gelfand topology. A commutative Banach algebra $\mathcal{A}$ is said to be radical if $\rho(a) = 0$ for every $a \in \mathcal{A}$, which means that $\widehat{\mathcal{A}} = \emptyset$. In the other direction a commutative Banach algebra $\mathcal{A} \neq \{0\}$ is said to be semisimple if $\cap_{\chi\in\widehat{\mathcal{A}}} ker(\chi) = \{0\}$, which means that $\rho(a) > 0$ for every $a \in \mathcal{A} \setminus \{0\}$. It follows from Shilov's idempotent theorem [7], Theorem 2.4.33 that if $\mathcal{A}$ is semisimple and $\widehat{\mathcal{A}}$ is compact, then $\mathcal{A}$ is unital. If $\mathcal{A}$ is unital and $a \in \mathcal{A}$, $spec(a)$ denotes the spectrum of a.

Let $\mathcal{A}$ be a commutative Banach algebra, and let $S \subset \mathcal{A}$. The *hull* of S is defined by the formula

$$h_{\mathcal{A}}(S) := \{\chi \in \widehat{\mathcal{A}} \ : \chi(a) = 0 \ \forall a \in S\},$$

and we have $h_{\mathcal{A}}(S) = h_{\mathcal{A}}(\overline{\mathcal{I}(S)})$, where $\mathcal{I}(S)$ denotes the ideal of $\mathcal{A}$ generated by S. We will often write $h(S)$ instead of $h_{\mathcal{A}}(S)$ if there is no risk of confusion. If $\mathcal{A}$ is not radical, the Gelfand transform $\widehat{a} \in \mathcal{C}(\widehat{\mathcal{A}})$ of $a \in \mathcal{A}$ is then defined by the formula

$$\widehat{a}(\chi) := \chi(a) \quad (\chi \in \widehat{\mathcal{A}}).$$

We now introduce the classical notions of regularity and spectral synthesis.

Definition 3.1. A non radical commutative Banach algebra $\mathcal{A}$ is said to be regular if for every proper closed subset F of $\widehat{\mathcal{A}}$ and every $\chi_0 \in \widehat{\mathcal{A}} \setminus F$ there exists $a \in \mathcal{A}$ such that $\chi_0(a) = 1$ and $\chi(a) = 0$ for every $\chi \in F$.

A commutative non unital Banach algebra $\mathcal{A}$ is said to satisfy spectral synthesis if $h(\mathcal{I}) \neq \emptyset$ for every proper closed ideal $\mathcal{I}$ of $\mathcal{A}$.

We list below some standard properties of a commutative semisimple regular Banach algebra $\mathcal{A}$.

1. $\mathcal{A}$ is normal: for every closed subset F of $\widehat{\mathcal{A}}$ and every compact subset K of $\widehat{\mathcal{A}}$ disjoint from F there exists $a \in \mathcal{A}$ such that $\chi(a) = 0$ for every $\chi \in F$ and $\chi(a) = 1$ for every $\chi \in K$, see, for example, [7], Proposition 4.1.14.
2. Let $F \subset \widehat{\mathcal{A}}$ be closed and nonempty, set $\mathcal{I}_F := \{a \in \mathcal{A} \ : \chi(a) = 0 \ \forall \chi \in F\}$ and denote by $\mathcal{J}_F$ the set of all $a \in \mathcal{A}$ such that there exists an open subset $U_a \supset F$ of $\widehat{\mathcal{A}}$ satisfying $\chi(a) = 0$ for every $\chi \in U_a$. Then $h(\mathcal{J}_F) = h(\mathcal{I}_F) = F$, and every ideal $\mathcal{I}$ of $\mathcal{A}$ such that $h(\mathcal{I}) = F$ satisfies $\mathcal{J}_F \subset \mathcal{I} \subset \mathcal{I}_F$, see, for example, [28], Propositions 3.2.6 or 7.3.2. In particular, if $\mathcal{A}$ is not unital and $h(\mathcal{I}) = \emptyset$, then $a \in \mathcal{I}$ for every $a \in \mathcal{A}$ such that $\widehat{a}$ is supported by some compact subset of $\widehat{\mathcal{A}}$.
3. Let $(\mathcal{I}_\lambda)_{\lambda\in\Lambda}$ be a family of closed ideals of $\mathcal{A}$. Then

$$\cup\{h(\mathcal{I}_\lambda) \ : \ \lambda \in \Lambda\} \text{ is dense in } h\left(\cap\{\mathcal{I}_\lambda \ : \lambda \in \Lambda\}\right). \tag{3.1}$$

4. Let ϕ be a homomorphism from $\mathcal{A}$ into a commutative unital Banach algebra $\mathcal{B}$ of unit element $e_{\mathcal{B}}$. Set $\phi^*(\chi) = \chi \circ \phi$ for $\chi \in \widehat{\mathcal{B}}$. Then we have, see, for example, [7], Proposition 4.1.27.,

$$h(\ker(\phi)) \subset \phi^*(\widehat{\mathcal{B}}) \subset h(\ker(\phi)) \cup \{0\}, \tag{3.2}$$

$$\widehat{a}(h(\ker(\phi)) \subset \mathrm{spec}_{\mathcal{B}}(\phi(a)) \subset \widehat{a}(h(\ker(\phi)) \cup \{0\} \quad (a \in \mathcal{A}). \tag{3.3}$$

If $e_{\mathcal{B}} \in \overline{\phi(\mathcal{A})}$, then $e_{\mathcal{B}} \in \phi(\mathcal{A})$, $h(\ker(\phi)) = \phi^*(\widehat{\mathcal{B}})$ is compact, and we have

$$\mathrm{spec}_{\mathcal{B}}(\phi(a)) = \widehat{a}(h(\ker(\phi)) \quad (a \in \mathcal{A}). \tag{3.4}$$

Conversely if $h(\ker(\phi))$ is compact and nonempty, then $\phi(\mathcal{A})$ is unital. Moreover ϕ^* is a homeomorphism from $\widehat{\mathcal{B}}$ onto $h(\ker(\phi))$ if $\phi(\mathcal{A})$ is dense in $\mathcal{B}$.

Property 3 is well known and easy to prove: denote F the closure of $\cup\{h(\mathcal{I}_\lambda) : \lambda \in \Lambda\}$, so that $F \subset h(\cap\{\mathcal{I}_\lambda : \lambda \in \Lambda\})$. If $\chi \notin F$, let $U \subset \widehat{\mathcal{A}}$ be an open set such that $F \subset U$ and $\chi \notin \overline{U}$. There exists $a \in \mathcal{A}$ such that $\widehat{a}(\chi) = 1$ and $\widehat{a}(\overline{U}) = \{0\}$. Since $h(\mathcal{I}_\lambda) \subset U$, we have $a \in \mathcal{I}_\lambda$ for $\lambda \in \Lambda$, and $\chi \notin h(\cap\{\mathcal{I}_\lambda : \lambda \in \Lambda\})$.

Property 4 means that $\phi(\mathcal{A})$ is a "full subalgebra" of $\mathcal{B}$ if $\mathcal{A}$ is semisimple and if $\overline{\phi(\mathcal{A})}$, or, equivalently, $\mathcal{A}$, contains the unit element of $\mathcal{B}$, since in this case $inv(\phi(\mathcal{A})) = inv(\mathcal{B}) \cap \phi(\mathcal{A})$ (inv denotes the set of invertible elements). Since the group $\mathcal{G}$ of invertible elements of a unital Banach algebra is open, and since the map $x \mapsto x^{-1}$ is continuous on $\mathcal{G}$, this shows that in this situation the Banach algebra $\overline{\phi(\mathcal{A})}$ is also a full subalgebra of $\mathcal{B}$ (this property also follows from the regularity of $\overline{\phi(\mathcal{A})}$, see [19], Lemma 1).

The fact that $e_{\mathcal{B}} \in \overline{\phi(\mathcal{A})}$ implies that $e_{\mathcal{B}} \in \phi(\mathcal{A})$ when $\mathcal{A}$ is semisimple and regular is also standard. In this situation $h(\ker(\phi))$ is compact and nonempty. Let U be a compact subset of $\widehat{\mathcal{A}}$ the interior of which contains $h(\ker(\phi))$, and let $u \in \mathcal{A}$ be such that $\widehat{u}(U) = \{1\}$. Then $a - au \in \mathcal{J}_{h(\ker(\phi))} \subset \ker(\phi)$ for every $a \in \mathcal{A}$. So $\phi(u)$ is a unit element of $\phi(\mathcal{A})$, and $\phi(u) = e_{\mathcal{B}}$.

Notice that it may happen that $e_{\mathcal{B}} \in \overline{\phi(\mathcal{A})}$ and that $\phi(\mathcal{A})$ is not unital when ϕ is a homomorphism from a commutative semisimple Banach algebra $\mathcal{A}$ into a commutative unital Banach algebra: for $r > 0$ denote by $\mathcal{A}_r$ the Banach algebra of holomorphic functions on the open disc $D_r := D(0,r)$ which admit a holomorphic extension to the closed disc $\overline{D_r}$. Set $\mathcal{M}_1 := \{f \in \mathcal{A}_1 : f(1) = 0\}$. For $f \in \mathcal{M}_1$ denote by $\phi(f)$ the restriction of f to the closed disc $\overline{D_{1/2}}$. Set $e_n(z) = \frac{z-1}{z-1-1/n}$ for $|z| \le 1$. Then $e_n \in \mathcal{M}_1$, $\lim_{n\to+\infty} \sup_{|z|\le 1/2} |1 - e_n(z)| = 0$, an easy verification shows that $\phi(\mathcal{M}_1)$ is dense in the unital Banach algebra $\mathcal{A}_{1/2}$, but $\phi(\mathcal{M}_1)$ does not contain 1 and $\phi(e_n)^{-1} \notin \phi(\mathcal{M}_1)$ for $n \ge 1$.

The closed subalgebra $reg(\mathcal{A})$ of a commutative Banach algebra generated by the union of all closed regular subalgebras of $\mathcal{A}$ is itself a closed regular subalgebra of $\mathcal{A}$, called the maximal regular subalgebra of $\mathcal{A}$. This result goes back to Albrecht [1] in the semisimple case, see also the proof of [27] given in [7], Proposition 4.1.17. A very simple proof of this fact based on the hull-kernel topology was obtained by

Inoue and Takahasi in [18], see [28], Corollary 3.2.11. Notice that reg$(\mathcal{A})$ is unital if $\mathcal{A}$ has a unit element e, since $\mathbb{C}e$ is regular.

We now state an abstract version of the spectral mapping theorem. We will use below the fact that if $\mathcal{I}$ is a closed ideal of $\mathcal{A}$ not contained in the radical of $\mathcal{A}$, then every $\chi \in \widehat{\mathcal{I}}$ has a unique extension $\tilde{\chi}$ to $\mathcal{A}$ given by the formula $\tilde{\chi}(a) := \chi(au)$, where u is any element of $\mathcal{I}$ such that $\chi(u) = 1$. The map $\chi \mapsto \tilde{\chi}$ is clearly a homeomorphism from $\widehat{\mathcal{I}}$ onto $\widehat{\mathcal{A}} \setminus h(\mathcal{I})$.

Theorem 3.2. *Let $\mathcal{A}$ be a commutative unital Banach algebra, let X be a Banach space, let X_* be a subspace of X' such that $\sup_{l\in X_*,\|l\|\le 1} |\langle x, l\rangle| = \|x\|$ for $x \in X$, and let $\phi : \mathcal{A} \to \mathcal{B}(X)$ be a continuous unital homomorphism. Let $\mathcal{I}$ be a closed ideal of $\mathcal{A}$, and for $\chi \in \widehat{\mathcal{I}}$ denote by $\tilde{\chi}$ the unique extension of χ to $\mathcal{A}$.*

Assume that $\mathcal{I}$ is a semisimple regular Banach algebra which satisfies spectral synthesis, and that we have

$$\inf_{a\in\mathcal{I}} |\langle \phi(a)x - x, l\rangle| = 0 \quad (x \in X,\, l \in X_*). \tag{3.5}$$

Then the following properties hold:

(i) *The set $h_{\mathcal{I}}(\ker(\phi) \cap \mathcal{I})$ is not empty.*

(ii) *If $h_{\mathcal{I}}(\ker(\phi) \cap \mathcal{I})$ is compact, then $I \in \phi(\mathcal{I})$, $\phi(\mathcal{I}) = \phi(\mathcal{A})$, $h(\ker(\phi)) = \{\tilde{\chi} : \chi \in h_{\mathcal{I}}(\ker(\phi) \cap \mathcal{I})\}$, and*

$$\mathrm{spec}(\phi(a)) = \{\tilde{\chi}(a) : \chi \in h_{\mathcal{I}}(\ker(\phi) \cap \mathcal{I})\} \quad (a \in \mathcal{A}).$$

(iii) *If $h_{\mathcal{I}}(\ker(\phi) \cap \mathcal{I})$ is not compact, then the weak*-closure of $h_{\mathcal{I}}(\ker(\phi) \cap \mathcal{I})$ in the unit ball of the dual of $\mathcal{I}$ contains 0, $0 \in \overline{\widehat{a}(h_{\mathcal{I}}(\ker(\phi) \cap \mathcal{I}))}$, and $\mathrm{spec}(\phi(a)) \setminus \{0\} \subset \widehat{a}(h_{\mathcal{I}}(\ker(\phi) \cap \mathcal{I}))$ for $a \in \mathcal{I}$.*

(iv) *If $\mathcal{U}$ is a closed regular subalgebra of $\mathcal{A}$ containing $\mathcal{I}$, then the set $\{\tilde{\chi}_{|\mathcal{U}} : \chi \in h_{\mathcal{I}}(\ker(\phi) \cap \mathcal{I})\}$ is dense in $h_{\mathcal{U}}(\ker(\phi) \cap \mathcal{I})$, and the set $\{\tilde{\chi}(\phi(a)) : \chi \in h_{\mathcal{I}}(\ker(\phi) \cap \mathcal{I})\}$ is dense in $\mathrm{spec}(\phi(a))$ for every $a \in \mathcal{U}$.*

Proof. (i) Let $x \in X \setminus \{0\}$ and $l \in X_*$ be such that $\langle x, l\rangle \neq 0$. It follows from (3.5) that $\langle \phi(a)x, l\rangle \neq 0$ for some $a \in \mathcal{I}$, $\ker(\phi) \cap \mathcal{I}$ is a proper closed ideal of $\mathcal{I}$ and $h_{\mathcal{I}}(\ker(\phi) \cap \mathcal{I}) \neq \emptyset$.

(ii) If $h_{\mathcal{I}}(\ker(\phi) \cap \mathcal{I})$ is compact, then $\phi(\mathcal{I})$ is unital and it follows from (3.5) that $I \in \phi(\mathcal{I})$, and so $\phi(\mathcal{I}) = \phi(\mathcal{A})$ since $\phi(\mathcal{I})$ is an ideal of $\phi(\mathcal{A})$. Let $u \in \mathcal{I}$ such that $e - u \in \ker(\phi)$, where e denotes the unit element of $\mathcal{A}$. Then $\chi(u) = \chi(e) = 1$ for $\chi \in h(\ker(\phi))$, and $\chi \notin h(\mathcal{I})$. Hence $h(\ker(\phi)) = \{\tilde{\chi} : \chi \in h_{\mathcal{I}}(\ker(\phi) \cap \mathcal{I})\}$, since $\tilde{\chi} \in h(\ker(\phi))$ for every $\chi \in h_{\mathcal{I}}(\ker(\phi) \cap \mathcal{I})$. Let $\mathcal{B}$ be a maximal commutative subalgebra of $\mathcal{B}(X)$ containing $\phi(\mathcal{A})$. Then it follows from (3.4) applied to $\mathcal{B}$ that $\mathrm{spec}(\phi(a)) = \mathrm{spec}(\phi(au)) = \mathrm{spec}_{\mathcal{B}}(\phi(au)) = \{\chi(au) : \chi \in h_{\mathcal{I}}(\ker(\phi) \cap \mathcal{I})\} = \{\tilde{\chi}(a) : \chi \in h_{\mathcal{I}}(\ker(\phi) \cap \mathcal{I})\}$.

(iii) Now assume that $h_{\mathcal{I}}(\ker(\phi) \cap \mathcal{I})$ is not compact. Then $I \notin \phi(\mathcal{I})$, $0 \in \overline{h_{\mathcal{I}}(\ker(\phi) \cap \mathcal{I})}$, and $0 \in \overline{\widehat{a}(h_{\mathcal{I}}(\ker(\phi) \cap \mathcal{I}))}$. The fact that $\mathrm{spec}(\phi(a)) \setminus \{0\} \subset \widehat{a}(h_{\mathcal{I}}(\ker(\phi) \cap \mathcal{I}))$ for $a \in \mathcal{I}$ follows from (3.3) applied to $\mathcal{B}$.

(iv) Denote by $\mathcal{K}(\widehat{\mathcal{I}})$ the set of all nonempty compact subsets of $\widehat{\mathcal{I}}$, and for $K \in \mathcal{K}(\widehat{\mathcal{I}})$ denote by $\mathcal{I}_K$ the set of all $a \in \mathcal{I}$ such that $\chi(a) = 0$ for every $\chi \in K$. Then $\mathcal{I}_K$ is a closed ideal of $\mathcal{A}$. Set

$$X_K = \{x \in X : \phi(a)x = 0 \ \forall a \in \mathcal{I}_K\}.$$

Then X_K is a closed subspace of X, and $\phi(a)(X_K) \subset X_K$ for every $a \in \mathcal{A}$. Set $\phi_K(a) = \phi(a)_{|X_K}$ for $a \in \mathcal{A}$. Then $\phi_K : \mathcal{A} \to \mathcal{B}(X_K)$ is a unital homomorphism. Let $\chi_0 \in \widehat{\mathcal{I}} \setminus K$, and let $u \in \mathcal{I}$ be such that $\chi_0(u) = 1$ and $\chi(u) = 0$ for every $\chi \in K$. Then $u \in \mathcal{I}_K \subset \ker(\phi_K)$, and $\chi_0 \notin h_{\mathcal{I}}(\ker(\phi_K) \cap \mathcal{I})$. This shows that $h_{\mathcal{I}}(\ker(\phi_K) \cap \mathcal{I}) \subset K$ is compact. It follows then from (ii) that $h(\ker(\phi_K)) = \{\tilde{\chi} : \chi \in h(\ker(\phi_K) \cap \mathcal{I})\}$.

Let $\mathcal{U}$ be a closed regular subalgebra of $\mathcal{A}$ containing $\mathcal{I}$. We have that $h_{\mathcal{U}}(\ker(\phi_K) \cap \mathcal{U}) = \{\tilde{\chi}_{|\mathcal{U}} : \chi \in h_{\mathcal{I}}(\ker(\phi_K) \cap \mathcal{I})\} \subset \{\tilde{\chi}_{|\mathcal{U}} : \chi \in h_{\mathcal{I}}(\ker(\phi) \cap \mathcal{I})\}$.

Clearly, $\ker(\phi) \subset \cap\{\ker(\phi_K) : K \in \mathcal{K}(\widehat{\mathcal{I}})\}$. Conversely assume that $a \in \mathcal{A}$ and that $\phi_K(a) = 0$ for every $K \in \mathcal{K}(\widehat{\mathcal{I}})$. Set $\Delta_K = \mathcal{I}_{\widehat{\mathcal{I}} \setminus \mathring{K}}$, where $\mathring{K}$ denotes the interior of $K \in \mathcal{K}(\widehat{\mathcal{I}})$. Let $x \in X$, and let $l \in X_*$. Since $\mathcal{I}$ satisfies spectral synthesis, $\cup\{\Delta_K : K \in \mathcal{K}(\widehat{\mathcal{I}})\}$ is dense in $\mathcal{I}$, and there exists a sequence $(u_n)_{n \geq 1}$ of elements of $\cup\{\Delta_K : K \in \mathcal{K}(\widehat{\mathcal{I}})\}$ such that $\langle \phi(a)x, l\rangle = \lim_{n \to +\infty} \langle \phi(u_n)\phi(a)x, l\rangle = \lim_{n \to +\infty} \langle \phi(a)\phi(u_n)x, l\rangle$. Let $K_n \in \mathcal{K}(\widehat{\mathcal{I}})$ such that $u_n \in \Delta_{K_n}$. If $b \in \mathcal{I}_{K_n}$, then $bu_n \in \cap_{\chi \in \widehat{I}} \ker(\chi) = \{0\}$, and so $\phi(u_n)x \in X_{K_n}$ and $\phi(a)\phi(u_n)x = 0$. Hence $\langle \phi(a)x, l\rangle = 0$ and $a \in \ker(\phi)$.

Hence $\ker(\phi) = \cap\{\ker(\phi_K) : K \in \mathcal{K}(\widehat{\mathcal{I}})\}$, $\ker(\phi) \cap \mathcal{U} = \cap\{\ker(\phi_K) \cap \mathcal{U} : K \in \mathcal{K}(\widehat{\mathcal{I}})\}$, and it follows from (3.1) that $\{\tilde{\chi}_{|\mathcal{U}} : \chi \in h_{\mathcal{I}}(\ker(\phi) \cap \mathcal{I})\} \supset \cup\{h_{\mathcal{U}}(\ker(\phi_K) \cap \mathcal{U}) : K \in \mathcal{K}(\widehat{\mathcal{I}})\}$ is dense in $h_{\mathcal{U}}(\ker(\phi) \cap \mathcal{U})$.

It follows then from (3.4) that the set $\{\tilde{\chi}(a) : \chi \in h_{\mathcal{I}}(\ker(\phi) \cap \mathcal{I})\}$ is dense in $\operatorname{spec}(\phi(a))$ for every $a \in \mathcal{U}$. □

4. The weak spectral mapping theorem for representations of locally compact abelian groups

Consider again a locally compact abelian group G. A submultiplicative locally bounded measurable weight on G is a function $\omega : G \to (0, +\infty)$ which is measurable with respect to the Haar measure m on G and satisfies $\omega(g_1 + g_2) \leq \omega(g_1)\omega(g_2)$ for $g_1, g_2 \in G$ and $\sup_{g \in K} \omega(g) < +\infty$ for every compact subset K of G. In this situation the space $L^1_\omega(G)$ of all Haar measurable functions f on G satisfying $\|f\|_\omega := \int_G |f(g)|\omega(g)dm(g) < +\infty$ is a Banach algebra with respect to convolution. If $\lim_{n \to +\infty} \omega(ng)^{\frac{1}{n}} = 1$ for every $g \in G$, then $\omega(g) \geq 1$ for every $g \in G$, and the map $s \mapsto \chi_s$ is a homeomorphism from the dual group $\widehat{G}$ onto $\widehat{L^1_\omega(G)}$, where the character χ_s is defined by the formula

$$\chi_s(f) = \int_G f(g)\langle g, s\rangle dm(g) = \widehat{f}(s) \quad (f \in L^1_\omega(G)). \tag{4.1}$$

Definition 4.1. A submultiplicative measurable locally bounded weight on G is said to be nonquasianalytic when it satisfies the condition

$$\sum_{n=1}^{+\infty} \frac{\log(\omega(ng))}{1+n^2} < +\infty \quad (g \in G). \tag{4.2}$$

Notice that if ω is any submultiplicative weight on G then we have $\omega(g) \geq \lim_{n\to+\infty} \omega(ng)^{\frac{1}{n}}$ for $g \in G$. Condition (4.2) implies that $\lim_{n\to+\infty} \omega(ng)^{\frac{1}{n}} \leq 1$ for every $g \in G$. Since $\lim_{n\to+\infty} \omega(ng)^{\frac{1}{n}} \cdot \lim_{n\to+\infty} \omega(-ng)^{\frac{1}{n}} \geq 1$, we have in fact $\omega(g) \geq \lim_{n\to+\infty} \omega(ng)^{\frac{1}{n}} = 1$ for $g \in G$ if ω is nonquasianalytic. This allows us to identify the character space $\widehat{L^1_\omega(G)}$ with the dual group $\widehat{G}$ by using formula (4.1).

The following result goes back to Domar [8].

Theorem 4.2 ([8]). *Let G be a locally compact abelian group, let ω be a submultiplicative measurable locally bounded nonquasianalytic weight on G. Then the convolution algebra $L^1_\omega(G)$ is a regular Banach algebra which satisfies spectral synthesis.*

We now state a general version of the weak spectral mapping theorem.

Theorem 4.3. *Let G be a locally compact abelian group, let (X, X_*) be a dual pairing and let $\mathbf{T} = (T(g))_{g\in G}$ be a representation of G on X which is weakly continuous with respect to (X, X_*). Assume that the representation satisfies the condition*

$$\sum_{n=1}^{+\infty} \frac{\log \|T(ng)\|}{1+n^2} < +\infty \quad (g \in G). \tag{4.3}$$

Set $\mathcal{M}_{\omega_\mathbf{T}}(G) = \{\mu \in \mathcal{M}(G) : \int_G^* \|T(g)\| d|\mu|(g) < +\infty\}$, and let $\phi_\mathbf{T} : \mathcal{M}_{\omega_\mathbf{T}}(G) \to \mathcal{B}(X)$ be the unital homomorphism defined by the formula

$$\langle \phi_\mathbf{T}(\mu)x, l\rangle = \int_G \langle T(g)x, l\rangle d\mu(g) \quad (\mu \in \mathcal{M}_{\omega_\mathbf{T}}(G),\ x \in X,\ l \in X_*).$$

Let

$$\operatorname{spec}(\mathbf{T}) := \{s \in \widehat{G} : \hat{f}(s) = 0\ \forall f \in \ker(\phi_\mathbf{T}) \cap L^1_{\omega_\mathbf{T}}(G)\}$$

be the Arveson spectrum of the representation. Then the following properties hold.

(i) $\operatorname{spec}(\mathbf{T})$ *is nonempty.*

(ii) *If* $\operatorname{spec}(\mathbf{T})$ *is compact, then the representation is continuous with respect to the norm of* $\mathcal{B}(X)$, $\phi_\mathbf{T}(\mathcal{M}_{\omega_\mathbf{T}}(G)) = \phi_\mathbf{T}(L^1_{\omega_\mathbf{T}}(G))$ *and* $\operatorname{spec}(\phi_\mathbf{T}(\mu)) = \hat{\mu}(\operatorname{spec}(\mathbf{T}))$ *for every* $\mu \in \mathcal{M}_{\omega_\mathbf{T}}(G)$.

(iii) *If μ is contained in a regular subalgebra of $\mathcal{M}_{\omega_\mathbf{T}}(G)$, then the set* $\hat{\mu}(\operatorname{spec}(\mathbf{T}))$ *is dense in* $\operatorname{spec}(\phi_\mathbf{T}(\mu))$. *In particular the set* $\{\langle g, s\rangle : s \in \operatorname{spec}(\mathbf{T})\}$ *is dense in* $\operatorname{spec}(T(g))$ *for every* $g \in G$.

Proof. It follows from Proposition 2.4 that $\inf_{f\in L^1_{\omega_\mathbf{T}}(G)} |\langle \phi_\mathbf{T}(f)x - x, l\rangle| = 0$, and it follows from Theorem 4.2 that the convolution algebra $L^1_{\omega_\mathbf{T}}(G)$ is regular and satisfies spectral synthesis. Denote by δ_g the Dirac measure at $g \in G$. Then

$T(g) = \phi_{\mathbf{T}}(\delta_g)$. Since the discrete topology on G is locally compact, the convolution algebra $l^1_{\omega_{\mathbf{T}}}(G) = \overline{\text{span}\{\delta_g\}_{g\in G}}$ is also regular, and it follows from Theorem 3.2 that (i), (iii) and the last two assertions of (ii) hold. Since the map $g \mapsto f * \delta_g$ is continuous on G with respect to the norm of $Lq_{\omega_{\mathbf{T}}}(G)$ for every $f \in \mathcal{C}_c(G)$, and since $\phi_{\mathbf{T}}$ is continuous, a density argument shows that the map $g \mapsto \phi_{\mathbf{T}}(f)T(g)$ is continuous with respect to the norm of $\mathcal{B}(X)$ for every $f \in L^1_{\omega_{\mathbf{T}}}(G)$. Hence the representation is continuous with respect to the norm of $\mathcal{B}(X)$ if spec($\mathbf{T}$) is compact, since in this case $I = \phi_{\mathbf{T}}(f)$ for some $f \in L^1_{\omega_{\mathbf{T}}}(G)$. □

Notice that if τ is a locally compact group topology on G coarser than the given one, then $L^1_{\omega_{\mathbf{T}}}(G, m_\tau)$ is a regular subalgebra of $\mathcal{M}_{\omega_{\mathbf{T}}}(G)$, and so $\hat{\mu}(\text{spec}(\mathbf{T}))$ is dense in $\text{spec}(\phi_{\mathbf{T}}(\mu))$ for $\mu \in L^1_{\omega_{\mathbf{T}}}(G, m_\tau)$ if $\mathbf{T}$ is non-quasianalytic. The union of all these convolution algebras may be strictly contained in $reg(\mathcal{M}_{\omega_{\mathbf{T}}}(G))$, see [17].

5. Link between the Arveson spectrum and the spectrum of the generator of a $\mathcal{C}_0$-group

Consider the case $G = \mathbb{R}$, $X_* = X'$, T a $\mathcal{C}_0$-group on X with non-quasianalytic weight. Denote by A the infinitesimal generator of T; this is an unbounded linear operator on X with dense domain $\mathcal{D}_A$. We identify $\widehat{G}$ with $i\mathbb{R}$ and for $f \in L^1_{\omega_{\mathbf{T}}}(\mathbb{R})$ the Fourier transform is

$$\widehat{f}(is) = \int_{\mathbb{R}} f(t)e^{its}dt = \chi_{is}(f)\,.$$

For the convenience of the reader, we give a proof of the following well-known result (stated in a slightly less general form in Corollary 4.1 of [32]).

Theorem 5.1. $$\text{spec}(\mathbf{T}) = \sigma(A)$$

Proof. For the inclusion "⊃" we use the following two poperties:

i) For $f \in L^1_{\omega_{\mathbf{T}}}(\mathbb{R})$ with $\phi_{\mathbf{T}}(f) = 0$ we have $\sigma_{ap}(A) \subset \ker \widehat{f}$ (σ_{ap} is the approximate point spectrum of the operator).
ii) $\sigma(A) = \sigma_{ap}(A)$

Proof of the inclusion "⊃":

$$\sigma(A) = \sigma_{ap}(A) \subset \bigcap_{f\in L^1_{\omega_{\mathbf{T}}}(\mathbb{R}),\phi_{\mathbf{T}}(f)=0} \ker \widehat{f} = \text{spec}(\mathbf{T}).$$

Proof of i): Let $\lambda \in \sigma_{ap}(A)$, $x_n \in \mathcal{D}_A$ a sequence with $\|x_n\| = 1$, $\lambda x_n - Ax_n \to 0$. Then

$$e^{\lambda t}x_n - T(t)x_n = \int_0^t e^{\lambda(t-s)}T(s)(\lambda x_n - Ax_n)ds,$$

and $e^{\lambda t}x_n - T(t)x_n \to 0$. We compute

$$\begin{aligned}
\|\phi_{\mathbf{T}}(f)\| &= \sup_{\|x\|=1} \|\phi_{\mathbf{T}}(f)x\| \\
&\geq \overline{\lim}\|\phi_{\mathbf{T}}(f)x_n\| \\
&= \overline{\lim}\| \int_{-\infty}^{\infty} \left(T(t)x_n - e^{\lambda t}x_n + e^{\lambda t}x_n\right) f(t)dt\| \\
&\geq \overline{\lim}| \int_{-\infty}^{\infty} e^{\lambda t} f(t)dt| \, \|x_n\| - \overline{\lim}\| \int_{-\infty}^{\infty} \left(T(t)x_n - e^{\lambda t}x_n\right) f(t)dt\| \\
&= |\widehat{f}(\lambda)|.
\end{aligned}$$

So $\phi_{\mathbf{T}}(f) = 0$ implies $\widehat{f}(\lambda) = 0$.

Proof of ii): This follows from the non-quasianalyticity of the weight, since $\sigma(A) \subset i\mathbb{R}$, and therefore $\sigma(A) = \partial\sigma(A)$, the boundary.

For the opposite inclusion "$\subset$" we use the following two properties:

iii) For $f \in L^1_{\omega_{\mathbf{T}}}(\mathbb{R})$ such that $\widehat{f}$ has compact support and vanishes on an open set containing $\sigma(A)$ we have $\phi_{\mathbf{T}}(f) = 0$.

iv) $L^1_{\omega_{\mathbf{T}}}(\mathbb{R})$ is a regular Banach algebra.

Proof of the inclusion "$\subset$": If $\lambda \in i\mathbb{R} \setminus \sigma(A)$, choose first a closed neighbourhood U of $\sigma(A)$ (in the Euclidean topology) such that $\lambda \notin U$. Then, by regularity, there exists $f \in L^1_{\omega_{\mathbf{T}}}(\mathbb{R})$ such that $\widehat{f} = 0$ on U and $\widehat{f}(\lambda) = 1$, and there exists $g \in L^1_{\omega_{\mathbf{T}}}(\mathbb{R})$ such that $\widehat{g}$ has compact support and $\widehat{g}(\lambda) = 1$. Set $h = f * g$. Then by iii) we have $\phi_{\mathbf{T}}(h) = 0$, with $\widehat{h}(\lambda) = 1$. Hence $\lambda \notin \operatorname{spec}(\mathbf{T})$.

Proof of iii): By Lebesgue and by inverse Fourier transform theorems we have

$$\phi_{\mathbf{T}}(f)x = \lim_{\delta \to 0+} \int_{-\infty}^{\infty} e^{-\delta|t|} T(t)xf(t)dt, \qquad f(t) = \frac{1}{2\pi} \int_{-\infty}^{\infty} \widehat{f}(-is)e^{ist}ds.$$

Furthermore, for $\Re\lambda > 0$ we have

$$(\lambda - A)^{-1}x = \int_0^{\infty} e^{-\lambda t}T(t)xdt, \quad (\lambda + A)^{-1}x = \int_{-\infty}^0 e^{\lambda t}T(t)xdt.$$

We compute, using Fubini and Lebesgue theorems:

$$\begin{aligned}
\phi_{\mathbf{T}}(f)x &= \frac{1}{2\pi} \lim_{\delta \to 0+} \int_{-\infty}^{\infty} \left(\int_{-\infty}^0 + \int_0^{\infty} \right) e^{ist} e^{-\delta|t|} T(t)xdt \widehat{f}(-is)ds \\
&= \frac{1}{2\pi} \lim_{\delta \to 0+} \int_{-\infty}^{\infty} \left((\delta - is - A)^{-1}x - (-\delta - is - A)^{-1}x\right) \widehat{f}(-is)ds \\
&= 0.
\end{aligned}$$

Proof of iv): This follows from the non-quasianaliticity of the weight (4.3).

The theorem is proved. □

References

[1] E. Albrecht, *Decomposable systems of operators in harmonic analysis*, Toeplitz Centennial (I. Golberg (ed.), Birkhäuser, Basel, 1982), 19–35.

[2] W. Arveson, *On group of automorphisms of operator algebras*, J. Funct. An. 15 (1974), 217–243.

[3] A.G. Baskakov, *Inequalities of Bernstein type in abstract harmonic analysis*, Sibirskij Mat. Zhurnal 20 (1979), English translation 665–672.

[4] N. Bourbaki, Intégration, Hermann, Paris, 1959.

[5] A. Connes, *Une classification des facteurs de type III*, Ann. scient. Éc. Norm. Sup. 4e série t. 6 (1973), 133–252.

[6] C. d'Antoni, R. Longo and L. Zsidó, *A spectral mapping theorem for locally compact group algebras*, Pac. J. Math. 103 (1981), 17–24.

[7] H.G. Dales, Banach algebras and automatic continuity, London Mathematical Society Monographs, vol. 24, The Clarendon Press, Oxford, 2000.

[8] Y. Domar, *Harmonic analysis based on certain commutative Banach algebras*, Acta Math. 96 (1956), 1–66.

[9] K.-J. Engel and R. Nagel, One-Parameter Semigroups for Linear Evolution Equations, Graduate Texts in Mathematics 194, Springer-Verlag, New York, 2000.

[10] D.E. Evans, *On the spectrum of a one-parameter strongly continuous representation*, Math. Scand. 39 (1976), 80–82.

[11] E. Fašangová, *A Banach algebra approach to the weak spectral mapping theorem for C_0-groups*, Ulmer Seminare Heft 5 (2000), 174–181.

[12] E. Fašangová, *Spectral mapping theorems and spectral space-independence*, Progress Nonlin. Diff. Eqns. Appl. 55 (2003), 157–168.

[13] E. Fašangová and P.J. Miana, *Spectral mapping inclusions for the Phillips functional calculus in Banach spaces and algebras*, Studia Math. 167 (2005), 219–226.

[14] M. Gerlach and M. Kunze, *Mean ergodic theorems on norming dual pairs.* Ergodic Theory Dynam. Systems 34 (2014), 1210–1229.

[15] S.-Z. Huang, *An equivalent description of non-quasianalyticity through spectral theory of C_0-groups*, J. Operator Theory 33 (1994), 299–309.

[16] S.-Z. Huang, *Spectral theory for non-quasianalytic representations of locally compact abelian groups*, Ph.D. Thesis, Tübingen, 1995.

[17] J. Inoue, *Some closed subalgebras of measure algebras and a generalization of P.J. Cohen's theorem*, J. Math. Soc. Japan 23 (1971), 278–294.

[18] J. Inoue and S.-E. Takahasi, *A note on the largest regular subalgebra of a Banach algebra*, Proc. Amer. Math. Soc. 116 (1992), 961–962.

[19] R. Kantrowitz and M.M. Neumann, *The greatest regular subalgebra of certain Banach algebras of vector-valued functions*, Rend. Circ. Mat. Palermo, Ser. 2, 43 (1994), 435–446.

[20] M. Kunze, *Continuity and equicontinuity of semigroups on norming dual pairs*, Semigroup Forum 79 (2009), 540–560.

[21] Y.I. Lyubich and V.I. Matsaev, *Operators with separable spectrum*, (Russian), Mat. Sb. (1962), 433–468.

[22] Y.I. Lyubich, V.I. Matsaev and G.M. Feldman, *The separability of the spectrum of a representation of a locally compact abelian group*, (Russian), translation in Dokl. Akad. Nauk SSSR 201 (1971), 1282–1284, Soviet Math. Dokl. 12 (1971), 1824–1827.

[23] Y.I. Lyubich, V.I. Matsaev and G.M. Feldman, *On representations with a separable spectrum*, (Russian), Funkcional. Anal. i Prilozen. 7 (1973), no. 2, 52–61.

[24] E. Marschall, *A spectral mapping theorem for local multipliers*, Math. Ann. 260 (1982), 143–150.

[25] E. Marschall, *On the functional calculus of nonquasianalytic groups of operators and cosine functions*, Rend. Circ. Mat. Palermo 35 (1986), no. 1, 58–81.

[26] R. Nagel and S.-Z. Huang, *Spectral Mapping Theorems for C_0-groups satisfying Non-quasianalytic growth conditions*, Math. Nach. 169 (1994), 207-218.

[27] M.M. Neumann, *Commutative Banach algebras and decomposable operators*, Monatsh. für Math. 113 (1992), 227–243.

[28] T.W. Palmer, Banach algebras and the general theory of $*$-algebras, Volume 1, Algebras and Banach algebras, Encyclopedia of Mathematics and its Applications, vol. 49, Cambridge University Press, 1993.

[29] H. Seferoğlu, *Spectral mapping theorem for representations of measure algebras*, Proc. Edinburgh Math. Soc. 40 (1997), 261–266.

[30] H. Seferoğlu, *Spectral mapping theorem for Banach modules*, Studia Math. 156 (2003), 99–103.

[31] H. Seferoğlu, *A spectral mapping theorem for representations of one-parameter groups*, Proc. Amer. Math. Soc. 134 (2006), 2457–2463.

[32] H. Seferoğlu, *Generators of C_0-groups with smallest spectra*, Math. Nachr. 280 (2007), 924–931.

[33] S.-E. Takahasi and J. Inoue, *A spectral mapping theorem for some representations of compact abelian groups*, Proc. Edinburgh Math. Soc. 35 (1992), 47–52.

[34] J.M.A.M. van Neerven, *Elementary operator-theoretic proof of Wiener's Tauberian theorem*, Rendic. Istit. Matem. Univ. Trieste., Suppl. Vol. 28 (1997), 281–286.

[35] K.F. Vu and Y.I. Lyubich, *On the spectral mapping theorem for one-parameter groups of operators*, Zap. Nauchn. Sem. Leningrad. Otdel. Mat. Inst. Steklov (LOMI) 178 (1989), 146–150, translation in J. Soviet Math. 61 (1992), 2035–2037.

Jean Esterle
Université de Bordeaux
IMB, UMR 5251
351, cours de la Libération
F-33405-Talence-Cedex, France
e-mail: esterle@math.u-bordeaux1.fr

Eva Fašangová
Technische Universität Dresden
Fachrichtung Mathematik
Institut für Analysis
D-01062 Dresden, Germany
e-mail: eva.fasangova@tu-dresden.de

Operator Theory:
Advances and Applications, Vol. 250, 171–197

Regularity Properties of Sectorial Operators: Counterexamples and Open Problems

Stephan Fackler

Dedicated to Professor Charles Batty on the occasion of his 60th birthday

Abstract. We give a survey on the different regularity properties of sectorial operators on Banach spaces. We present the main results and open questions in the theory and then concentrate on the known methods to construct various counterexamples.

Mathematics Subject Classification (2010). Primary 47D06; Secondary 35K90, 47A60, 47A20.

Keywords. Counterexample, sectorial operator, maximal regularity, $\mathcal{R}$-sectorial operator, bounded imaginary powers, H^∞-calculus, dilations, Schauder multiplier.

1. Introduction

By now sectorial operators play a central role in the study of abstract evolution equations. Moreover, in the past decades certain sectorial operators with additional properties have become important both from the point of view of operator theory and partial differential equations. We call these additional properties *regularity properties* of sectorial operators. Very important examples are the boundedness of the H^∞-calculus or the imaginary powers, $\mathcal{R}$-sectoriality and – in the case that the sectorial operator generates a semigroup – the property of having a dilation to a group. This survey is intended as a quick guide to these properties and the main results and open questions in this area. A particular emphasis is thereby given to the presentation of various methods to construct counterexamples.

In the first section we introduce all aforementioned properties and list the main results. In particular we will see that on L_p for $p \in (1, \infty)$ and on more

The author was supported by a scholarship of the "Landesgraduiertenförderung Baden-Württemberg".

general Banach spaces the following implications hold:

$$\text{loose dilation} \quad \Rightarrow \quad \text{bounded } H^\infty\text{-calculus} \quad \Rightarrow \quad \text{BIP} \quad \Rightarrow \quad \mathcal{R}\text{-sectorial}$$

and all of them imply sectoriality by their mere definitions. Our main goal in the sections thereafter is to give explicit counterexamples which show that for each of the above properties the converse implication $\Leftarrow$ does not hold. We present different approaches to construct such counterexamples. The first one is well known and the most far-reaching and uses Schauder multipliers. In [17] and [18] this approach has been developed further to give the first explicit example of a sectorial operator on L_p which is not $\mathcal{R}$-sectorial. The second approach uses a theorem of S. Monniaux to give examples of sectorial operators with BIP which do not have a bounded H^∞-calculus. Finally, we study the regularity properties on exotic Banach spaces and show how Pisier's counterexample to the Halmos problem can be used to give an example of a sectorial operator with a bounded $H^\infty(\Sigma_{\frac{\pi}{2}+})$-calculus which does not have a dilation. Moreover, we meet and motivate open problems in the theory and formulate them separately whenever they arise.

2. Main definitions and fundamental results

In this section we give the definitions of the regularity properties to be considered later. Further, we present the main results for these regularity classes. Our leitmotif is to present all results in the most general form that does not involve the introduction of new concepts apart from the main ones. We hope that this allows the reader to see the main ideas clearly without getting himself lost in details. For further information we refer to [38], [12] and [26]. Furthermore we make the following convention.

Convention 2.1. All Banach spaces are assumed to be complex.

2.1. Sectorial operators

We begin our journey with sectorial operators. For $\omega \in (0, \pi)$ we denote by

$$\Sigma_\omega := \{z \in \mathbb{C} \setminus \{0\} : |\arg(z)| < \omega\}$$

the open sector in the complex plane with opening angle ω, where our convention is that $\arg z \in (-\pi, \pi]$.

Definition 2.2 (Sectorial operator). A closed densely defined operator A with dense range on a Banach space X is called *sectorial* if there exists an $\omega \in (0, \pi)$ such that

$$\sigma(A) \subset \overline{\Sigma_\omega} \qquad \text{and} \qquad \sup_{\lambda \notin \overline{\Sigma_{\omega+\varepsilon}}} \|\lambda R(\lambda, A)\| < \infty \quad \forall \varepsilon > 0. \tag{S_ω}$$

One defines the *sectorial angle of A* as $\omega(A) := \inf\{\omega : (S_\omega) \text{ holds}\}$.

Remark 2.3. The above definition automatically implies that A is injective. The definition of sectorial operators varies in the literature. Some authors do not require a sectorial operator to be injective or to have dense range. Others even omit the density of the domain. We give this strict definition to reduce technical difficulties when dealing with bounded imaginary powers and bounded H^∞-calculus. For a very general treatment avoiding unnecessary restrictions in the development as far as possible see the monograph [26].

2.2. $\mathcal{R}$-sectorial operators

In the study of L_p-maximal parabolic regularity culminating in the work [60] an equivalent characterization of maximal L_p-regularity in terms of a stronger sectoriality condition has become very useful both for theory and applications. This condition is called $\mathcal{R}$-sectoriality. We will exclusively treat this condition from an operator theoretic point of view and refer to [38] and [12] for the connection with non-linear parabolic partial differential equations.

Let $r_k(t) := \operatorname{sign} \sin(2^k \pi t)$ be the kth *Rademacher function*. Then on the probability space $([0,1], \mathcal{B}([0,1]), \lambda)$, where $\mathcal{B}([0,1])$ is the Borel σ-algebra on $[0,1]$ and λ denotes the Lebesgue measure, the Rademacher functions form an independent identically distributed family of random variables satisfying $\mathbb{P}(r_k = \pm 1) = \frac{1}{2}$.

Definition 2.4 ($\mathcal{R}$-boundedness). A family of operators $\mathcal{T} \subseteq \mathcal{B}(X)$ on a Banach space X is called $\mathcal{R}$-*bounded* if for one $p \in [1, \infty)$ (equiv. all $p \in [1, \infty)$ by the Khintchine inequality) there exists a finite constant $C_p \geq 0$ such that for each finite subset $\{T_1, \ldots, T_n\}$ of $\mathcal{T}$ and arbitrary $x_1, \ldots, x_n \in X$ one has

$$\left\| \sum_{k=1}^{n} r_k T_k x_k \right\|_{L_p([0,1];X)} \leq C_p \left\| \sum_{k=1}^{n} r_k x_k \right\|_{L_p([0,1];X)}. \tag{2.1}$$

The best constant C_p such that (2.1) holds is called the $\mathcal{R}$-*bound* of $\mathcal{T}$ and is denoted (for an implicitly fixed p) by $\mathcal{R}(\mathcal{T})$.

Furthermore we denote by $\operatorname{Rad}(X)$ the closed span of the functions of the form $\sum_{k=1}^n r_k x_k$ in $L_1([0,1]; X)$. The $\mathcal{R}$-bound behaves in many ways similar to a classical norm. For example, if $\mathcal{S}$ is a second family of operators, one sees that (if the operations make sense)

$$\mathcal{R}(\mathcal{T} + \mathcal{S}) \leq \mathcal{R}(\mathcal{T}) + \mathcal{R}(\mathcal{S}), \qquad \mathcal{R}(\mathcal{T}\mathcal{S}) \leq \mathcal{R}(\mathcal{T})\mathcal{R}(\mathcal{S}).$$

Note that by the orthogonality of the Rademacher functions in $L_2([0,1])$ a family $\mathcal{T} \subseteq \mathcal{B}(H)$ for some Hilbert space H is $\mathcal{R}$-bounded if and only if $\mathcal{T}$ is bounded in operator norm. In fact, an $\mathcal{R}$-bounded subset $\mathcal{T} \subseteq \mathcal{B}(X)$ for a Banach space X is clearly always norm-bounded and one can show that the converse holds if and only if X is isomorphic to a Hilbert space [2, Proposition 1.13].

Now, if one replaces norm-boundedness by $\mathcal{R}$-boundedness, one obtains the definition of an $\mathcal{R}$-sectorial operator.

Definition 2.5 ($\mathcal{R}$-sectorial operator). A sectorial operator on a Banach space X is called $\mathcal{R}$-*sectorial* if for some $\omega \in (\omega(A), \pi)$ one has

$$\mathcal{R}\{\lambda R(\lambda, A) : \lambda \notin \overline{\Sigma_\omega}\} < \infty. \tag{$\mathcal{R}_\omega$}$$

One defines the *$\mathcal{R}$-sectorial angle* of A as $\omega_R(A) := \inf\{\omega : (\mathcal{R}_\omega) \text{ holds}\}$. If A is not $\mathcal{R}$-sectorial, we set $\omega_R(A) := \infty$.

By definition one has $\omega(A) \le \omega_R(A)$. In Hilbert spaces an operator is sectorial if and only if it is $\mathcal{R}$-sectorial. In this case the equality $\omega(A) = \omega_R(A)$ holds. There are examples of sectorial operators A on Banach spaces for which one has the strict inequalities $\omega(A) < \omega_R(A) < \infty$. For this see the examples cited in Section 2.3 and use the fact that $\omega_R(A) = \omega_{H^\infty}(A)$ on UMD-spaces. However, the following problem seems to be open.

Problem 1. Let A be an $\mathcal{R}$-sectorial operator on L_p for $p \in (1,\infty)$. Does one have $\omega(A) = \omega_R(A)$ (if A generates a positive / contractive / positive contractive analytic C_0-semigroup)?

In general Banach spaces $\mathcal{R}$-sectorial operators clearly are sectorial, the converse question whether every sectorial operator is $\mathcal{R}$-sectorial will be explicitly answered negatively in Theorem 3.18.

2.3. Bounded H^∞-calculus for sectorial operators

In complete analogy to the Dunford functional calculus for bounded operators one can define a holomorphic functional calculus for sectorial operators. This goes back to the work [46] in the Hilbert space case and to [10] in the Banach space case. We start by introducing the necessary function spaces.

Definition 2.6. For $\sigma \in (0,\pi)$ we define

$$H_0^\infty(\Sigma_\sigma) := \left\{ f : \Sigma_\sigma \to \mathbb{C} \text{ analytic} : |f(\lambda)| \le C \frac{|\lambda|^\varepsilon}{(1+|\lambda|)^{2\varepsilon}} \text{ on } \Sigma_\sigma \text{ for } C, \varepsilon > 0 \right\},$$
$$H^\infty(\Sigma_\sigma) := \{ f : \Sigma_\sigma \to \mathbb{C} \text{ analytic and bounded} \}.$$

Now let A be a sectorial operator on a Banach space X and $\sigma > \omega(A)$. Then for $f \in H_0^\infty(\Sigma_\sigma)$ one can define

$$f(A) = \frac{1}{2\pi i} \int_{\partial \Sigma_{\sigma'}} f(\lambda) R(\lambda, A)\, d\lambda \qquad (\omega(A) < \sigma' < \sigma).$$

This is well defined by the growth estimate on f and by the invariance of the contour integral and induces an algebra homomorphism $H_0^\infty(\Sigma_\sigma) \to \mathcal{B}(X)$.

One can show that this homomorphism can be extended to a bounded homomorphism on $H^\infty(\Sigma_\sigma)$ satisfying a continuity property similar to the one in Lebesgue's dominated convergence theorem if and only if the homomorphism $H_0^\infty(\Sigma_\sigma) \to \mathcal{B}(X)$ is bounded. This leads us to the next definition.

Definition 2.7 (Bounded H^∞-calculus). A sectorial operator A is said to have a *bounded $H^\infty(\Sigma_\sigma)$-calculus* for some $\sigma \in (\omega(A),\pi)$ if the homomorphism $f \mapsto f(A)$ from $H_0^\infty(\Sigma_\sigma)$ to $\mathcal{B}(X)$ is bounded. The infimum of the σ for which this homomorphism is bounded is denoted by $\omega_{H^\infty}(A)$. We say that A has a *bounded H^∞-calculus* if A has a bounded $H^\infty(\Sigma_\sigma)$-calculus for some $\sigma \in (0,\pi)$. If A does not have a bounded H^∞-calculus, we let $\omega_{H^\infty}(A) := \infty$.

One can extend the functional calculus to the broader class of holomorphic functions on Σ_σ with polynomial growth [38, Appendix B]. Of course, the so obtained operators cannot be bounded in general. Note that it follows directly from the definition that one always has $\omega(A) \le \omega_{H^\infty}(A)$ for a sectorial operator A. Moreover, there exist examples of sectorial operators A for which the strict inequalities $\omega(A) < \omega_{H^\infty}(A) < \infty$ hold: in [32] N.J. Kalton gives an example on a uniformly convex space and in the unpublished manuscript [35] there is an example on a subspace of an L_p-space by the same author.

There is a close connection to $\mathcal{R}$-boundedness and $\mathcal{R}$-sectoriality. A Banach space X is said to have *Pisier's property* (α) (as introduced in [51]) if there is a constant $C \ge 0$ such that for all $n \in \mathbb{N}$, all $n \times n$-matrices $[x_{ij}] \in M_n(X)$ of elements in X and all choices of scalars $[\alpha_{ij}] \in M_n(\mathbb{C})$ one has

$$\int_{[0,1]^2} \left\| \sum_{i,j=1}^{n} \alpha_{ij} r_i(s) r_j(t) x_{ij} \right\| ds\, dt \le C \sup_{i,j} |\alpha_{ij}| \int_{[0,1]^2} \left\| \sum_{i,j=1}^{n} r_i(s) r_j(t) x_{ij} \right\| ds\, dt.$$

We remark that L_p-spaces have Pisier's property (α) for $p \in (1, \infty)$. A proof of the following theorem can be found in [38, Theorem 12.8].

Theorem 2.8. *Let X be a Banach space with Pisier's property (α) and A a sectorial operator on X with a bounded $H^\infty(\Sigma_\sigma)$-calculus for some $\sigma \in (0, \pi)$. Then for all $\sigma' \in (\sigma, \pi)$ and all $C \ge 0$ the set*

$$\{f(A) : \|f\|_{H^\infty(\Sigma_{\sigma'})} \le C\}$$

is $\mathcal{R}$-bounded.

Note that this also implies under the above assumptions that a sectorial operator with a bounded H^∞-calculus is $\mathcal{R}$-sectorial. This can also be proved under the following weaker assumption on the Banach space [37, Theorem 5.3]. A Banach space X *has property* (Δ) if there is a constant $C \ge 0$ such that for all $n \in \mathbb{N}$ and all $n \times n$-matrices $[x_{ij}] \in M_n(X)$ one has

$$\int_{[0,1]^2} \left\| \sum_{i=1}^{n} \sum_{j=1}^{i} r_i(s) r_j(t) x_{ij} \right\| ds\, dt \le C \int_{[0,1]^2} \left\| \sum_{i,j=1}^{n} r_i(s) r_j(t) x_{ij} \right\| ds\, dt.$$

Theorem 2.9. *Let X be a Banach space with property (Δ). Further let A be a sectorial operator on X with a bounded H^∞-calculus. Then A is $\mathcal{R}$-sectorial with $\omega_{\mathcal{R}}(A) = \omega_{H^\infty}(A)$.*

The above theorem can be seen as a generalization of the result that a sectorial operator with a bounded H^∞-calculus on a Hilbert space satisfies $\omega(A) = \omega_{H^\infty}(A)$. In particular, the example for the strict inequality $\omega_{H^\infty}(A) > \omega(A)$ on a subspace of L_p gives the same strict inequality for the $\mathcal{R}$-sectorial angle $\omega_{\mathcal{R}}(A)$.

It is an important and natural question to ask which classes of sectorial operators have a bounded H^∞-calculus. In the following a *contractive analytic semigroup* is an analytic semigroup $(T(z))$ with $\|T(t)\| \le 1$ for all $t \ge 0$. In the Hilbert space case one has the following characterization.

Theorem 2.10. *Let A be a sectorial operator on a Hilbert space such that $-A$ generates a contractive analytic C_0-semigroup. Then A has a bounded H^∞-calculus with $\omega_{H^\infty}(A) = \omega(A) < \frac{\pi}{2}$.*

Conversely, if A has a bounded H^∞-calculus with $\omega_{H^\infty}(A) < \frac{\pi}{2}$, then there exists an invertible $S \in \mathcal{B}(H)$ such that $-S^{-1}AS$ generates a contractive analytic C_0-semigroup.

The first implication follows from the existence of a dilation to a C_0-group as discussed in Section 3 and the fact $\omega_{H^\infty}(A) = \omega(A)$, the second implication is a result of C. Le Merdy [40, Theorem 1.1]. There is an analogue in the L_p-case.

Theorem 2.11. *Let $p \in (1, \infty)$ and A be a sectorial operator on an L_p-space $L_p(\Omega)$ such that $-A$ generates a contractive positive analytic C_0-semigroup. Then A has a bounded H^∞-calculus with $\omega_{H^\infty}(A) = \omega_R(A) < \frac{\pi}{2}$.*

Conversely, if A has a bounded H^∞-calculus with $\omega_{H^\infty}(A) < \frac{\pi}{2}$, then there exists a sectorial operator B on a second L_p-space $L_p(\tilde{\Omega})$ with $\omega_{H^\infty}(B) < \frac{\pi}{2}$ such that $-B$ generates a positive contractive analytic C_0-semigroup, a quotient of a subspace E of $L_p(\tilde{\Omega})$ and an invertible $S \in \mathcal{B}(L_p(\Omega), E)$ with $A = S^{-1}BS$.

The first implication is due to L. Weis (see [60, Remark 4.9c)] and [59, Section 4d)]), the second one was obtained by the author in [19]. There are some open questions regarding generalizations of Weis' result.

Problem 2. Let A be a sectorial operator on some UMD-Banach lattice and suppose that $-A$ is the generator of a positive contractive C_0-semigroup. Does A have a bounded H^∞-calculus (bounded imaginary powers / is $\mathcal{R}$-analytic)?

Problem 3. Let A be a sectorial operator on some L_p-space for $p \in (1, \infty)$ and suppose that $-A$ is the generator of a contractive C_0-semigroup. Does A have a bounded H^∞-calculus (bounded imaginary powers / is $\mathcal{R}$-analytic)?

Problem 4. Let A be a sectorial operator on some L_p-space for $p \in (1, \infty)$ and suppose that $-A$ is the generator of a positive C_0-semigroup. Does A have a bounded H^∞-calculus (bounded imaginary powers / is $\mathcal{R}$-analytic)?

Problem 5. Find a similar characterization as in Theorem 2.10 or Theorem 2.11 in the case $\omega_{H^\infty(A)} = \frac{\pi}{2}$.

It was observed by C. Le Merdy in [41, p. 33] that a counterexample to Problem 3 on L_p would also provide a negative answer to a (largely) open conjecture by Matsaev. For an introduction to the problem, its noncommutative analogue and further references we refer to the recent article [4]. We note that there exists a 2×2-matrix counterexample to Matsaev's conjecture for the case $p = 4$ which was obtained with the help of numerics [14], but an analytic approach is missing.

2.4. Bounded imaginary powers (BIP)

Sectorial operators with bounded imaginary powers have been studied before the first appearance of the H^∞-calculus. They play an important role in the Dore–

Venni theorem [13, Theorem 2.1] and in the interpolation of fractional domain spaces [61].

Definition 2.12 (Bounded Imaginary Powers (BIP)). A sectorial operator on a Banach space X is said to have *bounded imaginary powers* (*BIP*) if for all $t \in \mathbb{R}$ the operator A^{it} associated to the functions $\lambda \mapsto \lambda^{it}$ via the holomorphic functional calculus is bounded.

In this case $(A^{it})_{t\in\mathbb{R}}$ is a C_0-group on X with generator $i \log A$ [26, Corollary 3.5.7]. The growth of the C_0-group $(A^{it})_{t\in\mathbb{R}}$ is used to define the BIP-angle.

Definition 2.13. For a sectorial operator A on some Banach space with bounded imaginary powers one defines

$$\omega_{\mathrm{BIP}}(A) := \inf\{\omega \geq 0 : \|A^{it}\| \leq Me^{\omega|t|} \text{ for all } t \in \mathbb{R} \text{ for some } M \geq 0\}.$$

If A does not have bounded imaginary powers, we set $\omega_{\mathrm{BIP}}(A) := \infty$.

Let A be a sectorial operator with a bounded $H^\infty(\Sigma_\sigma)$-calculus for some $\sigma \in (0, \pi)$. Then one has

$$|\lambda^{it}| \leq \exp(\mathrm{Re}(it \log \lambda)) \leq \exp(|t|\, \sigma)$$

for all $\lambda \in \Sigma_\sigma$. This shows that the boundedness of the H^∞-calculus for A implies that A has bounded imaginary powers with $\omega_{\mathrm{BIP}}(A) \leq \omega_{H^\infty}(A)$. A less obvious fact is that BIP implies $\mathcal{R}$-sectoriality on UMD-spaces [12, Theorem 4.5]. A Banach space is called a UMD-space if the vector-valued Hilbert transform is bounded on $L_2(\mathbb{R}; X)$. There are more equivalent definitions of UMD-spaces. For details we refer to [8] and [54]. We only note the following: if X is a UMD-space, then so is $L_p(\Omega; X)$ for all measure spaces Ω and $p \in (1, \infty)$. In particular, $L_p(\Omega)$ is UMD. Moreover, every UMD-space has property (Δ), but not every UMD-space has Pisier's property (α).

Theorem 2.14. *Let A be a sectorial operator with bounded imaginary powers on a UMD-space. Then A is $\mathcal{R}$-sectorial with $\omega_{\mathcal{R}}(A) \leq \omega_{\mathrm{BIP}}(A)$.*

In particular this implies that a sectorial operator A on a UMD-space with a bounded H^∞-calculus satisfies $\omega_{\mathcal{R}}(A) = \omega_{\mathrm{BIP}}(A) = \omega_{H^\infty}(A)$. The first example showing that the strict inequality $\omega(A) < \omega_{\mathrm{BIP}}(A)$ can hold was found by M. Haase [25, Corollary 5.3] (see also Remark 4.3).

2.5. Sectorial operators which have a dilation

A further regularity property which is not so inherent to sectorial operators but nevertheless very important for their study is the existence of group dilations. This powerful concept goes back to B. Sz.-Nagy. For a detailed treatment of dilation theory on Hilbert spaces see [57]. In particular one has the following result [57, Theorem 8.1].

Theorem 2.15. *Let $(T(t))_{t\geq 0}$ be a contractive C_0-semigroup on a Hilbert space H. Then there exists a second Hilbert space K, an embedding $J\colon H \to K$, an orthogonal projection $P\colon K \to H$ and a unitary group $(U(t))_{t\geq 0}$ on K with*

$$T(t) = PU(t)J \qquad \text{for all } t \geq 0.$$

It follows from the spectral theory of normal operators that the negative generator of $(U(t))_{t\in\mathbb{R}}$ and therefore also the negative generator of $(T(t))_{t\geq 0}$ has a bounded H^∞-calculus for all angles bigger than $\frac{\pi}{2}$. Hence, using the fact that $\omega(A) = \omega_{H^\infty}(A)$ we have found a proof of the first part of Theorem 2.10. We have seen the following.

Corollary 2.16. *Let A be a sectorial operator on a Hilbert space such that $-A$ generates a contractive C_0-semigroup. Then A has a bounded H^∞-calculus with $\omega_{H^\infty}(A) = \omega(A) \leq \frac{\pi}{2}$.*

It is now time to give a precise definition of semigroup dilations on general Banach spaces. We follow the terminology used in [5].

Definition 2.17. Let $(T(t))_{t\geq 0}$ be a C_0-semigroup on some Banach space X. Further let $\mathcal{X}$ denote a class of Banach spaces. We say that

(i) $(T(t))_{t\geq 0}$ has a *strict dilation* in $\mathcal{X}$ if for some Y in $\mathcal{X}$ there are contractive linear operators $J\colon X \to Y$ and $Q\colon Y \to X$ and an isometric C_0-group $(U(t))_{t\in\mathbb{R}}$ on Y such that

$$T(t) = QU(t)J \qquad \text{for all } t \geq 0.$$

(ii) $(T(t))_{t\geq 0}$ has a *loose dilation* in $\mathcal{X}$ if for some Y in $\mathcal{X}$ there are bounded linear operators $J\colon X \to Y$ and $Q\colon Y \to X$ and a bounded C_0-group $(U(t))_{t\in\mathbb{R}}$ on Y such that

$$T(t) = QU(t)J \qquad \text{for all } t \geq 0.$$

Note that in the above terminology Theorem 2.15 shows that every contractive C_0-semigroup on a Hilbert space has a strict dilation in the class of all Hilbert spaces. The main connection with the other regularity properties is the following observation.

Proposition 2.18. *Let A be a sectorial operator on a Banach space X such that $-A$ generates a C_0-semigroup which has a loose dilation in the class of all UMD-Banach spaces. Then A has a bounded H^∞-calculus with $\omega_{H^\infty}(A) \leq \frac{\pi}{2}$.*

This follows from the transference principle of R.R. Coifman and G. Weis developed in [9] which reduces the assertion to the case of the vector-valued shift group on $L_p(\mathbb{R}; Y)$ for some UMD-space Y which can be shown directly with the help of the vector-valued Mikhlin multiplier theorem [62, Proposition 3].

On L_p-spaces for $p \in (1, \infty)$ one has the following characterization of strict dilations. A bounded linear operator $T\colon L_p(\Omega) \to L_p(\Omega')$ is called a *subpositive contraction* if there exists a positive contraction $S\colon L_p(\Omega) \to L_p(\Omega')$, that is $\|S\| \leq 1$ and $f \geq 0 \Rightarrow Sf \geq 0$, such that $|Tf| \leq S\,|f|$ for all $f \in L_p(\Omega)$.

Theorem 2.19. *Let $(T(t))_{t\geq 0}$ be a C_0-semigroup on some σ-finite L_p-space for $p \in (1,\infty)\setminus\{2\}$. Then $(T(t))_{t\geq 0}$ has a strict dilation in the class of all σ-finite L_p-spaces if and only if $(T(t))_{t\geq 0}$ is a semigroup consisting of subpositive contractions.*

Every C_0-semigroup of subpositive contractions on L_p for $p \in (1,\infty)$ has a strict dilation by Fendler's dilation theorem [20]. For the converse it suffices to show that for a strict dilation $T(t) = QU(t)J$ all the operators $U(t)$, J and Q are subpositive contractions (notice that J and Q^* are isometries). For the first two this essentially follows from the Banach–Lamperti theorem [21, Theorem 3.2.5] on the structure of isometries on L_p-spaces, for the third as well if applied to the adjoint Q^*. However, there is no characterization of semigroups on L_p with a loose dilation.

Problem 6. Characterize those semigroups on L_p which have a loose dilation in the class of all L_p-spaces.

For a more concrete discussion in the setting of discrete semigroups see [5, Section 5]. We also do not know whether the following extension of Fendler's dilation theorem to UMD-Banach lattices holds.

Problem 7. Does every C_0-semigroup of positive contractions on a UMD-Banach lattice have a strict / loose dilation in the class of all UMD-spaces?

In the negative direction one knows the following: there exists a completely positive contraction, i.e., a discrete semigroup, on a noncommutative L_p-space which does not have a strict dilation in the class of all noncommutative L_p-spaces [30, Corollary 4.4]. For a weak discrete counterexample in the setting of $L_p(L_q)$-spaces see [24, Contre exemple 6.1].

Recall that by Proposition 2.18 a C_0-semigroup $(T(t))_{t\geq 0}$ with generator $-A$ that has a loose dilation in the class of all UMD-spaces has a bounded H^∞-calculus with $\omega_{H^\infty}(A) \leq \frac{\pi}{2}$. The following theorem by A. Fröhlich and L. Weis [22, Corollary 5.4] is a partial converse. Its proof uses square function techniques which we do not cover here, for an overview we refer to [42].

Theorem 2.20. *Let A be a sectorial operator on a UMD-space X with $\omega_{H^\infty}(A) < \frac{\pi}{2}$. Then the semigroup $(T(t))_{t\geq 0}$ generated by $-A$ has a loose dilation to the space $L_2([0,1];X)$.*

This shows that on UMD-spaces the existence of loose dilations in the class of UMD-spaces and of a bounded H^∞-calculus are equivalent under the restriction $\omega_R(A) < \frac{\pi}{2}$. However, we will see in Section 5 that there exists a semigroup generator $-A$ on a Hilbert space with $\omega_R(A) = \omega(A) = \frac{\pi}{2}$ that does not have a loose dilation in the class of all Hilbert spaces. So in general the existence of a dilation is a strictly stronger property than the existence of a bounded H^∞-calculus.

3. Counterexamples I: The Schauder multiplier method

In this section we develop the most fruitful known method to construct systematically counterexamples: the Schauder multiplier method. This method was first used in [6] and [58] in the context of sectorial operators to give examples of sectorial operators without bounded imaginary powers. After dealing with H^∞-calculus and bounded imaginary powers, we present a self-contained example of a sectorial operator on L_p which is not $\mathcal{R}$-sectorial.

3.1. Schauder multipliers

We start our journey by giving the definition of Schauder multipliers and by studying its fundamental properties. After that we show how Schauder multipliers can be used to construct (analytic) semigroups. From now on we need some background from Banach space theory. We refer to [1], [16], [43] and [55].

Definition 3.1 (Schauder multiplier). Let $(e_m)_{m\in\mathbb{N}}$ be a Schauder basis for a Banach space X. For a sequence $(\gamma_m)_{m\in\mathbb{N}} \subset \mathbb{C}$ the operator A defined by

$$D(A) = \left\{ x = \sum_{m=1}^{\infty} a_m e_m : \sum_{m=1}^{\infty} \gamma_m a_m e_m \text{ exists} \right\}$$

$$A\left(\sum_{m=1}^{\infty} a_m e_m\right) = \sum_{m=1}^{\infty} \gamma_m a_m e_m$$

is called the *Schauder multiplier* associated to $(\gamma_m)_{m\in\mathbb{N}}$.

3.1.1. Basic properties of Schauder multipliers. We now discuss some properties of Schauder multipliers whose proofs can be found in [26, Section 9.1.1] and [58].

Proposition 3.2. *The Schauder multiplier A associated to a sequence $(\gamma_m)_{m\in\mathbb{N}}$ is a densely defined closed linear operator.*

A central problem in the theory of Schauder multipliers is to determine for a given Schauder basis $(e_m)_{m\in\mathbb{N}}$ the set of all sequences $(\gamma_m)_{m\in\mathbb{N}}$ for which the associated Schauder multiplier is bounded. In general, it is an extremely difficult problem to determine this space exactly. For example, the trigonometric basis is a Schauder basis for $L_p([0,1])$ for $p \in (1,\infty)$. In this particular case the above problem asks for a characterization of all bounded Fourier multipliers on L_p.

However, some elementary general properties of this sequence space can be obtained easily. In what follows let BV be the Banach space of all sequences with bounded variation.

Proposition 3.3. *Let $(e_m)_{m\in\mathbb{N}}$ be a Schauder basis for a Banach space X. Then there exists a constant $K \geq 0$ such that for every $(\gamma_m)_{m\in\mathbb{N}} \in BV$ the Schauder multiplier A associated to $(\gamma_m)_{m\in\mathbb{N}}$ with respect to $(e_m)_{m\in\mathbb{N}}$ is bounded and satisfies*

$$\|A\| \leq K \left\|(\gamma_m)_{m\in\mathbb{N}}\right\|_{BV}.$$

Conversely, if A is a bounded Schauder multiplier associated to some sequence $(\gamma_m)_{m\in\mathbb{N}}$, then $(\gamma_m)_{m\in\mathbb{N}}$ is bounded.

Remark 3.4. In general the above result is optimal. For if $X = BV$, then $(e_m)_{m\in\mathbb{N}_0}$ defined by e_0 as the constant sequence $\mathbb{1}$ and $e_m = (\delta_{mn})_{n\in\mathbb{N}}$ form a conditional basis of BV and the multiplier associated to a sequence $(\gamma_m)_{m\in\mathbb{N}_0}$ is bounded if and only if $(\gamma_m) \in BV$.

3.1.2. Schauder multipliers as generators of analytic semigroups. Given an arbitrary Banach space X, it is difficult to guarantee, roughly spoken, the existence of interesting strongly continuous semigroups on this space. Of course, every bounded operator generates such a semigroup by means of exponentiation. Such an argument does in general not work to show the existence of C_0-semigroups with an unbounded generator. Indeed, on $L_\infty([0,1])$ a result by H.P. Lotz [45, Theorem 3] shows that every generator of a strongly continuous semigroup is already bounded.

One therefore has to make additional assumptions on the Banach space. A very convenient and rather general assumption for separable Banach spaces is to require the existence of a Schauder basis for that space. Indeed, all classical separable Banach spaces have a Schauder basis. Moreover, for a long time it has been an open problem whether all separable Banach spaces have a Schauder basis (this was solved negatively by P. Enflo [15]).

The next proposition shows that Schauder bases allow us to construct systematically strongly continuous semigroups (with unbounded generators) on the underlying Banach spaces.

Proposition 3.5. *Let $(e_m)_{m\in\mathbb{N}}$ be a Schauder basis for a Banach space X. Further let $(\gamma_m)_{m\in\mathbb{N}}$ be an increasing sequence of positive real numbers. Then the Schauder multiplier associated to $(\gamma_m)_{m\in\mathbb{N}}$ with respect to $(e_m)_{m\in\mathbb{N}}$ is a sectorial operator with $\omega(A) = 0$. In particular, $-A$ generates an analytic C_0-semigroup $(T(z))_{z\in\Sigma_{\frac{\pi}{2}}}$.*

3.2. Sectorial operators without a bounded H^∞-calculus

In this subsection we apply the so far developed methods to give examples of sectorial operators without a bounded H^∞-calculus. The first example was given in [47]. The elegant approach of this section goes back to [39] and [41].

One can easily show that one cannot obtain examples of sectorial operators without a bounded H^∞-calculus by using Schauder multipliers with respect to an unconditional basis. However, one can produce counterexamples from Schauder multipliers with respect to a conditional basis.

Theorem 3.6. *Let $(e_m)_{m\in\mathbb{N}}$ be a conditional Schauder basis for a Banach space X. Then the Schauder multiplier A associated to the sequence $(2^m)_{m\in\mathbb{N}}$ is a sectorial operator with $\omega(A) = 0$ which does not have a bounded H^∞-calculus.*

Proof. By Proposition 3.5 everything is already shown except for the fact that A does not have a bounded H^∞-calculus. For this observe that for each $f \in H^\infty(\Sigma_\sigma)$ for some $\sigma \in (0,\pi)$ the operator $f(A)$ is given by the Schauder multiplier associated to the sequence $(f(\gamma_m))_{m\in\mathbb{N}}$. Now assume that A has a bounded $H^\infty(\Sigma_\sigma)$-calculus for some $\sigma \in (0,\pi)$. By [26, Corollary 9.1.6] on the interpolation of sequences by holomorphic functions, for every element in ℓ_∞ there exists an $f \in H^\infty(\Sigma_\sigma)$

such that $(f(2^m))_{m\in\mathbb{N}}$ is the desired sequence. This means that every element in ℓ_∞ defines a bounded Schauder multiplier. However, this means that $(e_m)_{m\in\mathbb{N}}$ is unconditional in contradiction to our assumption. □

Corollary 3.7. *Let X be a Banach space that admits a Schauder basis. Then there exists a sectorial operator A with $\omega(A) = 0$ that does not have a bounded H^∞-calculus.*

Proof. Every Banach space which admits a Schauder basis does also admit a conditional Schauder basis [1, Theorem 9.5.6]. Then the result follows directly from Theorem 3.6. □

Next we give a concrete example of a sectorial operator of the above form which has boundary imaginary powers but no bounded H^∞-calculus. This goes back to G. Lancien [39] (see also [41]).

Example 3.8. We consider the trigonometric system $(e^{imz})_{m\in\mathbb{Z}}$ for the enumeration $(0,-1,1,-2,\ldots)$ which is a conditional basis of $L_p([0,2\pi])$ for $p\in(1,\infty)\setminus\{2\}$ [44, Theorem 2.c.16]. We can then consider the Schauder multiplier A associated to the sequence $(2^m)_{m\in\mathbb{Z}}$. As a consequence of the boundedness of the Hilbert transform on L_p one can consider the operator separately on the two complemented parts with respect to the decomposition

$$L_p([0,2\pi]) = \overline{\operatorname{span}}\{e^{imz} : m < 0\} \oplus \overline{\operatorname{span}}\{e^{imz} : m \geq 0\}.$$

Observe that A has a bounded H^∞-calculus if and only if both parts have a bounded H^∞-calculus. It then follows from Proposition 3.3 and Proposition 3.5 that A is a sectorial operator with $\omega(A) = 0$ which by Theorem 3.6 (applied to the second part) does not have a bounded H^∞-calculus. We now show that A has bounded imaginary powers with $\omega_{\mathrm{BIP}}(A) = 0$. For this we observe that

$$\begin{aligned} A^{it}\Big(\sum_{m\in\mathbb{Z}} a_m e^{imz}\Big) &= \sum_{m\in\mathbb{Z}} 2^{mit} a_m e^{imz} = \sum_{m\in\mathbb{Z}} a_m \exp(imt\log 2) e^{imz} \\ &= \sum_{m\in\mathbb{Z}} a_m \exp(im(t\log 2 + z)) = S(t\log 2)\Big(\sum_{m\in\mathbb{Z}} a_m e^{imz}\Big), \end{aligned}$$

where $(S(t))_{t\in\mathbb{R}}$ is the periodic shift group on $L_p([0,2\pi])$.

We will study examples of the above type more systematically in Section 4.

3.3. Sectorial operators without BIP

Similarly to the case of the bounded H^∞-calculus one can use Schauder multipliers to construct sectorial operators which do not have bounded imaginary powers. We start with a weighted version of Example 3.8 which gives an example of an $\mathcal{R}$-sectorial operator without bounded imaginary powers, a discrete version of the counterexample [38, Example 10.17]. However, before we need to state some facts on harmonic analysis and A_p-weights.

It is a natural question to ask for which weights w the trigonometric system is a Schauder basis for the space $L_p([0,2\pi],w)$. Indeed, a complete characterization of these weights is known. We identify the torus $\mathbb{T}$ with the interval $[0,2\pi)$ on the real line and functions in $L_p([0,2\pi])$ with their periodic extensions or with L_p-functions on the torus.

Definition 3.9 (A_p-weight). Let $p \in (1,\infty)$. A function $w\colon \mathbb{R} \to [0,\infty]$ with $w(t) \in (0,\infty)$ almost everywhere is called an *A_p-weight* if there exists a constant $K \geq 0$ such that for every compact interval $I \subset \mathbb{R}$ with positive length one has

$$\left(\frac{1}{|I|}\int_I w(t)\,dt\right)\left(\frac{1}{|I|}\int_I w(t)^{-1/(p-1)}\,dt\right)^{p-1} \leq K.$$

The set of all A_p-weights is denoted by $\mathcal{A}_p(\mathbb{R})$. Moreover, we set in the periodic case

$$\mathcal{A}_p(\mathbb{T}) := \{w \in \mathcal{A}_p(\mathbb{R}) : w \text{ is } 2\pi\text{-periodic}\}.$$

For a detailed treatment of these weights and their applications in harmonic analysis we refer to the monograph [56, Chapter V]. As an example the 2π-periodic extension of the function $t \mapsto |t|^\alpha$ for $\alpha \in \mathbb{R}$ lies in $\mathcal{A}_p(\mathbb{T})$ if and only if $\alpha \in (-1,p-1)$ [7, Example 2.4]. The characterization below can be found in [49, Proposition 2.3] and essentially goes back to methods developed by R. Hunt, B. Muckenhoupt and R. Wheeden in [29].

Theorem 3.10. *Let $w\colon \mathbb{R} \to [0,\infty]$ with $w(t) \in (0,\infty)$ almost everywhere be a 2π-periodic weight and $p \in (1,\infty)$. Then the trigonometric system is a Schauder basis for $L_p([0,2\pi],w)$ with respect to the enumeration $(0,-1,1,-2,2,\ldots)$ of $\mathbb{Z}$ if and only if $w \in \mathcal{A}_p(\mathbb{T})$.*

Now we are ready to give the example.

Example 3.11. Let $p \in (1,\infty)$ and $w \in \mathcal{A}_p(\mathbb{T})$ be an A_p-weight. Then the trigonometric system $(e^{imz})_{m\in\mathbb{Z}}$ is a Schauder basis for $L_p([0,2\pi],w)$ by Theorem 3.10. Let A again be the Schauder multiplier associated to the sequence $(2^m)_{m\in\mathbb{Z}}$. One sees as in Example 3.8 that A is a sectorial operator. It remains to show that A is $\mathcal{R}$-sectorial. Notice that for $\lambda = a2^l e^{i\theta} \in \mathbb{C} \setminus [0,\infty)$ with $|a| \in [1,2]$ one has for $x = \sum_{m\in\mathbb{Z}} a_m e^{imz}$

$$\begin{aligned}\lambda R(\lambda,A)x &= \sum_{m\in\mathbb{Z}} \frac{\lambda}{\lambda - 2^m} a_m e^{imz} = \sum_{m\in\mathbb{Z}} \frac{ae^{i\theta}}{ae^{i\theta} - 2^{m-l}} a_m e^{imz} \\ &= \sum_{m\in\mathbb{Z}} \frac{ae^{i\theta}}{ae^{i\theta} - 2^m} a_{m+l} e^{i(m+l)z} = ae^{i\theta} R(ae^{i\theta},A)\Bigl(\sum_{m\in\mathbb{Z}} a_{m+l} e^{imz}\Bigr) e^{ilz} \\ &= e^{ilz} ae^{i\theta} R(ae^{i\theta},A)(x \cdot e^{-ilz})\end{aligned}$$

Consequently for $\lambda_k = a2^{l_k}e^{i\theta}$ with $k \in \{1, \ldots, n\}$ and $x_1, \ldots, x_n \in L_p([0, 2\pi], w)$ one has

$$\left\| \sum_{k=1}^{n} r_k \lambda_k R(\lambda_k, A) x_k \right\| = \left\| \sum_{k=1}^{n} r_k e^{il_k z} a e^{i\theta} R(ae^{i\theta}, A)(e^{-il_k z} x_k) \right\|$$
$$\leq 2\left|a\right| \left\| R(ae^{i\theta}, A) \right\| \left\| \sum_{k=1}^{n} r_k e^{-il_k z} x_k \right\| \leq 8 \left\| R(ae^{i\theta}, A) \right\| \left\| \sum_{k=1}^{n} r_k x_k \right\|$$

by Kahane's contraction principle. Now it is easy to check that for every $\theta_0 > 0$ the sequences $(\frac{ae^{i\theta}}{ae^{i\theta} - 2^{\pm m}})_{m \in \mathbb{N}}$ satisfy the assumptions of Proposition 3.3 uniformly in $\theta \in [\theta_0, 2\pi)$ for $\theta_0 > 0$ and in $|a| \in [1, 2]$. By [60, Theorem 4.2 2)] and the boundedness of the Hilbert transform on $L_p([0, 2\pi], w)$ this shows that A is $\mathcal{R}$-analytic with $\omega_{\mathcal{R}}(A) = 0$.

By the same calculation as in Example 3.8 the operator A^{it} for $t \in \mathbb{R}$ is given by $S(t \log 2)$ on the dense set of trigonometric polynomials, where $(S(t))_{t \in \mathbb{R}}$ is the periodic shift group. Notice however that for example for $w(t) = |t|^{\alpha}$ for a suitable chosen $\alpha \in \mathbb{R}$ such that $w \in \mathcal{A}_p(\mathbb{T})$ this group obviously does not leave $L_p([0, 2\pi], w)$ invariant. Hence, A does not have bounded imaginary powers.

3.4. Sectorial operators which are not $\mathcal{R}$-sectorial

We now present a self-contained example of a sectorial operator on L_p which is not $\mathcal{R}$-sectorial based on [17]. In order to do that we need to study some geometric properties of L_p-spaces.

A key role in what follows is played by L_p-functions which stay away from zero in a sufficiently large set. More precisely, for $p \in [1, \infty)$ and $\varepsilon > 0$ we consider

$$M_\varepsilon^p := \left\{ f \in L_p([0,1]) : \lambda\left(\left\{ x \in [0,1] : |f(x)| \geq \varepsilon \left\| f \right\|_p \right\} \right) \geq \varepsilon \right\}.$$

Functions in these sets have a very important summability property which is comparable to the L_2-case. For the proofs of the next two lemmata we follow closely the main ideas in [55, §21].

Lemma 3.12. *For $p \in [2, \infty)$ and $\varepsilon > 0$ let $(f_m)_{m \in \mathbb{N}} \subset L_p([0,1])$ be a sequence in M_ε^p such that $\sum_{m=1}^{\infty} f_m$ converges unconditionally in $L_p([0,1])$. Then one has $\sum_{m=1}^{\infty} \left\| f_m \right\|_p^2 < \infty$.*

Proof. Since $p \in [2, \infty)$, it follows from Hölder's inequality that for all $f \in L_p([0,1])$ one has $\left\| f \right\|_2 \leq \left\| f \right\|_p$. This shows that the series $\sum_{m=1}^{\infty} f_m$ converges unconditionally in $L_2([0,1])$ as well. By the unconditionality of the series there exists a $K \geq 0$ such that $\left\| \sum_{m=1}^{\infty} \varepsilon_m f_m \right\|_2 \leq K$ for all $(\varepsilon_m)_{m \in \mathbb{N}} \in \{-1, 1\}^{\mathbb{N}}$. Now, for all $N \in \mathbb{N}$ one has

$$\sum_{m=1}^{N} \left\| f_m \right\|_2^2 = \int_0^1 \left\| \sum_{m=1}^{N} r_m(t) f_m \right\|_2^2 dt \leq K^2.$$

Hence, $\sum_{m=1}^{\infty} \|f_m\|_2^2 < \infty$. Notice that the assumption $f_m \in M_\varepsilon^p$ implies that for all $m \in \mathbb{N}$

$$\|f_m\|_2^2 \geq \int_{|f_m| \geq \varepsilon \|f_m\|_p} |f_m(x)|^2 \, dx \geq \varepsilon^3 \|f_m\|_p^2 .$$

Together with the summability shown above this yields $\sum_{m=1}^{\infty} \|f_m\|_p^2 < \infty$. □

The next lemma shows that unconditional basic sequences formed out of elements in M_ε^p behave like Hilbert space bases.

Lemma 3.13. *For $p \in [2, \infty)$ let $(e_m)_{m \in \mathbb{N}}$ be an unconditional normalized basic sequence in $L_p([0,1])$ for which there exists an $\varepsilon > 0$ such that $e_m \in M_\varepsilon^p$ for all $m \in \mathbb{N}$. Then*

$$\sum_{m=1}^{\infty} a_m e_m \quad \textit{converges} \qquad \Leftrightarrow \qquad (a_m)_{m \in \mathbb{N}} \in \ell_2.$$

Proof. Assume that the expansion $\sum_{m=1}^{\infty} a_m e_m$ converges. Since $(e_m)_{m \in \mathbb{N}}$ is an unconditional basic sequence, the series $\sum_{m=1}^{\infty} a_m e_m$ converges unconditionally in $L_p([0,1])$. By Lemma 3.12, one has

$$\sum_{m=1}^{\infty} |a_m|^2 = \sum_{m=1}^{\infty} \|a_m e_m\|_p^2 < \infty.$$

Conversely, we have to show that the expansion converges for all $(a_m)_{m \in \mathbb{N}} \in \ell_2$. One has $\left\|\sum_{m=1}^{N} a_m e_m\right\| \leq K \left\|\sum_{m=1}^{N} \varepsilon_m a_m e_m\right\|$ for all $(\varepsilon_m)_{m \in \mathbb{N}} \in \{-1,1\}^N$ and all $N \in \mathbb{N}$, where $K \geq 0$ denotes the unconditional basis constant of $(e_m)_{m \in \mathbb{N}}$. Now, since for $p \geq 2$ the space $L_p([0,1])$ has type 2, we have for all $N, M \in \mathbb{N}$

$$\left\|\sum_{m=M}^{N} a_m e_m\right\|_p \leq K \int_0^1 \left\|\sum_{m=M}^{N} r_m(t) a_m e_m\right\|_p dt \leq KC \Big(\sum_{m=M}^{N} |a_m|^2\Big)^{1/2}$$

for some constant $C > 0$. From this it is immediate that the sequence of partial sums $(\sum_{m=1}^{N} a_m e_m)_{N \in \mathbb{N}}$ is Cauchy in $L_p([0,1])$. □

For the following counterexample on L_p-spaces our starting point is a particular basis given by the Haar system.

Definition 3.14. The *Haar system* is the sequence $(h_n)_{n \in \mathbb{N}}$ of functions defined by $h_1 = 1$ and for $n = 2^k + s$ (where $k = 0, 1, 2, \ldots$ and $s = 1, 2, \ldots, 2^k$) by

$$h_n(t) = \mathbb{1}_{[\frac{2s-2}{2^{k+1}}, \frac{2s-1}{2^{k+1}})}(t) - \mathbb{1}_{[\frac{2s-1}{2^{k+1}}, \frac{2s}{2^{k+1}})}(t) = \begin{cases} 1 & \text{if } t \in [\frac{2s-2}{2^{k+1}}, \frac{2s-1}{2^{k+1}}) \\ -1 & \text{if } t \in [\frac{2s-1}{2^{k+1}}, \frac{2s}{2^{k+1}}) \\ 0 & \text{otherwise} \end{cases}.$$

The Haar basis is an unconditional Schauder basis for $L_p([0,1])$ for $p \in (1, \infty)$ (see [1, Proposition 6.1.3 & Theorem 6.1.6]).

Remark 3.15. Note that the Haar system is not normalized in $L_p([0,1])$ for $p \in [1,\infty)$. Of course, we can always work with $(h_m/\|h_m\|_p)_{m\in\mathbb{N}}$ instead which is a normalized basis. It is however important to note that the normalization constant $\|h_m\|_p = 2^{-k/p}$ depends on p and we can therefore not simultaneously normalize $(h_m)_{m\in\mathbb{N}}$ on the L_p-scale. This crucial point was overlooked in [17].

The following proposition is used to transfer the $\mathcal{R}$-boundedness of a sectorial operator to the boundedness of a single operator. This approach is closely motivated by the work [3].

Proposition 3.16. *Let A be an $\mathcal{R}$-sectorial operator. Then there exists a constant $C \geq 0$ such that for all $(q_n)_{n\in\mathbb{N}} \subset \mathbb{R}_-$ the associated operator*

$$\mathcal{R}\colon \sum_{n=1}^{N} r_n x_n \mapsto \sum_{n=1}^{N} r_n q_n R(q_n, A) x_n$$

defined on the finite Rademacher sums extends to a bounded operator on Rad(X) *with operator norm at most C.*

Proof. If A is $\mathcal{R}$-sectorial, one has $C := \mathcal{R}\{\lambda R(\lambda, A) : \lambda \in \mathbb{R}_-\} < \infty$. Hence, for all finite Rademacher sums we have by the definition of $\mathcal{R}$-boundedness

$$\left\| \sum_{n=1}^{N} r_n q_n R(q_n, A) x_n \right\| \leq C \left\| \sum_{n=1}^{N} r_n x_n \right\|. \qquad \square$$

One now uses the freedom in the choice of the sequence $(q_n)_{n\in\mathbb{N}}$. This is done in the following elementary lemma. We will see its usefulness very soon.

Lemma 3.17. *For $\gamma_m > \gamma_{m-1} > 0$ consider the function $d(t) := t[(t+\gamma_{m-1})^{-1} - (t+\gamma_m)^{-1}]$ on $\mathbb{R}_+$. Then d has a maximum bigger than $\frac{1}{2}\frac{\gamma_m - \gamma_{m-1}}{\gamma_m + \gamma_{m-1}}$.*

Proof. By the mean value theorem we have for some $\xi \in (\gamma_{m-1}, \gamma_m)$ and all $t > 0$ that

$$\frac{1}{t+\gamma_{m-1}} - \frac{1}{t+\gamma_m} = (\gamma_m - \gamma_{m-1})\frac{1}{(t+\xi)^2} \geq (\gamma_m - \gamma_{m-1})\frac{1}{(t+\gamma_m)^2}.$$

One now easily verifies that the function $t \mapsto (\gamma_m - \gamma_{m-1})\frac{t}{(t+\gamma_m)^2}$ has a unique maximum for $t = \gamma_m$. In particular one has

$$\max_{t>0} d(t) \geq d(\gamma_m) = \frac{1}{2}\frac{\gamma_m - \gamma_{m-1}}{\gamma_m + \gamma_{m-1}}. \qquad \square$$

We can now give examples of sectorial operators on L_p which are not $\mathcal{R}$-sectorial.

Theorem 3.18. *For $p \in (2,\infty)$ there exists a sectorial operator A on $L_p([0,1])$ with $\omega(A) = 0$ which is not $\mathcal{R}$-sectorial.*

Proof. Until the rest of the proof let $(h_m)_{m\in\mathbb{N}}$ denote the normalized Haar system on $L_p([0,1])$. Choose a subsequence $(m_k)_{k\in\mathbb{N}} \subset 2\mathbb{N}$ such that the functions h_{m_k} have pairwise disjoint supports. Then $(h_{m_k})_{k\in\mathbb{N}}$ is an unconditional basic sequence equivalent to the standard basis of ℓ_p. Indeed, for any finite sequence $a_1, \ldots, a_N$ we have by the disjointness of the supports

$$\left\| \sum_{k=1}^{N} a_k h_{m_k} \right\|_p^p = \sum_{k=1}^{N} \| a_k h_{m_k} \|_p^p = \sum_{k=1}^{N} |a_k|^p .$$

Choose a permutation of the even numbers such that $\pi(4k) = m_k$. We now define a new system $(f_m)_{m\in\mathbb{N}}$ as

$$f_m := \begin{cases} h_{\pi(m)} \\ h_{\pi(m)} + h_{\pi(m-1)} \end{cases} = \begin{cases} h_m & m \text{ odd} \\ h_{\pi(m)} + h_{m-1} & m \text{ even.} \end{cases}$$

Notice that by the unconditionality of the Haar basis $(h_{\pi(m)})_{m\in\mathbb{N}}$ is a Schauder basis of $L_p([0,1])$ as well. As a block perturbation of the normalized basis $(h_{\pi(m)})_{m\in\mathbb{N}}$ the sequence $(f_m)_{m\in\mathbb{N}}$ is a basis for $L_p([0,1])$ as well [55, Ch. I, § 4, Proposition 4.4]. Further, let A be the closed linear operator on $L_p([0,1])$ given by

$$D(A) = \left\{ x = \sum_{m=1}^{\infty} a_m f_m : \sum_{m=1}^{\infty} 2^m a_m f_m \text{ exists} \right\}$$

$$A\Big(\sum_{m=1}^{\infty} a_m f_m \Big) = \sum_{m=1}^{\infty} 2^m a_m f_m .$$

Proposition 3.5 shows that A is sectorial with $\omega(A) = 0$. The basic sequences $(h_{\pi(4m)})_{m\in\mathbb{N}}$ and $(h_{4m+1})_{m\in\mathbb{N}}$ are not equivalent: assume that the two basic sequences are equivalent. Then on the one hand for $(h_{4m+1})_{m\in\mathbb{N}}$ the block basic sequence

$$b_k = \sum_{\substack{m:4m+1 \\ \in[2^k+1,2^{k+1}]}} h_{4m+1}$$

satisfies for $k \geq 2$ by the disjointness of the summands

$$\|b_k\|_p^p = \sum_{\substack{m:4m+1 \\ \in[2^k+1,2^{k+1}]}} \|h_{4m+1}\|_p^p = \sum_{\substack{m:4m+1 \\ \in[2^k+1,2^{k+1}]}} 1 = \frac{1}{4} \cdot 2^k = 2^{k-2}.$$

Moreover, on the non-vanishing part b_k satisfies $|b_k(t)| = 2^{k/p}$ for $k \geq 2$. Hence, for the normalized block basic sequence $(\tilde{b}_k)_{k\geq 2} = (\frac{b_k}{\|b_k\|_p})_{k\geq 2}$ one has $|\tilde{b}_k(t)| = 2^{2/p}$. Therefore we have

$$\lambda\left(\left\{ t \in [0,1] : |\tilde{b}_k(t)| \geq \varepsilon \|\tilde{b}_k\|_p \right\}\right) = \lambda\left(\left\{ t \in [0,1] : |\tilde{b}_k(t)| \geq \varepsilon \right\}\right) = \frac{1}{4}$$

for $\varepsilon \leq 2^{2/p}$. In particular for $\varepsilon \leq \frac{1}{4}$ we have $\tilde{b}_k \in M_\varepsilon^p$ for all $k \geq 2$. By Lemma 3.13 this implies that $(\tilde{b}_k)_{k\geq 2}$ is equivalent to the standard basis in ℓ_2.

Since we have assumed that the basic sequence $(h_{\pi(4k)})_{k\in\mathbb{N}}$ is equivalent to $(h_{4k+1})_{k\in\mathbb{N}}$, the block basic sequence $(c_k)_{k\geq 2}$ defined by

$$c_k = \|b_k\|_p^{-1} \sum_{\substack{m:4m+1 \\ \in[2^k+1,2^{k+1}]}} h_{\pi(4m)}$$

is seminormalized. Recall that $(h_{\pi(4m)})_{m\in\mathbb{N}}$ is equivalent to the standard basis of ℓ_p. Since all semi-normalized block basic sequences of ℓ_p are equivalent to the standard basis of ℓ_p [1, Lemma 2.1.1], the sequence $(c_k)_{k\geq 2}$ is equivalent to the standard basis of ℓ_p. Altogether we have shown that the standard basic sequences of ℓ_p and ℓ_2 are equivalent, which is obviously wrong.

In particular, the above arguments show that there is a sequence $(a_m)_{m\in\mathbb{N}}$ which converges with respect to $(h_{\pi(2m)})_{m\in\mathbb{N}}$ but not with respect to $(h_{2m+1})_{m\in\mathbb{N}}$. Now assume that A is $\mathcal{R}$-sectorial. Let $(q_m)_{m\in\mathbb{N}} \subset \mathbb{R}_-$ be a sequence to be chosen later. It follows from Proposition 3.16 that the operator $\mathcal{R}$: $\operatorname{Rad}(L_p([0,1])) \to \operatorname{Rad}(L_p([0,1]))$ associated to the sequence $(q_n)_{n\in\mathbb{N}}$ is bounded. We now show that

$$x = \sum_{m=1}^{\infty} a_m h_{\pi(2m)} r_m \tag{3.1}$$

converges in $\operatorname{Rad}(L_p([0,1]))$. Indeed, for some fixed $\omega \in [0,1]$ the infinite series

$$\sum_{m=1}^{\infty} a_m r_m(\omega) h_{\pi(2m)}$$

converges by the unconditionality of the basic sequence $(h_{\pi(2m)})_{m\in\mathbb{N}}$ as $r_m(\omega) \in \{-1,1\}$. Hence, the above series defines a measurable function as the pointwise limit of measurable functions. Moreover, if K denotes the unconditional constant of $(h_{\pi(2m)})_{m\in\mathbb{N}}$, one has for each $\omega \in [0,1]$

$$\left\| \sum_{m=1}^{\infty} r_m(\omega) a_m h_{\pi(2m)} \right\| \leq K \left\| \sum_{m=1}^{\infty} a_m h_{\pi(2m)} \right\|. \tag{3.2}$$

This shows that the series (3.1) is in $L_1([0,1]; L_p([0,1]))$. Using an analogous estimate as (3.2) one sees that the sequence of partial sums $\sum_{m=1}^{N} a_m h_{\pi(2m)} r_m$ converges to $\sum_{m=1}^{\infty} a_m h_{\pi(2m)} r_m$ in $\operatorname{Rad}(L_p([0,1]))$. We now apply $\mathcal{R}$ to x. Because of $h_{\pi(2m)} = f_{2m} - f_{2m-1}$ we obtain

$$\begin{aligned} g := \mathcal{R}(x) &= \mathcal{R}\left(\sum_{m=1}^{\infty} a_m (f_{2m} - f_{2m-1}) r_m \right) \\ &= \sum_{m=1}^{\infty} \frac{a_m q_m}{q_m - \gamma_{2m}} f_{2m} - \frac{a_m q_m}{q_m - \gamma_{2m-1}} f_{2m-1} \\ &= \sum_{m=1}^{\infty} \frac{a_m q_m}{q_m - \gamma_{2m}} (h_{\pi(2m)} + h_{2m-1}) - \frac{a_m q_m}{q_m - \gamma_{2m-1}} h_{2m-1} \end{aligned}$$

$$= \sum_{m=1}^{\infty} \frac{a_m q_m}{q_m - \gamma_{2m}} h_{\pi(2m)} + a_m q_m \left(\frac{1}{q_m - \gamma_{2m}} - \frac{1}{q_m - \gamma_{2m-1}} \right) h_{2m-1}.$$

We now want to choose $(q_m)_{m\in\mathbb{N}}$ in such a way that the last term in the bracket is big. Notice that if we set $\gamma_m = 2^m$, then by Lemma 3.17 for $t = \gamma_{2m}$ one has

$$t[(t+\gamma_{2m-1})^{-1} - (t+\gamma_{2m})^{-1}] = \frac{1}{6}.$$

Hence, for the choice $q_m = -\gamma_{2m}$ we obtain

$$\mathcal{R}(x) = \sum_{m=1}^{\infty} \frac{1}{2} a_m h_{\pi(2m)} - \frac{1}{6} a_m h_{2m-1}.$$

Then after choosing a subsequence (N_k) there exists a set $N \subset [0,1]$ of measure zero such that

$$\sum_{m=1}^{N_k} \frac{1}{2} a_m r_m(\omega) h_{\pi(2m)} - \frac{1}{6} a_m r_m(\omega) h_{2m-1} \xrightarrow[k\to\infty]{} g(\omega) \quad \text{for all } \omega \in N^c. \tag{3.3}$$

Applying the coordinate functionals for $(h_m)_{m\in\mathbb{N}}$ to (3.3) shows that for $\omega \in N^c$ the unique coefficients $(h_m^*(g(\omega)))$ of the expansion of $g(\omega)$ with respect to $(h_m)_{m\in\mathbb{N}}$ satisfy $h_{2m-1}^*(g(\omega)) = -\frac{a_m}{6} r_m(\omega)$. Since $(h_m)_{m\in\mathbb{N}}$ is unconditional

$$\sum\nolimits_{m=1}^{\infty} a_m r_m(\omega) h_{2m-1} \quad \text{and therefore} \quad \sum\nolimits_{m=1}^{\infty} a_m h_{2m-1} \quad \text{converge.}$$

This contradicts the choice of $(a_m)_{m\in\mathbb{N}}$ and therefore A cannot be $\mathcal{R}$-sectorial. □

Note that by taking the adjoint operators A^* of the above counterexamples one obtains counterexamples on the range $p \in (1,2)$. Further, the above argument works for every Banach space that admits an unconditional normalized non-symmetric basis [18]. This allows one to prove the following result by N.J. Kalton & G. Lancien [33].

Theorem 3.19. *Let X be a Banach space that admits an unconditional basis. Then every negative generator of an analytic semigroup is $\mathcal{R}$-sectorial if and only if X is isomorphic to a Hilbert space.*

Note that on $L_\infty([0,1])$ by a result of H.P. Lotz [45, Theorem 3] every negative generator of a C_0-semigroup is already bounded and therefore $\mathcal{R}$-sectorial. However, the following questions are open [31, p. 68].

Problem 8. Does Theorem 3.19 hold in the bigger class of all Banach spaces admitting a Schauder basis / of all separable Banach spaces?

For partial results in this direction see [34].

4. Counterexamples II: Using Monniaux' theorem

In this section we present an alternative method to construct counterexamples. This method is based on a theorem of S. Monniaux. We consider the following straightforward analogue of sectorial operators on strips. For details see [26, Ch. 4].

Definition 4.1. For $\omega > 0$ let $H_\omega := \{z \in \mathbb{C} : |\operatorname{Im} z| < \omega\}$ be the *horizontal strip* of height 2ω. A closed densely defined operator B is called a *strip type operator* of height $\omega > 0$ if $\sigma(B) \subset \overline{H_\omega}$ and

$$\sup\{\|R(\lambda, B)\| : |\operatorname{Im} \lambda| \geq \omega + \varepsilon\} < \infty \qquad \text{for all } \varepsilon > 0. \tag{H_ω}$$

Further, we define the *spectral height* of B as $\omega_{st}(B) := \inf\{\omega > 0 : (H_\omega) \text{ holds}\}$.

Recall that if A is a sectorial operator with bounded imaginary powers, then $t \mapsto A^{it}$ is a strongly continuous group. Conversely, one may ask which C_0-groups can be written in this form. The following theorem of S. Monniaux [48] gives a very satisfying answer to this question (for an alternative proof see [27, Section 4]).

Theorem 4.2. *Let X be a UMD-space. Then there is a one-to-one correspondence*

$$\left\{\begin{matrix} A \text{ sectorial operator with BIP and} \\ \omega_{\mathrm{BIP}}(A) < \pi \end{matrix}\right\} \underset{e^B}{\overset{\log A}{\longleftrightarrow}} \left\{\begin{matrix} B \text{ strip type operator with} \\ iB \sim C_0\text{-group of type} < \pi \end{matrix}\right\}.$$

Proof. For the surjectivity let B be a strip type operator such that iB generates a C_0-group $(U(t))_{t\in\mathbb{R}}$ of type $< \pi$. Then by Monniaux' theorem [48, Theorem 4.3] there exists a sectorial operator A with bounded imaginary powers such that $A^{it} = U(t)$ for all $t \in \mathbb{R}$. Moreover, $(U(t))_{t\in\mathbb{R}}$ is generated by $i \log A$. It then follows from the uniqueness of the generator that $B = \log A$.

For the injectivity assume that $\log A = \log B$ for two sectorial operators from the left-hand side. Then by [26, Corollary 4.2.5] one has $A = e^{\log A} = e^{\log B} = B$. ☐

Remark 4.3. In [25] M. Haase shows that for every strip type operator B with $\omega_{st}(B) < \pi$ such that iB generates a C_0-group $(U(t))_{t\in\mathbb{R}}$ of arbitrary type there exists a sectorial operator A with $A^{it} = U(t)$ for all $t \in \mathbb{R}$. If one chooses B as above such that $(U(t))_{t\in\mathbb{R}}$ has group type bigger than π (which is possible on some UMD-spaces) one sees that there exists a sectorial operator A with $\omega_{\mathrm{BIP}}(A) > \pi$. By taking suitable fractional powers of A one then obtains a sectorial operator $\tilde{A}$ with $\omega(\tilde{A}) < \omega_{\mathrm{BIP}}(\tilde{A}) < \pi$.

Because of the above results, for a moment, we restrict our attention to a UMD-space X. A particular class of sectorial operators with bounded imaginary powers are those with a bounded H^∞-calculus. Recall that a sectorial operator A on X with a bounded H^∞-calculus satisfies $\omega_R(A) = \omega_{\mathrm{BIP}}(A) = \omega_{H^\infty}(A)$ by Theorem 2.9. In particular one has $\omega_{\mathrm{BIP}}(A) < \pi$. For sectorial operators with a bounded H^∞-calculus one can formulate an analogous correspondence which essentially follows from an unpublished result of N.J. Kalton & L. Weis.

In the following for a C_0-group $(U(t))_{t\in\mathbb{R}}$ on some Banach space we call the infimum of those $\omega > 0$ for which $\mathcal{R}\{e^{-\omega|t|}U(t) : t \in \mathbb{R}\} < \infty$ the $\mathcal{R}$-group type of $(U(t))_{t\in\mathbb{R}}$.

Theorem 4.4. *Let X be a Banach space with Pisier's property (α). Then there is a one-to-one correspondence*

$$\left\{\begin{array}{c} A \text{ sectorial operator with bounded} \\ H^\infty\text{-calculus} \end{array}\right\} \underset{e^B}{\overset{\log A}{\longleftrightarrow}} \left\{\begin{array}{c} B \text{ strip type operator with } iB \sim \\ C_0\text{-group of } \mathcal{R}\text{-type} < \pi \end{array}\right\}.$$

Proof. Let A be a sectorial operator with a bounded H^∞-calculus. Then it follows from Theorem 2.8 and the fact that the norm of $\lambda \mapsto \lambda^{it}$ in $H^\infty(\Sigma_\sigma)$ is bounded by $\exp(|t|\,\sigma)$ for $t \in \mathbb{R}$ that $\{e^{-|t|\sigma}A^{it} : t \in \mathbb{R}\}$ is $\mathcal{R}$-bounded for all $\sigma \in (\omega_{H^\infty}(A), \pi)$. In particular $(A^{it})_{t\in\mathbb{R}}$ is of $\mathcal{R}$-type $< \pi$.

Conversely, let B be from the right-hand side. We now use results developed in [36]. It follows from [28, Theorem 6.5] (note that $\mathcal{R}$- is stronger than γ-boundedness) that the $\mathcal{R}$-type assumption implies that B has a bounded H^∞-calculus on some strip of height smaller than π. By [26, Proposition 5.3.3], the operator e^B is sectorial and has a bounded H^∞-calculus.

The one-to-one correspondence then follows as in Theorem 4.2. □

From the above theorems it follows immediately that on L_p for $p \in (1,\infty)\backslash\{2\}$ there exist sectorial operators with bounded imaginary powers which do not have a bounded H^∞-calculus.

Corollary 4.5. *Let $p \in (1,\infty) \setminus \{2\}$. Then there exists a sectorial operator A on $L_p(\mathbb{R})$ with $\omega(A) = \omega_{\mathrm{BIP}}(A) = 0$ which does not have a bounded H^∞-calculus.*

Proof. Let $(U(t))_{t\in\mathbb{R}}$ be the shift group on $L_p(\mathbb{R})$. It follows from the Khintchine inequality that $\{U(t) : t \in [0,1]\}$ is not $\mathcal{R}$-bounded [38, Example 2.12]. By Theorem 4.2 there exists a sectorial operator A with bounded imaginary powers such that $A^{it} = U(t)$ for all $t \in \mathbb{R}$. Then one has $\omega(A) \le \omega_{\mathrm{BIP}}(A) = 0$. However, by construction, A^{it} is not $\mathcal{R}$-bounded on $[0,1]$ and therefore Theorem 4.4 implies that A cannot have a bounded H^∞-calculus. □

Note that the constructed counterexample is exactly the same as in Example 3.8 which was obtained by different methods except for the fact that we worked in Example 3.8 with the periodic shift. Of course, we could have started with the same periodic shift in Corollary 4.5.

4.1. Some results on exotic Banach spaces

In this subsection we want to investigate shortly sectorial operators on exotic Banach spaces. In the past twenty years Banach spaces were constructed whose algebra of operators has an extremely different structure from those of the well-known classical Banach spaces. The most prominent examples are probably the hereditarily indecomposable Banach spaces.

Definition 4.6 (Hereditarily indecomposable Banach space (H.I.)). A Banach space X is called *indecomposable* if it cannot be written as the sum of two closed infinite-dimensional subspaces. Further X is called *hereditarily indecomposable* (*H.I.*) if every infinite-dimensional closed subspace of X is indecomposable.

It is a deep result of B. Maurey and T. Gowers that such (separable) spaces do actually exist [23]. We are now interested in the properties of C_0-semigroups on such spaces. We will use the following theorem proved in [53, Theorem 2.3].

Theorem 4.7. *Let X be a H.I. Banach space. Then every C_0-group on X has a bounded generator.*

The above result can be directly used to show the following result on operators with bounded imaginary powers.

Corollary 4.8. *Let A be a sectorial operator with bounded imaginary powers on a H.I. Banach space. Then A is bounded.*

Proof. Let A be as in the assertion. Note that $(A^{it})_{t\in\mathbb{R}}$ is a C_0-group with generator $i\log A$. By Theorem 4.7 $\log A$ is a bounded operator. This implies that $e^{\log A} = A$ is bounded. □

In particular on H.I. Banach spaces the structure of sectorial operators with a bounded H^∞-calculus is rather trivial.

Corollary 4.9. *Let A be an invertible sectorial operator on a H.I. Banach space. Then the following assertions are equivalent.*

(i) *A is a bounded operator.*
(ii) *A has bounded imaginary powers.*
(iii) *A has a bounded H^∞-calculus.*

Proof. The implication (i) $\Rightarrow$ (iii) can easily directly be verified and holds for every Banach space, (iii) $\Rightarrow$ (ii) also holds on every Banach space as discussed before and (ii) $\Rightarrow$ (i) follows from Corollary 4.8. □

Note that since every Banach space contains a basic sequence [1, Corollary 1.5.3], there exist H.I. Banach spaces that admit Schauder bases. Then by Proposition 3.5 on these spaces there exist semigroups with unbounded generators which cannot have bounded imaginary powers. In particular the structure of semigroups on these spaces is not trivial. We do not know how $\mathcal{R}$-sectoriality behaves in these spaces.

5. Counterexamples III: Pisier's counterexample to the Halmos problem

We now present a counterexample to the last implication left open, namely that there exists a C_0-semigroup with generator $-A$ and $\omega_{H^\infty}(A) = \frac{\pi}{2}$ which does not have a loose dilation. The key ingredient here is Pisier's counterexample to the Halmos problem [52] (for a more elementary approach see [11]). He constructed a

Hilbert space H and an operator $T \in \mathcal{B}(H)$ that is polynomially bounded, i.e., for some $K \geq 0$ one has

$$\|p(T)\| \leq K \sup_{|z|\leq 1} |p(z)|$$

for all polynomials p, but is not similar to a contraction, i.e., there does not exist any invertible $S \in \mathcal{B}(H)$ such that $S^{-1}TS$ is a contraction.

Theorem 5.1. *There exists a generator $-A$ of a C_0-semigroup $(T(t))_{t\geq 0}$ on some Hilbert space with $\omega_{H^\infty}(A) = \frac{\pi}{2}$ such that $(T(t))_{t\geq 0}$ does not have a loose dilation in the class of all Hilbert spaces.*

Proof. Let T and H be as above from Pisier's counterexample to the Halmos problem. It is explained in [40, Proposition 4.8] that the concrete structure of T allows one to define $A = (I+T)(I-T)^{-1}$ which turns out to be a sectorial operator with $\omega(A) = \frac{\pi}{2}$. Moreover, it is shown that $-A$ generates a bounded C_0-semigroup $(T(t))_{t\geq 0}$ on H. Further, it follows from the polynomial boundedness of T with a conformal mapping argument that A has a bounded H^∞-calculus with $\omega_{H^\infty}(A) = \frac{\pi}{2}$ [40, Remark 4.4]. Now assume that $(T(t))_{t\geq 0}$ has a loose dilation in the class of all Hilbert spaces. Then it follows from Dixmier's unitarization theorem [50, Theorem 9.3] that $(T(t))_{t\geq 0}$ has a loose dilation to a unitary C_0-group $(U(t))_{t\in\mathbb{R}}$ on some Hilbert space K, i.e., there exist bounded operators $J\colon H \to K$ and $Q\colon K \to H$ such that

$$T(t) = QU(t)J \qquad \text{for all } t \geq 0.$$

Now let $\mathcal{A}$ be the unital subalgebra of $L_\infty([0,\infty))$ generated by the functions $x \mapsto e^{-itx}$ for $t \geq 0$, where we identify elements in $L_\infty([0,\infty))$ with multiplication operators on the Hilbert space $L_2([0,\infty))$. This gives $\mathcal{A}$ the structure of an operator space. We now show that the algebra homomorphism

$$u\colon \mathcal{A} \to \mathcal{B}(H), \qquad e^{-it\cdot} \mapsto T(t)$$

is completely bounded with respect to this operator space structure for $\mathcal{A}$. Indeed, observe that by Stone's theorem on unitary groups and the spectral theorem for self-adjoint operators there exists a measure space Ω and a measurable function $m\colon \Omega \to \mathbb{R}$ such that after unitary equivalence $U(t)$ is the multiplication operator with respect to the function e^{-itm} for all $t \in \mathbb{R}$. Now for $n \in \mathbb{N}$ let $[f_{ij}] \in M_n(\mathcal{A})$ with $f_{ij} = \sum_{k=1}^N a_k^{(ij)} e^{-it_k\cdot}$. Then one has

$$\begin{aligned}\|u_n([f_{ij}])\|_{M_n(\mathcal{B}(X))} &= \left\| \Big[\sum_{k=1}^N a_k^{(ij)} T(t_k)\Big] \right\|_{M_n(\mathcal{B}(X))} \\ &= \left\| \Big[Q \sum_{k=1}^N a_k^{(ij)} U(t_k) J\Big] \right\|_{M_n(\mathcal{B}(X))} \leq \|Q\|\,\|J\| \left\| \Big[\sum_{k=1}^N a_k^{(ij)} e^{-it_k m}\Big] \right\|_{M_n(\mathcal{B}(L_2(\Omega)))} \\ &\leq \|J\|\,\|Q\| \sup_{x\in\mathbb{R}} \left\| \Big[\sum_{k=1}^N a_k^{(ij)} e^{-it_k x}\Big] \right\|_{M_n} = \|J\|\,\|Q\|\,\|[f_{ij}]\|_{M_n(L_\infty[0,\infty))}.\end{aligned}$$

Here we have used the identification of the C^*-algebras $M_n(L_\infty(\Omega)) \simeq L^\infty(\Omega; M_n)$ for all $n \in \mathbb{N}$. We deduce from Theorem [50, Theorem 9.1] that $(T(t))_{t\geq 0}$ is similar to a semigroup of contractions. However, since by construction T is the cogenerator of $(T(t))_{t\geq 0}$, this holds if and only if T is similar to a contraction [57, III,8]. This is a contradiction to our choice of T. $\square$

References

[1] F. Albiac and N.J. Kalton, *Topics in Banach space theory*, Graduate Texts in Mathematics, vol. 233, Springer, New York, 2006.

[2] W. Arendt and S. Bu, *The operator-valued Marcinkiewicz multiplier theorem and maximal regularity*, Math. Z. **240** (2002), no. 2, 311–343.

[3] ———, *Tools for maximal regularity*, Math. Proc. Cambridge Philos. Soc. **134** (2003), no. 2, 317–336.

[4] C. Arhancet, *On Matsaev's conjecture for contractions on noncommutative L^p-spaces*, J. Operator Theory **69** (2013), no. 2, 387–421.

[5] C. Arhancet and C. Le Merdy, *Dilation of Ritt operators on L^p-spaces*, Israel J. Math. **201** (2014), no. 1, 373–414.

[6] J.-B. Baillon and P. Clément, *Examples of unbounded imaginary powers of operators*, J. Funct. Anal. **100** (1991), no. 2, 419–434.

[7] E. Berkson and T.A. Gillespie, *On restrictions of multipliers in weighted settings*, Indiana Univ. Math. J. **52** (2003), no. 4, 927–961.

[8] D.L. Burkholder, *Martingales and singular integrals in Banach spaces*, Handbook of the geometry of Banach spaces, vol. I, North-Holland, Amsterdam, 2001, pp. 233–269.

[9] R.R. Coifman and G. Weiss, *Transference methods in analysis*, American Mathematical Society, Providence, R.I., 1976, Conference Board of the Mathematical Sciences Regional Conference Series in Mathematics, no. 31.

[10] M. Cowling, I. Doust, A. McIntosh, and A. Yagi, *Banach space operators with a bounded H^∞ functional calculus*, J. Austral. Math. Soc. Ser. A **60** (1996), no. 1, 51–89.

[11] K.R. Davidson and V.I. Paulsen, *Polynomially bounded operators*, J. Reine Angew. Math. **487** (1997), 153–170.

[12] R. Denk, M. Hieber, and J. Prüss, *$\mathcal{R}$-boundedness, Fourier multipliers and problems of elliptic and parabolic type*, Mem. Amer. Math. Soc. **166** (2003), no. 788, viii+114.

[13] G. Dore and A. Venni, *On the closedness of the sum of two closed operators*, Math. Z. **196** (1987), no. 2, 189–201.

[14] S.W. Drury, *A counterexample to a conjecture of Matsaev*, Linear Algebra Appl. **435** (2011), no. 2, 323–329.

[15] P. Enflo, *A counterexample to the approximation problem in Banach spaces*, Acta Math. **130** (1973), 309–317.

[16] M. Fabian, P. Habala, P. Hájek, V. Montesinos, and V. Zizler, *Banach space theory*, CMS Books in Mathematics/Ouvrages de Mathématiques de la SMC, Springer, New York, 2011, The basis for linear and nonlinear analysis.

[17] S. Fackler, *An explicit counterexample for the L^p-maximal regularity problem*, C. R. Math. Acad. Sci. Paris **351** (2013), no. 1-2, 53–56.

[18] ______, *The Kalton–Lancien theorem revisited: Maximal regularity does not extrapolate*, J. Funct. Anal. **266** (2014), no. 1, 121–138.

[19] ______, *On the structure of semigroups on L_p with a bounded H^∞-calculus*, Bull. Lond. Math. Soc. **46** (2014), no. 5, 1063–1076.

[20] G. Fendler, *Dilations of one parameter semigroups of positive contractions on L^p spaces*, Canad. J. Math. **49** (1997), no. 4, 736–748.

[21] R.J. Fleming and J.E. Jamison, *Isometries on Banach spaces: function spaces*, Chapman & Hall/CRC Monographs and Surveys in Pure and Applied Mathematics, vol. 129, Chapman & Hall/CRC, Boca Raton, FL, 2003.

[22] A.M. Fröhlich and L. Weis, *H^∞ calculus and dilations*, Bull. Soc. Math. France **134** (2006), no. 4, 487–508.

[23] W.T. Gowers and B. Maurey, *The unconditional basic sequence problem*, J. Amer. Math. Soc. **6** (1993), no. 4, 851–874.

[24] S. Guerre and Y. Raynaud, *Sur les isométries de $L^p(X)$ et le théorème ergodique vectoriel*, Canad. J. Math. **40** (1988), no. 2, 360–391.

[25] M. Haase, *Spectral properties of operator logarithms*, Math. Z. **245** (2003), no. 4, 761–779.

[26] ______, *The functional calculus for sectorial operators*, Operator Theory: Advances and Applications, vol. 169, Birkhäuser Verlag, Basel, 2006.

[27] ______, *Functional calculus for groups and applications to evolution equations*, J. Evol. Equ. **7** (2007), no. 3, 529–554.

[28] ______, *Transference principles for semigroups and a theorem of Peller*, J. Funct. Anal. **261** (2011), no. 10, 2959–2998.

[29] R. Hunt, Benjamin Muckenhoupt, and Richard Wheeden, *Weighted norm inequalities for the conjugate function and Hilbert transform*, Trans. Amer. Math. Soc. **176** (1973), 227–251.

[30] M. Junge and C. Le Merdy, *Dilations and rigid factorisations on noncommutative L^p-spaces*, J. Funct. Anal. **249** (2007), no. 1, 220–252.

[31] N.J. Kalton, *Applications of Banach space theory to sectorial operators*, Recent progress in functional analysis (Valencia, 2000), North-Holland Math. Stud., vol. 189, North-Holland, Amsterdam, 2001, pp. 61–74.

[32] ______, *A remark on sectorial operators with an H^∞-calculus*, Trends in Banach spaces and operator theory (Memphis, TN, 2001), Contemp. Math., vol. 321, Amer. Math. Soc., Providence, RI, 2003, pp. 91–99.

[33] N.J. Kalton and G. Lancien, *A solution to the problem of L^p-maximal regularity*, Math. Z. **235** (2000), no. 3, 559–568.

[34] ______, *L^p-maximal regularity on Banach spaces with a Schauder basis*, Arch. Math. (Basel) **78** (2002), no. 5, 397–408.

[35] N.J. Kalton and L. Weis, *Euclidean structures and their applications to spectral theory*, unpublished manuscript.

[36] ______, *The H^∞-Functional Calculus and Square Function Estimates*, unpublished manuscript on arXiv: 1411.0472.

[37] ———, *The H^∞-calculus and sums of closed operators*, Math. Ann. **321** (2001), no. 2, 319–345.

[38] P.C. Kunstmann and L. Weis, *Maximal L_p-regularity for parabolic equations, Fourier multiplier theorems and H^∞-functional calculus*, Functional analytic methods for evolution equations, Lecture Notes in Math., vol. 1855, Springer, Berlin, 2004, pp. 65–311.

[39] G. Lancien, *Counterexamples concerning sectorial operators*, Arch. Math. (Basel) **71** (1998), no. 5, 388–398.

[40] C. Le Merdy, *The similarity problem for bounded analytic semigroups on Hilbert space*, Semigroup Forum **56** (1998), no. 2, 205–224.

[41] ———, *H^∞-functional calculus and applications to maximal regularity*, Semigroupes d'opérateurs et calcul fonctionnel (Besançon, 1998), Publ. Math. UFR Sci. Tech. Besançon, vol. 16, Univ. Franche-Comté, Besançon, 1999, pp. 41–77.

[42] ———, *Square functions, bounded analytic semigroups, and applications*, Perspectives in operator theory, Banach Center Publ., vol. 75, Polish Acad. Sci., Warsaw, 2007, pp. 191–220.

[43] J. Lindenstrauss and L. Tzafriri, *Classical Banach spaces. I*, Springer-Verlag, Berlin, 1977, Sequence spaces, Ergebnisse der Mathematik und ihrer Grenzgebiete, vol. 92.

[44] ———, *Classical Banach spaces. II*, Ergebnisse der Mathematik und ihrer Grenzgebiete [Results in Mathematics and Related Areas], vol. 97, Springer-Verlag, Berlin, 1979, Function spaces.

[45] H.P. Lotz, *Uniform convergence of operators on L^∞ and similar spaces*, Math. Z. **190** (1985), no. 2, 207–220.

[46] A. McIntosh, *Operators which have an H_∞ functional calculus*, Miniconference on operator theory and partial differential equations (North Ryde, 1986), Proc. Centre Math. Anal. Austral. Nat. Univ., vol. 14, Austral. Nat. Univ., Canberra, 1986, pp. 210–231.

[47] A. McIntosh and A. Yagi, *Operators of type ω without a bounded H_∞ functional calculus*, Miniconference on Operators in Analysis (Sydney, 1989), Proc. Centre Math. Anal. Austral. Nat. Univ., vol. 24, Austral. Nat. Univ., Canberra, 1990, pp. 159–172.

[48] S. Monniaux, *A new approach to the Dore–Venni theorem*, Math. Nachr. **204** (1999), 163–183.

[49] M. Nielsen, *Trigonometric quasi-greedy bases for $L^p(\mathbb{T}; w)$*, Rocky Mountain J. Math. **39** (2009), no. 4, 1267–1278.

[50] V.I. Paulsen, *Completely bounded maps and operator algebras*, Cambridge Studies in Advanced Mathematics, vol. 78, Cambridge University Press, Cambridge, 2002.

[51] G. Pisier, *Some results on Banach spaces without local unconditional structure*, Compositio Math. **37** (1978), no. 1, 3–19.

[52] ———, *A polynomially bounded operator on Hilbert space which is not similar to a contraction*, J. Amer. Math. Soc. **10** (1997), no. 2, 351–369.

[53] F. Räbiger and W.J. Ricker, *C_0-groups and C_0-semigroups of linear operators on hereditarily indecomposable Banach spaces*, Arch. Math. (Basel) **66** (1996), no. 1, 60–70.

[54] J.L. Rubio de Francia, *Martingale and integral transforms of Banach space-valued functions*, Probability and Banach spaces (Zaragoza, 1985), Lecture Notes in Math., vol. 1221, Springer, Berlin, 1986, pp. 195–222.

[55] I. Singer, *Bases in Banach spaces. I*, Springer-Verlag, New York, 1970, Die Grundlehren der mathematischen Wissenschaften, Band 154.

[56] E.M. Stein, *Harmonic analysis: real-variable methods, orthogonality, and oscillatory integrals*, Princeton Mathematical Series, vol. 43, Princeton University Press, Princeton, NJ, 1993, With the assistance of Timothy S. Murphy, Monographs in Harmonic Analysis, III.

[57] B. Sz.-Nagy, C. Foias, H. Bercovici, and L. Kérchy, *Harmonic analysis of operators on Hilbert space*, enlarged ed., Universitext, Springer, New York, 2010.

[58] A. Venni, *A counterexample concerning imaginary powers of linear operators*, Functional analysis and related topics, 1991 (Kyoto), Lecture Notes in Math., vol. 1540, Springer, Berlin, 1993, pp. 381–387.

[59] L. Weis, *A new approach to maximal L_p-regularity*, Evolution equations and their applications in physical and life sciences (Bad Herrenalb, 1998), Lecture Notes in Pure and Appl. Math., vol. 215, Dekker, New York, 2001, pp. 195–214.

[60] ______, *Operator-valued Fourier multiplier theorems and maximal L_p-regularity*, Math. Ann. **319** (2001), no. 4, 735–758.

[61] A. Yagi, *Coïncidence entre des espaces d'interpolation et des domaines de puissances fractionnaires d'opérateurs*, C. R. Acad. Sci. Paris Sér. I Math. **299** (1984), no. 6, 173–176.

[62] F. Zimmermann, *On vector-valued Fourier multiplier theorems*, Studia Math. **93** (1989), no. 3, 201–222.

Stephan Fackler
Institute of Applied Analysis
University of Ulm
Helmholtzstr. 18
D-89069 Ulm, Germany
e-mail: stephan.fackler@uni-ulm.de

Operator Theory:
Advances and Applications, Vol. 250, 199–211

Global Existence Results for the Navier–Stokes Equations in the Rotational Framework in Fourier–Besov Spaces

Daoyuan Fang, Bin Han and Matthias Hieber

Dedicated to Charles Batty on the occasion of his 60th Birthday

Abstract. Consider the equations of Navier–Stokes in $\mathbb{R}^3$ in the rotational setting, i.e., with Coriolis force. It is shown that this set of equations admits a unique, global mild solution provided the initial data is small with respect to the norm of the Fourier–Besov space $\dot{FB}^{2-3/p}_{p,r}(\mathbb{R}^3)$, where $p \in (1,\infty]$ and $r \in [1,\infty]$. In the two-dimensional setting, a unique, global mild solution to this set of equations exists for *non-small* initial data $u_0 \in L^p_\sigma(\mathbb{R}^2)$ for $p \in [2,\infty)$.

Mathematics Subject Classification (2010). 35Q35,76D03,76D05.

Keywords. Navier–Stokes, rotational framework, global existence.

1. Introduction and main results

Consider the flow of an incompressible, viscous fluid in $\mathbb{R}^3$ in the rotational framework which is described by the following set of equations:

$$\begin{cases} \partial_t u + u \cdot \nabla u - \Delta u + \Omega e_3 \times u + \nabla \pi = 0, & \text{in } \mathbb{R}^3 \times (0,\infty), \\ \operatorname{div} u = 0, & \text{in } \mathbb{R}^3 \times (0,\infty), \\ u(0) = u_0, & \text{in } \mathbb{R}^3. \end{cases} \tag{1.1}$$

Here, u and π represent the velocity and pressure of the fluid, respectively, and $\Omega \in \mathbb{R}$ denotes the speed of rotation around the unit vector $e_3 = (0,0,1)$ in x_3-direction. If $\Omega = 0$, the system reduces to the classical Navier–Stokes system.

This set of equations recently gained quite some attention due to its importance in applications to geophysical flows. In particular, large scale atmospheric and oceanic flows are dominated by rotational effects, see, e.g., [23] or [11].

This work is supported partially by NSFC 11271322, 11331005.

If $\Omega = 0$, the classical Navier–Stokes equations have been considered by many authors in various scaling invariant spaces, in particular in

$$\dot{H}^{\frac{1}{2}}(\mathbb{R}^3) \hookrightarrow L^3(\mathbb{R}^3) \hookrightarrow B_{p,\infty}^{-1+\frac{3}{p}}(\mathbb{R}^3) \hookrightarrow BMO^{-1}(\mathbb{R}^3) \hookrightarrow B_{\infty,\infty}^{-1}(\mathbb{R}^3),$$

where $3 < p < \infty$. The space $BMO^{-1}(\mathbb{R}^3)$ is the largest scaling invariant space known for which equation (1.1) with $\Omega = 0$ is well posed.

It is a very remarkable fact that the equation (1.1) allows a global, mild solution for arbitrary large data in the L^2-setting provided the speed Ω of rotation is fast enough, see [2], [3] and [11]. More precisely, it was proved by Chemin, Desjardins, Gallagher and Grenier in [11] that for initial data $u_0 \in L^2(\mathbb{R}^2)^3 + H^{1/2}(\mathbb{R}^3)^3$ satisfying $\operatorname{div} u_0 = 0$, there exists a constant $\Omega_0 > 0$ such that for every $\Omega \geq \Omega_0$ the equation (1.1) admits a unique, global mild solution. The case of periodic initial data was considered before by Babin, Mahalov and Nicolaenko in the papers [2] and [3].

It is now a natural question to ask whether, for given and fixed $\Omega > 0$, there exists a unique, global mild solution to (1.1) provided the initial data is sufficiently small with respect to the above or related norms. In this context it is natural to extend the classical Fujita–Kato approach for the Navier–Stokes equations to the rotational setting. Hieber and Shibata considered in [18] the case of initial data belonging to $H^{\frac{1}{2}}(\mathbb{R}^3)$ and proved a global well-posedness result for (1.1) for initial data being small with respect to $H^{\frac{1}{2}}(\mathbb{R}^3)$. Generalizations of this result to the case of Fourier–Besov spaces are due to Konieczny and Yoneda [21] and Iwabuchi and Takada [19] and Koh, Lee and Takada [20].

More precisely, Konieczy and Yoneda proved the existence of a unique global mild solution to (1.1) for initial data u_0 being small with respect to the norm of $F\dot{B}_{p,\infty}^{2-\frac{3}{p}}(\mathbb{R}^3)$, where $1 < p \leq \infty$. For the case $p = 1$ considered in [19], the existence of a unique global mild solution was proved provided the initial data u_0 are small with respect to $F\dot{B}_{1,2}^{-1}(\mathbb{R}^3)$. Moreover, it was shown in [19] that the space $F\dot{B}_{1,2}^{-1}(\mathbb{R}^3)$ is critical for the well-posedness of system (1.1). In fact, it was shown in [19] that equation (1.1) is ill posed in $F\dot{B}_{1,q}^{-1}(\mathbb{R}^3)$ whenever $2 < q \leq \infty$ and $\Omega \in \mathbb{R}$.

Giga, Inui, Mahalov and Saal considered in [15] the problem of non-decaying initial data and obtained the uniform global solvability of (1.1) in the scaling invariant space $FM_0^{-1}(\mathbb{R}^3)$. For details, see [15] and [16]. Note that all of these results rely on good mapping properties of the Stokes–Coriolis semigroups on these function spaces.

It seems to be unknown, whether global existence results are also true for initial data u_0 being small with respect to $L^p(\mathbb{R}^3)$ for $p \geq 3$. The main difficulty here is that Mikhlin's theorem applied to the Stokes–Coriolis semigroup T yields an estimate of the form

$$\|T(t)f\|_{L^q} \leq M_p \Omega^2 t^2 \|f\|_{L^p}, \quad t \geq 1, \quad f \in L^p_\sigma(\mathbb{R}^3),$$

which is not suitable for fixed point arguments. For this and the definition of T we refer to Section 2 and [18]. Nevertheless, a global existence result for equation (1.1) was recently proved by Chen, Miao and Zhang in [12] for highly oscillating initial data in certain hybrid Besov spaces.

The aim of this paper is twofold: first we prove the existence of a unique, global mild solution to the above problem for initial data u_0 being *small* in the space $\dot{FB}^{2-\frac{3}{p}}_{p,r}(\mathbb{R}^3)$, where $1 < p \leq \infty$ and $1 \leq r \leq \infty$, hereby generalizing the result in [21] for $r = \infty$ to the case $1 \leq r \leq \infty$. We note that Iwabuchi and Takada [19] recently proved the well-posedness of (1.1) for data being small with respect to the norm of $\dot{FB}^{-1}_{1,2}(\mathbb{R}^3)$.

Secondly, considering the two-dimensional situation in the L^p-setting, we prove that (1.1) admits a unique, global mild solution $u \in C([0,\infty); L^p(\mathbb{R}^2))$ for arbitrary $u_0 \in L^p_\sigma(\mathbb{R}^2)$ provided $2 \leq q < \infty$. Our argument is based on applying the curl operator to equation (1.1). The resulting *vorticity equation* allows then for a global estimate in two dimensions which can used to control the term ∇u in the L^p-norm.

In order to formulate our first result, let us recall the definition of Fourier–Besov spaces. To this end, let φ be a C^∞ function satisfying supp $\varphi \subset \{3/4 \leq |\xi| \leq 8/3\}$ and

$$\sum_{k\in\mathbb{Z}} \varphi(2^{-k}\xi) = 1, \quad \xi \in \mathbb{R}^n \backslash \{0\}.$$

For $k \in \mathbb{Z}$, set $\varphi_k(\xi) = \varphi(2^{-k}\xi)$ and $h_k = \mathscr{F}^{-1}\varphi_k$. For $s \in \mathbb{R}$ and $1 \leq p, r \leq \infty$, the space $\dot{FB}^{s}_{p,r}(\mathbb{R}^3)$ is defined to be the set of all $f \in \mathcal{S}'(\mathbb{R}^3)$ such that $\hat{f} \in L^1_{\mathrm{loc}}(\mathbb{R}^3)$ and

$$\|f\|_{\dot{FB}^s_{p,r}} := \left\| \{2^{js} \|\varphi_j \hat{f}\|_{L^p(\mathbb{R}^3)}\}_{j\in\mathbb{Z}} \right\|_{l^r} < \infty.$$

Given $1 \leq q \leq \infty$ and $T \in (0,\infty]$, we also make use of Chemin–Lerner type spaces $\tilde{L}^q([0,T); \dot{FB}^{s}_{p,r}(\mathbb{R}^3))$, which are defined to be the completion of $C([0,T]; \mathcal{S}(\mathbb{R}^3))$ with respect to the norm

$$\|f\|_{\tilde{L}^q([0,T);\dot{FB}^s_{p,r}(\mathbb{R}^3))} := \left\| \{2^{js} \|\varphi_j \hat{f}\|_{L^q([0,T];L^p(\mathbb{R}^3))}\}_{j\in\mathbb{Z}} \right\|_{l^r}.$$

We are now in the position to state our first result.

Theorem 1.1. *Let $\Omega \in \mathbb{R}$ and $1 < p \leq \infty$, $1 \leq r \leq \infty$. Then there exist constants $C > 0$ and $\varepsilon > 0$, independent of Ω, such that for every $u_0 \in \dot{FB}^{2-\frac{3}{p}}_{p,r}(\mathbb{R}^3)$ satisfying div $u_0 = 0$ and $\|u_0\|_{\dot{FB}^{2-\frac{3}{p}}_{p,r}} \leq \varepsilon$, the equation (1.1) admits a unique, global mild solution $u \in X$, where X is given by*

$$X = \{u \in C([0,\infty); \dot{FB}^{2-\frac{3}{p}}_{p,r}(\mathbb{R}^3)) : \|u\|_X \leq C\varepsilon, \operatorname{div} u = 0\}$$

with

$$\|u\|_X = \|u\|_{\tilde{L}^\infty([0,\infty);\dot{FB}^{2-\frac{3}{p}}_{p,r}(\mathbb{R}^3))} + \|u\|_{\tilde{L}^1([0,\infty);\dot{FB}^{4-\frac{3}{p}}_{p,r}(\mathbb{R}^3))}.$$

Remarks 1.2.

a) Observe that due to the results in [19], the above system (1.1) is ill posed provided $p = 1$ and $r > 2$.
b) Note that the case $r = \infty$ coincides with the result of Konieczny and Yoneda in [21].
c) Iwabuchi and Takada [19] recently proved the existence of a unique, global mild solution to equation (1.1) for initial data small with respect to the norm of $F\dot{B}^{-1}_{1,2}$.
d) Note that neither $F\dot{B}^{-1}_{1,2}(\mathbb{R}^3) \subset F\dot{B}^{2-3/p}_{p,r}(\mathbb{R}^3)$ for $r \in [1,\infty]$ nor $F\dot{B}^{2-3/p}_{p,r}(\mathbb{R}^3) \subset F\dot{B}^{-1}_{1,2}(\mathbb{R}^3)$ for $r > 2$.

Our second results concerns the two-dimensional setting. The rotational term given above by $\Omega e_3 \times u$ is replaced in the two-dimensional situation by the term $\Omega u^\perp$. We hence consider the set of equations

$$\begin{cases} \partial_t u + u\cdot\nabla u - \Delta u + \Omega u^\perp + \nabla\pi = 0, & \text{in } \mathbb{R}^2 \times (0,\infty), \\ \operatorname{div} u = 0, & \text{in } \mathbb{R}^2 \times (0,\infty), \\ u(0) = u_0, & \text{in } \mathbb{R}^2. \end{cases} \tag{1.2}$$

We denote by $L^p_\sigma(\mathbb{R}^2)$ the solenoidal subspace of $L^p(\mathbb{R}^2)$. Our second result concerning *non-small data* in the $L^p(\mathbb{R}^2)$-setting reads as follows.

Theorem 1.3. *Let* $2 \le p < \infty$ *and* $u_0 \in L^p_\sigma(\mathbb{R}^2)$. *Then equation* (1.2) *admits a unique, global mild solution* $u \in C([0,\infty), L^p_\sigma(\mathbb{R}^2))$.

2. Linear and bilinear estimates

We start this section by considering the linear Stokes problem with Coriolis force

$$\begin{cases} \partial_t u - \Delta u + \Omega e_3 \times u + \nabla\pi = 0, & \text{in } \mathbb{R}^3 \times (0,\infty), \\ \operatorname{div} u = 0, & \text{in } \mathbb{R}^3 \times (0,\infty), \\ u(0) = u_0, & \text{in } \mathbb{R}^3. \end{cases} \tag{2.1}$$

It was shown in [18] that the solution of (2.1) is given by the Stokes–Coriolis semigroup T, which has the explicit representation

$$T(t)f := \mathcal{F}^{-1}\Big[\cos\Big(\Omega\frac{\xi_3}{|\xi|}t\Big)e^{-|\xi|^2 t}\mathrm{Id}\hat{f}(\xi) + \sin\Big(\Omega\frac{\xi_3}{|\xi|}t\Big)e^{-\xi|^2 t}R(\xi)\hat{f}(\xi)\Big], \quad t > 0, \tag{2.2}$$

for divergence free vector fields $f \in \mathcal{S}(\mathbb{R}^3)$. Here Id is the identity matrix in $\mathbb{R}^3$ and $R(\xi)$ is the skew symmetric matrix defined by

$$R(\xi) := \frac{1}{|\xi|}\begin{pmatrix} 0 & \xi_3 & -\xi_2 \\ -\xi_3 & 0 & \xi_1 \\ \xi_2 & -\xi_1 & 0 \end{pmatrix}, \quad \xi \in \mathbb{R}^3\backslash\{0\}.$$

For more information on C_0-semigroups we refer, e.g., to [1] or [5]. In order to solve equation (1.1), consider the integral equation

$$\Phi(u) := T(t)u_0 - \int_0^t T(t-\tau)\mathbb{P}\mathrm{div}(u\otimes u)(\tau)d\tau,$$

where $\mathbb{P} := (\delta_{ij} + R_iR_j)_{1\le i,j\le 3}$ denotes the Helmholtz projection from $L^p(\mathbb{R}^3)$ onto its divergence free vector fields. Here R_i denotes the Riesz transforms for $i = 1,2,3$. Since the Riesz transforms R_i are bounded operators on $\dot{FB}^s_{p,q}$ for all values of $p,q\in[1,\infty]$ and $s\in\mathbb{R}$, we see that $\mathbb{P}$ defines a bounded operators also on these spaces.

Our first estimate concerns the above convolution integral.

Lemma 2.1. *Let $1\le p,q,a,r\le\infty$, $s\in\mathbb{R}$ and $f\in L^a([0,\infty);\dot{FB}^s_{p,r}(\mathbb{R}^3))$. Then there exists a constant $C>0$ such that*

$$\Big\|\int_0^t T(t-\tau)f(\tau)d\tau\Big\|_{\widetilde{L}^q([0,\infty);\dot{FB}^s_{p,r}(\mathbb{R}^3))} \le C\|f\|_{\widetilde{L}^a([0,\infty);\dot{FB}^{s-2-\frac{2}{q}+\frac{2}{a}}_{p,r}(\mathbb{R}^3))}.$$

Proof. By the definition of the norm of $\widetilde{L}^q([0,\infty);\dot{FB}^s_{p,r}(\mathbb{R}^3))$, and by Young's inequality

$$\begin{aligned}
&\Big\|\int_0^t T(t-\tau)f(\tau)d\tau\Big\|_{\widetilde{L}^q([0,\infty);\dot{FB}^s_{p,r}(\mathbb{R}^3))}\\
&= \Big(\sum_k 2^{ksr}\Big(\int_0^\infty \Big\|\int_0^t \mathcal{F}(T(t-\tau)f)(\tau)d\tau\cdot\varphi_k\Big\|_{L^p}^q dt\Big)^{\frac{r}{q}}\Big)^{\frac{1}{r}}\\
&\le \Big(\sum_k 2^{ksr}\Big(\int_0^\infty \Big(\int_0^t e^{-(t-\tau)\cdot 2^{2k}}\|\hat{f}(\tau)\cdot\varphi_k\|_{L^p}d\tau\Big)^q dt\Big)^{\frac{r}{q}}\Big)^{\frac{1}{r}}\\
&\le \Big(\sum_k 2^{ksr}\Big(\int_0^\infty e^{-t\cdot 2^{2k}\tilde{q}}dt\Big)^{\frac{r}{\tilde{q}}}\Big(\int_0^\infty \|\hat{f}(\tau)\cdot\varphi_k\|_{L^p}^a\Big)dt\Big)^{\frac{r}{a}}\Big)^{\frac{1}{r}},
\end{aligned}$$

where $\tilde{q}$ satisfies $1+\frac{1}{q} = \frac{1}{\tilde{q}}+\frac{1}{a}$. We hence obtain

$$\begin{aligned}
&\Big\|\int_0^t T(t-\tau)f(\tau)d\tau\Big\|_{\widetilde{L}^q([0,\infty);\dot{FB}^s_{p,r}(\mathbb{R}^3))}\\
&\le C\Big(\sum_k 2^{ksr}2^{-2k(1+\frac{1}{q}-\frac{1}{a})r}\|\hat{f}\cdot\varphi_k\|^r_{L^a([0,\infty);L^p)}\Big)^{\frac{1}{r}}\\
&\le C\|f\|_{\widetilde{L}^a([0,\infty);\dot{FB}^{s-2-\frac{2}{q}+\frac{2}{a}}_{p,r}(\mathbb{R}^3))}.
\end{aligned}$$

□

Lemma 2.2. *Let $1<p\le\infty$ and $1\le p,r\le\infty$ and assume that $-1<s<3-\frac{3}{p}$. Set*

$$Y := \widetilde{L}^\infty([0,\infty);\dot{FB}^s_{p,r}(\mathbb{R}^3))\cap\widetilde{L}^1([0,\infty);\dot{FB}^{4-\frac{3}{p}}_{p,r}(\mathbb{R}^3)).$$

Then there exists a constant $C>0$ such that

$$\|uv\|_{\widetilde{L}^1([0,\infty);\dot{FB}^{s+1}_{p,r}(\mathbb{R}^3))} \le C\|u\|_Y\|v\|_Y.$$

Proof. Let φ and h_k be defined as in Section 1 for $k \in \mathbb{Z}$. Define the homogeneous dyadic blocks $\dot{\Delta}_k$ by

$$\dot{\Delta}_k u := \varphi(2^{-k}D)u = \int_{\mathbb{R}^N} h_k(y)u(x-y)dy, \quad k \in \mathbb{Z},$$

and for $j \in \mathbb{Z}$, set $\dot{S}_j u := \sum_{k=-\infty}^{j} \dot{\Delta}_k u$. We then obtain

$$\|uv\|_{\widetilde{L}^1([0,\infty);\dot{FB}^{s+1}_{p,r}(\mathbb{R}^3))} = \Big(\sum_j 2^{j(s+1)r}\Big(\int_0^\infty \|\widehat{\dot{\Delta}_j(uv)}\|_{L^p}dt\Big)^r\Big)^{\frac{1}{r}}. \tag{2.3}$$

Using the Bony decomposition (see, e.g., [6], [9] and [4, Section 2.8.1]), we rewrite $\dot{\Delta}_j(uv)$ as

$$\begin{aligned}\dot{\Delta}_j(uv) &= \sum_{|k-j|\le 4} \dot{\Delta}_j(\dot{S}_{k+1}u\dot{\Delta}_k v) + \sum_{|k-j|\le 4} \dot{\Delta}_j(\dot{S}_{k+1}v\dot{\Delta}_k u) + \sum_{k\ge j-2} \dot{\Delta}_j(\dot{\Delta}_k u\tilde{\dot{\Delta}}_k v)\\ &=: I + II + III. \end{aligned}\tag{2.4}$$

Then, by triangle inequalities in $L^p(\mathbb{R}^3)$ and $l^r(\mathbb{Z})$, we have

$$\begin{aligned}&\|uv\|_{\widetilde{L}^1([0,\infty);\dot{FB}^{s+1}_{p,r}(\mathbb{R}^3))}\\ &\le \Big(\sum_j 2^{j(s+1)r}\Big(\int_0^\infty \|\widehat{\dot{\Delta}_j I}\|_{L^p}dt\Big)^r\Big)^{\frac{1}{r}} + \Big(\sum_j 2^{j(s+1)r}\Big(\int_0^\infty \|\widehat{\dot{\Delta}_j II}\|_{L^p}dt\Big)^r\Big)^{\frac{1}{r}}\\ &\quad + \Big(\sum_j 2^{j(s+1)r}\Big(\int_0^\infty \|\widehat{\dot{\Delta}_j III}\|_{L^p}dt\Big)^r\Big)^{\frac{1}{r}}\\ &=: J_1 + J_2 + J_3.\end{aligned}$$

For the term J_1, we note

$$J_1 = \Big(\sum_j 2^{j(s+1)r}\Big(\int_0^\infty \|\sum_{|k-j|\le 4} \dot{\Delta}_j(\widehat{\dot{S}_{k+1}u\dot{\Delta}_k v})\|_{L^p}dt\Big)^r\Big)^{\frac{1}{r}}.$$

For fixed j, Lemma 2.1 yields

$$\begin{aligned}&2^{j(s+1)}\int_0^\infty \Big\|\sum_{|k-j|\le 4} \dot{\Delta}_j(\widehat{\dot{S}_{k+1}u\dot{\Delta}_k v})\Big\|_{L^p} dt\\ &\le 2^{j(s+1)}\int_0^\infty \sum_{|k-j|\le 4} \|\varphi_j(\chi_k\hat{u} * \varphi_k\hat{v})\|_{L^p}dt\\ &\le 2^{j(s+1)}\int_0^\infty \sum_{|k-j|\le 4} \|\chi_k\hat{u}\|_{L^1}\|\varphi_k\hat{v})\|_{L^p}dt\\ &\le 2^{j(s+1)}\int_0^\infty \sum_{|k-j|\le 4}\sum_{k'\le k} \|\varphi_{k'}\hat{u}\|_{L^p}2^{k'(3-\frac{3}{p})}\|\varphi_k\hat{v})\|_{L^p}dt\end{aligned}$$

$$\leq 2^{j(s+1)} \int_0^\infty \sum_{|k-j|\leq 4} \Big(\sum_{k'\leq k} \|\varphi_{k'}\hat{u}\|_{L^p}^r 2^{k'sr}\Big)^{\frac{1}{r}} \Big(\sum_{k'\leq k} 2^{k'r'(3-\frac{3}{p}-s)}\Big)^{\frac{1}{r'}} \|\varphi_k \hat{v})\|_{L^p} dt$$

$$\leq C\|u\|_{\widetilde{L}^\infty([0,\infty);\dot{FB}^s_{p,r})} 2^{j(s+1)} \int_0^\infty \sum_{|k-j|\leq 4} 2^{k(3-\frac{3}{p}-s)} \|\varphi_k \hat{v})\|_{L^p} dt.$$

Hence, by Young's inequality,

$$J_1 \leq C\|u\|_{\widetilde{L}^\infty([0,\infty);\dot{FB}^s_{p,r})} \Big(\sum_j \Big(\sum_{|k-j|\leq 4} 2^{k(4-\frac{3}{p})} 2^{(j-k)(s+1)} \|\varphi_k \hat{v})\|_{L^1([0,\infty);L^p)}\Big)^r\Big)^{\frac{1}{r}}$$

$$\leq C\|u\|_{\widetilde{L}^\infty([0,\infty);\dot{FB}^s_{p,r})} \|v\|_{\widetilde{L}^1([0,\infty);\dot{FB}^{4-\frac{3}{p}}_{p,r})}.$$

The term J_2 is estimated in the same way as J_1. In fact,

$$J_2 \leq C\|v\|_{\widetilde{L}^\infty([0,\infty);\dot{FB}^s_{p,r})} \Big(\sum_j \Big(\sum_{|k-j|\leq 4} 2^{k(4-\frac{3}{p})} 2^{(j-k)(s+1)} \|\varphi_k \hat{u})\|_{L^1([0,\infty);L^p)}\Big)^r\Big)^{\frac{1}{r}}$$

$$\leq C\|v\|_{\widetilde{L}^\infty([0,\infty);\dot{FB}^s_{p,r})} \|u\|_{\widetilde{L}^1([0,\infty);\dot{FB}^{4-\frac{3}{p}}_{p,r})}.$$

Finally, we focus on the third term J_3. As in the estimate to J_1, for fixed j, we obtain

$$2^{j(s+1)} \int_0^\infty \Big\| \sum_{k\geq j-2} \sum_{|k-k'|\leq 1} \varphi_j(\varphi_k \hat{u} * \varphi_{k'}\hat{v})\Big\|_{L^p} dt$$

$$\leq 2^{j(s+1)} \int_0^\infty \sum_{k\geq j-2} \sum_{|k-k'|\leq 1} \|\varphi_{k'}\hat{v})\|_{L^p} \|\varphi_k \hat{u}\|_{L^p} 2^{k(3-\frac{3}{p})} dt$$

$$\leq C 2^{j(s+1)} \int_0^\infty \sum_{k\geq j-2} \Big(\sum_{|k-k'|\leq 1} \|\varphi_{k'}\hat{v})\|_{L^p}^r 2^{k'rs}\Big)^{\frac{1}{r}} 2^{-ks} 2^{k(3-\frac{3}{p})} \|\varphi_k \hat{u}\|_{L^p} dt$$

$$\leq C\|v\|_{\widetilde{L}^\infty([0,\infty);\dot{FB}^s_{p,r})} \sum_{k\geq j-2} 2^{(j-k)(s+1)} 2^{k(4-\frac{3}{p})} \int_0^\infty \|\varphi_k \hat{u}\|_{L^p} dt.$$

Thus, by Young's inequality

$$J_3 \leq C\|v\|_{\widetilde{L}^\infty([0,\infty);\dot{FB}^s_{p,r})} \|u\|_{\widetilde{L}^1([0,\infty);\dot{FB}^{4-\frac{3}{p}}_{p,r})}$$

since $s > -1$.

Summing up, we see that

$$\|uv\|_{\widetilde{L}^1([0,\infty);\dot{FB}^{s+1}_{p,r}(\mathbb{R}^3))} \leq C\|u\|_Y \|v\|_Y. \qquad \square$$

We conclude this section with the following Lemma.

Lemma 2.3. [24, Section 1.3.2] *Let $1 \leq p \leq q \leq \infty$, $0 < r < R < \infty$, $j \in \mathbb{Z}$ and $n \in \mathbb{N}$. Let $f \in \mathcal{S}(\mathbb{R}^n)$ and denote by $\hat{f}$ its Fourier transform. Then, for any multiindex $\gamma \in \mathbb{N}^n$, there exists a constant $C > 0$ such that*

a) *If* $\mathrm{supp}\hat{f} \subset \{\xi \in \mathbb{R}^n : |\xi| \le R\,2^j\}$, *then*

$$\|(i\cdot)^\gamma \hat{f}\|_{L^q(\mathbb{R}^n)} \le C 2^{j|\gamma| + nj(\frac{1}{p}-\frac{1}{q})} \|\hat{f}\|_{L^p(\mathbb{R}^n)}.$$

b) *If* $\mathrm{supp}\hat{f} \subset \{\xi \in \mathbb{R}^n : r\,2^j \le |\xi| \le R\,2^j\}$, *then*

$$\|\hat{f}\|_{L^q(\mathbb{R}^n)} \le C 2^{-j|\gamma|} \sup_{|\beta|=|\gamma|} \|(i\cdot)^\beta \hat{f}\|_{L^q(\mathbb{R}^n)}.$$

3. Proof of Theorem 1.1

For the proof of Theorem 1.1 we make use of the following standard fixed point result, see, e.g., [7] [22],[10] or [8]. For a detailed proof, we refer, e.g., to [22, Theorem 13.2] or [7].

Proposition 3.1. *Let X be a Banach space and $B : X \times X \to X$ be a bounded bilinear form satisfying $\|B(x_1, x_2)\|_X \le \eta \|x_1\|_X \|x_2\|_X$ for all $x_1, x_2 \in X$ and a constant $\eta > 0$. Then, if $0 < \varepsilon < \frac{1}{4\eta}$ and if $a \in X$ such that $\|a\|_X \le \varepsilon$, the equation $x = a + B(x, x)$ has a solution in X such that $\|x\|_X \le 2\varepsilon$. This solution is the only one in the ball $\overline{B}(0, 2\varepsilon)$. Moreover, the solution depends continuously on a in the following sense: if $\|\tilde{a}\|_X \le \varepsilon$, $\tilde{x} = \tilde{a} + B(\tilde{x}, \tilde{x})$, and $\|\tilde{x}\|_X \le 2\varepsilon$, then*

$$\|x - \tilde{x}\|_X \le \frac{1}{1 - 4\eta\varepsilon} \|a - \tilde{a}\|_X.$$

In the following, we choose an underlying Banach space X given by

$$X := \widetilde{L}^\infty([0,\infty); \dot{FB}^{2-\frac{3}{p}}_{p,r}(\mathbb{R}^3)) \cap \widetilde{L}^1([0,\infty); \dot{FB}^{4-\frac{3}{p}}_{p,r}(\mathbb{R}^3)),$$

and recall that Φ was defined by

$$\Phi(u) = T(t)u_0 - \int_0^t T(t-\tau)\mathbb{P}\mathrm{div}(u \otimes u)(\tau)d\tau.$$

We estimate first the term $T(t)u_0$.

Lemma 3.2. *Let $p, r \in [1, \infty]$, $s = 2 - 3/p$ and $u_0 \in \dot{FB}^s_{p,r}(\mathbb{R}^3)$. Then there exists a constant $C > 0$ such that*

$$\|T(t)u_0\|_{\widetilde{L}^\infty([0,\infty); \dot{FB}^s_{p,r})} \le C \|u_0\|_{\dot{FB}^s_{p,r}}, \quad t > 0, \tag{3.1}$$

$$\|T(t)u_0\|_{\widetilde{L}^1([0,\infty); \dot{FB}^{s+2}_{p,r})} \le C \|u_0\|_{\dot{FB}^s_{p,r}}, \quad t > 0. \tag{3.2}$$

Proof. We prove first estimate (3.1). By the definition of the norm, we have

$$\begin{aligned} \|T(t)u_0\|_{\widetilde{L}^\infty([0,\infty); \dot{FB}^{2-\frac{3}{p}}_{p,r})} &\le \Big(\sum_k 2^{k(2-\frac{3}{p})r} \sup_{t\in[0,\infty)} \|\varphi_k \widehat{T(t)u_0}\|^r_{L^p} \Big)^{\frac{1}{r}} \\ &\le C \|u_0\|_{\dot{FB}^{2-\frac{3}{p}}_{p,r}}. \end{aligned}$$

In order to prove the second estimate (3.2) above, note that

$$\|T(t)u_0\|_{\widetilde{L}^1([0,\infty);\dot{FB}_{p,r}^{4-\frac{3}{p}})} \leq \Big(\sum_k 2^{k(4-\frac{3}{p})r}\Big(\int_0^\infty e^{-t2^{2k}}\|\varphi_k\hat{u}_0\|_{L^p}dt\Big)^r\Big)^{\frac{1}{r}}$$
$$\leq C\|u_0\|_{\dot{FB}_{p,r}^{2-\frac{3}{p}}}. \qquad \square$$

We next consider the bilinear operator B given by

$$B(u,v) := \int_0^t T(t-\tau)\mathbb{P}\mathrm{div}(u\otimes v)d\tau.$$

By Lemma 2.1, Lemma 2.3 and Lemma 2.2 with $s = 2-\frac{3}{p}$, we obtain

$$\begin{aligned}\|B(u,v)\|_{\widetilde{L}^1([0,\infty);\dot{FB}_{p,r}^{4-\frac{3}{p}}(\mathbb{R}^3))} &= \Big\|\int_0^t T(t-\tau)\mathbb{P}\mathrm{div}(u\otimes v)d\tau\Big\|_{\widetilde{L}^1([0,\infty);\dot{FB}_{p,r}^{4-\frac{3}{p}}(\mathbb{R}^3))}\\ &\leq C\|\mathrm{div}(u\otimes v)\|_{\widetilde{L}^1([0,\infty);\dot{FB}_{p,r}^{2-\frac{3}{p}})}\\ &\leq C\|uv\|_{\widetilde{L}^1([0,\infty);\dot{FB}_{p,r}^{3-\frac{3}{p}})}\\ &\leq C\|u\|_X\|v\|_X.\end{aligned}$$

Similarly,

$$\begin{aligned}\|B(u,v)\|_{\widetilde{L}^\infty([0,\infty);\dot{FB}_{p,r}^{2-\frac{3}{p}}(\mathbb{R}^3))} &= \Big\|\int_0^t T(t-\tau)\mathbb{P}\mathrm{div}(u\otimes v)d\tau\Big\|_{\widetilde{L}^\infty([0,\infty);\dot{FB}_{p,r}^{2-\frac{3}{p}}(\mathbb{R}^3))}\\ &\leq C\|\mathrm{div}(u\otimes v)\|_{\widetilde{L}^1([0,\infty);\dot{FB}_{p,r}^{2-\frac{3}{p}})}\\ &\leq C\|uv\|_{\widetilde{L}^1([0,\infty);\dot{FB}_{p,r}^{3-\frac{3}{p}})}\\ &\leq C\|u\|_X\|v\|_X.\end{aligned}$$

Thus, combining these estimates with Lemma 3.2 yields

$$\|\Phi(u)\|_X \leq C\|u_0\|_{\dot{FB}_{p,r}^{2-\frac{3}{p}}} + 4C\varepsilon^2,$$

as well as

$$\|\Phi(u)-\Phi(v)\|_X \leq C(\|u\|_X+\|v\|_X)\|u-v\|_X.$$

Choosing now $\varepsilon \leq \frac{1}{8C}$, for every $u_0 \in \dot{FB}_{p,r}^{2-\frac{3}{p}}(\mathbb{R}^3)$ with $\|u_0\|_{\dot{FB}_{p,r}^{2-\frac{3}{p}}} \leq \frac{\varepsilon}{C}$, we finally obtain

$$\|\Phi(u)\|_X \leq 2\varepsilon$$

and

$$\|\Phi(u)-\Phi(v)\|_X \leq \frac{1}{2}\|u-v\|_X.$$

Applying Proposition 3.1 to the given situation completes the proof of Theorem 1.1.

4. Global existence for non-small data in $L^p_\sigma(\mathbb{R}^2)$

In this section we consider equation (1.1) in the two-dimensional setting and in the case where the initial data u_0 belong to $L^p_\sigma(\mathbb{R}^2)$ for $p > 2$. To this end, we note first that the equations of Navier–Stokes with Coriolis force are equivalent to the Navier–Stokes equations with linearly growing initial data. Indeed, we may rewrite equation (1.1) for a two-dimensional rotating fluid as

$$\begin{cases} \partial_t u - \Delta u + u \cdot \nabla u - 2Mu + \nabla \pi = 0, & y \in \mathbb{R}^2,\ t > 0, \\ \text{div } u = 0, & y \in \mathbb{R}^2,\ t > 0, \\ u(0) = u_0, & y \in \mathbb{R}^2, \end{cases} \tag{4.1}$$

where M is given by

$$M = -\frac{\Omega}{2}\begin{pmatrix} 0 & -1 \\ 1 & 0 \end{pmatrix}.$$

Then, by the change of variables $x = e^{-tM}y$ and by setting

$$v(t,x) := e^{-tM}u(t, e^{tM}x), \qquad q(t,x) := \pi(t, e^{tM}x),$$

we obtain the following set of equations for v:

$$\begin{cases} \partial_t v - \Delta v + v \cdot \nabla v - Mx \cdot \nabla v - Mv + \nabla q = 0, & x \in \mathbb{R}^2,\ t > 0, \\ \text{div } v = 0, & x \in \mathbb{R}^2,\ t > 0, \\ v(0) = u_0, & x \in \mathbb{R}^2. \end{cases} \tag{4.2}$$

These are the usual equations of Navier–Stokes with linearly growing initial data. Indeed, setting $U = v - Mx$, we have

$$\begin{cases} \partial_t U - \Delta U + U \cdot \nabla U + \nabla \tilde{\pi} = 0, & x \in \mathbb{R}^2,\ t > 0, \\ \text{div } U = 0, & x \in \mathbb{R}^2,\ t > 0, \\ U(0) = u_0 - Mx, & x \in \mathbb{R}^2, \end{cases} \tag{4.3}$$

with $\nabla\tilde{\pi} = \nabla q - M^2 x$. For initial data $u_0 \in L^p_\sigma(\mathbb{R}^2)$, it was shown in [17] that there exists a unique, local mild solution v to equation (4.2) in the space $C([0,T_0); L^p_\sigma(\mathbb{R}^2))$, where $2 \le p < \infty$. We note that if $u_0 \in L^p_\sigma(\mathbb{R}^2)$, Theorem 2.1 in [17] implies that $t^{\frac{1}{2}}\nabla v \in C([0,T_0); L^p(\mathbb{R}^2))$. Thus there exists $t_1 \in (0,T_0)$ such that $\nabla v(t_1) \in L^p(\mathbb{R}^2)$ which implies that rot $v(t_1) \in L^p(\mathbb{R}^2)$. Hence, in order to prove Theorem 1.3, it suffices to show an a priori estimate of the following form. In the sequel, we set $w :=$ rot v.

Proposition 4.1. *Let $2 \le p < \infty$ and $v(t_1) \in L^p_\sigma(\mathbb{R}^2)$ such that rot $v(t_1) \in L^p(\mathbb{R}^2)$ for some $t_1 \in (0,T_0)$. Let v be the mild solution of (4.2). Then there exists a constant $C > 0$ such that*

$$\|v(t)\|_{L^p} \le C\|v(t_1)\|_{L^p} \exp\Big(Ct\|w(t_1)\|_{L^p}\Big), \quad t > t_1,$$

where $w(t_1) =$ rot $v(t_1)$.

Proof. Consider the operator A in $L^p_\sigma(\mathbb{R}^2)$ given by

$$Au := -\Delta u - \langle M\cdot, \nabla u\rangle + Mu$$

equipped with the domain $D(A) = \{u \in W^{2,p}(\mathbb{R}^2) : \langle M\cdot, \nabla u\rangle \in L^p(\mathbb{R}^2)\}$. By the results in [17], the mild solution of (4.2) is represented by

$$v(t) = e^{-tA}v(t_1) - \int_{t_1}^{t} e^{-(t-s)A}\mathbb{P}(v\cdot\nabla v)(s)ds + 2\int_{t_1}^{t} e^{-(t-s)A}\mathbb{P}(Mv)(s)ds,$$

for $t > t_1$. Applying Proposition 3.4 in [17] yields

$$\begin{aligned} \|e^{-(t-s)A}\mathbb{P}(v\cdot\nabla v)(s)\|_{L^p} &\le \frac{C}{(t-s)^{\frac{1}{p}}}\cdot\|v\cdot\nabla v(s)\|_{L^{\frac{p}{2}}} \\ &\le \frac{C}{(t-s)^{\frac{1}{p}}}\cdot\|v(s)\|_{L^p}\cdot\|\nabla v(s)\|_{L^p}, \quad t > s > t_1. \end{aligned} \tag{4.4}$$

Employing the well-known inequality $\|\nabla v(s)\|_{L^p} \le \frac{p^2}{p-1}\|w(s)\|_{L^p}$ based on Biot–Savart's law (see, e.g., [4, Proposition 7.7.5]), we see that

$$\begin{aligned} \|v(t)\|_{L^p} \le C\|v(t_1)\|_{L^p} &+ C\int_{t_1}^{t}\frac{C}{(t-s)^{\frac{1}{p}}}\cdot\|v(s)\|_{L^p}\cdot\|w(s)\|_{L^p}ds \\ &+ C\int_{t_1}^{t}\|v(s)\|_{L^p}ds, \quad t > t_1. \end{aligned}$$

Next, applying curl to equation (4.2), we verify that the vorticity $w = \text{rot } v$ satisfies the equation

$$\begin{cases} \partial_t w - \Delta w + v\cdot\nabla w - Mx\cdot\nabla w = 0, & x\in\mathbb{R}^2,\ t>0, \\ w(0) = \text{rot } u_0. \end{cases} \tag{4.5}$$

A standard energy estimate allows us to show that

$$\|w(t)\|_{L^p} \le C\|w(t_1)\|_{L^p}, \quad t > t_1.$$

Hence, we have

$$\|v(t)\|_{L^p} \le C\|v(t_1)\|_{L^p} + C\|w(t_1)\|_{L^p}\int_{t_1}^{t}\left(\frac{1}{(t-s)^{\frac{1}{p}}}+1\right)\|v(s)\|_{L^p}ds, \quad t > t_1.$$

Finally, Gronwall's inequality yields the desired estimate. This finishes the proof of Proposition 4.1. □

By Proposition 4.1, we obtain a unique, global solution $\widetilde{v}$ of (4.2) on $[t_1,\infty)$ for the initial data $v(t_1)$. A uniqueness argument ensures that $v(t) = \widetilde{v}(t)$ on $[t_1, T_0)$. Therefore, the local solution v on $[0, T_0)$ can be continued globally. This finishes the proof of Theorem 1.3.

References

[1] W. Arendt, Ch. Batty, M. Hieber and F. Neubrander, *Vector-Valued Laplace Transforms and Cauchy Problems*, Monographs in Math., Vol. 96, Birkhäuser, Basel, 2nd edition, (2011).

[2] A. Babin, A. Mahalov, and B. Nicolaenko, *Regularity and integrability of 3D Euler and Navier–Stokes equations for rotating fluids*, Asymptot. Anal. 15, (1997), 103–150.

[3] A. Babin, A. Mahalov, and B. Nicolaenko, *Global regularity of 3D rotating Navier–Stokes equations for resonant domains*, Indiana Univ. Math. J. 48, (1999), 1133–1176.

[4] H. Bahouri, J.Y. Chemin and R. Danchin, *Fourier Analysis and Nonlinear Partial Differential Equations*, Springer Grundlehren, 343, (2011).

[5] Ch. Batty, *Unbounded operators: functional calculus, generation, perturbations.* Extracta Math. 24, (2009), 99–133.

[6] J.M. Bony, *Calcul symbolique et propagation des singularités pour équations aux dérivées partielles nonlinéaires.* Ann. Sci. L'École Normale Supérieure 14 (1981), 209–246.

[7] M. Cannone, *Ondolettes, Paraproduits et Navier–Stokes*, Diderot, Paris, 2, 1995.

[8] M. Cannone, *Harmonic Analysis Tools for solving the imcompressible Navier–Stokes Equations*, Handbook of Math. Fluid Dynamics, vol. III, 161–244, North-Holland, Amsterdam, 2004.

[9] M. Cannone and Y. Meyer, *Littlewood–Paley decompositions and Navier–Stokes equations*, Meth. Appl. Anal. 2, (1997), 307–319.

[10] M. Cannone, G. Karch, *Smooth or singular solutions to the Navier–Stokes system?*, J. Diff. Equ. 197, (2004), 247–274.

[11] J.Y. Chemin, B. Desjardins, I. Gallagher, and E. Grenier, *Mathematical Geophysics*, Oxford Lecture Series in Mathematics and its Applications, Oxford University Press, Oxford, 2006.

[12] Q. Chen, C. Miao, Z. Zhang, *Global well-posedness for the 3D rotating Navier–Stokes equations with highly oscillating initial data*, Arxiv: 0910.3064v2(2010).

[13] D. Fang, S. Wang, T. Zhang, *Wellposedness for anisotropic rotating fluid equations*, Appl. Math. J. Chinese Univ., to appear.

[14] Y. Giga, K. Inui, A. Mahalov, S. Matsui, *Uniform local solvability for the Navier–Stokes equations with the Coriolis force*, Meth. Appl. Anal. 12, (2005), 381–394.

[15] Y. Giga, K. Inui, A. Mahalov, J. Saal, *Global solvabiliy of the Navier–Stokes equations in spaces based on sum-closed frequency sets*, Adv. Differtial Equations. 12, (2007), 721–736.

[16] Y. Giga, K. Inui, A. Mahalov, J. Saal, *Uniform global solvability of the rotating Navier–Stokes equations for nondecaying initial data*, Indiana Univ. Math. J. 57, (2008), 2775–2791.

[17] M. Hieber, O. Sawada, *The Navier–Stokes equations in* $\mathbb{R}^N$ *with linearly growing initial data*, Arch. Rational Mech. Anal. 175, (2005), 269–285.

[18] M. Hieber, Y. Shibata, *The Fujita–Kato approach to the Navier–Stokes equations in the rotational framework*, Math. Z. 265, (2010), 481–491.

[19] T. Iwabuchi, R. Takada, *Global well-posedness for the Navier–Stokes equations with the Coriolis force in function spaces of Besov type*, J. Func. Anal. 5, (2014), 1321–1337.

[20] Y. Koh, S. Lee, R. Takada, *Dispersive estimates for the Navier–Stokes equations in the rotational framework*, Adv. Differential Equations 19, (2014), 857–878.

[21] P. Konieczny, T. Yoneda, *On dispersive effect of the Coriolis force for the stationary Navier–Stokes equations*, J. Diff. Equ. 250, (2011), 3859–3873.

[22] P.G. Lemarié-Rieusset, *Recent Develpoments in the Navier–Stokes Problem*, Chapman–Hall/CRC Press, Boca Raton, 2002.

[23] A. Majda, *Introduction to PDEs and Waves for the Atmosphere and Ocean.* Courant Lecture Notes in Math., 2003.

[24] H. Triebel, *Theory of Function Spaces*, Monographs in Mathematics, 78. Birkhäuser, Basel, 1983.

Daoyuan Fang and Bin Han
Department of Mathematics
Zhejiang University
Hangzhou 310027, China
e-mail: dyf@zju.edu.cn
hanbinxy@163.com

Matthias Hieber
Technische Universität Darmstadt
Fachbereich Mathematik
Schlossgartenstr. 7
D-64289 Darmstadt, Germany
and
University of Pittsburgh
Benedum Hall
3700 O'Hara Street
Pittsburgh, PA 15261, USA
e-mail: hieber@mathematik.tu-darmstadt.de

Operator Theory:
Advances and Applications, Vol. 250, 213–239

Some Operator Bounds Employing Complex Interpolation Revisited

Fritz Gesztesy, Yuri Latushkin, Fedor Sukochev and Yuri Tomilov

Dedicated with great pleasure to Charles Batty on the occasion of his 60th birthday.

Abstract. We revisit and extend known bounds on operator-valued functions of the type

$$T_1^{-z} S T_2^{-1+z}, \quad z \in \overline{\Sigma} = \{z \in \mathbb{C} \,|\, \mathrm{Re}(z) \in [0,1]\},$$

under various hypotheses on the linear operators S and T_j, $j = 1,2$. We particularly single out the case of self-adjoint and sectorial operators T_j in some separable complex Hilbert space $\mathcal{H}_j$, $j = 1,2$, and suppose that S (resp., S^*) is a densely defined closed operator mapping $\mathrm{dom}(S) \subseteq \mathcal{H}_1$ into $\mathcal{H}_2$ (resp., $\mathrm{dom}(S^*) \subseteq \mathcal{H}_2$ into $\mathcal{H}_1$), relatively bounded with respect to T_1 (resp., T_2^*). Using complex interpolation methods, a generalized polar decomposition for S, and (a variant of) the Loewner–Heinz inequality, the bounds we establish lead to inequalities of the following type: Given $k \in (0,\infty)$,

$$\begin{aligned}\big\|\overline{T_2^{-z} S T_1^{-1+z}}\big\|_{\mathcal{B}(\mathcal{H}_1,\mathcal{H}_2)} &\le N_1 N_2 e^{k(\mathrm{Im}(z))^2 + k\,\mathrm{Re}(z)[1-\mathrm{Re}(z)] + (4k)^{-1}(\theta_1+\theta_2)^2} \\ &\quad \times \big\|S T_1^{-1}\big\|_{\mathcal{B}(\mathcal{H}_1,\mathcal{H}_2)}^{1-\mathrm{Re}(z)} \big\|S^*(T_2^*)^{-1}\big\|_{\mathcal{B}(\mathcal{H}_2,\mathcal{H}_1)}^{\mathrm{Re}(z)}, \quad z \in \overline{\Sigma},\end{aligned}$$

which also implies,

$$\begin{aligned}\big\|\overline{T_2^{-x} S T_1^{-1+x}}\big\|_{\mathcal{B}(\mathcal{H}_1,\mathcal{H}_2)} &\le N_1 N_2 e^{(\theta_1+\theta_2)[x(1-x)]^{1/2}} \\ &\quad \times \big\|S T_1^{-1}\big\|_{\mathcal{B}(\mathcal{H}_1,\mathcal{H}_2)}^{1-x} \big\|S^*(T_2^*)^{-1}\big\|_{\mathcal{B}(\mathcal{H}_2,\mathcal{H}_1)}^{x}, \quad x \in [0,1],\end{aligned}$$

assuming that T_j have bounded imaginary powers, that is, for some $N_j \geqslant 1$ and $\theta_j \geqslant 0$,

$$\big\|T_j^{is}\big\|_{\mathcal{B}(\mathcal{H})} \le N_j e^{\theta_j |s|}, \quad s \in \mathbb{R},\ j = 1,2.$$

We also derive analogous bounds with $\mathcal{B}(\mathcal{H}_1,\mathcal{H}_2)$ replaced by trace ideals, $\mathcal{B}_p(\mathcal{H}_1,\mathcal{H}_2)$, $p \in [1,\infty)$. The methods employed are elementary, predominantly relying on Hadamard's three-lines theorem and the Loewner–Heinz inequality.

Partially supported by the NSF grant DMS-1067929, by the Research Board and the Research Council of the University of Missouri, by the ARC, the NCN grant DEC-2011/03/B/ST1/00407, and by the EU Marie Curie IRSES program, project "AOS", No. 318910.

Mathematics Subject Classification (2010). Primary 47A57, 47B10, 47B44; Secondary 47A30, 47B25.

Keywords. Complex interpolation, generalized Loewner–Heinz-type inequalities, operator and trace norm inequalities.

1. Introduction

This paper was inspired by an interesting result proved by Lesch in Appendix A to his 2005 paper [22], dealing with uniqueness of spectral flow on spaces of unbounded Fredholm operators. More precisely, upon a close inspection of the proof of [22, Proposition A.1], we derived the following interpolation result in [6] (cf. [6, Theorem 4.1]):

Theorem 1.1. *Let $\mathcal{H}$ be a separable Hilbert space and $T \geq 0$ be a self-adjoint operator with $T^{-1} \in \mathcal{B}(\mathcal{H})$. Assume that S is closed and densely defined in $\mathcal{H}$, with $\big(\operatorname{dom}(S) \cap \operatorname{dom}(S^*)\big) \supseteq \operatorname{dom}(T)$, implying $ST^{-1} \in \mathcal{B}(\mathcal{H})$ and $S^*T^{-1} \in \mathcal{B}(\mathcal{H})$. If, in addition, $ST^{-1} \in \mathcal{B}_1(\mathcal{H})$ and $S^*T^{-1} \in \mathcal{B}_1(\mathcal{H})$, then*

$$T^{-1/2}ST^{-1/2} \in \mathcal{B}_1(\mathcal{H}), \quad (T^{-1/2}ST^{-1/2})^* = T^{-1/2}S^*T^{-1/2} \in \mathcal{B}_1(\mathcal{H}). \tag{1.1}$$

Moreover,

$$\big\|T^{-1/2}ST^{-1/2}\big\|_{\mathcal{B}_1(\mathcal{H})} = \big\|T^{-1/2}S^*T^{-1/2}\big\|_{\mathcal{B}_1(\mathcal{H})} \leqslant \big\|ST^{-1}\big\|_{\mathcal{B}_1(\mathcal{H})}^{1/2} \big\|S^*T^{-1}\big\|_{\mathcal{B}_1(\mathcal{H})}^{1/2}. \tag{1.2}$$

Theorem 1.1 was used repeatedly in [6] (in Section 4 and especially, in Section 6). We then announced the present paper in 2010, but due to a variety of reasons, finishing it was delayed for quite a while. We should also mention that in the meantime we became aware of a paper by Huang [14], who proved, in fact, extended, some parts of Lesch's Proposition A.1 in [22] already in 1988 (we will return to this in Sections 2 and 3).

Given Theorem 1.1, we became interested in extensions of it of the following three types:

- The case of fractional powers of T different from $1/2$.
- General trace ideals $\mathcal{B}_p(\mathcal{H})$, $p \in (1, \infty)$.
- Classes of non-self-adjoint operators T, especially, sectorial operators T having bounded imaginary powers.

While interpolation theory has long been raised to a high art, we emphasize that the methods we use are entirely elementary, being grounded in complex interpolation, particularly, in Hadamard's three-lines theorem as pioneered by Kato [18, Sect. 3], and the Loewner–Heinz inequality. In fact, Kato [18, Sect. 3] presents a proof of the generalized Loewner–Heinz inequality applying Hadamard's three-lines theorem, and hence the latter is the ultimate ingredient in our proofs.

We continue with a brief summary of the content of each section. One of the principal results proven in Section 2 reads as follows: Assume that T_j are self-adjoint operators in separable complex Hilbert spaces $\mathcal{H}_j$ with $T_j^{-1} \in \mathcal{B}(\mathcal{H}_j)$, $j = 1, 2$, and suppose that S is a closed operator mapping $\operatorname{dom}(S) \subseteq \mathcal{H}_1$ into $\mathcal{H}_2$ satisfying $\operatorname{dom}(S) \supseteq \operatorname{dom}(T_1)$ and $\operatorname{dom}(S^*) \supseteq \operatorname{dom}(T_2)$. Then $T_2^{-z} S T_1^{-1+z}$ defined on $\operatorname{dom}(T_1)$, $z \in \overline{\Sigma}$, is closable, and given $k \in (0, \infty)$, one obtains

$$\big\|\overline{T_2^{-z} S T_1^{-1+z}}\big\|_{\mathcal{B}(\mathcal{H}_1,\mathcal{H}_2)} \leq \big\|S T_1^{-1}\big\|_{\mathcal{B}(\mathcal{H}_1,\mathcal{H}_2)}^{1-\operatorname{Re}(z)} \big\|S^* T_2^{-1}\big\|_{\mathcal{B}(\mathcal{H}_2,\mathcal{H}_1)}^{\operatorname{Re}(z)} \times \begin{cases} e^{k|\operatorname{Im}(z)|^2 + k\operatorname{Re}(z)[1-\operatorname{Re}(z)] + k^{-1}\pi^2}, \\ e^{k|\operatorname{Im}(z)|^2 + k\operatorname{Re}(z)[1-\operatorname{Re}(z)] + (4k)^{-1}\pi^2}, \text{ if } T_1 \geq 0, \text{ or } T_2 \geq 0, \\ 1, \text{ if } T_j \geq 0,\ j = 1, 2, \end{cases} \quad z \in \overline{\Sigma}, \tag{1.3}$$

as well as

$$\big\|\overline{T_2^{-x} S T_1^{-1+x}}\big\|_{\mathcal{B}(\mathcal{H}_1,\mathcal{H}_2)} \leq \big\|S T_1^{-1}\big\|_{\mathcal{B}(\mathcal{H}_1,\mathcal{H}_2)}^{1-x} \big\|S^* T_2^{-1}\big\|_{\mathcal{B}(\mathcal{H}_2,\mathcal{H}_1)}^{x} \times \begin{cases} e^{2\pi[x(1-x)]^{1/2}}, \\ e^{\pi[x(1-x)]^{1/2}}, \text{ if } T_1 \geq 0, \text{ or } T_2 \geq 0, \\ 1, \text{ if } T_j \geq 0,\ j = 1, 2, \end{cases} \quad x \in [0, 1]. \tag{1.4}$$

In Section 3 we turn to trace ideals $\mathcal{B}_p(\mathcal{H})$, $p \in (1, \infty)$. In addition to the hypotheses imposed on T_j, $j = 1, 2$, and S mentioned in the paragraph preceding (1.3), let $p \in [1, \infty)$ and $S T_1^{-1} \in \mathcal{B}_p(\mathcal{H}_1, \mathcal{H}_2)$ and $S^* T_2^{-1} \in \mathcal{B}_p(\mathcal{H}_2, \mathcal{H}_1)$. Then given $k \in (0, \infty)$, the principal result in Section 3 derives the analog of (1.3) and (1.4) in the form,

$$\big\|\overline{T_2^{-z} S T_1^{-1+z}}\big\|_{\mathcal{B}_p(\mathcal{H}_1,\mathcal{H}_2)} \leq \big\|S T_1^{-1}\big\|_{\mathcal{B}_p(\mathcal{H}_1,\mathcal{H}_2)}^{1-\operatorname{Re}(z)} \big\|S^* T_2^{-1}\big\|_{\mathcal{B}_p(\mathcal{H}_2,\mathcal{H}_1)}^{\operatorname{Re}(z)} \times \begin{cases} e^{k|\operatorname{Im}(z)|^2 + k\operatorname{Re}(z)[1-\operatorname{Re}(z)] + k^{-1}\pi^2}, \\ e^{k|\operatorname{Im}(z)|^2 + k\operatorname{Re}(z)[1-\operatorname{Re}(z)] + (4k)^{-1}\pi^2}, \text{ if } T_1 \geq 0, \text{ or } T_2 \geq 0, \\ 1, \text{ if } T_j \geq 0,\ j = 1, 2, \end{cases} \quad z \in \overline{\Sigma}, \tag{1.5}$$

as well as

$$\big\|\overline{T_2^{-x} S T_1^{-1+x}}\big\|_{\mathcal{B}_p(\mathcal{H}_1,\mathcal{H}_2)} \leq \big\|S T_1^{-1}\big\|_{\mathcal{B}_p(\mathcal{H}_1,\mathcal{H}_2)}^{1-x} \big\|S^* T_2^{-1}\big\|_{\mathcal{B}_p(\mathcal{H}_2,\mathcal{H}_1)}^{x} \times \begin{cases} e^{2\pi[x(1-x)]^{1/2}}, \\ e^{\pi[x(1-x)]^{1/2}}, \text{ if } T_1 \geq 0, \text{ or } T_2 \geq 0, \\ 1, \text{ if } T_j \geq 0,\ j = 1, 2, \end{cases} \quad x \in [0, 1]. \tag{1.6}$$

In our final Section 4 we discuss the extension of (1.3) and (1.4) from self-adjoint to sectorial operators T_j, $j = 1, 2$. One of our principal results there reads as follows: Assume that T_j are sectorial operators in $\mathcal{H}_j$ such that $T_j^{-1} \in \mathcal{B}(\mathcal{H}_j)$, and that for some $\theta_j \geq 0$, $N_j \geq 1$, $\big\|T_j^{is}\big\|_{\mathcal{B}(\mathcal{H}_j)} \leq N_j e^{\theta_j |s|}$, $s \in \mathbb{R}$, $j = 1, 2$. In addition, suppose that S is a closed operator mapping $\operatorname{dom}(S) \subseteq \mathcal{H}_1$ into $\mathcal{H}_2$, satisfying $\operatorname{dom}(S) \supseteq \operatorname{dom}(T_1)$ and $\operatorname{dom}(S^*) \supseteq \operatorname{dom}(T_2^*)$. Then $T_2^{-z} S T_1^{-1+z}$ defined on $\operatorname{dom}(T_1)$, $z \in \overline{\Sigma}$, is closable, and given $k \in (0,\infty)$, one obtains

$$\begin{aligned} \big\|\overline{T_2^{-z} S T_1^{-1+z}}\big\|_{\mathcal{B}(\mathcal{H}_1,\mathcal{H}_2)} &\leq N_1 N_2 e^{k(\operatorname{Im}(z))^2 + k\operatorname{Re}(z)[1-\operatorname{Re}(z)] + (4k)^{-1}(\theta_1+\theta_2)^2} \\ &\quad \times \big\|S T_1^{-1}\big\|_{\mathcal{B}(\mathcal{H}_1,\mathcal{H}_2)}^{1-\operatorname{Re}(z)} \big\|S^*(T_2^*)^{-1}\big\|_{\mathcal{B}(\mathcal{H}_2,\mathcal{H}_1)}^{\operatorname{Re}(z)}, \quad z \in \overline{\Sigma}, \end{aligned} \tag{1.7}$$

as well as

$$\begin{aligned} \big\|\overline{T_2^{-x} S T_1^{-1+x}}\big\|_{\mathcal{B}(\mathcal{H}_1,\mathcal{H}_2)} &\leq N_1 N_2 e^{(\theta_1+\theta_2)[x(1-x)]^{1/2}} \\ &\quad \times \big\|S T_1^{-1}\big\|_{\mathcal{B}(\mathcal{H}_1,\mathcal{H}_2)}^{1-x} \big\|S^*(T_2^*)^{-1}\big\|_{\mathcal{B}(\mathcal{H}_2,\mathcal{H}_1)}^{x}, \quad x \in [0,1]. \end{aligned} \tag{1.8}$$

Moreover, in addition to the hypotheses on T_j, $j = 1,2$, and S mentioned in the paragraph preceding (1.7), let $p \in [1,\infty)$ and $ST_1^{-1} \in \mathcal{B}_p(\mathcal{H}_1,\mathcal{H}_2)$, $S^*(T_2^*)^{-1} \in \mathcal{B}_p(\mathcal{H}_2,\mathcal{H}_1)$. Then given $k \in (0,\infty)$, one obtains the analog of (1.5) and (1.6) in the form,

$$\begin{aligned} \big\|\overline{T_2^{-z} S T_1^{-1+z}}\big\|_{\mathcal{B}_p(\mathcal{H}_1,\mathcal{H}_2)} &\leq N_1 N_2 e^{k(\operatorname{Im}(z))^2 + k\operatorname{Re}(z)[1-\operatorname{Re}(z)] + (4k)^{-1}(\theta_1+\theta_2)^2} \\ &\quad \times \big\|S T_1^{-1}\big\|_{\mathcal{B}_p(\mathcal{H}_1,\mathcal{H}_2)}^{1-\operatorname{Re}(z)} \big\|S^*(T_2^*)^{-1}\big\|_{\mathcal{B}_p(\mathcal{H}_2,\mathcal{H}_1)}^{\operatorname{Re}(z)}, \quad z \in \overline{\Sigma}, \end{aligned} \tag{1.9}$$

as well as

$$\begin{aligned} \big\|\overline{T_2^{-x} S T_1^{-1+x}}\big\|_{\mathcal{B}_p(\mathcal{H}_1,\mathcal{H}_2)} &\leq N_1 N_2 e^{(\theta_1+\theta_2)[x(1-x)]^{1/2}} \\ &\quad \times \big\|S T_1^{-1}\big\|_{\mathcal{B}_p(\mathcal{H}_1,\mathcal{H}_2)}^{1-x} \big\|S^*(T_2^*)^{-1}\big\|_{\mathcal{B}_p(\mathcal{H}_2,\mathcal{H}_1)}^{x}, \quad x \in [0,1]. \end{aligned} \tag{1.10}$$

We note that we permit operators T in (1.3)–(1.6) to have spectrum covering $\mathbb{R}$ except for a neighborhood of zero. Thus, our results in Section 4 for sectorial operators T do not cover the results (1.3)–(1.6).

In conclusion, we briefly summarize the basic notation used in this paper: Let $\mathcal{H}$ be a separable complex Hilbert space, $(\cdot,\cdot)_{\mathcal{H}}$ the scalar product in $\mathcal{H}$ (linear in the second factor), and $I_{\mathcal{H}}$ the identity operator in $\mathcal{H}$. Limits in the norm topology on $\mathcal{H}$ (also called strong limits in $\mathcal{H}$) will be denoted by s-lim. If T is a linear operator mapping (a subspace of) a Hilbert space into another, $\operatorname{dom}(T)$ denotes the domain of T. The closure of a closable operator S is denoted by $\overline{S}$. The spectrum and resolvent set of a closed linear operator in $\mathcal{H}$ will be denoted by $\sigma(\cdot)$ and $\rho(\cdot)$, respectively. The Banach spaces of bounded and compact linear operators in $\mathcal{H}$ are denoted by $\mathcal{B}(\mathcal{H})$ and $\mathcal{B}_\infty(\mathcal{H})$, respectively; in the context of two separable complex Hilbert spaces, $\mathcal{H}_j$, $j = 1,2$, we use the analogous abbreviations $\mathcal{B}(\mathcal{H}_1,\mathcal{H}_2)$ and $\mathcal{B}_\infty(\mathcal{H}_1,\mathcal{H}_2)$. Similarly, the usual ℓ^p-based Schatten–von Neumann (trace) ideals are denoted by $\mathcal{B}_p(\mathcal{H})$, $p \in [1,\infty)$.

2. Interpolation and some operator norm bounds revisited

In this section we revisit and extend a number of bounds collected by Lesch in [22, Proposition A.1].

Through most of this section we will make the following assumptions:

Hypothesis 2.1. *Assume that T is a self-adjoint operator in $\mathcal{H}$ with $T^{-1} \in \mathcal{B}(\mathcal{H})$. In addition, suppose that S is a closed operator in $\mathcal{H}$ satisfying*

$$\operatorname{dom}(S) \cap \operatorname{dom}(S^*) \supseteq \operatorname{dom}(T). \tag{2.1}$$

In particular, Hypothesis 2.1 (and the closed graph theorem) implies that

$$ST^{-1} \in \mathcal{B}(\mathcal{H}), \quad S^*T^{-1} \in \mathcal{B}(\mathcal{H}). \tag{2.2}$$

Remark 2.2.

(i) In the sequel we will adhere to the following convention: Operator products AB of two linear operators A and B in $\mathcal{H}$ are always assumed to be maximally defined, that is,

$$\operatorname{dom}(AB) = \{f \in \mathcal{H} \,|\, f \in \operatorname{dom}(B),\, Bf \in \operatorname{dom}(A)\}, \tag{2.3}$$

unless explicitly stated otherwise. The same convention is of course applied to products of three or more linear operators in $\mathcal{H}$.

(ii) We recall the following useful facts (see, e.g., [42, Theorem 4.19]): Suppose T_j, $j = 1, 2$, are two densely defined linear operators in $\mathcal{H}$ such that T_2T_1 is also densely defined in $\mathcal{H}$. Then,

$$(T_2T_1)^* \supseteq T_1^*T_2^*. \tag{2.4}$$

If in addition $T_2 \in \mathcal{B}(\mathcal{H})$, then

$$(T_2T_1)^* = T_1^*T_2^*. \tag{2.5}$$

Theorem 2.3. *Assume Hypothesis 2.1. Then the following facts hold:*

(i) *The operators $T^{-1}ST$ and $T^{-1}S^*T$ are well defined on $\operatorname{dom}(T^2)$, and hence densely defined in $\mathcal{H}$,*

$$\operatorname{dom}\left(T^{-1}ST\right) \cap \operatorname{dom}\left(T^{-1}S^*T\right) \supseteq \operatorname{dom}\left(T^2\right). \tag{2.6}$$

(ii) *The relations*

$$\left(T^{-1}ST\right)^* = TS^*T^{-1}, \quad \left(T^{-1}S^*T\right)^* = TST^{-1}, \tag{2.7}$$

hold, and hence TS^*T^{-1} and TST^{-1} are closed in $\mathcal{H}$.

(iii) *One infers that*

$$T^{-1}ST \text{ is bounded if and only if } (T^{-1}ST)^* = TS^*T^{-1} \in \mathcal{B}(\mathcal{H}). \tag{2.8}$$

In case $T^{-1}ST$ is bounded, then

$$\overline{T^{-1}ST} = (TS^*T^{-1})^*, \quad \left\|\overline{T^{-1}ST}\right\|_{\mathcal{B}(\mathcal{H})} = \left\|TS^*T^{-1}\right\|_{\mathcal{B}(\mathcal{H})}. \tag{2.9}$$

Analogously, one concludes that

$$T^{-1}S^*T \text{ is bounded if and only if } \left(T^{-1}S^*T\right)^* = TST^{-1} \in \mathcal{B}(\mathcal{H}). \tag{2.10}$$

In case $T^{-1}S^*T$ is bounded, then

$$\overline{T^{-1}S^*T} = \left(TST^{-1}\right)^*, \quad \left\|\overline{T^{-1}S^*T}\right\|_{\mathcal{B}(\mathcal{H})} = \left\|TST^{-1}\right\|_{\mathcal{B}(\mathcal{H})}. \tag{2.11}$$

Proof. We start by recalling that

$$\operatorname{dom}\big(T^{-1}ST\big) = \big\{g \in \operatorname{dom}(T) \,\big|\, Tg \in \operatorname{dom}(S)\big\}, \tag{2.12}$$

$$\operatorname{dom}\big(TS^*T^{-1}\big) = \big\{f \in \mathcal{H} \,\big|\, S^*T^{-1}f \in \operatorname{dom}(T)\big\}. \tag{2.13}$$

(i) Suppose that $g \in \operatorname{dom}\big(T^2\big)$. Then $Tg \in \operatorname{dom}(T) \subseteq \operatorname{dom}(S)$ and hence

$$\operatorname{dom}\big(T^{-1}ST\big) \supseteq \operatorname{dom}\big(T^2\big). \tag{2.14}$$

Since T and hence T^2 are self-adjoint in $\mathcal{H}$ and hence necessarily densely defined, $T^{-1}ST$ is densely defined in $\mathcal{H}$. The same applies to $T^{-1}S^*T$.

(ii) Applying Remark 2.2 (ii), one obtains

$$\big(T^{-1}ST\big)^* = \big(T^{-1}[ST]\big)^* = [ST]^*T^{-1} \supseteq TS^*T^{-1}, \tag{2.15}$$

since $T^{-1}ST$ is densely defined in $\mathcal{H}$ by item (i). To prove the converse inclusion in (2.15), we now assume that $f \in \operatorname{dom}\big(\big(T^{-1}ST\big)^*\big)$ and $g \in \operatorname{dom}\big(T^2\big) \subseteq \operatorname{dom}\big(T^{-1}ST\big)$. Then

$$\big(\big(T^{-1}ST\big)^*f, g\big)_{\mathcal{H}} = \big(f, T^{-1}STg\big)_{\mathcal{H}} = \big(T^{-1}f, STg\big)_{\mathcal{H}} = \big(S^*T^{-1}f, Tg\big)_{\mathcal{H}}. \tag{2.16}$$

Since $\operatorname{dom}\big(T^2\big)$ is an operator core for T (e.g., upon applying the spectral theorem), (2.16) extends to all $g \in \operatorname{dom}(T)$, that is, one has

$$\big(\big(T^{-1}ST\big)^*f, g\big)_{\mathcal{H}} = \big(S^*T^{-1}f, Tg\big)_{\mathcal{H}}, \quad f \in \operatorname{dom}\big(\big(T^{-1}ST\big)^*\big),\ g \in \operatorname{dom}(T). \tag{2.17}$$

Consequently, $S^*T^{-1}f \in \operatorname{dom}(T)$ and

$$\begin{aligned} \big(\big(T^{-1}ST\big)^*f, g\big)_{\mathcal{H}} = \big(S^*T^{-1}f, Tg\big)_{\mathcal{H}} = \big(TS^*T^{-1}f, g\big)_{\mathcal{H}}&, \\ f \in \operatorname{dom}\big(\big(T^{-1}ST\big)^*\big),\ g \in \operatorname{dom}(T)&, \end{aligned} \tag{2.18}$$

implying

$$\big(T^{-1}ST\big)^*f = TS^*T^{-1}f, \quad f \in \operatorname{dom}\big(\big(T^{-1}ST\big)^*\big), \tag{2.19}$$

and hence,

$$\big(T^{-1}ST\big)^* \subseteq TS^*T^{-1}. \tag{2.20}$$

Then (2.15) and (2.20) yield the first relation in (2.7). Replacing S by S^* yields the second relation in (2.7).

(iii) Since $T^{-1}ST$ is densely defined by (2.6), an application of [42, Theorem 4.14(a)] yields (2.8). Equation (2.9) is an immediate consequence of (2.8).

Again, replacing S by S^* implies (2.10) and (2.11). □

In the following we denote by $\Sigma \subset \mathbb{C}$ the open strip

$$\Sigma = \{z \in \mathbb{C} \,|\, \operatorname{Re}(z) \in (0,1)\}, \tag{2.21}$$

and by $\overline{\Sigma}$ its closure.

To state additional results we will have to apply a version of Hadamard's three-lines theorem and hence recall the following general result:

Theorem 2.4 ([13, p. 211]). *Suppose $\phi(\cdot)$ is an analytic function on Σ, continuous on $\overline{\Sigma}$, and satisfying for some fixed $C \in \mathbb{R}$ and $a \in [0, \pi)$,*

$$\sup_{z \in \overline{\Sigma}} \left[e^{-a|\operatorname{Im}(z)|} \ln(|\phi(z)|) \right] \leq C. \tag{2.22}$$

Then

$$|\phi(z)| \leq \exp\left\{ \frac{\sin(\pi \operatorname{Re}(z))}{2} \int_{\mathbb{R}} dy \left[\frac{\ln(|\phi(iy)|)}{\cosh(\pi(y - \operatorname{Im}(z))) - \cosh(\pi \operatorname{Re}(z))} + \frac{\ln(|\phi(1+iy)|)}{\cosh(\pi(y - \operatorname{Im}(z))) + \cosh(\pi \operatorname{Re}(z))} \right] \right\}, \quad z \in \Sigma. \tag{2.23}$$

If in addition, for some $C_0, C_1 \in (0, \infty)$,

$$|\phi(iy)| \leq C_0, \quad |\phi(1+iy)| \leq C_1, \quad y \in \mathbb{R}, \tag{2.24}$$

then

$$|\phi(z)| \leq C_0^{1-\operatorname{Re}(z)} C_1^{\operatorname{Re}(z)}, \quad z \in \overline{\Sigma}. \tag{2.25}$$

For a recent detailed exposition of such results we refer to [10, Sects. 1.3.2, 1.3.3] (see also [8, Sect. III.13]). A classical application of Theorem 2.4 to linear operators appeared in [35] (see also [3, Sect. 4.3]).

The growth condition (2.22) is of course familiar from Phragmen–Lindelöf-type arguments applied to the strip Σ (see, e.g., [32, Theorem 12.9]).

In the sequel, complex powers T^z, $z \in \overline{\Sigma}$, of a self-adjoint operator T in $\mathcal{H}$, with $T^{-1} \in \mathcal{B}(\mathcal{H})$, are defined in terms of the spectral representation of T,

$$T = \int_{\sigma(T)} \lambda \, dE_T(\lambda), \tag{2.26}$$

with $\{E_T(\lambda)\}_{\lambda \in \mathbb{R}}$ denoting the family of spectral projections of T, as follows: Since by hypothesis, $(-\varepsilon, \varepsilon) \cap \sigma(T) = \emptyset$ for some $\varepsilon > 0$, one defines

$$T^z = \int_{\sigma(T)} \lambda^z \, dE_T(\lambda), \quad z \in \overline{\Sigma}, \tag{2.27}$$

where

$$\lambda^z = \lambda^{\operatorname{Re}(z)} e^{i \operatorname{Im}(z) \ln(|\lambda|)} \left[\theta(\lambda) + e^{-\pi \operatorname{Im}(z)} \theta(-\lambda) \right], \quad \lambda \in \mathbb{R} \backslash \{0\}, \; z \in \overline{\Sigma}, \tag{2.28}$$

and

$$\theta(x) = \begin{cases} 1, & x > 0, \\ 0, & x < 0. \end{cases} \tag{2.29}$$

Consequently, one obtains the estimate

$$\left\| T^{iy} \right\|_{\mathcal{B}(\mathcal{H})} \leq \max\left(1, e^{-\pi y}\right) \leq e^{\pi |y|}, \quad y \in \mathbb{R}, \tag{2.30}$$

and

$$\text{if } T \geq \varepsilon I_{\mathcal{H}}, \text{ for some } \varepsilon > 0, \text{ then } T^{iy} \text{ is unitary, } \left\| T^{iy} \right\|_{\mathcal{B}(\mathcal{H})} = 1, \quad y \in \mathbb{R}. \tag{2.31}$$

Theorem 2.5. *Assume Hypothesis* 2.1 *and suppose that* $TST^{-1} \in \mathcal{B}(\mathcal{H})$ *as well as* $TS^*T^{-1} \in \mathcal{B}(\mathcal{H})$. *Then* $S \in \mathcal{B}(\mathcal{H})$ (*and hence* $S^* \in \mathcal{B}(\mathcal{H})$) *and*

$$\|S\|_{\mathcal{B}(\mathcal{H})} = \|S^*\|_{\mathcal{B}(\mathcal{H})} \leq \left\|TST^{-1}\right\|_{\mathcal{B}(\mathcal{H})}^{1/2} \left\|TS^*T^{-1}\right\|_{\mathcal{B}(\mathcal{H})}^{1/2} \begin{cases} e^{2\pi}, \\ 1, \ \text{if } T \geq 0. \end{cases} \tag{2.32}$$

Proof. Introducing

$$\varphi_k(z) = e^{kz(z-1)}\left(T^2 f, T^{2z-3} S T^{-1-2z} T^2 g\right)_{\mathcal{H}}, \ f, g \in \operatorname{dom}\left(T^2\right), \ z \in \overline{\Sigma}, \ k \in [0, \infty), \tag{2.33}$$

one infers that φ_k is analytic on Σ. (We note that the idea to exploit the factor $e^{kz(z-1)}$, $k > 0$, can already be found in the proof of [18, Theorem 6]. This factor is used in (2.34)–(2.36) below to neutralize factors of the type $e^{4\pi|y|}$ and $e^{4\pi|\operatorname{Im}(z)|}$.)

In the following we focus on the general case where T is self-adjoint and $k > 0$; in this case we will employ the bound (2.30).

Assuming $k > 0$, (2.30) yields the estimates

$$\begin{aligned} |\varphi_k(iy)| &= e^{-ky^2}\left|\left(f, T^{2iy-1} S T^{1-2iy} g\right)_{\mathcal{H}}\right| = e^{-ky^2}\left|\left(T^{-2iy} f, [T^{-1}ST] T^{-2iy} g\right)_{\mathcal{H}}\right| \\ &\leq e^{-ky^2 + 4\pi|y|} \left\|\overline{T^{-1}ST}\right\|_{\mathcal{B}(\mathcal{H})} \|f\|_{\mathcal{H}} \|g\|_{\mathcal{H}} \\ &= e^{-ky^2 + 4\pi|y|} \left\|TS^*T^{-1}\right\|_{\mathcal{B}(\mathcal{H})} \|f\|_{\mathcal{H}} \|g\|_{\mathcal{H}} \\ &\leq e^{k^{-1}4\pi^2} \left\|TS^*T^{-1}\right\|_{\mathcal{B}(\mathcal{H})} \|f\|_{\mathcal{H}} \|g\|_{\mathcal{H}}, \quad f, g \in \operatorname{dom}\left(T^2\right), \ y \in \mathbb{R}, \end{aligned} \tag{2.34}$$

using (2.9), and similarly,

$$\begin{aligned} |\varphi_k(1+iy)| &= e^{-ky^2}\left|\left(f, T^{1+2iy} S T^{-1-2iy} g\right)_{\mathcal{H}}\right| \\ &= e^{-ky^2}\left|\left(T^{-2iy} f, [TST^{-1}] T^{-2iy} g\right)_{\mathcal{H}}\right| \\ &\leq e^{k^{-1}4\pi^2} \left\|TST^{-1}\right\|_{\mathcal{B}(\mathcal{H})} \|f\|_{\mathcal{H}} \|g\|_{\mathcal{H}}, \quad f, g \in \operatorname{dom}\left(T^2\right), \ y \in \mathbb{R}, \end{aligned} \tag{2.35}$$

again employing (2.9) and (2.30). In addition, one obtains

$$\begin{aligned} |\varphi_k(z)| &= e^{-k|\operatorname{Im}(z)|^2 + k\operatorname{Re}(z)[\operatorname{Re}(z)-1]}\left|\left(T^2 f, T^{2z-3} S T^{-1-2z} T^2 g\right)_{\mathcal{H}}\right| \\ &\leq e^{-k|\operatorname{Im}(z)|^2 + k\operatorname{Re}(z)[\operatorname{Re}(z)-1] + 4\pi|\operatorname{Im}(z)|} \\ &\quad \times \left\|T^{2\operatorname{Re}(z)-3}\right\|_{\mathcal{B}(\mathcal{H})} \left\|ST^{-1}\right\|_{\mathcal{B}(\mathcal{H})} \left\|T^{-2\operatorname{Re}(z)}\right\|_{\mathcal{B}(\mathcal{H})} \left\|T^2 f\right\|_{\mathcal{H}} \left\|T^2 g\right\|_{\mathcal{H}} \\ &\leq e^{k^{-1}4\pi^2 + k\operatorname{Re}(z)[\operatorname{Re}(z)-1]} \\ &\quad \times \left\|T^{2\operatorname{Re}(z)-3}\right\|_{\mathcal{B}(\mathcal{H})} \left\|ST^{-1}\right\|_{\mathcal{B}(\mathcal{H})} \left\|T^{-2\operatorname{Re}(z)}\right\|_{\mathcal{B}(\mathcal{H})} \left\|T^2 f\right\|_{\mathcal{H}} \left\|T^2 g\right\|_{\mathcal{H}} \\ &\leq C, \quad f, g \in \operatorname{dom}\left(T^2\right), \ z \in \overline{\Sigma}, \end{aligned} \tag{2.36}$$

for some finite constant $C = C(f, g, S, T) > 0$, independent of $z \in \overline{\Sigma}$.

Applying the Hadamard three-lines estimate (2.25) to $\varphi_k(\cdot)$ then yields

$$\begin{aligned} |\varphi_k(z)| \leq e^{k^{-1}4\pi^2} \left\|TS^*T^{-1}\right\|_{\mathcal{B}(\mathcal{H})}^{1-\operatorname{Re}(z)} \left\|TST^{-1}\right\|_{\mathcal{B}(\mathcal{H})}^{\operatorname{Re}(z)} \|f\|_{\mathcal{H}} \|g\|_{\mathcal{H}}, \\ f, g \in \operatorname{dom}\left(T^2\right), \ z \in \overline{\Sigma}. \end{aligned} \tag{2.37}$$

Taking $z = 1/2$ in (2.37) implies

$$|(f, Sg)_{\mathcal{H}}| \leq e^{4^{-1}k+k^{-1}4\pi^2} \big\|TS^*T^{-1}\big\|_{\mathcal{B}(\mathcal{H})}^{1/2} \big\|TST^{-1}\big\|_{\mathcal{B}(\mathcal{H})}^{1/2} \|f\|_{\mathcal{H}} \|g\|_{\mathcal{H}}, \quad f, g \in \operatorname{dom}\big(T^2\big). \tag{2.38}$$

Optimizing with respect to $k > 0$ yields

$$|(f, Sg)_{\mathcal{H}}| \leq e^{2\pi} \big\|TS^*T^{-1}\big\|_{\mathcal{B}(\mathcal{H})}^{1/2} \big\|TST^{-1}\big\|_{\mathcal{B}(\mathcal{H})}^{1/2} \|f\|_{\mathcal{H}} \|g\|_{\mathcal{H}}, \quad f, g \in \operatorname{dom}\big(T^2\big). \tag{2.39}$$

Since $\operatorname{dom}\big(T^2\big)$ is dense in $\mathcal{H}$, this yields that S is a bounded operator in $\mathcal{H}$. Employing that S is closed in $\mathcal{H}$ finally proves $S \in \mathcal{B}(\mathcal{H})$ and hence the first estimate in (2.32).

If in addition, $T \geq 0$, we choose $k = 0$ in (2.33) and then rely on equality (2.31) (as opposed to (2.30)), which slightly simplifies the estimates (2.34)–(2.39), implying the second inequality in (2.32). □

Remark 2.6. In the special case where T is self-adjoint and $T \geq \varepsilon I_{\mathcal{H}}$ for some $\varepsilon > 0$, there exists an alternative way of deriving the bound (2.32) by means of Proposition A.1 (2) proved by Lesch [22] in the context of closed, symmetric operators S. Indeed, an application of [22, Proposition A.1 (2)] with S replaced by the symmetric, in fact, self-adjoint, S^*S yields

$$\begin{aligned} \|S^*S\|_{\mathcal{B}(\mathcal{H})} &\leq \big\|TS^*ST^{-1}\big\|_{\mathcal{B}(\mathcal{H})} = \big\|\big(TS^*T^{-1}\big)\big(TST^{-1}\big)\big\|_{\mathcal{B}(\mathcal{H})} \\ &\leq \big\|TS^*T^{-1}\big\|_{\mathcal{B}(\mathcal{H})} \big\|TST^{-1}\big\|_{\mathcal{B}(\mathcal{H})}. \end{aligned} \tag{2.40}$$

Thus, S^*S is bounded, so both S and S^* are bounded, and hence,

$$\|S\|_{\mathcal{B}(\mathcal{H})} = \|S^*\|_{\mathcal{B}(\mathcal{H})} = \|S^*S\|_{\mathcal{B}(\mathcal{H})}^{1/2} \leq \big\|TST^{-1}\big\|_{\mathcal{B}(\mathcal{H})}^{1/2} \big\|TS^*T^{-1}\big\|_{\mathcal{B}(\mathcal{H})}^{1/2}. \tag{2.41}$$

Thus, in this special case one needs no additional arguments to prove Theorem 2.5. However, this type of argument does not apply to the remaining statements in this section.

Theorem 2.5 allows us to derive the following result.

Theorem 2.7. *In addition to Hypothesis* 2.1 *suppose that* $T \geq 0$. *Then*

$$T^{-1/2}ST^{-1/2} \in \mathcal{B}(\mathcal{H}), \quad T^{-1/2}S^*T^{-1/2} \in \mathcal{B}(\mathcal{H}), \tag{2.42}$$

and

$$\big(T^{-1/2}ST^{-1/2}\big)^* = T^{-1/2}S^*T^{-1/2}, \quad \big(T^{-1/2}S^*T^{-1/2}\big)^* = T^{-1/2}ST^{-1/2}. \tag{2.43}$$

Moreover,

$$\big\|T^{-1/2}ST^{-1/2}\big\|_{\mathcal{B}(\mathcal{H})} = \big\|T^{-1/2}S^*T^{-1/2}\big\|_{\mathcal{B}(\mathcal{H})} \leq \big\|ST^{-1}\big\|_{\mathcal{B}(\mathcal{H})}^{1/2} \big\|S^*T^{-1}\big\|_{\mathcal{B}(\mathcal{H})}^{1/2}. \tag{2.44}$$

Proof. Applying the spectral theorem, the combined assumptions on T actually yield $T \geq \varepsilon I_{\mathcal{H}}$ for some $\varepsilon > 0$. (The condition $T \geq 0$ has inadvertently been omitted in [6, Theorem 4.1] and [22, Proposition A.1 (3)].) Introduce the operators $\widehat{S}$ and $\widehat{T}$ in $\mathcal{H}$ by

$$\widehat{S} = T^{-1/2} S T^{-1/2}, \quad \widehat{T} = T^{1/2}. \tag{2.45}$$

Then $\widehat{T}$ is self-adjoint and $\operatorname{dom}(\widehat{S}) \supseteq \operatorname{dom}(T^{1/2})$, that is, $\operatorname{dom}(\widehat{S}) \supseteq \operatorname{dom}(\widehat{T})$ yields that $\widehat{S}$ is densely defined. Next, we note that by Remark 2.2 (ii),

$$(\widehat{S})^* = \big(T^{-1/2}\big[ST^{-1/2}\big]\big)^* = \big[ST^{-1/2}\big]^* T^{-1/2} \supseteq T^{-1/2} S^* T^{-1/2}, \tag{2.46}$$

and hence $\operatorname{dom}\big((\widehat{S})^*\big) \supseteq \operatorname{dom}(\widehat{T})$. Moreover, since

$$\widehat{T}\widehat{S}(\widehat{T})^{-1} = ST^{-1} \in \mathcal{B}(\mathcal{H}), \quad \widehat{T}(\widehat{S})^*(\widehat{T})^{-1} \supseteq S^*T^{-1} \in \mathcal{B}(\mathcal{H}), \tag{2.47}$$

by Hypothesis 2.1 (resp., (2.2)), one also infers

$$\widehat{T}(\widehat{S})^*(\widehat{T})^{-1} = S^*T^{-1} \in \mathcal{B}(\mathcal{H}). \tag{2.48}$$

Thus, Theorem 2.5 applies to $\widehat{S}$ and $\widehat{T}$ and hence $\widehat{S}$, $(\widehat{S})^* \in \mathcal{B}(\mathcal{H})$, as well as

$$\begin{aligned} \|\widehat{S}\|_{\mathcal{B}(\mathcal{H})} = \|(\widehat{S})^*\|_{\mathcal{B}(\mathcal{H})} &\leq \big\|\widehat{T}\widehat{S}(\widehat{T})^{-1}\big\|_{\mathcal{B}(\mathcal{H})}^{1/2} \big\|\widehat{T}(\widehat{S})^*(\widehat{T})^{-1}\big\|_{\mathcal{B}(\mathcal{H})}^{1/2} \\ &= \big\|ST^{-1}\big\|_{\mathcal{B}(\mathcal{H})}^{1/2} \big\|S^*T^{-1}\big\|_{\mathcal{B}(\mathcal{H})}^{1/2}. \end{aligned} \tag{2.49}$$

Since our hypotheses are symmetric with respect to S and S^*, interchanging S and S^*, repeating (2.45)–(2.49) with $\widehat{S}$ replaced by

$$\widetilde{S} = T^{-1/2} S^* T^{-1/2}, \tag{2.50}$$

then also yields $(\widetilde{S})^* \supseteq T^{-1/2} S T^{-1/2} = \widehat{S}$. Since $\widehat{S} \in \mathcal{B}(\mathcal{H})$, one concludes that $(\widetilde{S})^* = \widehat{S}$. Applying Theorem 2.5 to $\widetilde{S}$ and $\widehat{T}$ then yields $\widetilde{S} \in \mathcal{B}(\mathcal{H})$ and hence also $(\widehat{S})^* = \widetilde{S}$, completing the proof. □

In the special case where S is symmetric in $\mathcal{H}$, $S \subseteq S^*$, and $T \geq 0$ (actually, $T \geq \varepsilon I_{\mathcal{H}}$ for some $\varepsilon > 0$ as also the condition $T^{-1} \in \mathcal{B}(\mathcal{H})$ is involved), nearly all the results of this section up to now (as well as the basic strategy of proofs employed), appeared in Lesch [22, Appendix A]. We emphasize, however, that some of these results, especially, Theorem 2.7, were previously derived in 1988 by Huang [14, Lemma 2.1.(b)]. In fact, combining the spectral theorem for T and the three-lines theorem, Huang arrives at an extension of Theorem 2.7 involving fractional powers of T^{α}, $\alpha \in [1/2, 1]$, on the right-hand side of (2.44).

Next, we recall the generalized polar decomposition for densely defined closed operators S in $\mathcal{H}$ derived in [7],

$$S = |S^*|^{\alpha} U_S |S|^{1-\alpha}, \quad \alpha \in [0,1], \tag{2.51}$$

where U_S denotes the partial isometry in $\mathcal{H}$ associated with the standard polar decomposition $S = U_S|S|$, and $|S| = (S^*S)^{1/2}$ (and we interpret $|S|^0 = I_{\mathcal{H}}$ in this particular context).

We will employ (2.51) (and its analog for S^*) to prove the following result:

Theorem 2.8. *Assume Hypothesis* 2.1. *Then* $T^{-z}ST^{-1+z}$, $z \in \overline{\Sigma}$, *defined on* $\mathrm{dom}(T)$, *is closable in* $\mathcal{H}$, *and*

$$\overline{T^{-z}ST^{-1+z}} = T^{-i\,\mathrm{Im}(z)}\big[|S^*|^{\mathrm{Re}(z)}T^{-\,\mathrm{Re}(z)}\big]^* U_S \times |S|^{1-\mathrm{Re}(z)}T^{-1+\mathrm{Re}(z)}T^{i\,\mathrm{Im}(z)} \in \mathcal{B}(\mathcal{H}), \quad z \in \overline{\Sigma}. \tag{2.52}$$

In addition, given $k \in (0,\infty)$, *one obtains*

$$\big\|\overline{T^{-z}ST^{-1+z}}\big\|_{\mathcal{B}(\mathcal{H})} \leq \big\|ST^{-1}\big\|_{\mathcal{B}(\mathcal{H})}^{1-\mathrm{Re}(z)} \big\|S^*T^{-1}\big\|_{\mathcal{B}(\mathcal{H})}^{\mathrm{Re}(z)} \times \begin{cases} e^{k|\,\mathrm{Im}(z)|^2+k\,\mathrm{Re}(z)[1-\mathrm{Re}(z)]+k^{-1}\pi^2}, \\ 1, \ \text{if } T \geq 0, \end{cases} \quad z \in \overline{\Sigma}, \tag{2.53}$$

and

$$\big\|\overline{T^{-x}ST^{-1+x}}\big\|_{\mathcal{B}(\mathcal{H})} \leq \big\|ST^{-1}\big\|_{\mathcal{B}(\mathcal{H})}^{1-x} \big\|S^*T^{-1}\big\|_{\mathcal{B}(\mathcal{H})}^{x} \begin{cases} e^{2\pi[x(1-x)]^{1/2}}, \\ 1, \ \text{if } T \geq 0, \end{cases} \quad x \in [0,1]. \tag{2.54}$$

In particular, assuming $T \geq 0$ *and taking* $x = 1/2$ *in* (2.54) *one recovers the estimate* (2.44) (*in this particular case the operator closure sign in* (2.53) *is superfluous since* $T^{-1/2}ST^{-1/2} \in \mathcal{B}(\mathcal{H})$ *by* (2.42)).

Proof. We start by noting that $\mathrm{dom}(S) \supseteq \mathrm{dom}(T)$ (together with S and T closed by hypothesis) implies that S is relatively bounded with respect to T and hence there exist $a > 0$ and $b > 0$ such that

$$\begin{aligned} \||S|f\|_{\mathcal{H}}^2 &= \|Sf\|_{\mathcal{H}}^2 \leq a^2\|Tf\|_{\mathcal{H}}^2 + b^2\|f\|_{\mathcal{H}}^2 = a^2\||T|f\|_{\mathcal{H}}^2 + b^2\|f\|_{\mathcal{H}}^2 \\ &= \big\|\big[a^2|T|^2+b^2\big]^{1/2}f\big\|_{\mathcal{H}}^2, \quad f \in \mathrm{dom}(T) = \mathrm{dom}(|T|). \end{aligned} \tag{2.55}$$

Thus, applying the Loewner–Heinz inequality (cf. [5, Sect. 3.2.1], [12, Theorem 3], [15], [21, Theorem IV.1.11], [23], [27]), one infers that

$$\mathrm{dom}\big(|S|^\alpha\big) \supseteq \mathrm{dom}\big(\big(a^2|T|^2+b^2\big)^{\alpha/2}\big) = \mathrm{dom}\big(|T|^\alpha\big), \quad \alpha \in [0,1], \tag{2.56}$$

and

$$\big\||S|^\alpha\big[a^2|T|^2+b^2\big]^{-\alpha/2}h\big\|_{\mathcal{H}} \leq \|h\|_{\mathcal{H}}^2, \quad h \in \mathcal{H}, \ \alpha \in [0,1]. \tag{2.57}$$

Similarly, interchanging S and S^*, one obtains

$$\mathrm{dom}\big(|S^*|^\alpha\big) \supseteq \mathrm{dom}\big(|T|^\alpha\big), \quad \alpha \in [0,1], \tag{2.58}$$

and

$$\big\||S^*|^\alpha\big[\widetilde{a}^2|T|^2+\widetilde{b}^2\big]^{-\alpha/2}h\big\|_{\mathcal{H}} \leq \|h\|_{\mathcal{H}}^2, \quad h \in \mathcal{H}, \ \alpha \in [0,1], \tag{2.59}$$

for appropriate $\widetilde{a} > 0$, $\widetilde{b} > 0$. Applying (2.51) to S (with $\alpha = \mathrm{Re}(z)$), and using (2.56) and (2.58), one concludes (cf. also Remark 2.2 (ii)) that

$$\begin{aligned} T^{-z}ST^{-1+z} &= T^{-z}|S^*|^{\mathrm{Re}(z)}U_S|S|^{1-\mathrm{Re}(z)}T^{-1+z} \\ &\subseteq \big[|S^*|^{\mathrm{Re}(z)}T^{-\overline{z}}\big]^*U_S\big[|S|^{1-\mathrm{Re}(z)}T^{-1+z}\big] \qquad (2.60) \\ &= T^{-i\,\mathrm{Im}(z)}\big(|S^*|^{\mathrm{Re}(z)}T^{-\,\mathrm{Re}(z)}\big)^*U_S|S|^{1-\mathrm{Re}(z)}T^{-1+\mathrm{Re}(z)}T^{i\,\mathrm{Im}(z)} \in \mathcal{B}(\mathcal{H}), \quad z \in \overline{\Sigma}, \end{aligned}$$

proving (2.52).

Next, one defines (repeatedly employing below the fact that for a closable operator A, $\overline{A}$ is an extension of A)

$$\begin{aligned} \phi_k(z) = e^{kz(z-1)}(Tf, T^{-1-z}ST^{-2+z}Tg)_{\mathcal{H}} = e^{kz(z-1)}\big(f, \overline{T^{-z}ST^{-1+z}}g\big)_{\mathcal{H}}, \\ f,g \in \mathrm{dom}(T),\ z \in \overline{\Sigma},\ k \in [0,\infty). \end{aligned} \qquad (2.61)$$

Again we primarily focus on the case where T is merely self-adjoint and hence choose $k > 0$ and employ the estimate (2.30) in the following.

One estimates

$$\begin{aligned} |\phi_k(iy)| &= e^{-ky^2}\big|(T^{iy}f, ST^{-1}T^{iy}g)_{\mathcal{H}}\big| \leq e^{-ky^2+2\pi|y|}\big\|ST^{-1}\big\|_{\mathcal{B}(\mathcal{H})}\|f\|_{\mathcal{H}}\|g\|_{\mathcal{H}} \\ &\leq e^{k^{-1}\pi^2}\big\|ST^{-1}\big\|_{\mathcal{B}(\mathcal{H})}\|f\|_{\mathcal{H}}\|g\|_{\mathcal{H}}, \quad y \in \mathbb{R}, \qquad (2.62) \\ |\phi_k(1+iy)| &= e^{-ky^2}\big|(T^{iy}f, \overline{T^{-1}S}T^{iy}g)_{\mathcal{H}}\big| \leq e^{-ky^2+2\pi|y|}\big\|\overline{T^{-1}S}\big\|_{\mathcal{B}(\mathcal{H})}\|f\|_{\mathcal{H}}\|g\|_{\mathcal{H}} \\ &\leq e^{k^{-1}\pi^2}\big\|S^*T^{-1}\big\|_{\mathcal{B}(\mathcal{H})}\|f\|_{\mathcal{H}}\|g\|_{\mathcal{H}}, \quad y \in \mathbb{R}, \qquad (2.63) \\ |\phi_k(z)| &= e^{-k|\mathrm{Im}(z)|^2+k\,\mathrm{Re}(z)[\mathrm{Re}(z)-1]}\big|(Tf, T^{-1-z}ST^{-2+z}Tg)_{\mathcal{H}}\big| \\ &\leq e^{-k|\mathrm{Im}(z)|^2+k\,\mathrm{Re}(z)[\mathrm{Re}(z)-1]+2\pi|\mathrm{Im}(z)|}\big\|T^{-1-\mathrm{Re}(z)}\big\|_{\mathcal{B}(\mathcal{H})} \\ &\quad \times \big\|ST^{-1}\big\|_{\mathcal{B}(\mathcal{H})}\big\|T^{-1+\mathrm{Re}(z)}\big\|_{\mathcal{B}(\mathcal{H})}\|Tf\|_{\mathcal{H}}\|Tg\|_{\mathcal{H}} \\ &\leq C, \quad f,g \in \mathrm{dom}(T),\ z \in \overline{\Sigma}, \qquad (2.64) \end{aligned}$$

where $C = C(f,g,S,T) > 0$ is a finite constant, independent of $z \in \overline{\Sigma}$.

Applying the Hadamard three-lines estimate (2.25) to $\phi_k(\cdot)$ then yields the first estimate in (2.53) since $\mathrm{dom}(T)$ is dense in $\mathcal{H}$ and $\overline{T^{-z}ST^{-1+z}} \in \mathcal{B}(\mathcal{H})$, $z \in \overline{\Sigma}$, by (2.52).

If in addition $T \geq 0$, one chooses $k = 0$ in (2.61) and uses (2.31) (instead of (2.30)) to arrive at the second estimate in (2.53). □

We emphasize that the case $T \geq 0$ (actually, $T \geq \varepsilon I_{\mathcal{H}}$ for some $\varepsilon > 0$) in the estimate (2.54) was also derived by Huang [14, Lemma 2.1.(a)].

The results provided thus far naturally extend to the situation where $T^{-z}ST^{-1+z}$ is replaced by $T_2^{-z}ST_1^{-1+z}$ for two self-adjoint operators T_j in $\mathcal{H}_j$, $j = 1, 2$. As an example, we now illustrate this in the context of Theorem 2.8.

Hypothesis 2.9. *Assume that T_j are self-adjoint operators in $\mathcal{H}_j$ with $T_j^{-1} \in \mathcal{B}(\mathcal{H}_j)$, $j = 1, 2$. In addition, suppose that S is a closed operator mapping* $\operatorname{dom}(S) \subseteq \mathcal{H}_1$ *into $\mathcal{H}_2$ satisfying*

$$\operatorname{dom}(S) \supseteq \operatorname{dom}(T_1) \ \textit{and} \ \operatorname{dom}(S^*) \supseteq \operatorname{dom}(T_2). \tag{2.65}$$

In particular, Hypothesis 2.9 implies that

$$ST_1^{-1} \in \mathcal{B}(\mathcal{H}_1, \mathcal{H}_2), \quad S^*T_2^{-1} \in \mathcal{B}(\mathcal{H}_2, \mathcal{H}_1). \tag{2.66}$$

Assuming Hypothesis 2.9, one obtains the following corollary of Theorem 2.8:

Corollary 2.10. *Assume Hypothesis 2.9. Then $T_2^{-z}ST_1^{-1+z}$ defined on* $\operatorname{dom}(T_1)$, *$z \in \overline{\Sigma}$, is closable, and*

$$\begin{aligned} \overline{T_2^{-z}ST_1^{-1+z}} &= T_2^{-i\operatorname{Im}(z)}\big[|S^*|^{\operatorname{Re}(z)}T_2^{-\operatorname{Re}(z)}\big]^* U_S \\ &\quad \times |S|^{1-\operatorname{Re}(z)}T_1^{-1+\operatorname{Re}(z)}T_1^{i\operatorname{Im}(z)} \in \mathcal{B}(\mathcal{H}_1, \mathcal{H}_2), \quad z \in \overline{\Sigma}. \end{aligned} \tag{2.67}$$

In addition, given $k \in (0, \infty)$, one obtains

$$\begin{aligned} &\big\|\overline{T_2^{-z}ST_1^{-1+z}}\big\|_{\mathcal{B}(\mathcal{H}_1,\mathcal{H}_2)} \leq \big\|ST_1^{-1}\big\|_{\mathcal{B}(\mathcal{H}_1,\mathcal{H}_2)}^{1-\operatorname{Re}(z)} \big\|S^*T_2^{-1}\big\|_{\mathcal{B}(\mathcal{H}_2,\mathcal{H}_1)}^{\operatorname{Re}(z)} \\ &\quad \times \begin{cases} e^{k|\operatorname{Im}(z)|^2 + k\operatorname{Re}(z)[1-\operatorname{Re}(z)] + k^{-1}\pi^2}, \\ e^{k|\operatorname{Im}(z)|^2 + k\operatorname{Re}(z)[1-\operatorname{Re}(z)] + (4k)^{-1}\pi^2}, \ \textit{if } T_1 \geq 0, \ \textit{or } T_2 \geq 0, \\ 1, \ \textit{if } T_j \geq 0, \ j = 1, 2, \end{cases} \qquad z \in \overline{\Sigma}, \end{aligned} \tag{2.68}$$

and

$$\begin{aligned} &\big\|\overline{T_2^{-x}ST_1^{-1+x}}\big\|_{\mathcal{B}(\mathcal{H}_1,\mathcal{H}_2)} \leq \big\|ST_1^{-1}\big\|_{\mathcal{B}(\mathcal{H}_1,\mathcal{H}_2)}^{1-x} \big\|S^*T_2^{-1}\big\|_{\mathcal{B}(\mathcal{H}_2,\mathcal{H}_1)}^{x} \\ &\quad \times \begin{cases} e^{2\pi[x(1-x)]^{1/2}}, \\ e^{\pi[x(1-x)]^{1/2}}, \ \textit{if } T_1 \geq 0, \ \textit{or } T_2 \geq 0, \\ 1, \ \textit{if } T_j \geq 0, \ j = 1, 2, \end{cases} \qquad x \in [0, 1]. \end{aligned} \tag{2.69}$$

Proof. Consider $\boldsymbol{\mathcal{H}} := \mathcal{H}_1 \oplus \mathcal{H}_2$ and introduce

$$\mathbf{S} = \begin{pmatrix} 0 & 0 \\ S & 0 \end{pmatrix}, \quad \operatorname{dom}(\mathbf{S}) = \operatorname{dom}(S) \oplus \mathcal{H}_2, \tag{2.70}$$

$$\mathbf{S}^* = \begin{pmatrix} 0 & S^* \\ 0 & 0 \end{pmatrix}, \quad \operatorname{dom}(\mathbf{S}^*) = \mathcal{H}_1 \oplus \operatorname{dom}(S^*), \tag{2.71}$$

and

$$\mathbf{T} = \begin{pmatrix} T_1 & 0 \\ 0 & T_2 \end{pmatrix}, \quad \operatorname{dom}(\mathbf{T}) = \operatorname{dom}(T_1) \oplus \operatorname{dom}(T_2). \tag{2.72}$$

Then $\mathbf{S}$ is a closed operator in $\boldsymbol{\mathcal{H}}$ and $\mathbf{T}$ is a self-adjoint operator in $\boldsymbol{\mathcal{H}}$ with bounded inverse given by

$$\mathbf{T}^{-1} = \begin{pmatrix} T_1^{-1} & 0 \\ 0 & T_2^{-1} \end{pmatrix} \in \mathcal{B}(\boldsymbol{\mathcal{H}}). \tag{2.73}$$

Moreover,

$$\operatorname{dom}(\mathbf{S})\cap\operatorname{dom}(\mathbf{S}^*) = \operatorname{dom}(S)\oplus\operatorname{dom}(S^*) \supseteq \operatorname{dom}(T_1)\oplus\operatorname{dom}(T_2) = \operatorname{dom}(\mathbf{T}), \quad (2.74)$$

that is, the pair $(\mathbf{S},\mathbf{T})$ satisfies Hypothesis 2.1.

Thus,

$$\mathbf{S}\mathbf{T}^{-1} = \begin{pmatrix} 0 & 0 \\ ST_1^{-1} & 0 \end{pmatrix} \in \mathcal{B}(\boldsymbol{\mathcal{H}}), \quad \mathbf{S}^*\mathbf{T}^{-1} = \begin{pmatrix} 0 & S^*T_2^{-1} \\ 0 & 0 \end{pmatrix} \in \mathcal{B}(\boldsymbol{\mathcal{H}}), \quad (2.75)$$

and

$$\|\mathbf{S}\mathbf{T}^{-1}\|_{\mathcal{B}(\boldsymbol{\mathcal{H}})} = \|ST_1^{-1}\|_{\mathcal{B}(\mathcal{H}_1,\mathcal{H}_2)}, \quad \|\mathbf{S}^*\mathbf{T}^{-1}\|_{\mathcal{B}(\boldsymbol{\mathcal{H}})} = \|S^*T_2^{-1}\|_{\mathcal{B}(\mathcal{H}_2,\mathcal{H}_1)}. \quad (2.76)$$

Applying Theorem 2.8 to the pair $(\mathbf{S},\mathbf{T})$, one obtains (2.68) and (2.69). □

In the special case, where $\mathcal{H}_1 = \mathcal{H}_2 = \mathcal{H}$, S and $T_1, T_2 \geqslant 0$ are bounded linear operators in $\mathcal{H}$, the third estimate (2.69) recovers Lemma 25 in [4]. On the other hand, the case $x \in [0,1]$ in the first estimate in (2.69) is a special case of (4.32), which in turn is recorded in [43, Lemma 16.3].

For connections with the generalized Loewner–Heinz inequality and bounds on operators of the type $T_2^{-x}ST_1^{-x}$, $x \in [0,1]$, under various conditions on S and T_j, $j = 1,2$, we also refer to [14, Lemma 2.1], [16, Theorem 3], [18, Theorem 6], and the references cited therein. In particular, [14, Lemma 2.1] predates and extends [22, Proposition A.1.(3)]. For additional variants on the Loewner–Heinz inequality we refer, for instance, to [28, Lemma 11, Remark 12], and especially, to [5, Sect. 3.12] and the detailed bibliography collected therein. For interesting extensions of the Loewner–Heinz inequality employing operator monotone functions we also refer to [38], [40], [41]. (An exhaustive list of references on (extensions of) the Loewner–Heinz inequality is beyond the scope of this short paper due to the enormous amount of literature on this subject.)

3. Interpolation and trace ideals revisited

In this section we recall a powerful result on interpolation theory in connection with linear operators in the trace ideal spaces $\mathcal{B}_p(\mathcal{H})$, $p \in [1,\infty)$, originally due to I.C. Gohberg, M.G. Krein, and S.G. Krein (cf. [8, p. 139]) and then apply it to the types of operators studied in Section 2. For background on trace ideals we refer to [8, Ch. I–III], [33], [34, Chs. 1–3].

In particular, we will dwell a bit on a particular case omitted in the discussion of Gohberg and Krein [8, Theorem III.13.1] (see also [9, Theorem III.5.1]):

Theorem 3.1 ([8], Theorem III.13.1 (see also [9], Theorem III.5.1)). *Let $p_0, p_1 \in [1,\infty)$, $p_0 \leq p_1$, and suppose that $A(z) \in \mathcal{B}(\mathcal{H})$, $z \in \overline{\Sigma}$, and that $A(\cdot)$ is analytic on Σ. Assume that for some $C_0, C_1 \in (0,\infty)$,*

$$\|A(iy)\|_{\mathcal{B}_{p_0}(\mathcal{H})} \leq C_0, \quad \|A(1+iy)\|_{\mathcal{B}_{p_1}(\mathcal{H})} \leq C_1, \quad y \in \mathbb{R}, \quad (3.1)$$

and suppose that for all $f, g \in \mathcal{H}$, *there exist* $C_{f,g} \in \mathbb{R}$ *and* $a_{f,g} \in [0, \pi)$, *such that*

$$\sup_{z\in\Sigma}\left[e^{-a_{f,g}|\operatorname{Im}(z)|}\ln(|(f, A(z)g)_{\mathcal{H}}|)\right] \leq C_{f,g}. \tag{3.2}$$

Then

$$A(z) \in \mathcal{B}_{p_z}(\mathcal{H}), \quad \frac{1}{p_z} = \frac{1-\operatorname{Re}(z)}{p_0} + \frac{\operatorname{Re}(z)}{p_1}, \quad z \in \overline{\Sigma}, \tag{3.3}$$

and

$$\|A(z)\|_{\mathcal{B}_{p_z}(\mathcal{H})} \leq C_0^{1-\operatorname{Re}(z)} C_1^{\operatorname{Re}(z)}, \quad z \in \overline{\Sigma}. \tag{3.4}$$

The estimate (3.4) *remains valid for* $p_0 \in [1, \infty)$ *and* $p_1 = \infty$ *in* (3.1) *and* (3.3). *In particular, if* $p_0 = p_1 \in [1, \infty)$, *then*

$$\|A(z)\|_{\mathcal{B}_{p_0}(\mathcal{H})} \leq C_0^{1-\operatorname{Re}(z)} C_1^{\operatorname{Re}(z)}, \quad z \in \overline{\Sigma}. \tag{3.5}$$

Proof. Theorem 3.1 and its proof is presented by Gohberg and Krein in [8, Theorem III.13.1] under the additional assumption that $p_0 < p_1$. (Moreover, this additional restriction is repeated in [9, Theorem III.5.1], where the result is stated without proof.) However, their proof extends to special case where $p_0 = p_1 \in [1, \infty)$ without any difficulties, in fact, it even simplifies a bit. For the convenience of the reader we now present the proof in this particular situation.

Let $F \in \mathcal{B}(\mathcal{H})$ be a finite-rank operator and suppose that

$$\begin{aligned} &\|F\|_{\mathcal{B}_{p_0'}(\mathcal{H})} = 1, \quad p_0^{-1} + (p_0')^{-1} = 1, \;\text{ if } p_0 \in (1, \infty), \\ &\|F\|_{\mathcal{B}(\mathcal{H})} = 1, \quad \text{if } p_0 = 1, \end{aligned} \tag{3.6}$$

and consider the function

$$\varphi(z) = \operatorname{tr}(A(z)F), \quad z \in \overline{\Sigma}. \tag{3.7}$$

By the assumptions on $A(\cdot)$, $\varphi(\cdot)$ is analytic on Σ and

$$\sup_{z\in\Sigma}\left[e^{-a|\operatorname{Im}(z)|}\ln(|\varphi(z)|)\right] \leq C \tag{3.8}$$

for some $a = a(F) \in [0, \pi)$ and $C = C(F) \in \mathbb{R}$. In addition, one estimates

$$|\varphi(iy)| \leq \|A(iy)F\|_{\mathcal{B}_1(\mathcal{H})} \leq \|A(iy)\|_{\mathcal{B}_{p_0}(\mathcal{H})}\, \|F\|_{\mathcal{B}_{p_0'}(\mathcal{H})} \leq C_0, \quad y \in \mathbb{R}, \tag{3.9}$$

$$|\varphi(1+iy)| \leq \|A(1+iy)F\|_{\mathcal{B}_1(\mathcal{H})} \leq \|A(1+iy)\|_{\mathcal{B}_{p_0}(\mathcal{H})}\, \|F\|_{\mathcal{B}_{p_0'}(\mathcal{H})} \leq C_1, \quad y \in \mathbb{R}. \tag{3.10}$$

Applying Hadamard's three-lines estimate (2.25) to $\varphi(\cdot)$ then yields

$$|\varphi(z)| = |\operatorname{tr}(A(z)F)| \leq C_0^{1-\operatorname{Re}(z)} C_1^{\operatorname{Re}(z)}, \quad z \in \overline{\Sigma}. \tag{3.11}$$

Next, denoting by $\mathcal{F}(\mathcal{H})$ the set of all finite-rank operators in $\mathcal{H}$, we recall that $B \in \mathcal{B}_{p_0}(\mathcal{H})$ if and only if the number $\|B\|_{\mathcal{B}_{p_0}(\mathcal{H})}$ is finite, where

$$\|B\|_{\mathcal{B}_{p_0}(\mathcal{H})} = \begin{cases} \sup_{0\neq F\in\mathcal{F}(\mathcal{H})} |\operatorname{tr}(BF)|/\|F\|_{\mathcal{B}_{p_0'}(\mathcal{H})}, & p_0 \in (1, \infty), \\ \sup_{0\neq F\in\mathcal{F}(\mathcal{H})} |\operatorname{tr}(BF)|/\|F\|_{\mathcal{B}(\mathcal{H})}, & p_0 = 1 \end{cases} \tag{3.12}$$

(cf. [8, Lemma III.12.1]). Thus, (3.11) implies (3.5) due to the normalization in (3.6). □

Alternatively, Theorem 3.1 follows from combining Theorems IX.20, IX.22, and Proposition 8 in [29].

Next we prove a trace ideal analog of Theorem 2.8, using Theorem 3.1:

Theorem 3.2. *In addition to Hypothesis* 2.1, *let* $p \in [1, \infty)$, *and assume that*

$$ST^{-1} \in \mathcal{B}_p(\mathcal{H}), \quad S^*T^{-1} \in \mathcal{B}_p(\mathcal{H}). \tag{3.13}$$

Then $T^{-z}ST^{-1+z}$, $z \in \overline{\Sigma}$, *defined on* $\operatorname{dom}(T)$, *is closable in* $\mathcal{H}$, *and*

$$\overline{T^{-z}ST^{-1+z}} \in \mathcal{B}_p(\mathcal{H}), \quad z \in \overline{\Sigma}. \tag{3.14}$$

In addition, given $k \in (0, \infty)$, *one obtains*

$$\begin{aligned}\big\|\overline{T^{-z}ST^{-1+z}}\big\|_{\mathcal{B}_p(\mathcal{H})} &\leq \big\|ST^{-1}\big\|_{\mathcal{B}_p(\mathcal{H})}^{1-\operatorname{Re}(z)} \big\|S^*T^{-1}\big\|_{\mathcal{B}_p(\mathcal{H})}^{\operatorname{Re}(z)} \\ &\quad \times \begin{cases} e^{k|\operatorname{Im}(z)|^2 + k\operatorname{Re}(z)[1-\operatorname{Re}(z)] + k^{-1}\pi^2}, \\ 1, \text{ if } T \geq 0, \end{cases} \quad z \in \overline{\Sigma},\end{aligned} \tag{3.15}$$

and

$$\big\|\overline{T^{-x}ST^{-1+x}}\big\|_{\mathcal{B}_p(\mathcal{H})} \leq \big\|ST^{-1}\big\|_{\mathcal{B}_p(\mathcal{H})}^{1-x} \big\|S^*T^{-1}\big\|_{\mathcal{B}_p(\mathcal{H})}^{x} \begin{cases} e^{2\pi[x(1-x)]^{1/2}}, \\ 1, \text{ if } T \geq 0, \end{cases} \quad x \in [0,1]. \tag{3.16}$$

In particular, assuming $T \geq 0$ *and taking* $x = 1/2$ *in* (3.16) *one obtains*

$$T^{-1/2}ST^{-1/2} = \big(T^{-1/2}S^*T^{-1/2}\big)^* \in \mathcal{B}_p(\mathcal{H}), \tag{3.17}$$

and

$$\big\|T^{-1/2}ST^{-1/2}\big\|_{\mathcal{B}_p(\mathcal{H})} = \big\|T^{-1/2}S^*T^{-1/2}\big\|_{\mathcal{B}_p(\mathcal{H})} \leq \big\|ST^{-1}\big\|_{\mathcal{B}_p(\mathcal{H})}^{1/2} \big\|S^*T^{-1}\big\|_{\mathcal{B}_p(\mathcal{H})}^{1/2}. \tag{3.18}$$

Proof. First we note that Theorem 2.8 applies and hence (2.52), (2.53) are at our disposal. Next, we introduce

$$A_k(z) = e^{kz(z-1)}\overline{T^{-z}ST^{-1+z}}, \quad z \in \overline{\Sigma}, \ k \in [0, \infty), \tag{3.19}$$

and focus again on $k > 0$ first.

Employing (2.30) one estimates

$$\begin{aligned}\|A_k(iy)\|_{\mathcal{B}_p(\mathcal{H})} &= e^{-ky^2}\big\|T^{-iy}ST^{-1}T^{iy}\big\|_{\mathcal{B}_p(\mathcal{H})} \leq e^{-ky^2+2\pi|y|}\big\|ST^{-1}\big\|_{\mathcal{B}_p(\mathcal{H})} \\ &\leq e^{k^{-1}\pi^2}\big\|ST^{-1}\big\|_{\mathcal{B}_p(\mathcal{H})}, \quad y \in \mathbb{R},\end{aligned} \tag{3.20}$$

$$\begin{aligned}\|A_k(1+iy)\|_{\mathcal{B}_p(\mathcal{H})} &= e^{-ky^2}\big\|\overline{T^{-1-iy}ST^{iy}}\big\|_{\mathcal{B}_p(\mathcal{H})} = e^{-ky^2}\big\|T^{-iy}(S^*T^{-1})^*T^{iy}\big\|_{\mathcal{B}_p(\mathcal{H})} \\ &= e^{k^{-1}\pi^2}\big\|S^*T^{-1}\big\|_{\mathcal{B}_p(\mathcal{H})}, \quad y \in \mathbb{R},\end{aligned} \tag{3.21}$$

$$\begin{aligned}\|A_k(z)\|_{\mathcal{B}(\mathcal{H})} &= e^{-k|\operatorname{Im}(z)|^2+k\operatorname{Re}(z)[\operatorname{Re}(z)-1]} \\ &\quad\times \left\|T^{-i\operatorname{Im}(z)}\left(|S^*|^{\operatorname{Re}(z)}T^{-\operatorname{Re}(z)}\right)^*U_S|S|^{1-\operatorname{Re}(z)}T^{-1+\operatorname{Re}(z)}T^{i\operatorname{Im}(z)}\right\|_{\mathcal{B}(\mathcal{H})} \\ &\leq e^{-k|\operatorname{Im}(z)|^2+k\operatorname{Re}(z)[\operatorname{Re}(z)-1]+2\pi|\operatorname{Im}(z)|} \\ &\quad\times \left\||S^*|^{\operatorname{Re}(z)}T^{-\operatorname{Re}(z)}\right\|_{\mathcal{B}(\mathcal{H})}\left\||S|^{1-\operatorname{Re}(z)}T^{-1+\operatorname{Re}(z)}\right\|_{\mathcal{B}(\mathcal{H})} \\ &\leq C, \quad z\in\overline{\Sigma}, \end{aligned} \tag{3.22}$$

where $C = C(S,T) > 0$ is a finite constant, independent of $z \in \overline{\Sigma}$, employing (2.57) and (2.59). Here again we used the generalized polar decomposition (2.51) for S (with $\alpha = \operatorname{Re}(z)$).

Applying the Hadamard three-lines estimate (3.5) to $A_k(\cdot)$ then yields the first relation in (3.14) and the estimate (3.15). ☐

In the special case where $T \geq 0$ and $S = S^* \in \mathcal{B}(\mathcal{H})$, the second estimate (3.16) recovers the result [36, Lemma 15] (see also [24, Lemma 5.10]). For applications of (3.16) to scattering theory we refer, for instance, to [30, Appendix 1].

Corollary 3.3. *In addition to Hypothesis* 2.9, *let* $p \in [1,\infty)$ *and assume that*

$$ST_1^{-1} \in \mathcal{B}_p(\mathcal{H}_1,\mathcal{H}_2), \quad S^*T_2^{-1} \in \mathcal{B}_p(\mathcal{H}_2,\mathcal{H}_1). \tag{3.23}$$

Then $T_2^{-z}ST_1^{-1+z}$ *defined on* $\operatorname{dom}(T_1)$, $z \in \overline{\Sigma}$, *is closable, and*

$$\overline{T_2^{-z}ST_1^{-1+z}} \in \mathcal{B}_p(\mathcal{H}_1,\mathcal{H}_2), \quad z \in \overline{\Sigma}. \tag{3.24}$$

In addition, given $k \in (0,\infty)$, *one obtains*

$$\begin{aligned}\left\|\overline{T_2^{-z}ST_1^{-1+z}}\right\|_{\mathcal{B}_p(\mathcal{H}_1,\mathcal{H}_2)} &\leq \left\|ST_1^{-1}\right\|_{\mathcal{B}_p(\mathcal{H}_1,\mathcal{H}_2)}^{1-\operatorname{Re}(z)}\left\|S^*T_2^{-1}\right\|_{\mathcal{B}_p(\mathcal{H}_2,\mathcal{H}_1)}^{\operatorname{Re}(z)} \\ &\quad\times\begin{cases} e^{k|\operatorname{Im}(z)|^2+k\operatorname{Re}(z)[1-\operatorname{Re}(z)]+k^{-1}\pi^2}, \\ e^{k|\operatorname{Im}(z)|^2+k\operatorname{Re}(z)[1-\operatorname{Re}(z)]+(4k)^{-1}\pi^2}, & \textit{if } T_1 \geq 0, \textit{ or } T_2 \geq 0, \\ 1, \ \textit{if } T_j \geq 0,\ j=1,2, \end{cases} \quad z \in \overline{\Sigma}, \end{aligned} \tag{3.25}$$

and

$$\begin{aligned}\left\|\overline{T_2^{-x}ST_1^{-1+x}}\right\|_{\mathcal{B}_p(\mathcal{H}_1,\mathcal{H}_2)} &\leq \left\|ST_1^{-1}\right\|_{\mathcal{B}_p(\mathcal{H}_1,\mathcal{H}_2)}^{1-x}\left\|S^*T_2^{-1}\right\|_{\mathcal{B}_p(\mathcal{H}_2,\mathcal{H}_1)}^{x} \\ &\quad\times\begin{cases} e^{2\pi[x(1-x)]^{1/2}}, \\ e^{\pi[x(1-x)]^{1/2}}, \ \textit{if } T_1 \geq 0, \textit{ or } T_2 \geq 0, \\ 1, \ \textit{if } T_j \geq 0,\ j=1,2, \end{cases} \quad x \in [0,1]. \end{aligned} \tag{3.26}$$

Proof. One can follow the proof of Corollary 2.10 step by step replacing $\mathcal{B}(\mathcal{H})$ and $\mathcal{B}(\mathcal{H}_j)$ by $\mathcal{B}_p(\mathcal{H})$ and $\mathcal{B}_p(\mathcal{H}_j)$, $j = 1,2$, respectively, applying Theorem 3.2 instead of Theorem 2.8. ☐

Finally, we recall the following known result in connection with the ideal $\mathcal{B}_\infty(\mathcal{H})$:

Theorem 3.4 ([31], pp. 115–116). *Suppose that $A(z) \in \mathcal{B}(\mathcal{H})$, $z \in \overline{\Sigma}$, that $A(\cdot)$ is analytic on Σ, weakly continuous on $\overline{\Sigma}$. Assume that for some $C_0, C_1 \in (0,\infty)$,*

$$\|A(iy)\|_{\mathcal{B}(\mathcal{H})} \leq C_0, \quad \|A(1+iy)\|_{\mathcal{B}(\mathcal{H})} \leq C_1, \quad y \in \mathbb{R}, \tag{3.27}$$

and suppose that for all $f, g \in \mathcal{H}$, there exist $C_{f,g} \in \mathbb{R}$ and $a_{f,g} \in [0,\pi)$, such that

$$\sup_{z\in\Sigma} \Big[e^{-a_{f,g}|\operatorname{Im}(z)|} \ln(|(f, A(z)g)_{\mathcal{H}}|) \Big] \leq C_{f,g}. \tag{3.28}$$

In addition, suppose that

$$\text{either } A(iy) \in \mathcal{B}_\infty(\mathcal{H}), \text{ or } A(1+iy) \in \mathcal{B}_\infty(\mathcal{H}), \quad y \in \mathbb{R}. \tag{3.29}$$

Then,

$$A(z) \in \mathcal{B}_\infty(\mathcal{H}), \quad z \in \Sigma. \tag{3.30}$$

A condition of the type (3.28) has inadvertently been omitted in [31, pp. 115–116].

4. Extensions to sectorial operators

In this section we revisit Theorems 2.3, 2.8, 3.2, and Corollaries 2.10, 3.3, and replace the self-adjointness hypothesis on T by appropriate sectoriality assumptions.

We start by recalling the definition of a sectorial operator and refer, for instance, to [11, Chs. 2, 3, 7] and [43, Chs. 2, 16] for a detailed treatment.

Definition 4.1. Let T be a densely defined, closed, linear operator in $\mathcal{H}$ and denote by $S_\omega \subset \mathbb{C}$, $\omega \in [0,\pi)$, the open sector

$$S_\omega = \begin{cases} \{z \in \mathbb{C} \,|\, z \neq 0, \, |\arg(z)| < \omega\}, & \omega \in (0,\pi), \\ (0,\infty), & \omega = 0, \end{cases} \tag{4.1}$$

with vertex at $z = 0$ along the positive real axis and opening angle 2ω. The operator T is called *sectorial of angle* $\omega \in [0,\pi)$, denoted by $T \in \operatorname{Sect}(\omega)$, if

$$\begin{aligned} &(\alpha)\ \sigma(T) \subseteq \overline{S_\omega}, \\ &(\beta)\ \text{for all } \omega' \in (\omega,\pi), \quad \sup_{z\in\mathbb{C}\setminus\overline{S_{\omega'}}} \|z(T - zI_{\mathcal{H}})^{-1}\|_{\mathcal{B}(\mathcal{H})} < \infty. \end{aligned} \tag{4.2}$$

One calls

$$\omega_T = \min\{\omega \in [0,\pi] \,|\, T \in \operatorname{Sect}(\omega)\}, \tag{4.3}$$

the *angle of sectoriality of* T.

For the remainder of this section we assume that T is sectorial (that is, $T \in \operatorname{Sect}(\omega)$ for some $\omega \in [0,\pi)$) and that $T^{-1} \in \mathcal{B}(\mathcal{H})$.

Then fractional powers T^{-z}, with $\operatorname{Re}(z) > 0$, of T can be defined by a standard Dunford integral in $\mathcal{B}(\mathcal{H})$ (cf., e.g., [43, Sect. 2.7.1]),

$$T^{-z} = (2\pi i)^{-1} \oint_\Gamma d\zeta\, \zeta^{-z} (T - \zeta I_{\mathcal{H}})^{-1}, \quad \operatorname{Re}(z) > 0, \tag{4.4}$$

using the principal branch of ζ^{-z}, $\{\zeta \in \mathbb{C} \,|\, |\arg(\zeta)| < \pi\}$, by excluding the negative real axis, with Γ surrounding $\sigma(T)$ clockwise in $(\mathbb{C}\backslash(-\infty, 0]) \cap \rho(T)$ (cf. [43, p. 92] for precise details). An important property of T^{-z} is that

$$T^{-z} \text{ is a } \mathcal{B}(\mathcal{H})\text{-valued analytic semigroup on } \{z \in \mathbb{C} \,|\, \operatorname{Re}(z) > 0\}. \tag{4.5}$$

Defining imaginary powers of T requires a bit more care. Following [43, p. 105], we introduce the imaginary powers T^{is}, $s \in \mathbb{R}$, of T as follows:

$$\begin{aligned} T^{is} f &= \operatorname*{s-lim}_{z \to is,\, \operatorname{Re}(z)>0} T^{-z} f, \\ f \in \operatorname{dom}(T^{is}) &= \Big\{ g \in \mathcal{H} \,\Big|\, \operatorname*{s-lim}_{z \to is,\, \operatorname{Re}(z)>0} T^{-z} g \text{ exists} \Big\}. \end{aligned} \tag{4.6}$$

We note that one can define imaginary powers of T also more explicitly as follows: for $s \in \mathbb{R}$, one sets as in [1, p. 153],

$$T^{is} f := \frac{\sin(\pi i s)}{\pi i s} \int_0^\infty t^{is} (T + t I_{\mathcal{H}})^{-2} T f \, dt, \quad f \in \operatorname{dom}(T). \tag{4.7}$$

Then the operator T^{is} is closable for every $s \in \mathbb{R}$ and one defines

$$T^{is} := \overline{T^{is}|_{\operatorname{dom}(T)}}, \quad s \in \mathbb{R}. \tag{4.8}$$

We also note that there are several definitions of the fractional (and imaginary) powers in the literature, see, for instance, [11, Section 3.2 and Proposition 3.5.5], [20], [25, Section 4], [26], [39, Section 1], [1, Section 4]. In our setting, all of these definitions coincide (cf. [2]), and we provided the most straightforward one.

To be able to argue as in previous sections one needs to deal with sectorial operators having *bounded imaginary powers* (*BIP*).

Definition 4.2. If T is a sectorial operator on $\mathcal{H}$ such that $T^{-1} \in \mathcal{B}(\mathcal{H})$, then T is said to have *bounded imaginary powers* if $T^{is} \in \mathcal{B}(\mathcal{H})$ for all $s \in \mathbb{R}$. This is then denoted by $T \in \operatorname{BIP}(\mathcal{H})$.

We recall that if T admits bounded imaginary powers then $\{T^{is}\}_{s\in\mathbb{R}}$ is a C_0-group on $\mathcal{H}$ (cf. [11, Corollary 3.5.7]). Hence, there exist $\theta \geq 0$ and $N \geq 1$ such that

$$\|T^{is}\|_{\mathcal{B}(\mathcal{H})} \leq N e^{\theta |s|}, \quad s \in \mathbb{R}, \tag{4.9}$$

and we write $T \in \operatorname{BIP}(N, \theta)$ in this case. Clearly,

$$\operatorname{BIP}(\mathcal{H}) = \bigcup_{N \geqslant 1,\, \theta \geqslant 0} \operatorname{BIP}(N, \theta). \tag{4.10}$$

We also define the type θ_T of the C_0-group $\{T^{is}\}_{s\in\mathbb{R}}$ by

$$\theta_T := \inf \big\{ \theta \geq 0 \,\big|\, \text{there exists } N_\theta \geq 1 \text{ such that } \big\|T^{is}\big\|_{\mathcal{B}(\mathcal{H})} \leq N_\theta e^{\theta|s|},\ s \in \mathbb{R} \big\}. \tag{4.11}$$

The standard example of operators T satisfying $T \in \mathrm{BIP}(\mathcal{H})$ (in addition to the situation described in (2.31)) are provided by strictly positive self-adjoint operators bounded from below (in this case $T \in \mathrm{BIP}(1,0)$) and boundedly invertible, m-accretive operators T (in this case $T \in \mathrm{BIP}(1,\pi/2)$). One recalls that T is said to be *m-accretive* (cf. [11, Sect. C.7], [17], [19, Sect. V.3.10], [25, Sect. 4.3], [37, Ch. 2]) if and only if

$$\overline{\mathrm{dom}(T)} = \mathcal{H}, \quad (-\infty,0) \subset \rho(T), \quad \big\|(T+\lambda I_{\mathcal{H}})^{-1}\big\|_{\mathcal{B}(\mathcal{H})} \le \lambda^{-1}, \ \lambda > 0. \tag{4.12}$$

The following extension of (4.5) will be vital for the remainder of this section:

Theorem 4.3 (See, e.g., [1], Theorem 4.7.1).
If $T \in \mathrm{BIP}(\mathcal{H})$ then $\{T^{-z} \,|\, \mathrm{Re}(z) \ge 0\}$ is a strongly continuous semigroup in the closed right half-plane $\{z \in \mathbb{C} \,|\, \mathrm{Re}(z) \ge 0\}$.

We note that by [11, Proposition 7.0.1] (or [43, p. 101]), $T \in \mathrm{Sect}(\omega)$ if and only if $T^* \in \mathrm{Sect}(\omega)$. Moreover, $(T^z)^* = (T^*)^{\bar{z}}$ and thus

$$T \in \mathrm{BIP}(N,\theta) \quad \text{if and only if} \quad T^* \in \mathrm{BIP}(N,\theta). \tag{4.13}$$

Theorem 4.3 together with (4.13) permits us to use the three-lines theorem in the present, more general setting of sectorial operators.

In the special case where T is self-adjoint and strictly positive in $\mathcal{H}$, that is, $T \ge \varepsilon I_{\mathcal{H}}$ for some $\varepsilon > 0$, T^α, $\alpha \in \mathbb{C}$, defined on one hand as sectorial operators as above, and on the other by the spectral theorem, coincide (cf., e.g., [25, Sect. 4.3.1], [39, Sect. 1.18.10]). In particular,

$$\mathrm{dom}(T^\alpha) = \bigg\{ f \in \mathcal{H} \,\bigg|\, \|T^\alpha f\|_{\mathcal{H}}^2 = \int_{[\varepsilon,\infty]} \lambda^{2\,\mathrm{Re}(\alpha)} d\|E_T(\lambda) f\|_{\mathcal{H}}^2 < \infty \bigg\}, \quad \alpha \in \mathbb{C}, \tag{4.14}$$

in this case. Here $\{E_T(\lambda)\}_{\lambda\in\mathbb{R}}$ denotes the family of spectral projections of T.

In the remainder of this section, we will use the following set of assumptions:

Hypothesis 4.4. *Assume that T is a sectorial operator in $\mathcal{H}$ such that $T^{-1} \in \mathcal{B}(\mathcal{H})$. In addition, we assume that S is a closed operator in $\mathcal{H}$ satisfying*

$$\mathrm{dom}(S) \supseteq \mathrm{dom}(T), \quad \mathrm{dom}(S^*) \supseteq \mathrm{dom}(T^*). \tag{4.15}$$

We start with the analog of Theorem 2.3:

Theorem 4.5. *Assume Hypothesis 4.4. Then the following facts hold:*

(i) *The operator $T^{-1}ST$ is well defined on $\mathrm{dom}(T^2)$, and hence densely defined in $\mathcal{H}$,*
$$\mathrm{dom}\big(T^{-1}ST\big) \supseteq \mathrm{dom}\big(T^2\big). \tag{4.16}$$

(ii) *The relation*
$$\big(T^{-1}ST\big)^* = T^*S^*(T^*)^{-1} \tag{4.17}$$
holds, and hence $T^*S^*(T^*)^{-1}$ is closed in $\mathcal{H}$.

(iii) *One infers that*
$$T^{-1}ST \text{ is bounded if and only if } (T^{-1}ST)^* = T^*S^*(T^*)^{-1} \in \mathcal{B}(\mathcal{H}). \tag{4.18}$$

If one of the assertions in (4.18) *hold, then*

$$\overline{T^{-1}ST} = (T^* S^* (T^*)^{-1})^*, \quad \big\|\overline{T^{-1}ST}\big\|_{\mathcal{B}(\mathcal{H})} = \big\|T^* S^* (T^*)^{-1}\big\|_{\mathcal{B}(\mathcal{H})}. \tag{4.19}$$

Proof. Since $\operatorname{dom}(T^2)$ is an operator core for T (cf. [11, Theorem 3.1.1]), one can follow the proof of Theorem 2.3 line by line. To illustrate this claim we just mention, for instance, the analog of (2.16) which now turns into

$$\begin{gathered}\big(\big(T^{-1}ST\big)^* f, g\big)_{\mathcal{H}} = \big(f, T^{-1}STg\big)_{\mathcal{H}} = \big((T^*)^{-1}f, STg\big)_{\mathcal{H}} = \big(S^*(T^*)^{-1}f, Tg\big)_{\mathcal{H}}, \\ f \in \operatorname{dom}\big(\big(T^{-1}ST\big)^*\big),\ g \in \operatorname{dom}\big(T^2\big) \subseteq \operatorname{dom}\big(T^{-1}ST\big),\end{gathered} \tag{4.20}$$

and hence once again extends to all $g \in \operatorname{dom}(T)$ as before in (2.17). □

Next, we turn to the analog of Theorem 2.8 and recall the notation used in (4.9):

Theorem 4.6. *Assume Hypothesis* 4.4. *If* $T \in \mathrm{BIP}(N, \theta)$, *then* $T^{-z}ST^{-1+z}$, $z \in \overline{\Sigma}$, *defined on* $\operatorname{dom}(T)$, *is closable in* $\mathcal{H}$, *and*

$$\begin{aligned}\overline{T^{-z}ST^{-1+z}} &= T^{-i\operatorname{Im}(z)}\big[|S^*|^{\operatorname{Re}(z)}(T^*)^{-\operatorname{Re}(z)}\big]^* U_S \\ &\quad \times |S|^{1-\operatorname{Re}(z)} T^{-1+\operatorname{Re}(z)} T^{i\operatorname{Im}(z)} \in \mathcal{B}(\mathcal{H}), \quad z \in \overline{\Sigma}.\end{aligned} \tag{4.21}$$

In addition, given $k \in (0,\infty)$, *one obtains*

$$\begin{aligned}\big\|\overline{T^{-z}ST^{-1+z}}\big\|_{\mathcal{B}(\mathcal{H})} &\le N^2 e^{k(\operatorname{Im}(z))^2 + k\operatorname{Re}(z)[1-\operatorname{Re}(z)] + k^{-1}\theta^2} \\ &\quad \times \big\|ST^{-1}\big\|_{\mathcal{B}(\mathcal{H})}^{1-\operatorname{Re}(z)} \big\|S^*(T^*)^{-1}\big\|_{\mathcal{B}(\mathcal{H})}^{\operatorname{Re}(z)}, \quad z \in \overline{\Sigma},\end{aligned} \tag{4.22}$$

and

$$\big\|\overline{T^{-x}ST^{-1+x}}\big\|_{\mathcal{B}(\mathcal{H})} \le N^2 e^{2\theta[x(1-x)]^{1/2}} \big\|ST^{-1}\big\|_{\mathcal{B}(\mathcal{H})}^{1-x} \big\|S^*(T^*)^{-1}\big\|_{\mathcal{B}(\mathcal{H})}^{x}, \quad x \in [0,1]. \tag{4.23}$$

Proof. Closely examining the first part of the proof of Theorem 2.8 based on the Loewner–Heinz inequality, one notes that everything up to (2.60) goes through without any change, implying the closability of $T^{-z}ST^{-1+z}$ and the validity of (4.21).

Next, one defines

$$\begin{gathered}\phi_k(z) = e^{kz(z-1)}\big(T^*f, T^{-1-z}ST^{-2+z}Tg\big)_{\mathcal{H}} = e^{kz(z-1)}\big(f, \overline{T^{-z}ST^{-1+z}}g\big)_{\mathcal{H}}, \\ f \in \operatorname{dom}(T^*),\ g \in \operatorname{dom}(T),\ z \in \overline{\Sigma},\ k \in (0,\infty).\end{gathered} \tag{4.24}$$

Then, employing (4.9) and (4.13), one estimates

$$\begin{aligned}|\phi_k(iy)| &= e^{-ky^2}\big|\big((T^*)^{iy}f, ST^{-1}T^{iy}g\big)_{\mathcal{H}}\big| \\ &\le e^{-ky^2} N^2 e^{2\theta|y|}\big\|ST^{-1}\big\|_{\mathcal{B}(\mathcal{H})}\|f\|_{\mathcal{H}}\|g\|_{\mathcal{H}} \\ &\le N^2 e^{k^{-1}\theta^2}\big\|ST^{-1}\big\|_{\mathcal{B}(\mathcal{H})}\|f\|_{\mathcal{H}}\|g\|_{\mathcal{H}}, \quad y \in \mathbb{R},\end{aligned} \tag{4.25}$$

$$
\begin{aligned}
|\phi_k(1+iy)| &= e^{-ky^2}\big|\big((T^*)^{iy}f, \overline{T^{-1}ST^{iy}}g\big)_{\mathcal{H}}\big| \\
&\leq e^{-ky^2}N^2e^{2\theta|y|}\big\|\overline{T^{-1}S}\big\|_{\mathcal{B}(\mathcal{H})}\|f\|_{\mathcal{H}}\|g\|_{\mathcal{H}} \\
&= e^{-ky^2}N^2e^{2\theta|y|}\big\|S^*(T^*)^{-1}\big\|_{\mathcal{B}(\mathcal{H})}\|f\|_{\mathcal{H}}\|g\|_{\mathcal{H}} \\
&\leq N^2e^{k^{-1}\theta^2}\big\|S^*(T^*)^{-1}\big\|_{\mathcal{B}(\mathcal{H})}\|f\|_{\mathcal{H}}\|g\|_{\mathcal{H}}, \quad y\in\mathbb{R}, \qquad (4.26)\\
|\phi_k(z)| &= e^{-k(\operatorname{Im}(z))^2+k\operatorname{Re}(z)[\operatorname{Re}(z)-1]}\big|\big(T^*f, T^{-1-z}ST^{-2+z}Tg\big)_{\mathcal{H}}\big| \\
&\leq e^{-k(\operatorname{Im}(z))^2+k\operatorname{Re}(z)[\operatorname{Re}(z)-1]}\big\|T^{-1-\operatorname{Re}(z)-i\operatorname{Im}(z)}\big\|_{\mathcal{B}(\mathcal{H})}\big\|ST^{-1}\big\|_{\mathcal{B}(\mathcal{H})} \\
&\quad\times\big\|T^{-1+\operatorname{Re}(z)+i\operatorname{Im}(z)}\big\|_{\mathcal{B}(\mathcal{H})}\|T^*f\|_{\mathcal{H}}\|Tg\|_{\mathcal{H}} \\
&\leq e^{-k(\operatorname{Im}(z))^2+k\operatorname{Re}(z)[\operatorname{Re}(z)-1]}N^2e^{2\theta|\operatorname{Im}(z)|}\big\|T^{-1-\operatorname{Re}(z)}\big\|_{\mathcal{B}(\mathcal{H})} \\
&\quad\times\big\|ST^{-1}\big\|_{\mathcal{B}(\mathcal{H})}\big\|T^{-1+\operatorname{Re}(z)}\big\|_{\mathcal{B}(\mathcal{H})}\|T^*f\|_{\mathcal{H}}\|Tg\|_{\mathcal{H}} \\
&\leq C_k, \quad f\in\operatorname{dom}(T^*),\ g\in\operatorname{dom}(T),\ z\in\overline{\Sigma}, \qquad (4.27)
\end{aligned}
$$

where $C_k = C_k(f,g,S,T) > 0$ is a finite constant, independent of $z \in \overline{\Sigma}$.

Applying the Hadamard three-lines estimate (2.25) to $\phi_k(\cdot)$ then yields (4.22) since $\operatorname{dom}(T)$ and $\operatorname{dom}(T^*)$ are dense in $\mathcal{H}$ and $\overline{T^{-z}ST^{-1+z}} \in \mathcal{B}(\mathcal{H})$, $z \in \overline{\Sigma}$, by (4.21). If $\operatorname{Im}(z) = 0$, optimizing (4.22) with respect to $k > 0$ implies (4.23). □

Remark 4.7. We recall that by McIntosh's theorem (cf. [11, Corollary 4.3.5]), one has

$$\theta_T = \omega_T, \tag{4.28}$$

where ω_T and θ_T are defined by (4.3) and (4.11), respectively. Thus, in principle, one can use ω_T to get estimates cruder than (4.22), (4.23), but then in *a priori* terms associated with T. However, we decided not to pursue this here. The same remark also concerns the statements in the remainder of this section.

In the special case where $T \geq 0$ and $S \in \mathcal{B}(\mathcal{H})$, the estimate (4.23) recovers [36, Lemma 15].

Again, these results naturally extend to the situation where $T^{-z}ST^{-1+z}$ is replaced by $T_2^{-z}ST_1^{-1+z}$ for two sectorial operators T_j in $\mathcal{H}_j$, $j = 1, 2$, having bounded imaginary powers, and once more we now illustrate this in the context of Theorem 4.6.

Hypothesis 4.8. *Assume that T_j are sectorial operators in $\mathcal{H}_j$ such that $T_j^{-1} \in \mathcal{B}(\mathcal{H}_j)$, $j = 1, 2$. In addition, suppose that S is a closed operator mapping* $\operatorname{dom}(S) \subseteq \mathcal{H}_1$ *into $\mathcal{H}_2$, satisfying*

$$\operatorname{dom}(S) \supseteq \operatorname{dom}(T_1) \ \textit{and} \ \operatorname{dom}(S^*) \supseteq \operatorname{dom}(T_2^*). \tag{4.29}$$

Then the analog of Corollary 2.10 reads as follows:

Corollary 4.9. *Assume Hypothesis* 4.8. *If* $T_j \in \mathrm{BIP}(N_j, \theta_j)$, $j = 1, 2$, *then* $T_2^{-z} S T_1^{-1+z}$ *defined on* $\mathrm{dom}(T_1)$, $z \in \overline{\Sigma}$, *is closable, and*

$$\begin{aligned}\overline{T_2^{-z} S T_1^{-1+z}} &= T_2^{-i\,\mathrm{Im}(z)} \big[|S^*|^{\mathrm{Re}(z)} (T_2^*)^{-\,\mathrm{Re}(z)}\big]^* U_S \\ &\quad \times |S|^{1-\mathrm{Re}(z)} T_1^{-1+\mathrm{Re}(z)} T_1^{i\,\mathrm{Im}(z)} \in \mathcal{B}(\mathcal{H}_1, \mathcal{H}_2), \quad z \in \overline{\Sigma}.\end{aligned} \tag{4.30}$$

In addition, given $k \in (0, \infty)$, *one obtains*

$$\begin{aligned}&\big\|\overline{T_2^{-z} S T_1^{-1+z}}\big\|_{\mathcal{B}(\mathcal{H}_1,\mathcal{H}_2)} \leq N_1 N_2 e^{k(\mathrm{Im}(z))^2 + k\,\mathrm{Re}(z)[1-\mathrm{Re}(z)] + (4k)^{-1}(\theta_1+\theta_2)^2} \\ &\quad \times \big\|S T_1^{-1}\big\|_{\mathcal{B}(\mathcal{H}_1,\mathcal{H}_2)}^{1-\mathrm{Re}(z)} \big\|S^* (T_2^*)^{-1}\big\|_{\mathcal{B}(\mathcal{H}_2,\mathcal{H}_1)}^{\mathrm{Re}(z)}, \quad z \in \overline{\Sigma},\end{aligned} \tag{4.31}$$

and

$$\begin{aligned}\big\|\overline{T_2^{-x} S T_1^{-1+x}}\big\|_{\mathcal{B}(\mathcal{H}_1,\mathcal{H}_2)} &\leq N_1 N_2 e^{(\theta_1+\theta_2)[x(1-x)]^{1/2}} \\ &\quad \times \big\|S T_1^{-1}\big\|_{\mathcal{B}(\mathcal{H}_1,\mathcal{H}_2)}^{1-x} \big\|S^* (T_2^*)^{-1}\big\|_{\mathcal{B}(\mathcal{H}_2,\mathcal{H}_1)}^{x}, \quad x \in [0, 1].\end{aligned} \tag{4.32}$$

Proof. Again, the 2×2 block operator formalism introduced in the proof of Corollary 2.10 applies to the case at hand. □

We emphasize that (4.31) is not new, it can be found in [43, Lemma 16.3]. Our proof, however, is slightly different.

Finally, we turn to the analogs of Theorem 3.2 and Corollary 3.3.

Theorem 4.10. *Assume Hypothesis* 4.4. *Moreover, let* $p \in [1, \infty)$, *and suppose that*

$$S T^{-1} \in \mathcal{B}_p(\mathcal{H}), \quad S^* (T^*)^{-1} \in \mathcal{B}_p(\mathcal{H}). \tag{4.33}$$

If $T \in \mathrm{BIP}(N, \theta)$, *then* $T^{-z} S T^{-1+z}$, $z \in \overline{\Sigma}$, *defined on* $\mathrm{dom}(T)$, *is closable in* $\mathcal{H}$, *and*

$$\overline{T^{-z} S T^{-1+z}} \in \mathcal{B}_p(\mathcal{H}), \quad z \in \overline{\Sigma}. \tag{4.34}$$

In addition, given $k \in (0, \infty)$, *one obtains*

$$\begin{aligned}\big\|\overline{T^{-z} S T^{-1+z}}\big\|_{\mathcal{B}_p(\mathcal{H})} &\leq N^2 e^{k(\mathrm{Im}(z))^2 + k\,\mathrm{Re}(z)[1-\mathrm{Re}(z)] + k^{-1}\theta^2} \\ &\quad \times \big\|S T^{-1}\big\|_{\mathcal{B}_p(\mathcal{H})}^{1-\mathrm{Re}(z)} \big\|S^* (T^*)^{-1}\big\|_{\mathcal{B}_p(\mathcal{H})}^{\mathrm{Re}(z)}, \quad z \in \overline{\Sigma},\end{aligned} \tag{4.35}$$

and

$$\begin{aligned}\big\|\overline{T^{-x} S T^{-1+x}}\big\|_{\mathcal{B}_p(\mathcal{H})} &\leq N^2 e^{2\theta[x(1-x)]^{1/2}} \\ &\quad \times \big\|S T^{-1}\big\|_{\mathcal{B}_p(\mathcal{H})}^{1-x} \big\|S^* (T^*)^{-1}\big\|_{\mathcal{B}_p(\mathcal{H})}^{x}, \quad x \in [0, 1].\end{aligned} \tag{4.36}$$

Proof. First we note that Theorem 4.6 applies and hence (4.21)–(4.23) are at our disposal. Next, one introduces

$$A_k(z) = e^{kz(z-1)} \overline{T^{-z} S T^{-1+z}}, \quad z \in \overline{\Sigma}, \ k \in (0, \infty), \tag{4.37}$$

and estimates

$$\|A_k(iy)\|_{\mathcal{B}_p(\mathcal{H})} = e^{-ky^2}\big\|T^{-iy}ST^{-1}T^{iy}\big\|_{\mathcal{B}_p(\mathcal{H})} \leq e^{-ky^2}N^2e^{2\theta|y|}\big\|ST^{-1}\big\|_{\mathcal{B}_p(\mathcal{H})}$$
$$\leq N^2e^{k^{-1}\theta^2}\big\|ST^{-1}\big\|_{\mathcal{B}_p(\mathcal{H})}, \quad y \in \mathbb{R}, \tag{4.38}$$
$$\|A_k(1+iy)\|_{\mathcal{B}_p(\mathcal{H})} = e^{-ky^2}\big\|\overline{T^{-1-iy}ST^{iy}}\big\|_{\mathcal{B}_p(\mathcal{H})}$$
$$= e^{-ky^2}\big\|T^{-iy}(S^*(T^*)^{-1})^*T^{iy}\big\|_{\mathcal{B}_p(\mathcal{H})} \leq e^{-ky^2}N^2e^{2\theta|y|}\big\|S^*(T^*)^{-1}\big\|_{\mathcal{B}_p(\mathcal{H})}$$
$$\leq N^2e^{k^{-1}\theta^2}\big\|S^*(T^*)^{-1}\big\|_{\mathcal{B}_p(\mathcal{H})}, \quad y \in \mathbb{R}, \tag{4.39}$$
$$\|A_k(z)\|_{\mathcal{B}(\mathcal{H})} = e^{-k(\operatorname{Im}(z))^2+k\operatorname{Re}(z)[\operatorname{Re}(z)-1]}$$
$$\times \big\|T^{-i\operatorname{Im}(z)}\big(|S^*|^{\operatorname{Re}(z)}(T^*)^{-\operatorname{Re}(z)}\big)^*U_S|S|^{1-\operatorname{Re}(z)}T^{-1+\operatorname{Re}(z)}T^{i\operatorname{Im}(z)}\big\|_{\mathcal{B}(\mathcal{H})}$$
$$\leq e^{-k(\operatorname{Im}(z))^2+k\operatorname{Re}(z)[\operatorname{Re}(z)-1]}N^2e^{2\theta|\operatorname{Im}(z)|}\,\big\||S^*|^{\operatorname{Re}(z)}(T^*)^{-\operatorname{Re}(z)}\big\|_{\mathcal{B}(\mathcal{H})}$$
$$\times \big\||S|^{1-\operatorname{Re}(z)}T^{-1+\operatorname{Re}(z)}\big\|_{\mathcal{B}(\mathcal{H})} \leq C_k, \quad z \in \overline{\Sigma}, \tag{4.40}$$

where $C_k = C_k(S,T) > 0$ is a finite constant, independent of $z \in \overline{\Sigma}$, applying (2.57) and (2.59). Here we used again the generalized polar decomposition (2.51) for S (with $\alpha = \operatorname{Re}(z)$).

Applying the Hadamard three-lines estimate (3.5) to $A_k(\cdot)$ then yields relation (4.34) and the estimate (4.35). If $\operatorname{Im}(z) = 0$, optimizing (4.35) with respect to $k > 0$ implies (4.36). □

Corollary 4.11. *In addition to Hypothesis* 4.8, *let* $p \in [1,\infty)$ *and assume that*

$$ST_1^{-1} \in \mathcal{B}_p(\mathcal{H}_1,\mathcal{H}_2), \quad S^*(T_2^*)^{-1} \in \mathcal{B}_p(\mathcal{H}_2,\mathcal{H}_1). \tag{4.41}$$

If $T_j \in \operatorname{BIP}(N_j,\theta_j)$, $j = 1,2$, *then* $T_2^{-z}ST_1^{-1+z}$ *defined on* $\operatorname{dom}(T_1)$, $z \in \overline{\Sigma}$, *is closable, and*

$$\overline{T_2^{-z}ST_1^{-1+z}} \in \mathcal{B}_p(\mathcal{H}_1,\mathcal{H}_2), \quad z \in \overline{\Sigma}. \tag{4.42}$$

In addition, given $k \in (0,\infty)$, *one obtains*

$$\big\|\overline{T_2^{-z}ST_1^{-1+z}}\big\|_{\mathcal{B}_p(\mathcal{H}_1,\mathcal{H}_2)} \leq N_1N_2e^{k(\operatorname{Im}(z))^2+k\operatorname{Re}(z)[1-\operatorname{Re}(z)]+(4k)^{-1}(\theta_1+\theta_2)^2}$$
$$\times \big\|ST_1^{-1}\big\|_{\mathcal{B}_p(\mathcal{H}_1,\mathcal{H}_2)}^{1-\operatorname{Re}(z)}\,\big\|S^*(T_2^*)^{-1}\big\|_{\mathcal{B}_p(\mathcal{H}_2,\mathcal{H}_1)}^{\operatorname{Re}(z)}, \quad z \in \overline{\Sigma}, \tag{4.43}$$

and

$$\begin{aligned}\big\|\overline{T_2^{-x}ST_1^{-1+x}}\big\|_{\mathcal{B}_p(\mathcal{H}_1,\mathcal{H}_2)} &\leq N_1N_2e^{(\theta_1+\theta_2)[x(1-x)]^{1/2}}\\ &\times \big\|ST_1^{-1}\big\|_{\mathcal{B}_p(\mathcal{H}_1,\mathcal{H}_2)}^{1-x}\,\big\|S^*(T_2^*)^{-1}\big\|_{\mathcal{B}_p(\mathcal{H}_2,\mathcal{H}_1)}^{x}, \quad x \in [0,1].\end{aligned} \tag{4.44}$$

Proof. Applying Theorem 4.10, one can follow the proof of Corollary 4.9 (see also Corollary 2.10) step by step replacing $\mathcal{B}(\mathcal{H})$ and $\mathcal{B}(\mathcal{H}_j)$ by $\mathcal{B}_p(\mathcal{H})$ and $\mathcal{B}_p(\mathcal{H}_j)$, $j = 1,2$, respectively. □

Acknowledgment

We are indebted to Alexander Gomilko for very helpful discussions. We also thank Matthias Lesch for valuable correspondence.

F.G. is indebted to all organizers of the Herrnhut Symposium, "Operator Semigroups meet Complex Analysis, Harmonic Analysis and Mathematical Physics" (June 3–7, 2013), and particularly, to Wolfgang Arendt, Ralph Chill, and Yuri Tomilov, for fostering an extraordinarily stimulating atmosphere during the meeting.

References

[1] H. Amann, *Linear and Quasilinear Parabolic Problems, Vol. I. Abstract Linear Theory*, Monographs in Mathematics, vol. 89, Birkhäuser, Basel, 1995.

[2] C.J.K. Batty, A. Gomilko, and Yu. Tomilov, *Product formulas in functional calculi for sectorial operators*, Math. Z. **279**, 479–507 (2015).

[3] C. Bennett and R. Sharpley, *Interpolation of Operators*, Academic Press, Boston, 1988.

[4] A. Carey, D. Potapov, and F. Sukochev, *Spectral flow is the integral of one forms on the Banach manifold of self adjoint Fredholm operators*, Adv. Math. **222**, 1809–1849 (2009).

[5] T. Furuta, *Invitation to Linear Operators. From Matrices to Bounded Linear Operators in a Hilbert Space*, Taylor & Francis, London, 2002.

[6] F. Gesztesy, Y. Latushkin, K.A. Makarov, F. Sukochev, and Y. Tomilov, *The index formula and the spectral shift function for relatively trace class perturbations*, Adv. Math. **227**, 319–420 (2011).

[7] F. Gesztesy, M. Malamud, M. Mitrea, and S. Naboko, *Generalized polar decompositions for closed operators in Hilbert spaces and some applications*, Integral Eq. Operator Theory **64**, 83–113 (2009).

[8] I.C. Gohberg and M.G. Krein, *Introduction to the Theory of Linear Nonselfadjoint Operators*, Translations of Mathematical Monographs, vol. 18, Amer. Math. Soc., Providence, RI, 1969.

[9] I.C. Gohberg and M.G. Krein, *Theory and Applications of Volterra Operators in Hilbert Space*, Translations of Mathematical Monographs, vol. 24, Amer. Math. Soc., Providence, RI, 1970.

[10] L. Grafakos, *Classical Fourier Analysis*, 2nd ed., Graduate Texts in Mathematics, vol. 249, Springer, New York, 2008.

[11] M. Haase, *The Functional Calculus for Sectorial Operators*, Operator Theory: Advances and Applications, vol. 169, Birkhäuser, Basel, 2006.

[12] E. Heinz, *Beiträge zur Störungstheorie der Spektralzerlegung*, Math. Ann. **123**, 415–438 (1951).

[13] I.I. Hirschman, *A convexity theorem for certain groups of transformations*, J. Analyse Math. **2**, 209–218 (1952).

[14] F. Huang, *On the mathematical model for linear elastic systems with analytic damping*, SIAM J. Control Optim. **26**, 714–724 (1988).

[15] T. Kato, *Notes on some inequaliies for linear operators*, Math. Ann. **125**, 208–212 (1952).

[16] T. Kato, *A generalization of the Heinz inequality*, Proc. Japan Acad. **37**, 305–308 (1961).

[17] T. Kato, *Fractional powers of dissipative operators*, J. Math. Soc. Japan, **13**, 246–274 (1961).

[18] T. Kato, *Fractional powers of dissipative operators, II*, J. Math. Soc. Japan, **14**, 242–248 (1962).

[19] T. Kato, *Perturbation Theory for Linear Operators*, corr. printing of the 2nd ed., Springer, Berlin, 1980.

[20] M.A. Krasnoselskii, P.P. Zabreiko, E.I. Pustylnik, and P.E. Sobolevskii, *Integral Operators in Spaces of Summable Functions*, Noordhoff, Leyden, 1976.

[21] S.G. Krein, Ju.I. Petunin, and E.M. Semenov, *Interpolation of Linear Operators*, Transl. Math. Monographs, vol. 54, Amer. Math. Soc., Providence, RI, 1982.

[22] M. Lesch, *The uniqueness of the spectral flow on spaces of unbounded self-adjoint Fredholm operators*, in *Spectral Geometry of Manifolds with Boundary and Decomposition of Manifolds*, B. Boss-Bavnbek, G. Grubb, and K.P. Wojciechowski (eds.), Contemp. Math. **366**, 193–224 (2005).

[23] K. Löwner, *Über monotone Matrixfunktionen*, Math. Z. **38**, 177–216 (1934).

[24] S. Lord, D. Potapov, and F. Sukochev, *Measures from Dixmier traces and zeta functions*, J. Funct. Anal. **259**, 1915–1949 (2010).

[25] A. Lunardi, *Interpolation Theory*, Lecture Notes, vol. 9, Scuola Normale Superiore Pisa, 2009.

[26] C. Martinez Carracedo and M. Sanz Alix, *The Theory of Fractional Powers of Operators*, North-Holland Mathematics Studies, vol. 187, Elsevier, Amsterdam, 2001.

[27] A. McIntosh, *Heinz inequalities and perturbation of spectral families*, Macquarie Mathematics Reports, Report 79-006, revised, 1980.

[28] D. Potapov and F. Sukochev, *Double operator integrals and submajorization*, Math. Model. Nat. Phenom. **5**, 317–339 (2010).

[29] M. Reed and B. Simon, *Methods of Modern Mathematical Physics. II: Fourier Analysis, Self-Adjointness*, Academic Press, New York, 1975.

[30] M. Reed and B. Simon, *The scattering of classical waves from inhomogeneous media*, Math. Z. **155**, 163–180 (1977).

[31] M. Reed and B. Simon, *Methods of Modern Mathematical Physics. IV: Analysis of Operators*, Academic Press, New York, 1978.

[32] W. Rudin, *Real and Complex Analysis*, 3rd. ed., McGraw-Hill, New York, 1987.

[33] R. Schatten, *Norm Ideals of Completely Continuous Operators*, Ergebnisse der Mathematik und ihrer Grenzgebiete, vol. 27, Springer, Berlin, 1960.

[34] B. Simon, *Trace Ideals and Their Applications*, 2nd ed., Mathematical Surveys and Monographs, vol. 120, Amer. Math. Soc., Providence, RI, 2005.

[35] E.M. Stein, *Interpolation of linear operators*, Trans. Amer. Math. Soc. **83**, 482–492 (1956).

[36] F.A. Sukochev, *On a conjecture of A. Bikchentaev*, in "Spectral Analysis, Differential Equations and Mathematical Physics: A Festschrift in Honor of Fritz Gesztesy's 60th Birthday", H. Holden, B. Simon, and G. Teschl (eds.), Proc. Symposia Pure Math., vol. 87, Amer. Math. Soc., Providence, RI, 2013, pp. 327–339.

[37] H. Tanabe, *Equations of Evolution*, Monographs and Studies in Mathematics, vol. 6, Pitman, London, 1979.

[38] K. Tanahashi, A. Uchiyama, and M. Uchiyama, *On Schwarz type inequalities*, Proc. Amer. Math. Soc. **131**, 2549–2552 (2003).

[39] H. Triebel, *Interpolation Theory, Function Spaces, Differential Operators*, 2nd ed., Barth, Heidelberg, 1995.

[40] M. Uchiyama, *Further extension of the Heinz–Kato–Furuta inequality*, Proc. Amer. Math. Soc. **127**, 2899–2904 (1999).

[41] M. Uchiyama, *Operator monotone functions and operator inequalities*, Sugaku Expos. **18**, 39–52 (2005).

[42] J. Weidmann, *Linear Operators in Hilbert Spaces*, Graduate Texts in Mathematics, vol. 68, Springer, New York, 1980.

[43] A. Yagi, *Abstract Parabolic Evolution Equations and their Applications*, Springer Monographs in Mathematics, Springer, Berlin, 2010.

Fritz Gesztesy and Yuri Latushkin
Department of Mathematics
University of Missouri
Columbia, MO 65211, USA
e-mail: `gesztesyf@missouri.edu`
`latushkiny@missouri.edu`
URL: `http://www.math.missouri.edu/personnel/faculty/gesztesyf.html`
URL: `http://www.math.missouri.edu/personnel/faculty/latushkiny.html`

Fedor Sukochev
School of Mathematics and Statistics
University of New South Wales
Kensington, NSW 2052, Australia
e-mail: `f.sukochev@unsw.edu.au`

Yuri Tomilov
Institute of Mathematics
Polish Academy od Sciences
Śniadeckich str. 8
PL-00-956 Warsaw, Poland

and

Faculty of Mathematics and Computer Science
Nicholas Copernicus University
ul. Chopina 12/18
PL-87-100 Toruń, Poland
e-mail: `tomilov@mat.umk.pl`

Operator Theory:
Advances and Applications, Vol. 250, 241–252

Power-bounded Invertible Operators and Invertible Isometries on L^p Spaces

T.A. Gillespie

Abstract. It is shown that an invertible isometry on ℓ^p, where $1 \leq p < \infty$ and $p \neq 2$, is a scalar-type spectral operator provided its spectrum is a *proper* subset of the unit circle. A similar, though weaker, analysis is also considered for invertible isometries on more general L^p spaces. These results are used to give several examples of invertible operators U on L^p spaces, where $p \in (1, \infty)$ and $p \neq 2$, such that $\sup_{n \in \mathbb{Z}} \|U^n\| < \infty$ but U is not similar to an invertible isometry. This contrasts with the situation on Hilbert space, where the condition $\sup_{n \in \mathbb{Z}} \|U^n\| < \infty$ on an invertible operator U implies that U is similar to a unitary operator.

Mathematics Subject Classification (2010). Primary 47B38, 47B37, 47B40; Secondary 43A15, 46E30.

Keywords. Invertible isometry, power-bounded operator, L^p spaces, similarity.

1. Introduction

It is well known that, for an invertible operator U on a Hilbert space H, the following statements are equivalent.

(i) $\sup_{n \in \mathbb{Z}} \|U^n\| < \infty$.
(ii) U is similar to a unitary operator on H.
(iii) U is a scalar-type spectral operator with spectrum contained in the unit circle $\mathbb{T}$; that is, U has a representation of the form

$$U = \int_{\mathbb{T}} z\, F(dz),$$

where $F(\cdot)$ is a projection-valued function, defined on the Borel subsets of $\mathbb{T}$ and countably additive in the strong operator topology.

For Charles Batty, a friend and colleague for many years, on the occasion of his sixtieth birthday.

As usual, $\mathbb{Z}$ here denotes the set of all integers $\{0, \pm 1, \pm 2, \dots\}$. The equivalence of (i) and (ii) goes back to the work of B. Sz.-Nagy, whilst the equivalence of (i) and (iii) is due to J. Wermer (see, for instance, accounts of these matters in [4, Chapter XV, §6] and [3, Chapter 8]). It is also known (see [2, Theorem 4.8]) that, if X is an L^p space (or, more generally, a closed subspace of an L^p space), where $1 < p < \infty$, and U is an invertible operator on X, then the condition

$$\sup_{n\in\mathbb{Z}} \|U^n\| < \infty \tag{1.1}$$

implies that U has the weaker spectral representation

$$U = \int_{0-}^{2\pi} e^{i\lambda}\, dE(\lambda)\,. \tag{1.2}$$

Here, $E(\cdot)$ is a (uniquely determined) projection-valued function from the reals $\mathbb{R}$ to $B(X)$, the algebra of all bounded linear operators on X, such that

(a) $E(\lambda)E(\mu) = E(\lambda \wedge \mu)$ for $\lambda, \mu \in \mathbb{R}$;
(b) $\lim_{\mu\to\lambda-} E(\mu)$ exists and $\lim_{\mu\to\lambda+} E(\mu) = E(\lambda)$ in the strong operator topology for $\lambda \in \mathbb{R}$;
(c) $E(\lambda) = 0$ if $\lambda < 0$ and $E(\lambda) = I$ (the identity operator) if $\lambda \geq 2\pi$;
(d) $\lim_{\mu\to 2\pi-} E(\mu) = I$,

and the integral in (1.2) exists as a Riemann–Stieltjes integral in the strong operator topology. Furthermore, the integral

$$\int_{0-}^{2\pi} \lambda\, dE(\lambda) \tag{1.3}$$

exists strongly as a Riemann–Stieltjes integral and defines a bounded linear operator A on X with spectrum contained in $[0, 2\pi]$ such that $\exp(iA) = U$.

The existence of a logarithm and a representation of the form (1.2) had been obtained earlier ([7, Theorem 2], [8, Theorem 1]) for translation operators on $L^p(G)$, where G is a locally compact abelian group and $1 < p < \infty$. More precisely, let U_s be the operator on $L^p(G)$ given by translation by $s \in G$ (that is, $U_s f(t) = f(t-s)$, $t \in G$ a.e.). Then U_s has a spectral representation of the form (1.2) and can be written as $\exp(iA_s)$, where the A_s has a representation of the form (1.3). It was also shown in [8, Theorem 2] that, if $p \neq 2$ and s has infinite order in G, then U_s is not a scalar-type spectral operator (nor, for that matter, a spectral operator).

Since such translation operators are invertible isometries, these results show that, although a weakened version of the implication (i) $\Rightarrow$ (iii) is valid on reflexive L^p spaces, the full implication fails in general. However, there remains the question as to whether, for invertible operators U on reflexive L^p spaces, (1.1) implies that U is similar to an invertible isometry or whether (as seems more likely) this fails as well. This question was raised at a lecture to the North British Functional Analysis Seminar in 2011 by Yuri Tomilov. The aim of this note is to show that this implication does indeed fail in the more general context of reflexive L^p spaces when $p \neq 2$. Examples will be given to illustrate this both on the sequence space ℓ^p and more generally on $L^p(G)$ for G a locally compact abelian group. The examples rely on spectral properties of invertible isometries on ℓ^p and on L^p spaces (see Theorems 2.4 and 3.1 below).

In addition to the notation already introduced, $\mathbb{N}$ will as usual denote the set of positive integers, $\mathbb{C}$ the complex numbers and $\mathbb{T}$ the unit circle in $\mathbb{C}$. It is convenient to take ℓ^p to be the space of doubly infinite p-summable sequences $x = \{x_n\}_{n\in\mathbb{Z}}$ with complex terms. Also, for $n \in \mathbb{N}$, ℓ^p_n will denote $\mathbb{C}^n$ endowed with the standard p-norm. Each $n \times n$ complex matrix A acts on the elements of ℓ^p_n when written as column vectors; the norm of the resulting linear operator will be denoted by $\|A\|_p$. In this context, we allow the value $p = \infty$. The symbol Sp will denote the spectrum of a linear operator (or the set of eigenvalues of an $n \times n$ matrix).

2. Invertible isometries on ℓ^p

An important property of a linear isometry U on an arbitrary L^p space when $1 \leq p < \infty$ and $p \neq 2$ is that it is *separation-preserving*; in other words, if f and g are functions in the underlying L^p space with disjoint supports, then Uf and Ug also have disjoint supports (i.e., $(Uf).(Ug) = 0$ almost everywhere whenever $f.g = 0$ almost everywhere). This result goes back to Banach ([1, Chapitre XI]), where it is stated for the real spaces $L^p[0,1]$ and ℓ^p, and discussed in fuller generality, allowing for complex scalars and arbitrary measure spaces, in [11]. It follows that the invertible isometries on ℓ^p, where $1 \leq p < \infty$ and $p \neq 2$, are precisely the operators $U : \ell^p \to \ell^p$ of the form

$$U\{x_n\}_{n\in\mathbb{Z}} = \{\alpha_n x_{\tau(n)}\}_{n\in\mathbb{Z}}, \tag{2.1}$$

where $\alpha_n \in \mathbb{C}$ with $|\alpha_n| = 1$ for all $n \in \mathbb{Z}$ and $\tau : \mathbb{Z} \to \mathbb{Z}$ is a bijection, a result that appeared in [1, pp. 178–180], though with real scalars.

Although this description of the invertible isometries on ℓ^p shows that there is no general restriction on their spectra beyond being closed subsets of $\mathbb{T}$ (just take τ to be the identity mapping and $\{\alpha_n\}_{n\in\mathbb{Z}}$ a dense subset of any required closed set), nevertheless additional assumptions on the spectrum $\mathrm{Sp}(U)$ of a given invertible isometry U can sometimes yield more detailed structural information about U. In particular, it will be shown that, if $\mathrm{Sp}(U)$ is a *proper* subset of $\mathbb{T}$, then U is a scalar-type spectral operator.

To prove this, it is convenient to establish several lemmas. The first provides an estimate for the norm of the inverse of a Vandermonde matrix and appears in [6]. There, the norm of an $n \times n$ complex matrix $A = (a_{jk})$ is taken as

$$\|A\| = \max_{1\leq j\leq n} \sum_{k=1}^{n} |a_{jk}|,$$

see [6, (1.4)]. In the present notation, this is the norm $\|A\|_\infty$ of A considered as a linear mapping on ℓ_n^∞. The relevant result [6, Theorem 1] is as follows.

Lemma 2.1. *Let $n \in \mathbb{N}$ with $n \geq 2$, let $\lambda_1, \lambda_2, \ldots, \lambda_n$ be distinct complex numbers, and let V_n denote the $n \times n$ Vandermonde matrix*

$$V_n = \begin{pmatrix} 1 & 1 & \cdots & 1 \\ \lambda_1 & \lambda_2 & \cdots & \lambda_n \\ \lambda_1^2 & \lambda_2^2 & \cdots & \lambda_n^2 \\ \vdots & \vdots & \vdots & \vdots \\ \lambda_1^{n-1} & \lambda_2^{n-1} & \cdots & \lambda_n^{n-1} \end{pmatrix}.$$

Then

$$\|V_n^{-1}\|_\infty \leq \max_{1\leq j\leq n} \prod_{\substack{k=1\\k\neq j}}^{n} \frac{1+|\lambda_k|}{|\lambda_k - \lambda_j|}. \tag{2.2}$$

Comment. As is well known, the condition that the λ_j's are distinct is necessary and sufficient for V_n to be invertible.

Lemma 2.2. *Let $\lambda_1, \ldots, \lambda_n$ and V_n be as in Lemma 2.1, and suppose further that $|\lambda_j| = 1$ for each j. Then*

$$\|V_n\|_p \leq n \quad \text{and} \quad \|V_n^{-1}\|_p \leq 2^{n-1} n^{1/p} \max_{1\leq j\leq n} \prod_{\substack{k=1\\k\neq j}}^{n} \frac{1}{|\lambda_k - \lambda_j|} \tag{2.3}$$

for $1 \leq p < \infty$.

Proof. When $p = 1$, the first inequality in (2.3) is clear since, for any $n \times n$ matrix A, $\|A\|_1$ equals the maximum ℓ^1 norm of the columns of A. Suppose then that $1 < p < \infty$ and that p' is the index conjugate to p. Given $x \in \mathbb{C}^n$, the jth coordinate of $V_n x$ is $\sum\limits_{k=1}^{n} \lambda_k^{j-1} x_k$ and, since $|\lambda_j| = 1$ for each j, Hölder's inequality then gives

$$\begin{aligned} \|V_n x\|_p^p &= \sum_{j=1}^{n} \left| \sum_{k=1}^{n} \lambda_k^{j-1} x_k \right|^p \\ &\leq \sum_{j=1}^{n} n^{p/p'} \|x\|_p^p = n^{1+p/p'} \|x\|_p^p. \end{aligned}$$

It follows, taking pth roots, that $\|V_n x\|_p \leq n\|x\|_p$ and so $\|V_n\|_p \leq n$ as required.

For the second inequality in (2.3), note that

$$\|V_n^{-1}x\|_p \leq n^{1/p}\|V_n^{-1}x\|_\infty \leq n^{1/p}\|V_n^{-1}\|_\infty\|x\|_\infty \leq n^{1/p}\|V_n^{-1}\|_\infty\|x\|_p$$

for $x \in \mathbb{C}^n$ and then use (2.2). □

Lemma 2.3. *Let $n \geq 2$, let $\beta_1, \ldots, \beta_n \in \mathbb{C}$ with $|\beta_j| = 1$ for each j, let A be the $n \times n$ matrix defined as*

$$A = \begin{pmatrix} 0 & \beta_1 & 0 & \cdots & 0 \\ 0 & 0 & \beta_2 & \cdots & 0 \\ \vdots & \vdots & \ddots & \ddots & \vdots \\ 0 & 0 & \cdots & 0 & \beta_{n-1} \\ \beta_n & 0 & \cdots & 0 & 0 \end{pmatrix}, \tag{2.4}$$

and let $1 \leq p < \infty$. Then there exists an invertible $n \times n$ matrix W with $\|W\|_p \leq n$ and $\|W^{-1}\|_p \leq n^{1/p}(\operatorname{cosec} \frac{\pi}{n})^n$ such that $W^{-1}AW$ is a diagonal matrix with unimodular diagonal entries spaced equally round $\mathbb{T}$.

Proof. Let S be the $n \times n$ diagonal matrix with diagonal entries $\eta_1, \eta_2, \eta_3, \ldots, \eta_n$, where

$$\eta_1 = \beta_1\beta_2 \ldots \beta_{n-1}, \quad \eta_2 = \beta_2 \ldots \beta_{n-1}, \ldots, \eta_{n-1} = \beta_{n-1}, \quad \eta_n = 1.$$

Then $\|S\|_p = \|S^{-1}\|_p = 1$ and $S^{-1}AS$ is the matrix $\tilde{A}$ given by

$$\tilde{A} = \begin{pmatrix} 0 & 1 & 0 & \cdots & 0 \\ 0 & 0 & 1 & \cdots & 0 \\ \vdots & \vdots & \ddots & \ddots & \vdots \\ 0 & 0 & \cdots & 0 & 1 \\ \beta & 0 & \cdots & 0 & 0 \end{pmatrix},$$

where $\beta = \beta_1\beta_2 \ldots \beta_n$. Write β as $e^{i\theta}$. The characteristic polynomial of $\tilde{A}$ is $\lambda^n - \beta$ and this has the n distinct roots

$$\lambda_k = e^{i(\theta+2\pi k)/n} \qquad (k = 0, 1, 2, \ldots, n-1),$$

which are spaced equally round $\mathbb{T}$. It is easy to check that $(1, \lambda_k, \lambda_k^2, \ldots, \lambda_k^{n-1})$ (written as a column vector) is an eigenvector corresponding to the eigenvalue λ_k. Hence the Vandermonde matrix

$$V = \begin{pmatrix} 1 & 1 & 1 & \cdots & 1 \\ \lambda_1 & \lambda_2 & \lambda_3 & \cdots & \lambda_n \\ \lambda_1^2 & \lambda_2^2 & \lambda_3^2 & \cdots & \lambda_n^2 \\ \vdots & \vdots & \vdots & \vdots & \vdots \\ \lambda_1^{n-1} & \lambda_2^{n-1} & \lambda_3^{n-1} & \cdots & \lambda_n^{n-1} \end{pmatrix}$$

diagonalizes $\tilde{A}$, with $V^{-1}\tilde{A}V$ the diagonal matrix with (unimodular) diagonal entries $\lambda_1, \lambda_2, \ldots, \lambda_n$. Since the points $\lambda_1, \lambda_2, \ldots, \lambda_n$ are spaced equally round $\mathbb{T}$,

$$|\lambda_k - \lambda_j| \geq 2\left(\sin \frac{\pi}{n}\right)$$

for $1 \le j, k \le n$ with $j \ne k$. An application of the result of Lemma 2.2 gives $\|V\|_p \le n$ and

$$\|V^{-1}\|_p \le 2^{n-1} n^{1/p} \max_{1\le j\le n} \prod_{\substack{k=1\\k\ne j}}^{n} \frac{1}{|\lambda_k - \lambda_j|}$$

$$\le 2^{n-1} n^{1/p} \max_{1\le j\le n} \prod_{\substack{k=1\\k\ne j}}^{n} \frac{1}{2\sin(\pi/n)} = n^{1/p}\left(\operatorname{cosec}\frac{\pi}{n}\right)^{n-1}.$$

Setting $W = SV$, $W^{-1}AW$ is the diagonal matrix with diagonal entries $\lambda_1, \lambda_2, \dots, \lambda_n$, $\|W\|_p = \|V\|_p \le n$ and $\|W^{-1}\|_p = \|V^{-1}\|_p \le n^{1/p}(\operatorname{cosec}\frac{\pi}{n})^{n-1}$ as required. □

We are now in a position to establish the promised result concerning an invertible isometry on ℓ^p with spectrum a proper subset of $\mathbb{T}$.

Theorem 2.4. *Let $1 \le p < \infty$ with $p \ne 2$ and let U be an invertible isometry on ℓ^p such that* $\mathrm{Sp}(U) \ne \mathbb{T}$. *Then U is similar to a diagonal operator on ℓ^p (that is, to an operator with diagonal matrix relative to the standard basis of ℓ^p) and hence is a scalar-type spectral operator.*

Proof. Let $\{e_n\}_{n\in\mathbb{Z}}$ be the standard basis of ℓ^p and let U have the form given by (2.1).

We first show that the mapping τ does not have an infinite orbit. Suppose it did and let $\{\tau^n(k_0) : n \in \mathbb{Z}\}$ be such an orbit for some fixed $k_0 \in \mathbb{Z}$. The subspace $X = \overline{\mathrm{span}}\{e_{\tau^n(k_0)} : n \in \mathbb{Z}\}$ is invariant under U and U^{-1}. Furthermore, X can be identified in a natural way with a copy Y of ℓ^p (for $n \in \mathbb{Z}$, $e_{\tau^n(k_0)}$ in X corresponds to e_n in Y) and, under this identification, the restriction $U|X$ of U to X corresponds to the backward bilateral weighted shift $S_\beta : \{y_n\}_{n\in\mathbb{Z}} \to \{\beta_n y_{n+1}\}_{n\in\mathbb{Z}}$ on Y, where $\beta_n = \alpha_{\tau^n(k_0)}$. The spectrum of a weighted shift on ℓ^p is rotationally invariant (in the present context, zS_β is isometrically equivalent to S_β for each $z \in \mathbb{T}$) and hence $\mathrm{Sp}\,(U|X) = \mathrm{Sp}\,(S_\beta) = \mathbb{T}$. Since the spectrum of an invertible isometry equals its approximate point spectrum, the spectrum of the restriction $U|X$ is contained in $\mathrm{Sp}(U)$ and the equality $\mathrm{Sp}\,(U|X) = \mathbb{T}$ contradicts the assumption that $\mathrm{Sp}(U) \ne \mathbb{T}$. Thus the mapping τ does not have an infinite orbit.

We next show that there is in fact a bound on the cardinalities of the (necessarily finite) orbits of τ. To do this, fix $\gamma > 0$ such that $\mathbb{T}\setminus \mathrm{Sp}(U)$ contains a closed arc Γ of length γ and consider an orbit with cardinality $n \ge 2$, say $\{\tau^m(k_0) : 0 \le m \le n-1\}$ for some $k_0 \in \mathbb{Z}$ with $\tau^n(k_0) = k_0$. Here, the subspace $Z = \mathrm{span}\{e_{k_0}, e_{\tau(k_0)}, \dots, e_{\tau^{n-1}(k_0)}\}$ of ℓ^p is invariant under U and the matrix of $U|Z$ with respect to the basis $\{e_{k_0}, e_{\tau(k_0)}, \dots, e_{\tau^{n-1}(k_0)}\}$ of Z is the matrix A in (2.4) with $\beta_m = \alpha_{\tau^{m-1}(k_0)}$ for $1 \le m \le n$. Applying the result of Lemma 2.3, it follows that $\mathrm{Sp}\,(U|Z)$ consists of n points spaced equally round $\mathbb{T}$. Since $\mathrm{Sp}\,(U|Z) \subseteq \mathrm{Sp}\,(U) \subseteq \mathbb{T}\setminus\Gamma$, it follows that $\frac{2\pi}{n} > \gamma$. Thus $\frac{2\pi}{\gamma}$ is an upper bound for the cardinalities of the orbits of τ.

Fix a positive integer N_0 with $N_0 \geq \frac{2\pi}{\gamma}$ and let $\{\Lambda_r\}_{r\in\mathbb{N}}$ be the set of distinct orbits of τ, so that $\mathbb{Z}$ is the disjoint union

$$\mathbb{Z} = \bigcup_{r=1}^{\infty} \Lambda_r.$$

For each $r \in \mathbb{N}$, fix $k_r \in \Lambda_r$, let n_r denote the cardinality of Λ_r, so that

$$\Lambda_r = \{k_r,\, \tau(k_r),\, \ldots\, ,\, \tau^{n_r-1}(k_r)\},$$

and let

$$Z_r = \operatorname{span}\{e_{k_r},\, e_{\tau(k_r)},\, \ldots\, ,\, e_{\tau^{n_r-1}(k_r)}\}.$$

Note that ℓ^p is the ℓ^p-direct sum of the subspaces $\{Z_r\}_{r\in\mathbb{N}}$, each of which is invariant under U. If $n_r = 1$, then $Z_r = \operatorname{span}\{e_{k_r}\}$, e_{k_r} is an eigenvector of U with corresponding eigenvalue α_{k_r}, and $\|U|Z_r\| = 1$. If $n_r \geq 2$, the properties of the matrix W in Lemma 2.3 imply that there is an invertible linear mapping $T_r : Z_r \to Z_r$ with

$$\|T_r\| \leq n_r < N_0 \tag{2.5}$$

and

$$\|T_r^{-1}\| \leq n_r^{1/p} \left(\operatorname{cosec} \frac{\pi}{n_r}\right)^{n_r} < N_0^{1/p} \left(\operatorname{cosec} \frac{\pi}{N_0}\right)^{N_0} \tag{2.6}$$

such that each basis element $\{e_{k_r},\, e_{\tau(k_r)},\, \ldots\, ,\, e_{\tau^{n_r-1}(k_r)}\}$ is an eigenvector for $T_r^{-1}(U|Z_r)T_r$. Take T_r to be the identity operator on Z_r when $n_r = 1$. Since ℓ^p is the ℓ^p-direct sum of the Z_r's, it follows from (2.5) there is a bounded linear operator $S : \ell^p \to \ell^p$ with $\|S\| \leq N_0$ such that $S|Z_r = T_r$ for each r. Furthermore, (2.6) implies that S is invertible, with $S^{-1}|Z_r = T_r^{-1}$ and $\|S^{-1}\| \leq N_0^{1/p}(\operatorname{cosec} \frac{\pi}{N_0})^{N_0}$. The basis elements $\{e_n : n \in \Lambda_r\}$ of Z_r are eigenvectors of $S^{-1}US$ for each r and hence each of the standard basis elements of ℓ^p is an eigenvector of $S^{-1}US$ since $\mathbb{Z} = \bigcup_{r=1}^{\infty} \Lambda_r$. It is well known and easy to prove that, for $1 \leq p < \infty$, such operators on ℓ^p (that is, operators having a diagonal matrix with diagonal entries, say $\{\lambda_n\}_{n\in\mathbb{Z}}$, with respect to the standard basis) are always scalar-type spectral operators; the spectral measure of a Borel subset σ of $\mathbb{C}$ is the projection onto the coordinate positions $\{n : \lambda_n \in \sigma\}$. Thus $S^{-1}US$ is a scalar-type spectral operator and hence so is U since this latter property is invariant under similarity. □

3. Invertible isometries on L^p spaces

We now consider spectral properties of invertible isometries on more general L^p spaces and obtain a result that is comparable to, though somewhat weaker than, the result of Theorem 2.4.

Let (Ω, Σ, μ) be a σ-finite measure space and let U be an invertible isometry on $L^p(\mu)$, where $1 \leq p < \infty$ and $p \neq 2$. A general structure theorem for U was obtained by Lamperti ([11, Theorem 3.1]) and by Kan ([10, Theorem 4.1 and

Proposition 4.1]); it is a consequence of the fact that U is necessarily separation-preserving.

The structure theorem takes the following form. There is an isomorphism Φ_0 of the measure algebra associated with (Ω, Σ, μ) and a measurable function $h : \Omega \to \mathbb{C}$ with $h \neq 0$ μ-a.e. such that Uf is given by the pointwise product

$$Uf = h.\Phi(f) \tag{3.1}$$

μ-a.e. for $f \in L^p$. Here Φ is the isomorphism of the algebra of all complex-valued measurable functions on Ω induced by Φ_0; in particular, $\Phi(\chi_\sigma) = \chi_{\Phi_0(\sigma)}$ for $\sigma \in \Sigma$, where χ denotes characteristic function. Furthermore,

$$|h|^p = \frac{d(\mu \circ \Phi_0^{-1})}{d\mu}. \tag{3.2}$$

Note that from, (3.1),

$$U^n f = h_n.\Phi^n(f) \tag{3.3}$$

for $n \in \mathbb{N}$, where h_n is the pointwise product $h.\Phi(h) \dots \Phi^{n-1}(h)$.

Theorem 3.1. *Let U, Φ_0, Φ and h be as above.*

(i) *If Φ_0 is the identity mapping, then U is a scalar-type spectral operator.*

(ii) *Suppose on the other hand that Φ_0 is not the identity mapping. Then there exists $\sigma \in \Sigma$ with $0 < \mu(\sigma) < \infty$ such that σ and $\Phi_0(\sigma)$ are disjoint.*

Furthermore, if M denotes the set of $m \in \mathbb{N}$ for which there exists $\sigma_m \in \Sigma$ with $0 < \mu(\sigma_m) < \infty$ such that

$$\sigma_m,\ \Phi_0(\sigma_m),\ \dots,\ \Phi_0^m(\sigma_m)$$

are mutually disjoint, then either (a) $M = \mathbb{N}$ *and* $\mathrm{Sp}(U) = \mathbb{T}$, *or* (b) M *is bounded above and* $\mathrm{Sp}(U)$ *contains* m_0+1 *distinct points spaced equally round* $\mathbb{T}$, *where* $m_0 = \max M$.

Comment. Since Φ_0 is defined on the measure algebra associated with (Ω, Σ, μ), set theoretic notions such as containment, disjointness and so on are as usual in this context to be interpreted as holding to within a μ-null set. This convention will be adopted in the following proof.

Proof. Firstly, suppose that Φ_0 is the identity mapping. Then the extension Φ of Φ_0 to the space of all measurable functions on Ω is also the identity mapping. Hence $\mu \circ \Phi_0^{-1} = \mu$, $|h| = 1$ a.e. by (3.2) and $Uf = h.f$ for $f \in L^p(\mu)$ by (3.1). Such multiplication operators are always scalar-type spectral operators (here the spectral measure of a measurable subset τ of $\mathbb{T}$ is given by multiplication by the characteristic function of $h^{-1}(\tau)$). Thus (i) holds in this situation.

Suppose now that Φ_0 is not the identity mapping, so that there exists $\tau \in \Sigma$ with $\tau \neq \Phi_0(\tau)$. Then either $\sigma = \tau \backslash \Phi_0(\tau)$ or $\sigma = \Phi_0(\tau) \backslash \tau$ has positive measure. If $\mu(\tau \backslash \Phi_0(\tau)) > 0$ and $\sigma = \tau \backslash \Phi_0(\tau)$, then $\Phi_0(\sigma) \subseteq \Phi_0(\tau)$ and so σ and $\Phi_0(\sigma)$ are disjoint. On the other hand, if $\mu(\Phi_0(\tau) \backslash \tau) > 0$ and $\sigma = \Phi_0(\tau) \backslash \tau$, then $\sigma \subseteq \Phi_0(\tau)$ and $\Phi_0(\sigma) \subseteq \Phi_0^2(\tau) \backslash \Phi_0(\tau)$, and so σ and $\Phi_0(\sigma)$ are disjoint in this case also. We thus have shown that there exists $\sigma \in \Sigma$ with $\mu(\sigma) > 0$ such that σ and $\Phi_0(\sigma)$ are disjoint. Since μ is σ-finite and Φ_0 preserves containment, by passing to a subset if necessary, we can assume that $0 < \mu(\sigma) < \infty$. This establishes the first part of (ii) and shows that the set M is non-empty. Notice that M has the property that, if $m \in M$ and $n \in \mathbb{N}$ with $n < m$, then $n \in M$. Thus either $M = \mathbb{N}$ or $M = \{1, 2, \ldots, m_0\}$ for some $m_0 \in \mathbb{N}$.

(a) Suppose that $M = \mathbb{N}$ and let $\lambda \in \mathbb{T}$. For $m \in \mathbb{N}$, let $\sigma_m \in \Sigma$ with $0 < \mu(\sigma_m) < \infty$ and

$$\sigma_m,\ \Phi_0(\sigma_m),\ \ldots,\ \Phi_0^m(\sigma_m) \tag{3.4}$$

mutually disjoint. Now χ_{σ_m} belongs to $L^p(\mu)$ and the functions

$$\{\chi_{\sigma_m}, U\chi_{\sigma_m}, \ldots, U^m\chi_{\sigma_m}\}$$

have disjoint support (up to a null set) by (3.3) and (3.4). Hence, with

$$f_m = \sum_{n=0}^{m} \lambda^{-n} U^n \chi_{\sigma_m},$$

we have

$$\|f_m\|_p^p = \sum_{n=0}^{m} \|U^n\chi_{\sigma_m}\|_p^p = \sum_{n=0}^{m} \|\chi_{\sigma_m}\|_p^p = (m+1)\mu(\sigma_m),$$

whilst $(\lambda I - U)f_m = \lambda\chi_{\sigma_m} - \lambda^{-m}U^{m+1}\chi_{\sigma_m}$, and

$$\|(\lambda I - U)f_m\|_p \leq \|\chi_{\sigma_m}\|_p + \|U^{m+1}\chi_{\sigma_m}\|_p = 2\|\chi_{\sigma_m}\|_p = 2\mu(\sigma_m)^{1/p}.$$

Thus

$$\frac{\|(\lambda I - U)f_m\|_p}{\|f_m\|_p} \leq \frac{2}{(m+1)^{1/p}} \to 0$$

as $m \to \infty$. Hence λ belongs to the approximate point spectrum of U and $Sp(U) = \mathbb{T}$ as required.

(b) Now suppose that $M = \{1, 2, \ldots, m_0\}$ for some $m_0 \in \mathbb{N}$ and let $\sigma \in \Sigma$ with $0 < \mu(\sigma) < \infty$ and

$$\sigma,\ \Phi_0(\sigma),\ \ldots,\ \Phi_0^{m_0}(\sigma) \tag{3.5}$$

mutually disjoint. We claim that $\Phi_0^{m_0+1}(\sigma) = \sigma$. Suppose not, so that either $\sigma\backslash\Phi_0^{m_0+1}(\sigma)$ or $\Phi_0^{m_0+1}(\sigma)\backslash\sigma$ has positive measure.

Consider first the case when $\tau = \sigma\backslash\Phi_0^{m_0+1}(\sigma)$ has positive measure. Then $0 < \mu(\tau) < \infty$ and, from the disjointness of sets in (3.5), the sets

$$\tau,\ \Phi_0(\tau),\ \ldots,\ \Phi_0^{m_0}(\tau) \tag{3.6}$$

are mutually disjoint (since $\tau \subseteq \sigma$), as are the sets

$$\Phi_0(\tau),\ \ldots,\ \Phi_0^{m_0}(\tau),\ \Phi_0^{m_0+1}(\tau).$$

(Apply Φ_0 to the sets in (3.6).) Further, $\Phi_0^{m_0+1}(\tau)$ does not meet τ from the definition of τ and so the sets

$$\tau,\ \Phi_0(\tau),\ \ldots,\ \Phi_0^{m_0}(\tau),\ \Phi_0^{m_0+1}(\tau)$$

are mutually disjoint, contradicting the maximality of m_0 in M.

Suppose now that $\tau = \Phi_0^{m_0+1}(\sigma)\backslash\sigma$ has positive measure. A similar argument, in this case applying $\Phi_0^{m_0+1}$ to (3.5) and noting that $\Phi_0^{m_0+1}(\tau)$ does not meet $\Phi_0^{m_0+1}(\sigma)$ and hence does not meet τ in a set of positive measure, again gives that the sets τ, $\Phi_0(\tau)$, ..., $\Phi_0^{m_0}(\tau)$, $\Phi_0^{m_0+1}(\tau)$ are mutually disjoint. Passing to a subset of τ of finite positive measure if necessary, the maximality of m_0 in M is again contradicted. Thus $\Phi_0^{m_0+1}(\sigma) = \sigma$ as claimed.

Let $\Omega_0 = \bigcup_{n=0}^{m_0} \Phi_0^n(\sigma)$, where σ is as in (3.5), and let X be the (closed) subspace of $L^p(\mu)$ consisting of the functions in $L^p(\mu)$ vanishing a.e. on $\Omega\backslash\Omega_0$. Since $\Phi_0^{m_0+1}(\sigma) = \sigma$, (3.1) implies that X is U-invariant. Let $\lambda \in \mathbb{T}$ with $\lambda^{m_0+1} = 1$ and define $S : X \to X$ by

$$Sf = \sum_{n=0}^{m_0} \lambda^{-n} \chi_{\Phi_0^n(\sigma)} f \qquad (f \in X).$$

Then

$$\begin{aligned} USf &= \sum_{n=0}^{m_0} \lambda^{-n} U(\chi_{\Phi_0^n(\sigma)} f) = \sum_{n=0}^{m_0} \lambda^{-n} h.\Phi(\chi_{\Phi_0^n(\sigma)} f) \\ &= \sum_{n=0}^{m_0} \lambda^{-n} h.\Phi(\chi_{\Phi_0^n(\sigma)})\Phi(f) = \sum_{n=0}^{m_0} \lambda^{-n} h.(\chi_{\Phi_0^{n+1}(\sigma)})\Phi(f) \\ &= \lambda \sum_{n=0}^{m_0} \lambda^{-n} \chi_{\Phi_0^n(\sigma)} h.\Phi(f) = \lambda SUf \end{aligned}$$

since Φ is multiplicative, $\lambda^{m_0+1} = 1$ and $\Phi_0^{m_0+1}(\sigma) = \sigma$. Noting that S is an invertible isometry on X, we then have

$$S^{-1}US = \lambda U.$$

Hence $\mathrm{Sp}(U|X)$ is invariant under multiplication by λ. In particular, if $\mu \in \mathrm{Sp}(U|X)$ and $\lambda_0 = e^{2\pi i/(m_0+1)}$, then $\{\mu, \lambda_0\mu, \dots, \lambda_0^{m_0}\mu\}$ are $m_0 + 1$ points in $\mathrm{Sp}(U|X)$ spaced equally round $\mathbb{T}$. Since, as noted in the proof of Theorem 2.4, $\mathrm{Sp}(U|X)$ consists of points in the approximate point spectrum of $U|X$ and hence is contained in $\mathrm{Sp}(U)$, this completes the proof of (ii). □

Remark 1. As it stands, Theorem 3.1 does not imply that an invertible isometry on $L^p(\mu)$ with spectrum a proper subset of $\mathbb{T}$ is given by multiplication by a measurable function (the analogue of a diagonal operator on ℓ^p) and hence is a scalar-type spectral operator. This result may be true, but it is not clear how to adapt the proof of Theorem 2.4 to this more general context, in part because the measure algebra isomorphism Φ_0 may be somewhat more complicated than the bijection τ appearing in (2.1). In particular, τ is measure-preserving whilst Φ_0 need not be.

Remark 2. Some aspects of the proof of Theorem 3.1 bear resemblance to the concepts of periodic and aperiodic sets introduced in [9]. However, the context there is rather different, involving measure-preserving transformations and associated L^2 spaces.

4. Examples of invertible power-bounded operators on L^p spaces

In this final section, examples of invertible power-bounded operators on non-Hilbert reflexive L^p spaces are constructed and the results of Theorems 2.4 and 3.1 are then used to show that these operators are not similar to any invertible isometry.

An example on ℓ^p. Let U_1 be the bilateral shift on ℓ^p, where $1 < p < \infty$ and $p \neq 2$. Then, from the discussion of translation operators in §1, U_1 can be written as

$$U_1 = \exp(iA_1),$$

where $A_1 \in B(\ell^p)$ has spectrum contained in the interval $[0, 2\pi]$. In this case, $\mathrm{Sp}\, A_1 = [0, 2\pi]$ since $\mathrm{Sp}\, U_1 = \mathbb{T}$. In fact, A_1 is the operator $\pi I + iH$, where H is the discrete Hilbert transform (see [7, Remark 2, p. 1044]). Further, as discussed in §1, since the element 1 has infinite order in the group $\mathbb{Z}$, U_1 is not a spectral operator (a fact that had been noted earlier in [5]).

Let $U = \exp(iA_1/2)$. Then $U^2 = U_1$ and so

$$\|U^{2k}\| = \|U_1^2\| = 1 \qquad \text{and} \qquad \|U^{2k+1}\| = \|U_1^k U\| = \|U\|$$

for $k \in \mathbb{Z}$. Thus U is power-bounded. Further,

$$\mathrm{Sp}\,(U) = \{e^{i\theta} : 0 \leq \theta \leq \pi\}$$

since $\mathrm{Sp}\,(A_1) = [0, 2\pi]$ and, since $U^2 = U_1$ is not a spectral operator, neither is U. Suppose that $\tilde{U}$ is similar to U. Then $\tilde{U}$ is not spectral and has spectrum equal to the proper subset $\{e^{i\theta} : 0 \leq \theta \leq \pi\}$ of $\mathbb{T}$. Thus, by the result of Theorem 2.4, $\tilde{U}$ is not an invertible isometry and so U provides an example of a power-bounded invertible operator on ℓ^p that is not similar to an invertible isometry.

Examples on $L^p(G)$. Let G be any locally compact abelian group and, in order to apply the results in §3, assume that Haar measure on G is σ-finite. Let $s \in G$ have infinite order and let U_s be translation by s on $L^p(G)$, where $1 < p < \infty$. Then, as discussed in §1, U_s can be written as

$$U_s = \exp(iA_s)$$

for some $A_s \in B(L^p(G))$ with $\mathrm{Sp}(A_s) \subseteq [0, 2\pi]$. Since $\mathrm{Sp}(U_s) = \mathbb{T}$ by [7, Theorem 1], in fact $\mathrm{Sp}(A_s) = [0, 2\pi]$. Now suppose that $p \neq 2$ and let $U = \exp(iA_s/3)$. Then $U^3 = U_s$ is isometric,

$$\sup_{n\in\mathbb{Z}} \|U^n\| = \max\{1, \|U\|, \|U^2\|\}$$

and so U is invertible and power-bounded. Furthermore,

$$\mathrm{Sp}(U) = \{e^{i\theta} : 0 \leq \theta \leq \pi/3\}.$$

Since in this case $\mathrm{Sp}(U)$ does not contain at least two points spaced equally round $\mathbb{T}$, U is not similar to an invertible isometry by Theorem 3.1(ii)b.

As a specific example, let $\alpha \in \mathbb{R}$ be irrational, take $s = e^{-2\pi i\alpha}$, and let $\mu_n = 2\pi(n\alpha - [n\alpha])$, where $[\cdot]$ denotes integer part. Then U is the mapping on $L^p(\mathbb{T})$ given by

$$U : \sum_{n\in\mathbb{Z}} a_n e^{int} \to \sum_{n\in\mathbb{Z}} a_n e^{i\mu_n/3} e^{int}.$$

References

[1] S. Banach, *Théorie des Opérations Linéaires*, Warsaw, 1932; Second Edition reprinted by Chelsea Publ. Co., New York, 1963.

[2] E. Berkson and T.A. Gillespie, *Stečkin's theorem, transference and spectral decompositions*, J. Functional Analysis 70 (1987), 140–170.

[3] H.R. Dowson, *Spectral Theory of Linear Operators*, London Math. Soc. Monographs 12, Academic Press, London, 1978.

[4] N. Dunford and J.T. Schwartz, *Linear Operators*, Part III: *Spectral Operators*, Wiley, New York, 1971.

[5] U. Fixman, *Problems in spectral operators*, Pacific J. Math. 9 (1959), 1029–1051.

[6] W. Gautschi, *On inverses of Vandermonde and confluent Vandermonde matrices*, Numerische Mathematik 4 (1962), 117–123.

[7] T.A. Gillespie, *Logarithms of L^p translations*, Indiana Univ. Math. J. 24 (1975), 1037–1045.

[8] T.A. Gillespie, *A spectral theorem for L^p translations*, J. London Math. Soc. (2) 11 (1975), 499–508.

[9] A. Ionescu Tulcea, *Random series and spectra of measure-preserving transformations*, in *Ergodic Theory* (*Proc. Internat. Symp., Tulane Univ.*, 1961), 273–292, Academic Press, New York, 1963.

[10] C.-H. Kan, *Ergodic properties of Lamperti operators*, Canadian J. Math. 30 (1978), 1206–1214.

[11] J. Lamperti, *On the isometries of certain function-spaces*, Pacific J. Math. 8 (1958), 459–466.

T.A. Gillespie
School of Mathematics and
Maxwell Institute for Mathematical Sciences
University of Edinburgh
James Clerk Maxwell Building
Mayfield Road
Edinburgh EH9 3JZ, Scotland
e-mail: t.a.gillespie@ed.ac.uk

Operator Theory:
Advances and Applications, Vol. 250, 253–271

Generation of Subordinated Holomorphic Semigroups via Yosida's Theorem

Alexander Gomilko and Yuri Tomilov

To Charles Batty, colleague and friend, on the occasion of his sixtieth anniversary with admiration

Abstract. Using functional calculi theory, we obtain several estimates for $\|\psi(A)g(A)\|$, where ψ is a Bernstein function, g is a bounded completely monotone function and $-A$ is the generator of a holomorphic C_0-semigroup on a Banach space, bounded on $[0,\infty)$. Such estimates are of value, in particular, in approximation theory of operator semigroups. As a corollary, we obtain a new proof of the fact that $-\psi(A)$ generates a holomorphic semigroup whenever $-A$ does, established recently in [8] by a different approach.

Mathematics Subject Classification (2010). Primary 47A60, 65J08, 47D03; Secondary 46N40, 65M12.

Keywords. Holomorphic C_0-semigroup, Bernstein functions, functional calculus.

1. Introduction

Bernstein functions play an important role in analysis, and in particular, in the study of Lévy processes in probability theory. Recently they found a number of applications in operator and ergodic theories, mainly in issues related to rates of convergence of semigroups and related operator families. At a core of many applications of Bernstein functions is an abstract subordination principle going back to Bochner, Nelson and Phillips (see [19, p. 171] for more on its historical background). Given a Bernstein function ψ and a generator $-A$ of a bounded C_0-semigroup on a Banach space X, the principle allows one to define the operator $-\psi(A)$ which again is the generator of a bounded C_0-semigroup on X. Thus, it is natural to ask whether Bernstein functions preserve other classes of (bounded)

This work was completed with partial support by the NCN grant DEC-2011/03/B/ST1/00407 and by the EU Marie Curie IRSES program, project AOS, No. 318910.

semigroups relevant for applications such as holomorphic semigroups, differentiable semigroups or any of their subclasses. This paper treats the permanence of the class of holomorphic semigroups under Bernstein functions.

Recall that a positive function $g \in \mathrm{C}^\infty(0,\infty)$ is called *completely monotone* if

$$(-1)^n g^{(n)}(\tau) \geq 0, \qquad \tau > 0,$$

for each $n \in \mathbb{N}$.

A positive function $\psi \in \mathrm{C}^\infty(0,\infty)$ is called a *Bernstein function* if its derivative is completely monotone.

A basic property of Bernstein functions is that their exponentials arise as Laplace transforms of uniquely defined convolution semigroups of subprobability measures. This property is a core of the notion of subordination discussed below.

Recall that a family of positive Borel measures $(\mu_t)_{t\geq 0}$ on $[0,\infty)$ is called a vaguely continuous convolution semigroup of subprobability measures if for all $t \geq 0$, $s \geq 0$,

$$\mu_t([0,\infty)) \leq 1, \quad \mu_{t+s} = \mu_t * \mu_s, \quad \text{and} \quad \text{vague} - \lim_{t\to 0+} \mu_t = \delta_0,$$

where δ_0 stands for the Dirac measure at zero. Such a semigroup is often called a subordinator. The next classical characterization of Bernstein functions goes back to Bochner and can be found, e.g., in [19, Theorem 5.2].

Theorem 1.1. *A function $\psi : (0,\infty) \to (0,\infty)$ is Bernstein if and only if there exists a vaguely continuous convolution semigroup $(\mu_t)_{t\geq 0}$ of subprobability measures on $[0,\infty)$ such that*

$$\widehat{\mu}_t(\tau) := \int_0^\infty e^{-s\tau}\, \mu_t(ds) = e^{-t\psi(\tau)}, \quad \tau > 0, \tag{1.1}$$

for all $t \geq 0$.

Theorem 1.1 has its operator-theoretical counterpart. One of the most natural ways to construct a new C_0-semigroup from a given one is to use subordinators. Recall that if $(e^{-tA})_{t\geq 0}$ is a bounded C_0-semigroup on a Banach space X and $(\mu_t)_{t\geq 0}$ is a vaguely continuous convolution semigroup of bounded Radon measures on $[0,\infty)$ then the formula

$$e^{-t\mathcal{A}} := \int_0^\infty e^{-sA}\, \mu_t(ds), \qquad t \geq 0, \tag{1.2}$$

defines a bounded C_0-semigroup $(e^{-t\mathcal{A}})_{t\geq 0}$ on X whose generator $-\mathcal{A}$ can be considered as $-\psi(A)$, thus we will write $\psi(A)$ instead of $\mathcal{A}$ (see the next subsection for more on that). The C_0-semigroup $(e^{-t\psi(A)})_{t\geq 0}$ is called subordinated to the C_0-semigroup $(e^{-tA})_{t\geq 0}$ via the subordinator $(\mu_t)_{t\geq 0}$ (or the corresponding Bernstein function ψ).

Despite the construction of subordination being very natural and appearing often in various contexts, some of its permanence properties have not been made precise so far. In this note, we show that subordination preserves the class of holomorphic C_0-semigroups. In particular, we present a positive answer to the following open question posed in [12, p. 63], see also [3]: suppose that $-A$ generates a bounded holomorphic C_0-semigroup on a Banach space X and ψ is a Bernstein function. Does $-\psi(A)$ also generate a (bounded) holomorphic C_0-semigroup?

A partial answer to a strengthened version of this question was given in [3, Proposition 7.4]: for any Bernstein function ψ the operator $-\psi(A)$ generates a sectorially bounded holomorphic C_0-semigroup of angle at least θ if $-A$ generates a sectorially bounded holomorphic C_0-semigroup of angle θ greater then $\pi/4$. Moreover, it was proved in [3, Theorem 6.1 and Remark 6.2] that the above claim is true with no restrictions on $\theta \in (0, \pi/2]$ if the Bernstein function ψ is, in addition, *complete*. (See [19, Chapters 6–7] concerning the definition and properties of complete Bernstein functions.) It was asked in [3] whether this additional assumption can, in fact, be removed.

If X is a uniformly convex Banach space, e.g., if X is a Hilbert space, then a positive answer to Kishimoto–Robinson's question was obtained in [14, Theorem 1] using Kato–Pazy's criteria for holomorphicity of semigroup.

Recently, based on the machinery of functional calculi, positive answers to both questions in their full generality, were provided in [8]. In particular, it was proved in [8] that if $-A$ generates a sectorially bounded holomorphic C_0-semigroup of angle θ, then for any Bernstein function ψ the operator $-\psi(A)$ also generate a sectorially bounded holomorphic C_0-semigroup of angle at least θ.

The aim of this note is to present an alternative and comparatively simple argument providing positive answers to the questions from [12] and [3] apart from the permanence of sectors of holomorphy. (The permanence property requires additional arguments going beyond the scope of the paper, see [8] for its proof.) Our approach has merits of being self-contained, transparent and much less technical in a sense of using only elementary properties of functional calculi theory.

The proof arises as a byproduct of estimates for $\|\psi(A)e^{-t\varphi(A)}\|$, $t > 0$, where ψ, φ are Bernstein functions satisfying appropriate conditions. In turn such estimates appeared to be crucial in putting approximation theory of operator semigroups into the framework of Bernstein functions of semigroup generators, see [7]. In fact, the technique developed in [7] is basic in this paper.

It is not clear whether the permanence of sectors of holomorphy can be proved by the methods of present note. See however [2] where still another, direct approach to subordination was worked out in details.

2. Preliminaries

2.1. Function theory

Let us recall some basic facts on completely monotone and Bernstein functions from [19] relevant for the following.

First, note that by Bernstein's theorem [19, Theorem 1.4] a real-valued function $g \in \mathrm{C}^\infty(0, \infty)$ is completely monotone if and only if it is the Laplace transform of a (necessarily unique) positive Laplace-transformable Radon measure ν

on $[0,\infty)$:

$$g(\tau)=\widehat{\nu}(\tau)=\int_0^\infty e^{-\tau s}\nu(ds) \qquad \text{for all} \quad \tau>0. \tag{2.1}$$

In particular, (2.1) implies that a completely monotone function extends holomorphically to the open right half-plane $\mathbb{C}_+ := \{z\in\mathbb{C} : \operatorname{Re} z>0\}$. The set of completely monotone functions will be denoted by $\mathcal{CM}$, and the set of bounded complete monotone functions will be denoted by $\mathcal{BCM}$. The standard examples of completely monotone functions include $e^{-t\tau}, \tau^{-\alpha}$, for fixed $t>0$ and $\alpha\ge 0$, and $(\log(1+\tau))^{-1}$.

Bernstein functions constitute a class "dual" in a sense to the class of completely monotone functions. A representation similar in a sense to (2.1) holds also for Bernstein functions. Indeed, by [19, Thm. 3.2], a function ψ is a Bernstein function if and only if there exist $a,b\ge 0$ and a positive Radon measure γ on $(0,\infty)$ satisfying

$$\int_{0+}^\infty \frac{s}{1+s}\,\gamma(\mathrm{d}s)<\infty$$

such that

$$\psi(\tau)=a+b\tau+\int_{0+}^\infty (1-e^{-s\tau})\gamma(ds), \qquad \tau>0. \tag{2.2}$$

The formula (2.2) is called the Lévy–Khintchine representation of ψ. The triple (a,b,γ) is uniquely determined by ψ and is called the Lévy–Khintchine triple. Thus we will write occasionally $\psi\sim(a,b,\gamma)$. Note that a Bernstein function $\psi\sim(a,b,\gamma)$ is increasing, and it satisfies

$$a=\psi(0+) \qquad \text{and} \qquad b=\lim_{t\to\infty}\frac{\psi(t)}{t}.$$

Moreover, by (2.2), ψ extends holomorphically to $\mathbb{C}_+$ and continuously to the closure $\overline{\mathbb{C}}_+$ of $\mathbb{C}_+$. Thus we identify ψ with its continuous extension to $\overline{\mathbb{C}}_+$. Note that ψ growths at most linearly in $\overline{\mathbb{C}}_+$. The Bernstein function ψ is bounded if and only if $b=0$ and $\gamma((0,\infty))<\infty$, see [19, Corollary 3.7].

In the sequel, we will denote the set of Bernstein functions by $\mathcal{BF}$. As examples of Bernstein functions we mention $1-e^{-t\tau}$, τ^α, for fixed $t>0$ and $\alpha\in[0,1]$, and $\log(1+\tau)$.

Now we introduce a functional J which will be an important tool in getting operator norm estimates for the products of functions of a negative semigroup generator A.

For $g\in\mathcal{CM}$ and $\psi\in\mathcal{BF}$ let us define

$$J[g,\psi]:=\int_0^\infty g(s)\psi'(s)\,ds. \tag{2.3}$$

Note that J is well defined if we allow $J[g,\psi]$ to be ∞.

The following choice of g and ψ will be of particular importance. Observe that if $t>0$ is fixed, φ is a Bernstein function, and $g=e^{-t\varphi}$ then $g\in\mathcal{BCM}$ by

Theorem 1.1 and

$$J[e^{-t\varphi}, \psi] = \int_0^\infty e^{-t\varphi(s)}\psi'(s)\, ds. \tag{2.4}$$

Let us note several conditions on g and ψ guaranteeing that $J[g,\psi]$ is finite.

Example 2.1. a) Let $g \in \mathcal{CM}$ and $\psi \in \mathcal{BF}$. If there exists a continuous function $q : (0,\infty) \mapsto (0,\infty)$ such that

$$\int_0^\infty q(s)\, ds < \infty, \quad \text{and} \quad g(s) \le q(\psi(s)), \qquad s > 0, \tag{2.5}$$

then

$$J[g,\psi] \le \int_0^\infty q(\psi(s))\psi'(s)\, ds = \int_{\psi(0)}^{\psi(\infty)} q(s)\, ds \le \int_0^\infty q(s)\, ds < \infty.$$

On the other hand, if $g \in \mathcal{CM}$, $\psi \in \mathcal{BF}$ is such that $\psi \not\equiv \text{const}$, and $J[g,\psi] < \infty$, then

$$g(\tau) = q(\psi(\tau)), \quad \tau > 0, \quad q(s) := g(\psi^{-1}(s)), \quad s \in (\psi(0), \psi(\infty)),$$

and

$$\int_{\psi(0)}^{\psi(\infty)} q(s)\, ds = \int_{\psi(0)}^{\psi(\infty)} g(\psi^{-1}(s))\, dt = \int_0^\infty g(s)\psi'(s)\, ds < \infty.$$

Thus, (2.5) is also necessary (in a sense described above) for $J[g,\psi] < \infty$.

b) Let $g \in \mathcal{BCM}$ be such that $g(0) \le 1$ and $g(\infty) = 0$, and let $\psi \in \mathcal{BF}$. Suppose that there exists a continuous function $f : (0,1) \mapsto (0,\infty)$ such that

$$\int_0^1 f(s)\, ds < \infty, \quad \text{and} \quad \psi(s) \le f(g(s)), \qquad s > 0. \tag{2.6}$$

Then

$$J[g,\psi] \le \int_0^1 f(s)\, ds. \tag{2.7}$$

Indeed, note that $g'(s) \le 0$, $s > 0$. Then, by (2.6), for all $\tau > 1 > \epsilon > 0$,

$$\begin{aligned} \int_\epsilon^\tau g(s)\psi'(s)\, ds &= g(\tau)\psi(\tau) - g(\epsilon)\psi(\epsilon) - \int_\epsilon^\tau g'(s)\psi(s)\, ds \\ &\le g(\tau)f(g(\tau)) - \int_\epsilon^\tau g'(s)f(g(s))\, ds \\ &= g(\tau)f(g(\tau)) + \int_{g(\tau)}^{g(\epsilon)} f(s)\, ds \\ &\le g(\tau)f(g(\tau)) + \int_0^1 f(s)\, ds. \end{aligned} \tag{2.8}$$

Note that $g(\tau)$ decreases to zero monotonically as $\tau \to \infty$. Since $f \in L^1(0,1)$ there exists $(\tau_k)_{k\ge1} \subset (1,\infty)$ such that

$$\lim_{k\to\infty} \tau_k = \infty, \qquad \text{and} \qquad \lim_{k\to\infty} g(\tau_k)f(g(\tau_k)) = 0.$$

Since g and ψ' are positive, setting $\tau = \tau_k, k \in \mathbb{N}$, in (2.8) and letting $k \to \infty$ and $\epsilon \to 0$, we obtain (2.7).

We proceed with several estimates for $J[g, \psi]$, where g is of the form $e^{-t\varphi}, t > 0$, for a Bernstein function φ. They will be important for exploring holomorphicity of $(e^{-t\varphi(A)})_{t\geq 0}$ in the next section.

Example 2.2. a) For any $\psi \in \mathcal{BF}$, we have

$$J[e^{-t\psi}, \psi] = \int_0^\infty e^{-t\psi(s)}\psi'(s)\,ds = t^{-1}[e^{-t\psi(0)} - e^{-t\psi(\infty)}] \leq t^{-1}, \qquad t > 0. \tag{2.9}$$

b) If $\psi \in \mathcal{BF}$ and $\varphi_\alpha(\tau) := \tau^\alpha$, $\alpha \in (0, 1]$, then using monotonicity of ψ and the fact that

$$\psi(c\tau) \leq c\psi(\tau), \qquad \tau \geq 0, \quad c \geq 1, \tag{2.10}$$

see, e.g., [11, p. 205], it follows that

$$\begin{aligned} J[e^{-t\varphi_\alpha}, \psi] + \psi(0) &= t\alpha \int_0^\infty e^{-ts^\alpha} s^{\alpha-1}\psi(s)\,ds \\ &= \alpha \int_0^\infty e^{-s^\alpha} s^{\alpha-1}\psi(s/t^{1/\alpha})\,ds \\ &\leq \psi(1/t^\alpha) \int_0^\infty e^{-s} \max\{1, s^{1/\alpha}\}\,ds \\ &\leq \left(1 + \frac{1}{\alpha e}\right)\psi(1/t^\alpha), \qquad t > 0. \end{aligned} \tag{2.11}$$

Let now $\psi \sim (a, b, \gamma)$ and $\alpha = 1$ so that $\varphi_1(\tau) = \tau$. Then using (2.2), the inequality

$$\frac{s}{t+s} = \frac{s/t}{1+s/t} \leq 1 - e^{-s/t}, \qquad s, t > 0,$$

and Fubini's theorem, we infer that

$$\begin{aligned} J[e^{-t\varphi_1}, \psi] &= \int_0^\infty e^{-ts}\psi'(s)\,ds = \frac{b}{t} + \int_{0+}^\infty \frac{s}{t+s}\,\gamma(ds) \\ &\leq \frac{b}{t} + \int_{0+}^\infty (1 - e^{-s/t})\,\gamma(ds) = \psi(1/t) - \psi(0) \\ &\leq \psi(1/t), \qquad t > 0. \end{aligned} \tag{2.12}$$

The following estimate for J generalizes the one in a).

c) Let ψ be a bounded Bernstein function satisfying

$$\psi(0) = 0, \qquad \psi'(0+) < \infty, \tag{2.13}$$

and let φ be a Bernstein function. Then,

$$J[e^{-t\varphi}, \psi] = \int_0^\infty e^{-t\varphi(s)}\psi'(s)\,ds \leq \psi(\infty), \qquad t > 0.$$

On the other hand, if we are interested in asymptotics of $J[e^{-t\varphi},\psi]$ for big t and $\varphi \not\equiv \text{const}$, then a better estimate is available. Since

$$\varphi(\tau) = \int_0^\tau \varphi'(s)\,ds + \varphi(0) \geq \varphi'(1)\tau, \qquad \tau \in (0,1),$$

it follows that

$$\begin{aligned} J[e^{-t\varphi},\psi] &= \int_0^1 e^{-t\varphi(s)}\psi'(s)\,ds + \int_1^\infty e^{-t\varphi(s)}\psi'(s)\,ds \\ &\leq \psi'(0+)\int_0^1 e^{-t\varphi'(1)s}\,ds + e^{-t\varphi(1)}\int_1^\infty \psi'(s)\,ds \\ &\leq \left[\frac{\psi'(0+)}{\varphi'(1)} + \frac{\psi(\infty)-\psi(1)}{\varphi(1)}\right]\frac{1}{t}, \qquad t>0. \end{aligned}$$

We finish this subsection with several estimates playing a light on the interplay between the functional $J[g,\psi]$ and the product $g\cdot\psi$. They will be needed as an illustration of our main statement.

The following estimate is well known for so-called complete Bernstein functions. However, it seems, it has not been noted for the whole class of Bernstein functions. In the proof, we use an idea from the proof of [4, Theorem 4].

Proposition 2.3. *Let $\psi \in \mathcal{BF}$. Then*

$$|\psi(z)| \leq 2\sigma^{-1}\psi(|z|), \quad \operatorname{Re} z \geq 0, \quad \sigma = 1-e^{-1}. \tag{2.14}$$

Proof. Recall that

$$|1-e^{-z}| \leq \min(|z|,2) \leq 2\min(|z|,1), \quad \operatorname{Re} z \geq 0,$$

and

$$1-e^{-s} \geq \sigma \min(s,1), \quad s\geq 0, \quad \sigma = 1-e^{-1},$$

see [11, Lemma 2.1.2]. Therefore,

$$|1-e^{-z}| \leq 2\sigma^{-1}(1-e^{-|z|}), \quad \operatorname{Re} z \geq 0. \tag{2.15}$$

Let $\psi \in \mathcal{BF}$ be given by (2.2). Then, using (2.15) and noting that $1 < 2\sigma^{-1}$, we obtain

$$\begin{aligned} |\psi(z)| &\leq a + b|z| + \int_{0+}^\infty |1-e^{-sz}|\,\gamma(ds) \leq a + b|z| + 2\sigma^{-1}\int_{0+}^\infty (1-e^{-|z|s})\,\gamma(ds) \\ &\leq 2\sigma^{-1}\psi(|z|), \qquad \operatorname{Re} z \geq 0. \end{aligned}$$ □

In the following result, we show that for $g \in \mathcal{BCM}$ and $\psi \in \mathcal{BF}$ the assumption $J[g,\psi] < \infty$ implies that $g\cdot\psi$ is bounded in any sector

$$\Sigma_\beta := \{z \in \mathbb{C} : |\arg z| < \beta\}, \qquad \beta \in (0,\pi/2).$$

Corollary 2.4. *Let $\psi \in \mathcal{BF}$. Then the following statements hold.*

(i) *For every $g \in \mathcal{CM}$ and every $\beta \in (0,\pi/2)$,*

$$|g(z)\psi(z)| \leq \frac{2}{\sigma\cos\beta}\, g(|z|\cos\beta)\psi(|z|\cos\beta), \quad z \in \Sigma_\beta. \tag{2.16}$$

(ii) *Let $g \in \mathcal{BCM}$ and $J[g,\psi] < \infty$. Then for every $\beta \in (0, \pi/2)$,*

$$|g(z)\psi(z)| \le \frac{2}{\sigma \cos\beta} \{g(0+)\psi(0) + J[g,\psi]\}, \qquad z \in \Sigma_\beta. \tag{2.17}$$

Proof. To prove (i) suppose that g is given by (2.1) and $z \in \Sigma_\beta$. Then

$$|g(z)| \le \int_0^\infty e^{-s\mathrm{Re}\, z}\, \nu(ds) \le \int_0^\infty e^{-s|z|\cos\beta}\, \nu(ds) = g(|z|\cos\beta).$$

Using Proposition 2.3 and the inequality (2.10), we then obtain

$$|g(z)\psi(z)| \le 2\sigma^{-1} g(|z|\cos\beta)\psi(|z|) \le \frac{2}{\sigma\cos\beta}\, g(|z|\cos\beta)\psi(|z|\cos\beta),$$

and the proof is complete.

If $g \in \mathcal{BCM}$ and $J[g,\psi] < \infty$, then, since g is decreasing, for every $\tau > 0$,

$$\begin{aligned} g(\tau)\psi(\tau) &= g(0+)\psi(0) + \int_0^\tau (g'(s)\psi(s) + g(s)\psi'(s))\, ds \\ &\le g(0+)\psi(0) + \int_0^\tau g(s)\psi'(s)\, ds \le g(0+)\psi(0) + J[g,\psi]. \end{aligned}$$

Hence, by (i),

$$|g(z)\psi(z)| \le \frac{2}{\sigma\cos\beta} \{g(0+)\psi(0) + J[g,\psi]\}, \qquad z \in \Sigma_\beta,$$

so that (ii) holds. □

2.2. Functional calculus and holomorphic semigroups

In this subsection we will set up the extended Hille–Phillips functional calculus. The calculus will enable us to define Bernstein functions of a negative semigroup generator and to establish some of their basic properties including operator counterparts of the formulas (1.1) and (2.2). As we will see below, the formulas remain essentially the same upon replacement of the independent variable by an operator A.

Let $\mathrm{M_b}(\mathbb{R}_+)$ be the Banach algebra of bounded Radon measures on $\mathbb{R}_+ := [0,\infty)$ with the standard, total variation norm $\|\mu\|_{\mathrm{M_b}(\mathbb{R}_+)} := |\mu|(\mathbb{R}_+)$. Note that

$$\mathrm{A}^1_+(\mathbb{C}_+) := \{\widehat{\mu} : \mu \in \mathrm{M}_b(\mathbb{R}_+)\}$$

is a commutative Banach algebra with pointwise multiplication and with the norm inherited from $\mathrm{A}^1_+(\mathbb{C}_+)$:

$$\|\widehat{\mu}\|_{\mathrm{A}^1_+(\mathbb{C}_+)} := \|\mu\|_{\mathrm{M_b}(\mathbb{R}_+)}. \tag{2.18}$$

Let $-A$ be the generator of a bounded C_0-semigroup $(e^{-tA})_{t\ge 0}$ on X, and let $\mathcal{L}(X)$ be the Banach space of bounded linear operators on X. Define an algebra homomorphism $\Phi : \mathrm{A}^1_+(\mathbb{C}_+) \mapsto \mathcal{L}(X)$ by the formula

$$\Phi(\widehat{\mu})x := \int_0^\infty e^{-sA}x\, \mu(ds), \qquad x \in X.$$

Since

$$\|\Phi(\hat{\mu})\| \le \sup_{t\ge 0} \|e^{-tA}\| |\mu|(\mathbb{R}_+), \tag{2.19}$$

Φ is clearly continuous. The homomorphism Φ is called the *Hille–Phillips* (HP-) functional calculus for A. If $g \in \mathrm{A}^1_+(\mathbb{C}_+)$ so that $g = \widehat{\mu}$ for $\mu \in \mathrm{M_b}(\mathbb{R}_+)$, we then put

$$g(A) = \Phi(\widehat{\mu}).$$

Basic properties of the Hille–Phillips functional calculus can be found in [10, Chapter XV] and in [9, Chapter 3.3]. It is crucial to note that if $g \in \mathcal{BCM}$, then $g \in A^1_+(\mathbb{C}_+)$ by Fatou's theorem, so that $g(A)$ is defined in the HP-calculus and $g(A) \in \mathcal{L}(X)$.

Let now $O(\mathbb{C}_+)$ be the algebra of functions holomorphic in $\mathbb{C}_+$. Denote by $\mathrm{A}^1_{+,r}(\mathbb{C}_+)$ the set of $f \in O(\mathbb{C}_+)$ such that there exists $e \in \mathrm{A}^1_+(\mathbb{C}_+)$ with $ef \in \mathrm{A}^1_+(\mathbb{C}_+)$ and the operator $e(A)$ is injective. Then for any $f \in \mathrm{A}^1_{+,r}(\mathbb{C}_+)$ one defines a closed operator $f(A)$ as

$$f(A) := (e(A))^{-1}(ef)(A). \tag{2.20}$$

The above definition does not depend on the choice of a regularizer e, and thus the mapping $f \to f(A)$ is well defined. We will call this mapping the *extended Hille–Phillips calculus* for A.

The extended HP-calculus satisfies, in particular, the following, natural sum and product rules, see, e.g., [9, Chapter 1].

Proposition 2.5. *Let f and g belong to $\mathrm{A}^1_{+,r}(\mathbb{C}_+)$, and let $-A$ be the generator of a bounded C_0-semigroup. Then*

(i) $f(A)g(A) \subset (fg)(A)$;

(ii) $f(A) + g(A) \subset (f + g)(A)$;

If $g(A)$ is bounded then the inclusions above are, in fact, equalities.

Recall that, as it was shown in [6, Lemma 2.5], Bernstein functions are regularisable by $e(z) = 1/(1 + z)$, that is $e\psi \in \mathrm{A}^1_+(\mathbb{C})$ for every Bernstein function ψ, and then, in particular, by the HP-calculus,

$$[\psi(z)(1 + z)^{-1}](A) \in \mathcal{L}(X). \tag{2.21}$$

Thus, according to (2.20), for any $\psi \in \mathcal{BF}$,

$$\psi(A) = (1 + A)[\psi(z)(1 + z)^{-1}](A). \tag{2.22}$$

While Bernstein functions can formally be defined in the extended *HP*-calculus by (2.20), this definition can hardly be used for practical purposes. However, following analogy to the scalar-valued case, one can derive representations for operator Bernstein functions similar to (1.1) and (2.2), see, e.g., [6, Corollary 2.6] and [19, Proposition 2.1 and Theorem 12.6].

Theorem 2.6. *Let $-A$ generate a bounded C_0-semigroup $(e^{-tA})_{t\ge 0}$ on X, and let ψ be a Bernstein function with the corresponding Lévy–Hintchine triple (a, b, γ). Then the following statements hold.*

(i) *For every* $x \in \mathrm{dom}(A)$,

$$\psi(A)x = ax + bAx + \int_{0+}^{\infty} (1 - e^{-sA})x\,\gamma(ds), \tag{2.23}$$

where the integral is understood as a Bochner integral. Moreover, $\mathrm{dom}(A)$ *is a core for* $\psi(A)$.

(ii) *The operator* $-\psi(A)$ *generates a bounded* C_0*-semigroup* $(e^{-t\psi(A)})_{t\geq 0}$ *on* X *given by*

$$e^{-t\psi(A)} = \int_0^{\infty} e^{-sA}\,\mu_t(ds), \qquad t \geq 0, \tag{2.24}$$

where $(\mu_t)_{t\geq 0}$ *is a vaguely continuous convolution semigroup of subprobability measures on* $[0,\infty)$ *corresponding to* ψ *by* (1.1).

Thus, the operator Bernstein function $\psi(A)$ can be recovered from its restriction to $\mathrm{dom}(A)$ by means of (2.23). Moreover, $-\psi(A)$ generates a bounded C_0-semigroup if $-A$ does, and this fact motivates further study of the permanence properties for the mapping $-A \to -\psi(A)$, e.g., preservation of the class of generators of holomorphic semigroups on X.

It will be crucial to note that subordination does not increase the norm. Indeed, as an immediate consequence of Theorem 2.6, (ii), one obtains

$$\sup_{t\geq 0} \|e^{-t\psi(A)}\| \leq \sup_{t\geq 0} \|e^{-tA}\|. \tag{2.25}$$

While the relations (2.23) and (2.24) hold for any bounded C_0-semigroup, in this note we will concentrate on bounded C_0-semigroups which are, in addition, holomorphic. Recall that a C_0-semigroup $(e^{-tA})_{t\geq 0}$ is said to be holomorphic if it extends holomorphically to a sector Σ_β for some $\beta \in (0, \frac{\pi}{2}]$ and the extension is bounded on $\Sigma_\theta \cap \{z \in \mathbb{C} : |z| \leq 1\}$ for any $\theta \in (0,\beta)$. If $e^{-\cdot A}$ is bounded in Σ_θ whenever $0 < \theta < \beta$, then $(e^{-tA})_{t\geq 0}$ is said to be a sectorially bounded holomorphic semigroup of angle β.

It is well known that sectorially bounded holomorphic semigroups can be described by means of their asymptotics on the real axis. Namely, $-A$ is the generator of a sectorially bounded holomorphic C_0-semigroup $(e^{-tA})_{t\geq 0}$ on a Banach space X if and only if $e^{-tA}(X) \subset \mathrm{dom}(A)$ for every $t > 0$, and $\sup_{t\geq 0} \|e^{-tA}\|$ and $\sup_{t>0} \|tAe^{-tA}\|$ are finite, see, e.g., [5, Theorem 4.6].

It is often useful to omit the assumption of sectorial boundedness and to consider C_0-semigroups bounded on $\mathbb{R}_+$ and having a holomorphic extension to a sector around the real axis. This situation can also be characterized in terms of behavior of $(e^{-tA})_{t\geq 0}$ on the positive half-axis.

By a classical theorem due to Yosida, a C_0-semigroup $(e^{-tA})_{t\geq 0}$ on X is holomorphic if and only if

$$e^{-tA}(X) \subset \mathrm{dom}(A), \quad t > 0, \quad \text{and} \quad \limsup_{t\to 0} \|tAe^{-tA}\| < \infty. \tag{2.26}$$

Since it is not easy to find this statement in the literature, we sketch its proof below. Note that by [1, Proposition 3.7.2 *b*)] a C_0-semigroup $(e^{-tA})_{t\geq 0}$ on X is holomorphic if and only if there exists $a > 0$ such that $(e^{-t(A+a)})_{t\geq 0}$ is a sectorially bounded holomorphic C_0-semigroup. Then, by [5, Theorem 4.6] mentioned above, the latter property is equivalent to $e^{-tA}(X) \subset \mathrm{dom}(A)$ for every $t > 0$, and

$$\sup_{t>0} (e^{-at}\|e^{-tA}\| + \|te^{-at}Ae^{-tA}\|) < \infty. \tag{2.27}$$

Thus, in particular, (2.26) holds. Conversely, if (2.26) is true, then (2.27) is satisfied for certain $a > 0$, and the sectorial boundedness of $(e^{-t(A+a)})_{t\geq 0}$ yields the holomorphicity of $(e^{-tA})_{t\geq 0}$. (Concerning Yosida's theorem and its proof see also [20] and [13, Remark, p. 332].)

Note that if $(e^{-tA})_{t\geq 0}$ is holomorphic and bounded, then for all $\delta > 0$ and $t > \delta$,

$$\|Ae^{-tA}\| \leq \left(\sup_{t\geq 0} \|e^{-tA}\|\right) \sup_{t\in(\delta/2,\delta)} \|Ae^{-tA}\|.$$

In other words, if $(e^{-tA})_{t\geq 0}$ is bounded, then the Yosida condition (2.26) can be given the equivalent form

$$\|Ae^{-tA}\| \leq c_0 + \frac{c_1}{t}, \qquad t > 0, \tag{2.28}$$

with some constants $c_0 \geq 0$ and $c_1 > 0$ which will be crucial in the estimates below. Thus, if $(e^{-tA})_{t\geq 0}$ satisfies (2.28), then we say that $(e^{-tA})_{t\geq 0}$ satisfies *the Yosida condition* $Y(c_0, c_1)$ (which is just an explicit form of the classical Yosida condition (2.26) above).

It will be convenient to rewrite (2.28) in terms of only $(e^{-tA})_{t\geq 0}$. To this aim, we first prove the following simple proposition.

Proposition 2.7. *Let $-A$ be the generator of a bounded C_0-semigroup on a Banach space X, and let*

$$M := \sup_{t\geq 0} \|e^{-tA}\|. \tag{2.29}$$

Suppose that $e^{-tA}(X) \subset \mathrm{dom}(A)$, $t > 0$, and there exists an increasing function $r : (0,\infty) \mapsto (0,\infty)$ such that

$$\sup_{t>0} r(t)\|Ae^{-tA}\| \leq 1. \tag{2.30}$$

Then

$$\|(1 - e^{-sA})\, e^{-tA}\| \leq \frac{4M\, s}{2Mr(t) + s}, \qquad s, t > 0. \tag{2.31}$$

Proof. By (2.29), for all $s, t > 0$,

$$\|(1 - e^{-sA})\, e^{-tA}\| \leq 2M.$$

On the other hand, since

$$(1 - e^{-sA})\, e^{-tA} = \int_t^{t+s} Ae^{-\tau A}\, d\tau, \tag{2.32}$$

we infer by (2.30) that

$$\|(1-e^{-sA})\,e^{-tA}\| \quad \le \quad \int_t^{t+s} \frac{d\tau}{r(\tau)} \le \frac{s}{r(t)}, \qquad s,t>0.$$

Then, since

$$\min\{a,b\} \le \frac{2ab}{a+b}, \qquad a,b>0,$$

it follows that

$$\|(1-e^{-sA})\,e^{-tA}\| \le \min\{2M, s/r(t)\} \le \frac{4M\,s}{2Mr(t)+s}. \qquad \square$$

Now we are ready to recast (2.28) in semigroup terms, and the following corollary of Proposition 2.7 is almost immediate.

Corollary 2.8. *Let $-A$ be the generator of a C_0-semigroup on X satisfying (2.29) and the Yosida condition $Y(c_0,c_1)$. Then*

$$\|(1-e^{-sA})\,e^{-tA}\| \le 2s\left\{\frac{2Mc_0}{1+c_0 s} + \frac{\max(2M,c_1)}{t+s}\right\}, \qquad s,t>0. \tag{2.33}$$

Conversely, if $(e^{-tA})_{t\ge 0}$ satisfies (2.33), then $(e^{-tA})_{t\ge 0}$ satisfies the Yosida condition $Y(4Mc_0, 2\max(2M,c_1))$.

Proof. By Proposition 2.7 applied to

$$r(t) := \frac{t}{c_0 t + c_1}, \qquad t>0,$$

we obtain that

$$\begin{aligned}
\|(1-e^{-sA})\,e^{-tA}\| &\le 4Ms\frac{(c_0 t + c_1)}{2Mt+(c_0 t + c_1)s} \\
&= 4Ms\left\{\frac{c_0 t}{2Mt+(c_0 t + c_1)s} + \frac{c_1}{2Mt+(c_0 t + c_1)s}\right\} \\
&\le 4Ms\left\{\frac{c_0}{2M+c_0 s} + \frac{c_1}{2Mt+c_1 s}\right\} \\
&\le 2s\left\{\frac{2Mc_0}{1+c_0 s} + \frac{\max(2M,c_1)}{t+s}\right\}.
\end{aligned}$$

If, conversely, (2.33) is true, then dividing both sides of it by s, using (2.32) and passing to the limit as $s\to 0+$ for a fixed $t>0$, we get

$$\|Ae^{-tA}\| \le 4Mc_0 + 2\frac{\max(2M,c_1)}{t},$$

that is $Y(4Mc_0, 2\max(2M,c_1))$ holds. $\square$

The elementary estimate (2.33) will play a key role in the subsequent arguments.

3. Main results

To obtain a positive answer to Kishimoto–Robinson's question, we need to show that if $(e^{-tA})_{t\geq 0}$ is a bounded C_0-semigroup satisfying Yosida's condition, then for any Bernstein function ψ one has $e^{-t\psi(A)}(X) \subset \operatorname{dom}(\psi(A)), t > 0$, and the function $t \mapsto \|t\psi(A)e^{-t\psi(A)}\|$ is bounded in an appropriate neighborhood of zero. This will be derived as a simple consequence of the following operator norm estimate for $\psi(A)g(A)$ where $\psi \in \mathcal{BF}$ and $g \in \mathcal{BCM}$. In a different context, a related estimate was obtained in [16, Theorem 1].

For the rest of the paper, if $(e^{-tA})_{t\geq 0}$ is a bounded C_0-semigroup on a Banach space X then we let

$$M(A) := \sup_{t\geq 0} \|e^{-tA}\|.$$

Theorem 3.1. *Let $\psi \in \mathcal{BF}$ and $g \in \mathcal{BCM}$ be such that $J[g,\psi] < \infty$. Let $-A$ be the generator of a bounded C_0-semigroup satisfying the Yosida condition $Y(c_0, c_1)$. Then $\psi(A)g(A) \in \mathcal{L}(X)$ and*

$$\|\psi(A)g(A)\| \leq \psi(0)\|g(A)\| + 2\max(2M(A), c_1)J[g,\psi] + 4M(A)g(0+)C[c_0;\psi], \tag{3.1}$$

where

$$C[c_0;\psi] := \int_0^\infty e^{-s/c_0}\psi'(s)\,ds, \qquad c_0 > 0, \qquad C[0;\psi] := 0. \tag{3.2}$$

Proof. By assumption and Bernstein's theorem, there exists a finite Radon measure ν on $[0,\infty)$ such that

$$g(s) = \int_0^\infty e^{-\tau s}\,\nu(d\tau), \quad s > 0, \qquad g(0+) = \nu([0,\infty)) < \infty. \tag{3.3}$$

Let $\psi \sim (a, b, \gamma)$ so that the representation (2.2) holds. Then (3.2) takes the form

$$C[c_0;\psi] = bc_0 + \int_{0+}^\infty \frac{c_0 s}{1 + c_0 s}\,\gamma(ds).$$

Note that it suffices to prove (3.1) for a Bernstein function ψ with $a = \psi(0) = 0$.

Suppose first that $a = b = 0$ in (2.2). Let $x \in \operatorname{dom}(A)$ be fixed. Then, by (2.21) and Proposition 2.5,

$$g(A)x \in \operatorname{dom}(A) \subset \operatorname{dom}\psi(A).$$

Hence, by Fubini's theorem, we have

$$\begin{aligned}\psi(A)g(A)x = g(A)\psi(A)x &= \int_0^\infty e^{-\tau A}\,\nu(d\tau)\int_{0+}^\infty [1 - e^{-sA}]x\,\gamma(ds)\\ &= \int_0^\infty \int_{0+}^\infty [1 - e^{-sA}]e^{-\tau A}x\,\gamma(ds)\,\nu(d\tau).\end{aligned}$$

Using (2.33) and (3.3), from here it follows that

$$\begin{aligned}\|\psi(A)g(A)x\| \qquad\qquad (3.4)\\
&\le 2\|x\| \int_0^\infty \int_{0+}^\infty \left\{ \frac{2M(A)c_0 s}{1+c_0 s} + \frac{\max(2M(A),c_1)s}{\tau+s} \right\} \gamma(ds)\,\nu(d\tau)\\
&= 2\|x\| \left\{ 2g(0+)M(A)C[c_0;\psi] + \max(2M(A),c_1) \int_0^\infty \int_{0+}^\infty \frac{s}{\tau+s}\gamma(ds)\,\nu(d\tau) \right\}.\end{aligned}$$

Again, by applying Fubini's theorem twice, we obtain that (as in (2.12))

$$\int_0^\infty e^{-\tau t}\psi'(t)\,dt = \int_{0+}^\infty \frac{s\,\gamma(ds)}{s+\tau}, \qquad \tau > 0.$$

and

$$\begin{aligned}\int_0^\infty \int_{0+}^\infty \frac{s\,\gamma(ds)}{s+\tau}\,\nu(d\tau) &= \int_0^\infty \int_0^\infty e^{-\tau t}\psi'(t)\,dt\,\nu(d\tau) \qquad (3.5)\\
&= \int_0^\infty \int_0^\infty e^{-\tau t}\,\nu(d\tau)\,\psi'(t)\,dt = \int_0^\infty g(t)\psi'(t)\,dt = J[g,\psi].\end{aligned}$$

So, (3.4) yields

$$\|\psi(A)g(A)x\| \le 2\|x\|\{\max(2M(A),c_1)J[g,\psi] + 2M(A)g(0+)C[c_0;\psi]\}. \qquad (3.6)$$

From (3.6), since $\psi(A)g(A)$ is closed as a product of closed and bounded operators and $\mathrm{dom}(A)$ is dense in X, we conclude that

$$\mathrm{ran}(g(A)) \subset \mathrm{dom}(\psi(A)), \qquad (3.7)$$

and (3.1) holds. This finishes the proof in the case $a = b = 0$.

Let now $a = 0$ and $b > 0$. Arguing as above, if $x \in \mathrm{dom}(A)$ is fixed, then

$$\begin{aligned}\psi(A)g(A)x &= g(A)\psi(A)x\\
&= b\int_0^\infty Ae^{-\tau A}x\,\nu(d\tau) + \int_0^\infty \int_{0+}^\infty [1-e^{-sA}]\,e^{-\tau A}x\,\gamma(ds)\,\nu(d\tau).\end{aligned}$$

Note that $\psi'(s) \ge b$, $s > 0$, and

$$\int_0^\infty \tau^{-1}\nu(d\tau) = \int_0^\infty g(s)\,ds \le b^{-1}\int_0^\infty g(s)\psi'(s)\,ds = b^{-1}J[g,\psi] < \infty.$$

Therefore,

$$\|Ag(A)x\| \le \int_0^\infty \|Ae^{-\tau A}x\|\,\nu(d\tau) \le \|x\| \int_0^\infty (c_0 + c_1/\tau)\,\nu(d\tau). \qquad (3.8)$$

Now using (3.5) for the Bernstein function $\psi(t) - bt$, and taking into account (3.8), we obtain that

$$\begin{aligned}\|\psi(A)g(A)x\| &\le b\|x\| \int_0^\infty (c_0 + c_1\tau^{-1})\,\nu(d\tau)\\
&\quad + 2\|x\| \int_0^\infty \int_{0+}^\infty \left\{ \frac{2M(A)c_0 s}{1+c_0 s} + \frac{\max(2M(A),c_1)s}{\tau+s} \right\} \gamma(ds)\,\nu(d\tau)\end{aligned}$$

$$\begin{aligned}
&\le g(0+)bc_0\|x\| + \|x\|b\int_0^\infty g(s)\,ds \\
&\quad + 4M(A)g(0+)\|x\|\int_{0+}^\infty \frac{c_0 s}{1+c_0 s}\gamma(ds) \\
&\quad + 2\max(M(A),c_1)\|x\|\int_0^\infty g(s)[\psi'(s)-b]\,ds \\
&\le 4M(A)g(0+)\|x\|\left\{bc_0 + \int_{0+}^\infty \frac{c_0 s}{1+c_0 s}\gamma(ds)\right\} \\
&\quad + 2\max(2M(A),c_1)\|x\|\int_0^\infty g(s)\psi'(s)\,ds \\
&= 2\|x\|\left\{\max(2M(A),c_1)J[g,\psi] + 2M(A)g(0+)C[c_0;\psi]\right\}.
\end{aligned}$$

Since the operator $\psi(A)g(A)$ is closed and $\mathrm{dom}(A)$ is dense, the last inequality implies (3.7) and (3.1). □

Remark 3.2. The assumption $J[g,\psi] < \infty$ is not necessary to ensure the boundedness of $\psi(A)g(A)$. To see this, it is enough to consider the Bernstein function $\psi(\tau) = \tau + 1$ and the bounded completely monotone function $g(\tau) = 1/(\tau+1)$. However, the assumption implies the boundedness of $\psi \cdot g$ in any sector Σ_β with $\beta \in (0, \pi/2)$, see Corollary 2.4. If $-A$ generates a sectorially bounded holomorphic C_0-semigroup and admits, in addition, a bounded H^∞-calculus on a sector Σ_θ, the boundedness of $\psi \cdot g$ in Σ_β, $\beta > \theta$, implies also the boundedness of $\psi(A)g(A)$.

For a choice of g as $e^{-t\varphi}$, where φ is a Bernstein function, Theorem 3.1 yields immediately the following corollaries.

Corollary 3.3. *Let ψ and φ be Bernstein functions such that $J[e^{-t\varphi},\psi] < \infty$ for every $t > 0$. Let $-A$ be the generator of a bounded C_0-semigroup on X satisfying the Yosida condition $Y(c_0, c_1)$. Then for every $t > 0$,*

$$\begin{aligned}
\|\psi(A)e^{-t\varphi(A)}\| &\le \psi(0)\|e^{-t\varphi(A)}\| \\
&+ 2\max(2M(A),c_1)J[e^{-t\varphi},\psi] + 4M(A)e^{-t\varphi(0)}C[c_0;\psi].
\end{aligned}$$

Corollary 3.4. *Let ψ be a Bernstein function and let $-A$ be the generator of a bounded C_0-semigroup $(e^{-tA})_{t\ge 0}$ on X satisfying the Yosida condition $Y(c_0, c_1)$. Then for every $t > 0$,*

$$\|\psi(A)e^{-tA}\| \le 2\max(2M(A),c_1)\psi(1/t) + 4M(A)C[c_0;\psi]. \tag{3.9}$$

In particular, if $-A$ generates a sectorially bounded holomorphic C_0-semigroup, then

$$\|\psi(A)e^{-tA}\| \le 2\max(2M(A),c_1)\psi(1/t), \qquad t > 0. \tag{3.10}$$

Proof. By (2.12) and Corollary 3.3 applied to the Bernstein function $\varphi_1(\tau) = \tau$,

$$\begin{aligned}\|\psi(A)e^{-tA}\| &\le \psi(0)M(A) + 2\max(2M(A), c_1)J[e^{-t\varphi_1}, \psi] + 4M(A)C[c_0;\psi]\\ &\le 2\max(2M(A), c_1)\{J[e^{-t\varphi_1}, \psi] + \psi(0)\} + 4M(A)C[c_0:\psi]\\ &\le 2\max(2M(A), c_1)\psi(1/t) + 4M(A)C[c_0;\psi].\end{aligned}$$

□

As we explained in the beginning of this section, Corollary 3.3 leads to a positive answer to Kishimoto–Robinson's question which is contained in the next statement. Incidentally, it also partially answers the question from [3] and shows that Bernstein functions map the class of generators of sectorially bounded holomorphic C_0-semigroups into itself. The statement was proved in [8] by a different technique.

Corollary 3.5. *Let ψ be a Bernstein function and let $-A$ be the generator of a bounded C_0-semigroup satisfying the Yosida condition $Y(c_0, c_1)$. Then $-\psi(A)$ generates a bounded C_0-semigroup on X satisfying the following Yosida condition:*

$$\|\psi(A)e^{-t\psi(A)}\| \le M(A)(\psi(0) + 4)C[c_0;\psi])e^{-t\psi(0)} + 2\max(2M(A), c_1)t^{-1} \quad (3.11)$$

for every $t > 0$. If $-A$ generates a sectorially bounded C_0-semigroup on X, then the same is true for $-\psi(A)$.

Proof. Note that $\psi = \psi(0) + \psi_0$, $\psi_0 \in \mathcal{BF}$, and then

$$\|e^{-t\psi(A)}\| \le e^{-\psi(0)t}\|e^{-t\psi_0(A)}\| \le M(A)e^{-\psi(0)t}, \qquad t > 0. \quad (3.12)$$

Now Corollary 3.3 and Example 2.2, a) yield (3.11). If $(e^{-tA})_{t\ge 0}$ is sectorially bounded, then $c_0 = 0$ and, by definition, $C[c_0;\psi] = 0$ as well. In this case, (3.11) implies that $t\psi(A)e^{-t\psi(A)}$ is bounded on $(0,\infty)$. Since $(e^{-t\psi(A)})_{t\ge 0}$ is bounded, it is moreover sectorially bounded. □

Next we turn to other applications of Theorem 3.1 arising in a general framework for approximation theory of operator semigroups developed in [7]. Note that Corollary 3.3 and Example 2.2, c) imply directly the next statement (cf. [7, Theorem 6.8]).

Theorem 3.6. *Let ψ be a bounded Bernstein function satisfying* (2.13), *and let $\varphi \not\equiv const$ be a Bernstein function. Let $-A$ be the generator of a sectorially bounded holomorphic C_0-semigroup $(e^{-tA})_{t\ge 0}$ on X. Then*

$$\sup_{t>0} \|t\psi(A)e^{-t\varphi(A)}\| \le 2\max(2M(A), c_1)\left[\frac{\psi'(0+)}{\varphi'(1)} + \frac{\psi(\infty) - \psi(1)}{\varphi(1)}\right]. \quad (3.13)$$

The following corollary of Theorem 3.6 was obtained in [7, Corollary 6.9].

Corollary 3.7. *Let φ be a Bernstein function such that*

$$\varphi'(0+) = 1, \quad |\varphi''(0+)| < \infty. \quad (3.14)$$

Let $-A$ be the generator of a sectorially bounded holomorphic C_0-semigroup $(e^{-tA})_{t\geq 0}$ on X. Then

$$\|(1-\varphi'(A))e^{-t\varphi(A)}\| \leq \frac{2\max(2M(A),c_1)}{t}\left[\frac{|\varphi''(0+)|}{\varphi'(1)}+\frac{\varphi'(1)}{\varphi(1)}\right],$$

for all $t>0$.

Proof. Note that by (3.14) the Bernstein function $\psi(\tau)=1-\varphi'(\tau), \tau>0$, is bounded and satisfies (2.13). Applying Theorem 3.6 to the Bernstein function φ and the bounded Bernstein function ψ and taking into account the relations $\psi'(0+)=-\varphi''(0+)=|\varphi''(0+)|$ and

$$\psi(\infty)-\psi(1)=\varphi'(1)-\varphi'(\infty)\leq\varphi'(1),$$

we get the assertion. □

Remark 3.8. Note that in [7, Theorem 6.8] the second term $\frac{\psi(\infty)-\psi(1)}{\varphi(1)}$ in the right-hand of (3.13) has a wrong form $\psi(1)/\varphi(1)$ due to incorrect evaluation of $\|\psi'\|_{L^1([a,\infty))}=\int_a^\infty \psi'(s)\,ds$ in the last line of the proof. Thus [7, Eq. (6.12)] should take the form of (3.13). However, [7, Corollary 6.9] (i.e., Corollary 3.7 here) which was a basis for subsequent estimates in [7, Section 6] remains unchanged.

We finish with relating our estimates to the following generalization of the moment inequality for generators of bounded C_0-semigroups given in [19, Corollary 12.8]. As proved in [19], if $-A$ is the generator of a bounded C_0-semigroup on X and $\psi\in\mathcal{BF}$, then

$$\|\psi(A)x\| \leq \frac{2e}{e-1}M(A)\psi\left(\frac{\|Ax\|}{2\|x\|}\right)\|x\|, \qquad x\neq 0, \qquad x\in \operatorname{dom}(A). \tag{3.15}$$

If $\psi(\tau)=\tau^\alpha, \alpha\in(0,1)$, then (3.15) reduces to the classical moment inequality for fractional powers of A. It is instructive to note the following corollary of (3.15).

Corollary 3.9. *Let $-A$ be the generator of a bounded C_0-semigroup such that*

$$\|tAe^{-tA}\|\leq M_a, \qquad t\in(0,a], \qquad a\leq\infty, \tag{3.16}$$

and $\psi\in\mathcal{BF}$. Then

$$\|\psi(A)e^{-tA}\| \leq \frac{e}{e-1}M(A)\max\{2M(A),M_a\}\,\psi(1/t), \qquad t\in(0,a]. \tag{3.17}$$

Proof. Setting in (3.15) $x=e^{-tA}y$, $y\in X$, $t\in(0,a]$ and using (3.16) and (2.10), we obtain that

$$\begin{aligned}\|\psi(A)e^{-tA}y\| &\leq \frac{2e}{e-1}M(A)\|e^{-tA}y\|\psi\left(\frac{M_a\|y\|}{2t\|e^{-tA}y\|}\right)\\ &\leq \frac{2e}{e-1}M(A)\|e^{-tA}y\|\max\left\{1,\frac{M_a\|y\|}{2\|e^{-tA}y\|}\right\}\psi(1/t)\end{aligned}$$

$$= \frac{e}{e-1} M(A) \max\left\{2\|e^{-tA}y\|, M_a\|y\|\right\} \psi(1/t)$$
$$\leq \frac{e}{e-1} M(A) \max\{2M(A), M_a\} \psi(1/t)\|y\|,$$

that is (3.17) holds. □

As an illustration of Corollary 3.9, note that if $\psi(\tau) = \log(1+\tau)$ then Corollary 3.9 yields the estimate

$$\sup_{t\in(0,1/e]} \frac{\|\log(1+A)e^{-tA}\|}{\log(1/t)} < \infty.$$

proved originally in [17, Proposition 2.7].

Finally, we note that it is possible to develop an approach to the permanence problems from [12] and [3] different from the ones in [8] and in the present note. This approach based on direct resolvent estimates for Bernstein functions of semigroup generators is worked out in [2]. While it allows one to get sharp estimates for subordinated semigroups (and their sectors of holomorphy), it is much more involved than the arguments in this article.

References

[1] W. Arendt, C.J.K. Batty, M. Hieber, and F. Neubrander, *Vector-valued Laplace Transforms and Cauchy Problems*, 2nd ed., Monographs in Mathematics, vol. 96, Birkhäuser, Basel, 2011.

[2] C.J.K. Batty, A. Gomilko and Yu. Tomilov, *Bernstein functions of semigroup generators: a resolvent approach*, preprint.

[3] C. Berg, K. Boyadzhiev and R. deLaubenfels, *Generation of generators of holomorphic semigroups.* J. Austral. Math. Soc. Ser. A **55** (1993), 246–269.

[4] A. Carasso and T. Kato, *On subordinated holomorphic semigroups.* Trans. Amer. Math. Soc. **327** (1991), 867–878.

[5] K.-J. Engel and R. Nagel, *One-Parameter Semigroups for Linear Evolution Equations*, Graduate Texts in Math., **194**, Springer, New York, 2000.

[6] A. Gomilko, M. Haase, and Yu. Tomilov, *Bernstein functions and rates in mean ergodic theorems for operator semigroups*, J. d'Analyse Math. **118** (2012), 545–576.

[7] A. Gomilko and Yu. Tomilov, *On rates in approximation theory for operator semigroups*, J. Funct. Anal. **266** (2014), 3040–3082.

[8] A. Gomilko and Yu. Tomilov, *On subordination of holomorphic semigroups*, Adv. in Math., to appear.

[9] M. Haase, *The Functional Calculus for Sectorial Operators.* Operator Theory: Advances and Applications **169**, Birkhäuser, Basel, 2006.

[10] E. Hille and R.S. Phillips, *Functional Analysis and Semi-Groups*, 3rd printing of rev. ed. of 1957, Colloq. Publ. **31**, AMS, Providence, RI, 1974.

[11] N. Jacob, *Pseudo differential operators and Markov processes. Vol. I. Fourier analysis and semigroups*, Imperial College Press, London, 2001.

[12] A. Kishimoto and D.W. Robinson, *Subordinate semigroups and order properties.* J. Austral. Math. Soc. Ser. A **31** (1981), 59–76.

[13] H. Komatsu, *Fractional powers of operators*, Pacific J. Math. **19** (1966), 285–346.

[14] A.R. Mirotin, *On the T-calculus of generators of C_0-semigroups*, Sibirsk. Mat. Zh. **39** (1998), 571–582 (in Russian); transl. in Siberian Math. J. **39** (1998), 493–503.

[15] A.R. Mirotin, *The multidimensional T-calculus of generators of C_0-semigroups.* (Russian) Algebra i Analiz **11** (1999), 142–169 (in Russian); transl. in St. Petersburg Math. J. **11** (2000), no. 2, 315–335.

[16] A.R. Mirotin, *Criteria for analyticity of subordinate semigroups*, Semigroup Forum **78** (2009), 262–275.

[17] N. Okazawa, *Logarithms and imaginary powers of closed linear operators.* Integral Eq. Operator Theory **38** (2000), 458–500.

[18] R.S. Phillips, *On the generation of semigroups of linear operators*, Pacific J. Math. **2** (1952), 343-369.

[19] R. Schilling, R. Song, and Z. Vondraček, *Bernstein functions*, de Gruyter Studies in Mathematics, **37**, Walter de Gruyter, Berlin, 2010.

[20] K. Yosida, *On the differentiability of semigroups of linear operators.* Proc. Japan Acad. **34** (1958), 337–340.

Alexander Gomilko
Faculty of Mathematics and Computer Science
Nicolas Copernicus University
ul. Chopina 12/18
PL-87-100 Toruń, Poland

and

Institute of Telecommunications and Global Information Space
National Academy of Sciences of Ukraine
13, Chokolivsky Blvd
03186 Kiev, Ukraine
e-mail: gomilko@mat.umk.pl

Yuri Tomilov
Institute of Mathematics
Polish Academy of Sciences
Śniadeckich 8
PL-00-956 Warsaw, Poland
e-mail: tomilov@mat.umk.pl

Operator Theory:
Advances and Applications, Vol. 250, 273–279

A Quantitative Coulhon–Lamberton Theorem

Tuomas P. Hytönen

Dedicated to Professor Charles Batty on the occasion of his 60th birthday.

Abstract. Let X be a Banach space and $p \in (1, \infty)$. We denote the L^p-norms of several operators on X-valued functions as follows: the norm of martingale transforms (i.e., the UMD constant) by $\beta_{p,X}$, the norm of the Hilbert transform by $\hbar_{p,X}$, and the norm of the maximal regularity operator for the Poisson semigroup by $\mathfrak{m}_{p,X}$. Qualitatively, all three are known to be finite or infinite simultaneously. We prove the quantitative relation

$$\frac{1}{2}\max(\beta_{p,X}, \hbar_{p,X}) \leq \mathfrak{m}_{p,X} \leq \beta_{p,X} + \hbar_{p,X}.$$

Mathematics Subject Classification (2010). Primary 42B15; Secondary 47D06.

Keywords. Maximal regularity, Poisson semigroup, Fourier multiplier, UMD space, Hilbert transform.

1. Introduction

Let X be a Banach space and $p \in (1, \infty)$. An analytic semigroup $(P_t)_{t\geq 0}$ (or its generator $-A$, or the Cauchy problem $\dot{u} + Au = f \in L^p(\mathbb{R}_+; X)$, $u(0) = 0$) has *maximal L^p-regularity* if the related singular integral operator

$$\mathfrak{M}f(t) := \int_0^t AP_s f(t-s)\,\mathrm{d}s = \int_0^\infty AP_s f(t-s)\,\mathrm{d}s$$

is bounded on $L^p(\mathbb{R}_+; X) \eqsim \{f \in L^p(\mathbb{R}; X) : \operatorname{supp} f \subseteq [0, \infty)\}$, or equivalently, by translation invariance, on all of $L^p(\mathbb{R}; X)$. It is of considerable interest to determine whether a given semigroup possesses this property. This frequently requires the UMD property of the underlying Banach space X, as illustrated by the following prototypical result:

The author is supported by the European Union through the ERC Starting Grant "Analytic-probabilistic methods for borderline singular integrals". He is a member of the Finnish Centre of Excellence in Analysis and Dynamics Research.

Theorem 1.1 (Coulhon, Lamberton [5]). *Let $p \in (1,\infty)$. The Poisson semigroup*

$$P_t\phi(x) = \frac{1}{\pi}\int_{\mathbb{R}} \frac{t}{t^2+(x-y)^2}\phi(y)\,\mathrm{d}y,$$

on $L^p(\mathbb{R};X)$ has maximal L^p-regularity if and only if X is a UMD space.

In this note we revisit Theorem 1.1 from a quantitative point of view. Let us denote by $\mathfrak{m}_{p,X}$ the L^p-norm of the above-defined maximal regularity operator for the Poisson semigroup on $L^p(\mathbb{R};X)$.

Recall that X is a UMD space if for one (or equivalently all) $p \in (1,\infty)$, there exists a constant C such that for all martingale difference sequences $(d_k)_{k=1}^n$ on $L^p(\Omega;X)$ over any probability space Ω, and all signs $\epsilon_k = \pm 1$, we have

$$\Big\|\sum_{k=1}^n \epsilon_k d_k\Big\|_{L^p(\Omega;X)} \le C\Big\|\sum_{k=1}^n d_k\Big\|_{L^p(\Omega;X)}.$$

We use Burkholder's notation $\beta_{p,X}$ for the best constant C in this inequality.

There is a well-known characterization [2, 3] of UMD spaces in terms of the Hilbert transform

$$H\phi(x) := \lim_{\epsilon\to 0}\frac{1}{\pi}\Big(\int_{-\infty}^{x-\epsilon} + \int_{x+\epsilon}^{\infty}\Big)\frac{\phi(y)}{x-y}\,\mathrm{d}y.$$

Namely, X is a UMD space if [2] and only if [3] for one (or equivalently all) $p \in (1,\infty)$, and all $\phi \in L^p(\mathbb{R};X)$, we have

$$\|H\phi\|_{L^p(\mathbb{R};X)} \le C\|\phi\|_{L^p(\mathbb{R};X)}.$$

We use the non-standard notation $\hbar_{p,X}$ for the best constant C in this inequality. (Apologies to physicists, but there is really no use for the Planck constant in this context.)

The constants $\beta_{p,X}$ and $\hbar_{p,X}$ are known to be finite or infinite simultaneously, but their precise quantitative relation remains a mystery. It is known that

$$1 \le \beta_{p,X} \le (\hbar_{p,X})^2 \quad [2], \qquad 1 \le \hbar_{p,X} \le (\beta_{p,X})^2 \quad [3],$$

and it is an open (and presumably difficult) problem to prove or disprove a linear estimate in either place. By Theorem 1.1, the L^p-norm of the maximal regularity operator of the Poisson semigroup on $L^p(\mathbb{R};X)$, denoted by $\mathfrak{m}_{p,X}$, is also finite at the same time as $\beta_{p,X}$ and $\hbar_{p,X}$. In this note we prove:

Theorem 1.2.

$$\frac{1}{2}\max(\beta_{p,X},\hbar_{p,X}) \le \mathfrak{m}_{p,X} \le \beta_{p,X} + \hbar_{p,X}.$$

Thus, while we are not able to bring new light to the possible linear dependence between the UMD and the Hilbert transform constants, we find that there is a linear dependence between their sum and the Poisson maximal regularity constant. The proof proceeds via the theory of Fourier multipliers, by interpreting the constants above in this framework and finding algebraic relations between the various symbols involved.

2. Facts about Fourier multipliers

2.1. Generalities

For the Fourier transform on $\mathbb{R}^n$ (we are mostly concerned with $n = 1, 2$), we use the normalization

$$\mathscr{F} f(\xi) := \hat{f}(\xi) := \int_{\mathbb{R}^n} f(x) e^{-i2\pi\xi\cdot x}\, \mathrm{d}x.$$

A Fourier multiplier with symbol $m \in L^\infty(\mathbb{R}^n)$ (or more generally, an operator-valued symbol $m \in L^\infty(\mathbb{R}^n; \mathscr{L}(X))$) is the operator defined (say, on sufficiently nice functions) by

$$T_m f := \mathscr{F}^{-1}(m\hat{f}).$$

We define the multiplier norm

$$\|m\|_{ML^p(\mathbb{R}^n;X)} := \|T_m\|_{\mathscr{L}(L^p(\mathbb{R}^n;X))}.$$

2.2. Orthogonal invariance

If O is an orthogonal transformation of $\mathbb{R}^n$, then

$$\|m(O\,\cdot\,)\|_{ML^p(\mathbb{R}^n;X)} = \|m\|_{ML^p(\mathbb{R}^n;X)}.$$

This follows easily from the corresponding invariance of the L^p norms.

2.3. Even and odd parts

Let $m_e(\xi) := \frac{1}{2}(m(\xi) + m(-\xi))$, $m_o(\xi) := \frac{1}{2}(m(\xi) - m(-\xi))$, so that $m(\xi) = m_e(\xi) + m_o(\xi)$. By 2.2, $\xi \mapsto m(-\xi)$ has the same multiplier norm as m. This and the triangle inequality imply that

$$\begin{aligned}\max(\|m_e\|_{ML^p(\mathbb{R}^n;X)}, \|m_o\|_{ML^p(\mathbb{R}^n;X)}) &\leq \|m\|_{ML^p(\mathbb{R}^n;X)} \\ &\leq \|m_e\|_{ML^p(\mathbb{R}^n;X)} + \|m_o\|_{ML^p(\mathbb{R}^n;X)}.\end{aligned}$$

2.4. Extension from a subspace

Let $V \eqsim \mathbb{R}^k \subseteq \mathbb{R}^n$ be a linear subspace, and P be the orthogonal projection of $\mathbb{R}^n$ onto V. If m is a Fourier multiplier on $L^p(V; X)$, then

$$\|m(P\,\cdot\,)\|_{ML^p(\mathbb{R}^n;X)} = \|m\|_{ML^p(V;X)}.$$

This is an easy consequence of Fubini's theorem.

2.5. Restriction to a subspace

Let $V \eqsim \mathbb{R}^k \subseteq \mathbb{R}^n$ be a linear subspace such that for a.e. (with respect to the k-dimensional Lebesgue measure) $v \in V$, the point $0 \in V^\perp$ is a Lebesgue point (with respect to the $(n-k)$-dimensional Lebesgue measure) of $w \in V^\perp \mapsto m(v + w)$. Then

$$\|v \mapsto m(v)\|_{ML^p(V;X)} \leq \|m\|_{ML^p(\mathbb{R}^n;X)}.$$

This is an X-valued extension of a classical theorem of de Leeuw [8]. It can also be seen as a consequence of (the proof of) a theorem of Clément and Prüss [4], which states that the essential range of an operator-valued Fourier multiplier is bounded (even R-bounded). Namely, one can (at least formally) identify $T_m \in \mathscr{L}(L^p(\mathbb{R}^n;X))$ with $T_{\tilde{m}} \in \mathscr{L}(L^p(V^\perp;L^p(V;X)))$, where

$$\tilde{m}(w) = T_{m(\cdot+w)} \in \mathscr{L}(L^p(V;X)),$$

in which case the multiplier restriction estimate takes the form

$$\|\tilde{m}(0)\|_{\mathscr{L}(Y)} \le \|\tilde{m}\|_{ML^p(V^\perp;Y)}, \quad Y = L^p(V;X).$$

2.6. The Hilbert transform

The Hilbert transform is a Fourier multiplier on $\mathbb{R}$ with symbol $-i\operatorname{sgn}(\xi)$. Thus

$$\|\operatorname{sgn}\|_{ML^p(\mathbb{R};X)} = \hbar_{p,X}.$$

A combination of 2.2, 2.4, and the previous display shows that

$$\|\xi \mapsto \operatorname{sgn}(\xi\cdot\theta)\|_{ML^p(\mathbb{R}^n;X)} = \hbar_{p,X} \qquad \forall\theta\in\mathbb{R}^n\setminus\{0\}.$$

2.7. Riesz transforms

The Riesz transform R_j on $\mathbb{R}^n$ is the Fourier multiplier operator

$$R_j = T_{m_j}, \qquad m_j = i\frac{\xi_j}{|\xi|}.$$

We are particularly concerned with second-order Riesz transforms, namely, compositions of two R_j's. These satisfy

$$\Big\|\sum_{j=1}^n \alpha_j R_j^2\Big\|_{\mathscr{L}(L^p(\mathbb{R}^n;X))} \le \beta_{p,X}, \qquad \alpha_j\in\{-1,0,+1\},$$

and moreover

$$\Big\|\sum_{j=1}^n \alpha_j R_j^2\Big\|_{\mathscr{L}(L^p(\mathbb{R}^n;X))} \ge \beta_{p,X}, \quad \text{if} \quad \{-1,+1\}\subseteq\{\alpha_j\}_{j=1}^n\subseteq\{-1,0,+1\}^n.$$

For $X=\mathbb{C}$, the upper bound is due to Bañuelos and Méndez-Hernández [1, (2.11)]. Their proof, based on the UMD property of $\mathbb{C}$, can be extended to the general case. Alternatively, although not explicitly stated, one may also extract this result from the paper of Geiss, Montgomery-Smith and Saksman [6]. The last-mentioned paper, [6, Proposition 3.4], contains the important reversal of the estimate for non-degenerate coefficients. (While simple to state, these precise relations between the second-order Riesz transforms and the UMD property are the deepest ingredients behind the rather short proof of Theorem 1.2 below!)

2.8. Multipliers for (Poisson) maximal regularity

The abstract maximal regularity operator $\mathfrak{M}$ on $L^p(\mathbb{R}; X)$ can be expressed as a Fourier multiplier operator T_M with an operator-valued symbol

$$M(\xi_1) = A(i2\pi\xi_1 + A)^{-1} \in \mathscr{L}(X), \qquad \xi_1 \in \mathbb{R}.$$

When X is of the form $L^p(\mathbb{R}; \tilde{X})$ and $A = \sqrt{-\Delta}$ is the negative of the generator of the Poisson semigroup, we may also view $\mathfrak{M}$ on $L^p(\mathbb{R}; X) = L^p(\mathbb{R}; L^p(\mathbb{R}; \tilde{X})) \eqsim L^p(\mathbb{R}^2; \tilde{X})$ as a two-dimensional Fourier multiplier T_m with scalar symbol

$$\begin{aligned} m(\xi) = m(\xi_1, \xi_2) = 2\pi|\xi_2|(i2\pi\xi_1 + 2\pi|\xi_2|)^{-1} &= \frac{|\xi_2|}{i\xi_1 + |\xi_2|} \\ &= \frac{|\xi_2|(-i\xi_1 + |\xi_2|)}{\xi_1^2 + \xi_2^2} = -i\frac{\xi_1|\xi_2|}{|\xi|^2} + \frac{\xi_2^2}{|\xi|^2} = m_o(\xi) + m_e(\xi). \end{aligned}$$

This is the point of view that we adopt for the rest of this note.

3. Proof of Theorem 1.2

3.1. The lower bound for $\mathfrak{m}_{p,X}$

From 2.3 we see that it is enough to bound from below the multiplier norms of m_e and m_o from 2.8.

Note that $T_{m_e} = -R_2^2$. Thus,

$$\begin{aligned} \beta_{p,X} = \|R_1^2 - R_2^2\|_{\mathscr{L}(L^p(\mathbb{R}^2;X))} &\le \|R_1^2\|_{\mathscr{L}(L^p(\mathbb{R}^2;X))} + \|R_2^2\|_{\mathscr{L}(L^p(\mathbb{R}^2;X))} \\ &= 2\|R_2^2\|_{\mathscr{L}(L^p(\mathbb{R}^2;X))} = 2\|m_e\|_{\mathcal{M}L^p(\mathbb{R}^n;X)}, \end{aligned}$$

where the first equality was based on 2.7 and the second on 2.2.

For the odd part, we have $m_o(\xi_1, \xi_1) = -\frac{i}{2}\operatorname{sgn}(\xi_1)$, so 2.6 and 2.5 imply

$$\hbar_{p,X} = \|\operatorname{sgn}\|_{\mathcal{M}L^p(\mathbb{R};X)} \le 2\|m_o\|_{\mathcal{M}L^p(\mathbb{R}^2;X)}.$$

The lower bound of Theorem 1.2 now follows from the above estimates and 2.3.

3.2. The upper bound for $\mathfrak{m}_{p,X}$

For m_e, it is immediate from 2.7 that

$$\|m_e\|_{\mathcal{M}L^p(\mathbb{R}^2;X)} = \|R_2^2\|_{\mathscr{L}(L^p(\mathbb{R}^2;X))} \le \beta_{p,X}.$$

So it remains to consider the odd symbol m_o. In order to apply 2.6, we wish to express it as an integral average

$$m_o(\xi) = -i \fint_{S^1} \operatorname{sgn}(\xi \cdot \theta)\Omega(\theta)\,\mathrm{d}\sigma(\theta),$$

or, writing $\xi = |\xi|(\cos t, \sin t)$, $\theta = (\cos u, \sin u)$, $\Omega(\theta) = \omega(u)$, as

$$\cos t|\sin t| = \frac{1}{2}\operatorname{sgn}(\sin t)\sin(2t) = \frac{1}{2\pi}\int_0^{2\pi} \operatorname{sgn}(\cos(t-u))\omega(u)\,\mathrm{d}u.$$

Rather than dwelling into a systematic study of such equations, we just pull out of the hat the following integral:

Lemma 3.3.

$$\int_0^{2\pi} \operatorname{sgn}(\cos(u-t)) \operatorname{sgn}(\cos u)\cos(2u)\,du = 2\operatorname{sgn}(\sin t)\sin(2t).$$

Proof. This is an elementary computation, which may be facilitated by the following observations: both sides are continuous in t and change sign on replacing t by $t+\pi$, so it suffices to verify the identity for $t \in (0,\pi)$. By periodicity, we may also replace the integration interval $[0,2\pi]$ by $[-\pi/2, 3\pi/2]$. In this case, we have

$$-\frac{\pi}{2} < t - \frac{\pi}{2} < \frac{\pi}{2} < t + \frac{\pi}{2} < \frac{3\pi}{2}.$$

By inspection of the sign factors, we find that our integral is equal to

$$\Big(-\int_{-\pi/2}^{t-\pi/2} + \int_{t-\pi/2}^{\pi/2} - \int_{\pi/2}^{t+\pi/2} + \int_{t+\pi/2}^{3\pi/2}\Big)\cos(2u)\,du,$$

and this is readily computed and found to be equal to $2\sin(2t)$.

Note that $\operatorname{sgn}(\sin t) = 1$ in the considered range $t \in (0,\pi)$, so we are done. □

Thus, a solution of our integral equation is given by

$$\omega(u) = \frac{\pi}{2}\operatorname{sgn}(\cos u)\cos(2u).$$

It follows from 2.6 that

$$\begin{aligned}\|m_o\|_{ML^p(\mathbb{R}^2;X)} &\le \fint_{S^1} \|\xi \mapsto \operatorname{sgn}(\xi\cdot\theta)\|_{ML^p(\mathbb{R}^2;X)}|\Omega(\theta)|\,d\sigma(\theta)\\ &= \hbar_{p,X}\frac{1}{2\pi}\int_0^{2\pi}|\omega(u)|\,du = \hbar_{p,X}\frac{1}{4}\int_0^{2\pi}|\cos(2u)|\,du = \hbar_{p,X}.\end{aligned}$$

By 2.3, this completes the proof of Theorem 1.2.

4. Further questions

It might be of some interest to try to carry out a similar analysis for some other maximal regularity operators with a simple Fourier multiplier representation: in particular, for the heat semigroup $e^{t\Delta}$, whose maximal regularity is described by the multiplier

$$m(\xi_1,\xi_2) = \frac{\xi_2^2}{i\xi_1 + \xi_2^2}.$$

Compared to the case already treated, this has the added difficulty that the dilation invariance $m(t\xi) = m(\xi)$, $t > 0$, of the Poisson maximal regularity symbol is replaced by the more complicated anisotropic invariance $m(t^2\xi_1, t\xi_2) = m(\xi)$. Such multipliers have been studied in [7], but altogether less is known than in the isotropic case. In particular, the theory of Geiss, Montgomery-Smith and Saksman [6] is only available for isotropic multipliers. Extending their results to the anisotropic situation could be of independent interest.

Acknowledgment

Thanks to the referee for detailed corrections.

References

[1] R. Bañuelos and P.J. Méndez-Hernández. *Space-time Brownian motion and the Beurling-Ahlfors transform.* Indiana Univ. Math. J., 52(4):981–990, 2003.

[2] J. Bourgain. *Some remarks on Banach spaces in which martingale difference sequences are unconditional.* Ark. Mat., 21(2):163–168, 1983.

[3] D.L. Burkholder. *A geometric condition that implies the existence of certain singular integrals of Banach-space-valued functions.* In *Conference on harmonic analysis in honor of Antoni Zygmund, vol. I, II (Chicago, Ill.,* 1981), Wadsworth Math. Ser., pages 270–286. Wadsworth, Belmont, CA, 1983.

[4] P. Clément and J. Prüss. *An operator-valued transference principle and maximal regularity on vector-valued L_p-spaces.* In *Evolution equations and their applications in physical and life sciences (Bad Herrenalb,* 1998), volume 215 of *Lecture Notes in Pure and Appl. Math.*, pages 67–87. Dekker, New York, 2001.

[5] T. Coulhon and D. Lamberton. *Régularité L^p pour les équations d'évolution.* In *Séminaire d'Analyse Fonctionelle* 1984/1985, volume 26 of *Publ. Math. Univ. Paris VII*, pages 155–165. Univ. Paris VII, Paris, 1986.

[6] S. Geiss, S. Montgomery-Smith, and E. Saksman. *On singular integral and martingale transforms.* Trans. Amer. Math. Soc., 362(2):553–575, 2010.

[7] T.P. Hytönen. *Anisotropic Fourier multipliers and singular integrals for vector-valued functions.* Ann. Mat. Pura Appl. (4), 186(3):455–468, 2007.

[8] K. de Leeuw. *On L_p multipliers.* Ann. of Math. (2), 81:364–379, 1965.

Tuomas P. Hytönen
Department of Mathematics and Statistics
P.O. Box 68 (Gustaf Hällströmin katu 2b)
FI-00014 University of Helsinki, Finland
e-mail: `tuomas.hytonen@helsinki.fi`

Operator Theory:
Advances and Applications, Vol. 250, 281–285

An Analytic Family of Contractions Generated by the Volterra Operator

L. Khadkhuu, D. Tsedenbayar and J. Zemánek

Dedicated to Charles Batty on his 60th anniversary

Abstract. Let V denote the classical Volterra operator on $L_2(0,1)$, and let z_1, z_2 be complex numbers. We prove that $||(I - z_1V)(I + z_2V)^{-1}|| = 1$ if and only if $z_1 + z_2 \geq 0$ and $|\text{Re}z_1| \leq |\text{Re}z_2|$. In particular, this generalizes the Cayley transform case, for the operator V, and also provides a simple way of showing that $||I - aV|| > 1$ for all complex $a \neq 0$.

Mathematics Subject Classification (2010). 47A10, 47B44, 47G10.

Keywords. Volterra operator, contraction, power-bounded operator, accretive operator, Cayley transform.

1. Introduction

Let H be a complex Hilbert space equipped with the inner product $\langle \cdot, \cdot \rangle$, which induces the norm $|| \cdot ||$. Denote by $B(H)$ the Banach algebra of bounded linear operators acting on H with the operator norm defined by

$$||B|| = \sup_{||x||=1} \{||Bx|| : x \in H\}, \quad B \in B(H).$$

A bounded linear operator B on H is *power-bounded*, if $\sup_{n\geq 0} ||B^n|| < +\infty$, and is a *contraction*, if $||B|| \leq 1$. In particular, all contractions are power-bounded.

A bounded linear operator B on H is *accretive*, if

$$\text{Re}B = \frac{B + B^*}{2} \geq 0.$$

The classical *Volterra operator* V on the Hilbert space $L_2(0,1)$ is defined by

$$(Vf)(t) := \int_0^t f(s)ds, \quad f \in L_2(0,1).$$

It is well known that V is quasi-nilpotent, compact, and accretive. Both quasi-nilpotency and accretivity of V will play a role in our results.

In [LT, page 183], a question was posed to describe the class of analytic functions $\varphi(z)$ such that $\varphi(0) = 1$ and $||\varphi(V)|| = 1$. Continuing the paper [KT], we present here a family of contractions (of norm 1) induced by V via a generalized Cayley transform. Other examples, including some polynomials in V, can be found in [EZ]. A complete characterization is not yet known, but the present paper and [EZ] show that the class in question is quite large. Namely, our main result is the following.

2. The results

Theorem 2.1. *Let z_1 and z_2 be complex numbers. We have*

$$||(I - z_1V)(I + z_2V)^{-1}|| = 1$$

if and only if $z_1 + z_2 \geq 0$ and $|\mathrm{Re} z_1| \leq |\mathrm{Re} z_2|$.

Proof. Since V is quasi-nilpotent, we have

$$||(I - z_1V)(I + z_2V)^{-1}|| \geq 1$$

for all complex z_1 and z_2. To characterize the property

$$||(I - z_1V)(I + z_2V)^{-1}x|| \leq ||x|| \tag{2.1}$$

for all $x \in L_2(0, 1)$, we proceed as follows. Put $y = (I + z_2V)^{-1}x$. Then (2.1) is equivalent to

$$||(I - z_1V)y|| \leq ||(I + z_2V)y|| \tag{2.2}$$

for all $y \in L_2(0, 1)$. By direct calculation, we see that property (2.2) is equivalent to

$$(|z_1|^2 - |z_2|^2)||Vy||^2 \leq \alpha\langle Py, y\rangle - 2\beta\langle(\mathrm{Im}V)y, y\rangle \tag{2.3}$$

for all $y \in L_2(0, 1)$, where

$$\alpha := \mathrm{Re}(z_1 + z_2), \; \beta := \mathrm{Im}(z_1 + z_2),$$

and $P := V + V^*$ is the orthogonal projection of $L_2(0, 1)$ onto the constant functions, that is,

$$(Pf)(t) = \int_0^1 f(s)ds, \quad f \in L_2(0, 1).$$

As in [KT], observe that for the function

$$y_*(t) := t - \frac{1}{2}, \quad t \in [0, 1]$$

we have $\langle Vy_*, y_*\rangle = 0 = \langle V^*y_*, y_*\rangle$, so that the right-hand side in (2.3) is equal to zero, while $||Vy_*|| \neq 0$. Consequently,

$$|z_1| \leq |z_2|. \tag{2.4}$$

Now, we intend to prove that $\beta = 0$. To this end, consider the functions

$$y_k(t) := e^{2\pi kit}, \quad k \in \mathbb{Z}, \quad t \in [0, 1].$$

Then $Py_k = 0$ for all $k \neq 0$, so that (2.3) reduces to

$$(|z_1|^2 - |z_2|^2)||Vy_k||^2 \leq -2\beta\langle(\mathrm{Im}V)y_k, y_k\rangle,$$

that is,

$$\left(|z_1|^2 - |z_2|^2\right)\frac{1}{2\pi k^2} \leq \frac{\beta}{k}$$

for all $k \in \mathbb{Z}\backslash\{0\}$.

If $\beta > 0$, then for $k \to -\infty$ we get

$$\left(|z_1|^2 - |z_2|^2\right)\frac{1}{2\pi k} \geq \beta > 0,$$

a contradiction, since the left-hand side tends to zero.

Similarly, if $\beta < 0$, then for $k \to +\infty$ we get

$$\left(|z_1|^2 - |z_2|^2\right)\frac{1}{2\pi k} \leq \beta < 0,$$

a contradiction again, since the left-hand side tends to zero.

Note that in the preceding argument we did not need (2.4).

Thus, we conclude that $\beta = 0$, as claimed. A similar argument also appears in [EZ], in a more general context.

Next, we have to show that $\alpha \geq 0$. Indeed, if $\alpha < 0$, then (2.3) yields

$$(\mathrm{Re}z_1 - \mathrm{Re}z_2)||Vy||^2 \geq \langle Py, y\rangle$$

for all $y \in L_2(0, 1)$. However, this is impossible for the functions f_k, in place of y, where

$$f_k(t) := t^k, \quad k \in \mathbb{N}, \quad t \in [0, 1],$$

since

$$\frac{\langle Pf_k, f_k\rangle}{||Vf_k||^2} = 2k + 3 \to +\infty.$$

Hence, $\alpha \geq 0$, as claimed. Also in this step, (2.4) was not used.

The conclusions that $\beta = 0$ and $\alpha \geq 0$ can also be obtained from the mere power boundedness of the operator considered, by using the deep characterization [L, Theorem 1.1]. However, under the stronger assumption of contractivity, the above proof is elementary. As mentioned above, a similar idea also occurs in [EZ].

So, $z_1 + z_2 = \alpha$ is a real number, and hence

$$|z_1|^2 - |z_2|^2 = (\mathrm{Re}z_1)^2 - (\mathrm{Re}z_2)^2 \leq 0$$

by (2.4). It follows that

$$|\mathrm{Re}z_1| \leq |\mathrm{Re}z_2|.$$

Until now, we have shown the necessity of both conditions in Theorem 2.1.

In the converse direction, our elementary reasoning leads to a stronger conclusion than [L, Theorem 1.1], namely, that the operator in question is actually a contraction, not merely power-bounded.

Indeed, if $z_1 + z_2 = \alpha \geq 0$ and $|\mathrm{Re}z_1| \leq |\mathrm{Re}z_2|$, then (2.3) holds, since $\beta = 0$ and $\langle Py, y\rangle \geq 0$ for all y. Then we can proceed back to (2.2) and (2.1). □

In view of [L, Theorem 1.1], Theorem 2.1 says that the operator $(I-z_1V)(I+z_2V)^{-1}$ is a contraction if and only if it is power-bounded and $|\mathrm{Re} z_1| \le |\mathrm{Re} z_2|$.

It is curious to note that the preceding theorem yields an elementary proof, even simpler than those mentioned in [EZ] and [KT], of the following fact, originally established in [LT, Corollary 2.5].

Corollary 2.2. *We have* $||I - aV|| > 1$ *for all complex* $a \neq 0$.

Proof. Suppose that $||I - z_1V|| = 1$ for some $z_1 := a \neq 0$. Then for all $z_2 \geq 0$ we have

$$||(I - z_1V)(I + z_2V)^{-1}|| = 1,$$

since $||(I + z_2V)^{-1}|| = 1$ by solution to [H, Problem 150]. Choosing here $z_2 = 0$, Theorem 2.1 yields a contradiction, if $z_1 \neq 0$. □

It is also surprising to see that both the factors $(I - z_1V)$ and $(I + z_2V)^{-1}$ may not be power-bounded, cf. [T, Theorem 1] and [L, Theorem 1.1], while their product can be a contraction (!), according to Theorem 2.1.

A more general context, involving certain accretive operators B in place of V, has been considered in [KT], however, under the a priori assumption that $z_1 + z_2 \geq 0$. This assumption is superfluous in the case where $B := V$; in fact, it becomes part of the result, which is the main point of the present paper.

Also [N, Theorem 1.4.2] deals with the Cayley transform of accretive operators, i.e., with the case where $z_1 = z_2 > 0$. Thus, our progress consists in splitting up the coefficients z_1 and z_2, and allowing them to be complex, in the Volterra case.

Example 2.3. The operator

$$(I - V)(I + V)^{-1}$$

is a contraction, by Theorem 2.1 or [N, Theorem 1.4.2]. However, its inverse

$$(I + V)(I - V)^{-1}$$

is not even power-bounded, by the classical Gelfand theorem [Z], or by [L, Theorem 1.1].

Remark 2.4. If $z_1 + z_2 = 0$, then $(I - z_1V)(I + z_2V)^{-1} = I$. On the other hand, if $z_1 + z_2 > 0$, then our second condition actually means that $|\mathrm{Re} z_1| \leq \mathrm{Re} z_2$, hence $\mathrm{Re} z_2 > 0$. The preceding example shows that the roles of z_1 and z_2 in the condition $|\mathrm{Re} z_1| \leq |\mathrm{Re} z_2|$ cannot be interchanged.

Corollary 2.5. *Let* $\mathrm{Re} z_2 < 0$. *Then*

$$||(I - z_1V)(I + z_2V)^{-1}|| > 1$$

for all complex $z_1 \neq -z_2$. *However, for all complex* z_1 *such that* $z_1 + z_2 \geq 0$, *this operator is power-bounded, by* [L, Theorem 1.1].

Acknowledgement

The second-named author was supported in part by the Warsaw Center of Mathematics and Computer Science during his stay at the Institute of Mathematics of the Polish Academy of Sciences in March 2014. We are grateful to the referee for careful reading of the paper.

References

[EZ] A.F.M. ter Elst and J. Zemánek, *Contractive polynomials in the Volterra operator*, submitted.

[H] P.R. Halmos, *A Hilbert Space Problem Book*, Van Nostrand, Princeton, New Jersey, 1967.

[KT] L. Khadkhuu and D. Tsedenbayar, *Some norm one functions of the Volterra operator*, Mathematica Slovaca, to appear.

[L] Yu. Lyubich, *The power boundedness and resolvent conditions for functions of the classical Volterra operator*, Studia Mathematica, **196** (2010), 41–63.

[LT] Yu. Lyubich and D. Tsedenbayar, *The norms and singular numbers of polynomials of the classical Volterra operator in* $L_2(0,1)$, Studia Mathematica, **199** (2010), 171–184.

[N] N.K. Nikolski, *Operators, Functions and Systems: An Easy Reading, vol. 2: Model Operators and Systems*, American Mathematical Society, Providence, Rhode Island, 2002.

[T] D. Tsedenbayar, *On the power boundedness of certain Volterra operator pencils*, Studia Mathematica, **156** (2003), 59–66.

[Z] J. Zemánek, *On the Gelfand–Hille theorems*, in: Functional Analysis and Operator Theory, Banach Center Publications, **30**, Institute of Mathematics of the Polish Academy of Sciences, Warsaw, 1994, 369–385.

L. Khadkhuu
National University of Mongolia
School of Arts and Sciences
Department of Mathematics
P.O.Box 46/145
Ulaanbaatar 210646, Mongolia
e-mail: `hadhuul@yahoo.com`

D. Tsedenbayar
Mongolian University of Science and Technology
School of Applied Sciences
Department of Mathematics
P.O.Box 46/520
Ulaanbaatar 210646, Mongolia
e-mail: `cdnbr@yahoo.com`

J. Zemánek
Institute of Mathematics
Polish Academy of Sciences
Śniadeckich 8
PL-00-656 Warsaw, Poland
e-mail: `zemanek@impan.pl`

Operator Theory:
Advances and Applications, Vol. 250, 287–296

Lattice Dilations of Bistochastic Semigroups

Leonard J. Konrad

Abstract. An alternative proof is given for Fendler's dilation result for bistochastic semigroups on L^p, $1 \leq p < \infty$, including the result for $p = 1$ as well as minimality and uniqueness of the dilation.

Mathematics Subject Classification (2010). 47A20.

Keywords. Lattice dilation; bistochastic semigroups.

1. Introduction

Dilation methods on Hilbert spaces introduced by Halmos ([13]) and Sz.-Nagy ([23]) in the 1950s have proved to be powerful instruments. Generalisations lead to power lattice dilation results on L^p-spaces for positive contractions by Akcoglu, Sucheston, Kopp ([1], [2], [3], [4], [5], [6]) and Kern, Nagel and Palm ([10, Appendix U], [14], [18]) in the 1970s. Neglecting the assumption of positivity, the dilatable operators on L^p, i.e., operators where a dilation to an isometrically invertible operator exists, have also been identified as the ones admitting a contractive majorant ([19]). Fendler ([11], [12]) used Akcoglu's lattice dilation construction and extended it to positive contraction semigroups on $L^p(\Omega, \Sigma, \mu)$, $1 < p < \infty$, as well as a variant for subpositive semigroups if $1 < p < \infty$ and $p \neq 2$.

In this paper, we use use a construction from [14] which is based on a construction by Rota ([20]) and the theory of Markov processes to obtain a more illustrative proof for bistochastic semigroups. We also extend Fendler's result to L^1 for such semigroups. This construction has been used to get similar results for C^*- and W^*-algebras (e.g., [7], [8], [9], [21], [25]) although there are some obstacles ([16, Section 2]).

We assume that $(T_t, t \geq 0)$ is a strongly continuous and bistochastic semigroup on $L^p(\Omega, \Sigma, \mu)$, $1 \leq p < \infty$, with a probability measure μ. This means that the map $t \mapsto T_t f$ is norm continuous for $t \geq 0$ and for all $f \in L^p(\Omega, \Sigma, \mu)$, $T_t T_s = T_{t+s}$ holds for all $t, s \geq 0$ as well as $T_t \mathbb{1} = \mathbb{1} = T_t' \mathbb{1}$ with positive operators

The author gratefully acknowledges the support of EPSRC and the Andrew Mullins Award.

T_t for $t \geq 0$. We usually drop the term strongly continuous if it is clear from the context and we note that a bistochastic operator is a contraction for all L^p, $1 \leq p \leq \infty$.

Our main goal in this paper is to obtain a lattice semigroup dilation.

Definition 1.1. We call $((\widehat{T}_t, t \geq 0), L^p(\widehat{\Omega}, \widehat{\Sigma}, \widehat{\mu}), J, Q)$ a lattice dilation of $((T_t, t \geq 0), L^p(\Omega, \Sigma, \mu))$ with a lattice isomorphism semigroup $(\widehat{T}_t, t \geq 0)$ on $L^p(\widehat{\Omega}, \widehat{\Sigma}, \widehat{\mu})$ if

$$Q\widehat{T}_t J = T_t$$

holds for all $t \geq 0$ where $J\colon L^p(\Omega, \Sigma, \mu) \to L^p(\widehat{\Omega}, \widehat{\Sigma}, \widehat{\mu})$ is an isometric lattice homomorphism and $Q\colon L^p(\widehat{\Omega}, \widehat{\Sigma}, \widehat{\mu}) \to L^p(\Omega, \Sigma, \mu)$ the corresponding positive contraction (i.e., $Q' = J$ on associated L^p spaces).

In this paper, we construct the semigroup dilation in Section 2. In Section 3, we prove that this construction satisfies all required properties, and we obtain the following dilation result.

Theorem 1.2. *Let $(T_t, t \geq 0)$ be a bistochastic semigroup on $L^p(\Omega, \Sigma, \mu)$, $1 \leq p < \infty$, with a probability measure μ. Then there exists a lattice dilation.*

We further prove the Markov property and the uniqueness of this dilation in Section 4 and we obtain the following result.

Theorem 1.3. *Let $((\tilde{T}_t, t \geq 0), L^p(\tilde{\Omega}, \tilde{\Sigma}, \tilde{\mu}), \tilde{J}, \tilde{Q})$ be a minimal lattice dilation of the bistochastic semigroup $((T_t, t \geq 0), L^p(\Omega, \Sigma, \mu))$, $1 \leq p < \infty$, where $(\tilde{T}_t, t \geq 0)$ satisfies the Markov property, $\tilde{T}_t \tilde{\mathbb{1}} = \tilde{\mathbb{1}}$ for all $t \in \mathbb{R}$ and $\tilde{J}\mathbb{1} = \tilde{\mathbb{1}}$. Then it is lattice isomorphic to the dilation in Section 2.*

2. Construction

In this section, we construct a dilation by introducing all required spaces and operators.

The following Gelfand type lemma simplifies the later considerations although its real value appears in particular in the technical details which we omit in this paper.

Lemma 2.1. *Let $(T_t, t \geq 0)$ be a bistochastic semigroup on $L^p(\Omega, \Sigma, \mu)$. Then there exists a compact Hausdorff measure space $(\tilde{\Omega}, \tilde{\Sigma}, \tilde{\mu})$, a bistochastic semigroup $(\tilde{T}_t, t \geq 0)$ on $L^p(\tilde{\Omega}, \tilde{\Sigma}, \tilde{\mu})$, where both $\tilde{T}_t$ and $\tilde{T}_t'$ leave $C(\tilde{\Omega})$ invariant for all $t \geq 0$, and an isometric lattice isomorphism $\iota\colon L^p(\Omega, \Sigma, \mu) \to L^p(\tilde{\Omega}, \tilde{\Sigma}, \tilde{\mu})$ such that*

$$T_t = \iota^{-1} \circ \tilde{T}_t \circ \iota$$

for all $t \geq 0$.

Proof. There exists a unit preserving lattice isomorphism $\iota : L^\infty(\Omega, \Sigma, \mu) \to C(\tilde{\Omega})$ with some compact Hausdorff space $\tilde{\Omega}$ by the Kakutani representation theorem ([17, Theorem 2.1.3]). We define a measure $\tilde{\mu} := \mu \circ \iota^{-1}$ on $\tilde{\Omega}$ and ι extends as a lattice isomorphism to all L^p, $1 \leq p < \infty$. We set $\tilde{T}_t := \iota \circ T_t \circ \iota^{-1}$ which satisfies the required properties. □

In view of this lemma, we can assume without loss of generality that the semigroups and their adjoints leave $C(\Omega)$ invariant, and that Ω is a compact Hausdorff space. Also, $T_t\mathbb{1} = \mathbb{1}$ and positivity imply that T_t is contractive on $C(\Omega)$.

We define a space $\widehat{\Omega} := \prod_{t\in\mathbb{R}} \Omega$, which is a compact Hausdorff space by Tychonoff's theorem, and we denote its elements by either $\widehat{\omega}$ or $(\omega_t)_{t\in\mathbb{R}}$. On $\widehat{\Omega}$, we define $\widehat{\Sigma}$ as the product σ-algebra on $\widehat{\Omega}$, i.e., the σ-algebra generated by sets of the form $\prod_{t\in\mathbb{R}} A_t$ with only finitely many $\Omega \neq A_t \in \Sigma$.

We introduce a tensor product notation by

$$\left(f_{t_{-n}} \otimes \cdots \otimes f_{t_n}\right)(\widehat{\omega}) := f_{t_{-n}}(\omega_{t_{-n}}) \cdots f_{t_n}(\omega_{t_n})$$

for time steps $t_{-n} < \cdots < t_n \in \mathbb{R}$ and $f_{t_{-n}}, \ldots, f_{t_n} \in C(\Omega)$. Two functions $f_{t_{-n}} \otimes \cdots \otimes f_{t_n}$ and $g_{t_{-m}} \otimes \cdots \otimes g_{t_m}$ are said to be equivalent if they denote the same function in $C(\widehat{\Omega})$. Hence, we do not change the function $f_{t_{-n}} \otimes \cdots \otimes f_{t_n}$ if we remove $\mathbb{1}$-elements or add them at certain additional times $t_{i_1}, \ldots, t_{i_m}$.

We set

$$\bigotimes_{t\in\mathbb{R}} C(\Omega) := \operatorname{lin}\{f_{t_{-n}} \otimes \cdots \otimes f_{t_n} : n \in \mathbb{N}_0, t_m \in \mathbb{R}, f_i \in C(\Omega)\}.$$

We note that $\bigotimes_{t\in\mathbb{R}} C(\Omega)$ is dense in $C(\widehat{\Omega})$ in the topology of uniform convergence by the Stone–Weierstraß theorem, and it is therefore legitimate to work with these functions in view of linearity and continuity of the operators being involved.

We use the notation and spaces just introduced and without loss of generality, we always assume $t_0 = 0$. By Lemma 2.1, we can assume that all T_t and T_t' leave $C(\Omega)$ invariant. We define a map $Q_\sharp \colon \{f_{t_{-n}} \otimes \cdots \otimes f_{t_n} : n \in \mathbb{N}_0, f_i \in C(\Omega)\} \to C(\Omega)$ by

$$\begin{aligned} Q_\sharp(f_{t_{-n}} \otimes \cdots \otimes f_{t_n}) := f_0 \cdot T_{t_1}(f_{t_1} T_{t_2-t_1}(f_{t_2} T_{t_3-t_2}(\ldots T_{t_n-t_{n-1}}(f_{t_n})) \\ \cdot T'_{-t_{-1}}(f_{t_{-1}} T'_{t_{-1}-t_{-2}}(f_{t_{-2}} T'_{t_{-2}-t_{-3}}(\ldots T'_{t_{-n+1}-t_{-n}}(f_{t_{-n}})) \end{aligned}$$

for a function $f_{t_{-n}} \otimes \cdots \otimes f_{t_n} \in \bigotimes_{t\in\mathbb{R}} C(\Omega)$, and we set $f_0 = \mathbb{1}$ if $0 \notin \{t_{-n}, \ldots, t_n\}$. We then extend $Q_\sharp$ to $\bigotimes_{t\in\mathbb{R}} C(\Omega)$ by linearity and continuity to a positive contraction Q^c on $C(\widehat{\Omega})$. We remark that some technical issues are required to show that this extension is well defined.

We define a positive unital functional $\widehat{\mu}$ on $C(\widehat{\Omega})$ by

$$\widehat{\mu} := \mu \circ Q^c,$$

which then induces a probability measure on $\widehat{\Omega}$ by the Riesz representation theorem. We hence have the space $L^p(\widehat{\Omega}, \widehat{\Sigma}, \widehat{\mu})$.

Now that we have defined the space $L^p(\widehat{\Omega},\widehat{\Sigma},\widehat{\mu})$, we define the embedding operator J, the contraction Q and the lattice isomorphisms $(\widehat{T}_t, t\in\mathbb{R})$.

We define the lattice homomorphism $J\colon L^p(\Omega,\Sigma,\mu)\to L^p(\widehat{\Omega},\widehat{\Sigma},\widehat{\mu})$ by

$$Jf := f_0$$

and we note that J is multiplicative on $C(\Omega)$. This is hence a positive isometry since

$$\|Jf\|_p^p = \int_{\widehat{\Omega}} |Jf|^p \, d\widehat{\mu} = \int_{\widehat{\Omega}} J(|f|)^p \, d\widehat{\mu} = \int_{\Omega} |f|^p \, d\mu = \|f\|_p^p$$

for $f\in C(\Omega)$ and hence for $f\in L^p(\Omega,\Sigma,\mu)$, $1\le p<\infty$ and we clearly have $\|Jf\|_\infty = \|f\|_\infty$ for $f\in L^\infty(\Omega,\Sigma,\mu)$. For $1<p<\infty$, we define the operator $Q = Q^p\colon L^p(\widehat{\Omega},\widehat{\Sigma},\widehat{\mu})\to L^p(\Omega,\Sigma,\mu)$ by $Q := J'$ where J is interpreted as the isometry on $L^q(\Omega,\Sigma,\mu)$ with $\frac{1}{p}+\frac{1}{q}=1$. For $p=1$, we define $Q = Q^1 := J_*$ where J_* is the preadjoint of J which exists since the operators Q^p for $p>1$ are consistent which each other and the spaces $L^p(\widehat{\Omega},\widehat{\Sigma},\widehat{\mu})$ are dense in $L^1(\widehat{\Omega},\widehat{\Sigma},\widehat{\mu})$. We have

$$\begin{aligned}
\langle Q\widehat{f}, g\rangle = \langle \widehat{f}, Jg\rangle &= \int_{\widehat{\Omega}} \widehat{f} Jg \, d\widehat{\mu} \\
&= \int_{\Omega} f_0 \cdot g \cdot T_{t_1}(f_{t_1}\cdots(f_{t_n}))\cdot T'_{-t_{-1}}(f_{t_{-1}}\cdots(f_{t_{-n}}))\, d\mu \\
&= \int_{\Omega} Q_\sharp \widehat{f}\cdot g\, d\mu = \langle Q_\sharp\widehat{f}, g\rangle
\end{aligned}$$

for all $\widehat{f} = f_{t_{-n}}\otimes\cdots\otimes f_{t_n}$ and all $g\in C(\Omega)$, hence, Q and $Q_\sharp$ coincide on $\bigotimes_{t\in\mathbb{R}} C(\Omega)$, and we exploit this fact in later calculations.

We define the operator $\widehat{T}_t$ as translation, i.e.,

$$\left(\widehat{T}_t\widehat{f}\right)((\omega_s)_{s\in\mathbb{R}}) := \widehat{f}\left((\omega_{s+t})_{s\in\mathbb{R}}\right) \tag{2.1}$$

for $\widehat{f}\in L^p(\widehat{\Omega},\widehat{\Sigma},\widehat{\mu})$, or as $\widehat{T}_t\widehat{f} = \widehat{f}\circ\tau_t$ where $\tau_t(\omega_s)_{s\in\mathbb{R}} = (\omega_{s+t})_{s\in\mathbb{R}}$. Clearly, it satisfies $\widehat{T}_t\widehat{T}_s = \widehat{T}_{t+s}$ for all $t,s\ge 0$ and $T_0 = \mathrm{Id}$.

3. The dilation result

We now show that the operators and spaces defined in Section 2 satisfy the dilation property and that $(\widehat{T}_t, t\ge 0)$ is indeed a strongly continuous lattice isomorphism semigroup.

Proposition 3.1. *The operator $\widehat{T}_t$ defined by* (2.1) *is a lattice isomorphism on $L^p(\widehat{\Omega},\widehat{\Sigma},\widehat{\mu})$ for all $t\in\mathbb{R}$.*

Proof. The operator $\widehat{T}_t$ is clearly positive, linear and invertible for all $t \in \mathbb{R}$. Let $\widehat{f} = f_{t_{-n}} \otimes \cdots \otimes f_{t_n}$ with $f_{t_i} \in C(\Omega)$ and $t \in \mathbb{R}$ with $-t \in [t_{-i-1}, t_{-i}]$ be given. If $t_{-i-1} \neq -t \neq t_{-i}$, we set $f_{-t} = \mathbb{1}$. We then have

$$\begin{aligned}
\int_{\widehat{\Omega}} \widehat{f} \circ \tau_t \, d\widehat{\mu} &= \int_{\widehat{\Omega}} \left(f_{t_n}(\omega_{t_n}) \cdots f_{-t}(\omega_{-t}) \cdot f_{t_{-i}}(\omega_{t_{-i}}) \cdots f_{t_n}(\omega_{t_n})\right) \circ \tau_t \, d\widehat{\mu}(\widehat{\omega}) \\
&= \int_{\widehat{\Omega}} f_{t_n}(\omega_{t_n+t}) \cdots f_{-t}(\omega_0) \cdot f_{t_{-i}}(\omega_{t_{-i}+t}) \cdots f_{t_n}(\omega_{t_n+t}) \, d\widehat{\mu}(\widehat{\omega}) \\
&= \int_\Omega Q\left(f_{t_n} \otimes \cdots \otimes f_{t_{-i-1}} \otimes f_{-t} \otimes f_{t_{-i}} \otimes \cdots \otimes f_{t_n}\right) d\mu \\
&= \int_\Omega f_{-t} \cdot T_{t_{-i}+t}(f_{t_{-i}} T_{t_{-i+1}-t_{-i}}(\ldots T_{t_n - t_{n-1}}(f_{t_n})) \\
&\qquad \cdot T'_{-t-t_{i-1}}(f_{t_{-i-1}} T'_{t_{-i-1}-t_{-i-2}}(\ldots T'_{t_{-n+1}-t_{-n}}(f_{t_{-n}})) \, d\mu \\
&= \int_\Omega f_{t_{-i}} T_{t_{-i+1}-t_{-i}}(\ldots T_{t_n - t_{n-1}}(f_{t_n})) \\
&\qquad \cdot T'_{t_{-i}-(-t)}(f_{-t} T'_{-t-t_{i-1}}(f_{t_{-i-1}} T'_{t_{-i-1}-t_{-i-2}}(\ldots (f_{t_{-n}})) \, d\mu.
\end{aligned}$$

We iterate the last step and obtain

$$\int_{\widehat{\Omega}} \left(\widehat{f} \circ \tau_t\right) d\widehat{\mu} = \int_{\widehat{\Omega}} \widehat{f} \, d\widehat{\mu}$$

for all $\widehat{f} = f_{t_{-n}} \otimes \cdots \otimes f_{t_n} \in \bigotimes_{t \in \mathbb{R}} C(\Omega)$ and hence, by linearity and continuity, for all $\widehat{f} \in L^p(\widehat{\Omega}, \widehat{\Sigma}, \widehat{\mu})$. In particular, it holds for $\widehat{f} = |\widehat{g}|^p$ where $\widehat{g} \in L^p(\widehat{\Omega}, \widehat{\Sigma}, \widehat{\mu})$. Since $|\widehat{g} \circ \tau_t|^p = |\widehat{g}|^p \circ \tau_t$, we obtain

$$\begin{aligned}
\|\widehat{T}_t \widehat{g}\|_p^p &= \int_{\widehat{\Omega}} |\widehat{T}_t \widehat{g}|^p \, d\widehat{\mu} = \int_{\widehat{\Omega}} |\widehat{g} \circ \tau_t|^p \, d\widehat{\mu} = \int_{\widehat{\Omega}} |\widehat{g}|^p \circ \tau_t \, d\widehat{\mu} = \int_{\widehat{\Omega}} \left(\widehat{f} \circ \tau_t\right) d\widehat{\mu} \\
&= \int_{\widehat{\Omega}} \widehat{f} \, d\widehat{\mu} = \int_{\widehat{\Omega}} |\widehat{g}|^p \, d\widehat{\mu} = \|\widehat{g}\|_p^p
\end{aligned}$$

for all $\widehat{g} \in L^p(\widehat{\Omega}, \widehat{\Sigma}, \widehat{\mu})$. Hence, the operator $\widehat{T}_t$ is an isometry for all $L^p(\widehat{\Omega}, \widehat{\Sigma}, \widehat{\mu})$ and all $t \in \mathbb{R}$. □

Proposition 3.2. *The construction of Section 2 satisfies*

$$Q\widehat{T}_t J = T_t$$

for all $t \geq 0$.

Proof. We have

$$\begin{aligned}
\langle Q\widehat{T}_t Jf, g\rangle = \langle \widehat{T}_t Jf, Jg\rangle = \langle f_t, g_0\rangle &= \int_{\widehat{\Omega}} f_t g_0 \, d\widehat{\mu} \\
&= \int_\Omega Q(f_t g_0) \, d\mu = \int_\Omega g_0 \cdot T_t f_t \, d\mu = \langle T_t f, g\rangle
\end{aligned}$$

for all $t \geq 0$ and for all $f, g \in C(\Omega)$. Hence, we can conclude the claim by the continuity of the operators. □

Proposition 3.3. *The semigroup constructed in Section 2 is strongly continuous on* $L^p(\widehat{\Omega}, \widehat{\Sigma}, \widehat{\mu})$, $1 \leq p < \infty$.

Proof. We first note that the semigroup $(\widehat{T}_t, t \geq 0)$ is uniformly bounded by 1 since it consists of isometries. It therefore suffices to show strong continuity for a dense subspace in $L^p(\widehat{\Omega}, \widehat{\Sigma}, \widehat{\mu})$ such as characteristic functions depending on finitely many coordinates.

We also note that the shift operator constructed in Section 2 is multiplicative on $L^\infty(\widehat{\Omega}, \widehat{\Sigma}, \widehat{\mu})$, strong continuity on indicator functions depending only on one coordinate of Ω can therefore be extended to strong continuity on $L^p(\widehat{\Omega}, \widehat{\Sigma}, \widehat{\mu})$.

Now let $\widehat{f} = \mathbb{1}_{A_i}$ for some $i \in \mathbb{R}$ with $A_i = A \times \prod_{t \neq i} \Omega \subseteq \widehat{\Omega}$ and $A \in \Sigma$ be given. We show that

$$\|\widehat{T}_t \widehat{f} - \widehat{f}\|_p^p = \int_\Omega Q|\widehat{T}_t \widehat{f} - \widehat{f}|^p \, d\mu \longrightarrow 0$$

as $t \to 0$. We have

$$|\widehat{T}_t \widehat{f} - \widehat{f}|^p = |\mathbb{1}_{A_{i+t}} - \mathbb{1}_{A_i}|^p = |\mathbb{1}_{A_{i+t}} \mathbb{1}_{A_i^c} - \mathbb{1}_{A_{i+t}^c} \mathbb{1}_{A_i}|^p = \mathbb{1}_{A_{i+t}} \mathbb{1}_{A_i^c} + \mathbb{1}_{A_{i+t}^c} \mathbb{1}_{A_i}$$

since $A_{i+t} \cap A_i^c$ and $A_{i+t}^c \cap A_i$ are disjoint. Integrating yields

$$\begin{aligned} \|\widehat{T}_t \widehat{f} - \widehat{f}\|_p^p &= \int_{\widehat{\Omega}} (\mathbb{1}_{A_{i+t}} \mathbb{1}_{A_i^c} + \mathbb{1}_{A_{i+t}^c} \mathbb{1}_{A_i}) \, d\widehat{\mu} \\ &= \int_\Omega Q(\mathbb{1}_{A_{i+t}} \mathbb{1}_{A_i^c}) \, d\mu + \int_\Omega Q(\mathbb{1}_{A_{i+t}^c} \mathbb{1}_{A_i}) \, d\mu \\ &= \int_\Omega \mathbb{1}_{A^c} T_t(\mathbb{1}_A) \, d\mu + \int_\Omega \mathbb{1}_A T_t(\mathbb{1}_{A^c}) \, d\mu \longrightarrow 0 \end{aligned}$$

as $t \to 0$ by the strong continuity of $(T_t, t \geq 0)$. Therefore, $(\widehat{T}_t, t \geq 0)$ is strongly continuous. □

Combining the results of Propositions 3.1, 3.2 and 3.3, we obtain the following dilation result.

Theorem 3.4. *Let* $(T_t, t \geq 0)$ *be a bistochastic semigroup on* $L^p(\Omega, \Sigma, \mu)$, $1 \leq p < \infty$, *with a probability measure* μ. *Then there exists a lattice dilation.*

By construction, we obtain also a minimality result.

Corollary 3.5. *The lattice dilation constructed in Section 2 is minimal in the sense that*

$$\operatorname{lin} \bigcup_{t \in (-\infty, \infty)} \widehat{T}_t J(L^p(\Omega, \Sigma, \mu))$$

is a dense sublattice in $L^p(\widehat{\Omega}, \widehat{\Sigma}, \widehat{\mu})$.

Remark 3.6. The construction in this paper heavily relies on the assumption of bistochastic semigroups. However, it is possible to modify and extend the construction in the discrete case ([10, Appendix U], [14], [18]).

4. Markov property and uniqueness

An important property of a dilation is uniqueness (up to isometric isomorphisms). The dilation in Section 2 satisfies both minimality and uniqueness, hence, it gives essentially *the* lattice dilation.

An additional feature of dilations is the Markov property which enables us to show uniqueness. The Markov property plays an important role in stochastic processes which are closely connected with dilations, and it is an important tool to show the existence of dilations for W^*-algebras (see, e.g., [15]). This section is an adaption of [14, Section 4] to our setting. The proofs and observations correspond with the proofs for power dilations in [14], and we refer there for all technical details which can be shown in an analogous way.

Let E_I be the closed sublattice of $L^p(\widehat{\Omega}, \widehat{\Sigma}, \widehat{\mu})$ generated by

$$\bigcup_{t\in I} \widehat{T}_t J(L^p(\Omega, \Sigma, \mu))$$

for some index set I. By [22, III.11.2], there is a unique positive contractive projection $Q_I\colon L^p(\widehat{\Omega}, \widehat{\Sigma}, \widehat{\mu}) \to E_I = L^p(\widehat{\Omega}, \Sigma_I, \widehat{\mu}) \subseteq L^p(\widehat{\Omega}, \widehat{\Sigma}, \widehat{\mu})$ for some σ-algebra Σ_I, and its adjoint is given by the corresponding embedding $J_I\colon L^p(\widehat{\Omega}, \Sigma_I, \widehat{\mu}) \to L^p(\widehat{\Omega}, \widehat{\Sigma}, \widehat{\mu})$. We note that finite tensors depending on Ω_I are dense in $L^p(\widehat{\Omega}, \Sigma_I, \widehat{\mu})$, $1 \le p < \infty$, and $Q_{\{0\}} = JQ$. We also note that $Q_I \widehat{f} = \widehat{f}$ for all $\widehat{f} \in L^p(\widehat{\Omega}, \Sigma_I, \widehat{\mu})$.

Definition 4.1. A dilation satisfies the Markov property if

$$Q_{[t,0]} \widehat{f} = Q_{\{0\}} \widehat{f}$$

holds for all $t \le 0$ and for all $\widehat{f} \in L^p(\Omega, \Sigma_{[0,\infty)}, \mu)$.

Proposition 4.2. *The dilation constructed in Section* 2 *satisfies the Markov property.*

Proof. We have

$$\begin{aligned}
&\langle Q_{[t,0]}(f_0 \otimes \cdots \otimes f_{t_n}), g_{t_{-m}} \otimes \cdots \otimes g_0\rangle \\
&\quad = \langle (f_0 \otimes \cdots \otimes f_{t_n}), J_{[t,0]} g_{t_{-m}} \otimes \cdots \otimes g_0\rangle \\
&\quad = \int_\Omega f_0 \cdot g_0 \cdot T_{t_1}(f_{t_1} T_{t_2 - t_1}(\dots (f_{t_n})) \cdot T'_{-t_{-1}}(g_{t_{-1}} T'_{t_{-1}-t_{-2}}(\dots (g_{t_{-m}})) \, d\mu \\
&\quad = \int_\Omega Q(f_0 \otimes \cdots \otimes f_{t_n}) \cdot Q(g_{t_{-m}} \otimes \cdots \otimes g_0) \, d\mu \\
&\quad = \int_{\widehat{\Omega}} JQ(f_0 \otimes \cdots \otimes f_{t_n}) \cdot g_{t_{-m}} \otimes \cdots \otimes g_0 \, d\widehat{\mu} \\
&\quad = \langle Q_{\{0\}}(f_0 \otimes \cdots \otimes f_{t_n}), g_{t_{-m}} \otimes \cdots \otimes g_0\rangle
\end{aligned}$$

for $t \le 0$, for all $f_0 \otimes \cdots \otimes f_{t_n} \in L^p(\Omega, \Sigma_{[0,\infty)}, \mu)$ and for all $g_{t_{-m}} \otimes \cdots \otimes g_0 \in L^q(\widehat{\Omega}, \Sigma_{[t,0]}, \widehat{\mu})$, hence, for all $\widehat{g} \in L^q(\widehat{\Omega}, \Sigma_{[t,0]}, \widehat{\mu})$. □

Following from the definitions and since $\widehat{T}_t$ is a lattice isomorphism for all $t \in \mathbb{R}$, we have

$$Q_I = \widehat{T}_{-t} Q_{I+t} \widehat{T}_t$$

for all index sets I. It is therefore not essential whether we consider the Markov property at $t = 0$ or at any other time.

Lemma 4.3. *The following are equivalent.*

(i) *The Markov property is satisfied at $t = 0$.*

(ii) *The Markov property is satisfied at $t = t_0$.*

We now assume that we have a lattice dilation $((\tilde{T}_t, t \geq 0), L^p(\tilde{\Omega}, \tilde{\Sigma}, \tilde{\mu}), \tilde{J}, \tilde{Q})$ satisfying the Markov property, $\tilde{T}_t \tilde{\mathbb{1}} = \tilde{\mathbb{1}}$ for all $t \in \mathbb{R}$ and $\tilde{J}\mathbb{1} = \tilde{\mathbb{1}}$ (and hence by the dilation property and $T_t \mathbb{1} = \mathbb{1}$ for all $t \geq 0$ also $\tilde{Q}\tilde{\mathbb{1}} = \mathbb{1}$). We show in Theorem 4.5 that this dilation is essentially the one we constructed in Section 2.

The following identity (4.1) plays an important role in the proof of Theorem 4.5. We note that a unit preserving lattice homomorphism is multiplicative on $C(K)$ ([22, Theorem III.9.1]) and that $T_{-t} = T_t'$ as well as $Q = J'$ hold on corresponding L^p-spaces.

Lemma 4.4. *Let $((\tilde{T}_t, t \geq 0), L^p(\tilde{\Omega}, \tilde{\Sigma}, \tilde{\mu}), \tilde{J}, \tilde{Q})$ be a minimal lattice dilation of $((T_t, t \geq 0), L^p(\Omega, \Sigma, \mu))$ with $\tilde{T}_t \tilde{\mathbb{1}} = \tilde{\mathbb{1}}$ for all $t \in \mathbb{R}$ and $\tilde{J}\mathbb{1} = \tilde{\mathbb{1}}$. Then*

$$\tilde{Q}_{\{t\}} \tilde{J} f = \tilde{T}_t \tilde{J} T_{-t} f \tag{4.1}$$

for all $t \leq 0$ and $f \in L^p(\Omega, \Sigma, \mu)$.

Proof. We have

$$\begin{aligned}
\langle \tilde{Q}_{\{t\}} \tilde{J} f, \tilde{T}_t \tilde{J} g \rangle &= \langle \tilde{J} f, \tilde{J}_t \tilde{T}_t \tilde{J} g \rangle = \langle \tilde{J} f, \tilde{T}_t \tilde{J} g \rangle = \langle \tilde{T}_{-t} \tilde{J} f, \tilde{J} g \rangle \\
&= \langle \tilde{Q} \tilde{T}_{-t} \tilde{J} f, g \rangle = \langle T_{-t} f, g \rangle = \int_\Omega T_{-t} f \cdot g \, d\mu = \int_\Omega (T_{-t} f \cdot g) \cdot \tilde{Q} \tilde{\mathbb{1}} \, d\mu \\
&= \int_{\tilde{\Omega}} \tilde{J}(T_{-t} f \cdot g) \cdot \tilde{\mathbb{1}} \, d\tilde{\mu} = \langle \tilde{J}(g \cdot T_{-t} f), \tilde{\mathbb{1}} \rangle = \langle \tilde{J}(g \cdot T_{-t} f), \tilde{T}_{-t} \tilde{\mathbb{1}} \rangle \\
&= \langle \tilde{T}_t \tilde{J}(g \cdot T_{-t} f), \tilde{\mathbb{1}} \rangle = \int_{\tilde{\Omega}} \tilde{T}_t \tilde{J}(g \cdot T_{-t} f) \, d\tilde{\mu} = \int_{\tilde{\Omega}} \tilde{T}_t \tilde{J} T_{-t} f \cdot \tilde{T}_t \tilde{J} g \, d\tilde{\mu} \\
&= \langle \tilde{T}_t \tilde{J} T_{-t} f, \tilde{T}_t \tilde{J} g \rangle
\end{aligned}$$

for all $f, g \in C(\Omega)$, and we conclude the claim by continuity and minimality since linear combinations of $\tilde{T}_t \tilde{J} g$ for $g \in C(\Omega)$ form a dense sublattice in $L^p(\tilde{\Omega}, \tilde{\Sigma}, \tilde{\mu})$. □

Theorem 4.5. *Let $((\tilde{T}_t, t \geq 0), L^p(\tilde{\Omega}, \tilde{\Sigma}, \tilde{\mu}), \tilde{J}, \tilde{Q})$ be a minimal lattice dilation of the bistochastic semigroup $((T_t, t \geq 0), L^p(\Omega, \Sigma, \mu))$, $1 \leq p < \infty$, where $(\tilde{T}_t, t \geq 0)$ satisfies the Markov property, $\tilde{T}_t \tilde{\mathbb{1}} = \tilde{\mathbb{1}}$ for all $t \in \mathbb{R}$ and $\tilde{J}\mathbb{1} = \tilde{\mathbb{1}}$. Then it is lattice isomorphic to the dilation in Section 2.*

Proof. Let $((\tilde{T}_t, t \geq 0), L^p(\tilde{\Omega}, \tilde{\Sigma}, \tilde{\mu}), \tilde{J}, \tilde{Q})$ be such a lattice dilation of $((T_t, t \geq 0), L^p(\Omega, \Sigma, \mu))$.

We define an operator $\Phi\colon \bigotimes_{t\in\mathbb{R}} C(\Omega) \to L^p(\tilde{\Omega}, \tilde{\Sigma}, \tilde{\mu})$ by

$$\Phi(f_{t_{-n}} \otimes \cdots \otimes f_{t_n}) = \Phi\left(\prod_{k=-n}^{n} \widehat{T}_{t_k} J f_{t_k}\right) := \prod_{k=-n}^{n} \tilde{T}_{t_k} \tilde{J} f_{t_k}$$

for $f_{t_k} \in C(\Omega)$ and we note that Φ is well defined by minimality and that the definition of Φ does not depend on the equivalence class of $f_{t_{-n}} \otimes \cdots \otimes f_{t_n}$ since $\tilde{J}\mathbb{1} = \tilde{\mathbb{1}}$ and $\tilde{T}_t\tilde{\mathbb{1}} = \tilde{\mathbb{1}}$ for all $t \in \mathbb{R}$.

We remark that Φ is linear and multiplicative. It also satisfies $\Phi(\widehat{\mathbb{1}}) = \tilde{\mathbb{1}}$ and $|\Phi(\widehat{f})| = \Phi(|\widehat{f}|)$ on $\bigotimes_{t\in\mathbb{R}} C(\Omega)$ ([22, Theorem III.9.1]). It can be shown that Φ is $\|\cdot\|_p$-isometric and hence bounded by using the Markov property and Lemma 4.4 (cf. [14, Corollary 4.6]). We then extend Φ by linearity and continuity to $L^p(\widehat{\Omega}, \widehat{\Sigma}, \widehat{\mu})$. Since $\tilde{T}_t$ is multiplicative for all $t \in \mathbb{R}$, we finally have

$$\begin{aligned}\tilde{T}_t\Phi(f_{t_{-n}} \otimes \cdots \otimes f_{t_n}) &= \tilde{T}_t\left(\prod_{k=-n}^{n} \tilde{T}_{t_k}\tilde{J} f_{t_k}\right) = \prod_{k=-n}^{n}\left(\tilde{T}_{t_k+t}\tilde{J} f_{t_k}\right) \\ &= \Phi\left(\prod_{k=-n}^{n} \widehat{T}_{t_k+t} J f_{t_k}\right) = \Phi\widehat{T}_t(f_{t_{-n}} \otimes \cdots \otimes f_{t_n})\end{aligned}$$

for all $f_{t_{-n}} \otimes \cdots \otimes f_{t_n} \in \bigotimes_{t\in\mathbb{R}} C(\Omega)$, hence,

$$\Phi \circ \widehat{T}_t = \tilde{T}_t \circ \Phi$$

for all $t \in \mathbb{R}$. □

References

[1] M.A. Akcoglu, *Positive Contractions of L^1-Spaces.* Math. Z. **143** (1975), 5–13.

[2] M.A. Akcoglu, *A Pointwise Ergodic Theorem in Lp-spaces.* Canad. J. Math. **27** (1975), 1075–1082.

[3] M.A. Akcoglu and P.E. Kopp, *Construction of Dilations of Positive Lp-Contractions.* Math. Z. **155** (1977), 119–127.

[4] M.A. Akcoglu and L. Sucheston, *On Convergence of Iterates of Positive Contractions in Lp Spaces.* J. Approximation Theory **13** (1975), 348–362.

[5] M.A. Akcoglu and L. Sucheston, *On Positive Dilations to Isometries in Lp Spaces.* Springer Lecture Notes in Math. **541** (1976), 389–401.

[6] M.A. Akcoglu and L. Sucheston, *Dilations of Positive Contractions on Lp Spaces.* Canad. Math. Bull. **20** (1977), 285–92.

[7] B.V.R. Bhat, *An Index Theory for Quantum Dynamical Semigroups.* Trans. Amer. Math. Soc. **348** (1996), 561–583.

[8] B.V.R. Bhat, *Minimal Dilations of Quantum Dynamical Semigroups to Semigroups of Endomorphisms of C^*-Algebras.* J. Ramanujan Math. Soc. **14** (1999), 109–124.

[9] B.V.R. Bhat and K.R. Parthasarathy, *Markov Dilations of Nonconservative Dynamical Semigroups and a Quantum Boundary Theory.* Ann. Inst. H. Poincaré Probab. Statist. **31** (1995), 601–651.

[10] R. Derndinger, R. Nagel and G. Palm, *Ergodic Theory in the Perspective of Functional Analysis.* Unpublished Manuscript.

[11] G. Fendler, *Dilation of One Parameter Semigroups of Positive Contractions on L^p spaces.* Can. J. Math. **49** (1997), 736–748.

[12] G. Fendler, *On Dilations and Transference for Continuous One-Parameter Semigroups of Positive Contractions on Lp-Spaces.* Ann. Univ. Sarav. Ser. Math. **9** (1998).

[13] P.R. Halmos, *Normal Dilations and Extensions of Operators.* Summa Brasil. Math. **2** (1950), 125–134.

[14] M. Kern, R. Nagel and G. Palm, *Dilations of Positive Operators: Construction and Ergodic Theory.* Math. Z. **156** (1977), 265–277.

[15] B. Kümmerer, *Markov Dilations on W^*-Algebras.* J. Funct. Anal. **63** (1985), 139–177.

[16] B. Kümmerer, *On the Structure of Markov Dilations on W^*-algebras.* Springer Lecture Notes in Math. **1136** (1985), 318–331.

[17] P. Meyer-Nieberg, *Banach Lattices.* Springer-Verlag, 1991.

[18] R. Nagel and G. Palm, *Lattice Dilations of Positive Contractions on L^p-Spaces.* Canad. Math. Bull. **25** (1982), 371–374.

[19] V.V. Peller, *Analogue of J. von Neumann's Inequality, Isometric Dilation of Contractions and Approximation by Isometries in Spaces of Measurable Functions.* Trudy Mat. Inst. Steklov **155** (1981), 103–150.

[20] G.-C. Rota, *An "Alternierende Verfahren" for General Positive Operators.* Bull. Amer. Math. Soc. **68** (1962), 95–102.

[21] J.-L. Sauvageot, *Markov Quantum Semigroups Admit Covariant Markov C^*-Dilations.* Comm. Math. Phys. **106** (1986), 91–103.

[22] H.H. Schaefer, *Banach Lattices and Positive Operators.* Springer-Verlag, 1974.

[23] B. Sz.-Nagy, *Sur les Contractions de l'Espace de Hilbert.* Acta Sci. Math. Szeged **15** (1953), 87–92.

[24] B. Sz.-Nagy and C. Foiaş, *Harmonic Analysis of Operators on Hilbert Spaces.* North-Holland Publishing Co., 1970.

[25] G.F. Vincent-Smith, *Dilation of a Dissipative Quantum Dynamical System to a Quantum Markov Process.* Proc. London Math. Soc. **49** (1984), 58–72.

Leonard J. Konrad
Trinity College, Broad Street
OX13BH Oxford, UK
e-mail: Leonard.Konrad@trinity.oxon.org

Operator Theory:
Advances and Applications, Vol. 250, 297–309

Domains of Fractional Powers of Matrix-valued Operators: A General Approach

Irena Lasiecka and Roberto Triggiani

Abstract. We present a general approach for identifying explicitly domains of fractional powers of matrix-valued operators. Such strategy will be illustrated by revisiting the topic of domains of fractional powers of 'strongly damped' abstract elastic equations, and recovering in this case established results [10]. These were obtained instead by use of the classical Balakrishnan formula [3], [21, p. 69], which is based on knowledge of the resolvent operator. A virtue of the present approach – which results evident in the present illustrative case – is that it is more conceptual and less computationally intensive than the former approach. In particular, it is resolvent independent.

Mathematics Subject Classification (2010). Primary 47F05; Secondary 35.

Keywords. Domains of fractional powers, matrix-valued operators.

0. Introduction

In this paper we present a general approach that permits to obtain a precise and explicit identification of domains of fractional powers of matrix-valued operators arising in a differential equation context. This method – which is critically based on Baiocchi's result [2], [20, Section 14.3, pp. 96–98] on interpolating subspaces – was actually first employed in [15] to re-obtain the domains of fractional powers of single elliptic operators, subject to appropriate boundary conditions, originally due to [13] for second-order operators. A more recent application of this method was employed in [18, Appendix A, p. 255]. For the sake of concreteness and space constraints, we shall illustrate and employ the present strategy by revisiting the topic of "strongly damped elastic systems" [7], [8]–[11], [4], reported also in [16, Appendix 3B, pp. 285–296]. In [10], a precise and explicit characterization was given by a radically different approach; namely by use of the classical resolvent-based Balakrishnan formula [3], [21, Section 2.6]. These results will be reproved in the present paper by the new described approach using interpolation of subspaces. It may be said that the approach of the present paper is more conceptual and much less computational than the original one [10]. In particular, it is resolvent-independent. Thus, in general, one may expect to be able to apply this approach to characterize domains of fractional powers of generators modeling complicated and coupled PDE systems which describe various interactive phenomena in the natural sciences. In such cases, the expression for the resolvent operator is likely to be cumbersome and/or unhelpful. Thus, an approach based on the Balakrishnan formula is likely to be out of question. A most recent application of the resolvent-free approach of the present paper is given in a forthcoming paper [17] dealing with a fluid-strongly damped structure model.

1. Setting of the Problem on the Energy Space E. Results [16, Appendix 3B]

1.1. The original model

We return to the setting of damped elastic operators reported in [16, Appendix 3B, pp. 285–296]. Throughout this paper, H is a separable Hilbert space. On it, we consider two operators $\mathcal{A}$ and $\mathcal{B}$ subject to the following assumptions:

(H.1) $\mathcal{A}$ (the elastic operator): $H \supset \mathcal{D}(\mathcal{A}) \to H$, with domain $\mathcal{D}(\mathcal{A})$ dense in H, is a strictly positive, self-adjoint operator.

(H.2) $\mathcal{B}$ (the dissipation operator: $H \supset \mathcal{D}(\mathcal{B}) \to H$, with domain $\mathcal{D}(\mathcal{B})$ dense in H, is a positive, self-adjoint operator.

(H.3) There exists a constant $0 < \alpha \leq 1$ and two constants $0 < \rho_1 < \rho_2 < \infty$, such that

$$\rho_1(\mathcal{A}^\alpha x, x)_H \leq (\mathcal{B}x, x)_H \leq \rho_2(\mathcal{A}^\alpha x, x)_H, \quad x \in \mathcal{D}(\mathcal{A}^{\frac{\alpha}{2}}) \subset \mathcal{D}(\mathcal{B}^{\frac{1}{2}}). \tag{1.1}$$

Equivalent (as well as sufficient) versions of (H.3) = (1.1) and a number of related considerations are given in [16, Remark 3B.0, p. 286, Remark 3B.1, p. 288] and will not be repeated. The object of our interest is the second-order abstract equation

$$\ddot{x} + \mathcal{B}\dot{x} + \mathcal{A}x = 0 \text{ on } H. \tag{1.2}$$

On the Energy Space $E \equiv \mathcal{D}(\mathcal{A}^{\frac{1}{2}}) \times H$. We rewrite (1.2) as a first-order equation on the space E:

$$\frac{d}{dt}\begin{bmatrix} x \\ \dot{x} \end{bmatrix} = A_B \begin{bmatrix} x \\ \dot{x} \end{bmatrix}, \quad A_B = \begin{bmatrix} 0 & I \\ -\mathcal{A} & -\mathcal{B} \end{bmatrix},$$

$$\text{with domain } \mathcal{D}(A_B) \text{ containing } \mathcal{D}(\mathcal{A}) \times \mathcal{D}(\mathcal{B}). \tag{1.3}$$

A_B is dissipative, hence closable (but not necessarily closed with domain $\mathcal{D}(\mathcal{A}) \times \mathcal{D}(\mathcal{B})$). To make it closed, one needs to enlarge the domain to a 'maximal' domain. This is done [16, (3B.5a), p. 287], after which one shows that with such an enlarged domain, A_B is closed [16, Claim, p. 287]. These results will not be strictly needed in the present paper and hence will be only referred to [16]. Henceforth, A_B denotes such a closed operator: $E \supset \mathcal{D}(A_B) \to E$.

Theorem 1.1 ([8], [9], [11], [16, Theorem 3B.1, p. 288]).

(a) *(generation) Assume the standing hypotheses* (H.1), (H.2), *and* (H.3). *Then, the operator* $A_{\mathcal{B}}$ *is maximal dissipative, and thus (by the Lumer–Phillips Theorem) it generates a s.c.* (C_0-) *contraction semigroup* $e^{A_{\mathcal{B}}t}$ *on the energy space* $E = \mathcal{D}(\mathcal{A}^{\frac{1}{2}}) \times H$.

(b) ($\frac{1}{2} \le \alpha \le 1$*: analyticity) If the parameter* α *in* (H.3) *satisfies* $\frac{1}{2} \le \alpha \le 1$, *then the s.c. semigroup* $e^{A_{\mathcal{B}}t}$ *of part* (a) *is, moreover, analytic on* E.

(c) *The range of analyticity in* (b) *is optimal. For* $0 < \alpha < \frac{1}{2}$ *the s.c. semigroup is generally not analytic (counterexample in* [9]). *It is, however, of Gevrey class* $\delta > 1/(2\alpha)$, *(see* [11], [16] *for more details), hence differentiable for all* $t > 0$ *on* E.

(d) *For all* $0 < \alpha \le 1$, *the s.c. semigroup* $e^{A_{\mathcal{B}}t}$ *is uniformly stable on* E*: There exist constants* $M \ge 1$ *and* $a > 0$ [*indeed* $-a = \sup\ Re\ \sigma(A_{\mathcal{B}})$] *such that*

$$\left\| e^{A_{\mathcal{B}}t} \right\|_{\mathcal{L}(E)} \le M e^{-at}, \quad t \ge 0. \tag{1.4}$$

1.2. The model case $\mathcal{B} = \rho\mathcal{A}^\alpha$, $\rho > 0$

In this subsection, we specialize to the canonical case:

$$\mathcal{B} = \rho\mathcal{A}^\alpha, \text{ hence } \ddot{x} + \mathcal{A}^\alpha\dot{x} + \rho\mathcal{A}x = 0 \text{ on } H, \text{ or} \tag{1.5}$$

$$\frac{d}{dt}\begin{bmatrix} x \\ \dot{x} \end{bmatrix} = A_{\rho\alpha}\begin{bmatrix} x \\ \dot{x} \end{bmatrix}, \quad A_{\rho\alpha} = \begin{bmatrix} 0 & I \\ -\mathcal{A} & -\rho\mathcal{A}^\alpha \end{bmatrix} : E \supset \mathcal{D}(A_{\rho\alpha}) \to E; \tag{1.6}$$

$$\begin{aligned} \mathcal{D}(A_{\rho\alpha}) = \Bigg\{ \begin{bmatrix} x_1 \\ x_2 \end{bmatrix} \in E : x_1 \in \mathcal{D}(\mathcal{A}^{1-\alpha}); \quad x_2 \in \mathcal{D}(\mathcal{A}^{\frac{1}{2}}) : \\ \mathcal{A}^\alpha[\mathcal{A}^{1-\alpha}x_1 + \rho x_2] \in H \Bigg\}. \end{aligned} \tag{1.7}$$

Thus, according to Theorem 1.1, $A_{\rho\alpha}$ generates a contraction s.c. semigroup $e^{A_{\rho\alpha}t}$ on E which, moreover, in the range $\frac{1}{2} \le \alpha \le 1$ is analytic. Thus, the corresponding dynamics has a 'parabolic' behavior. It is therefore important to determine the domains $\mathcal{D}((-A_{\rho\alpha})^\theta)$ of fractional power, $0 \le \theta \le 1$, $\frac{1}{2} \le \alpha \le 1$, $\rho > 0$, of its generator $A_{\rho\alpha}$. This was done in [10] and is reported also in [16, Theorem 3B.2, p. 290].

Remark 1.1. Model (1.5) is mostly, but not exclusively, of mathematical value to test the validity of the results. This is so since only partial differential equations (on an arbitrary domain) with 'special' boundary conditions can be accommodated under model (1.5). See examples in [16, Section 3.4, p. 204, Section 3.6, p. 211, for $\alpha = \frac{1}{2}$; Section 3.5, p. 208, for $\alpha = 1$]. More realistic, physically significant boundary conditions escape model (1.5) and instead are captured by model (1.2). See, e.g., a PDE example of a plate with clamped boundary conditions [16, Section 3.7, p. 214].

Domains of fractional powers $\mathcal{D}((-A_{\rho\alpha})^\theta)$, $0 \le \theta \le 1$, $\frac{1}{2} \le \alpha \le 1$, $\rho > 0$. The next result characterizes the domains of fractional powers $\mathcal{D}((-A_{\rho\alpha})^\theta)$ of the generator $A_{\rho\alpha}$ in (1.6), (1.7), in the specialized range $\frac{1}{2} \leq \alpha \leq 1$ of analyticity for $e^{A_{\rho\alpha}t}$ on E, in which case [16, (3B.15), p. 290] the domain (1.7) can be rewritten as

$$\mathcal{D}(A_{\rho\alpha}) = \left\{ \begin{bmatrix} x_1 \\ x_2 \end{bmatrix} \in E : x_1 \in \mathcal{D}\left(\mathcal{A}^{\frac{1}{2}+1-\alpha}\right),\ x_2 \in \mathcal{D}(\mathcal{A}^{\frac{1}{2}}), \right.$$
$$\left. \mathcal{A}^{1-\alpha}x_1 + \rho x_2 \in \mathcal{D}(\mathcal{A}^{\alpha}) \right\}, \quad \frac{1}{2} \leq \alpha \leq 1, \tag{1.8}$$

since now $x_1 \in \mathcal{D}(\mathcal{A}^{\frac{1}{2}}) \subset \mathcal{D}(\mathcal{A}^{1-\alpha})$, $x_2 \in \mathcal{D}(\mathcal{A}^{\frac{1}{2}})$ and $\mathcal{A}^{1-\alpha}x_1 + \rho x_2 \in \mathcal{D}(\mathcal{A}^{\alpha}) \subset \mathcal{D}(\mathcal{A}^{\frac{1}{2}})$ for $\alpha \geq \frac{1}{2}$ implies $\mathcal{A}^{1-\alpha}x_1 \in \mathcal{D}(\mathcal{A}^{\frac{1}{2}})$, as desired. For $\alpha = \frac{1}{2}$, we obtain

$$\mathcal{D}(A_{\rho,\alpha=\frac{1}{2}}) = \mathcal{D}(\mathcal{A}) \times \mathcal{D}(\mathcal{A}^{\frac{1}{2}}). \tag{1.9}$$

Theorem 1.2 ([10], [16, Thm. 3B.2, p. 290]). *Consider the generator* $A_{\rho\alpha} : E \supset \mathcal{D}(A_{\rho\alpha}) \to E$ *in* (1.6) *with the domain given by* (1.8), *for* $\rho > 0$, $\frac{1}{2} \leq \alpha \leq 1$.

(i) *Let* $\frac{1}{2} \leq \theta \leq 1$. *Then*

$$\mathcal{D}((-A_{\rho\alpha})^{\theta}) = \mathcal{D}\left(\mathcal{A}^{\frac{1}{2}+\theta(1-\alpha)}\right) \times \mathcal{D}(\mathcal{A}^{\alpha\theta}). \tag{1.10}$$

(ii) *Let* $0 \leq \theta \leq \frac{1}{2}$. *Then*

$$\mathcal{D}((-A_{\rho\alpha})^{\theta}) = \left\{ \begin{bmatrix} x_1 \\ x_2 \end{bmatrix} \in E : x_1 \in \mathcal{D}\left(\mathcal{A}^{\frac{1}{2}+\theta(1-\alpha)}\right);\ x_2 \in \mathcal{D}\left(\mathcal{A}^{\alpha-\frac{1}{2}+\theta(1-\alpha)}\right); \right.$$
$$\left. \mathcal{A}^{1-\alpha}x_1 + \rho x_2 \in \mathcal{D}(\mathcal{A}^{\alpha\theta}) \right\}. \tag{1.11}$$

For $\theta = \frac{1}{2}$ *or* $\alpha = \frac{1}{2}$, *the third requirement in* (1.10) *is automatically satisfied.*

(iii) *For* $\alpha = 1$, $A_{\rho\alpha}$ *does not have compact resolvent on* E, *even when* $\mathcal{A}$ *has compact resolvent on* H.

The proof in [10] was based on the classical Balakrishnan formula [3], [21, p. 69], in fact for the inverse operator. More precisely, in [10], it was used that $\mathcal{D}((-A_{\rho\alpha})^{\theta})$
$= (-A_{\rho\alpha})^{-\theta}E$ and hence $[x_1, x_2] \in \mathcal{D}((-A_{\rho\alpha})^{\theta})$ if and only if

$$\begin{bmatrix} x_1 \\ x_2 \end{bmatrix} = (-A_{\rho\alpha})^{-\theta}w = \frac{\sin \pi\theta}{\pi} \int_0^{\infty} \lambda^{-\theta}(\lambda I - A_{\rho\alpha})^{-1}w\, d\lambda,\ w \in E. \tag{1.12}$$

In the next section – the core of the present paper – we shall give a radically different resolvent-free proof, in a sense much less computational and more conceptual. It is critically based on a (specialization, in the present Hilbert setting, of a) very general result due to C. Baiocchi [2], and reported in [20, pp. 96–93]. It is intrinsically a result on interpolation between subspaces (which holds true also in the Banach space setting).

Domains of fractional powers $\mathcal{D}((-A_{\mathcal{B}})^{\theta})$. The next result extends the usefulness of Theorem 1.2 to obtain information on the domains of fractional powers $\mathcal{D}((-A_{\mathcal{B}})^{\theta})$ of $(-A_{\mathcal{B}})$ in (1.3) with maximal domain [16, Eqn. (3B.5a), p. 287].

Theorem 1.3 ([16, Corollary 3B.4, p. 291]). *Assume the above operator $A_{\mathcal{B}}$ in* (1.3) *with maximal domain generates a s.c. analytic semigroup in the situation of Theorem* 1.1(b) *for $\frac{1}{2} \leq \alpha \leq 1$. Then, for $0 < \theta_2 < \theta_1 < \theta < 1$, we have*

$$\mathcal{D}((-A_{\rho\alpha})^{\theta}) \subset \mathcal{D}((-A_{\mathcal{B}})^{\theta_1}) \subset \mathcal{D}((-A_{\rho\alpha})^{\theta_2}). \tag{1.13}$$

2. New proof of Theorem 1.2 using interpolation of subspaces

Step 1. We shall deliberately use the notation of [20, pp. 96–98] in invoking Baiocchi's result. Our first step is to obtain the following unifying form, for all $0 \leq \theta \leq 1$, $\frac{1}{2} \leq \alpha \leq 1$, of $\mathcal{D}((-A_{\rho\alpha})^{\theta})$.

Proposition 2.1. *Let $\rho > 0$, $0 \leq \theta \leq 1$, $\frac{1}{2} \leq \alpha \leq 1$, so that $\frac{1}{2} + \theta(1-\alpha) \geq (1-\alpha)$. Then the following representation for $\mathcal{D}((-A_{\rho\alpha})^{\theta})$ can be given:*

$$\mathcal{D}((-A_{\rho\alpha})^{\theta}) = \left\{ \begin{bmatrix} x_1 \\ x_2 \end{bmatrix} \in E : x_1 \in \mathcal{D}\left(\mathcal{A}^{\frac{1}{2}+\theta(1-\alpha)}\right); \right.$$

$$\left. x_2 \in \mathcal{D}\left(\mathcal{A}^{\frac{\theta}{2}}\right); \, \mathcal{A}^{1-\alpha}x_1 + \rho x_2 \in \mathcal{D}(\mathcal{A}^{\alpha\theta}) \right\}. \tag{2.1}$$

Proof. Step (i). With reference to the case $\theta = 1$, $\frac{1}{2} \leq \alpha \leq 1$ given by (1.8), define the space X, the (constraint) map ∂ and the corresponding space $\mathcal{X}$, as follows:

$$X \equiv \begin{bmatrix} \mathcal{D}(\mathcal{A}^{\frac{3}{2}-\alpha}) \\ \mathcal{D}(\mathcal{A}^{\frac{1}{2}}) \end{bmatrix} : \partial(X) \;=\; \partial \begin{bmatrix} x_1 \in \mathcal{D}(\mathcal{A}^{\frac{3}{2}-\alpha}) \\ x_2 \in \mathcal{D}(\mathcal{A}^{\frac{1}{2}}) \end{bmatrix} \tag{2.2}$$

$$\stackrel{\text{def}}{\equiv} \mathcal{A}^{1-\alpha}x_1 + \rho x_2 \in \mathcal{D}(\mathcal{A}^{\alpha}) \equiv \mathcal{X}. \tag{2.3}$$

Then, via (2.2), (2.3), we can rewrite $\mathcal{D}(A_{\rho\alpha})$ in (1.8) as follows:

$$\mathcal{D}(-A_{\rho\alpha}) = (X)_{\partial,\mathcal{X}} = \text{subspace of } X \text{ mapped into } \mathcal{X} \text{ by map } \partial$$

$$= \{x : x \in X, \, \partial x \in \mathcal{X}\}. \tag{2.4}$$

Next, we rewrite the space $E = \mathcal{D}(\mathcal{A}^{\frac{1}{2}}) \times H$ as follows by means of the spaces Y ($\equiv E$) and $\mathcal{Y}$ ($\equiv Y$) via the same (constraint) map ∂:

$$Y \equiv E \equiv \begin{bmatrix} \mathcal{D}(\mathcal{A}^{\frac{1}{2}}) \\ H \end{bmatrix}; \, \partial(Y) = \partial \begin{bmatrix} y_1 \in \mathcal{D}(\mathcal{A}^{\frac{1}{2}}) \\ y_2 \in H \end{bmatrix} : \; \mathcal{A}^{1-\alpha}y_1 + \rho y_2 \in H \equiv \mathcal{Y}, \tag{2.5}$$

where we note that, for $\frac{1}{2} \leq \alpha$ as assumed, $1 - \alpha \leq \frac{1}{2}$, hence $y_1 \in \mathcal{D}(\mathcal{A}^{\frac{1}{2}}) \subset \mathcal{D}(\mathcal{A}^{1-\alpha})$, and thus the map ∂ actually imposes no constraint on Y. In other words,

$$E \equiv (Y)_{\partial,\mathcal{Y}} = \{y : y \in Y, \partial y \in \mathcal{Y}\}. \tag{2.6}$$

Step (ii). Since $A_{\rho\alpha}$ is maximal dissipative, and $A_{\rho\alpha}^{-1} \in \mathcal{L}(E)$ (e.g., by (1.4)), then by the results reported in [16, p. 5, in particular case c] and [5, Prop. 6.1, p. 171] the following intermediate/interpolation result holds true (this fact was already used in [16, (3B.20), p. 291]

$$\mathcal{D}((-A_{\rho\alpha})^{\theta}) = [\mathcal{D}(A_{\rho\alpha}), E]_{1-\theta} = [(X)_{\partial,\mathcal{X}}, (Y)_{\partial,\mathcal{Y}}]_{1-\theta}, \tag{2.7}$$

where in the second identification we have invoked (2.4) and (2.6).

It is at this point that we appeal to Baiocchi's result [20, pp. 96–98].

The setting, in the notation of [20, pp. 96–97] is as follows with $\frac{1}{2} \le \alpha \le 1$:

$$\Phi = \begin{bmatrix} \mathcal{D}(\mathcal{A}^{1-\alpha}) \\ H \end{bmatrix}; \quad X \equiv \begin{bmatrix} \mathcal{D}(\mathcal{A}^{\frac{3}{2}-\alpha}) \\ \mathcal{D}(\mathcal{A}^{\frac{1}{2}}) \end{bmatrix} \subset \Phi; \quad Y \equiv E = \begin{bmatrix} \mathcal{D}(\mathcal{A}^{\frac{1}{2}}) \\ H \end{bmatrix} \subset \Phi;$$

$$\mathcal{X} \equiv \mathcal{D}(\mathcal{A}^{\alpha}) \subset \tilde{\mathcal{X}} \equiv \mathcal{D}(\mathcal{A}^{\frac{1}{2}}) \subset H \equiv \Psi; \quad r \equiv 0;$$

$$\mathcal{Y} \equiv \tilde{\mathcal{Y}} \equiv H \equiv \Psi; \; \mathcal{G}x = \begin{bmatrix} \mathcal{A}^{-(1-\alpha)}x \\ 0 \end{bmatrix} \subset \Phi, \text{ for } x \in \tilde{\mathcal{Y}} \equiv H \supset \tilde{\mathcal{X}},$$

so that

$$\partial : \Phi \to \Psi \text{ means: } \partial \begin{bmatrix} x_1 \in \mathcal{D}(\mathcal{A}^{1-\alpha}) \\ x_2 \in H \end{bmatrix} = \mathcal{A}^{1-\alpha}x_1 + \rho x_2 \in H \equiv \Psi;$$

$$\partial : X \to \tilde{\mathcal{X}} \text{ means: } \partial \begin{bmatrix} x_1 \in \mathcal{D}(\mathcal{A}^{\frac{3}{2}-\alpha}) \\ x_2 \in \mathcal{D}(\mathcal{A}^{\frac{1}{2}}) \end{bmatrix} = \mathcal{A}^{-\frac{1}{2}}(\mathcal{A}^{\frac{3}{2}-\alpha})x_1 + \rho x_2 \in \mathcal{D}(\mathcal{A}^{\frac{1}{2}}) \equiv \tilde{\mathcal{X}};$$

$$\partial : Y \to \tilde{\mathcal{Y}} \text{ means: } \partial \begin{bmatrix} x_1 \in \mathcal{D}(\mathcal{A}^{\frac{1}{2}}) \\ x_2 \in H \end{bmatrix} = \mathcal{A}^{-(\alpha-\frac{1}{2})}\mathcal{A}^{\frac{1}{2}}x_1 + \rho x_2 \in H \equiv \tilde{\mathcal{Y}};$$

$$\partial\mathcal{G}x = \partial \begin{bmatrix} \mathcal{A}^{-(1-\alpha)}x \\ 0 \end{bmatrix} = \mathcal{A}^{1-\alpha}\mathcal{A}^{-(1-\alpha)}x + \rho 0 = x, \quad x \in \tilde{\mathcal{Y}} \equiv H \equiv \Psi,$$

and thus the assumptions of [20, Thm. 14.3, p. 97] are all satisfied.

This result essentially interchanges the operation of a "restriction to a subspace" with the operation of "interpolation." Thus, "first restriction ∂ followed by interpolation" coincides with "first interpolation followed by restriction ∂." Technically, we then can write starting from (2.7):

$$\mathcal{D}((-A_{\rho\alpha})^{\theta}) = [(X)_{\partial,\mathcal{X}}, (Y)_{\partial,\mathcal{Y}}]_{1-\theta} \tag{2.8a}$$

$$= ([X, Y]_{1-\theta})_{\partial,[\mathcal{X},\mathcal{Y}]_{1-\theta}}. \tag{2.8b}$$

This step from (2.8a) to (2.8b) is critical in our proof. We shall now employ the RHS of (2.8b) to compute $\mathcal{D}((-A_{\rho\alpha})^{\theta})$.

Step (iii). Since $\mathcal{A}$ is positive self-adjoint on H, the usual (Hilbert) interpolation formulas apply [20, p. 10]. By (2.2) for X and (2.5) for Y, we obtain

$$[X,Y]_{1-\theta} = \left[\begin{bmatrix} \mathcal{D}\left(\mathcal{A}^{\frac{3}{2}-\alpha}\right) \\ \mathcal{D}\left(\mathcal{A}^{\frac{1}{2}}\right) \end{bmatrix}, \begin{bmatrix} \mathcal{D}\left(\mathcal{A}^{\frac{1}{2}}\right) \\ H \end{bmatrix}\right]_{1-\theta}$$

$$= \begin{bmatrix} \mathcal{D}\left(\mathcal{A}^{\theta(\frac{3}{2}-\alpha)+\frac{1}{2}(1-\theta)}\right) \\ \mathcal{D}\left(\mathcal{A}^{\theta\frac{1}{2}+0(1-\theta)}\right) \end{bmatrix} = \begin{bmatrix} \mathcal{D}\left(\mathcal{A}^{\frac{1}{2}+\theta(1-\alpha)}\right) \\ \mathcal{D}\left(\mathcal{A}^{\frac{\theta}{2}}\right) \end{bmatrix}. \tag{2.9}$$

Step (iv). Similarly, recalling (2.3) for $\mathcal{X}$ and (2.5) for $\mathcal{Y}$, we obtain

$$[\mathcal{X},\mathcal{Y}]_{1-\theta} = [\mathcal{D}(\mathcal{A}^{\alpha}), H]_{1-\theta} = \mathcal{D}(\mathcal{A}^{\alpha\theta}). \tag{2.10}$$

Step (v). The identification (2.8b) says that:

$$\mathcal{D}((-A_{\rho\alpha})^{\theta}) = \{\text{all elements of } [X,Y]_{1-\theta} \text{ which are mapped into } [\mathcal{X},\mathcal{Y}]_{1-\theta} \text{ by the map } \partial\} = \{u \in [X,Y]_{1-\theta} : \partial u \in [\mathcal{X},\mathcal{Y}]_{1-\theta}\}. \tag{2.11}$$

Interpreting (2.11) by virtue of the spaces in (2.9) and (2.10) and recalling the map ∂, we obtain

$$\mathcal{D}((-A_{\rho\alpha})^{\theta}) = \left\{\begin{bmatrix} x_1 \\ x_2 \end{bmatrix} \in E : x_1 \in \mathcal{D}(\mathcal{A}^{\frac{1}{2}+\theta(1-\alpha)});\ x_2 \in \mathcal{D}(\mathcal{A}^{\frac{\theta}{2}})\ ; \right.$$
$$\left. \mathcal{A}^{1-\alpha}x_1 + \rho x_2 \in \mathcal{D}(\mathcal{A}^{\alpha\theta})\right\}. \tag{2.12}$$

This establishes (2.1), and Proposition 2.1 is proved. □

Step 2. In this step we restrict to the range $0 \le \theta \le \frac{1}{2}$ and find an equivalent more explicit version for $\mathcal{D}((-A_{\rho\alpha})^{\theta})$ in (2.1), actually proving Theorem 1.2(i) = (1.10).

Proposition 2.2. *Assume* $\rho > 0$, $\frac{1}{2} \le \alpha \le 1$, $0 \le \theta \le \frac{1}{2}$. *Then we may rewrite* (2.1) *more explicitly as*

$$\mathcal{D}((-A_{\rho\alpha})^{\theta}) = \mathcal{D}(\mathcal{A}^{\frac{1}{2}+\theta(1-\alpha)}) \times \mathcal{D}(\mathcal{A}^{\alpha\theta}). \tag{2.13}$$

Proof. Step (i). We with reference to the third constraint in (2.1), seek to show that

$$\mathcal{A}^{1-\alpha}x_1 \in \mathcal{D}(\mathcal{A}^{\alpha\theta}), \text{ or } x_1 \in \mathcal{D}(\mathcal{A}^{1-\alpha+\alpha\theta}), \tag{2.14}$$

after which, then, the third constraint in (2.1) would yield

$$x_2 \in \mathcal{D}(\mathcal{A}^{\alpha\theta}) \subset \mathcal{D}(\mathcal{A}^{\frac{\theta}{2}}) \text{ for } \alpha \ge \tfrac{1}{2} \text{ as assumed.} \tag{2.15}$$

This way, characterization (2.1) would then be reduced to the desired form (2.13).
Step (ii). We shall now show (2.14). The first constraint for x_1 in (2.1) is that $x_1 \in \mathcal{D}(\mathcal{A}^{\frac{1}{2}+\theta(1-\alpha)})$. We then seek to show that, more precisely,

$$x_1 \in \mathcal{D}(\mathcal{A}^{\frac{1}{2}+\theta(1-\alpha)}) \subset \mathcal{D}(\mathcal{A}^{1-\alpha+\alpha\theta}), \tag{2.16}$$

which is true just in case

$$\frac{1}{2} + \theta(1-\alpha) \geq 1 - \alpha + \alpha\theta. \tag{2.17}$$

Equation (2.17) holds true as an equality for $\alpha = \frac{1}{2}$. For $\frac{1}{2} < \alpha \leq 1$, we rewrite it as $\theta(2\alpha - 1) \leq \frac{1}{2}(2\alpha - 1) \iff \theta \leq \frac{1}{2}$. Then, for $0 \leq \theta \leq \frac{1}{2}$, $\frac{1}{2} \leq \alpha \leq 1$, (2.14) is shown, as desired, and Proposition 2.2 is established. □

Step 3. In this step we restrict to the range $\frac{1}{2} \leq \theta \leq 1$ and find an equivalent, more explicit expression for $\mathcal{D}((-A_{\rho\alpha})^\theta)$ in (2.1), actually proving Theorem 1.2(ii) = (1.11).

Proposition 2.3. *Assume* $\rho > 0$, $\frac{1}{2} \leq \alpha \leq 1$, $\frac{1}{2} \leq \theta \leq 1$. *Then we may rewrite* (2.1) *more explicitly as*

$$\mathcal{D}((-A_{\rho\alpha})^\theta) = \left\{ \begin{bmatrix} x_1 \\ x_2 \end{bmatrix} \in E : x_1 \in \mathcal{D}\left(\mathcal{A}^{\frac{1}{2}+\theta(1-\alpha)}\right); \, x_2 \in \mathcal{D}\left(\mathcal{A}^{\alpha-\frac{1}{2}+\theta(1-\alpha)}\right); \right.$$
$$\left. \mathcal{A}^{1-\alpha} x_1 + \rho x_2 \in \mathcal{D}(\mathcal{A}^{\alpha\theta}) \right\}. \tag{2.18}$$

Proof. Step (i). By comparing expression (2.1) with expression (2.18), we see that we need to show that the three conditions in (2.1) actually imply the second condition for x_2 in (2.18); more precisely that

$$x_2 \in \mathcal{D}(\mathcal{A}^{\alpha-\frac{1}{2}+\theta(1-\alpha)}) \subset \mathcal{D}(\mathcal{A}^{\frac{\theta}{2}}). \tag{2.19}$$

First, the containment in (2.19) does hold true, since

$$\left(\alpha - \frac{1}{2}\right) + \theta(1-\alpha) - \frac{\theta}{2} = \left(\alpha - \frac{1}{2}\right) + \theta\left(\frac{1}{2} - \alpha\right) \geq \left(\alpha - \frac{1}{2}\right) + \left(\frac{1}{2} - \alpha\right) = 0, \tag{2.20}$$

in the present case $\theta \leq 1$, $\frac{1}{2} \leq \alpha$. Next we now present the strategy to show (2.19) in the present range $\frac{1}{2} \leq \theta \leq 1$. We rewrite the third constraint in (2.1) = (2.12) as follows:

$$\left[\mathcal{A}^{\alpha\theta-(\alpha-\frac{1}{2})-\theta(1-\alpha)}\right] \mathcal{A}^{(1-\alpha)+(\alpha-\frac{1}{2})+\theta(1-\alpha)} x_1$$
$$+ \left[\mathcal{A}^{\alpha\theta-(\alpha-\frac{1}{2})-\theta(1-\alpha)}\right] \mathcal{A}^{(\alpha-\frac{1}{2})+\theta(1-\alpha)} \rho x_2 = \mathcal{A}^{\alpha\theta}\left\{\mathcal{A}^{1-\alpha}x_1 + \rho x_2\right\} \in H. \tag{2.21}$$

The idea is to begin by expressing x_2 as acted upon by the desired operator $\mathcal{A}^{(\alpha-\frac{1}{2})+\theta(1-\alpha)}$. This results in producing the operator $\mathcal{A}^{\alpha\theta-(\alpha-\frac{1}{2})-\theta(1-\alpha)}$, which we then want to be a common component also acting on x_1. This idea yields (2.21). Setting, with $\theta \geq \frac{1}{2}$, $\alpha \geq \frac{1}{2}$:

$$\beta \equiv \alpha\theta - \left(\alpha - \frac{1}{2}\right) - \theta(1-\alpha) = \theta(2\alpha - 1) - \alpha + \frac{1}{2} \geq \frac{1}{2}(2\alpha - 1) - \alpha + \frac{1}{2} = 0, \tag{2.22}$$

and noticing that

$$(1-\alpha)+\left(\alpha-\frac{1}{2}\right)+\theta(1-\alpha)=\frac{1}{2}+\theta(1-\alpha), \tag{2.23}$$

we rewrite (2.21) as

$$\mathcal{A}^{\beta}\left\{\mathcal{A}^{\frac{1}{2}+\theta(1-\alpha)}x_1+\mathcal{A}^{(\alpha-\frac{1}{2})+\theta(1-\alpha)}\rho x_2\right\}\in H, \tag{2.24}$$

or, since $\beta \geq 0$ by (2.22), we obtain *a fortiori*

$$\mathcal{A}^{\frac{1}{2}+\theta(1-\alpha)}x_1+\rho\mathcal{A}^{(\alpha-\frac{1}{2})+\theta(1-\alpha)}x_2\in H. \tag{2.25}$$

But the first term $\mathcal{A}^{\frac{1}{2}+\theta(1-\alpha)}x_1 \in H$ by the first constraint in (2.1). Then (2.25) holds if and only if

$$\mathcal{A}^{(\alpha-\frac{1}{2})+\theta(1-\alpha)}x_2\in H,\ \text{ or } x_2\in\mathcal{D}(\mathcal{A}^{(\alpha-\frac{1}{2})+\theta(1-\alpha)}, \tag{2.26}$$

and the LHS containment in (2.19) is proved. The full (2.19) now shows that expression (2.1) can be rewritten more explicitly as in (2.18). Proposition 2.3 is established. Theorem 1.2 is thus proved. □

3. Setting of the problem on the product space $H \times H$

In this section, we return to the operator $A_{\rho\alpha}$ in (1.6), however, now viewed in the state space $H \times H$:

$$A_{\rho\alpha}=\begin{bmatrix}0 & I\\ -\mathcal{A} & -\rho\mathcal{A}^{\alpha}\end{bmatrix}: H\times H\supset\mathcal{D}(A_{\rho\alpha})\to H\times H; \tag{3.1a}$$

$$\mathcal{D}(A_{\rho\alpha})=\left\{\begin{bmatrix}x_1\\ x_2\end{bmatrix}: x_2\in H;\ x_1\in\mathcal{D}(\mathcal{A}^{1-\alpha});\ \mathcal{A}^{1-\alpha}x_1+\rho x_2\in\mathcal{D}(\mathcal{A}^{\alpha})\right\}$$

$$\supset\mathcal{D}(\mathcal{A})\times\mathcal{D}(\mathcal{A}^{\alpha}). \tag{3.1b}$$

Such $A_{\rho\alpha}$ is densely defined and closed.

Here $\mathcal{A}$ is the strictly positive self-adjoint operator in (H.1). We again wish to revisit the issue of the domains of fractional powers $\mathcal{D}((-A_{\rho\alpha})^{\theta})$ in the case of interest for the corresponding dynamics: $\ddot{x}+\rho\mathcal{A}^{\alpha}\dot{x}+\mathcal{A}x=0$ as in (1.5). Before doing this, we need to recall the following negative and positive results in the present case.

Case #1. $\rho = 0$; or else $\rho > 0$ and $0 \leq \alpha < 1$. In this case we have a negative result on $H \times H$.

Theorem 3.1 ([4]). *Let $\mathcal{A}$ satisfy* (H.1).

(a) *Let $\rho = 0$, or else $\rho > 0$ and $0 \leq \alpha < 1$. Let $\lambda > 0$. Then the resolvent operator $R(\lambda, A_{\rho\alpha})$ of $A_{\rho\alpha}$ in* (3.1) *satisfies the lower bound*

$$\|R(\lambda, A_{\rho\alpha}\|_{\mathcal{L}(H\times H)}\geq\frac{1}{2+\rho},\ \forall\,\lambda>0. \tag{3.2}$$

A fortiori, (3.2) *violates the necessary and sufficient condition of generation of a s.c. semigroup by* $A_{\rho\alpha}$ *on* $H \times H$ [12], [21]. *Thus,* $A_{\rho\alpha}$ *does not generate a s.c. semigroup on* $H \times H$.

(b) *Let* $\rho > 0$, $\frac{1}{2} \le \alpha \le 1$. *Then*

$$\|R(\lambda, A_{\rho\alpha})\|_{\mathcal{L}(H\times H)} \le \text{const}_{\rho\alpha r_0}, \ \forall\, \lambda \ \textit{with}\ Re\,\lambda > r_0 > 0, \tag{3.3}$$

and so [1, p. 341] $A_{\rho\alpha}$ *is the generator of a so-called integrated semigroup (or distribution semigroup* [19]) *in* $H \times H$.

Case #2. $\rho > 0$ and $\alpha = 1$. In this case, we have a positive result: $A_{\rho\alpha}$ does generate a s.c. even analytic semigroup on $H \times H$, though not contractive. The following result provides also the domain $\mathcal{D}((-A_{\rho\alpha})^{\theta})$, $\alpha = 1$, of fractional powers.

Theorem 3.2. *Assume* (H.1) *for* $\mathcal{A}$. *Let* $\rho > 0$ *and* $\alpha = 1$. *Consider the operator* $A_{\rho,\alpha=1} : H \times H \supset \mathcal{D}(A_{\rho,\alpha=1}) \to H \times H$, *as in* (3.1b) *for* $\alpha = 1$:

$$\mathcal{D}(A_{\rho\alpha}) = \{x_1, x_2 \in H : x_1 + \rho x_2 \in \mathcal{D}(\mathcal{A})\}. \tag{3.4}$$

Then:

(a) [6], [25, p. 314] *The operator* $A_{\rho,\alpha=1}$ *generates a s.c. (non-contractive) semigroup on* $H \times H$ *which, moreover, satisfies the estimate*

$$\|(R(\lambda, A_{\rho,\alpha=1})\|_{\mathcal{L}(H\times H)} \le \frac{C}{|\lambda|}, \ Re\,\lambda > 0. \tag{3.5}$$

Hence, such s.c. semigroup is, moreover, analytic on $H \times H$, $t > 0$.

(b) [10, p. 292], *For* $0 \le \theta \le 1$, *the domains of fractional powers* $\mathcal{D}((-A_{\rho\alpha})^{\theta})$, $\alpha = 1$, *are given by*

$$\mathcal{D}((-A_{\rho\alpha})^{\theta}) = \{x_1, x_2 \in H :\ x_1 + \rho x_2 \in \mathcal{D}(\mathcal{A}^{\theta})\},\ \rho > 0,\ \alpha = 1. \tag{3.6}$$

The goal of the present section is to re-prove part (b), Equation (3.6), by using the ideas and approach of Section 2, critically relying on the interpolation result of subspaces in [2].

Proof of (3.6). *Step* 1. With reference to (3.4), we define the space X and the (constraint) map ∂ as follows:

$$X = \begin{bmatrix} H \\ H \end{bmatrix};\ \partial(X) = \partial \begin{bmatrix} x_1 \in H \\ x_2 \in H \end{bmatrix} \stackrel{\text{def}}{=} x_1 + \rho x_2 \in \mathcal{D}(\mathcal{A}) \equiv \mathcal{X}. \tag{3.7}$$

Then, according to (3.1b), we can rewrite the domain $\mathcal{D}(A_{\rho\alpha})$ of $A_{\rho\alpha}$ (Case $\theta = 1$, $\alpha = 1$) as follows:

$$\begin{aligned} \mathcal{D}(A_{\rho\alpha}) = (X)_{\partial,\mathcal{X}} &= \text{subspace of } X \text{ mapped into } \mathcal{X} \text{ by the map } \partial \\ &= \{x : x \in X;\ \partial x \in \mathcal{X}\}. \end{aligned} \tag{3.8}$$

Next, we rewrite the space X in (3.7), and the map ∂ on it, as follows:

$$Y \equiv X = \begin{bmatrix} H \\ H \end{bmatrix};\ \partial(Y) = \partial \begin{bmatrix} x_1 \in H \\ x_2 \in H \end{bmatrix} \stackrel{\text{def}}{=} x_1 + \rho x_2 \in H \equiv \mathcal{Y}. \tag{3.9}$$

Of course, (3.9) is automatically satisfied and imposes no constraint on the map δ. We can write

$$Y \equiv \left| \begin{array}{c} H \\ H \end{array} \right| = (Y)_{\partial,\mathcal{Y}} = \{y \in Y : \partial y \in \mathcal{Y}\}. \tag{3.10}$$

Step 2. In the present case, $A_{\rho,\alpha=1}$ is no longer maximal dissipative on the space $H \times H$, while it is still boundedly invertible on $H \times H$. Thus, we cannot use the reason of Step (ii), Proposition 2.1 to justify the critical relation

$$\mathcal{D}((-A_{\rho\alpha})^{\theta}) = \left[\mathcal{D}(A_{\rho\alpha}), \begin{bmatrix} H \\ H \end{bmatrix}\right]_{1-\theta}, \quad 0 \leq \theta \leq 1, \ \alpha = 1. \tag{3.11}$$

However, (3.11) continues to be true for two reasons. The first reason is that the closed operator $(-A_{\rho\alpha})$, $\alpha = 1$ is positive in the sense of [22, Definition 1.14.1, p. 91] (as, in particular, $A_{\rho\alpha}$, $\alpha = 1$, is the generator of a s.c. semigroup of negative type), and the required resolvent condition in [22, Definition 1.14.1] holds true. The second reason is that $(-A_{\rho\alpha})$, $\alpha = 1$, has locally bounded imaginary powers: There exist two positive numbers ϵ and C such that $\|(-A_{\rho\alpha})^{it}\|_{\mathcal{L}(X\times X)} \leq C$ for $-\epsilon \leq t \leq \epsilon$. This property can be verified in our case by using the spectral properties of $(-A_{\rho\alpha})$ on $X\times X$: $(-A_{\rho\alpha})$, $\alpha = 1$, is the sum of two normal operators [24], [25], [9, Appendix], [22, Theorem 1.15.3, p. 103] applies and yields (3.11).

Step 3. Thus, starting from (3.11) and recalling (3.8) for $\mathcal{D}(A_{\rho\alpha})$ and (3.10) for $[H, H]$, we can write

$$\mathcal{D}((-A_{\rho\alpha})^{\theta}) = [(X)_{\partial,\mathcal{X}}, (Y)_{\partial,\mathcal{Y}}]_{1-\theta} \tag{3.12a}$$

$$= ([X, Y]_{1-\theta})_{\partial,[\mathcal{X},\mathcal{Y}]_{1-\theta}}, \tag{3.12b}$$

where in going from (3.12a) to (3.12b) we have again critically invoked Baiocchi's result [2], [20, pp. 96–98]. The setting in the notation of [20, pp. 96–98] is as follows:

$$\Phi \equiv \begin{bmatrix} H \\ H \end{bmatrix} \equiv X \equiv Y; \quad \mathcal{X} \equiv \mathcal{D}(\mathcal{A}) \subset \tilde{\mathcal{X}} \equiv H \equiv \Psi;$$

$$\mathcal{Y} \equiv \tilde{\mathcal{Y}} \equiv H \equiv \Psi; \quad r \equiv 0; \quad \mathcal{G}x \equiv \begin{bmatrix} x \\ 0 \end{bmatrix} \subset \Phi, \text{ for } x \in H \equiv \tilde{\mathcal{X}} \equiv \tilde{\mathcal{Y}} = \Psi,$$

so that

$$\partial : \Phi \equiv X \to \tilde{\mathcal{X}} \equiv \Psi \text{ means: } \partial \begin{bmatrix} x_1 \in H \\ x_2 \in H \end{bmatrix} = x_1 + \rho x_2 \in H \equiv \tilde{\mathcal{X}} = \Psi;$$

$$\partial : Y \to \tilde{\mathcal{Y}} \text{ means: } \partial \begin{bmatrix} x_1 \in H \\ x_2 \in H \end{bmatrix} = x_1 + \rho x_2 \in H \equiv \tilde{\mathcal{Y}};$$

$$\partial \mathcal{G}x = \partial \begin{bmatrix} x \\ 0 \end{bmatrix} = x + \rho 0 = x, \quad x \in \tilde{\mathcal{X}} \equiv \tilde{\mathcal{Y}},$$

and thus the assumptions of [20, Thm. 14.3, p. 97] are all satisfied. This justifies the passage from (3.12a) to (3.12b).

Step 4. From (3.9), we obtain

$$[X,Y]_{1-\theta} = \left[\begin{bmatrix} H \\ H \end{bmatrix}, \begin{bmatrix} H \\ H \end{bmatrix} \right]_{1-\theta} = \begin{bmatrix} H \\ H \end{bmatrix}, \tag{3.13}$$

while $\mathcal{A}$ being positive, self-adjoint, one obtains via (3.7) for $\mathcal{X}$ and (3.9) for $\mathcal{Y}$:

$$[\mathcal{X},\mathcal{Y}]_{1-\theta} = [\mathcal{D}(\mathcal{A}), H]_{1-\theta} = \mathcal{D}(\mathcal{A}^{\theta}). \tag{3.14}$$

Step 5. The identification formula (3.12b) says for $\alpha = 1$:

$$\mathcal{D}((-A_{\rho\alpha})^{\theta}) = \{\text{all elements of } [X,Y]_{1-\theta} \text{ which are mapped into } [\mathcal{X},\mathcal{Y}]_{1-\theta} \text{ by the map } \partial\} = \{u \in [X,Y]_{1-\theta} : \ \partial u \in [\mathcal{X},\mathcal{Y}]_{1-\theta}\}. \tag{3.15}$$

Thus, invoking (3.13) and (3.14) in (3.15) and recalling the definition of the map ∂, we obtain

$$\mathcal{D}((-A_{\rho\alpha})^{\theta}) = \{x_1 \in H, \, x_2 \in H : \ x_1 + \rho x_2 \in \mathcal{D}(\mathcal{A}^{\theta})\}, \tag{3.16}$$

and (3.6) is established. Theorem 3.2(b) is proved, as desired. □

Acknowledgment

The authors wish to thank a referee for much appreciated comments.

Research partially supported by the National Science Foundation under grant DMS-1108871 and by the Air Force Office of Scientific Research under grant FA9550-12-1-0354.

References

[1] W. Arendt, *Vector-valued Laplace transforms and Cauchy problems*, Israel J. Math. 59 (1987), 327–352.

[2] C. Baiocchi, *Un teorema di interpolazione; applicazioni ai problemi ai limiti per le equazioni a derivate parziali*, Ann. Math. Pura Appl. 4 LXXIII (1966), 235–252.

[3] A.V. Balakrishnan, *Fractional powers of closed operators and the semigroups generated by them*, Pacific J. Math. 10 (1960), 419–437.

[4] A.V. Balakrishnan and R. Triggiani, *Lack of generation of strongly continuous semigroups by the damped wave operator on $H \times H$*, Appl. Math. Letters 6 (1993), 33–37.

[5] A. Bensousson, G. Da Prato, M. Delfour, and S. Mitter, *Representation and Control of Infinite-Dimensional Systems.*

[6] F. Bucci, *A Dirichlet boundary control problem for the strongly damped wave equation*, SIAM J. Control & Optimiz. 30(5) (1992), 1092–1100.

[7] G. Chen and D.L. Russel, *A mathematical model for linear elastic systems with structural damping*, Quart. Appl. Math. 1982, 433–454.

[8] S. Chen and R. Triggiani, Proof of two conjectures of G. Chen and D.L. Russell on structural damping for elastic systems: The case $\alpha = 1/2$ (with S. Chen), Springer-Verlag Lecture Notes in Mathematics 1354 (1988), 234–256. *Proceedings of Seminar on Approximation and Optimization*, University of Havana, Cuba (January 1987).

[9] S. Chen and R. Triggiani, *Proof of extensions of two conjectures on structural damping for elastic systems: The case $1/2 \le \alpha \le 1$*, Pacific J. Math. 136 (1989), 15–55.

[10] S. Chen and R. Triggiani, *Characterization of domains of fractional powers of certain operators arising in elastic systems, and applications*, J. Diff. Eqns. 88 (1990), 279–293.

[11] S. Chen and R. Triggiani, *Gevrey class semigroups arising from elastic systems with gentle perturbation*, Proceedings Amer. Math. Soc. 110 (1990), 401–415.

[12] H. O. Fattorini, *The Cauchy Problem*, Encyclopedia of Mathematics and its Applications, Addison-Wesley, 1983.

[13] D. Fujiwara, *Concrete characteriations of domains of fractional powers of some elliptic differential power of some elliptic differential operators of the second order*, Proc. Acad. Japan 43 (1967), 82–86.

[14] P. Grisvard, *Caracterizaiton de qualques espaces d'interpolation*, Arch. Rat. Mech. Anal. 25 (1967), 40–63.

[15] I. Lasiecka, *Unified theory for abstract parabolic boundary problems – a semigroup approach*, Appl. Math. & Optimiz. 6 (1980), 31–62.

[16] I. Lasiecka and R. Triggiani, Control Theory for Partial Differential Equations: Continuous and Approximation Theories, *Encyclopedia of Mathematics and Its Applications Series*, Cambridge University Press, January 2000.

[17] I. Lasiecka and R. Triggiani, A fluid-strongly damped structure interaction: Analyticity, spectral analysis, exponential stability, preprint 2014.

[18] C. Lebiedzik and R. Triggiani, The optimal interior regularity for the critical case of a clamped thermoelastic system with point control revisited, *Modern Aspects of the Theory of PDEs.* Vol. 216 of Operator Theory: Advances and Applications, 243–259, Birkhäuser/Springer, Basel, 2011. M. Ruzhansky and J. Wirth, eds.

[19] J.L. Lions, *Les semigroupe distributions*, Portug. Math. 19(1960), 141–164.

[20] J.L. Lions and E. Magenes, Nonhomogeneous Boundary Value Propblems and Applications, vol. I, Springer-Verlag (1972), 357 pp.

[21] A. Pazy, Semigroups of linear operators and applications to partial dfferential equations, 1983, 279 pp.

[22] H. Triebel, Interpolation Theory and Function Spaces, Differential Operators, North-Holland, 1978.

[23] R. Triggiani, *On the stabilizability problem in Banach space*, J. Math. Anal. Appl. 52 (1975), 383–403.

[24] R. Triggiani, *Improving stability properties of hyperbolic damped equations by boundary feedback*, Springer-Verlag Lecture Notes 75 (1985), 400–409.

[25] R. Triggiani, *Regularity of structurally damped systems with point/boundary control*, J. Math. Anal. Appl. 161(2) (1991), 299–331.

Irena Lasiecka
Department of Mathematical Sciences
University of Memphis
Memphis, TN, 38152, USA
e-mail: lasiecka@memphis.edu

and

IBS, Polish Academy of Sciences
ul. Newelska 6, Warsaw, Poland

Roberto Triggiani
Department of Mathematical Sciences
University of Memphis
Memphis, TN, 38152, USA
e-mail: rtrggani@memphis.edu

Operator Theory:
Advances and Applications, Vol. 250, 311–342

General Mazur–Ulam Type Theorems and Some Applications

Lajos Molnár

Dedicated to Professor Charles J.K. Batty on the occasion of his 60th birthday.

Abstract. Recently we have presented several structural results on certain isometries of spaces of positive definite matrices and on those of unitary groups. The aim of this paper is to put those previous results into a common perspective and extend them to the context of operator algebras, namely, to that of von Neumann factors.

Mathematics Subject Classification (2010). Primary: 47B49. Secondary: 46L40.

Keywords. Mazur–Ulam type theorems, generalized distance measures, positive definite cone, unitary group, Jordan triple map, inverted Jordan triple map, operator algebras.

1. Introduction and statement of the results

The famous Mazur–Ulam theorem states that every surjective isometry (i.e., surjective distance preserving map) from a normed real linear space onto another one is automatically affine, in other words, it is necessarily an isomorphism with respect to the operation of convex combinations.

Recently we have extensively investigated how this fundamental theorem can be generalized to more general settings. In [12] we have obtained results in the context of groups (and some of their substructures) which state that under certain conditions surjective distance preserving transformations between such structures necessarily preserve locally the operation of the so-called inverted Jordan product. This means that also in that general setting the surjective isometries necessarily have a particular algebraic property. This property opens the way for employing algebraic ideas, techniques and computations to get more information about the

The author was supported by the "Lendület" Program (LP2012-46/2012) of the Hungarian Academy of Sciences and by the Hungarian Scientific Research Fund (OTKA) Reg.No. K81166 NK81402.

isometries under considerations. In some of our latter papers we have successfully used that approach to describe explicitly the isometries of different non-linear structures of matrices and operators.

In [13] we have determined the surjective isometries of the unitary group over a Hilbert space equipped with the metric of the operator norm. In [21] we described the surjective isometries of the space of all positive definite operators on a Hilbert space relative to the so-called Thompson part metric. In [14] we have presented generalizations of the latter two results for the setting of C^*-algebras. It has turned out that the corresponding surjective isometries are closely related to Jordan *-isomorphisms between the underlying full algebras. In [25], [22] we have proceeded further and described the structure of surjective isometries of the unitary group with respect to complete symmetric norms (see the definition later) both in the infinite- and in the finite-dimensional cases. Furthermore, in [22] we have also determined the isometries relative to the elements of a recently introduced collection of metrics [8] (having connections to quantum information science) on the group of unitary matrices. In [23] we have revealed the structure of surjective isometries of the space of positive definite matrices relative to certain metrics of differential geometric origin (they are common generalizations of the Thompson part metric and the natural Riemannian metric on positive definite matrices) as well as to a new metric obtained from the Jensen–Shannon symmetrization of the important divergence called Stein's loss. In [26] we have made an important step toward further generality. Namely, we have described the structure of those surjective maps on the space of all positive definite matrices which leave invariant a given element of a large collection of certain so-called generalized distance measures. In that way we could present a common generalization of the mentioned results in [23] and also provide structural information on a large class of transformations preserving other particular important distance measures including Stein's loss itself. We also mention that by the help of appropriate modifications in our general results in [12] we have managed to determine the surjective isometries of Grassmann spaces of projections of a fixed rank on a Hilbert space relative to the gap metric [5].

In this paper we develop even further the ideas and approaches we have worked out and used in the papers [22], [23], [26], and extend our previous results concerning distance measure preserving maps on matrix algebras for the case of operator algebras, especially, von Neumann factors. We obtain results which show that if the positive definite cones or the unitary groups in those algebras equipped with a sort of very general distance measures are "isometric", then the underlying full algebras are Jordan *-isomorphic (either *-isomorphic or *-antiisomorphic).

We begin the presentation with the case of positive definite cones. As the starting point of the route leading to our corresponding result we exhibit a Mazur–Ulam type theorem for a certain very general structure called point-reflection geometry equipped with a generalized distance measure. In fact, we believe that with this result we have found in some sense the most general version of the Mazur–Ulam theorem that one can obtain using the approach followed in [12] in the setting of groups. We point out that many of the arguments below use ideas that have already appeared in our previous papers [22], [23], [26]. In several cases only small changes need to be performed while in other cases we really have to work to find solutions for particular problems that emerge from the fact that instead of matrix algebras here we consider much more complicated objects, namely operator algebras. In order to make the material readable we present the results with complete proofs.

For our new general Mazur–Ulam type result we need the following concept that has been defined by Manara and Marchi in [19] (also see [16], [18]).

Definition 1. Let X be a set equipped with a binary operation $\diamond$ which satisfies the following conditions:

(a1) $a \diamond a = a$ holds for every $a \in X$;
(a2) $a \diamond (a \diamond b) = b$ holds for any $a, b \in X$;
(a3) the equation $x \diamond a = b$ has a unique solution $x \in X$ for any given $a, b \in X$.

In this case the pair $(X, \diamond)$ (or X itself) is called a point-reflection geometry.

Observe that from the property (a1) above we easily obtain that the equation $a \diamond x = b$ also has unique solution $x \in X$ for any given $a, b \in X$.

As for our present purposes, the most important example of such a structure is given as follows. In the rest of the paper by a C^*-algebra we always mean a unital C^*-algebra with unit I. Let $\mathcal{A}$ be such an algebra. We denote by $\mathcal{A}_s$ the self-adjoint part of $\mathcal{A}$ and $\mathcal{A}_+$ stands for the cone of all positive elements of $\mathcal{A}$ (self-adjoint elements with non-negative spectrum). The set of all invertible elements in $\mathcal{A}_+$ is denoted by $\mathcal{A}_+^{-1}$. Sometimes $\mathcal{A}_+^{-1}$ is called positive definite cone and its elements are said positive definite. For any $A, B \in \mathcal{A}_+^{-1}$ define $A \diamond B = AB^{-1}A$. In that way $\mathcal{A}_+^{-1}$ becomes a point-reflection geometry. Indeed, the conditions (a1), (a2) above are trivial to check. Concerning (a3) we recall that for any given $A, B \in \mathcal{A}_+^{-1}$, the so-called Ricatti equation $XA^{-1}X = B$ has a unique solution $X = A\#B$ which is just the geometric mean of A and B defined by

$$A\#B = A^{1/2}(A^{-1/2}BA^{-1/2})^{1/2}A^{1/2}.$$

This assertion is usually called the Anderson–Trapp theorem (for the original source see [1]).

In our general Mazur–Ulam type theorem that we are going to present we do not need to confine the considerations to true metrics, the theorem works also for so-called generalized distance measures.

Definition 2. Given an arbitrary set X, the function $d : X \times X \to [0, \infty[$ is called a generalized distance measure if it has the property that for an arbitrary pair $x, y \in X$ we have $d(x, y) = 0$ if and only if $x = y$.

Hence, in the definition above we require only the definiteness property of a metric but neither the symmetry nor the triangle inequality is assumed. Our new general Mazur–Ulam type theorem reads as follows.

Theorem 3. *Let X, Y be sets equipped with binary operations $\diamond, \star$, respectively, with which they form point-reflection geometries. Let $d : X \times X \to [0, \infty[$, $\rho : Y \times Y \to [0, \infty[$ be generalized distance measures. Pick $a, b \in X$, set*

$$L_{a,b} = \{x \in X \,:\, d(a, x) = d(x, b \diamond a) = d(a, b)\}$$

and assume the following:

(b1) *$d(b \diamond x, b \diamond x') = d(x', x)$ holds for all $x, x' \in X$;*
(b2) *$\sup\{d(x, b) \,:\, x \in L_{a,b}\} < \infty$;*
(b3) *there exists a constant $K > 1$ such that $d(x, b \diamond x) \geq K d(x, b)$ holds for every $x \in L_{a,b}$.*

Let $\phi : X \to Y$ be a surjective map such that

$$\rho(\phi(x), \phi(x')) = d(x, x'), \quad x, x' \in X$$

and also assume that

(b4) *for the element $c \in Y$ with $c \star \phi(a) = \phi(b \diamond a)$ we have $\rho(c \star y, c \star y') = \rho(y', y)$ for all $y, y' \in Y$.*

Then we have

$$\phi(b \diamond a) = \phi(b) \star \phi(a).$$

The maps ϕ appearing in the theorem may be called "generalized isometries". Moreover, observe that the above result trivially includes the original Mazur–Ulam theorem. To see this, take normed real linear spaces X, Y and a surjective isometry $\phi : X \to Y$. Define the operation $\diamond$ by $x \diamond x' = 2x - x'$, $x, x' \in X$ and the operation $\star$ similarly. Let d, ρ be the metrics corresponding to the norms on X and Y. Selecting any pair a, b of points in X, it is apparent that all conditions in the theorem are fulfilled and hence we have $\phi(2b - a) = 2\phi(b) - \phi(a)$. It easily implies that ϕ respects the operation of the arithmetic mean from which it follows that ϕ respects all dyadic convex combinations and finally, by the continuity of ϕ, we conclude that ϕ is affine.

The above result shows that maps which conserve the "distances" with respect to a pair of generalized distance measures respect a pair of algebraic operations in some sense. We emphasize that in the result above as well as in our other general Mazur–Ulam type results that appeared in [12], the isometries respect or, in other words, preserve algebraic operations only locally, for certain pairs a, b of elements. In fact, in that generality nothing more can be expected. To see this, one may refer to groups equipped with the discrete metrics: any bijection between them is a surjective isometry but clearly not necessarily an isomorphism in any adequate sense. Nonetheless, even if only locally, surjective distance measure preserver transformations appearing in the above theorem do have a certain algebraic property. And in the cases that we consider in the present paper it turns out that they in fact have this property globally. Therefore, the problem of describing those distance measure preserver transformations can be transformed to the problem of describing certain algebraic isomorphisms. This is exactly the strategy we are going to follow below.

Let us proceed toward the first group of our results which concern transformations between the positive definite cones of C^*-algebras. Before presenting the results we need to make some preparations. By a symmetric norm on a C^*-algebra $\mathcal{A}$ we mean a norm N for which $N(AXB) \leq \|A\|N(X)\|B\|$ holds for all $A, X, B \in \mathcal{A}$. Here and in what follows $\|.\|$ stands for the original norm on $\mathcal{A}$ which we sometimes call operator norm. Whenever we speak about topological properties (convergence, continuity, etc.) without specifying the topology we always mean the norm topology of $\|.\|$. We call a norm N on $\mathcal{A}$ unitarily invariant if $N(UAV) = N(A)$ holds for all $A \in \mathcal{A}$ and unitary $U, V \in \mathcal{A}$. Furthermore, a norm N on $\mathcal{A}$ is said to be unitary similarity invariant if we have $N(UAU^*) = N(A)$ for all $A \in \mathcal{A}$ and unitary $U \in \mathcal{A}$. It is easy to see that any symmetric norm is unitarily invariant and it is trivial that every unitarily invariant norm is unitary similarity invariant. For several examples of complete symmetric norms on $B(H)$, the algebra of all bounded linear operators on a complex Hilbert space H, we refer to [7]. They include the so-called (c,p)-norms and, in particular, the Ky Fan k-norms. We mention that in that paper the authors use the expression "uniform norm" for symmetric norms. Apparently, the above examples provide examples of complete symmetric norms on any C^*-subalgebra of $B(H)$ and hence on von Neumann algebras, too.

Now, we recall that in [23] we have described the structure of isometries of the space $\mathbb{P}_n$ of all positive definite $n \times n$ complex matrices with respect to the metric defined by

$$d_N(A, B) = N(\log A^{-1/2}BA^{-1/2}), \quad A, B \in \mathbb{P}_n, \tag{1.1}$$

where N is a unitarily invariant norm on the space $\mathbb{M}_n$ of all $n \times n$ complex matrices. (It is a well-known fact that on matrix algebras a norm is symmetric if and only if it is unitarily invariant, see Proposition IV.2.4 in [3].) The importance of that metric comes from its differential geometric background (it is a shortest path distance in a Finsler-type structure on $\mathbb{P}_n$ which generalizes its fundamental natural Riemann structure, for references see [23]). In the recent paper [26] we have presented a substantial extension of that result for the case where the logarithmic function in (1.1) is replaced by any continuous function $f\colon]0,\infty[\to \mathbb{R}$ that satisfies

(c1) $f(y) = 0$ holds if and only if $y = 1$;
(c2) there exists a number $K > 1$ such that

$$|f(y^2)| \geq K|f(y)|, \qquad y \in]0,\infty[.$$

We must point out that with this replacement we usually get not a true metric, only a generalized distance measure. However in that way we cover the cases of many important concepts of matrix divergences whose preserver transformations could hence be explicitly described, for details see [26].

We now define that new class of generalized distance measures in the context of C^*-algebras. Let $\mathcal{A}$ be a C^*-algebra, N a norm on $\mathcal{A}$, $f\colon]0,\infty[\to \mathbb{R}$ a given continuous function with property (c1). Define $d_{N,f} : \mathcal{A}_+^{-1} \times \mathcal{A}_+^{-1} \to [0, \infty[$ by

$$d_{N,f}(A, B) = N(f(A^{1/2}B^{-1}A^{1/2})), \qquad A, B \in \mathcal{A}_+^{-1}. \tag{1.2}$$

It is apparent that $d_{N,f}$ is a generalized distance measure.

We also need the following notions. If $\mathcal{A}$ is a C^*-algebra and $A, B \in \mathcal{A}$, then ABA is called the Jordan triple product of A and B while $AB^{-1}A$ is said to be their inverted Jordan triple product. If $\mathcal{B}$ is another C^*-algebra and $\phi : \mathcal{A}_+^{-1} \to \mathcal{B}_+^{-1}$ is a map which satisfies

$$\phi(ABA) = \phi(A)\phi(B)\phi(A), \qquad A, B \in \mathcal{A}_+^{-1},$$

then it is called a Jordan triple map. If $\phi : \mathcal{A}_+^{-1} \to \mathcal{B}_+^{-1}$ fulfills

$$\phi(AB^{-1}A) = \phi(A)\phi(B)^{-1}\phi(A), \qquad A, B \in \mathcal{A}_+^{-1},$$

then ϕ is said to be an inverted Jordan triple map. A bijective Jordan triple map is called a Jordan triple isomorphism and a bijective inverted Jordan triple map is said to be an inverted Jordan triple isomorphism.

Applying Theorem 3 we shall prove the following result.

Theorem 4. *Let $\mathcal{A}, \mathcal{B}$ be C^*-algebras with complete symmetric norms N, M, respectively. Assume N satisfies $N(|A|) = N(A)$ for all $A \in \mathcal{A}$. Suppose $f, g\colon]0, \infty[\to \mathbb{R}$ are continuous functions both satisfying* (c1) *and f also fulfilling* (c2). *Let $\phi : \mathcal{A}_+^{-1} \to \mathcal{B}_+^{-1}$ be a surjective map which respects the pair $d_{N,f}, d_{M,g}$ of generalized distance measures in the sense that*

$$d_{M,g}(\phi(A), \phi(B)) = d_{N,f}(A, B), \qquad A, B \in \mathcal{A}_+^{-1}.$$

Then ϕ is a continuous inverted Jordan triple isomorphism, i.e., a continuous bijective map that satisfies

$$\phi(AB^{-1}A) = \phi(A)\phi(B)^{-1}\phi(A), \qquad A, B \in \mathcal{A}_+^{-1}.$$

Having this result, the next natural step is to try to describe the structure of all continuous inverted Jordan triple isomorphisms between positive definite cones. This is exactly what we do. Observe that the inverted Jordan triple isomorphisms are closely related to Jordan triple isomorphisms which are much more common, they appear, e.g., in pure ring theory, too (though there they are considered between full rings and usually assumed to be additive which is definitely not the case here). Indeed, if $\phi : \mathcal{A}_+^{-1} \to \mathcal{B}_+^{-1}$ is an inverted Jordan triple isomorphism, then elementary computation shows that the transformation $\psi(.) = \phi(I)^{-1/2}\phi(.)\phi(I)^{-1/2}$ is a unital inverted Jordan triple isomorphism which can easily be seen to be a Jordan triple isomorphism. Recall a map is called unital if it sends the identity to the identity.

In the following theorem we describe the structure of continuous Jordan triple isomorphisms between the positive definite cones of von Neumann factors. A linear functional $l : \mathcal{A} \to \mathbb{C}$ on an algebra $\mathcal{A}$ is said to be tracial if it has the property $l(AB) = l(BA)$, $A, B \in \mathcal{A}$.

Theorem 5. *Assume $\mathcal{A}, \mathcal{B}$ are von Neumann algebras and $\mathcal{A}$ is a factor not of type I_2. Let $\phi : \mathcal{A}_+^{-1} \to \mathcal{B}_+^{-1}$ be a continuous Jordan triple isomorphism. Then there is either an algebra *-isomorphism or an algebra *-antiisomorphism $\theta : \mathcal{A} \to \mathcal{B}$, a number $c \in \{-1, 1\}$, and a continuous tracial linear functional $l : \mathcal{A} \to \mathbb{C}$ which is real valued on $\mathcal{A}_s$ and $l(I) \neq -c$ such that*

$$\phi(A) = e^{l(\log A)}\theta(A^c), \qquad A \in \mathcal{A}_+^{-1}. \tag{1.3}$$

*Conversely, for any algebra *-isomorphism or algebra *-antiisomorphism $\theta : \mathcal{A} \to \mathcal{B}$, number $c \in \{-1, 1\}$, and continuous tracial linear functional $l : \mathcal{A} \to \mathbb{C}$ which is real valued on $\mathcal{A}_s$ and $l(I) \neq -c$, the above displayed formula* (1.3) *defines a continuous Jordan triple isomorphism between $\mathcal{A}_+^{-1}$ and $\mathcal{B}_+^{-1}$.*

As for the tracial linear functional l appearing above we mention the following. It is proven in [11] that in a properly infinite von Neumann algebra, every element is the sum of two commutators. This gives us that if $\mathcal{A}$ in the theorem is of one of the types I_∞, II_∞, III, then the functional l above vanishes. However, if $\mathcal{A}$ is of type I_n or type II_1, then due to the existence of a normalized trace, it really shows up.

After this we shall easily obtain our theorem on the structure of surjective maps between the positive definite cones of von Neumann factors which respect pairs of generalized distance measures. The statement reads as follows.

Theorem 6. *Let $\mathcal{A}, \mathcal{B}$ be von Neumann algebras with complete symmetric norms N, M, respectively. Assume $f, g \colon]0, \infty[\to \mathbb{R}$ are continuous functions both satisfying* (c1) *and f also fulfilling* (c2). *Suppose that $\mathcal{A}$ is a factor not of type I_2. Let $\phi : \mathcal{A}_+^{-1} \to \mathcal{B}_+^{-1}$ be a surjective map which respects the pair $d_{N,f}, d_{M,g}$ of generalized distance measures in the sense that*

$$d_{M,g}(\phi(A), \phi(B)) = d_{N,f}(A, B), \qquad A, B \in \mathcal{A}_+^{-1}. \tag{1.4}$$

*Then there is either an algebra *-isomorphism or an algebra *-antiisomorphism $\theta : \mathcal{A} \to \mathcal{B}$, a number $c \in \{-1, 1\}$, an element $T \in \mathcal{B}_+^{-1}$ and a continuous tracial linear functional $l : \mathcal{A} \to \mathbb{C}$ which is real valued on $\mathcal{A}_s$ and $l(I) \neq -c$ such that*

$$\phi(A) = e^{l(\log A)}T\theta(A^c)T, \qquad A \in \mathcal{A}_+^{-1}. \tag{1.5}$$

In the case where $\mathcal{A}$ is an infinite factor, the linear functional l is in fact missing.

Let us emphasize the interesting consequence of the above theorem that if the positive definite cones of two von Neumann factors (not of type I_2) are "isometric" in a very general sense (with respect a pair of generalized distance measures), then the underlying algebras are necessarily isomorphic or antiisomorphic as algebras.

The second part of our results concerns transformations between unitary groups. For any C^*-algebra $\mathcal{A}$, we denote its unitary group by $\mathcal{A}_u$. Similarly to the case of the positive definite cone, we are going to consider certain generalized distance measures on $\mathcal{A}_u$. Let N be a norm on $\mathcal{A}$ and $f : \mathbb{T} \to \mathbb{C}$ a continuous function having zero exactly at 1 and define

$$d_{N,f}(U, V) = N(f(UV^{-1})), \quad U, V \in \mathcal{A}_u. \tag{1.6}$$

Clearly, $d_{N,f}$ is a generalized distance measure on $\mathcal{A}_u$. In particular, when $f(z) = z-1$, $z \in \mathbb{T}$, and N is unitarily invariant, we obtain $d_{N,f}(U,V) = N(U-V)$, $U,V \in \mathcal{A}_u$, i.e., the usual norm distance with respect to N.

In paper [22] we have considered a recently defined collection of metrics on the group of $n\times n$ unitary matrices. To the definition we recall that for any unitary matrix U we have a unique Hermitian matrix H with spectrum in $]-\pi,\pi]$ such that $U=\exp(iH)$. This H is called the angular matrix of U. Now, for a given unitarily invariant norm N on the algebra $\mathbb{M}_n$ of all $n\times n$ complex matrices the distance $d_N(U,V)$ between unitary matrices U and V is defined by $d_N(U,V) = N(H)$, where H is the angular matrix of UV^{-1}. These metrics have been introduced and studied in [8] and the corresponding isometries have been determined in Theorem 4 in [22]. Observe that these metrics "almost" fit into the general framework we have presented in (1.6) above. Indeed, there f should be the argument function on $\mathbb{T}$, but the problem is that this function is not continuous. Apparently, it did not cause any problem in [22] since there we have considered a finite-dimensional setting where the spectrum is finite and on such a set all functions can be viewed continuous.

As in the case of the positive definite cone, in order to obtain reasonable structural results on transformations between unitary groups that respect generalized distance measures, we need to require certain conditions on the generating continuous function $f:\mathbb{T}\to\mathbb{C}$. These are the following:

(d1) $f(y)=0$ holds if and only if $y=1$;
(d2) there exists a number $K>1$ such that

$$|f(y^2)| \geq K|f(y)|$$

holds for all $y\in\mathbb{T}$ from a neighborhood of 1.

One may ask what is the reason for the locality in (d2). The trivial answer is that we want to cover the case of the function $f(z) = z-1$, $z\in\mathbb{T}$ which corresponds to the usual norm distance.

Our result parallel to Theorem 4 which states that the "generalized isometries" between unitary groups of von Neumann algebras are continuous inverted Jordan triple isomorphisms is formulated below. To this we note that having a look at the concepts relating to maps on the positive definite cone which are given before Theorem 4, the notions of Jordan triple maps, inverted Jordan triple maps, Jordan triple isomorphisms, inverted Jordan triple isomorphisms between unitary groups should be self-explanatory.

Theorem 7. *Let $\mathcal{A},\mathcal{B}$ be von Neumann algebras with complete symmetric norms N,M, respectively. Assume $f,g:\mathbb{T}\to\mathbb{C}$ are continuous functions having the property* (d1) *and f also satisfies* (d2). *Let $\phi:\mathcal{A}_u\to\mathcal{B}_u$ be a surjective map that respects the pair $d_{N,f}, d_{M,g}$ of generalized distance measures in the sense that*

$$d_{M,g}(\phi(U),\phi(V)) = d_{N,f}(U,V), \qquad U,V\in\mathcal{A}_u. \tag{1.7}$$

Then ϕ is a continuous inverted Jordan triple isomorphism, i.e., a continuous bijective map which satisfies

$$\phi(UV^{-1}U) = \phi(U)\phi(V)^{-1}\phi(U), \qquad U, V \in \mathcal{A}_u.$$

Just as in the first part of this section, one can easily see that inverted Jordan triple maps between unitary groups are closely related to Jordan triple maps. In particular, if $\phi : \mathcal{A}_u \to \mathcal{B}_u$ is an inverted Jordan triple isomorphism, then it is apparent that the map $\psi(.) = \phi(I)^{-1}\phi(.)$ is again an inverted Jordan triple isomorphism which is unital and hence it is a Jordan triple isomorphism. The following theorem describes the structure of continuous such isomorphisms in the case of von Neumann factors. In the course of its proof we employ an argument involving one-parameter unitary groups.

Theorem 8. *Assume $\mathcal{A}, \mathcal{B}$ are von Neumann algebras $\mathcal{A}$ is a factor. Let $\phi : \mathcal{A}_u \to \mathcal{B}_u$ be a continuous Jordan triple isomorphism, i.e., a continuous bijective map which satisfies*

$$\phi(UVU) = \phi(U)\phi(V)\phi(U), \qquad U, V \in \mathcal{A}_u.$$

*Then there is either an algebra *-isomorphism or an algebra *-antiisomorphism $\theta : \mathcal{A} \to \mathcal{B}$ and scalars $c, d \in \{-1, 1\}$ such that*

$$\phi(A) = d\theta(A^c), \qquad A \in \mathcal{A}_u.$$

Combining the above two results we can readily obtain our theorem on generalized distance measure preserving maps between unitary groups of von Neumann factors which reads as follows.

Theorem 9. *Let $\mathcal{A}, \mathcal{B}$ be von Neumann algebras with complete symmetric norms N, M, respectively. Suppose that $\mathcal{A}$ is a factor. Assume $f, g : \mathbb{T} \to \mathbb{C}$ are continuous functions having the property* (d1) *and f also satisfies* (d2). *Let $\phi : \mathcal{A}_u \to \mathcal{B}_u$ be a surjective map that respects the pair $d_{N,f}, d_{M,g}$ of generalized distance measures in the sense that*

$$d_{M,g}(\phi(U), \phi(V)) = d_{N,f}(U, V), \qquad U, V \in \mathcal{A}_u. \tag{1.8}$$

*Then there is either an algebra *-isomorphism or an algebra *-antiisomorphism $\theta : \mathcal{A} \to \mathcal{B}$, a unitary element $W \in \mathcal{B}_u$ and a number $c \in \{-1, 1\}$ such that*

$$\phi(U) = W\theta(U^c), \qquad A \in \mathcal{A}_u.$$

Just as in the case of the positive definite cone, we point out the interesting consequence of the above theorem that if the unitary groups of two von Neumann factors are "isometric" in a very general sense (with respect to some pair of generalized distance measures), then the underlying algebras are necessarily isomorphic or antiisomorphic as algebras.

2. Proofs

This section is devoted to the proofs of our results.

We begin with the proof of our new general Mazur–Ulam type theorem, Theorem 3. In fact, the argument we use here follows closely the ideas given in the proofs of Proposition 9, Lemma 10 and Proposition 11 in [26] which statements have been formulated in the context of so-called twisted subgroups of groups. The main novelty here is that we have found in a sense the most general structure (point-reflection geometry) for which that argument can be employed.

The first step toward the proof of Theorem 3 is the following lemma that appeared in [26] as Lemma 8. The proof is so short that for the sake of completeness we repeat it here.

Lemma 10. *Let X be a set and $d : X \times X \to [0, \infty[$ an arbitrary function. Assume $\varphi : X \to X$ is a bijective map satisfying*

$$d(\varphi(x), \varphi(x')) = d(x', x), \qquad x, x' \in X. \tag{2.1}$$

Assume further that we have $b \in X$ for which $\sup\{d(x, b) | x \in X\} < \infty$, and there is a constant $K > 1$ such that

$$d(x, \varphi(x)) \geq K d(x, b), \qquad x \in X.$$

Then for every bijective map $f : X \to X$ satisfying

$$d(f(x), f(x')) = d(x', x), \qquad x, x' \in X \tag{2.2}$$

we have $d(f(b), b) = 0$.

Proof. For temporary use we call a map $f : X \to X$ d-reversing if it satisfies (2.2). Let

$$\lambda = \sup\{d(f(b), b) | f : X \to X \text{ is a bijective } d\text{-reversing map}\}.$$

Then $0 \leq \lambda < \infty$. For an arbitrary bijective d-reversing map $f : X \to X$, consider $\tilde{f} = f^{-1} \circ \varphi \circ f$. Then $\tilde{f}$ is also a bijective d-reversing transformation and

$$\lambda \geq d(\tilde{f}(b), b) = d(f(b), \varphi(f(b))) \geq K d(f(b), b).$$

By the definition of λ we get $\lambda \geq K\lambda$ which implies that $\lambda = 0$ and this completes the proof. □

The next proposition in the case where $X = Y$, $d = \rho$ appeared as Proposition 9 in [26].

Proposition 11. *Let X be a set and $d : X \times X \to [0, \infty[$ any function. Let $a, b \in X$ and assume that $\varphi : X \to X$ is a bijective map which satisfies (2.1). Moreover, assume that $\varphi(b) = b$ and the composition map $\varphi \circ \varphi$ equals the identity on X. Set*

$$L = \{x \in X | d(a, x) = d(x, \varphi(a)) = d(a, b)\}.$$

Suppose that $\sup\{d(x, b) | x \in L\} < \infty$ and there exists a constant $K > 1$ such that

$$d(x, \varphi(x)) \geq K d(x, b), \qquad x \in L.$$

Let Y be another set and $\rho : Y \times Y \to [0, \infty[$ any function. Assume $\psi \colon Y \to Y$ is a bijective map such that

$$\rho(\psi(y), \psi(y')) = d(y', y), \qquad y, y' \in Y.$$

If $T \colon X \to Y$ is a bijective map satisfying

$$\rho(T(x), T(x')) = d(x, x'), \qquad x, x' \in X, \tag{2.3}$$

and

$$\psi(T(a)) = T(\varphi(a)), \quad \psi(T(\varphi(a))) = T(a), \tag{2.4}$$

then we have

$$\rho(\psi(T(b)), T(b)) = 0.$$

Proof. Since $\varphi(b) = b$ and φ satisfies (2.1), we have

$$d(a, b) = d(\varphi(b), \varphi(a)) = d(b, \varphi(a)),$$

which implies that $b \in L$. Let

$$L' = \{y \in Y | \rho(T(a), y) = \rho(y, T(\varphi(a))) = d(a, b)\}.$$

By the bijectivity and the property (2.3) of T one can easily check that $T(L) = L'$. Furthermore, using corresponding properties of the maps φ, ψ as well as the intertwining properties (2.4), we obtain that $\varphi(L) = L$ and $\psi(L') = L'$. Consider now the transformation $\tilde{T} = T^{-1} \circ \psi \circ T$. Plainly, the restriction of this map onto L is a self-bijection of L and it satisfies

$$\rho(\tilde{T}(x), \tilde{T}(x')) = d(x', x), \qquad x, x' \in L.$$

Since $\sup\{d(x, b) | x \in L\} < \infty$, we can apply Lemma 10 and deduce that

$$0 = d(\tilde{T}(b), b) = d(\psi(T(b)), T(b)).$$

□

After this preparation we can present the proof of our general Mazur–Ulam type result.

Proof of Theorem 3. First observe that by the definiteness of generalized distance measures the surjective "generalized isometry" ϕ is also injective. Let $\varphi(x) = b \diamond x$ for every $x \in X$ and define $\psi \colon Y \to Y$ by $\psi(y) = c \star y$, $y \in Y$. By the properties of point-reflection geometries and the assumptions in the theorem, φ is a bijective map on X and ψ is a bijective map on Y, moreover all conditions appearing in Proposition 11 are easily seen to be satisfied with ϕ in the place of T. In fact, we obviously have $\psi(\phi(a)) = \phi(\varphi(a))$ which, by taking into account that ψ is an involution, implies that $\psi(\phi(\varphi(a))) = \phi(a)$. Applying Proposition 11 we get that $\rho(\psi(\phi(b)), \phi(b)) = 0$ which implies $\phi(b) = \psi(\phi(b)) = c \star \phi(b)$. By the properties (a1), (a3) of point-reflection geometries we infer that $c = \phi(b)$ implying

$$\phi(b \diamond a) = \phi(b) \star \phi(a).$$

□

In the next proposition on which the proof of Theorem 4 relies we shall need the following lemma about the monotonicity of symmetric norms. We believe its content is well known but we could not find it in the literature. Therefore, we present it with a short proof.

Lemma 12. *Let N be a symmetric norm on a C^*-algebra $\mathcal{A}$ and assume $A, B \in \mathcal{A}_+$ are such that $A \leq B$. Then we have $N(A) \leq N(B)$.*

Proof. Suppose first that B is invertible. Let D be the geometric mean of B^{-1} and A, i.e.,

$$D = B^{-1} \# A = B^{-1/2}(B^{1/2}AB^{1/2})^{1/2}B^{-1/2}.$$

Clearly, we have $A = DBD$ and observe that $D \leq I$ (this can be proven directly using the operator monotonicity of the square-root function, or referring to the monotonicity property of general operator means of Kubo–Ando sense). It follows that

$$N(A) = N(DBD) \leq \|D\|N(B)\|D\| \leq N(B).$$

For non-invertible B, consider $B + \epsilon I$ for positive numbers ϵ tending to 0. □

Before presenting the next result we point out the easy fact that for any function $f :]0, \infty[\to \mathbb{R}$ with the properties (c1) and (c2) we necessarily have

$$\lim_{y \to 0} |f(y)| = \lim_{y \to \infty} |f(y)| = \infty.$$

We also remark the following. If f is a continuous scalar-valued function on the set of positive real numbers, then for any $A \in \mathcal{A}_+^{-1}$ and unitary $U \in \mathcal{A}$ we have $f(UAU^*) = Uf(A)U^*$. This is obviously true if f is a polynomial and then one can refer to the fact that any continuous function on a compact interval can be uniformly approximated by polynomials to obtain the general statement. For any unitary similarity invariant norm N on $\mathcal{A}$, it follows that $N(f(UAU^*)) = N(f(A))$ holds for all $A \in \mathcal{A}_+^{-1}$ and unitary $U \in \mathcal{A}$.

Finally, we admit that if N is a complete symmetric norm on the C^*-algebra $\mathcal{A}$, then it is necessarily equivalent to the operator norm. Indeed, we have $N(A) \leq N(I)\|A\|$, $A \in \mathcal{A}$ and then the equivalence follows from the completeness of N and $\|.\|$.

We are now in a position to prove the following proposition.

Proposition 13. *Let $\mathcal{A}$ be a C^*-algebra with complete symmetric norm N such that $N(|A|) = N(A)$ holds for all $A \in \mathcal{A}$. Assume $f \colon]0, \infty[\to \mathbb{R}$ is a continuous function satisfying* (c1), (c2). *Consider the standard point-reflection geometry operation on $\mathcal{A}_+^{-1}$, i.e., let $A \diamond B = AB^{-1}A$, $A, B \in \mathcal{A}_+^{-1}$. Define $d_{N,f}$ as in* (1.2). *Then for the structure $\mathcal{A}_+^{-1}$ equipped with this operation $\diamond$ and generalized distance measure $d_{N,f}$ the assumptions* (b1)–(b3) *in Theorem 3 are satisfied for every pair $A, B \in \mathcal{A}_+^{-1}$. Moreover, for a sequence (X_n) in $\mathcal{A}_+^{-1}$ and element $X \in \mathcal{A}_+^{-1}$ we have $X_n \to X$ in the operator norm topology if and only if $d_{N,f}(X, X_n) \to 0$.*

Proof. Essentially, we follow the argument given in the proof of Theorem 1 in [26].

Pick $A, B \in \mathcal{A}_+^{-1}$ and consider the polar decomposition

$$B^{-1/2}A^{1/2} = U|B^{-1/2}A^{1/2}|.$$

Observe that by the invertibility of A, B such a unitary $U \in \mathcal{A}$ does exist. We see that

$$B^{-1/2}AB^{-1/2} = U|B^{-1/2}A^{1/2}|^2U^* = UA^{1/2}B^{-1}A^{1/2}U^*. \tag{2.5}$$

Now, take an arbitrary invertible element $T \in \mathcal{A}$. Set

$$X = A^{-1/2}BT^*(TAT^*)^{-1/2}.$$

We deduce

$$XX^* = A^{-1/2}BA^{-1}BA^{-1/2} = (A^{-1/2}BA^{-1/2})^2.$$

Consider the polar decomposition $X = V|X|$, $V \in \mathcal{A}$ being unitary. We compute

$$\begin{aligned}
&(TAT^*)^{-1/2}TBT^*(TAT^*)^{-1/2}\\
&= ((TAT^*)^{-1/2}TBT^*(TAT^*)^{-1}TBT^*(TAT^*)^{-1/2})^{1/2}\\
&= ((TAT^*)^{-1/2}TBA^{-1}BT^*(TAT^*)^{-1/2})^{1/2}\\
&= (X^*X)^{1/2} = |X| = V^*|X^*|V = V^*(A^{-1/2}BA^{-1/2})V.
\end{aligned} \tag{2.6}$$

Recall that the generalized distance measure $d_{N,f}$ is defined by

$$d_{N,f}(A, B) = N(f(A^{1/2}B^{-1}A^{1/2})), \quad A, B \in \mathcal{A}_+^{-1}.$$

Pick $B, X, Y \in \mathcal{A}_+^{-1}$. It follows from the content of (2.6) that for some unitary $V \in \mathcal{A}$ we have

$$\begin{aligned}
(B \diamond X)^{1/2}(B \diamond Y)^{-1}(B \diamond X)^{1/2} &= (BX^{-1}B)^{1/2}(BY^{-1}B)^{-1}(BX^{-1}B)^{1/2}\\
&= (B^{-1}XB^{-1})^{-1/2}(B^{-1}YB^{-1})(B^{-1}XB^{-1})^{-1/2} = V(X^{-1/2}YX^{-1/2})V^*.
\end{aligned}$$

By (2.5), $X^{-1/2}YX^{-1/2}$ is unitarily similar to $Y^{1/2}X^{-1}Y^{1/2}$ and hence we obtain the unitary similarity of $(B \diamond X)^{1/2}(B \diamond Y)^{-1}(B \diamond X)^{1/2}$ to $Y^{1/2}X^{-1}Y^{1/2}$. As we have mentioned above $N(f(UAU^*)) = N(f(A))$ holds for all $A \in \mathcal{A}_+^{-1}$ and unitary $U \in \mathcal{A}$. These observations imply that

$$d_{N,f}(B \diamond X, B \diamond Y) = d_{N,f}(Y, X)$$

holds for any $B, X, Y \in \mathcal{A}_+^{-1}$. Hence condition (b1) in Theorem 3 is fulfilled.

As for condition (b2), let us consider the set $\mathcal{H}$ of those elements $X \in \mathcal{A}_+^{-1}$ for which we have

$$\begin{aligned}
d_{N,f}(A, X) &= N(f(A^{1/2}X^{-1}A^{1/2}))\\
&= N(f(A^{1/2}B^{-1}A^{1/2})) = d_{N,f}(A, B).
\end{aligned}$$

(With the notation of Theorem 3 we clearly have $L_{A,B} \subset \mathcal{H}$.) We show that the corresponding set of numbers

$$d_{N,f}(X, B) = N(f(X^{1/2}B^{-1}X^{1/2})) = N(f(B^{-1/2}XB^{-1/2}))$$

is bounded. Indeed, since $N(f(A^{1/2}X^{-1}A^{1/2}))$ is constant on $\mathcal{H}$ and N is equivalent to the operator norm $||.||$, the set

$$\{\|f(A^{1/2}X^{-1}A^{1/2})\| : X \in \mathcal{H}\}$$

is bounded. We have already mentioned that $|f(y)| \to \infty$ as $y \to 0$ or $y \to \infty$. It follows easily that there are positive numbers m, M such that $mI \leq A^{1/2}X^{-1}A^{1/2} \leq MI$ holds for all $X \in \mathcal{H}$. Clearly, we then have another pair m', M' of positive numbers such that $m'I \leq X \leq M'I$ and finally another one m'', M'' such that $m''I \leq B^{-1/2}XB^{-1/2} \leq M''I$ holds for all $X \in \mathcal{H}$. Equivalently, $m''I \leq X^{1/2}B^{-1}X^{1/2} \leq M''I$ holds for each $X \in \mathcal{H}$. By continuity, f is bounded on the interval $[m'', M'']$ and this implies that the set

$$\{N(f(X^{1/2}B^{-1}X^{1/2})) : X \in \mathcal{H}\}$$

is bounded. We conclude that condition (b2) is also fulfilled.

Concerning condition (b3) we first note that, by Lemma 12,

$$N(f(C^2)) = N(|f(C^2)|) \geq KN(|f(C)|) = KN(f(C))$$

holds for every $C \in \mathcal{A}_+^{-1}$.

Now, selecting any $X \in \mathcal{A}_+^{-1}$ and setting $Y = X^{1/2}B^{-1}X^{1/2}$ we easily deduce that

$$\begin{aligned} d_{N,f}(X, B \diamond X) &= N(f(X^{1/2}(BX^{-1}B)^{-1}X^{1/2})) \\ &= N(f(X^{1/2}B^{-1}XB^{-1}X^{1/2})) = N(f(Y^2)) \geq KN(f(Y)) \\ &= KN(f(X^{1/2}B^{-1}X^{1/2})) = Kd_{N,f}(X, B). \end{aligned}$$

This means that condition (b3) is also satisfied. Therefore, all assumptions (b1)–(b3) are fulfilled for any pair $A, B \in \mathcal{A}_+^{-1}$.

Let us show now that for the sequence (X_n) in $\mathcal{A}_+^{-1}$ and element $X \in \mathcal{A}_+^{-1}$ we have the convergence $X_n \to X$ in the operator norm topology if and only if $d_{N,f}(X, X_n) \to 0$. To see this, assume $X_n \to X$ in the operator norm topology. Then $X^{1/2}X_n^{-1}X^{1/2} \to I$ which implies $f(X^{1/2}X_n^{-1}X^{1/2}) \to f(I) = 0$. By the equivalence of N to the operator norm, we obtain

$$d_{N,f}(X, X_n) = N(f(X^{1/2}X_n^{-1}X^{1/2})) \to 0.$$

Conversely, if the above convergence holds, then we have

$$f(X^{1/2}X_n^{-1}X^{1/2}) \to 0$$

in the operator norm. By the continuity of f and the property (c1), it is easy to verify that we necessarily have

$$X^{1/2}X_n^{-1}X^{1/2} \to I,$$

which implies that $X_n \to X$ in the operator norm. □

Remark 14. In the above proposition we have supposed that the complete symmetric norm N on $\mathcal{A}$ satisfies $N(|A|) = N(A)$ for all $A \in \mathcal{A}$. We do not know if it is really necessary to assume this or any complete symmetric norm automatically has this property. Nevertheless we suspect a negative answer.

Proof of Theorem 4. By Proposition 13 the conditions (b1)–(b3) are satisfied in Theorem 3 for all $A, B \in \mathcal{A}_+^{-1}$. The argument used there to verify (b1) shows that (b4) is fulfilled, too. Therefore, the conclusion in Theorem 3 holds for any pair $A, B \in \mathcal{A}_+^{-1}$ which gives us that the transformation ϕ is an inverted Jordan triple isomorphism. Its continuity follows from the last statement in Proposition 13 which is valid for the generalized distance measure $d_{M,g}$, too. □

In accordance with our original plan, the next step we make is to prove Theorem 5 which describes the structure of continuous Jordan triple isomorphisms between the positive definite cones of von Neumann factors. Our idea of how to do it comes from the paper [23]. Namely, we first verify that any such map is automatically Lipschitz in a small neighborhood of the identity I. This is the content of the next lemma. Its proof relies on some appropriate modifications in the proof of Lemma 5 in [23].

Lemma 15. *Let $\mathcal{A}, \mathcal{B}$ be C^*-algebras. Let $\phi : \mathcal{A}_+^{-1} \to \mathcal{B}_+^{-1}$ be a continuous Jordan triple map. Then ϕ is a Lipschitz function in a neighborhood of the identity.*

Proof. We begin with the following important observation. For any $A \in \mathcal{A}_+^{-1}$ that is close enough to I, we have

$$\frac{1}{2}\|A - I\| \leq \|\log A\| \leq 2\|A - I\|. \tag{2.7}$$

Indeed, this follows easily from the inequalities

$$\|e^H - I\| \leq e^{\|H\|} - 1, \quad H \in \mathcal{A}_s,$$

and

$$\|\log A\| \leq -\log(1 - \|A - I\|), \quad A \in \mathcal{A}_+^{-1} \text{ with } \|A - I\| < 1,$$

and from elementary properties of the exponential and logarithm functions of a real variable.

For temporary use, let $\mathcal{G}_r$ denote the closed ball in $\mathcal{A}_+^{-1}$ with center I and radius $0 < r < 1$. We assert that there exists an r with $0 < r < 1$ and another positive number L for which $\|\phi(A) - I\| \leq L\|A - I\|$ holds for all $A \in \mathcal{G}_r$. Assume on the contrary that there is a sequence (A_k) of elements of $\mathcal{A}_+^{-1}$ such that

$$\|A_k - I\| < 1/k \quad \text{and} \quad \|\phi(A_k) - I\| > k\|A_k - I\| \tag{2.8}$$

hold for every $k \in \mathbb{N}$. Since $A_k \to I$, it follows that $\phi(A_k) \to \phi(I) = I$. (Observe that $\phi(I)^3 = \phi(I^3) = \phi(I)$ implies $\phi(I) = I$.) Clearly, we have

$$\|\phi(A_k) - I\| = \epsilon_k, \quad \epsilon_k < 1$$

for large enough $k \in \mathbb{N}$ and in what follows we consider only such indexes k. Choose positive integers l_k such that

$$1/(l_k+1) \leq \epsilon_k < 1/l_k.$$

By (2.7), for large enough k we have $\|\log A_k\| \leq 2\|A_k - I\|$ and hence obtain

$$\|\log A_k^{l_k}\| = l_k\|\log A_k\| \leq 2l_k\|A_k - I\| < 2l_k(\epsilon_k/k) < 2/k \to 0$$

as $k \to \infty$. It follows that $A_k^{l_k} \to I$ and hence $A_k^{l_k+1} \to I$. Therefore, we infer $\phi(A_k^{l_k+1}) \to \phi(I) = I$. However, using (2.7) again, for large enough k we also have

$$\begin{aligned}\frac{1}{2} \leq \frac{\epsilon_k(l_k+1)}{2} &= \frac{l_k+1}{2}\|\phi(A_k) - I\| \\ &\leq (l_k+1)\|\log\phi(A_k)\| = \|\log\phi(A_k^{l_k+1})\| \to 0\end{aligned}$$

which is a contradiction. Consequently, there do exist positive real numbers $r(<1)$ and L such that $\|\phi(A)-I\| \leq L\|A-I\|$ holds for all $A \in \mathcal{G}_r$. Clearly, ϕ is necessarily bounded on $\mathcal{G}_r$.

To complete the proof, let s be a not yet specified positive number with $s < r$ and pick arbitrary $C, D \in \mathcal{G}_s$. Let $B = \sqrt{C}$ and $A = B^{-1}DB^{-1}$. Considering the inequality

$$\begin{aligned}\|A - I\| &= \|B^{-1}DB^{-1} - I\| \\ &\leq \|B^{-1} - I\|\|D\|\|B^{-1}\| + \|D - I\|\|B^{-1}\| + \|B^{-1} - I\|\end{aligned}$$

we see that choosing small enough $s > 0$ we have $\|A - I\| < r$ and $\|B - I\| < r$. Assuming $C \neq D$ we can compute

$$\begin{aligned}\frac{\|\phi(D) - \phi(C)\|}{\|D - C\|} = \frac{\|\phi(B)\phi(A)\phi(B) - \phi(B)^2\|}{\|BAB - B^2\|} &\leq \frac{\|B^{-1}\|^2\|\phi(B)\|^2\|\phi(A) - I\|}{\|A - I\|} \\ = \|C^{-1}\|\|\phi(C)\|\frac{\|\phi(A) - I\|}{\|A - I\|} &\leq L\|C^{-1}\|\|\phi(C)\|.\end{aligned}$$

Clearly, the function $C \mapsto \|C^{-1}\|\|\phi(C)\|$ is bounded on $\mathcal{G}_s$ (recall that $s < r$) and thus we obtain the desired Lipschitz property of ϕ in a neighborhood of I. □

The next lemma shows that every continuous Jordan triple map from $\mathcal{A}_+^{-1}$ into $\mathcal{B}_+^{-1}$ is the exponential of a commutativity preserving linear map from $\mathcal{A}_s$ to $\mathcal{B}_s$ composed by the logarithmic function. Similarly to the case of matrix algebras treated in [23], this plays an essential role in the proof of Theorem 5. We say that a linear transformation $f : \mathcal{A}_s \to \mathcal{B}_s$ preserves commutativity (more precisely preserves commutativity in one direction) if for any pair $T, S \in \mathcal{A}_s$ of commuting elements we have that $f(T), f(S) \in \mathcal{B}_s$ commute, too. The proof of the next lemma follows the proof of Lemma 6 in [23] (presented for matrices) except its last paragraph.

Lemma 16. *Let $\mathcal{A}, \mathcal{B}$ be C^*-algebras. Assume $\phi : \mathcal{A}_+^{-1} \to \mathcal{B}_+^{-1}$ is a continuous Jordan triple map. Then there exists a commutativity preserving linear transformation $f : \mathcal{A}_s \to \mathcal{B}_s$ such that*

$$\phi(A) = e^{f(\log A)}, \quad A \in \mathcal{A}_+^{-1}. \tag{2.9}$$

Proof. We define $f : \mathcal{A}_s \to \mathcal{B}_s$ by

$$f(T) = \log \phi(e^T), \quad T \in \mathcal{A}_s.$$

We clearly have (2.9) and need only to show that f is linear and preserves commutativity.

Pick arbitrary $A \in \mathcal{A}_+^{-1}$. Observe that since ϕ is a Jordan triple map, we have

$$\phi(A^n) = \phi(A)^n \tag{2.10}$$

for all integers $n = 0, 1, 2, \ldots$. This easily implies that $\phi(A^{1/n}) = \phi(A)^{1/n}$. From $\phi(A)\phi(A^{-2})\ \phi(A) = \phi(I) = I$ we have $\phi(A^{-1})^2 = (\phi(A)^{-1})^2$ implying that ϕ preserves the inverse operation and hence (2.10) holds for all integers n. Therefore, $\phi(A^r) = \phi(A)^r$ is valid for all $A \in \mathcal{A}_+^{-1}$ and rational number r. By the continuity of ϕ we obtain that $\phi(A^t) = \phi(A)^t$ is true for every real number t, too. This obviously implies that f is homogeneous.

We next prove that $f : \mathcal{A}_s \to \mathcal{B}_s$ is additive. Pick $T, S, H \in \mathcal{A}_s$. We compute

$$\begin{aligned} &\frac{e^{(t/2)T} e^{tS} e^{(t/2)T} - e^{tH}}{t} \\ &= \frac{(e^{(t/2)T} - I)e^{tS} e^{(t/2)T} + (e^{tS} - I)e^{(t/2)T} + (e^{(t/2)T} - I) - (e^{tH} - I)}{t} \\ &\to T/2 + S + T/2 - H = T + S - H \end{aligned} \tag{2.11}$$

as $t \to 0$. It follows that

$$\lim_{t \to 0} \frac{e^{(t/2)T} e^{tS} e^{(t/2)T} - e^{tH}}{t} = 0 \Longleftrightarrow H = T + S.$$

If $H = T + S$, then using (2.9) and the Lipschitz property of ϕ in a neighborhood of I that has been proven in Lemma 15 we have

$$\begin{aligned} \frac{e^{(t/2)f(T)} e^{tf(S)} e^{(t/2)f(T)} - e^{tf(H)}}{t} &= \frac{\phi(e^{(t/2)T})\phi(e^{tS})\phi(e^{(t/2)T}) - \phi(e^{tH})}{t} \\ &= \frac{\phi(e^{(t/2)T} e^{tS} e^{(t/2)T}) - \phi(e^{tH})}{t} \to 0 \end{aligned}$$

as $t \to 0$. On the other hand, as in (2.11) we infer

$$\frac{e^{(t/2)f(T)} e^{tf(S)} e^{(t/2)f(T)} - e^{tf(H)}}{t} \to f(T) + f(S) - f(H).$$

This gives us that $f(T) + f(S) - f(T + S) = 0$, i.e., f is additive.

To verify the commutativity preserving property of f first observe that we have

$$\phi(\sqrt{A} B \sqrt{A}) = \phi(\sqrt{A})\phi(B)\phi(\sqrt{A}) = \sqrt{\phi(A)}\phi(B)\sqrt{\phi(A)}$$

for every $A, B \in \mathcal{A}_+^{-1}$. We now recall the following notion and fact. Given positive elements D, F of a unital C^*-algebra define their so-called sequential product by $\sqrt{D}F\sqrt{D}$. It is an interesting fact that commutativity of D, F with respect to this product is equivalent to the commutativity of D, F with respect to the usual product. A short proof of this fact has been given in Proposition 1 in [2]. It is then clear that ϕ preserves commutativity which apparently implies the commutativity preserving property of f, too. □

The next lemma that we shall need for the proof of Theorem 5 provides a characterization of tracial continuous linear functionals on von Neumann algebras in terms of their behavior with respect to the Jordan triple product. It says that the continuous linear functional l which is real valued on self-adjoint elements is tracial if and only if the (non-linear) functional $\exp \circ l \circ \log$ is a Jordan triple map on the positive definite cone.

Lemma 17. *Let $\mathcal{A}$ be a von Neumann algebra and $l : \mathcal{A} \to \mathbb{C}$ a continuous linear functional which has real values on $\mathcal{A}_s$. Then l satisfies*

$$l(\log ABA) = l(\log A) + l(\log B) + l(\log A), \quad A, B \in \mathcal{A}_+^{-1} \tag{2.12}$$

if and only if l is tracial.

Proof. Assume that l satisfies (2.12). We first follow an argument similar to the one given in the proof of Theorem 2 in [20]. Pick projections P, Q in $\mathcal{A}$. Let

$$A = I + tP, \quad B = I + tQ,$$

where $t > -1$ is any real number. Easy computation shows that

$$\begin{aligned} ABA &= (I + tP)(I + tQ)(I + tP) \\ &= I + t(2P + Q) + t^2(P + PQ + QP) + t^3(PQP). \end{aligned}$$

Recall that in an arbitrary unital Banach algebra, for any element a with $\|a\| < 1$ we have

$$\log(1 + a) = \sum_{n=1}^{\infty} \frac{(-1)^{n+1} a^n}{n}.$$

This shows that for a suitable positive ϵ, the elements $\log(ABA)$, $\log A$, $\log B$ of $\mathcal{A}$ can be expressed by power series of t ($|t| < \epsilon$) with algebra coefficients. In particular, considering the coefficients of t^3 on both sides of the equality (2.12) and using their uniqueness, we obtain the equation

$$\begin{aligned} l\Big(PQP &- \frac{1}{2}((2P + Q)(P + PQ + QP) \\ &+ (P + PQ + QP)(2P + Q)) + \frac{1}{3}(2P + Q)^3 \Big) = l\Big(\frac{1}{3}(P + Q + P) \Big). \end{aligned}$$

Executing the operations and subtracting those terms which appear on both sides of this equation, we arrive at the equality

$$l\Big(\frac{1}{3}(PQP) - \frac{1}{3}(QPQ) \Big) = 0.$$

Therefore,

$$l(PQP) = l(QPQ)$$

holds for all projections $P, Q \in \mathcal{A}$. We assert that this implies that l is tracial, i.e., $l(XY) = l(YX)$, $X, Y \in \mathcal{A}$. To verify this, we apply an idea from the proof of Lemma 1 in [4]. Namely, select an arbitrary pair P, Q of projections in $\mathcal{A}$, define $S = I - 2P$ and compute

$$\begin{aligned} l(Q + SQS) &= \frac{1}{2} l((I - S)Q(I - S) + (I + S)Q(I + S)) \\ &= \frac{1}{2} l(4PQP + 4(I - P)Q(I - P)) = 2l(PQP + (I - P)Q(I - P)) \\ &= 2l(QPQ + Q(I - P)Q) = 2l(Q). \end{aligned}$$

Since the symmetries (i.e., self-adjoint unitaries) in $\mathcal{A}$ are exactly the elements of the form $S = I - 2P$ with some projection $P \in \mathcal{A}$, we obtain that $l(Q) = l(SQS)$ holds for every symmetry S and every projection Q in $\mathcal{A}$. By the continuity of the linear functional l and using the spectral theorem, we infer that $l(X) = l(SXS)$ holds for any $X \in \mathcal{A}$ and symmetry $S \in \mathcal{A}$. This implies that

$$l(SX) = l(S(XS)S) = l(XS)$$

for all $X \in \mathcal{A}$ and symmetry $S \in \mathcal{A}$. Plainly, this gives us that $l(PX) = l(XP)$ holds for every projection $P \in \mathcal{A}$. Finally, we conclude that $l(XY) = l(YX)$ for all $X, Y \in \mathcal{A}$.

Conversely, if $l : \mathcal{A} \to \mathbb{C}$ is a continuous linear functional which has real values on $\mathcal{A}_s$ and tracial, we need to prove that

$$e^{l(\log ABA)} = e^{l(\log A) + l(\log B) + l(\log A)}$$

holds for all $A, B \in \mathcal{A}_+^{-1}$. This can be proven following the proofs of Lemma 2 and Lemma 3 in [9]. □

In the proofs of our theorems on the structures of continuous Jordan triple isomorphisms we shall also need a particular case of the following result which has appeared in [24].

Proposition 18. *Let $\mathcal{A}$ be a C^*-algebra. If $c \notin \{-1, 0, 1\}$ is a real number with the property that for any pair $A, B \in \mathcal{A}_+^{-1}$ we have a real number λ such that*

$$(ABA)^c = \lambda A^c B^c A^c,$$

then the algebra $\mathcal{A}$ is commutative. Similarly, if m is an integer $m \notin \{-1, 0, 1\}$ and for any pair $U, V \in \mathcal{A}_u$ we have a scalar μ such that

$$(UVU)^m = \mu U^m V^m U^m,$$

then $\mathcal{A}$ is commutative.

We now have all preliminary information to present the proof of Theorem 5.

Proof of Theorem 5. Let $\phi : \mathcal{A}_+^{-1} \to \mathcal{B}_+^{-1}$ be a continuous Jordan triple isomorphism. Applying Lemma 16 we have a bijective linear transformation $f : \mathcal{A}_s \to \mathcal{B}_s$

such that $\phi(A) = \exp f(\log A)$, $A \in \mathcal{A}_+^{-1}$ and f preserves commutativity. Since ϕ^{-1} is also a Jordan triple isomorphism, it follows from the last part of the proof of Lemma 16 that ϕ^{-1} also preserves commutativity implying that f preserves commutativity in both directions. It follows that $\mathcal{B}$ is a factor von Neumann algebra, too.

If $\mathcal{A}$ is of type I_1, then so is $\mathcal{B}$ and the theorem reduces to the description of all continuous multiplicative bijections of the positive real line. These maps are well known to be exactly the power functions corresponding to nonzero exponents. So, in this case the assertion is trivial and hence in what follows we assume that $\mathcal{A}$ is not of type I_1 and not of type I_2.

We extend f from $\mathcal{A}_s$ to a linear transformation onto $\mathcal{A}$ in the following trivial way:

$$F(A + iB) = f(A) + if(B), \qquad A, B \in \mathcal{A}_s.$$

Clearly, $F : \mathcal{A} \to \mathcal{B}$ is a bijective linear transformation. Since an operator is normal if and only if its real and imaginary parts commute, we see that F sends the normal elements of $\mathcal{A}$ to normal elements of $\mathcal{B}$. There are structural results concerning such maps. We refer to Theorem 4.1 in [6] on the form of normal preserving linear transformations between centrally closed prime algebras satisfying some additional conditions which can be applied here (see the introduction of that paper for the explanation of the necessary concepts). We obtain that there is a nonzero complex number c, an algebra *-isomorphism or an algebra *-antiisomorphism $\theta : \mathcal{A} \to \mathcal{B}$, and a linear functional $l : \mathcal{A} \to \mathbb{C}$ such that

$$F(A) = c\theta(A) + l(A)I, \qquad A \in \mathcal{A}.$$

We claim that c is real and l maps $\mathcal{A}_s$ into $\mathbb{R}$. In order to see this, let P be a nontrivial projection in $\mathcal{A}$. Then $c\theta(P) + l(P)I$ is a self-adjoint element of $\mathcal{B}$ and $\theta(P)$ is a nontrivial projection. This easily gives us first that $l(P)$ and then that c are real numbers. Next, for any $A \in \mathcal{A}_s$ we have that $c\theta(A) + l(A)I$ and $c\theta(A)$ are both self-adjoint implying that $l(A) \in \mathbb{R}$.

Clearly, c is not zero. We have

$$\phi(A) = e^{c\theta(\log A) + l(\log A)I} = e^{l(\log A)}\theta(A^c), \qquad A \in \mathcal{A}_+^{-1}.$$

Since ϕ is a Jordan triple isomorphism from $\mathcal{A}_+^{-1}$ onto $\mathcal{B}_+^{-1}$ and θ is an algebra *-isomorphism or an algebra *-antiisomorphism from $\mathcal{A}$ onto $\mathcal{B}$, it readily follows that

$$e^{l(\log ABA)}\theta((ABA)^c) = e^{l(\log A) + l(\log B) + l(\log A)}\theta(A^c B^c A^c), \quad A, B \in \mathcal{A}_+^{-1}$$

from which we obtain

$$e^{l(\log ABA)}(ABA)^c = e^{l(\log A) + l(\log B) + l(\log A)} A^c B^c A^c, \quad A, B \in \mathcal{A}_+^{-1}.$$

In particular, $(ABA)^c$ and $A^c B^c A^c$ are scalar multiples of each other for all $A, B \in \mathcal{A}_+^{-1}$. By Proposition 18 it follows that c is either 1 or -1. Therefore, we have $e^{l(\log ABA)} = e^{l(\log A) + l(\log B) + l(\log A)}$ implying

$$l(\log ABA) = l(\log A) + l(\log B) + l(\log A), \quad A, B \in \mathcal{A}_+^{-1}.$$

By the continuity of ϕ and using the fact that θ is necessarily isometric, we have that l is a continuous linear functional. Applying Lemma 17 we have that l is tracial. By the injectivity of ϕ we have $\phi(eI) \neq \phi(I) = I$ which gives $l(I) + c \neq 0$. This proves the necessity part of our theorem.

Conversely, if $l : \mathcal{A} \to \mathbb{C}$ is a continuous linear functional which is real valued on $\mathcal{A}_s$ and tracial, then by Lemma 17

$$e^{l(\log ABA)} = e^{l(\log A)+l(\log B)+l(\log A)}$$

holds for all $A, B \in \mathcal{A}_+^{-1}$. If $c \in \{-1, 1\}$ and $l(I) \neq -c$ is also true, then one can readily verify that for any algebra *-isomorphism or algebra *-antiisomorphism $\theta : \mathcal{A} \to \mathcal{B}$, the map $\phi : \mathcal{A}_+^{-1} \to \mathcal{B}_+^{-1}$ defined by

$$\phi(A) = e^{l(\log A)}\theta(A^c), \qquad A \in \mathcal{A}_+^{-1}$$

is a continuous Jordan triple isomorphism. □

After this the proof of Theorem 6 is very simple. We note that on a von Neumann algebra $\mathcal{A}$ any symmetric norm N has the property $N(|A|) = N(A)$, $A \in \mathcal{A}$ which follows easily from the fact that the components of the polar decomposition of any element in $\mathcal{A}$ belong to $\mathcal{A}$.

Proof of Theorem 6. Let $\phi : \mathcal{A}_+^{-1} \to \mathcal{B}_+^{-1}$ be a surjective function which satisfies

$$d_{M,g}(\phi(A), \phi(B)) = d_{N,f}(A, B), \qquad A, B \in \mathcal{A}_+^{-1}.$$

From Theorem 4 we obtain that ϕ is a continuous inverted Jordan triple isomorphism. As mentioned in between the formulations of Theorems 4 and 5, the map $\psi(.) = \phi(I)^{-1/2}\,\phi(.)\phi(I)^{-1/2}$ is a continuous Jordan triple isomorphism and hence the latter result applies and we obtain the form (1.5). As for the last statement in the theorem on the disappearance of l in the case of infinite factors, we refer the remark given after Theorem 5. The proof is complete. □

To see cases where the tracial linear functional in (1.5) really shows up we refer to Theorem 3 in [23].

Remark 19. We present a sort of application of Theorems 4 and 6. In the paper [15] Honma and Nogawa considered metrics on the positive definite cone of a C^*-algebra $\mathcal{A}$ of the form

$$d_\alpha(A, B) = \| \log(A^{-\alpha/2}B^\alpha A^{-\alpha/2})^{1/\alpha}\|, \qquad A, B \in \mathcal{A}_+^{-1},$$

where α is a given nonzero real number. They described the structure of surjective isometries between two such spaces.

Observe that the problem can also be treated in the framework that we have presented above. Indeed, let $\mathcal{A}, \mathcal{B}$ be C^*-algebras, α, β given nonzero real numbers and $\phi : \mathcal{A}_+^{-1} \to \mathcal{B}_+^{-1}$ a surjective map which is an isometry with respect to the pair d_α, d_β of metrics, i.e., which satisfies

$$\| \log(\phi(A)^{-\beta/2}\phi(B)^\beta\phi(A)^{-\beta/2})^{1/\beta}\| = \| \log(A^{-\alpha/2}B^\alpha A^{-\alpha/2})^{1/\alpha}\|$$

for all $A, B \in \mathcal{A}_+^{-1}$. Define $\psi(A) = \phi(A^{1/a})^\beta$, $A \in \mathcal{A}_+^{-1}$. Hence, $\psi : \mathcal{A}_+^{-1} \to \mathcal{B}_+^{-1}$ is a bijective map for which

$$(1/|\beta|)\| \log(\psi(A)^{-1/2}\psi(B)\psi(A)^{-1/2})\| = (1/|\alpha|)\| \log(A^{-1/2}BA^{-1/2})\|$$

holds for all $A, B \in \mathcal{A}_+^{-1}$. Now, apply Theorem 4 for $N(.) = (1/|\beta|)\|.\|$, $f(y) = -\log y$, $y > 0$ and $M(.) = (1/|\alpha|)\|.\|$, $g(y) = -\log y$, $y > 0$. We infer that $\psi : \mathcal{A}_+^{-1} \to \mathcal{B}_+^{-1}$ is an inverted Jordan triple isomorphism. This means that

$$\phi(A^{1/\alpha})^\beta \phi(B^{1/\alpha})^{-\beta} \phi(A^{1/\alpha})^\beta = \phi((AB^{-1}A)^{1/\alpha})^\beta, \qquad A, B \in \mathcal{A}_+^{-1}.$$

Replacing here A by A^α and B by B^α we immediately get

$$\phi(A)^\beta \phi(B)^{-\beta} \phi(A)^\beta = \phi((A^\alpha B^{-\alpha} A^\alpha)^{1/\alpha})^\beta$$

and hence that

$$(\phi(A)^\beta \phi(B)^{-\beta} \phi(A)^\beta)^{1/\beta} = \phi((A^\alpha B^{-\alpha} A^\alpha)^{1/\alpha}), \qquad A, B \in \mathcal{A}_+^{-1}.$$

This is just the conclusion formulated in Proposition 2 in [15] on the algebraic behavior of the isometries with respect to the pair d_α, d_β of metrics which statement plays important role in that paper. Using the same ideas and applying Theorem 6 it should not be difficult to derive Corollary 9 in [15]. Furthermore, observe that by our general results one can treat the cases of more general distances. Indeed, in [15] also metrics on the set of all positive definite matrices of the form

$$d_{|||\,.\,|||,\alpha}(A, B) = ||| \log(A^{-\alpha/2} B^\alpha A^{-\alpha/2})^{1/\alpha} |||$$

where $|||\,.\,|||$ is a unitarily invariant norm on $\mathbb{M}_n$ have been mentioned but structural result on the corresponding isometries was not obtained. Clearly, our approach above applies also in that situation and using it one can easily describe the corresponding isometries. We omit the details.

We now turn to the proofs of our results on transformations between unitary groups which respect a pair of generalized distance measures. Our approach is quite similar to the one we have followed in the case of the positive definite cone. We first present an appropriate general Mazur–Ulam type result, then show that under certain conditions the surjective "generalized isometries" that we consider are continuous inverted Jordan triple isomorphisms. Next we describe the structure of those isomorphisms between von Neumann factors and finally, after gathering the necessary information, we prove our result on the structure of surjective "generalized isometries" between unitary groups of von Neumann factors.

The general Mazur–Ulam type theorem that we need here reads as follows.

Proposition 20. *Suppose that G and H are groups equipped with generalized distance measures d and ρ, respectively. Pick $a, b \in G$, set*

$$L_{a,b} = \{x \in G : d(a, x) = d(x, ba^{-1}b) = d(a, b)\},$$

and assume the following:

(e1) *$d(bx^{-1}b, bx'^{-1}b) = d(x', x)$ holds for all $x, x' \in G$;*

(e2) *$\sup\{d(x, b) : x \in L_{a,b}\} < \infty$;*

(e3) *there exists a constant* $K > 1$ *such that*

$$d(x, bx^{-1}b) \geq Kd(x, b), \quad x \in L_{a,b};$$

(e4) $\rho(cy^{-1}c', cy'^{-1}c') = \rho(y', y)$ *holds for all* $c, c', y, y' \in H$.

Then for any surjective map $\phi : G \to H$ *which satisfies*

$$\rho(\phi(x), \phi(x')) = d(x, x'), \quad x, x' \in G$$

we have

$$\phi(ba^{-1}b) = \phi(b)\phi(a)^{-1}\phi(b).$$

Proof. First observe that ϕ is also injective and hence bijective. Let $\varphi(x) = bx^{-1}b$, $x \in G$, define $\psi(y) = \phi(a)y^{-1}\phi(ba^{-1}b)$, $y \in H$ and let $T = \phi$. One can readily check that Proposition 11 applies and results in $\psi(\phi(b)) = \phi(b)$. This immediately implies

$$\phi(ba^{-1}b) = \phi(b)\phi(a)^{-1}\phi(b)$$

and we are done. □

We proceed with the proof of Theorem 7.

Proof of Theorem 7. We apply Proposition 20 in the following setting: $G = \mathcal{A}_u$, $H = \mathcal{B}_u$, $d = d_{N,f}$, $\rho = d_{M,g}$. We observe that conditions (e1), (e2), (e4) are satisfied for all $A, B, X, X' \in \mathcal{A}_u$ and $C, D, Y, Y' \in \mathcal{B}_u$. Indeed, to see (e1) we can compute

$$\begin{aligned} f(BX^{-1}X'B^{-1}) &= Bf(X^{-1}X')B^{-1} = Bf(X^{-1}(X'X^{-1})X)B^{-1} \\ &= BX^{-1}f(X'X^{-1})XB^{-1} \end{aligned}$$

and hence, by the unitary similarity invariance of N, it follows that

$$d_{N,f}(BX^{-1}B', BX'^{-1}B') = d_{N,f}(X', X)$$

holds for all $B, B', X, X' \in \mathcal{A}_u$. As for (e2), the set of all values of $d_{N,f}(.,.)$ is bounded which is a consequence of the boundedness of f and the equivalence of N to the operator norm.

By (d2) we have that $N(f(U^2)) \geq KN(f(U))$ if $U \in \mathcal{A}_u$ is close enough (in the operator norm) to I. In what follows we show that (e3) is also satisfied provided $A, B \in \mathcal{A}_u$ are close enough to each other in the operator norm. We warn the reader that we are going to argue rather vaguely avoiding the precise "$\epsilon - \delta$ technique" which would make the proof much more lengthy. So, let $A, B \in \mathcal{A}_u$ be unitaries which are close to each other in the operator norm. Then picking X from $L_{A,B}$, referring to the property (d1) of f and the equivalence of N to the operator norm, we have that $N(f(AX^{-1})) = N(f(AB^{-1}))$ is small. This gives us that AX^{-1} is close to the identity, i.e., X is close to A in the operator norm. Hence we obtain that X is close also to B implying that XB^{-1} is close to the identity in the operator norm. By (d2) we have

$$d_{N,f}(X, BX^{-1}B) = N(f((XB^{-1})^2)) \geq KN(f(XB^{-1})) = Kd_{N,f}(X, B).$$

This shows that (e3) really holds for $A, B \in \mathcal{A}_u$ which are close enough to each other in the operator norm. The fact that (e4) is also valid follows from the argument we have presented relating to condition (e1) above.

Therefore, by Proposition 20 we have that for $A, B \in \mathcal{A}_u$ which are close enough to each other in the operator norm, the equality

$$\phi(BA^{-1}B) = \phi(B)\phi(A)^{-1}\phi(B)$$

holds. We can now follow the argument presented in the proof of Theorem 8 in [13] to conclude that from the validity of that equality for close enough A, B we obtain that it necessarily holds globally, i.e., for all $A, B \in \mathcal{A}_u$, too. The continuity of ϕ can be shown in a way similar to the proof of the last statement in Proposition 13. This completes the proof. □

As we have mentioned after the formulation of Theorem 7, multiplying any inverted Jordan triple isomorphism ϕ between unitary groups by the element $\phi(I)^{-1}$ we obtain a Jordan triple isomorphism. Our next aim is to determine the structure of the continuous Jordan triple isomorphisms between unitary groups of von Neumann factors. As in the case of the positive definite cone, the first step is to show that any such map has Lipschitz property which in fact holds globally in the present situation.

Lemma 21. *Let $\mathcal{A}, \mathcal{B}$ be von Neumann algebras and $\phi : \mathcal{A}_u \to \mathcal{B}_u$ a continuous Jordan triple map, i.e., assume that*

$$\phi(UVU) = \phi(U)\phi(V)\phi(U), \quad U, V \in \mathcal{A}_u.$$

Then ϕ is a Lipschitz function.

Proof. Since ϕ is a Jordan triple map, it follows that $\phi(I)^3 = \phi(I)$, hence $\phi(I)^2 = I$ which means that $\phi(I)$ is a symmetry (i.e., self-adjoint unitary).

From $\phi(I)\phi(V)\phi(I) = \phi(V)$ we obtain $\phi(I)\phi(V) = \phi(V)\phi(I)$ for any $V \in \mathcal{A}_u$ which means that $\phi(I)$ commutes with the range of ϕ. It follows easily that $\phi(I)^{-1}\phi(.)$ is also a Jordan triple map. Therefore, we may and do assume that $\phi(I) = I$. It is easy to see that in that case we have $\phi(U^m) = \phi(U)^m$ for any positive integer m and $U \in \mathcal{A}_u$.

Next we follow the argument given in the proof of Lemma 6 in [22] but we need to make some necessary modifications due to the fact that here we consider operator algebras, not matrix algebras. We first assert that there exist positive real numbers r, L such that $\|\phi(U) - I\| \leq L\|U - I\|$ holds for all $U \in \mathcal{A}_u$ with $\|U - I\| < r$. Assume on the contrary that we have a sequence (U_k) in $\mathcal{A}_u$ such that $\|U_k - I\| < 1/k$ and

$$\|\phi(U_k) - I\| > k\|U_k - I\| \tag{2.13}$$

holds for every $k \in \mathbb{N}$. We have $U_k \to I$ and hence $\phi(U_k) \to I$ as $k \to \infty$. Denoting $\epsilon_k = \|\phi(U_k) - I\|$ we obviously have $\epsilon_k < 1/2$ for large enough k. Choose positive integers l_k such that

$$1/(l_k + 1) \leq \epsilon_k < 1/l_k.$$

By (2.13) we have

$$\epsilon_k/k > \|U_k - I\|$$

and hence

$$\|U_k^{l_k} - I\| \leq \|U_k - I\|\|U_k^{l_k-1} + \cdots + I\| \leq l_k\|U_k - I\| < (l_k\epsilon_k)/k < 1/k.$$

It follows that $U_k^{l_k+1} \to I$ and we infer $\phi(U_k^{l_k+1}) = \phi(U_k)^{l_k+1} \to I$.

We shall need the following simple observation. Let λ be a complex number of modulus 1 in the upper half-plane. Clearly, the length of arc from 1 to λ divided by the length of the corresponding chord is less than $\pi/2$. Geometrical considerations show that assuming n is a positive integer such that $n(\pi/2)|\lambda - 1| < \pi$, we necessarily have $n|\lambda - 1| < (\pi/2)|\lambda^n - 1|$. It follows that for any unitary $V \in \mathcal{A}_u$ and positive integer n, the inequality $n\|V - I\| < 2$ implies $n\|V - I\| < (\pi/2)\|V^n - I\|$.

We have

$$2\epsilon_k(l_k + 1) < 2(l_k + 1)/l_k < \pi,$$

where in the last inequality we have used $l_k \geq 2$. This gives us that $(l_k+1)\|\phi(U_k) - I\| < 2$. Therefore, we compute

$$1 = (1/\epsilon_k)\|\phi(U_k) - I\| \leq (l_k + 1)\|\phi(U_k) - I\| \leq (\pi/2)\|\phi(U_k)^{l_k+1} - I\|.$$

But this clearly contradicts the fact that $\phi(U_k)^{l_k+1} \to I$. Therefore, we do have positive real numbers r, L such that $\|\phi(U) - I\| \leq L\|U - I\|$ holds for every $U \in \mathcal{A}_u$ with $\|U - I\| < r$. Since ϕ is bounded, we have (probably with another constant L) that $\|\phi(U) - I\| \leq L\|U - I\|$ is valid for every $U \in \mathcal{A}_u$.

To complete the proof, pick arbitrary unitaries $W, W' \in \mathcal{A}_u$. Since every unitary in $\mathcal{A}$ is the exponential of a self-adjoint element multiplied by the imaginary unit i, we can choose $V \in \mathcal{A}_u$ such that $V^2 = W'$ and then find $U \in \mathcal{A}_u$ such that $VUV = W$. We compute

$$\begin{aligned}\|\phi(W) - \phi(W')\| &= \|\phi(VUV) - \phi(V^2)\| = \|\phi(V)\phi(U)\phi(V) - \phi(V)^2\| \\ &= \|\phi(U) - I\| \leq L\|U - I\| = L\|VUV - V^2\| = L\|W - W'\|.\end{aligned}$$

This proves that ϕ is a Lipschitz function. □

In the next lemma we show that every continuous Jordan triple map between unitary groups of von Neumann algebras gives rise to a certain commutativity preserving linear map between the self-adjoint parts of the underlying algebras. This result is parallel to Lemma 16. In the proof we follow the argument given in the proof of Lemma 7 in [22].

Lemma 22. *Let $\mathcal{A}, \mathcal{B}$ be von Neumann algebras and $\phi : \mathcal{A}_u \to \mathcal{B}_u$ a continuous Jordan triple map. Then $\phi(I)$ is a symmetry which commutes with the range of ϕ and we have a commutativity preserving linear transformation $f : \mathcal{A}_s \to \mathcal{B}_s$ such that*

$$\phi(e^{itA}) = \phi(I)e^{itf(A)}, \quad t \in \mathbb{R}, A \in \mathcal{A}_s.$$

Moreover, f satisfies

$$f(VAV) = \phi(V)f(A)\phi(V)$$

for every $A \in \mathcal{A}_s$ and symmetry (self-adjoint unitary) $V \in \mathcal{A}_u$.

Proof. The assertion concerning $\phi(I)$ has been verified in the first part of the proof of Lemma 21. Just as there we can assume that ϕ is a unital Jordan triple map, $\phi(I) = I$. We have also learned that $\phi(V^k) = \phi(V)^k$ holds for every $V \in \mathcal{A}_u$ and positive integer k. We now show that ϕ preserves the inverse operation. To prove this, let $W \in \mathcal{A}_u$ be such that $W^2 = V$. We compute

$$\phi(W)\phi(V^{-1})\phi(W) = \phi(WV^{-1}W) = \phi(I) = I$$

which implies that

$$\phi(V^{-1}) = \phi(W)^{-2} = \phi(W^2)^{-1} = \phi(V)^{-1}.$$

It follows that $\phi(V^k) = \phi(V)^k$ holds for every integer k and for every $V \in \mathcal{A}_u$. In the rest of the proof we shall use several times that, in particular, ϕ sends symmetries to symmetries.

In the next step we show that ϕ sends norm-continuous one-parameter unitary groups to norm-continuous one-parameter unitary groups. Pick an arbitrary self-adjoint element $T \in \mathcal{A}_s$ and define $S_T : \mathbb{R} \to \mathcal{B}_u$ by

$$S_T(t) = \phi(e^{itT}), \quad t \in \mathbb{R}.$$

We assert that S_T is a norm-continuous one-parameter unitary group in $\mathcal{B}_u$. Since ϕ is continuous, S_T is also continuous. We verify that $S_T(t+t') = S_T(t)S_T(t')$ holds for every pair t, t' of real numbers. First select rational numbers r and r' such that $r = \frac{k}{m}$ and $r' = \frac{k'}{m'}$ with integers k, k', m, m'. We compute

$$\begin{aligned} S_T(r + r') &= \phi(e^{i\frac{km'+k'm}{mm'}T}) = \phi(e^{i\frac{1}{mm'}T})^{km'+k'm} \\ &= \phi(e^{i\frac{1}{mm'}T})^{km'} \phi(e^{i\frac{1}{mm'}T})^{k'm} = S_T(r)S_T(r'). \end{aligned}$$

By the continuity of S_T, we deduce that $S_T(t+t') = S_T(t)S_T(t')$ holds for every pair t, t' of real numbers. By Stone's theorem we obtain that there exists a unique self-adjoint element $f(T) \in \mathcal{B}_s$, the generator of S_T, such that

$$\phi(e^{itT}) = S_T(t) = e^{itf(T)}, \quad t \in \mathbb{R}.$$

Observe that the generator belongs to $\mathcal{B}$ since in the present case (norm-continuous one-parameter unitary group) it can be obtained by differentiation, the limit of difference quotients taken in the norm topology.

We next prove that $f : \mathcal{A}_s \to \mathcal{B}_s$ is in fact a linear transformation. Pick $A, B, C \in \mathcal{A}_s$. Similarly to the proof of Lemma 16 we compute

$$\begin{aligned} &\frac{e^{i(t/2)A}e^{itB}e^{i(t/2)A} - e^{itC}}{it} \qquad (2.14) \\ &= \frac{(e^{i(t/2)A} - I)e^{itB}e^{i(t/2)A} + (e^{itB} - I)e^{i(t/2)A} + (e^{i(t/2)A} - I) - (e^{itC} - I)}{it} \\ &\to A/2 + B + A/2 - C = A + B - C \end{aligned}$$

as $t \to 0$. It follows that

$$\lim_{t \to 0} \frac{e^{i(t/2)A}e^{itB}e^{i(t/2)A} - e^{itC}}{it} = 0 \iff C = A + B.$$

If $C = A + B$, then using the Lipschitz property of ϕ proven in Lemma 21 we have

$$\begin{aligned}\frac{e^{i(t/2)f(A)}e^{itf(B)}e^{i(t/2)f(A)} - e^{itf(C)}}{it} &= \frac{\phi(e^{i(t/2)A})\phi(e^{itB})\phi(e^{i(t/2)A}) - \phi(e^{itC})}{it}\\ &= \frac{\phi(e^{i(t/2)A}e^{itB}e^{i(t/2)A}) - \phi(e^{itC})}{it} \to 0\end{aligned}$$

as $t \to 0$. On the other hand, just as in (2.14) above we have

$$\frac{e^{i(t/2)f(A)}e^{itf(B)}e^{i(t/2)f(A)} - e^{itf(C)}}{it} \to f(A) + f(B) - f(C).$$

This gives us that $f(A)+f(B)-f(A+B) = 0$, i.e., f is additive. The homogeneity of f is trivial to see. Indeed, we have

$$e^{it\lambda f(A)} = \phi(e^{it\lambda A}) = e^{itf(\lambda A)}$$

for every $t, \lambda \in \mathbb{R}$ which implies $\lambda f(A) = f(\lambda A)$. Consequently, $f : \mathcal{A}_s \to \mathcal{B}_s$ is a linear transformation.

To obtain the last statement of the lemma, we use the fact that for any symmetry $V \in \mathcal{A}_u$ the element $\phi(V)$ is also a symmetry and compute

$$\begin{aligned}e^{it\phi(V)f(A)\phi(V)} &= \phi(V)e^{itf(A)}\phi(V) = \phi(V)\phi(e^{itA})\phi(V)\\ &= \phi(Ve^{itA}V) = \phi(e^{itVAV}) = e^{itf(VAV)}.\end{aligned}$$

Since this holds for every $t \in \mathbb{R}$ we deduce the desired equality $f(VAV) = \phi(V)f(A)\phi(V)$ for every $A \in \mathcal{A}_s$ and symmetry $V \in \mathcal{A}_u$.

It remains to prove that f preserves commutativity. Pick commuting elements $A, B \in \mathcal{A}_s$. Then for every $t, s \in \mathbb{R}$ we have

$$e^{itA}e^{i2sB}e^{itA} = e^{isB}e^{i2tA}e^{isB}$$

implying

$$\phi(e^{itA})\phi(e^{i2sB})\phi(e^{itA}) = \phi(e^{isB})\phi(e^{i2tA})\phi(e^{isB})$$

and hence

$$e^{itf(A)}e^{i2sf(B)}e^{itf(A)} = e^{isf(B)}e^{i2tf(A)}e^{isf(B)}.$$

Fixing the real variable s and putting the complex variable z into the place of it we have that the equality

$$e^{zf(A)}e^{i2sf(B)}e^{zf(A)} = e^{isf(B)}e^{2zf(A)}e^{isf(B)}$$

between von Neumann algebra-valued holomorphic (entire) functions of the variable z holds along the imaginary axis in the complex plane. By the uniqueness theorem of holomorphic functions we infer that the above equality necessarily holds on the whole plane. Next, fixing z and inserting the complex variable w into the place of is, the same reasoning leads to the conclusion that the equality

$$e^{zf(A)}e^{2wf(B)}e^{zf(A)} = e^{wf(B)}e^{2zf(A)}e^{wf(B)}$$

holds for all values of the variables $z, w \in \mathbb{C}$. In particular, for arbitrary real numbers t, s setting $z = t/2, w = s/2$ we have

$$\sqrt{e^{tf(A)}}e^{sf(B)}\sqrt{e^{tf(A)}} = \sqrt{e^{sf(B)}}e^{tf(A)}\sqrt{e^{sf(B)}}. \tag{2.15}$$

Just as in the last paragraph of the proof of Lemma 16 we infer from (2.15) that

$$e^{tf(A)}e^{sf(B)} = e^{sf(B)}e^{tf(A)}$$

holds for all $t, s \in \mathbb{R}$. This immediately implies $f(A)f(B) = f(B)f(A)$ which verifies that f indeed preserves commutativity. □

We are now in a position to prove Theorem 8 on the structure of continuous Jordan triple isomorphisms between unitary groups of von Neumann factors.

Proof of Theorem 8. Let $\phi : \mathcal{A}_u \to \mathcal{B}_u$ be a continuous Jordan triple isomorphism. By the first statement in Lemma 22 we know that $\phi(I)$ is a symmetry in $\mathcal{B}$ which commutes with $\mathcal{B}_u$ implying that it is a central element in $\mathcal{B}$. Considering the map $\psi(.) = \phi(I)^{-1}\phi(.)$ we have a unital continuous Jordan triple isomorphism from $\mathcal{A}_u$ onto $\mathcal{B}_u$. Clearly this map and also its inverse send symmetries to symmetries. Apparently, for a symmetry S and unitary U, we have S, U commute if and only if $SUS = U$. This gives us that ψ, ψ^{-1} preserve commutativity between an arbitrary symmetry and an arbitrary unitary. Hence both transformations send central unitaries to central unitaries (in fact, by spectral theorem it is apparent that a unitary is central if and only if it commutes with all symmetries). Since $\mathcal{A}$ is a factor, it has only two central symmetries. Therefore, the same must hold for $\mathcal{B}$, too. This means that $\mathcal{B}$ is also a factor and concerning the central symmetry $\phi(I)$ in $\mathcal{B}$ we have $\phi(I) \in \{I, -I\}$. Without loss of generality we may and do assume that $\phi(I) = I$ (in particular, we have $\psi = \phi$).

Before proceeding further let us consider the case where $\mathcal{A}$ if of type I_1 (i.e., where $\mathcal{A}$ is isomorphic to $\mathbb{C}$). In that case all unitaries in $\mathcal{A}$ are central which implies that the same holds in $\mathcal{B}$, too. This means that $\mathcal{B}$ is commutative and hence it is also isomorphic to $\mathbb{C}$. The same argument applies when $\mathcal{B}$ is assumed to be of type I_1 and yields that $\mathcal{A}$ must be of the same type, too. The structure of all continuous automorphisms of the circle group is well known and one can trivially complete the proof in the particular case where one of $\mathcal{A}, \mathcal{B}$ is of type I_1. So, in what follows we assume that $\mathcal{A}$, $\mathcal{B}$ are not of that type.

Next, by Lemma 22 we have a commutativity preserving linear transformation $f : \mathcal{A}_s \to \mathcal{B}_s$ such that

$$\phi(e^{itA}) = e^{itf(A)}, \quad t \in \mathbb{R}, A \in \mathcal{A}_s. \tag{2.16}$$

We claim that f is bijective. Observe that this would be trivial if we knew that ϕ^{-1} is also a continuous Jordan triple isomorphism. Since continuity of the inverse has not been assumed, we have to find another way to show the bijectivity. By the injectivity of ϕ we easily obtain that f is also injective. Since ϕ sends central unitaries to central unitaries, we obtain that $\exp(itf(I))$ is a scalar (meaning that scalar times the identity) for every $t \in \mathbb{R}$ implying that $f(I)$ is also scalar which is nonzero by the injectivity of f. We show that every projection in $\mathcal{B}$ belongs to the range of f. To see this, pick an arbitrary projection Q from $\mathcal{B}$. Then $\exp(i\pi Q)$ is a symmetry in $\mathcal{B}$ and since the symmetries in $\mathcal{A}_u$ are bijectively mapped onto the symmetries in $\mathcal{B}_u$, it follows that there is a projection P in $\mathcal{A}$ such that

$$e^{i\pi Q} = \phi(e^{i\pi P}) = e^{i\pi f(P)}.$$

From this equality we infer that $f(P) = mI + nQ$ holds for some integers m, n. If Q is a nontrivial projection, then $f(P)$ is obviously not a scalar. Therefore, we have $n \neq 0$ which implies that Q is in the range of f. Therefore, all projections in $\mathcal{B}$ belong to that range. Since every element of a von Neumann factor is the finite linear combination of projections (e.g., see [10]), it follows that f maps $\mathcal{A}_s$ onto $\mathcal{B}_s$. Consequently, f is really bijective.

In the case where $\mathcal{A}$ is of type I_n ($1 < n$ is finite), referring to linear dimensions we obtain that $\mathcal{B}$ is of the same type and hence both algebras are isomorphic to $\mathbb{M}_n$. The statement of Theorem 8 for matrix algebras has been proven in [22], see Corollary 2. So in what follows we assume that none of $\mathcal{A}$, $\mathcal{B}$ is of type I_n, n being finite.

We continue as in the proof of Theorem 5. Namely, we extend f from $\mathcal{A}_s$ onto $\mathcal{A}$ by the formula

$$F(A + iB) = f(A) + if(B), \qquad A, B \in \mathcal{A}_s$$

and obtain a normal preserving bijective linear map $F : \mathcal{A} \to \mathcal{B}$. Just as there we can infer that there is a nonzero real number c, an algebra *-isomorphism or an algebra *-antiisomorphism $\theta : \mathcal{A} \to \mathcal{B}$, and a linear functional $l : \mathcal{A} \to \mathbb{C}$ which is real valued on $\mathcal{A}_s$ such that

$$F(A) = c\theta(A) + l(A)I, \qquad A \in \mathcal{A}.$$

We have

$$\phi(e^{itA}) = e^{it(c\theta(A)+l(A)I)} = e^{itl(A)}\theta(e^{itcA}), \quad t \in \mathbb{R}, A \in \mathcal{A}_s. \tag{2.17}$$

It follows that $\phi(\exp(itA))$ is a scalar multiple of $\theta(\exp(itcA))$ for every $t \in \mathbb{R}$ and $A \in \mathcal{A}_s$. We claim that $c = \pm 1$. To verify this, we again recall that ϕ sends symmetries to symmetries. Let P be a nontrivial projection in $\mathcal{A}$. Then $\exp(i\pi P)$ is a symmetry and it follows that the symmetry $\phi(\exp(i\pi P))$ is a scalar multiple of $\theta(\exp(i\pi cP)) = \exp(i\pi c\theta(P))$, where $\theta(P)$ is a nontrivial projection. It is easy to see that the scalar multiplier in question is necessarily ± 1 and then we obtain that the number $\exp(i\pi c)$ also equals ± 1. From this we get that c is an integer, say $c = m$. Since every element of $\mathcal{A}_u$ is of the form $\exp(iA)$ with some $A \in \mathcal{A}_s$, we thus obtain from (2.17) that $\phi(V)$ is a scalar multiple of $\theta(V^m)$ for every $V \in \mathcal{A}_u$.

By the Jordan triple multiplicativity of ϕ and θ this gives us that the nonzero integer m has the property that $(VWV)^m$ and $V^m W^m V^m$ are scalar multiples of each other whenever $V, W \in \mathcal{A}_u$. Applying Proposition 18 we infer that $m = \pm 1$.

Using (2.17) we can write

$$\phi(U) = \varphi(U)\theta(U^m), \quad U \in \mathcal{A}_u,$$

where the functional $\varphi : \mathcal{A}_u \to \mathbb{T}$ satisfies

$$\varphi(e^{itA}) = e^{itl(A)}, \quad t \in \mathbb{R}, A \in \mathcal{A}_s.$$

Clearly, φ is a continuous Jordan triple map with values in the circle group and we have $\varphi(I) = 1$. It follows that $\varphi(S) = \pm 1$ for any symmetry $S \in \mathcal{A}_u$. Applying the last assertion in Lemma 22 in the particular case where $\mathcal{B} = \mathbb{C}$, we obtain that $l(SAS) = l(A)$ holds for every $A \in \mathcal{A}_s$ and symmetry $S \in \mathcal{A}_u$. By linearity the same holds for any $A \in \mathcal{A}$, too. Then just as in the proof of Lemma 17 we deduce for any $X \in \mathcal{A}$ and symmetry $S \in \mathcal{A}_u$ that $l(SX) = l(S(XS)S) = l(XS)$ implying $l(PX) = l(XP)$ whenever $P \in \mathcal{A}$ is a projection and $X \in \mathcal{A}$. Using the fact that any factor as a linear space is generated by its projections, it follows that l is a tracial linear functional on $\mathcal{A}$. We have mentioned above that $\varphi(S) = \pm 1$ holds for any symmetry $S \in \mathcal{A}_u$. It follows that

$$\pm 1 = \varphi(e^{i\pi P}) = e^{i\pi l(P)}$$

which implies that the value $l(P)$ is an integer for every projection $P \in \mathcal{A}$. Since l is a tracial linear functional on $\mathcal{A}$, it takes equal values on equivalent projections. If $\mathcal{A}$ is of one of the types II_1, II_∞, III, then any nonzero projection P in $\mathcal{A}$ can be written as the sum of an arbitrary finite number of equivalent projections. It follows that the integer $l(P)$ is divisible by any positive integer and this implies that $l(P) = 0$. If $\mathcal{A}$ is of type I_∞, in the same way we obtain that $l(P) = 0$ holds for any infinite projection and then refer to the fact that any finite projection is the difference of two infinite ones. In all those cases we can infer that l vanishes on the set of all projections in $\mathcal{A}$ which then implies that l is zero everywhere.

Therefore, we have $\phi(U) = \theta(U^m)$, $U \in \mathcal{A}_u$ where $m = \pm 1$ and this completes the proof. □

We now can easily prove our last result Theorem 9.

Proof of Theorem 9. First observe that ϕ is also injective. We next define $\psi(U) = \phi(I)^{-1}\, \phi(U)$, $U \in \mathcal{A}_u$. It is apparent that this bijective map also satisfies (1.8). By Theorem 7 ψ is a continuous inverted Jordan triple isomorphism. But it is a unital map and hence easily follows that it is necessarily a continuous Jordan triple isomorphism. Applying Theorem 8 one can complete the proof readily. □

Remark 23. We conclude with a few remarks.

Notice that in our results on the structure of continuous Jordan triple isomorphisms as well as on that of the "generalized isometries" between positive definite cones we have assumed that the underlying algebras are not of type I_2 (while there has not been such an assumption relating to unitary groups). The reason is connected to the use of the structural result concerning normal preserving maps which does not hold in algebras of type I_2. In the case of unitary groups, referring to a result in [22] we could handle that situation but, unfortunately, we have not been able to do so in the case of the positive definite cone. So, it might be rather surprising, but we do not have a proof in the very particular case represented by

2×2 matrices. In fact, we believe the problem is far from being as simple as one might think at the first sight. We leave this as an open problem.[1]

Finally, we emphasize that our results on "generalized isometries" are not "if and only if" type results. Indeed, they assert that every invariance transformation under consideration is of a certain form but it is not necessary that all maps of those given forms have the corresponding invariance properties. This is due to the generality of the circumstances in those results. Hence, in concrete situations one needs to go further and select from the groups of transformations that appear in the conclusions of our results those ones which really have the actual invariance property.

References

[1] W.N. Anderson and G.E. Trapp, *Operator means and electrical networks*, Proc. 1980 IEEE International Symposium on Circuits and Systems (1980), 523–527.

[2] R. Beneduci and L. Molnár, *On the standard K-loop structure of positive invertible elements in a C^*-algebra*, J. Math. Anal. Appl. **420** (2014), 551–562.

[3] R. Bhatia, *Matrix Analysis*, Graduate Texts in Mathematics 169, New York, Springer, (1996).

[4] A.M. Bikchentaev, *On a property of L_p-spaces on semifinite von Neumann algebras*, Math. Notes **64** (1998), 159–163.

[5] F. Botelho, J. Jamison and L. Molnár, *Surjective isometries on Grassmann spaces*, J. Funct. Anal. **265** (2013), 2226–2238.

[6] K.I. Beidar, M. Brešar, M.A. Chebotar and Y. Fong, *Applying functional identities to some linear preserver problems*, Pacific J. Math. **204** (2002), 257–271.

[7] J.T. Chan, C.K. Li and C.C.N. Tu, *A class of unitarily invariant norms on $B(H)$*, Proc. Amer. Math. Soc. **129** (2001), 1065–1076.

[8] H.F. Chau, C.K. Li, Y.T Poon and N.S. Sze, *Induced metric and matrix inequalities on unitary matrices*, J. Phys. A: Math. Theor. **45** (2012), 095201, 8 pp.

[9] B. Fuglede and R.V. Kadison, *Determinant theory in finite factors*, Ann. of Math. **55** (1952), 520–530.

[10] S. Goldstein and A. Paszkiewicz, *Linear combinations of projections in von Neumann algebras*, Proc. Amer. Math. Soc. **116** (1992), 175–183.

[11] H. Halpern, *Commutators in properly infinite von Neumann algebras*, Trans. Amer. Math. Soc. **139** (1969), 55–73.

[12] O. Hatori, G. Hirasawa, T. Miura and L. Molnár, *Isometries and maps compatible with inverted Jordan triple products on groups*, Tokyo J. Math. **35** (2012), 385–410.

[13] O. Hatori and L. Molnár, *Isometries of the unitary group*, Proc. Amer. Math. Soc. **140** (2012), 2141–2154.

[1] Added in proof reading: The problem has recently been solved by the author and D. Virosztek and the solution will soon be published in the joint paper entitled "Continuous Jordan triple endomorphisms of $\mathbb{P}_2$".

[14] O. Hatori and L. Molnár, *Isometries of the unitary groups and Thompson isometries of the spaces of invertible positive elements in C^*-algebras*, J. Math. Anal. Appl. **409** (2014), 158–167.

[15] S. Honma and T. Nogawa, *Isometries of the geodesic distances for the space of invertible positive operators and matrices*, Linear Algebra Appl. **444** (2014), 152–164.

[16] H. Hotje, M. Marchi and S. Pianta, *On a class of point-reflection geometries*, Discrete Math. **129** (1994), 139–147.

[17] R.V. Kadison and J.R. Ringrose, *Fundamentals of the Theory of Operator Algebras, vol. II*, Academic Press, 1986.

[18] H. Karzel, M. Marchi and S. Pianta, *On commutativity in point-reflection geometries*, J. Geom. **44** (1992), 102–106.

[19] C.F. Manara and M. Marchi, *On a class of reflection geometries*, Istit. Lombardo Accad. Sci. Lett. Rend. A **125** (1991), 203–217.

[20] L. Molnár, *A remark to the Kochen–Specker theorem and some characterizations of the determinant on sets of Hermitian matrices*, Proc. Amer. Math. Soc. **134** (2006) 2839–2848.

[21] L. Molnár, *Thompson isometries of the space of invertible positive operators*, Proc. Amer. Math. Soc. **137** (2009), 3849–3859

[22] L. Molnár, *Jordan triple endomorphisms and isometries of unitary groups*, Linear Algebra Appl. **439** (2013), 3518–3531.

[23] L. Molnár, *Jordan triple endomorpisms and isometries of spaces of positive definite matrices*, Linear and Multilinear Alg. **63** (2015), 12–33.

[24] L. Molnár, *A few conditions for a C^*-algebra to be commutative*, Abstr. Appl. Anal. Volume 2014 (2014), Article ID 705836, 4 pages.

[25] L. Molnár and P. Šemrl, *Transformations of the unitary group on a Hilbert space*, J. Math. Anal. Appl. **388** (2012), 1205–1217.

[26] L. Molnár and P. Szokol, *Transformations on positive definite matrices preserving generalized distance measures*, Linear Algebra Appl. **466** (2015), 141–159.

Lajos Molnár
MTA-DE "Lendület" Functional Analysis Research Group
Institute of Mathematics
University of Debrecen
P.O. Box 12
H-4010 Debrecen, Hungary
e-mail: `molnarl@science.unideb.hu`
URL: `http://www.math.unideb.hu/~molnarl/`

Operator Theory:
Advances and Applications, Vol. 250, 343–351

Traces of Non-regular Vector Fields on Lipschitz Domains

Sylvie Monniaux

Abstract. In this note, for Lipschitz domains $\Omega \subset \mathbb{R}^n$, I propose to show the boundedness of the trace operator for functions from $H^1(\Omega)$ to $L^2(\partial\Omega)$ as well as for square integrable vector fields in L^2 with square integrable divergence and curl satisfying a half boundary condition. Such results already exist in the literature. The originality of this work lies on the control of the constants involved. The proofs are based on integration by parts formulas applied to the right expressions.

Mathematics Subject Classification (2010). Primary 35B65; Secondary 35J56.

Keywords. Traces, Lipschitz domains, integration by parts.

1. Introduction

It is well known that for a bounded Lipschitz domain $\Omega \subset \mathbb{R}^n$, the trace operator $\mathrm{Tr}_{\partial\Omega} : \mathscr{C}(\overline{\Omega}) \to \mathscr{C}(\partial\Omega)$ restricted to $\mathscr{C}(\overline{\Omega}) \cap H^1(\Omega)$ extends to a bounded operator from $H^1(\Omega)$ to $L^2(\partial\Omega)$ and the following estimate holds:

$$\|\mathrm{Tr}_{\partial\Omega}\, u\|_{L^2(\partial\Omega)} \leq C \left(\|u\|_{L^2(\Omega)} + \|\nabla u\|_{L^2(\Omega,\mathbb{R}^n)}\right) \qquad \text{for all } u \in H^1(\Omega), \tag{1.1}$$

where $C = C(\Omega) > 0$ is a constant depending on the domain Ω. This result can be proved via a simple integration by parts (see, e.g., [6]). If the domain is the upper graph of a Lipschitz function, i.e.,

$$\Omega = \left\{x = (x_h, x_n) \in \mathbb{R}^{n-1} \times \mathbb{R}; x_n > \omega(x_h)\right\} \tag{1.2}$$

where $\omega : \mathbb{R}^{n-1} \to \mathbb{R}$ is a globally Lipschitz function, the method presented here allows an explicit constant C in (1.1) to be given. We pass from domains of type (1.2) to bounded Lipschitz domains via a partition of unity.

The same question arises for vector fields instead of scalar functions. In dimension 3, Costabel [1] gave the following estimate for square integrable vector fields u in a bounded Lipschitz domain with square integrable rotational and divergence and either $\nu \times u$ or $\nu \times u$ square integrable on the boundary (ν denotes the outer unit normal of Ω):

$$\|\mathrm{Tr}_{\partial\Omega}\, u\|_{L^2(\partial\Omega)} \le C\Big(\|u\|_{L^2(\Omega)} + \|\mathrm{curl}\, u\|_{L^2(\Omega,\mathbb{R}^n)} + \|\mathrm{div}\, u\|_{L^2(\Omega)} + \min\big\{\|\nu\cdot u\|_{L^2(\partial\Omega)}, \|\nu\times u\|_{L^2(\partial\Omega,\mathbb{R}^n)}\big\}\Big). \tag{1.3}$$

This result was generalized to differential forms on Lipschitz domains of compact manifolds (and L^p for certain $p \neq 2$) by D. Mitrea, M. Mitrea and M. Taylor in [5, Theorem 11.2]. As for scalar functions on bounded Lipschitz domains (or special Lipschitz domains as (1.2)), we can prove a similar estimate for vector fields (see Theorem 4.2 and Theorem 4.3 below) using essentially integration by parts.

2. Tools and notations

2.1. About the domains

Let $\Omega \subset \mathbb{R}^n$ be a domain of the form (1.2). The exterior unit normal ν of Ω at a point $x = (x_h, \omega(x_h)) \in \Gamma$ on the boundary of Ω,

$$\Gamma := \big\{x = (x_h, x_n) \in \mathbb{R}^{n-1} \times \mathbb{R};\, x_n = \omega(x_h)\big\}, \tag{2.1}$$

is given by

$$\nu(x_h, \omega(x_h)) = \frac{1}{\sqrt{1 + |\nabla_h \omega(x_h)|^2}}\,(\nabla_h \omega(x_h), -1) \tag{2.2}$$

(∇_h denotes the "horizontal gradient" on $\mathbb{R}^{n-1}$ acting on the "horizontal variable" x_h). We denote by $\theta \in [0, \frac{\pi}{2})$ the angle

$$\theta = \arccos\left(\inf_{x_h \in \mathbb{R}^{n-1}} \frac{1}{\sqrt{1 + |\nabla_h \omega(x_h)|^2}}\right), \tag{2.3}$$

so that in particular for $e = (0_{\mathbb{R}^{n-1}}, 1)$ the "vertical" direction, we have

$$-e \cdot \nu(x_h, \omega(x_h)) = \frac{1}{\sqrt{1 + |\nabla_h \omega(x_h)|^2}} \ge \cos\theta, \quad \text{for all } x_h \in \mathbb{R}^{n-1}. \tag{2.4}$$

2.2. Vector fields

We assume here that $\Omega \subset \mathbb{R}^n$ is either a special Lipschitz domain of the form (1.2) or a bounded Lipschitz domain. Let $u : \Omega \to \mathbb{R}^n$ be an $\mathbb{R}^n$-valued distribution. We denote by $\mathrm{curl}\, u \in \mathscr{M}_n(\mathbb{R})$ the antisymmetric part of the Jacobian matrix of first-order partial derivatives considered in the sense of distributions, i.e., $\nabla u = (\partial_\ell u_\alpha)_{1 \le \ell, \alpha \le n}$:

$$\big(\mathrm{curl}\, u\big)_{\ell,\alpha} = \tfrac{1}{\sqrt{2}}\,(\partial_\ell u_\alpha - \partial_\alpha u_\ell) = \tfrac{1}{\sqrt{2}}\,\big(\nabla u - (\nabla u)^\top\big)_{\ell,\alpha}, \quad 1 \le \ell, \alpha \le n. \tag{2.5}$$

On $\mathscr{M}_n(\mathbb{R})$, we choose the following scalar product:

$$\langle v, w\rangle := \sum_{\ell,\alpha=1}^{n} v_{\ell,\alpha} w_{\ell,\alpha}, \quad v = (v_{\ell,\alpha})_{1\leq \ell,\alpha\leq n}, w = (w_{\ell,\alpha})_{1\leq \ell,\alpha\leq n} \in \mathscr{M}_n(\mathbb{R}). \quad (2.6)$$

We will use the notation $|\cdot|$ for the norm associated to the previous scalar product:

$$|w| = \langle w, w\rangle^{\frac{1}{2}}, \quad w \in \mathscr{M}_n(\mathbb{R}). \quad (2.7)$$

Remark 2.1. In dimension 3, if we denote by $\operatorname{rot} u$ the usual rotational of a smooth vector field u, i.e.,

$$\mathbb{R}^3 \ni \operatorname{rot} u = (\partial_2 u_3 - \partial_3 u_2, \partial_3 u_1 - \partial_1 u_3, \partial_1 u_2 - \partial_2 u_1),$$

it is immediate that $|\operatorname{rot} u|$, the Euclidian norm in $\mathbb{R}^3$ (also denoted by $|\cdot|$) of $\operatorname{rot} u$, is equal to $|\operatorname{curl} u|$.

To proceed, we define the wedge product of two vectors as follows:

$$e \wedge \varepsilon := \tfrac{1}{\sqrt{2}}\left(e_\ell \varepsilon_\alpha - e_\alpha \varepsilon_\ell\right)_{1\leq \ell,\alpha\leq n} \in \mathscr{M}_n(\mathbb{R}), \quad e, \varepsilon \in \mathbb{R}^n. \quad (2.8)$$

It is immediate that $e\wedge e = 0$, $e\wedge\varepsilon = -\varepsilon\wedge e$ and we obtain the higher-dimensional version of a well-known formula in $\mathbb{R}^3$:

$$|e|^2|\varepsilon|^2 = (e\cdot\varepsilon)^2 + |e\wedge\varepsilon|^2, \quad e, \varepsilon \in \mathbb{R}^n \quad (2.9)$$

as a consequence of the decomposition

$$\varepsilon = (e\cdot\varepsilon)\, e - \sqrt{2}\,(e\wedge\varepsilon)\, e, \quad e, \varepsilon \in \mathbb{R}^n. \quad (2.10)$$

One can also verify that for three vectors $e, \varepsilon, \nu \in \mathbb{R}^n$, the two following identities hold:

$$\langle e\wedge\varepsilon, \nu\wedge\varepsilon\rangle = (e\cdot\nu)|\nu\wedge\varepsilon|^2 + (\nu\cdot\varepsilon)\langle e\wedge\nu, \nu\wedge\varepsilon\rangle, \quad (2.11)$$

$$(e\cdot\varepsilon)(\nu\cdot\varepsilon) = (e\cdot\nu)(\nu\cdot\varepsilon)^2 - (\nu\cdot\varepsilon)\langle e\wedge\nu, \nu\wedge\varepsilon\rangle. \quad (2.12)$$

If $u : \Omega \to \mathbb{R}^n$ and $\varphi : \Omega \to \mathbb{R}$ are both smooth, the following holds:

$$\operatorname{curl}(\varphi u) = \varphi \operatorname{curl} u + \nabla\varphi \wedge u. \quad (2.13)$$

The (formal) transpose of the curl operator given by (2.5) acts on matrix-valued distributions $w = (w_{\ell,\alpha})_{1\leq \ell,\alpha\leq n}$ according to

$$\left(\operatorname{curl}^\top w\right)_\ell = \frac{1}{\sqrt{2}} \sum_{\alpha=1}^{n} \partial_\alpha (w_{\ell,\alpha} - w_{\alpha,\ell}), \quad 1 \leq \ell \leq n. \quad (2.14)$$

As usual, the divergence of a vector field $u : \Omega \to \mathbb{R}^n$ of distributions is denoted by $\operatorname{div} u$ and is the trace of the matrix ∇u:

$$\operatorname{div} u = \sum_{\ell=1}^{n} \partial_\ell u_\ell. \quad (2.15)$$

Let now $u : \Omega \to \mathbb{R}^n$ be a vector field of distributions and let $e \in \mathbb{R}^n$ be a fixed vector. Then the following formula holds:

$$\operatorname{curl}^\top (e \wedge u) = (\operatorname{div} u)\, e - (e \cdot \nabla)\, u \in \mathbb{R}^n, \tag{2.16}$$

where the notation $e \cdot \nabla$ stands for $\sum_{\ell=1}^n e_\ell \partial_\ell$. Next, for $\varphi : \overline{\Omega} \to \mathbb{R}$, $u : \overline{\Omega} \to \mathbb{R}^n$ and $w : \overline{\Omega} \to \mathscr{M}_n(\mathbb{R})$ smooth with compact supports in $\overline{\Omega}$, the following integration by parts formulas are easy to verify:

$$\int_\Omega \varphi\, (\operatorname{div} u)\, \mathrm{d}x = -\int_\Omega \nabla\varphi \cdot u\, \mathrm{d}x + \int_{\partial\Omega} \varphi\, (\nu \cdot u)\, \mathrm{d}\sigma, \tag{2.17}$$

$$\int_\Omega \langle w, \operatorname{curl} u\rangle\, \mathrm{d}x = \int_\Omega \operatorname{curl}^\top w \cdot u\, \mathrm{d}x + \int_{\partial\Omega} \langle w, \nu \wedge u\rangle\, \mathrm{d}\sigma, \tag{2.18}$$

where $\partial\Omega$ is the boundary of the Lipschitz domain Ω and $\nu(x)$ denotes the exterior unit normal of Ω at a point $x \in \partial\Omega$. The equation (2.17) corresponds to the well-known divergence theorem. The equation (2.18) generalizes in higher dimensions the more popular corresponding integration by parts in dimension 3 (see, e.g., [1, formula (2)]):

$$\int_\Omega w \cdot \operatorname{rot} u\, \mathrm{d}x = \int_\Omega \operatorname{rot} w \cdot u\, \mathrm{d}x + \int_{\partial\Omega} w \cdot (\nu \times u)\, \mathrm{d}\sigma, \quad u, w : \overline{\Omega} \to \mathbb{R}^3 \text{ smooth,}$$

where $\nu \times u = (\nu_2 u_3 - \nu_3 u_2, \nu_3 u_1 - \nu_1 u_3, \nu_1 u_2, \nu_2 u_1)$ denotes the usual 3D vector product. Combining the previous results, we are now in position to present our last formula which will be used in Section 4: for $e \in \mathbb{R}^n$ a fixed vector and $u : \overline{\Omega} \to \mathbb{R}^n$ a smooth vector field,

$$\begin{aligned} 2\int_\Omega \langle e \wedge u, \operatorname{curl} u\rangle\, \mathrm{d}x &- 2\int_\Omega (e \cdot u) \operatorname{div} u\, \mathrm{d}x \\ &= \int_{\partial\Omega} \langle e \wedge u, \nu \wedge u\rangle\, \mathrm{d}\sigma - \int_{\partial\Omega} (e \cdot u)(\nu \cdot u)\, \mathrm{d}\sigma. \end{aligned} \tag{2.19}$$

3. The scalar case

3.1. Special Lipschitz domains

We assume here that Ω is of the form (1.2). The following result is classical (see, e.g., [7, Theorem 1.2]). We will propose an elementary proof of it.

Theorem 3.1. *Let $\varphi : \Omega \to \mathbb{R}$ belong to the Sobolev space $H^1(\Omega)$. Then $\operatorname{Tr}_\Gamma \varphi \in L^2(\Gamma)$ and*

$$\|\operatorname{Tr}_\Gamma \varphi\|^2_{L^2(\Gamma)} \leq \frac{2}{\cos\theta}\, \|\varphi\|_{L^2(\Omega)} \|\nabla\varphi\|_{L^2(\Omega,\mathbb{R}^n)}, \tag{3.1}$$

where θ has been defined in (2.3). In other words, the trace operator originally defined on continuous functions $\operatorname{Tr}_\Gamma : \mathscr{C}_c(\overline{\Omega}) \to \mathscr{C}_c(\Gamma)$ extends to a bounded operator from $H^1(\Omega)$ to $L^2(\Gamma)$ with a norm controlled by the Lipschitz character of Ω.

Proof. Assume first that $\varphi : \overline{\Omega} \to \mathbb{R}$ is smooth, and apply the divergence theorem with $u = \varphi^2 e$ where $e = (0_{\mathbb{R}^{n-1}}, 1)$. Since $\operatorname{div}(\varphi^2 e) = 2\varphi(e \cdot \nabla\varphi)$, we obtain

$$\int_\Omega \operatorname{div}(\varphi^2 e)\,\mathrm{d}x = \int_\Omega 2\varphi(e \cdot \nabla\varphi)\,\mathrm{d}x = \int_\Gamma \nu \cdot (\varphi^2 e)\,\mathrm{d}\sigma.$$

Therefore using the definition of θ and Cauchy–Schwarz inequality, we get

$$\cos\theta \int_\Gamma \varphi^2\,\mathrm{d}\sigma \le -2\int_\Omega \varphi(e \cdot \nabla\varphi)\,\mathrm{d}x \le 2\,\|\varphi\|_{L^2(\Omega)}\|\nabla\varphi\|_{L^2(\Omega,\mathbb{R}^n)}, \tag{3.2}$$

since $|e| = 1$, which gives the estimate (3.1) for smooth functions φ. Since $\mathscr{C}_c(\overline{\Omega})$ is dense in $H^1(\Omega)$ (see, e.g., [2, Theorem 4.7, p. 248] or [7, §1.1.1]), we conclude easily that (3.1) holds for every $\varphi \in H^1(\Omega)$. □

3.2. Bounded Lipschitz domains

Let now Ω be a bounded Lipschitz domain. Then there exist $N \in \mathbb{N}$, a partition of unity $(\eta_k)_{1\le k\le N}$ of $\mathscr{C}_c^\infty(\mathbb{R}^n)$-functions and domains $(\Omega_k)_{1\le k\le N}$ such that

$$\overline{\Omega} \cap \left(\bigcup_{k=1}^N \Omega_k\right) = \overline{\Omega}, \quad \operatorname{supp}\eta_k \subset \Omega_k \ \ (1 \le k \le N),$$
$$0 \le \eta_k \le 1 \ \ (1 \le k \le N)$$

and

$$\sum_{k=1}^N \eta_k(x)^2 = 1 \quad \text{for all } x \in \Omega. \tag{3.3}$$

Matters can be arranged such that, for $1 \le k \le N$, there is a direction e_k and an angle $\theta_k \in [0, \frac{\pi}{2})$ such that $-e_k \cdot \nu(x) \ge \cos\theta_k$ for all $x \in \partial\Omega \cap \Omega_k$ (see, e.g., [7, §1.1.3]). We denote by γ the minimum of all $\cos\theta_k$, $1 \le k \le N$: γ depends only on the boundary of Ω. We are now in position to state the following result, analogue to Theorem 3.1 in the case of bounded Lipschitz domains.

Theorem 3.2. *Let $\Omega \subset \mathbb{R}^n$ be a bounded Lipschitz domain. Then there exists a constant $C = C(\Omega) > 0$ such that for all $\varphi \in H^1(\Omega)$, $\mathrm{Tr}_{\partial\Omega}\varphi \in L^2(\partial\Omega)$ and the following estimate holds:*

$$\|\mathrm{Tr}_{\partial\Omega}\varphi\|^2_{L^2(\partial\Omega)} \le \frac{1}{\gamma}\,\|\varphi\|_{L^2(\Omega)}\Big(2\,\|\nabla\varphi\|_{L^2(\Omega,\mathbb{R}^n)} + C(\Omega)\,\|\varphi\|_{L^2(\Omega)}\Big). \tag{3.4}$$

Remark 3.3. Compared to Theorem 3.1, the estimate (3.4) contains the extra term $\|\varphi\|^2_{L^2(\Omega)}$. An estimate of the form (3.1) cannot hold in bounded Lipschitz domains as the example of constant functions shows.

Proof. Let η_k, Ω_k, $1 \le k \le N$, as in (3.3), and let $\gamma := \min\{\cos\theta_k, 1 \le k \le N\}$. Using (3.2) for the functions $\eta_k\varphi$, $1 \le k \le N$, we obtain

$$\begin{aligned}\gamma \int_{\partial\Omega} \varphi^2 \,\mathrm{d}\sigma = \gamma \sum_{k=1}^{N} \int_{\partial\Omega} \eta_k^2 \varphi^2 \,\mathrm{d}\sigma \;&\le\; 2\Big|\sum_{k=1}^{N} \int_{\Omega} \eta_k\varphi\,(e_k \cdot \nabla(\eta_k\varphi))\,\mathrm{d}x\Big| \\ &\le 2\Big|\int_{\Omega} \varphi\,\nabla\varphi \cdot \big(\sum_{k=1}^{N} \eta_k^2 e_k\big)\,\mathrm{d}x\Big| + \Big|\int_{\Omega} \varphi^2 \sum_{k=1}^{N} (e_k \cdot \nabla(\eta_k^2))\,\mathrm{d}x\Big| \\ &\le 2\,\|\varphi\|_{L^2(\Omega)} \|\nabla\varphi\|_{L^2(\Omega,\mathbb{R}^n)} + \Big(\sum_{k=1}^{N} \|\nabla(\eta_k^2)\|_{L^\infty(\Omega,\mathbb{R}^n)}\Big)\,\|\varphi\|^2_{L^2(\Omega)}\end{aligned}$$

which proves the estimate (3.4) with

$$C(\Omega) = \sum_{k=1}^{N} \|\nabla(\eta_k^2)\|_{L^\infty(\Omega,\mathbb{R}^n)}.$$

□

4. The case of vector fields

We begin this section by a remark allowing us to make sense of values on the boundary of certain quantities involving vectors fields with minimal smoothness. See also [1, equations (2) and (3)].

Remark 4.1.

1. For $u \in L^2(\Omega;\mathbb{R}^n)$ such that $\operatorname{div} u \in L^2(\Omega)$, one can define $\nu \cdot u$ as a distribution on $\partial\Omega$ as follows: for any $\phi \in H^{\frac12}(\partial\Omega)$, we denote by Φ an extension of ϕ to Ω in $H^1(\Omega)$ (see, e.g., [4, Theorem 3, Chap. VII, §2, p. 197]) and we define, according to (2.17),
$$_{H^{-\frac12}(\partial\Omega)}\langle \nu \cdot u, \phi\rangle_{H^{\frac12}(\partial\Omega)} = \int_\Omega \Phi \operatorname{div} u \,\mathrm{d}x + \int_\Omega u \cdot \nabla\Phi \,\mathrm{d}x; \tag{4.1}$$
this definition is independent of the choice of the extension Φ of ϕ. See, e.g., [8, Theorem 1.2].
2. Following the same lines, for $u \in L^2(\Omega;\mathbb{R}^n)$ such that $\operatorname{curl} u \in L^2(\Omega;\mathscr{M}_n(\mathbb{R}))$, one can define $\nu \wedge u$ as a distribution in $H^{-\frac12}(\partial\Omega;\mathscr{M}_n(\mathbb{R}))$ as follows: for any $\psi \in H^{\frac12}(\partial\Omega;\mathscr{M}_n(\mathbb{R}))$, we denote by Ψ an extension of ψ to Ω in $H^1(\Omega;\mathscr{M}_n(\mathbb{R}))$ and we define, according to (2.18),
$$\begin{aligned}&_{H^{-\frac12}(\partial\Omega,\mathscr{M}_n(\mathbb{R}))}\langle \nu \wedge u, \psi\rangle_{H^{\frac12}(\partial\Omega;\mathscr{M}_n(\mathbb{R}))} \\ &\quad = \int_\Omega \langle \Psi, \operatorname{curl} u\rangle \,\mathrm{d}x - \int_\Omega \operatorname{curl}^\top \Psi \cdot u \,\mathrm{d}x;\end{aligned} \tag{4.2}$$
this definition is independent of the choice of the extension Ψ of ψ. See, e.g., [3, Theorem 2.5] for the case $n = 3$ and [5, Chap. 11] for the more general setting of differential forms.

4.1. Special Lipschitz domains

Theorem 4.2. *Let Ω be a special Lipschitz domain of the form* (1.2) *and let θ be defined by* (2.3). *Let $u \in L^2(\Omega, \mathbb{R}^n)$ such that* $\operatorname{div} u \in L^2(\Omega)$ *and* $\operatorname{curl} u \in L^2(\Omega, \mathscr{M}_n(\mathbb{R}))$. *If $\nu \cdot u \in L^2(\Gamma)$ or $\nu \wedge u \in L^2(\Gamma, \mathscr{M}_n(\mathbb{R}))$, then* $\operatorname{Tr}_\Gamma u \in L^2(\Gamma, \mathbb{R}^n)$ *and*

$$
\begin{aligned}
&\max\big\{\|\nu \cdot u\|^2_{L^2(\Gamma)}, \|\nu \wedge u\|^2_{L^2(\Gamma,\mathscr{M}_n(\mathbb{R}))}\big\} \\
&\quad\leq \frac{2}{\cos\theta}\Big(\frac{2}{\cos\theta} + 1\Big) \min\big\{\|\nu \cdot u\|^2_{L^2(\Gamma)}, \|\nu \wedge u\|^2_{L^2(\Gamma,\mathscr{M}_n(\mathbb{R}))}\big\} \\
&\qquad + \frac{4}{\cos\theta}\, \|u\|_{L^2(\Omega,\mathbb{R}^n)} \big(\|\operatorname{curl} u\|_{L^2(\Omega,\mathscr{M}_n(\mathbb{R}))} + \|\operatorname{div} u\|_{L^2(\Omega)}\big),
\end{aligned} \tag{4.3}
$$

and

$$
\begin{aligned}
&\|\operatorname{Tr}_\Gamma u\|^2_{L^2(\Gamma,\mathbb{R}^n)} \\
&\quad\leq \Big(\frac{4}{\cos^2\theta} + \frac{2}{\cos\theta} + 1\Big) \min\big\{\|\nu \cdot u\|^2_{L^2(\Gamma)}, \|\nu \wedge u\|^2_{L^2(\Gamma,\mathscr{M}_n(\mathbb{R}))}\big\} \\
&\qquad + \frac{4}{\cos\theta}\, \|u\|_{L^2(\Omega,\mathbb{R}^n)} \big(\|\operatorname{curl} u\|_{L^2(\Omega,\mathscr{M}_n(\mathbb{R}))} + \|\operatorname{div} u\|_{L^2(\Omega)}\big).
\end{aligned} \tag{4.4}
$$

Proof. Assume first that $u : \overline{\Omega} \to \mathbb{R}^n$ is smooth, and apply (2.19) together with (2.11) and (2.12):

$$
\begin{aligned}
&\int_\Gamma (e\cdot\nu)|\nu\wedge u|^2\, d\sigma + 2\int_\Gamma (\nu\cdot u)\langle e\wedge\nu, \nu\wedge u\rangle\, d\sigma - \int_\Gamma (e\cdot\nu)(\nu\cdot u)^2\, d\sigma \\
&\quad = 2\int_\Omega \langle e\wedge u, \operatorname{curl} u\rangle\, dx - 2\int_\Omega (e\cdot u)\operatorname{div} u\, dx.
\end{aligned} \tag{4.5}
$$

Denote now by M the maximum between $\|\nu \cdot u\|_{L^2(\Gamma)}$ and $\|\nu \wedge u\|_{L^2(\Gamma,\mathscr{M}_n(\mathbb{R}))}$ and by m the minimum between the same quantities, so that in particular

$$
Mm = \|\nu \cdot u\|_{L^2(\Gamma)} \|\nu \wedge u\|_{L^2(\Gamma,\mathscr{M}_n(\mathbb{R}))}. \tag{4.6}
$$

Taking into account that $|e\cdot\nu| \leq 1$ and $|e\wedge\nu| \leq 1$, the equation (4.5) together with the estimate (2.4) for $\cos\theta$ and Cauchy–Schwarz inequality yield

$$
M^2\cos\theta \leq m^2 + 2mM + 2\|u\|_{L^2(\Omega,\mathbb{R}^n)}\big(\|\operatorname{curl} u\|_{L^2(\Omega,\mathscr{M}_n(\mathbb{R}))} + \|\operatorname{div} u\|_{L^2(\Omega)}\big). \tag{4.7}
$$

The obvious inequality $2mM \leq \frac{\cos\theta}{2} M^2 + \frac{2}{\cos\theta} m^2$ then implies

$$
\tfrac{\cos\theta}{2} M^2 \leq \big(1 + \tfrac{2}{\cos\theta}\big) m^2 + 2\|u\|_{L^2(\Omega,\mathbb{R}^n)}\big(\|\operatorname{curl} u\|_{L^2(\Omega,\mathscr{M}_n(\mathbb{R}))} + \|\operatorname{div} u\|_{L^2(\Omega)}\big), \tag{4.8}
$$

which gives (4.3) from which (4.4) follows immediately thanks to (2.10) and (2.9) for smooth vector fields. As in the proof of Theorem 3.1, we conclude by density of smooth vector fields in the space

$$
\big\{u \in L^2(\Omega,\mathbb{R}^n), \operatorname{div} u \in L^2(\Omega), \operatorname{curl} u \in L^2(\Omega,\mathscr{M}_n(\mathbb{R})), \nu\cdot u \in L^2(\Gamma)\big\} \tag{4.9}
$$

or in the space

$$
\big\{u \in L^2(\Omega,\mathbb{R}^n), \operatorname{div} u \in L^2(\Omega), \operatorname{curl} u \in L^2(\Omega,\mathscr{M}_n(\mathbb{R})), \nu\wedge u \in L^2(\Gamma,\mathscr{M}_n(\mathbb{R}))\big\} \tag{4.10}
$$

endowed with their natural norms. □

4.2. Bounded Lipschitz domains

In the case of bounded Lipschitz domains, Theorem 4.2 becomes

Theorem 4.3. *Let $\Omega \subset \mathbb{R}^n$ be a bounded Lipschitz domain and let γ be defined as in § 3.2. Then there exists a constant $C = C(\Omega) > 0$ with the following significance: let $u \in L^2(\Omega, \mathbb{R}^n)$ such that $\operatorname{div} u \in L^2(\Omega)$ and $\operatorname{curl} u \in L^2(\Omega, \mathscr{M}_n(\mathbb{R}))$. If $\nu \cdot u \in L^2(\partial\Omega)$ or $\nu \wedge u \in L^2(\partial\Omega, \mathscr{M}_n(\mathbb{R}))$, then $\mathrm{Tr}_{\partial\Omega} u \in L^2(\partial\Omega, \mathbb{R}^n)$ and*

$$\begin{aligned}
&\max\bigl\{\|\nu\cdot u\|^2_{L^2(\partial\Omega)}, \|\nu\wedge u\|^2_{L^2(\partial\Omega,\mathscr{M}_n(\mathbb{R}))}\bigr\}\\
&\quad\le \frac{2}{\gamma}\left(\frac{2}{\gamma}+1\right)\min\bigl\{\|\nu\cdot u\|^2_{L^2(\partial\Omega)}, \|\nu\wedge u\|^2_{L^2(\partial\Omega,\mathscr{M}_n(\mathbb{R}))}\bigr\} \qquad (4.11)\\
&\qquad + \frac{2}{\gamma}\|u\|_{L^2(\Omega,\mathbb{R}^n)}\bigl(2\|\operatorname{curl} u\|_{L^2(\Omega,\mathscr{M}_n(\mathbb{R}))} + 2\|\operatorname{div} u\|_{L^2(\Omega)} + C(\Omega)\|u\|_{L^2(\Omega,\mathbb{R}^n)}\bigr),
\end{aligned}$$

and

$$\begin{aligned}
&\|\mathrm{Tr}_{\partial\Omega} u\|^2_{L^2(\partial\Omega,\mathbb{R}^n)}\\
&\quad\le \left(\frac{4}{\gamma^2}+\frac{2}{\gamma}+1\right)\min\bigl\{\|\nu\cdot u\|^2_{L^2(\partial\Omega)}, \|\nu\wedge u\|^2_{L^2(\partial\Omega,\mathscr{M}_n(\mathbb{R}))}\bigr\} \qquad (4.12)\\
&\qquad + \frac{2}{\gamma}\|u\|_{L^2(\Omega,\mathbb{R}^n)}\bigl(2\|\operatorname{curl} u\|_{L^2(\Omega,\mathscr{M}_n(\mathbb{R}))} + 2\|\operatorname{div} u\|_{L^2(\Omega)} + C(\Omega)\|u\|_{L^2(\Omega,\mathbb{R}^n)}\bigr).
\end{aligned}$$

Proof. As in the proof of Theorem 3.2, let η_k, Ω_k, $1 \le k \le N$ and denote by γ the minimum of all $\cos\theta_k$, $\gamma = \min\{\cos\theta_k, 1 \le k \le N\}$. Using the formula (2.13) and the fact that $\operatorname{div}(\varphi u) = \varphi \operatorname{div} u + \nabla\varphi \cdot u$ for (smooth) scalar functions φ, we apply (4.5) for the N vector fields $\eta_k u$, $1 \le k \le N$, and we obtain, summing over k,

$$\begin{aligned}
\gamma M \le m^2 + 2Mm &+ 2\|u\|_{L^2(\Omega;\mathbb{R}^n)}\bigl(\|\operatorname{curl} u\|_{L^2(\Omega,\mathscr{M}_n(\mathbb{R}))} + \|\operatorname{div} u\|_{L^2(\Omega)}\bigr)\\
&+ \left(\sum_{k=1}^N \|\nabla(\eta_k^2)\|_\infty\right)\|u\|^2_{L^2(\Omega;\mathbb{R}^n)}, \qquad (4.13)
\end{aligned}$$

where, as in the proof of Theorem 4.2,

$$M := \max\bigl\{\|\nu\cdot u\|_{L^2(\partial\Omega)}, \|\nu\wedge u\|_{L^2(\partial\Omega,\mathscr{M}_n(\mathbb{R}))}\bigr\}$$

and

$$m := \min\bigl\{\|\nu\cdot u\|_{L^2(\partial\Omega)}, \|\nu\wedge u\|_{L^2(\partial\Omega,\mathscr{M}_n(\mathbb{R}))}\bigr\}.$$

This gives (4.11) with

$$C(\Omega) = \sum_{k=1}^N \|\nabla(\eta_k^2)\|_\infty.$$

As before, (4.12) follows immediately thanks to (2.10) and (2.9) for smooth vector fields. We conclude by density of smooth vector fields in the spaces (4.9) and (4.10). □

References

[1] M. Costabel, *A remark on the regularity of solutions of Maxwell's equations on Lipschitz domains*, Math. Methods Appl. Sci. **12** (1990), no. 4, 365–368.

[2] D.E. Edmunds and W.D. Evans, *Spectral theory and differential operators*, Oxford Mathematical Monographs, The Clarendon Press, Oxford University Press, New York, 1987, Oxford Science Publications.

[3] V. Girault and P.-A. Raviart, *Finite element approximation of the Navier–Stokes equations*, Lecture Notes in Mathematics, vol. 749, Springer-Verlag, Berlin, 1979.

[4] A. Jonsson and H. Wallin, *Function spaces on subsets of* $\mathbf{R}^n$, Math. Rep. **2** (1984), no. 1, xiv+221.

[5] D. Mitrea, M. Mitrea, and M. Taylor, *Layer potentials, the Hodge Laplacian, and global boundary problems in nonsmooth Riemannian manifolds*, Mem. Amer. Math. Soc. **150** (2001), no. 713, x+120.

[6] S. Monniaux, *A three lines proof for traces of* H^1 *functions on special Lipschitz domains*, Ulmer Sem. **19**(2014), 339–340.

[7] J. Nečas, *Direct methods in the theory of elliptic equations*, Springer Monographs in Mathematics, Springer, Heidelberg, 2012, translated from the 1967 French original by Gerard Tronel and Alois Kufner, editorial coordination and preface by Šárka Nečasová and a contribution by Christian G. Simader.

[8] R. Temam, *Navier–Stokes equations*, revised ed., Studies in Mathematics and its Applications, vol. 2, North-Holland Publishing Co., Amsterdam, 1979, Theory and numerical analysis, with an appendix by F. Thomasset.

Sylvie Monniaux
Aix-Marseille Université
CNRS, Centrale Marseille
I2M, UMR 7373
F-13453 Marseille, France
e-mail: sylvie.monniaux@univ-amu.fr

Operator Theory:
Advances and Applications, Vol. 250, 353–368

The L^p-Poincaré Inequality for Analytic Ornstein–Uhlenbeck Semigroups

Jan van Neerven

Abstract. Consider the linear stochastic evolution equation

$$dU(t) = AU(t)\,dt + dW_H(t), \quad t \geqslant 0,$$

where A generates a C_0-semigroup on a Banach space E and W_H is a cylindrical Brownian motion in a continuously embedded Hilbert subspace H of E. Under the assumption that the solutions to this equation admit an invariant measure μ_∞ we prove that if the associated Ornstein–Uhlenbeck semigroup is analytic and has compact resolvent, then the Poincaré inequality

$$\|f - \overline{f}\|_{L^p(E,\mu_\infty)} \leqslant \|D_H f\|_{L^p(E,\mu_\infty)}$$

holds for all $1 < p < \infty$. Here $\overline{f}$ denotes the average of f with respect to μ_∞ and D_H the Fréchet derivative in the direction of H.

Mathematics Subject Classification (2010). Primary 47D07; Secondary: 35R15, 35R60.

Keywords. Analytic Ornstein–Uhlenbeck semigroups, Poincaré inequality, compact resolvent, joint functional calculus.

1. Introduction

Let E be a real Banach space and let H be a Hilbert subspace of E, with continuous embedding $i : H \hookrightarrow E$. Let A be the generator of a C_0-semigroup $S = (S(t))_{t\geqslant 0}$ on E and let W_H be a cylindrical Brownian motion in H. Under the assumption that the linear stochastic evolution equation

$$dU(t) = AU(t) + dW_H(t), \quad t \geqslant 0, \tag{1.1}$$

The author gratefully acknowledges financial support by VICI subsidy 639.033.604 of the Netherlands Organisation for Scientific Research (NWO).

has an invariant measure μ_∞, we wish to establish sufficient conditions for the validity of the Poincaré inequality

$$\|f - \overline{f}\|_{L^p(E,\mu_\infty)} \leqslant C\|D_H f\|_{L^p(E,\mu_\infty)}, \quad 1 < p < \infty.$$

Here $\overline{f}$ denotes the average of f with respect to μ_∞ and D_H the directional Fréchet derivative in the direction of H (see (2.4) below). To the best of our knowledge, this problem has been considered so far only for $p = 2$ and Hilbert spaces E. For this setting, Chojnowska-Michalik and Goldys [5] obtained various necessary and sufficient conditions for the inequality to be true. Here we show that these conditions are equivalent to another, formally weaker, condition and that these equivalent conditions imply the validity of the Poincaré inequality for all $1 < p < \infty$ (Theorem 2.4). Our proof depends crucially on the L^p-gradient estimates for analytic Ornstein–Uhlenbeck semigroups obtained in the recent papers [25, 26].

Related L^p-Poincaré inequalities have been proved in various other settings, e.g., for the classical Ornstein–Uhlenbeck semigroup (this corresponds to the case $A = -I$ of the setting considered here) [32, Eq. (2.5)], for the Walsh system [11], and in certain non-commutative situations [17, 35]. Poincaré inequalities are intimately related to other functional inequalities such as, log-Sobolev inequalities and transportation cost inequalities, and imply concentration-of-measure inequalities. For a comprehensive study of these topics we refer the reader to the recent monograph of Bakry, Gentil and Ledoux [1].

As an application of Theorem 2.4 we find that the L^p-Poincaré inequality holds if the Ornstein–Uhlenbeck semigroup P associated with (1.1) (see (2.1)) is analytic on $L^p(E,\mu_\infty)$ and has compact resolvent. In Section 3 we provide some examples in which the various assumptions are satisfied. In the final Section 4 we address the problem of compactness of certain tensor products of resolvents naturally associated with P.

All vector spaces are real. We will always identify Hilbert spaces with their dual via the Riesz representation theorem. The domain, kernel, and range of a linear operator A will be denoted by $\mathsf{D}(A)$, $\mathsf{N}(A)$, and $\mathsf{R}(A)$, respectively. We write $a \lesssim b$ to mean that there exists a constant C, independent of a and b, such that $a \leqslant Cb$.

2. The L^p-Poincaré inequality

Throughout this note we fix a Banach space E and a Hilbert subspace H of E, with continuous embedding $i : H \hookrightarrow E$, and make the following assumption.

Assumption 2.1. There exists a centred Gaussian Radon measure μ_∞ on E whose covariance operator $Q_\infty \in \mathscr{L}(E^*, E)$ is given by

$$\langle Q_\infty x^*, y^* \rangle = \int_0^\infty \langle QS^*(s)x^*, S^*(s)y^* \rangle \, ds, \quad x^*, y^* \in E^*.$$

Here $Q := i \circ i^*$; we identify H and its dual in the usual way. The convergence of the integrals on the right-hand side is part of the assumption. As is well known, Assumption 2.1 is equivalent to the existence of an invariant measure for the problem (1.1); we refer the reader to [10, 16] for the details. In fact, the measure μ_∞ is the minimal (in the sense of covariance domination) invariant measure for (1.1).

The formula

$$P(t)f(x) = \mathbb{E}(f(U(t,x))), \quad t \geqslant 0, \ x \in E, \tag{2.1}$$

where $U(t,x)$ denotes the unique mild solution of (1.1) with initial value x, defines a semigroup of linear contractions $P = (P(t))_{t\geqslant 0}$ on the space $B_{\mathrm{b}}(E)$ of bounded real-valued Borel functions on E. This semigroup is called the *Ornstein–Uhlenbeck semigroup* associated with the pair (A, H). By an easy application of Hölder's inequality, this semigroup extends uniquely to C_0-semigroup of contractions on $L^p(E, \mu_\infty)$, which we shall also denote by P. Its generator will be denoted by L.

By a result of Chojnowska-Michalik and Goldys [4, 5] (see [28] for the formulation of this result in its present generality), the reproducing kernel Hilbert space H_∞ associated with the measure μ_∞ is invariant under the semigroup S and the restriction of S is a C_0-semigroup of contractions on H_∞. We shall denote this restricted semigroup by S_∞ and its generator by A_∞. The inclusion mapping $H_\infty \hookrightarrow E$ will be denoted by i_∞; recall that $Q_\infty = i_\infty \circ i_\infty^*$ (see [16, 28]).

It has been shown in [4] (see also [28, 29]) that $P(t)$ is the so-called *second quantisation* of the adjoint semigroup $S_\infty^*(t)$. More precisely, the Wiener–Itô isometry establishes an isometric identification $L^2(E, \mu_\infty) = \bigoplus_{n\geqslant 0} H_\infty^{\circledS n}$, where $H_\infty^{\circledS n}$ is the n-fold symmetric tensor product of H_∞ (the so-called nth Wiener–Itô chaos), and under this isometry we have

$$P(t) = \bigoplus_{n\geqslant 0} S_\infty^{*\circledS n}(t).$$

We have $H_\infty^{\circledS 0} = \mathbb{R}\mathbf{1}$ (by definition) and $H_\infty^{\circledS 1} = H_\infty$. The latter identification allows us to deduce many properties of P from the corresponding properties of S_∞^* and vice versa and will be used freely in what follows.

Following [3, 16] we define $\mathscr{F}^k$ as the space of all functions $f : E \to \mathbb{R}$ of the form

$$f(x) = \phi(\langle x, x_1^* \rangle, \ldots, \langle x, x_d^* \rangle) \tag{2.2}$$

for some $d \geqslant 1$, with $x_j^* \in E^*$ for all $j = 1, \ldots, d$ with $\phi \in C_{\mathrm{b}}^k(\mathbb{R}^d)$. Let

$$\mathscr{F}_A^k = \{f \in \mathscr{F}^k : \ x_j^* \in \mathsf{D}(A^*) \text{ for all } j = 1, \ldots, d \text{ and } \langle \cdot\,, A^* Df(\cdot) \rangle \in C_{\mathrm{b}}(E)\}.$$

It follows from [3, 16] that $\mathscr{F}_A^2$ is a core for $\mathsf{D}(L)$ in each $L^p(E, \mu_\infty)$ and that for $f, g \in \mathscr{F}_A^2$ we have the identity

$$\langle Lf, g \rangle + \langle Lg, f \rangle = -\int_E \langle D_H f, D_H g \rangle \, d\mu_\infty. \tag{2.3}$$

Here D_H denotes the Fréchet derivative in the direction of H, defined on $\mathscr{F}^1$ by

$$D_H f(x) := \sum_{j=1}^{n} \frac{\partial \phi}{\partial x_j}(\langle x, x_1^*\rangle, \dots, \langle x, x_n^*\rangle)\, i^* x_j^* \tag{2.4}$$

with f and ϕ as in (2.2). It should be emphasized that D_H is not always closable; various conditions for closability as well as a counterexample are given in [15]. If P is analytic on $L^p(E, \mu_\infty)$ for some/all $1 < p < \infty$ (the equivalence being a consequence of the Stein interpolation theorem), then D_H is closable as an operator from $L^p(E, \mu_\infty)$ to $L^p(E, \mu_\infty; H)$ [16, Proposition 8.7].

The following necessary and sufficient condition for the L^2-Poincaré inequality is essentially due to Chojnowska-Michalik and Goldys [6] (see also [9, Proposition 10.5.2]). Since the present formulation is slightly more general, for the convenience we include the proof which follows the lines of [6].

Proposition 2.2 (Poincaré inequality, the case $p = 2$). *Let Assumption* 2.1 *hold and fix a number $\omega > 0$. If D_H is closable as a densely defined operator in $L^2(E, \mu_\infty)$, then the following assertions are equivalent:*

(1) $\|S_\infty(t)\| \leqslant e^{-\omega t}$ *for all* $t \geqslant 0$*;*

(2) *The Poincaré inequality*

$$\|f - \overline{f}\|_{L^2(E,\mu_\infty)} \leqslant \frac{1}{\sqrt{2\omega}} \|D_H f\|_{L^2(E,\mu_\infty)}, \quad f \in \mathsf{D}(D_H),$$

holds. Here, $\overline{f} = \int_E f\, d\mu_\infty$.

Proof. (1)⇒(2): Since $t \mapsto e^{\omega t} S_\infty^*(t)$ is a C_0-contraction semigroup, by second quantisation the same is true for the direct sum for $n \geqslant 1$ of their n-fold symmetric tensor products, $\bigoplus_{n \geqslant 1} e^{n\omega t} S_\infty^{*\circledS n}(t)$. Replacing $e^{n\omega t}$ by $e^{\omega t}$, the resulting direct sum $\bigoplus_{n \geqslant 1} e^{\omega t} S_\infty^{*\circledS n}(t)$ is contractive as well. This semigroup is generated by the part $L_0 + \omega$ of $L + \omega$ in $L_0^2(E, \mu_\infty) := L^2(E, \mu_\infty) \ominus \mathbb{R}\mathbf{1}$. Thus we obtain the dissipativity inequality

$$-\langle (L_0 + \omega) f, f\rangle \geqslant 0, \quad f \in \mathsf{D}(L_0).$$

In view of (2.3), this gives the inequality

$$\omega \|f\|_2^2 \leqslant -\langle L_0 f, f\rangle = \frac{1}{2}\|D_H f\|_2^2, \quad f \in \mathsf{D}(L_0) \cap \mathscr{F}_A^2.$$

As a consequence,

$$\omega \|f - \overline{f}\|_2^2 \leqslant \frac{1}{2}\|D_H f\|_2^2, \quad f \in \mathscr{F}_A^2. \tag{2.5}$$

It is routine (albeit somewhat tedious) to check that the inequality (2.5) extends to $f \in \mathscr{F}^1$, and since by definition this is a core for $\mathsf{D}(D_H)$ it extends to arbitrary elements $g \in \mathsf{D}(D_H)$.

(2)⇒(1): Every $x^* \in E^*$, when viewed as an element of $L^2(E, \mu_\infty)$, satisfies $D_H x^* = i^* x^*$. Moreover, if $x^* \in \mathsf{D}(A^*)$, then $A_\infty^* x^* \in \mathsf{D}(A_\infty^*)$, and therefore (identifying H_∞ with the first Wiener–Itô chaos) $x^* \in \mathsf{D}(L)$ as an element of $L^2(E,\mu_\infty)$.

By specialising the Poincaré inequality to functionals x^* we obtain the inequality

$$\|i_\infty^* x^*\| = \|x^*\|_{L^2(E,\mu_\infty)} \leqslant \frac{1}{\sqrt{2\omega}}\|i^*x^*\|, \quad x^* \in \mathsf{D}(A^*).$$

In the same way, (2.3) takes the form

$$\langle A_\infty^* i_\infty^* x^*, i_\infty^* x^*\rangle = -\frac{1}{2}\|i^*x^*\|^2, \quad x^* \in \mathsf{D}(A^*).$$

Combining these inequalities, we obtain

$$-\langle A_\infty^* i_\infty^* x^*, i_\infty^* x^*\rangle \geqslant \omega\|i_\infty^* x^*\|^2, \quad x^* \in \mathsf{D}(A^*).$$

Since the elements $i_\infty^* x^*$ with $x^* \in \mathsf{D}(A^*)$ form a core for $\mathsf{D}(A_\infty^*)$, this is equivalent to saying that $A_\infty^* + \omega$ is dissipative on H_∞. It follows that $\|S_\infty^*(t)\| \leqslant \exp(-\omega t)$ for all $t \geqslant 0$. □

The main result of this note (Theorem 2.4) asserts that if P is analytic and A_∞^* has closed range, then all conditions of Proposition 2.2 are satisfied and the Poincaré inequality extends to $L^p(E,\mu_\infty)$ for all $1 < p < \infty$. To prepare for the proof we need to recall some preliminary facts. We begin by imposing the following assumption, which will be in force for the rest of this section.

Assumption 2.3. For some (equivalently, for all) $1 < p < \infty$ the semigroup P extends to an analytic C_0-semigroup on $L^p(E,\mu_\infty)$.

The problem of analyticity of P has been studied in several articles [13, 14, 16, 24, 26]. In these, necessary and sufficient conditions for analyticity can be found. We have already mentioned the fact that if P is analytic on $L^p(E,\mu_\infty)$ for some/all $1 < p < \infty$, then D_H is closable as an operator from $L^p(E,\mu_\infty)$ to $L^p(E,\mu_\infty;H)$. In what follows, D_H will always denote this closure and $\mathsf{D}(D_H)$ its domain in $L^p(E,\mu)$. Note that there is a slight abuse of notation here, as $\mathsf{D}(D_H)$ obviously depends on p. The choice of p will always be clear from the context, and for this reason we prefer not to overburden notations. The same slight abuse of notation applies to the notation $\mathsf{D}(L)$ for the domain of L in $L^p(E,\mu_\infty)$.

From [24] we know that if P is analytic, then the generator L of P can be represented as

$$L = D_H^* B D_H \tag{2.6}$$

for a unique bounded operator B on H which satisfies

$$B + B^* = -I.$$

The rigorous interpretation of (2.6) is that for $p = 2$ the operator $-L$ is the sectorial operator associated with the closed continuous accretive form

$$(f,g) \mapsto -\langle BD_H f, D_H g\rangle.$$

In the sequel we will use the standard fact (which is proved by hypercontractivity arguments) that for each $n \geqslant 0$ the summand $H_\infty^{\circledS n}$ in the Wiener–Itô decomposition for $L^2(E,\mu_\infty)$ is contained as a closed subspace in $L^p(E,\mu_\infty)$ for all $1<p<\infty$. In view of this we will continue to refer to $H_\infty^{\circledS n}$ as the nth Wiener chaos. By an interpolating argument (see [29, Lemma 4.2]) we obtain the estimate $\|P(t)\|_p \leqslant \|S_\infty(t)\|^{n\theta_p}$ on each of these subspaces, with a constant $0<\theta_p<1$ depending only on p. Summing over $n \geqslant 1$ and passing to the closure of the linear span, we obtain the estimate

$$\|P(t)\|_p \leqslant \|S_\infty(t)\|^{\theta_p} \quad \text{on } L^p(E,\mu_\infty) \ominus \mathbb{R}\mathbf{1}. \tag{2.7}$$

Theorem 2.4 (L^p-Poincaré inequality). *Let Assumptions* 2.1 *and* 2.3 *hold. Then the following assertions are equivalent:*

(1) A_∞^* *has closed range;*
(2) *there exists* $\omega>0$ *such that* $\|S_\infty(t)\| \leqslant e^{-\omega t}$ *for all* $t \geqslant 0$*;*
(3) *there exist* $M \geqslant 1$ *and* $\omega>0$ *such that* $\|S_\infty(t)\| \leqslant Me^{-\omega t}$ *for all* $t \geqslant 0$*;*
(4) *there exist* $M \geqslant 1$ *and* $\omega>0$ *such that* $\|S_H(t)\| \leqslant Me^{-\omega t}$ *for all* $t \geqslant 0$*;*
(5) H_∞ *embeds continuously in* H*;*
(6) *for some* $1<p<\infty$ *there exists a finite constant* $C \geqslant 0$ *such that*
$$\|f-\overline{f}\|_{L^p(E,\mu_\infty)} \leqslant C_p\|D_H f\|_{L^p(E,\mu_\infty)}, \quad f \in \mathsf{D}(D_H);$$
(7) *for all* $1<p<\infty$ *there exists a finite constant* $C \geqslant 0$ *such that*
$$\|f-\overline{f}\|_{L^p(E,\mu_\infty)} \leqslant C_p\|D_H f\|_{L^p(E,\mu_\infty)}, \quad f \in \mathsf{D}(D_H).$$

In what follows we will say that *the L^p-Poincaré inequality holds* if condition (7) is satisfied.

Before we start with the proof we recall some further useful facts. Firstly, on the first Wiener chaos, (2.6) reduces to the identity

$$A_\infty^* = V^*BV,$$

where V is the closure of the mapping $i_\infty^* x^* \mapsto i^* x^*$; see [15, 25, 26]. Secondly, in [26] it is shown that Assumption 2.3 implies that S maps H into itself and that its restriction to H extends to a bounded analytic C_0-semigroup on H. We shall denote this semigroup by S_H and its generator by A_H.

Proof of Theorem 2.4. (1)⇒(3): Let us first observe that the strong stability of S_∞^* [16, Proposition 2.4] implies that $\mathsf{N}(A_\infty^*)=\{0\}$.

Suppose next that some $h \in H_\infty$ annihilates the range of A_∞^*. As $\langle A_\infty^* g, h\rangle = \langle V^*BVg, h\rangle = 0$ for all $g \in \mathsf{D}(A_\infty^*)$, it follows that $h \in \mathsf{D}(V)$ and $\langle BV^*g, Vh\rangle = 0$ for all $g \in \mathsf{D}(A_\infty^*)$. Using that $\mathsf{D}(A_\infty^*)$ is a core for $\mathsf{D}(V)$ (see [25]), it follows that $\langle BV^*g, Vh\rangle = 0$ for all $g \in \mathsf{D}(V)$. In particular, $\langle BV^*h, Vh\rangle = 0$. Since also $\langle BV^*h, Vh\rangle = -\frac{1}{2}\|Vh\|^2$ by the identity $B+B^*=-I$, it follows that $Vh=0$ and therefore $h \in \mathsf{N}(A_\infty^*) = \mathsf{N}(V)$. But we have already seen that $\mathsf{N}(A_\infty^*)=\{0\}$ and we conclude that $h=0$.

This argument proves that $\mathsf{R}(A_\infty^*)$ is dense. Since by assumption A_∞^* has closed range, it follows that A_∞^* is surjective. As we observed at the beginning of the proof, A^*_∞ is also injective, and therefore A^*_∞ is boundedly invertible by the closed graph theorem. Since A^*_∞ generates an analytic C_0-contraction semigroup, the spectral mapping theorem for analytic C_0-semigroups (see [12]) implies that S^*_∞ is uniformly exponentially stable.

(3)⇒(7): Fix an arbitrary $1 < p < \infty$. Fix a function $f \in \mathscr{F}^0$ and let $\frac{1}{p} + \frac{1}{q}$. Then

$$\|f - \overline{f}\| = \sup_{\substack{\|g\|_q \leqslant 1 \\ \overline{g}=0}} |\langle f - \overline{f}, g\rangle| = \sup_{\substack{\|g\|_q \leqslant 1 \\ \overline{g}=0}} |\langle f - \overline{f}, g - \overline{g}\rangle| = \sup_{\substack{\|g\|_q \leqslant 1 \\ \overline{g}=0}} |\langle f, g - \overline{g}\rangle|,$$

where it suffices to consider functions $g \in \mathscr{F}^0$. Next we observe that, by (2.7),

$$\langle f, g - \overline{g}\rangle = \lim_{t\to\infty} \langle f, g - P(t)g\rangle.$$

Following an argument in [22, Lemma 3] we have

$$\langle f, g - P(t)g\rangle = -\int_0^t \langle f, LP(s)g\rangle\, ds = -\int_0^t \langle D_H f, BD_H P(s)g\rangle\, ds.$$

If in addition $\overline{g} = 0$ (i.e., if $g \in L^p(E, \mu_\infty) \ominus \mathbb{R}\mathbf{1}$), then for all $t \geqslant 1$ we have

$$\begin{aligned} |\langle f, g - P(t)g\rangle| &\leqslant \|B\|\|D_H f\|_p \Big(\int_0^1 + \int_1^\infty\Big) \|D_H P(s)g\|_q\, ds \\ &\lesssim \|D_H f\|_p \Big(\int_0^1 \frac{1}{\sqrt{s}} \|g\|_q\, ds + \|D_H P(1)\| \int_0^\infty e^{-\omega\theta_q} \|g\|_q\, ds\Big). \end{aligned}$$

where we used the gradient estimates of [25] and (2.7). Taking the supremum over all $g \in \mathscr{F}^0$ of L^q-norm 1 with $\overline{g} = 0$, this gives

$$\|f - \overline{f}\|_p \lesssim \|D_H f\|_p.$$

Since $\mathscr{F}^0$ is a core for $\mathsf{D}(D_H)$ this concludes the proof of the implication.

(7)⇒(6): This implication is trivial.

(6)⇒(3): This follows from Proposition 2.2 along with the fact that H_∞ is isomorphic to the first Wiener–Itô chaos in $L^p(E, \mu_\infty)$.

(3)⇒(1): The uniform exponential stability of S^*_∞ implies that A^*_∞ is boundedly invertible.

(3)⇔(4)⇔(5): These equivalences have been proved in [16, Theorem 5.4].

(7)⇒(2): This follows from Proposition 2.2.

(2)⇒(3): Trivial. □

The equivalent conditions of the theorem do not in general imply the existence of an $\omega > 0$ such that $\|S_H(t)\| \leqslant e^{-\omega t}$ for all $t \geqslant 0$:

Example 2.5. Consider the Dirichlet Laplacian Δ on $E = L^2(-1, 1)$ and take $H = E$. Let S denote the heat semigroup generated by Δ on this space. Fix $\omega > 0$. As is well known and easy to check, Assumptions 2.1 and 2.3 are satisfied for the operator $\Delta - \omega$. Let us now replace the norm of $L^2(-1, 1)$ by the equivalent (Hilbertian) norm

$$\|f\|^2_{(r)} := \|f|_{(-1,0)}\|^2 + r^2 \|f|_{(0,1)}\|^2,$$

where $r > 0$ is a positive scalar. Starting from an initial condition with support in $(-1, 0)$, the semigroup $s_\omega(t) = e^{-\omega t}S(t)$ generated by $\Delta - \omega$ will instantaneously spread out the support of f over the entire interval $(-1, 1)$. Hence if we fix $t_0 > 0$ and $\omega > 0$ we may choose $r_0 > 0$ so large that

$$\|S_\omega(t_0)f\|_{(r)} > \|f\|_{(r)}.$$

As a result, the semigroup S_ω is uniformly exponentially stable but not contractive on $L^2(-1, 1)$ endowed with the norm $\|\cdot\|_{(r_0)}$.

One could object to this example that there is an equivalent Hilbertian norm (namely, the original norm of $L^2(-1, 1)$) on which we do have $\|S_\omega(t)\| \leqslant e^{-\omega t}$. There exist examples, however, of bounded analytic Hilbert space semigroups which are not similar to an analytic contraction semigroup. Such examples may be realised as multiplication semigroups on a suitable (pathological) Schauder basis (see, e.g., [21] and the references given there). For such examples, Assumptions 2.1 and 2.3 are again satisfied and we obtain a counterexample that cannot be repaired by a Hilbertian renorming.

As an application of Theorem 2.4 we have the following sufficient condition for the validity of the L^p-Poincaré inequality.

Theorem 2.6 (Compactness implies the L^p-Poincaré inequality). *Let Assumptions 2.1 and 2.3 hold and fix $1 < p < \infty$. The following assertions are equivalent:*

(1) *L has compact resolvent on $L^p(E, \mu_\infty)$;*
(2) *P is compact on $L^p(E, \mu_\infty)$;*
(3) *A_∞ has compact resolvent on H_∞;*
(4) *S_∞ is compact on H_∞;*
(5) *A_H has compact resolvent on H;*
(6) *S_H is compact on H.*

If these equivalent conditions are satisfied, then the L^p-Poincaré inequality holds for all $1 < p < \infty$.

Proof. The equivalences (1)$\Leftrightarrow$(2), (3)$\Leftrightarrow$(4), and (5)$\Leftrightarrow$(6) follow from [12, Theorem 4.29] since P, S_∞, and S_H are analytic semigroups.

We will prove next that (4) implies the validity of the L^p-Poincaré inequality. We will use some elementary facts from semigroup theory which can all be found in [12]. The strong stability of S_∞^* implies that 1 is not an eigenvalue of $S_\infty^*(t)$ for any $t > 0$. Since these operators are compact it follows that $1 \notin \sigma(S_\infty^*(t))$, which in turn implies that $0 \notin \sigma(A_\infty^*)$ by the spectral mapping theorem for eventually norm continuous semigroups. By the equality spectral bound and growth bound for such semigroups, it follows that S_∞^* (and hence also S_∞) is uniformly exponentially stable. We may now apply Theorem 2.4 to obtain the conclusion.

(2)$\Rightarrow$(4): This follows by restricting to the first Wiener–Itô chaos.

(4)$\Rightarrow$(2): We have already seen that (4) implies that S_∞^* is uniformly exponentially stable. Because of this, the compactness of $S_\infty^*(t)$ implies, by second quantisation, the compactness of $P(t)$ on $L^p(E, \mu_\infty)$ (cf. [29, Lemma 4.2]).

(4)⇒(6): By [16, Theorem 3.5] combined with [28, Proposition 1.3], for each $t > 0$ the operator $S(t)$ maps H into H_∞; we shall denote this operator by $S_{H,\infty}(t)$. Furthermore we have a continuous embedding $i_{\infty,H} : H_\infty \hookrightarrow H$ [16, Theorem 5.4] (this result can be applied here since, by what has already been proved, (2) implies the uniform exponential stability of S_∞). Now if S_∞ is compact, the compactness of S_H follows from the factorisation

$$S_H(t) = i_{\infty,H} \circ S_\infty(t/2) \circ S_{H,\infty}(t/2).$$

(6)⇒(4): We will show that (6) implies that H_∞ embeds into H. Once we know this, (4) follows from the factorisation $S_\infty(t) = S_{H,\infty}(t/2) \circ S_H(t/2) \circ i_{\infty,H}$.

This concludes the proof of the equivalences of the conditions (1)–(6). To complete the proof we will now show that these conditions imply the validity of the Poincaré inequality.

Suppose that $h \in H$ is a vector satisfying $S_H(t)h = h$ for all $t \geqslant 0$. Since $S(t)$ maps H into H_∞ (see [16, Proposition 2.3]) this means that $h \in H_\infty$. But then in E for all $t \geqslant 0$ we have $i_\infty S_\infty(t)h = i_H S_H(t)h = i_H h = i_\infty h$, so that in H_∞ we obtain $S_\infty(t)h = h$ for all $t \geqslant 0$. Hence, for all $h' \in H_\infty$,

$$\langle h, h'\rangle_{H_\infty} = \lim_{t\to\infty} \langle S_\infty(t)h, h'\rangle_{H_\infty} = \lim_{t\to\infty} \langle h, S_\infty^*(t)h'\rangle_{H_\infty} = 0$$

by the strong stability of S_∞^*. This being true for all $h' \in H_\infty$, it follows that $h = 0$. We have thus shown that 1 is not an eigenvalue of $S_H(t)$. Having arrived at this conclusion, the argument given above for S_∞ can now be repeated to conclude that S_H is uniformly exponentially stable. Now Theorem 2.4 implies that H_∞ embeds into H. □

Remark 2.7. The equivalence of (4) and (6) for symmetric Ornstein–Uhlenbeck semigroups follows from [8, Theorem 2.9].

Corollary 2.8. *Let $1 < p < \infty$. If the embedding $\mathsf{D}(D_H) \hookrightarrow L^p(E, \mu_\infty)$ is compact, then the L^p-Poincaré inequality holds.*

Recall our abuse of notation to denote by $\mathsf{D}(D_H)$ and $\mathsf{D}(L)$ the domains of closed operators D_H and L in $L^p(E, \mu_\infty)$. Necessary and sufficient conditions for the compactness of the embedding $\mathsf{D}(D_H) \hookrightarrow L^p(E, \mu_\infty)$ are stated in [15].

Proof. Since $\mathsf{D}(L)$ embeds into $\mathsf{D}(D_H)$ (see [25, Theorem 8.2]) this is immediate from the previous theorem. □

Our next aim is to show that also an L^p-inequality holds for the adjoint operator D_H^*. Here we view D_H as a closed densely defined operator from $L^q(E, \mu_\infty)$ into $\overline{\mathsf{R}(D_H)}$ and D_H^* a closed densely defined operator from $\overline{\mathsf{R}(D_H)}$ into $L^p(E, \mu_\infty)$, $\frac{1}{p} + \frac{1}{q} = 1$. The proof relies on some facts that have been proved in [25, 26]. We start by observing that if Assumptions 2.1 and 2.3 hold, then the semigroup

$$\underline{P}(t) := P(t) \otimes S_H^*(t)$$

extends to a bounded analytic C_0-semigroup on $L^p(E,\mu_\infty;H)$, $1<p<\infty$. We will need the fact that on $\overline{\mathsf{R}(D_H)}$ the generator $\underline{L}$ of this semigroup is given by

$$\underline{L} = D_H D_H^* B;$$

the proof as well as the rigorous interpretation of the right-hand side is given in the references just quoted.

Theorem 2.9 (L^p-Poincaré inequality for D_H^*). *Let Assumptions* 2.1 *and* 2.3 *hold. If the equivalent conditions of Theorem* 2.4 *are satisfied, then there exists a finite constant $C\geqslant 0$ such that for all $1<p<\infty$ we have*

$$\|f\|_{L^p(E,\mu_\infty;H)} \leqslant C_p \|D_H^* f\|_{L^p(E,\mu_\infty;H)}, \quad f\in \mathsf{D}(D_H^*),$$

where D_H^* is interpreted as explained above.

Proof. We can follow the proof of Theorem 2.4, this time using that for bounded cylindrical functions $f,g\in\overline{\mathsf{R}(D_H)}$ we have

$$\langle f, g-\underline{P}(t)g\rangle = -\int_0^t \langle f, \underline{LP}(s)g\rangle\,ds = -\int_0^t \langle D_H^* f, D_H^* B\underline{P}(s)g\rangle\,ds.$$

For $t\geqslant 1$ we then have

$$\begin{aligned}|\langle f, g-\underline{P}(t)g\rangle| &\leqslant \|D_H^* f\|_p\Big(\int_0^1 + \int_1^\infty\Big)\|D_H^* B\underline{P}(s)g\|_q\,ds\\ &\lesssim \|D_H^* f\|_p\Big(\int_0^1 \frac{1}{\sqrt{s}}\|g\|_q\,ds + \|D_H^* B\underline{P}(1)\|\int_0^\infty e^{-\omega\theta_q}\|g\|_q\,ds\Big),\end{aligned}$$

this time using the gradient estimates for $D_H^* B$ (cf. the proof of [25, Proposition 9.3] where resolvents are used instead of the semigroup operators) and the uniform exponential stability of $\underline{P}=P\otimes S_H^*$. The proof can be finished along the lines of Theorem 2.4; this time we use that $\lim_{t\to\infty}\langle f, g-\underline{P}(t)g\rangle = \langle f,g\rangle$. □

3. Examples

Example 3.1 (Finite dimensions and non-degenerate noise). Suppose that $H=E=\mathbb{R}^d$ and let Assumption 2.1 hold. Then $H^\infty=\mathbb{R}^d$. Under these assumptions, a result of Fuhrman [13, Theorem 3.6 and Corollary 3.8] implies that Assumption 2.3 holds. By finite-dimensionality, the conditions of Theorems 2.4 and 2.9 are satisfied. It follows that the L^p-Poincaré inequalities for D_H and D_H^* hold for $1<p<\infty$.

Example 3.2 (The self-adjoint case). Suppose that $H=E$ and S is self-adjoint on E. Then Assumption 2.1 holds if and only if S is uniformly exponentially stable. In this situation, by [16] also S_∞ is self-adjoint and uniformly exponentially stable, and P is self-adjoint on $L^2(E,\mu_\infty)$. In particular, Assumption 2.3 then holds and therefore the equivalent conditions of Theorem 2.4 are satisfied. It follows that the L^p-Poincaré inequality holds for $1<p<\infty$.

Example 3.3 (The strong Feller case). Suppose that Assumptions 2.1 and 2.3 hold, and that P is strongly Feller. As is well known, this is equivalent to the condition that for each $t > 0$ the semigroup operator $S(t)$ maps E into the reproducing kernel Hilbert space H_t associated with μ_t, the centred Gaussian Radon measure on E whose covariance operator $Q_t \in \mathscr{L}(E^*, E)$ is given by

$$\langle Q_t x^*, y^* \rangle = \int_0^t \langle QS^*(s)x^*, S^*(s)y^* \rangle \, ds, \quad x^*, y^* \in E^*.$$

These measures exist by a standard covariance domination argument (note that $\langle Q_t x^*, x^* \rangle \leqslant \langle Q_\infty x^*, x^* \rangle$). By [28] we have a contractive embedding $i_{i,\infty} : H_t \hookrightarrow H_\infty$. Then $S_\infty(t) = i_{t,\infty} \circ S(t) \circ i_\infty$, where $i_\infty : H_\infty \hookrightarrow E$ is the inclusion mapping. The compactness of $i_\infty : H_\infty \hookrightarrow E$ (this mapping being γ-radonifying; see [30]) implies that $S_\infty(t)$ is compact for all $t > 0$, and by a general result from semigroup theory this implies that the resolvent operators $R(\lambda, A_\infty)$ are compact. Similarly from $S_H(t) = i_{t,\infty} i_{\infty,H} \circ S(t) \circ i_\infty$, where $i_{\infty,H} : H_\infty \hookrightarrow H$ is the embedding mapping (see [16, Theorem 5.4] for the proof that this inclusion holds under the present assumptions) it follows that $S_H(t)$ is compact and therefore $R(\lambda, A_H)$ are compact. It follows that the L^p-Poincaré inequalities for D_H and D_H^* hold for $1 < p < \infty$.

Example 3.4 (The case $\mathsf{D}(A) \hookrightarrow H_\infty$). Suppose that Assumptions 2.1 and 2.3 hold, and that we have a continuous inclusion $\mathsf{D}(A) \hookrightarrow H_\infty$. Then $R(\lambda, A_\infty) = i_A R(\lambda, A) i_\infty$, where $i_\infty : H_\infty \hookrightarrow E$ and $i_A : \mathsf{D}(A) \hookrightarrow H_\infty$ are the inclusion mappings. The compactness of $i_\infty : H_\infty \hookrightarrow E$ implies that $R(\lambda, A_\infty)$ is compact. It follows that the L^p-Poincaré inequality for D_H holds for $1 < p < \infty$. A similar argument (using again that $H_\infty \hookrightarrow H$) shows that if the inclusion $H \hookrightarrow E$ is compact, then $R(\lambda, A_H)$ is compact as well and the L^p-Poincaré inequalities for D_H and D_H^* hold for $1 < p < \infty$.

In fact the same results hold if $\mathsf{D}(A^n) \hookrightarrow H_\infty$ for some large enough $n \geqslant 1$. We give the argument for $n = 2$; it is clear from this argument that we may proceed inductively to prove the general case. For $n = 2$ we repeat the above proof we now obtain $\mu R(\mu, A_\infty) R(\lambda, A_\infty) = \mu i_{A^2} R(\mu, A) R(\lambda, A) i_\infty$, where $i_\infty : H_\infty \hookrightarrow E$ and $i_{A^2} : \mathsf{D}(A^2) \hookrightarrow H_\infty$ are the inclusion mappings. It follows that $\mu R(\mu, A_\infty) R(\lambda, A_\infty)$ is compact for each $\mu \in \varrho(A_\infty)$. Passing to the limit $\mu \to \infty$, noting that by the resolvent identity we have

$$\begin{aligned}
&\Big\| \mu R(\mu, A_\infty) R(\lambda, A_\infty) - R(\lambda, A_\infty) \Big\| \\
&\quad = \Big\| \frac{\mu}{\mu - \lambda} (R(\lambda, A_\infty) - R(\mu, A_\infty)) - R(\lambda, A_\infty) \Big\| \\
&\quad \leqslant \Big\| \frac{\mu}{\mu - \lambda} R(\mu, A_\infty) \Big\| + \Big\| \Big(\frac{\mu}{\mu - \lambda} - 1 \Big) R(\lambda, A_\infty) \Big\|,
\end{aligned}$$

and using that $\|R(\nu, A_\infty)\| \leqslant 1/\nu$, it follows that $R(\lambda, A_\infty)$ is compact, being the uniform limit of compact operators.

4. Compactness results

In [5], a condition equivalent to the Poincaré inequality has been used to prove, under an additional Hilbert–Schmidt assumption, the compactness of the semigroup $P \otimes S_H^*$ on $L^p(E, \mu_\infty; H)$. The importance of this semigroup is apparent from the proof of Theorem 2.9 and the results in [5, 7, 25, 26] where this semigroup plays a crucial rôle in identifying the domains of $\sqrt{-L}$ and L. Here we wish to show that the compactness of this semigroup and its resolvent can be deduced under quite minimal assumptions.

We begin with a lemma which is based on the classical result of Paley [31] and Marcinkiewicz and Zygmund [27] (see also [33]) that if T is a bounded operator on a space $L^p(\nu)$ and if H is a Hilbert space, then $T \otimes I$ is bounded on $L^p(\nu; H)$ and $\|T \otimes I\| = \|T\|$. As a direct consequence, if S is a bounded operator on H, then $T \otimes S = (T \otimes I) \circ (I \otimes S)$ is bounded on $L^p(\nu; H)$ and $\|T \otimes S\| \leqslant \|T\|\|S\|$.

Lemma 4.1. *Let $1 \leqslant p < \infty$. If T is compact on $L^p(\nu)$ and S is compact on H, then $T \otimes S$ is compact on $L^p(\nu; H)$.*

Proof. Since compactness can be tested sequentially, there is no loss of generality in assuming that both $L^p(\nu)$ and H are separable. Since separable spaces $L^p(\nu)$ have the approximation property, by [23, Theorem 1.e.4] there is a finite rank operator T' on $L^p(\nu)$ such that $\|T - T'\| < \varepsilon$. Similarly there is a finite rank operator S' on H such that $\|S - S'\| < \varepsilon$. Then $T' \otimes S'$ is a finite rank operator on $L^p(\nu; H)$ and

$$\|T' \otimes S' - T \otimes S\| \leqslant \|T' \otimes (S' - S)\| + \|(T' - T) \otimes S\| \leqslant \varepsilon((\|T\| + \varepsilon) + \|S\|).$$

This proves that $T \otimes S$ can be uniformly approximated by finite rank operators. □

We now return to the setting of the previous section. Since a semigroup which is norm continuous for $t > 0$ is compact for $t > 0$ if and only if its resolvent operators are compact, Lemma 4.1 implies:

Proposition 4.2. *Let $1 < p < \infty$ and suppose that Assumptions 2.1 and 2.3 hold. If P has compact resolvent on $L^p(E, \mu_\infty)$, then $P \otimes S_H^*$ has compact resolvent on $L^p(E, \mu_\infty; H)$.*

The generator of $P \otimes S_H^*$ equals $L \otimes I + I \otimes A_H^*$. As we have seen, the compactness of the resolvent of L implies the compactness of the resolvent A_H^*. Thus the proposition suggests the more general problem whether $A \otimes I + I \otimes B$ has compact resolvent if A and B have compact resolvents. Our final result gives an affirmative answer for sectorial operators A and B of angle $< \frac{1}{2}\pi$. Recall that a densely defined closed linear operator A is said to be *sectorial operator of angle* $< \frac{1}{2}\pi$ if there exists an angle $0 < \theta < \frac{1}{2}\pi$ such that $\{|\arg z| > \theta\} \subseteq \varrho(A)$ and $\sup_{\{|\arg z| > \theta\}} \|zR(z, A)\| < \infty$.

Proposition 4.3. *Let $1 \leqslant p < \infty$ and suppose that A and B are sectorial operators of angle $< \frac{1}{2}\pi$ on $L^p(\nu)$ and H, respectively. If, for some $w_0 \in \varrho(A)$ and $z_0 \in \varrho(A)$, the operators $R(w_0, A)$ and $R(z_0, B)$ are compact, then $A \otimes I + I \otimes B$ has compact resolvent on $L^p(\nu; H)$.*

Proof. Fix numbers $\omega_A < \theta_A < \frac{1}{2}\pi$, $\omega_B < \theta_B < \frac{1}{2}\pi$, where ω_A and ω_B denote the angles of sectoriality of A and B. Fix $\lambda \in \mathbb{C}$ with $|\arg \lambda| > \theta$ and fix a number $0 < r < |\lambda|$.

Let $\gamma_{A,r}$ and $\gamma_{B,r}$ be the downwards oriented boundaries of

$$\{|z| < r\} \cup \{|\arg z| < \theta_A\} \quad \text{and} \quad \{|z| < r\} \cup \{|\arg z| < \theta_B\}.$$

It follows from [18, Formulas (2.2), (2.3)] and a limiting argument that

$$\begin{aligned} &R(\lambda, A \otimes I + B \otimes I) \\ &\quad = \frac{1}{(2\pi i)^2} \int_{\gamma_{B,r}} \int_{\gamma_{A,r}} \frac{1}{\lambda - (w+z)} R(w, A) \otimes R(z, B)\, dw\, dz; \end{aligned} \tag{4.1}$$

note that the double integral on the right-hand side converges absolutely.

Given $\varepsilon > 0$ fix $R > r$ so large that

$$\left\| \frac{1}{(2\pi i)^2} \int_{\gamma_{B,r} \cap \complement B_R} \int_{\gamma_{A,r} \cap \complement B_R} \frac{1}{\lambda - (w+z)} R(w, A) \otimes R(z, B)\, dw\, dz \right\| < \varepsilon,$$

where $B_R = \{z \in \mathbb{C} : |z| < \mathbb{R}\}$ and $\complement B_R$ is its complement. By Lemma 4.1 and [34, Theorem 1.3] the operator

$$\frac{1}{(2\pi i)^2} \int_{\gamma_{B,r} \cap B_R} \int_{\gamma_{A,r} \cap B_R} \frac{1}{\lambda - (w+z)} R(w, A) \otimes R(z, B)\, dw\, dz$$

is compact, as it is the strong integral over a finite measure space of an integrand with values in the space of compact operators. As a consequence, for each $\varepsilon > 0$ we obtain that $R(\lambda, A \otimes +I \otimes B) = K_\varepsilon + L_\varepsilon$ with K_ε compact and L_ε bounded with $\|L_\varepsilon\| < \varepsilon$. It follows that the range of the unit ball of $L^p(\nu; H)$ under $R(\lambda, A \otimes I + I \otimes B)$ is totally bounded and therefore relatively compact. $\square$

The formula (4.1) for the resolvent of the sum of two operators goes back to Bianchi and Favella [2] who considered bounded A and B. It can be viewed as a special instance of the so-called joint functional calculus for sectorial operators; see [20, Theorem 2.2], [19, Theorem 12.12].

Remark 4.4. The above proof easily extends to tensor products of C_0-semigroups on arbitrary Banach spaces, provided one makes appropriate assumptions on the boundedness of the tensor products of the various bounded operators involved.

Remark 4.5. The same proof may be used to see that if A and B are resolvent commuting sectorial operators of angle $< \frac{1}{2}\pi$ on a Banach space X and if, for some $w_0 \in \varrho(A)$ and $z_0 \in \varrho(A)$, the operator $R(w_0, A)R(z_0, B)$ is compact on X, then $A + B$ has compact resolvent on X.

Acknowledgment

I thank Ben Goldys and Jan Maas for providing helpful comments and the anonymous referee for suggesting some improvements.

References

[1] D. Bakry, I. Gentil, and M. Ledoux. *Analysis and geometry of Markov diffusion operators*, volume 348 of *Grundlehren der Mathmatischen Wissenschaften*. Springer-Verlag, 2013.

[2] L. Bianchi and L. Favella. *A convolution integral for the resolvent of the sum of two commuting operators*. Il Nuovo Cimento (1955–1965), 34(6):1825–1828, 1964.

[3] S. Cerrai and F. Gozzi. *Strong solutions of Cauchy problems associated to weakly continuous semigroups*. Differential Integral Equations, 8(3):465–486, 1995.

[4] A. Chojnowska-Michalik and B. Goldys. *Nonsymmetric Ornstein–Uhlenbeck semigroup as second quantized operator*. J. Math. Kyoto Univ., 36(3):481–498, 1996.

[5] A. Chojnowska-Michalik and B. Goldys. *On regularity properties of nonsymmetric Ornstein–Uhlenbeck semigroup in L^p spaces*. Stochastics Stochastics Rep., 59(3-4):183–209, 1996.

[6] A. Chojnowska-Michalik and B. Goldys. Nonsymmetric Ornstein–Uhlenbeck generators. In *Infinite-dimensional stochastic analysis (Amsterdam,* 1999), volume 52 of *Verh. Afd. Natuurkd. 1. Reeks. K. Ned. Akad. Wet.*, pages 99–116. Roy. Neth. Acad. Arts Sci., Amsterdam, 2000.

[7] A. Chojnowska-Michalik and B. Goldys. *Generalized Ornstein–Uhlenbeck semigroups: Littlewood–Paley–Stein inequalities and the P.A. Meyer equivalence of norms*. J. Funct. Anal., 182(2):243–279, 2001.

[8] A. Chojnowska-Michalik and B. Goldys. *Symmetric Ornstein–Uhlenbeck semigroups and their generators*. Probab. Theory Related Fields, 124(4):459–486, 2002.

[9] G. Da Prato and J. Zabczyk. *Second-order partial differential equations in Hilbert spaces*, volume 293 of *London Mathematical Society Lecture Note Series*. Cambridge University Press, Cambridge, 2002.

[10] G. Da Prato and J. Zabczyk. *Stochastic equations in infinite dimensions*, volume 44 of *Encyclopedia of Mathematics and its Applications*. Cambridge University Press, Cambridge, 2nd edition, 2014.

[11] L.B. Efraim and F. Lust-Piquard. *Poincaré type inequalities on the discrete cube and in the CAR algebra*. Probab. Theory Related Fields, 141(3-4):569–602, 2008.

[12] K.-J. Engel and R. Nagel. *One-parameter semigroups for linear evolution equations*, volume 194 of *Graduate Texts in Mathematics*. Springer-Verlag, New York, 2000.

[13] M. Fuhrman. *Analyticity of transition semigroups and closability of bilinear forms in Hilbert spaces*. Studia Math., 115(1):53–71, 1995.

[14] B. Goldys. *On analyticity of Ornstein–Uhlenbeck semigroups*. Atti Accad. Naz. Lincei Cl. Sci. Fis. Mat. Natur. Rend. Lincei (9) Mat. Appl., 10(3):131–140, 1999.

[15] B. Goldys, F. Gozzi, and J.M.A.M. van Neerven. *On closability of directional gradients*. Potential Analysis, 18(4):289–310, 2003.

[16] B. Goldys and J.M.A.M. van Neerven. *Transition semigroups of Banach space-valued Ornstein–Uhlenbeck processes.* Acta Appl. Math., 76(3):283–330, 2003.

[17] M. Junge and Q. Zeng. Noncommutative martingale deviation and Poincaré type inequalities with applications. Probability Theory and Related Fields, 161(3-4):449–507, 2015.

[18] N.J. Kalton and L.W. Weis. *The H^∞-calculus and sums of closed operators.* Math. Ann., 321(2):319–345, 2001.

[19] P.C. Kunstmann and L.W. Weis. Maximal L_p-regularity for parabolic equations, Fourier multiplier theorems and H^∞-functional calculus. In *Functional analytic methods for evolution equations*, volume 1855 of *Lecture Notes in Math.*, pages 65–311. Springer, Berlin, 2004.

[20] F. Lancien, G. Lancien, and C. Le Merdy. *A joint functional calculus for sectorial operators with commuting resolvents.* Proc. London Math. Soc. (3), 77(2):387–414, 1998.

[21] C. Le Merdy. A bounded compact semigroup on Hilbert space not similar to a contraction one. In *Semigroups of Operators: Theory and Applications*, volume 42 of *Progr. Nonlinear Differential Equations Appl.*, pages 213–216. Birkhäuser, Basel, 2000.

[22] M. Ledoux. *On improved Sobolev embedding theorems.* Math. Res. Lett., 10(5-6):659–669, 2003.

[23] J. Lindenstrauss and L. Tzafriri. *Classical Banach spaces. I.* Springer-Verlag, Berlin, 1977. Sequence spaces, Ergebnisse der Mathematik und ihrer Grenzgebiete, vol. 92.

[24] J. Maas and J.M.A.M. van Neerven. *On analytic Ornstein–Uhlenbeck semigroups in infinite dimensions.* Arch. Math. (Basel), 89(3):226–236, 2007.

[25] J. Maas and J.M.A.M. van Neerven. *Boundedness of Riesz transforms for elliptic operators on abstract Wiener spaces.* J. Funct. Anal., 257(8):2410–2475, 2009.

[26] J. Maas and J.M.A.M. van Neerven. Gradient estimates and domain identification for analytic Ornstein–Uhlenbeck operators. In *Parabolic problems*, volume 80 of *Progr. Nonlinear Differential Equations Appl.*, pages 463–477. Birkhäuser, Basel, 2011.

[27] J. Marcinkiewicz and A. Zygmund. *Quelques inégalités pour les opérations linéaires.* Fund. Math., 32(1):115–121, 1939.

[28] J.M.A.M. van Neerven. *Nonsymmetric Ornstein–Uhlenbeck semigroups in Banach spaces.* J. Funct. Anal., 155(2):495–535, 1998.

[29] J.M.A.M. van Neerven. *Second quantization and the L^p-spectrum of nonsymmetric Ornstein–Uhlenbeck operators.* Infin. Dimens. Anal. Quantum Probab. Relat. Top., 8(3):473–495, 2005.

[30] J.M.A.M. van Neerven. γ-Radonifying operators – a survey. In *The AMSI-ANU Workshop on Spectral Theory and Harmonic Analysis*, volume 44 of *Proc. Centre Math. Appl. Austral. Nat. Univ.*, pages 1–61. Austral. Nat. Univ., Canberra, 2010.

[31] R.E.A.C. Paley. *A remarkable series of orthogonal functions (I).* Proc. London Math. Soc., 2(1):241–264, 1932.

[32] G. Pisier. Probabilistic methods in the geometry of Banach spaces. In *Probability and analysis (Varenna, 1985)*, volume 1206 of *Lecture Notes in Math.*, pages 167–241. Springer, Berlin, 1986.

[33] J.L. Rubio de Francia. Weighted norm inequalities and vector-valued inequalities. In *Harmonic analysis (Minneapolis, Minn.*, 1981), volume 908 of *Lecture Notes in Math.*, pages 86–101. Springer, Berlin, 1982.

[34] J. Voigt. *On the convex compactness property for the strong operator topology.* Note Mat., 12:259–269, 1992.

[35] Q. Zeng. *Poincaré type inequalities for group measure spaces and related transportation cost inequalities.* J. Funct. Anal., 266(5):3236–3264, 2014.

Jan van Neerven
Fac. EWI/TWA
Technical University Delft
P.O.Box 5031
NL-2600 GA Delft, The Netherlands
e-mail: `J.M.A.M.vanNeerven@tudelft.nl`

Operator Theory:
Advances and Applications, Vol. 250, 369–395

A Murray–von Neumann Type Classification of C^*-algebras

Chi-Keung Ng and Ngai-Ching Wong

This paper is dedicated to Charles Batty on the occasion of his 60th birthday

Abstract. We define type $\mathfrak{A}$, type $\mathfrak{B}$, type $\mathfrak{C}$ as well as C^*-semi-finite C^*-algebras.

It is shown that a von Neumann algebra is a type $\mathfrak{A}$, type $\mathfrak{B}$, type $\mathfrak{C}$ or C^*-semi-finite C^*-algebra if and only if it is, respectively, a type I, type II, type III or semi-finite von Neumann algebra. Any type I C^*-algebra is of type $\mathfrak{A}$ (actually, type $\mathfrak{A}$ coincides with the discreteness as defined by Peligrad and Zsidó), and any type II C^*-algebra (as defined by Cuntz and Pedersen) is of type $\mathfrak{B}$. Moreover, any type $\mathfrak{C}$ C^*-algebra is of type III (in the sense of Cuntz and Pedersen). Conversely, any separable purely infinite C^*-algebra (in the sense of Kirchberg and Rørdam) with either real rank-zero or stable rank-one is of type $\mathfrak{C}$.

We also prove that type $\mathfrak{A}$, type $\mathfrak{B}$, type $\mathfrak{C}$ and C^*-semi-finiteness are stable under taking hereditary C^*-subalgebras, multiplier algebras and strong Morita equivalence. Furthermore, any C^*-algebra A contains a largest type $\mathfrak{A}$ closed ideal $J_{\mathfrak{A}}$, a largest type $\mathfrak{B}$ closed ideal $J_{\mathfrak{B}}$, a largest type $\mathfrak{C}$ closed ideal $J_{\mathfrak{C}}$ as well as a largest C^*-semi-finite closed ideal $J_{\mathfrak{sf}}$. Among them, we have $J_{\mathfrak{A}} + J_{\mathfrak{B}}$ being an essential ideal of $J_{\mathfrak{sf}}$, and $J_{\mathfrak{A}} + J_{\mathfrak{B}} + J_{\mathfrak{C}}$ being an essential ideal of A. On the other hand, $A/J_{\mathfrak{C}}$ is always C^*-semi-finite, and if A is C^*-semi-finite, then $A/J_{\mathfrak{B}}$ is of type $\mathfrak{A}$.

Mathematics Subject Classification (2010). 46L05, 46L35.

Keywords. C^*-algebra; open projections; Murray–von Neumann type classification.

The authors are supported by National Natural Science Foundation of China (11071126), and Taiwan NSC grant (99-2115-M-110-007-MY3).

1. Introduction

In their seminal works ([27], see also [26]), Murray and von Neumann defined three types of von Neumann algebras (namely, type I, type II and type III) according to the properties of their projections. They showed that any von Neumann algebra is a sum of a type I, a type II, and a type III von Neumann subalgebras. This classification was shown to be very important and becomes the basic theory for the study of von Neumann algebras (see, e.g., [20]). Since a C^*-algebra needs not have any projection, a similar classification for C^*-algebras seems impossible. There is, however, an interesting classification scheme for C^*-algebras proposed by Cuntz and Pedersen in [14], which captures some features of the classification of Murray and von Neumann.

The classification theme of C^*-algebras took a drastic turn after an exciting work of Elliott on the classification of AF-algebras through the ordered K-theory, in the sense that two AF-algebras are isomorphic if and only if they have the same ordered K-theory ([16]). Elliott then proposed an invariant consisting of the tracial state space and some K-theory datum of the underlying C^*-algebra (called the *Elliott invariant*) which could be a suitable candidate for a complete invariant for simple separable nuclear C^*-algebras. Although it is known recently that it is not the case (see [38]), this Elliott invariant still works for a very large class of such C^*-algebras (namely, those satisfying certain regularity conditions as described in [18]). Many people are still making progress in this direction in trying to find the biggest class of C^*-algebras that can be classified through the Elliott invariant (see, e.g., [17, 36]). Notice that this classification is very different from the classification in the sense of Murray and von Neumann.

In this article, we reconsider the classification of C^*-algebras through the idea of Murray and von Neumann. Instead of considering projections in a C^*-algebra A, we consider open projections and we twist the definition of the finiteness of projections slightly to obtain our classification scheme.

The notion of open projections was introduced by Akemann (in [1]). A projection p in the universal enveloping von Neumann algebra (i.e., the biduals) A^{**} of a C^*-algebra A (see, e.g., [37, §III.2]) is an *open* projection of A if there is an increasing net $\{a_i\}_{i\in\mathfrak{I}}$ of positive elements in A_+ with $\lim_i a_i = p$ in the $\sigma(A^{**}, A^*)$-topology. In the case when A is commutative, open projections of A are exactly characteristic functions of open subsets of the spectrum of A. In general, there is a bijective correspondence between open projections of A and hereditary C^*-subalgebras of A (where a hereditary C^*-subalgebra B corresponds to an open projection p such that $B = pA^{**}p \cap A$; see, e.g., [31, 3.11.10]). Characterisations and further developments of open projections can be found in, e.g., [2, 3, 4, 9, 15, 30, 33]. Since every element in a C^*-algebra is in the closed linear span of its open projections, it is reasonable to believe that the study of open projections will provide fruitful information about the underlying C^*-algebra. Moreover, because of the correspondence between open projections (respectively, central open projections) and hereditary C^*-subalgebras (respectively, closed ideals), the notion of strong

Morita equivalence as defined by Rieffel (see [34] and also [11, 35]) is found to be very useful in this scheme.

One might wonder why we do not consider the classification of the universal enveloping von Neumann algebras of C^*-algebras to obtain a classification of C^*-algebras. A reason is that for a C^*-algebra A, its bidual A^{**} always contains many minimum projections (see, e.g., [1, II.17]), and hence a reasonable theory of type classification cannot be obtained without serious modifications. Furthermore, A^{**} are usually very far away from A, and information of A might not always be respected very well in A^{**}; for example, c and c_0 have isomorphic biduals, but the structure of their open projections can be used to distinguish them (see, e.g., Example 2.1 and also Proposition 2.3(b)).

As in the case of von Neumann algebras, in order to give a classification of C^*-algebras, one needs, first of all, to consider a good equivalence relation among open projections. After some thoughts and considerations, we end up with the "spatial equivalence" as defined in Section 2, which is weaker than the one defined by Peligrad and Zsidó in [32] and stronger than the ordinary Murray–von Neumann equivalence. One reason for making this choice is that it is precisely the "hereditarily stable version of Murray–von Neumann equivalence" that one might want (see Proposition 2.7(a)(5)), and it also coincides with the "spatial isomorphism" of the hereditary C^*-subalgebras (see Proposition 2.7(a)(2)).

Using the spatial equivalence relation, we introduce in Section 3, the notion of C^*-finite C^*-algebras. It is shown that the sum of all C^*-finite hereditary C^*-subalgebra is a (not necessarily closed) ideal of the given C^*-algebra. In the case when the C^*-algebra is $\mathcal{B}(H)$ or $\mathcal{K}(H)$, this ideal is the ideal of all finite rank operators on H. Moreover, through C^*-finiteness, we define type $\mathfrak{A}$, type $\mathfrak{B}$, type $\mathfrak{C}$ as well as C^*-semi-finite C^*-algebras, and we study some properties of them. In particular, we will show that these properties are stable under taking hereditary C^*-subalgebras, multiplier algebras, unitalization (if the algebra is not unital) as well as strong Morita equivalence. We will also show that the notion of type $\mathfrak{A}$ coincides precisely with the discreteness as defined in [32].

In Section 4, we will compare these notions with some results in the literature and give some examples. In particular, we show that any type I C^*-algebra (see, e.g., [31]) is of type $\mathfrak{A}$; any type II C^*-algebra (as defined by Cuntz and Pedersen) is of type $\mathfrak{B}$; any semi-finite C^*-algebras (in the sense of Cuntz and Pedersen) is C^*-semi-finite; any purely infinite C^*-algebra (in the sense of Kirchberg and Rørdam) with real rank-zero and any separable purely infinite C^*-algebra with stable rank-one are of type $\mathfrak{C}$; and any type $\mathfrak{C}$ C^*-algebra is of type III (as introduced by Cuntz and Pedersen). Using our arguments for these results, we also show that any purely infinite C^*-algebra is of type III. Moreover, a von Neumann algebra M is a type $\mathfrak{A}$, a type $\mathfrak{B}$, a type $\mathfrak{C}$ or a C^*-semi-finite C^*-algebra if and only if M is, respectively, a type I, a type II, a type III, or a semi-finite von Neumann algebra.

In Section 5, we show that any C^*-algebra A contains a largest type $\mathfrak{A}$ closed ideal $J^A_{\mathfrak{A}}$, a largest type $\mathfrak{B}$ closed ideal $J^A_{\mathfrak{B}}$, a largest type $\mathfrak{C}$ closed ideal $J^A_{\mathfrak{C}}$ as well as a largest C^*-semi-finite closed ideal $J^A_{\mathfrak{sf}}$. It is further shown that $J^A_{\mathfrak{A}} + J^A_{\mathfrak{B}}$ is an essential ideal of $J^A_{\mathfrak{sf}}$, and $J^A_{\mathfrak{A}} + J^A_{\mathfrak{B}} + J^A_{\mathfrak{C}}$ is an essential ideal of A. On the other hand, $A/J^A_{\mathfrak{C}}$ is always a C^*-semi-finite C^*-algebra, while $B/J^B_{\mathfrak{B}}$ is always of type $\mathfrak{A}$ if one sets $B := A/J^A_{\mathfrak{C}}$. We also compare $J^{M(A)}_{\mathfrak{A}}$, $J^{M(A)}_{\mathfrak{B}}$, $J^{M(A)}_{\mathfrak{C}}$ and $J^{M(A)}_{\mathfrak{sf}}$ with $J^A_{\mathfrak{A}}$, $J^A_{\mathfrak{B}}$, $J^A_{\mathfrak{C}}$ and $J^A_{\mathfrak{sf}}$, respectively.

Notation 1.1. Throughout this paper, A is a non-zero C^*-algebra, $M(A)$ is the multiplier algebra of A, $\mathrm{Z}(A)$ is the center of A, and A^{**} is the bidual of A. Furthermore, $\mathrm{Proj}(A)$ is the set of all projections in A, while $\mathrm{OP}(A) \subseteq \mathrm{Proj}(A^{**})$ is the set of all open projections of A. All ideals in this paper are two-sided ideals (not assumed to be closed unless specified).

If $x, y \in A^{**}$ and E is a subspace of A^{**}, we set $xEy := \{xzy : z \in E\}$, and denote by $\overline{E}$ the norm closure of E. For any $x \in A^{**}$, we set $\mathrm{her}_A(x)$ to be the hereditary C^*-subalgebra $\overline{x^*A^{**}x} \cap A$ of A (note that if $u \in A^{**}$ is a partial isometry, then $\mathrm{her}_A(u) = u^*A^{**}u \cap A = \{x \in A : x = u^*uxu^*u\} = \mathrm{her}_A(u^*u)$). When A is understood, we will use the notation $\mathrm{her}(x)$ instead. Moreover, p_x is the right support projection of a norm one element $x \in A$, i.e., p_x is the $\sigma(A^{**}, A^*)$-limit of $\{(x^*x)^{1/n}\}_{n\in\mathbb{N}}$ and is the smallest open projection in A^{**} with $xp_x = x$.

2. Spatial equivalence of open projections

In this section, we will consider a suitable equivalence relation on the set of open projections of a C^*-algebra. Let us start with the following example, which shows that the structure of open projections is rich enough to distinguish c and c_0, while they have isomorphic biduals (see Proposition 2.3(b) below for a more general result).

Example 2.1. The sets of open projections of c_0 and c can be regarded as the collections $\mathcal{X}$ and $\mathcal{Y}$, of open subsets of $\mathbb{N}$ and of open subsets of the one point compactification of $\mathbb{N}$, respectively. As ordered sets, $\mathcal{X}$ and $\mathcal{Y}$ are not isomorphic. In fact, suppose on the contrary that there is an order isomorphism $\Psi : \mathcal{Y} \to \mathcal{X}$. Then $\Psi(\mathbb{N})$ is a proper open subset of $\mathbb{N}$. Let $k \notin \Psi(\mathbb{N})$ and $U \in \mathcal{Y}$ with $\Psi(U) = \{k\}$. As U is a minimal element, it is a singleton set. Thus, $U \subseteq \mathbb{N}$, which gives the contradiction that $\{k\} \subseteq \Psi(\mathbb{N})$.

Secondly, we give the following well-known remarks which says that open projections and the hereditary C^*-subalgebras they define, are "hereditarily invariant". These will clarify some discussions later on.

Remark 2.2. Let $B \subseteq A$ be a hereditary C^*-subalgebra and $e \in \mathrm{OP}(A)$ be the open projection with $\mathrm{her}_A(e) = B$.

(a) For any $p \in \mathrm{Proj}(B^{**})$, one has $\mathrm{her}_B(p) = \mathrm{her}_A(p)$.
(b) $\mathrm{OP}(B) = \mathrm{OP}(A) \cap B^{**}$. In fact, if $p \in \mathrm{OP}(A) \cap B^{**}$ and $\{a_i\}_{i\in\mathfrak{I}}$ is an approximate unit in $\mathrm{her}_A(p) = \mathrm{her}_B(p)$, then $\{a_i\}_{i\in\mathfrak{I}}$ will $\sigma(B^{**}, B^*)$-converge to p and $p \in \mathrm{OP}(B)$.

(c) If $z \in A$ satisfying $zz^*, z^*z \in B$, then $z \in B$. In fact, as $z^*z \in \mathrm{her}_A(e) = eA^{**}e \cap A$, by considering the polar decomposition of z, we see that $ze = z$. Similarly, we have $ez = z$.
(d) If $f \in \mathrm{OP}(A)$, the open projections corresponding to $\mathrm{her}(e) \cap \mathrm{her}(f)$ and the hereditary C^*-subalgebra generated by $\mathrm{her}(e) + \mathrm{her}(f)$ are $e \wedge f$ and $e \vee f$ respectively.

Let $j_A : M(A) \to A^{**}$ be the canonical *-monomorphism, i.e., $j_A(x)(f) = \tilde{f}(x)$ $(x \in M(A), f \in A^*)$, where $\tilde{f} \in M(A)^*$ is the unique strictly continuous extension of f. The proposition below can be regarded as a motivation behind the study of C^*-algebras through their open projections. It could be a known result (especially, part (a)). However, since we need it for the equivalence of (1) and (5) in Proposition 2.7(a), we give a proof here for completeness.

Proposition 2.3. *Suppose that A and B are C^*-algebras, and $\Phi : A^{**} \to B^{**}$ is a *-isomorphism.*
(a) *If $\Phi(j_A(M(A))) = j_B(M(B))$, then $\Phi(A) = B$.*
(b) *If $\Phi(\mathrm{OP}(A)) = \mathrm{OP}(B)$, then $\Phi(A) = B$.*

Proof. (a) Let $p_A \in \mathrm{OP}(M(A))$ such that $\mathrm{her}_{M(A)}(p_A) = A$. It is not hard to verify that p_A is the support of $\tilde{j}_A$, where $\tilde{j}_A : M(A)^{**} \to A^{**}$ is the *-epimorphism induced by j_A. Consider $\Psi := j_B^{-1} \circ \Phi_{|j_A(M(A))} \circ j_A : M(A) \to M(B)$ (which is well defined by the hypothesis). Since $j_B \circ \Psi = \Phi_{|j_A(M(A))} \circ j_A$, we see that $\tilde{j}_B \circ \Psi^{**} = \Phi \circ \tilde{j}_A$ (as Φ is automatically weak-*-continuous). Thus, $\tilde{j}_B(\Psi^{**}(p_A)) = 1_{B^{**}}$ which implies $\Psi^{**}(p_A) \geq p_B$. Similarly,

$$(\Psi^{**})^{-1}(p_B) = (j_A^{-1} \circ \Phi^{-1}{}_{|j_B(M(B))} \circ j_B)^{**}(p_B) \geq p_A$$

and we have $\Psi^{**}(p_A) = p_B$. Consequently, $\Psi(\mathrm{her}_{M(A)}(p_A)) = \mathrm{her}_{M(B)}(p_B)$ as required.
(b) If $a \in M(A)_{sa}$ and U is an open subset of $\sigma(a) = \sigma(\Phi(j_A(a)))$, then $\chi_U(\Phi(j_A(a))) = \Phi(\chi_U(j_A(a)))$ is an element of $\mathrm{OP}(B)$ (by [5, Theorem 2.2] and the hypothesis). Thus, by [5, Theorem 2.2] again, we have $\Phi(j_A(a)) \in j_B(M(B))$. A similar argument shows that $\Phi^{-1}(j_B(M(B))) \subseteq j_A(M(A))$. Now, we can apply part (a) to obtain the required conclusion. □

Remark 2.4. Note that if A and B are separable and $\Psi : M(A) \to M(B)$ is a *-isomorphism, then $\Psi(A) = B$, by a result of Brown in [10]. However, the same result is not true if one of them is not separable (e.g., take $A = M(B)$ and $\Psi = \mathrm{id}$, where B is non-unital). Proposition 2.3(a) shows that one has $\Psi(A) = B$ if (and only if) Ψ extends to a *-isomorphism from A^{**} to B^{**}.

We now consider a suitable equivalence relation on $\mathrm{OP}(A)$. A naive choice is to use the original "Murray–von Neumann equivalence" $\sim_{\mathrm{Mv}}$. However, this choice is not good because [23] tells us that two open projections that are Murray–von Neumann equivalent might define non-isomorphic hereditary C^*-subalgebras. On the other hand, one might define $p \sim_{\mathrm{her}} q$ $(p, q \in \mathrm{OP}(A))$ whenever $\mathrm{her}(p) \cong \mathrm{her}(q)$

as C^*-algebras. The problem of this choice is that two distinct open projections of $C([0,1])$ can be equivalent (if they correspond to homeomorphic open subsets of $[0,1]$), which means that the resulting classification, even if possible, will be very different from the Murray–von Neumann classification.

After some thoughts, we end up with an equivalence relation $\sim_{\rm sp}$ on $\mathrm{OP}(A)$: $p \sim_{\rm sp} q$ if there is a partial isometry $v \in A^{**}$ satisfying

$$v^* \operatorname{her}_A(p) v \;=\; \operatorname{her}_A(q) \quad \text{and} \quad v \operatorname{her}_A(q) v^* \;=\; \operatorname{her}_A(p).$$

Note that this relation is precisely the "hereditarily stable version" of the Murray–von Neumann equivalence (see Proposition 2.7(a)(5) below and the discussion following it).

In [32, Definition 1.1], Peligrad and Zsidó introduced another equivalence relation on $\mathrm{Proj}(A^{**})$: $p \sim_{\rm PZ} q$ if there is a partial isometry $v \in A^{**}$ such that

$$p = vv^*, \quad q = v^*v, \quad v^* \operatorname{her}_A(p) \subseteq A \quad \text{and} \quad v \operatorname{her}_A(q) \subseteq A. \tag{2.1}$$

It is not difficult to see that $\sim_{\rm PZ}$ is stronger than $\sim_{\rm sp}$, and a natural description of $\sim_{\rm PZ}$ on the set of range projections of positive elements of A is given in [29, Proposition 4.3]. Moreover, we also gave in [28, Proposition 3.1] an equivalent description of $\sim_{\rm PZ}$ that is similar to $\sim_{\rm sp}$ but use right ideals instead of hereditary C^*-subalgebras. However, it is now known that $\sim_{\rm PZ}$ and $\sim_{\rm sp}$ are actually different even for very simple kind of C^*-algebras (see [28, Theorem 5.3]). We decide to use $\sim_{\rm sp}$ as it seems to be more natural in the way of using open projections (see Proposition 2.7(a) below).

Let us start with an extension of $\sim_{\rm sp}$ to the whole of $\mathrm{Proj}(A^{**})$.

Definition 2.5. We say that $p, q \in \mathrm{Proj}(A^{**})$ are *spatially equivalent with respect to* A, denoted by $p \sim_{\rm sp} q$, if there exists a partial isometry $v \in A^{**}$ satisfying

$$p = vv^*, \quad q = v^*v, \quad v^* \operatorname{her}_A(p) v = \operatorname{her}_A(q) \quad \text{and} \quad v \operatorname{her}_A(q) v^* = \operatorname{her}_A(p). \tag{2.2}$$

In this case, we also say that the hereditary C^*-subalgebras $\operatorname{her}_A(p)$ and $\operatorname{her}_A(q)$ are *spatially isomorphic.*

It might happen that $\operatorname{her}(p) = 0$ but $p \neq 0$ and this is why we need to consider the first two conditions in (2.2). We will see in Proposition 2.7(a) that the first two conditions are redundant if p and q are both open projections.

Obviously, $\sim_{\rm sp}$ is stronger than $\sim_{\rm Mv}$ (for elements in $\mathrm{Proj}(A^{**})$). Moreover, if $p \sim_{\rm sp} q$, then $x \mapsto v^*xv$ is a *-isomorphism from $\operatorname{her}(p)$ to $\operatorname{her}(q)$, which means that $\sim_{\rm sp}$ is stronger than $\sim_{\rm her}$ in the context of open projections.

A good point of the spatial equivalence is that open projections are stable under $\sim_{\rm sp}$, as can be seen in part (b) of the following lemma.

Lemma 2.6.

(a) $\sim_{\rm sp}$ *is an equivalence relation in* $\mathrm{Proj}(A^{**})$.

(b) *Let $p, q \in \mathrm{Proj}(A^{**})$ and $u \in A^{**}$ be a partial isometry. If p is open, $u^*pu = q$, $\mathrm{her}_A(p) \subseteq u\,\mathrm{her}_A(q)u^*$ and $\mathrm{her}_A(q) \subseteq u^*\,\mathrm{her}_A(p)u$, then q is open and $p \sim_{\mathrm{sp}} q$. Consequently, if $p \sim_{\mathrm{sp}} q$ and p is open, then q is open.*

(c) *If $B \subseteq A$ is a hereditary C^*-subalgebra and $p, q \in \mathrm{Proj}(B^{**})$, then p and q are spatially equivalent with respect to B if and only if they are spatially equivalent with respect to A.*

Proof. (a) It suffices to verify the transitivity. Suppose that p, q and v are as in Definition 2.5. If $w \in A^{**}$ and $r \in \mathrm{Proj}(A^{**})$ satisfy that

$$p = w^*w, \quad r = ww^*, \quad w\,\mathrm{her}_A(p)w^* = \mathrm{her}_A(r) \quad \text{and} \quad w^*\,\mathrm{her}_A(r)w = \mathrm{her}_A(p),$$

then the partial isometry wv gives the equivalence $r \sim_{\mathrm{sp}} q$.

(b) As p is open and $\mathrm{her}_A(p)$ is contained in the weak-*-closed subspace $uA^{**}u^*$, one has $p \leq uu^*$. Let $v := pu$. Then $vv^* = p$ and $v^*v = u^*pu = q$. Moreover, it is clear that $\mathrm{her}_A(p) \subseteq v\,\mathrm{her}_A(q)v^*$ and $\mathrm{her}_A(q) \subseteq v^*\,\mathrm{her}_A(p)v$. Now, it is easy to see that the relations in (2.2) are satisfied. Furthermore, if $\{a_i\}_{i\in\mathfrak{I}}$ is an approximate unit in $\mathrm{her}_A(p)$, then $\{v^*a_iv\}$ is an increasing net in $\mathrm{her}_A(q)$ that weak-*-converges to $v^*pv = q$, and so q is open. The second statement follows directly from the first one.

(c) Suppose that p and q are spatially equivalent with respect to A and $v \in A^{**}$ satisfies the relations in (2.2). As $vv^*, v^*v \in B^{**}$, Remark 2.2(c) tells us that $v \in B^{**}$. Now the equivalence follows from Remark 2.2(a). □

Proposition 2.7.

(a) *If $p, q \in \mathrm{OP}(A)$, the following statements are equivalent.*
   (1) $p \sim_{\mathrm{sp}} q$.
   (2) $\mathrm{her}(q) = u^*\,\mathrm{her}(p)u$ *and* $\mathrm{her}(p) = u\,\mathrm{her}(q)u^*$ *for a partial isometry* $u \in A^{**}$.
   (3) $\mathrm{her}(q) \subseteq u^*\,\mathrm{her}(p)u$ *and* $\mathrm{her}(p) \subseteq u\,\mathrm{her}(q)u^*$ *for a partial isometry* $u \in A^{**}$.
   (4) $q \leq v^*v$ *and* $v\,\mathrm{her}(q)v^* = \mathrm{her}(p)$ *for a partial isometry* $v \in A^{**}$.
   (5) *There is a partial isometry* $w \in A^{**}$ *such that* $p = ww^*$ *and*

$$\{w^*rw : r \in \mathrm{OP}(A); r \leq p\} = \{s \in \mathrm{OP}(A) : s \leq q\}.$$

(b) *If M is a von Neumann algebra and $p, q \in \mathrm{Proj}(M)$, then $p \sim_{\mathrm{sp}} q$ if and only if $p \sim_{\mathrm{Mv}} q$ as elements in* $\mathrm{Proj}(M)$.

Proof. (a) The implications (1) ⇒ (2) ⇒ (3) and (1) ⇒ (4) are clear.

(3) ⇒ (1). Since q is open, one has $q \leq u^*u$. Thus, $(uq)^*uq = q$ and Statement (3) also holds when u is replaced by uq. As p is also open, a similar argument shows that $p \leq uqu^*$ and Statement (3) holds if we replace u by $v := puq$ and that $p = vv^*$. Furthermore, since $vqv^* = vv^* = p$, Lemma 2.6(b) tells us that $p \sim_{\mathrm{sp}} q$.

(4) ⇒ (2). This follows from $v^*\,\mathrm{her}(p)v = v^*v\,\mathrm{her}(q)v^*v = \mathrm{her}(q)$.

(1) ⇒ (5). Notice that $\mathrm{OP}(\mathrm{her}(p)) = \{r \in \mathrm{OP}(A) : r \leq p\}$ (see Remark 2.2(b)). Suppose that $v \in A^{**}$ satisfies (2.2) and $r \in \mathrm{OP}(\mathrm{her}(p))$. If $\{a_i\}_{i\in\mathfrak{I}}$ is an increasing

net in $\mathrm{her}(p)$ that $\sigma(A^{**}, A^*)$-converge to r, then $\{v^*a_iv\}_{i\in\mathfrak{I}}$ is an increasing net in $\mathrm{her}(q)$ that $\sigma(A^{**}, A^*)$-converge to v^*rv and hence $v^*rv \in \mathrm{OP}(\mathrm{her}(q))$. The argument for the other inclusion is similar.

(5) $\Rightarrow$ (1). By Statement (5), we have $q = w^*pw$, and the map $\Phi : x \mapsto w^*xw$ is a *-isomorphism from $\mathrm{her}(p)^{**}$ to $\mathrm{her}(q)^{**}$. By Proposition 2.3(b), we see that $\Phi(\mathrm{her}(p)) = \mathrm{her}(q)$ and Statement (4) holds.

(b) If $p \sim_{\mathrm{sp}} q$, then $p \sim_{\mathrm{Mv}} q$ as elements in $\mathrm{Proj}(M^{**})$, which implies that $p \sim_{\mathrm{Mv}} q$ as elements in $\mathrm{Proj}(M)$ (by considering the canonical *-homomorphism $\Lambda_M : M^{**} \to M$). Conversely, if $v \in M$ satisfying $p = vv^*$ and $q = v^*v$, then clearly $v^*\,\mathrm{her}(p)v = \mathrm{her}(q)$. □

One can reformulate Statement (5) of Proposition 2.7(a) in the following way.

> There is a partial isometry $w \in A^{**}$ that induces Murray–von Neumann equivalences between open subprojections of p (including p) and open subprojections of q (including q).

Therefore, one may regard $\sim_{\mathrm{sp}}$ as the "hereditarily stable version" of the Murray–von Neumann equivalence. Moreover, if $v \in A^{**}$ satisfies the relations in (2.2), then by Lemma 2.6(b), $r \sim_{\mathrm{sp}} v^*rv$ for all $r \in \mathrm{OP}(\mathrm{her}(p))$, which means that spatial equivalence is automatically "hereditarily stable".

Remark 2.8. (a) Let $p, q \in \mathrm{Proj}(A^{**})$. We call the unique $p_{\mathrm{int}} \in \mathrm{OP}(A)$ with $\mathrm{her}(p) = \mathrm{her}(p_{\mathrm{int}})$ the *interior* of p. By the bijective correspondence between hereditary C^*-subalgebras and open projections, p_{int} is the largest open projection dominated by p. As a direct consequence of Proposition 2.7(a), we know that $p_{\mathrm{int}} \sim_{\mathrm{sp}} q_{\mathrm{int}}$ if and only if

> $\mathrm{her}(q) \subseteq u^*\,\mathrm{her}(p)u$ and $\mathrm{her}(p) \subseteq u\,\mathrm{her}(q)u^*$ for a partial isometry $u \in A^{**}$.

(b) Suppose that $p, q \in \mathrm{OP}(A)$. One might attempt to define $p \precsim_{\mathrm{sp}} q$ if there is $q_1 \in \mathrm{OP}(A)$ with $p \sim_{\mathrm{sp}} q_1 \leq q$. However, unlike the Murray–von Neumann equivalence situation, $p \precsim_{\mathrm{sp}} q$ and $q \precsim_{\mathrm{sp}} p$ does not imply that $p \sim_{\mathrm{sp}} q$. This can be shown by using a result of Lin. More precisely, it was shown in [23, Theorem 9] that there exist a separable unital simple C^*-algebra A as well as $p \in \mathrm{Proj}(A)$ and $u \in A$ such that $uu^* = 1$ and $p_1 = u^*u \leq p$, but $\mathrm{her}(p)$ and A are not *-isomorphic. In particular, $p \nsim_{\mathrm{sp}} 1$. Now, we clearly have $p \precsim_{\mathrm{sp}} 1$. On the other hand, as $u \in A$, we have

$$u^*Au = \mathrm{her}(p_1) \quad \text{and} \quad u\,\mathrm{her}(p_1)u^* = A,$$

which implies that $1 \precsim_{\mathrm{sp}} p$.

This example also shows that the same problematic situation appears even if we replace $\sim_{\mathrm{sp}}$ with the stronger equivalence relation $\sim_{\mathrm{PZ}}$ as defined in (2.1) (because $u \in A$). Nevertheless, it was shown in [32, Theorem 1.13] that a weaker conclusion holds if one adds an extra assumption on either p or q, but we will not recall the details here.

Let us end this section with the following well-known example. We give an explicit argument here for future reference. Note that parts (a) and (b) of it mean that if $a, b \in A_+$ are equivalent in the sense of Blackadar (i.e., there exists $x \in A$ with $a = x^*x$ and $b = xx^*$; see, e.g., [29, Definition 2.1]), then their support projections are spatially equivalence (which is also a corollary of [29, Proposition 4.3], since $\sim_{\mathrm{PZ}}$ is stronger than $\sim_{\mathrm{sp}}$).

Example 2.9. Suppose that $x \in A$ with $\|x\| = 1$. Set $a = x^*x$ and $b = xx^*$. Let $x = ua^{1/2}$ be the polar decomposition.

(a) It is easy to see that $\overline{aAa} = u^*(\overline{xAx^*})u$ and $\overline{xAx^*} = u(\overline{aAa})u^*$, i.e., $\overline{xAx^*}$ is spatially isomorphic to $\overline{aAa}$ (by Proposition 2.7(a)).

(b) Notice that $u(\overline{aAa})u^* = \overline{xAx^*} \supseteq \overline{xx^*Axx^*} \supseteq \overline{xx^*xAx^*xx^*} \supseteq \overline{ua^{3/2}Aa^{3/2}u^*} = u(\overline{aAa})u^*$, and we have $\overline{xAx^*} = \overline{bAb}$. Similarly, $\overline{x^*Ax} = \overline{aAa}$ and $\overline{x^*A^{**}x} = \overline{aA^{**}a}$, which implies that $\mathrm{her}(x) = \mathrm{her}(a)$. On the other hand, as $\overline{aAa}$ is a hereditary C^*-subalgebra of $\mathrm{her}(a)$ and $\{a^{1/k}\}_{k\in\mathbb{N}}$ is a sequence in $\overline{aAa}$ which is an approximate unit for $\mathrm{her}(a)$, one has $\overline{aAa} = \mathrm{her}(a)$. Consequently, $\mathrm{her}(x) = \overline{x^*Ax}$.

(c) Suppose that $B \subseteq A$ is a hereditary C^*-subalgebra and $x \in B$. Since $\overline{aAa} = \overline{a^2Aa^2}$, we see that $\overline{aBa} = \overline{aAa}$. Therefore, $\mathrm{her}_B(x) = \mathrm{her}_A(x)$ by part (b).

3. C^*-semi-finiteness and three types of C^*-algebras

As in the case of von Neumann algebras ([27]), in order to define different "types" of C^*-algebras, we need to define "abelian" and "finite" open projections. "Abelian" open projections are defined in the same way as that of von Neumann algebras. However, in order to define "finite" open projections, we need to use our "hereditarily stable version" of Murray–von Neumann equivalence in Section 2. Note that one cannot go very far with the original Murray–von Neumann equivalence, because there exist $p, q \in \mathrm{OP}(A)$ with $p \sim_{\mathrm{Mv}} q$ but $\mathrm{her}(p)$ and $\mathrm{her}(q)$ are not isomorphic (see [23]). Moreover, one cannot use a direct verbatim translation of the Murray–von Neumann finiteness.

Definition 3.1.

(a) Let $q \in \mathrm{OP}(A)$ and $p \in \mathrm{Proj}(qA^{**}q)$. The *closure of p in q*, denoted by $\bar{p}^q$, is the smallest closed projection of $\mathrm{her}(q)$ that dominates p.

(b) Let $p, q \in \mathrm{OP}(A)$ with $p \leq q$. The projection p is said to be

  i. *dense in q* if $\bar{p}^q = q$;

  ii. *abelian* if $\mathrm{her}(p)$ is a commutative C^*-algebra;

  iii. *C^*-finite* if for any $r, s \in \mathrm{OP}(\mathrm{her}(p))$ with $r \leq s$ and $r \sim_{\mathrm{sp}} s$, one has $\bar{r}^s = s$.

If p is dense in q, we say that $\mathrm{her}(p)$ is *essential* in $\mathrm{her}(q)$. We denote by $\mathrm{OP}_{\mathfrak{C}}(A)$ and $\mathrm{OP}_{\mathfrak{F}}(A)$ the set of all abelian open projections and the set of all C^*-finite open projections of A, respectively.

The terminology "p is dense in q" is used in many places (e.g., [32]), while the terminology "essential" comes from [39].

Some people might wonder why we do not use the finiteness as defined in [14]. The reason is that we want to give a classification scheme for C^*-algebras using open projections (and the definition of finiteness in [14] seems not related to open projections).

Remark 3.2. Let $p \in \mathrm{OP}(A)$.

(a) Suppose that p is abelian. If $r, s \in \mathrm{OP}(\mathrm{her}(p))$ satisfying $r \leq s$ and $r \sim_{\mathrm{sp}} s$, then $r = s$. Thus, p is C^*-finite.

(b) If $\mathrm{her}(p)$ is finite dimensional, then p is C^*-finite.

(c) One might ask why we do not define C^*-finiteness of p in the following way: for any $r \in \mathrm{OP}(\mathrm{her}(p))$ with $r \sim_{\mathrm{sp}} p$, one has $\bar{r}^p = p$. The reason is that the stronger condition in Definition 3.1(b) can ensure every open subprojection of a C^*-finite projection being C^*-finite. Such a phenomena is automatic for von Neumann algebras.

(d) A hereditary C^*-subalgebra $B \subseteq A$ is essential in A if and only if for any non-zero hereditary C^*-subalgebra $C \subseteq A$, one has $B \cdot C \neq \{0\}$. Thus, a closed ideal $I \subseteq A$ is essential in the sense of Definition 3.1 if and only it is essential in the usual sense (i.e., any non-zero closed ideal of A intersects I non-trivially).

Definition 3.3. A C^*-algebra A is said to be:

i. *C^*-finite* if $1 \in \mathrm{OP}_{\mathfrak{F}}(A)$;
ii. *C^*-semi-finite* if every element in $\mathrm{OP}(A) \setminus \{0\}$ dominates an element in $\mathrm{OP}_{\mathfrak{F}}(A) \setminus \{0\}$;
iii. *of Type $\mathfrak{A}$* if every element in $\mathrm{OP}(A) \cap \mathrm{Z}(A^{**}) \setminus \{0\}$ dominates an element in $\mathrm{OP}_{\mathfrak{C}}(A) \setminus \{0\}$;
iv. *of Type $\mathfrak{B}$* if $\mathrm{OP}_{\mathfrak{C}}(A) = \{0\}$ but each element in $\mathrm{OP}(A) \cap \mathrm{Z}(A^{**}) \setminus \{0\}$ dominates an element in $\mathrm{OP}_{\mathfrak{F}}(A) \setminus \{0\}$;
v. *of Type $\mathfrak{C}$* if $\mathrm{OP}_{\mathfrak{F}}(A) = \{0\}$.

Let us give an equivalent form of the above abstract definition through the relation between open projections (respectively, central open projections) and hereditary C^*-subalgebras (respectively, ideals). A C^*-algebra A is

- C^*-finite if and only if for each hereditary C^*-subalgebra $B \subseteq A$, every hereditary C^*-subalgebra of B that is spatially isomorphic to B is essential in B;
- C^*-semi-finite if and only if every non-zero hereditary C^*-subalgebra of A contains a non-zero C^*-finite hereditary C^*-subalgebra;
- of type $\mathfrak{A}$ if and only if every non-zero closed ideal of A contains a non-zero abelian hereditary C^*-subalgebra;
- of type $\mathfrak{B}$ if and only if A does not contain any non-zero abelian hereditary C^*-subalgebra and every non-zero closed ideal of A contains a non-zero C^*-finite hereditary C^*-subalgebra;

- of type $\mathfrak{C}$ if and only if A does not contain any non-zero C^*-finite hereditary C^*-subalgebra.

Remark 3.4. Suppose that A is simple.

(a) A is either of type $\mathfrak{A}$, type $\mathfrak{B}$ or type $\mathfrak{C}$.

(b) We will see in Corollary 4.5 that A is of type $\mathfrak{A}$ if and only if A is of type I (see, e.g., [31, 6.1.1] for its definition). Moreover, if A is of type II (in the sense of [14]), then A is of type $\mathfrak{B}$ (by Proposition 4.7 below), while if A is purely infinite (in the sense of [13]), then A is of type $\mathfrak{C}$ (by Proposition 4.11(a) below and [40, Theorem 1.2(ii)]). However, we do not know if the converse of the last two statements hold.

A positive element $a \in A_+$ is said to be *C^*-finite* if $\mathrm{her}(a)$ (i.e., $\overline{aAa}$) is C^*-finite.

Proposition 3.5.

(a) *The sum, $\mathcal{C}(A)$, of all abelian hereditary C^*-subalgebras of A is a (not necessarily closed) ideal of A. If $\mathcal{C}(A)_+ := \mathcal{C}(A) \cap A_+$, then $\mathcal{C}(A)$ coincides with the vector space* $\operatorname{span}\mathcal{C}(A)_+$ *generated by $\mathcal{C}(A)_+$.*

(b) *The sum, $\mathcal{F}(A)$, of all C^*-finite hereditary C^*-subalgebras of A is a (not necessarily closed) ideal of A. If $\mathcal{F}(A)_+ := \mathcal{F}(A) \cap A_+$, then $\mathcal{F}(A) = \operatorname{span}\mathcal{F}(A)_+$.*

(c) *If $B \subseteq A$ is a hereditary C^*-subalgebra, then $\mathcal{C}(B)_+ = \mathcal{C}(A) \cap B_+$ and $\mathcal{F}(B)_+ = \mathcal{F}(A) \cap B_+$.*

Proof. Since parts (a) and (b) follow from the arguments of [31, Proposition 6.1.7], we will only give the proof for part (c). Moreover, we will only establish the second equality as the argument for the first one is similar. As K_A is a hereditary cone, the argument of part (b) tells us that $\mathcal{F}(A)_+ = K_A$. It is clear that $\mathcal{F}(B) \subseteq \mathcal{F}(A) \cap B$. Conversely, if $w \in K_A \cap B$ and $w_1, \ldots, w_n \in F_A$ such that $w = \sum_{i=1}^n w_i$, then $w_i \leq w \in B_+$, which implies that $w_i \in F_A \cap B = F_B$ (see Example 2.9(c)). Consequently, $w \in K_B$ as required. □

Clearly, $\mathcal{C}(A) \subseteq \mathcal{F}(A)$. We will see in Theorem 5.2(d) below that the closed ideal $\overline{\mathcal{C}(A)}$ is of type $\mathfrak{A}$, while $\overline{\mathcal{F}(A)}$ is C^*-semi-finite.

Example 3.6. (a) If A is commutative, then A is of type $\mathfrak{A}$ and is C^*-finite. Moreover, $\mathcal{C}(A) = \mathcal{F}(A) = A$.

(b) Let $p \in \mathrm{OP}(\mathcal{B}(\ell^2)) \subseteq \mathcal{B}(\ell^2)^{**}$ such that $\mathrm{her}(p) = \mathcal{K}(\ell^2)$ (the C^*-algebra of all compact operators). Then $p \neq 1$ but $\mathrm{her}(1-p) = (0)$. In fact, if $T \in \mathrm{her}(1-p)$, we have $pT = 0$ and $ST = SpT = 0$ for any $S \in \mathcal{K}(\ell^2)$, which gives $T = 0$. Moreover, p is dense in 1 because $\mathcal{K}(\ell^2)$ is an essential closed ideal of $\mathcal{B}(\ell^2)$ (see Remark 3.2(d)).

(c) If H is an infinite-dimensional Hilbert space, then $\mathcal{K}(H)$ is a C^*-algebra of type $\mathfrak{A}$, which is not C^*-finite but is C^*-semi-finite. In fact, as $\mathcal{K}(H)$ is simple and contains many rank-one projections, it is of type $\mathfrak{A}$. On the other hand, suppose that $e \in \mathrm{Proj}(\mathcal{K}(H))$ is a rank-one projection. Then $1 - e \in \mathrm{OP}(\mathcal{K}(H)) \subseteq \mathcal{B}(H)$ and there is an isometry $v \in \mathcal{B}(H)$ with $vv^* = 1 - e$. Thus,

$$v^* \operatorname{her}(1-e)v \;=\; \mathcal{K}(H) \quad \text{and} \quad 1 - e \;\sim_{\mathrm{sp}}\; 1.$$

Moreover, as $e \in \mathrm{Proj}(\mathcal{K}(H))$, we see that $1-e$ is also a closed projection and hence it is not dense in 1. Finally, as all hereditary C^*-subalgebras of $\mathcal{K}(H)$ are given by projections in $\mathcal{B}(H)$, they are of the form $\mathcal{K}(K)$ for some subspaces $K \subseteq H$. Hence, $\mathcal{K}(H)$ is C^*-semi-finite (see Remark 3.2(b)).

(d) Let H be a Hilbert space. Clearly, $\mathrm{Proj}(\mathcal{K}(H)) \subseteq \mathrm{OP}_{\mathcal{F}}(\mathcal{B}(H))$. Hence, if $\mathfrak{F}(H)$ is the set of all finite rank operators, then $\mathfrak{F}(H) \subseteq \mathcal{F}(\mathcal{B}(H))$. Suppose that $B \subseteq \mathcal{B}(H)$ is a C^*-finite hereditary C^*-subalgebra and $p \in \mathrm{Proj}(B)$. As p is C^*-finite and $pBp = p\mathcal{B}(H)p \cong \mathcal{B}(K)$ for a subspace $K \subseteq H$, we see that K is finite dimensional (see part (c)) and so $p \in \mathcal{K}(H)$. Since $B \subseteq \mathcal{B}(H)$ is a hereditary C^*-subalgebra, B is generated by its projections. Thus, B is a hereditary C^*-subalgebra of $\mathcal{K}(H)$, and $B \cong \mathcal{K}(H')$ for a subspace $H' \subseteq H$. The C^*-finiteness of B again implies that $\dim H' < \infty$, and $B \subseteq \mathfrak{F}(H)$. Consequently,

$$\mathcal{F}(\mathcal{B}(H)) \;=\; \mathfrak{F}(H).$$

On the other hand, since any finite rank projection is a sum of rank-one projections and any rank-one projection belongs to $\mathcal{C}(\mathcal{B}(H))$, we see that $\mathfrak{F}(H) = \mathcal{C}(\mathcal{B}(H)) = \mathcal{F}(\mathcal{B}(H))$. Furthermore, by Proposition 3.5(c), we also have $\mathcal{F}(\mathcal{K}(H)) = \mathcal{C}(\mathcal{K}(H)) = \mathfrak{F}(H)$.

Remark 3.7. Let $e \in \mathrm{OP}(A)$ and $z(e)$ be the central support of e in A^{**}.

(a) $z(e) = \sup_{u \in U_{M(A)}} ueu^*$ (see, e.g., [31, Lemma 2.6.3]), and $z(e)$ is an open projection (see Remark 2.2(d)) with $\operatorname{her}(z(e))$ being the smallest closed ideal containing $\operatorname{her}(e)$.

(b) Recall that $B := \operatorname{her}(e) \subseteq A$ is said to be *full* if $\operatorname{her}(z(e)) = A$. In this case, B is strongly Morita equivalent to A (see, e.g., [35]). Consequently, $\operatorname{her}(e)$ is always strongly Morita equivalent to $\operatorname{her}(z(e))$.

The following provides an important tool to us in this paper. An essential ingredient of its proof (in particular, part (b)) is a result of Peligrad and Zsidó in [32].

Proposition 3.8. *Let A and B be two strongly Morita equivalent C^*-algebras.*

(a) *A contains a non-zero abelian hereditary C^*-subalgebra if and only if B does.*
(b) *A contains a non-zero C^*-finite hereditary C^*-subalgebra if and only if B does.*

Proof. There exist a C^*-algebra D and $e \in \mathrm{Proj}(M(D))$ such that both A and B are full hereditary C^*-subalgebras of D and we have

$$A \cong eDe \quad \text{and} \quad B \cong (1-e)D(1-e)$$

(see, e.g., [8, Theorem II.7.6.9]). Thus, $z(e) = 1 = z(1-e)$.

(a) It suffices to show that A contains a non-zero abelian hereditary C^*-subalgebra whenever D does. Let $p \in \mathrm{OP}_{\mathfrak{C}}(D)\setminus\{0\}$. As $pz(e) = p \neq 0$, we see that $pueu^* \neq 0$ for some $u \in U_{M(D)}$. By replacing p with u^*pu, we may assume that $pe \neq 0$, and hence $e\,\mathrm{her}_D(p)e \neq (0)$. If $x, y \in \mathrm{her}_D(p)$ and $\{b_j\}_{j\in\mathfrak{J}}$ is an approximate unit of $\mathrm{her}_D(p)$, then $b_ieb_j \in \mathrm{her}_D(p)$ which implies that

$$xey \;=\; \lim xb_ieb_jy \;=\; \lim yb_ieb_jx \;=\; yex.$$

Consequently, $e\,\mathrm{her}_D(p)e$ is an abelian hereditary C^*-subalgebra of A.

(b) It suffices to show that if D contains a non-zero C^*-finite hereditary C^*-subalgebra, then so does A. Suppose that $p \in \mathrm{OP}_{\mathfrak{F}}(D) \setminus \{0\}$. By [32, Theorem 1.9], there exist $e_0, e_1 \in \mathrm{OP}(\mathrm{her}_D(e))$ and $p_0, p_1 \in \mathrm{OP}(\mathrm{her}_D(p))$ satisfying

$$\overline{e_0 + e_1}^{\,e} = e,\;\; \overline{p_0 + p_1}^{\,p} = p,\;\; z(e_0)z(p_0) = 0 \;\text{ and }\; e_1 \sim_{\mathrm{PZ}} p_1.$$

Suppose that $p_1 = 0$. Then $e_1 = 0$ and $z(e_0)$ is dense in $z(e) = 1$ (by [32, Lemma 1.8]). This implies that $z(p_0) = 0$, and we have a contradiction that $p_0 = 0$ is dense in the non-zero open projection p. Therefore, $p_1 \neq 0$ and is C^*-finite. Since $\mathrm{her}_D(e_1) \cong \mathrm{her}_D(p_1)$ (note that $\sim_{\mathrm{PZ}}$ is stronger than $\sim_{\mathrm{sp}}$), we see that $\mathrm{her}_D(e_1)$ is a non-zero C^*-finite hereditary C^*-subalgebra of $A = \mathrm{her}_D(e)$. □

One may also use the argument of part (b) to obtain part (a), but we keep the alternative argument since it is also interesting.

Suppose that E is a full Hilbert A-module implementing the strong Morita equivalence between A and B, i.e., $B \cong \mathcal{K}_A(E)$ (see, e.g., [22]). If I is a closed ideal of A, then EI is a full Hilbert I-module and $\mathcal{K}_I(EI)$ is a closed ideal of B.

We recall from [32, Definition 2.1] that A is said to be *discrete* if any non-zero open projection of A dominates a non-zero abelian open projection.

Theorem 3.9.

(a) *Let A and B be two strongly Morita equivalent C^*-algebras. Then A is of type $\mathfrak{A}$ (respectively, type $\mathfrak{B}$ or type $\mathfrak{C}$) if and only if B is of the same type.*

(b) *A C^*-algebra A is of type $\mathfrak{A}$ if and only if it is discrete.*

Proof. (a) Suppose that A is of type $\mathfrak{B}$. If $\mathrm{OP}_{\mathfrak{C}}(B) \neq \{0\}$, then $\mathrm{OP}_{\mathfrak{C}}(A) \neq \{0\}$ (because of Proposition 3.8(a)), which is a contradiction. Let J be a non-zero closed ideal of B. As in the paragraph above, the strong Morita equivalence of A and B gives a closed ideal J_0 of A that is strongly Morita equivalent to J. As J_0 contains a non-zero C^*-finite hereditary C^*-subalgebra, so is J (by Proposition 3.8(b)). This shows that B is of type $\mathfrak{B}$. The argument for the other two types are similar and easier.

(b) It suffices to show that if A is of type $\mathfrak{A}$, then it is discrete. Let $B \subseteq A$ be a non-zero hereditary C^*-subalgebra and $J \subseteq A$ be the closed ideal generated by B (which is strongly Morita equivalent to B; see Remark 3.7(b)). As J contains a non-zero abelian hereditary C^*-subalgebra, so does B (by Proposition 3.8(a)). □

The following result follows from Proposition 3.8(b) and the argument of Theorem 3.9.

Corollary 3.10.

(a) *A is C^*-semi-finite if and only if any non-zero closed ideal of A contains a non-zero C^*-finite hereditary C^*-subalgebra.*
(b) *If A is strongly Morita equivalent to a C^*-semi-finite C^*-algebra, then A is also C^*-semi-finite.*
(c) *A is of type $\mathfrak{B}$ if and only if it is C^*-semi-finite and anti-liminary (i.e., it does not contain any non-zero commutative hereditary C^*-subalgebra).*

Remark 3.11. (a) As in the case of von Neumann algebra, strong Morita equivalence does not preserve C^*-finiteness. In fact, for any C^*-algebra A, the algebra $A \otimes \mathcal{K}(\ell^2)$ is not C^*-finite (using the same argument as Example 3.6(c); note that $1 \otimes (1 - e)$ is both an open and a closed projection of $A \otimes \mathcal{K}(\ell^2)$). Consequently, any stable C^*-algebra is not C^*-finite.

(b) By Remark 3.7(b), Theorem 3.9(a) and Corollary 3.10(b), any type $\mathfrak{A}$, type $\mathfrak{B}$, type $\mathfrak{C}$ or C^*-semi-finite hereditary C^*-subalgebra is contained in a closed ideal of the same type.

Recall that a C^*-algebra A has *real rank-zero* in the sense of Brown and Pedersen if the set of elements in A_{sa} with finite spectrum is norm dense in A_{sa} (see, e.g., [12, Corollary 2.6]). The following result follows from Theorem 3.9(b), Corollary 3.10(c) as well as the fact that any hereditary C^*-subalgebra of a real rank-zero C^*-algebra is again of real rank-zero (see, e.g., [12, Corollary 2.8]).

Corollary 3.12. *Let A be a C^*-algebra with real rank-zero.*

(a) *A is of type $\mathfrak{A}$ if and only if every projection in* $\mathrm{Proj}(A) \setminus \{0\}$ *dominates an abelian projection in* $\mathrm{Proj}(A) \setminus \{0\}$.
(b) *A is of type $\mathfrak{B}$ if and only if every projection in* $\mathrm{Proj}(A) \setminus \{0\}$ *is non-abelian but dominates a C^*-finite projection in* $\mathrm{Proj}(A) \setminus \{0\}$.
(c) *A is of type $\mathfrak{C}$ if and only if A does not contain any non-zero C^*-finite projection.*
(d) *A is C^*-semi-finite if and only if every projection in* $\mathrm{Proj}(A)\setminus\{0\}$ *dominates a C^*-finite projection in* $\mathrm{Proj}(A) \setminus \{0\}$.

Remark 3.13. Suppose that A is a C^*-finite C^*-algebra with real rank-zero. If $r, p \in \mathrm{Proj}(A)$ such that $r \leq p$ and there exists $u \in A$ with $uu^* = r$ and $u^*u = p$, then $r \sim_{\mathrm{sp}} p$ and so, $r = \bar{r}^p = p$.

Corollary 3.14. *If A is of real rank-zero, then the closures of the ideals $\mathcal{C}(A)$ and $\mathcal{F}(A)$ (see Proposition 3.5) are the closed linear spans of abelian projections and of C^*-finite projections in* $\mathrm{Proj}(A)$, *respectively.*

Proof. If $B \subseteq A$ is a C^*-finite hereditary C^*-subalgebra, then B is the closed linear span of $\mathrm{Proj}(B) \cap \mathrm{OP}_{\mathcal{F}}(B)$. Thus, $\mathcal{F}(A)$ lies inside the closed linear span of $\mathrm{Proj}(A) \cap \mathrm{OP}_{\mathcal{F}}(A)$. Conversely, it is clear that $\mathrm{Proj}(A) \cap \mathrm{OP}_{\mathcal{F}}(A) \subseteq \mathcal{F}(A)$. The argument for the statement concerning $\mathcal{C}(A)$ is similar. □

Corollary 3.15. *Let A be of type $\mathfrak{A}$ (respectively, of type $\mathfrak{B}$, of type $\mathfrak{C}$ or C^*-semi-finite).*

(a) *If B is a hereditary C^*-subalgebra of A, then B is of type $\mathfrak{A}$ (respectively, of type $\mathfrak{B}$, of type $\mathfrak{C}$ or C^*-semi-finite).*

(b) *If A is a hereditary C^*-subalgebra of A_0 that generates an essential ideal $I \subseteq A_0$, then A_0 is of type $\mathfrak{A}$ (respectively, of type $\mathfrak{B}$, of type $\mathfrak{C}$ or C^*-semi-finite).*

Proof. (a) As any hereditary C^*-subalgebra of B is a hereditary C^*-subalgebra of A, this result follows directly from the definitions, Theorem 3.9(b) and Corollary 3.10(c).

(b) Note that A is strongly Morita equivalent to I and any hereditary C^*-subalgebra of A_0 intersects I non-trivially. Thus, this part follows from the definitions, Theorem 3.9 and Corollary 3.10. □

Consequently, we have the following result.

Corollary 3.16. *Suppose that A is non-unital, and $\tilde{A}$ is the unitalization of A. Then A is of type $\mathfrak{A}$ (respectively, of type $\mathfrak{B}$, of type $\mathfrak{C}$ or C^*-semi-finite) if and only if $\tilde{A}$ is of type $\mathfrak{A}$ (respectively, of type $\mathfrak{B}$, of type $\mathfrak{C}$ or C^*-semi-finite). The same is true when $\tilde{A}$ is replaced by $M(A)$.*

Our next lemma is probably well known, but we give a simple argument here for completeness.

Lemma 3.17. *Let $e, f \in \mathrm{OP}(A)$ and $p, q \in \mathrm{OP}(A) \cap \mathrm{Z}(A^{**})$.*

(a) *$ep \in \mathrm{OP}(A)$ and $\mathrm{her}(ep) = \mathrm{her}(e) \cap \mathrm{her}(p)$.*

(b) *If $e \neq 0$ and $\mathrm{her}(e) \subseteq \mathrm{her}(p) + \mathrm{her}(q)$, then $\mathrm{her}(e) \cap \mathrm{her}(p) \neq (0)$ or $\mathrm{her}(e) \cap \mathrm{her}(q) \neq (0)$.*

(c) *If $z(e)z(f) = 0$, then $\mathrm{her}(e) + \mathrm{her}(f) = \mathrm{her}(e + f)$.*

Proof. Parts (a) and (c) are obvious (see Remark 2.2(d)). To show part (b), note that as $\mathrm{her}(p) + \mathrm{her}(q) \subseteq \mathrm{her}(p+q-pq)$, we have $e \leq p+q-pq$. If $ep = 0 = eq$, one obtains a contradiction that $e = e(p + q - pq) = 0$. Thus, the conclusion follows from part (a). □

Lemma 3.18. *If $\{p_i\}_{i\in\mathfrak{I}}$ is a family in $\mathrm{OP}_{\mathcal{F}}(A)$ with $z(p_i)z(p_j) = 0$ for $i \neq j$, then $p := \sum_{i\in\mathfrak{I}} p_i \in \mathrm{OP}_{\mathcal{F}}(A)$.*

Proof. It is clear that p is an open projection and $z(p) = \sum_{i\in\mathfrak{I}} z(p_i)$. Suppose that $r, q \in \mathrm{OP}(\mathrm{her}(p))$ with $r \leq q$ and $r \sim_{\mathrm{sp}} q$. Let $u \in A^{**}$ with $q = u^*u$ and $u\,\mathrm{her}(q)u^* = \mathrm{her}(r)$. For any $i \in \mathfrak{I}$, we set $q_i := z(p_i)q, r_i := z(p_i)r \in \mathrm{OP}(A)$ and $u_i := z(p_i)u$. It is easy to see that $q = \sum_{i\in\mathfrak{I}} q_i$, $r = \sum_{i\in\mathfrak{I}} r_i$, $q_i = u_i^*u_i$ and $r_i \leq q_i \leq z(p_i)p = p_i$. By Lemma 3.17(c), we see that

$$z(p_i)\,\mathrm{her}(q) \;=\; z(p_i)\big(\mathrm{her}(q_i) + \mathrm{her}\,\big(\textstyle\sum_{j\in\mathfrak{I}\setminus\{i\}} q_j\big)\big) \;=\; \mathrm{her}(q_i).$$

Similarly, $z(p_i)\,\mathrm{her}(r) = \mathrm{her}(r_i)$ and we have $u_i\,\mathrm{her}(q_i)u_i^* = \mathrm{her}(r_i)$. By Proposition 2.7(a), we know that $r_i \sim_{\mathrm{sp}} q_i$ and the C^*-finiteness of p_i tells us that r_i is dense in q_i. If $e \in \mathrm{OP}(\mathrm{her}(q))$ with $re = 0$, then $e_i := z(p_i)e \in \mathrm{OP}(\mathrm{her}(q_i))$ with $r_ie_i = 0$, which means that $e_i = 0$ (because $\overline{r_i}^{q_i} = q_i$). Consequently, $e = \sum_{i\in\mathfrak{I}} e_i = 0$ and r is dense in q as required. □

Part (a) of the following result is the equivalence of statements (i) and (iii) in [32, Theorem 2.3], while part (b) follows from the proof of [32, Theorem 2.3], Lemma 3.18, Theorem 3.9(a) and Corollary 3.15(b).

Proposition 3.19.

(a) *A C^*-algebra A is of type $\mathfrak{A}$ if and only if there is an abelian hereditary C^*-subalgebra of A that generates an essential closed ideal of A.*
(b) *A C^*-algebra A is C^*-semi-finite if and only if there is a C^*-finite hereditary C^*-subalgebra of A that generates an essential closed ideal of A.*

4. Comparison with existing theories

In this section, we compare our "Murray–von Neumann type classification" with existing results in the literature. Through these comparisons, we obtain many examples of C^*-algebras of different types. Moreover, we will show that a von Neumann algebra is a type $\mathfrak{A}$, type $\mathfrak{B}$, type $\mathfrak{C}$ or C^*-semi-finite C^*-algebra if and only if it is, respectively, a type I, type II, type III or semi-finiteness von Neumann algebra.

4.1. Comparison with type I algebras

Recall that a C^*-algebra A is said to be of *type I* if for any irreducible representation (π, H) of A, one has $\mathcal{K}(H) \subseteq \pi(A)$. We have already seen in Theorem 3.9(b) that type $\mathfrak{A}$ is the same as discreteness. Thus, the following result is a direct consequence of [32, Theorem 2.3]. Note that one can also obtain it using Theorem 3.9(a) and [6, Theorems 1.8 and 2.2].

Corollary 4.1. *Any type I C^*-algebra is of type $\mathfrak{A}$.*

The converse of the above is not true even for real rank-zero C^*-algebras, as can be seen in the following example.

Example 4.2. Example 3.6(c) and Corollary 3.15(b) tell us that $\mathcal{B}(\ell^2)$ is of type $\mathfrak{A}$. However, $\mathcal{B}(\ell^2)$ is not a type I C^*-algebra (see, e.g., [31, 6.1.2]).

Proposition 4.3.

(a) *A is of type I if and only if every primitive quotient of A is of type $\mathfrak{A}$.*
(b) *If A is of type $\mathfrak{A}$ and contains no essential primitive ideal, then A is of type I.*

Proof. (a) Because of Corollary 4.1 and the fact that quotients of type I C^*-algebras are also of type I, we only need to show the "if" part. Let $\pi : A \to \mathcal{B}(H)$ be an irreducible representation and B be a non-zero abelian hereditary C^*-subalgebra of $A/\ker\pi$. If $\tilde{\pi} : A/\ker\pi \to \mathcal{B}(H)$ is the induced representation, the restriction $\tilde{\pi}_B : B \to \mathcal{B}(\tilde{\pi}(B)H)$ is non-zero and irreducible. Thus, $\dim\tilde{\pi}(B)H = 1$ and $\tilde{\pi}(b)$ is a rank-one operator (and hence is compact) for any $b \in B \setminus \{0\}$. This shows that $\tilde{\pi}(A/\ker\pi) \cap \mathcal{K}(H) \neq (0)$, and $\pi(A) \supseteq \mathcal{K}(H)$.

(b) Suppose that $\pi : A \to \mathcal{B}(H)$ is an irreducible representation and J is a non-zero closed ideal of A with $J \cap \ker\pi = (0)$. If $B \subseteq J$ is a non-zero abelian hereditary C^*-subalgebra, the restriction $\pi_B : B \to \mathcal{B}(\pi(B)H)$ is non-zero and irreducible. The same argument as in part (a) tells us that $\pi(A) \supseteq \mathcal{K}(H)$. □

Remark 4.4. (a) Proposition 4.3(a) actually shows that A is of type I if and only if any primitive quotient contains a non-zero abelian hereditary C^*-subalgebra, which is likely to be a known fact.

(b) If every quotient of $\mathcal{B}(\ell^2)$ were of type $\mathfrak{A}$, then Proposition 4.3(a) told us that $\mathcal{B}(\ell^2)$ were a type I C^*-algebra, which contradicted [31, 6.1.2]. Consequently, not every quotient of a type $\mathfrak{A}$ C^*-algebra is of type $\mathfrak{A}$.

If A is simple and of type $\mathfrak{A}$, then by Proposition 4.3(b), it is of type I. This, together with Example 3.6(c), gives the following.

Corollary 4.5. *If A is a simple C^*-algebra of type $\mathfrak{A}$, then $A = \mathcal{K}(H)$ for some Hilbert space H. If, in addition, A is C^*-finite, then $A = M_n$ for some positive integer n.*

4.2. Comparison with type II and (semi-)finite C^*-algebras

The following is a direct consequence of Remark 3.4(a) and Corollary 4.5.

Corollary 4.6. *Any infinite-dimensional C^*-finite simple C^*-algebra is of type $\mathfrak{B}$.*

In the following, we compare type $\mathfrak{B}$ and type $\mathfrak{C}$ with the notions of type *II* and type *III* as introduced by Cuntz and Pedersen in [14]. Let us recall from [14, p. 140] that $x \in A_+$ is said to be *finite* if for any sequence $\{z_k\}_{k\in\mathbb{N}}$ in A with $x = \sum_{k=1}^{\infty} z_k^* z_k$, the condition $\sum_{k=1}^{\infty} z_k z_k^* \leq x$ will imply $x = \sum_{k=1}^{\infty} z_k z_k^*$. We also recall that A is said to be *finite* (respectively, *semi-finite*) if every $x \in A_+ \setminus \{0\}$ is finite (respectively, x dominates a non-zero finite element). Furthermore, A is said to be of *type II* if it is anti-liminary and finite, while A is said to be of *type III* if it has no non-zero finite elements (see [14, p. 149]).

Let $T_s(A)$ be the set of all tracial states on A. It follows from [14, Theorem 3.4] that $T_s(A)$ separates points of A_+ if A is finite.

Proposition 4.7. *If $T_s(A)$ separates points of A_+, then A is C^*-finite. Consequently, if A is finite, then A is C^*-finite.*

Proof. Suppose on the contrary that there exist $r, q \in \mathrm{OP}(A)$ with $r \leq q$, $r \sim_{\mathrm{sp}} q$ but $\bar{r}^q \lneq q$. For any $\tau \in T_s(A)$, if $\tilde{\tau}$ is the normal tracial state on A^{**} extending

τ, then $\tilde{\tau}(r) = \tilde{\tau}(q)$ (because $r = vv^*$ and $q = v^*v$ for some $v \in A^{**}$). Moreover, if $\{a_i\}_{i\in\mathfrak{I}}$ is an approximate unit in $\mathrm{her}(r)$, one has $\tilde{\tau}(r) = \lim \tau(a_i)$. Since $\bar{r}^q \precsim q$, there exists $s \in \mathrm{OP}(\mathrm{her}(q))\setminus\{0\}$ with $rs = 0$. If $x \in \mathrm{her}(s)_+$ with $\|x\| = 1$, one can find $\tau_0 \in T_s(A)$ with $\tau_0(x) > 0$. Thus, we have $\tau_0(a_i)+\tau_0(x) \leq \tilde{\tau}_0(q)$ (as $a_i + x \leq q$ because $a_i x = 0$), which gives the contradiction that $\tilde{\tau}_0(r) + \tau_0(x) \leq \tilde{\tau}_0(q)$. □

As in [14], we denote by $\mathscr{F}^A$ the set of all finite elements in A_+. If $B \subseteq A$ is a hereditary C^*-subalgebra, then

$$\mathscr{F}^B = \mathscr{F}^A \cap B.$$

In fact, it is obvious that $\mathscr{F}^A \cap B \subseteq \mathscr{F}^B$. Conversely, suppose that $x \in \mathscr{F}^B$. Consider $y \in A_+$ and a sequence $\{z_k\}_{k\in\mathbb{N}}$ in A satisfying $y \leq x$, $y = \sum_{k=1}^{\infty} z_k z_k^*$ and $x = \sum_{k=1}^{\infty} z_k^* z_k$. Since B_+ is a hereditary cone of A_+, we have $y \in B_+$ and $z_k^* z_k, z_k z_k^* \in B_+$ ($k \in \mathbb{N}$). By Remark 2.2(c), we know that $z_k \in B$ and so, $y = x$ as required.

Corollary 4.8.

(a) *A is semi-finite if and only if every non-zero hereditary C^*-subalgebra of A contains a non-zero finite hereditary C^*-subalgebra.*

(b) *If A is semi-finite (respectively, of type II), then A is C^*-semi-finite (respectively, of type $\mathfrak{B}$).*

Proof. (a) For the necessity, let $B \subseteq A$ be a non-zero hereditary C^*-subalgebra. If $y \in B_+ \setminus \{0\}$, there is $x \in \mathscr{F}^A \setminus \{0\}$ with $x \leq y$. By [14, Lemma 4.1] and [14, Theorem 4.8] as well as their arguments, one can find a non-zero finite hereditary C^*-subalgebra of $\mathrm{her}(x)$. More precisely, let $f \in C(\sigma(x))_+$ such that f vanishes in a neighborhood of 0 and $f(t) \leq t \leq f(t) + \frac{\|x\|}{2}$ ($t \in \sigma(x)$). There exists $g \in C(\sigma(x))_+$ and $\lambda > 0$ such that $f = fg$ and $g(t) < \lambda t$ ($t \in \sigma(x)$). Then $g(x) \in \mathscr{F}^A$ and $f(x) = f(x)g(x)$, i.e.,

$$f(x) \in \mathscr{F}_0 := \{a \in A_+ : a = ay \text{ for some } y \in \mathscr{F}^A\} \subseteq \mathscr{F}^A.$$

For any $z \in \mathrm{her}(f(x))_+$, we have $zg(x) = z$ and $z \in \mathscr{F}_0 \cap \mathrm{her}(f(x)) \subseteq \mathscr{F}^A \cap \mathrm{her}(f(x)) = \mathscr{F}^{\mathrm{her}(f(x))}$. Thus, $\mathrm{her}(f(x))$ is a non-zero finite hereditary C^*-subalgebra of $\mathrm{her}(x)$.

For the sufficiency, let $y \in A_+ \setminus \{0\}$ and C be a non-zero finite hereditary C^*-subalgebra of $\mathrm{her}(y)$. Observe that $C_+ = \mathscr{F}^C = \mathscr{F}^A \cap C$. Take any $x \in C_+$ with $\|x\| = 1$. Since $x^{1/2}yx^{1/2} \leq \|y\|x \in \mathscr{F}^A$, we know, from [14, Lemma 4.1], that

$$y^{1/2}xy^{1/2} = y^{1/2}x^{1/2}(y^{1/2}x^{1/2})^* \in \mathscr{F}^A.$$

Moreover, as $y^{1/2}xy^{1/2} \leq y$, we see that A is semi-finite.

(b) This follows from part (a), Proposition 4.7 and Corollary 3.10(c). □

Example 4.9. (a) If A is an infinite-dimensional simple C^*-algebra with a faithful tracial state, then A is of type $\mathfrak{B}$ (by Corollary 4.6 and Proposition 4.7). In particular, if Γ is an infinite discrete group such that $C_r^*(\Gamma)$ is simple (see, e.g., [7] for some examples of such groups), then $C_r^*(\Gamma)$ is of type $\mathfrak{B}$.

(b) Every simple AF algebra which is not of the form $\mathcal{K}(H)$ is of type $\mathfrak{B}$ (because of [14, Proposition 4.11] as well as Corollaries 4.5 and 4.8(b)).

4.3. Comparison with type III and purely infinite C^*-algebras

If a C^*-algebra A contains a non-zero (positive) finite element x, the argument of the necessity of Corollary 4.8(a) tells us that there is a non-zero finite hereditary C^*-subalgebra of A, and hence A is not of type $\mathfrak{C}$, because of Proposition 4.7. This gives the following corollary.

Corollary 4.10. *If A is of type $\mathfrak{C}$, then it is of type III.*

In the following, we will also compare type $\mathfrak{C}$ with the notion of pure infinity as defined by Cuntz (in the case of simple C^*-algebras) and by Kirchberg and Rørdam (in the general case). Suppose that $a \in M_n(A)$ and $b \in M_m(A)$ $(m, n \in \mathbb{N})$. As in [21, Definition 2.1], we say that *$a \precsim b$ relative to $M_{m,n}(A)$* if there is a sequence $\{x_k\}_{k\in\mathbb{N}}$ in $M_{m,n}(A)$ such that $\|x_k^* b x_k - a\| \to 0$. An element $a \in A$ is said to be *properly infinite* if $a \oplus a \precsim a$ relative to $M_{1,2}(A)$. Moreover, A is said to be *purely infinite* if every element in A_+ is properly infinite (see [21, Theorem 4.16]). Note that if A is simple, this notion coincides with the one in [13], namely, every hereditary C^*-subalgebra of A contains a non-zero infinite projection (see, e.g., the work of Lin and Zhang in [24]).

Proposition 4.11.

(a) *If A has real rank-zero and is purely infinite, then it is of type $\mathfrak{C}$.*

(b) *If A is a separable purely infinite C^*-algebra with stable rank-one, then A is of type $\mathfrak{C}$.*

Proof. (a) By [21, Theorem 4.16], any element $p \in \mathrm{Proj}(A)\setminus\{0\}$ is properly infinite and hence is infinite, in the sense that there exist $q \in \mathrm{Proj}(A)$ and $v \in A$ such that $q \leq p$, $v^*v = p$ and $q = vv^*$ (see, e.g., [21, Lemma 3.1]). Thus, $p \sim_{\mathrm{sp}} q$ (as $v \in A$) but q is not dense in p (because $p - q \in \mathrm{Proj}(A) \setminus \{0\}$). Consequently, any non-zero projection in A is not C^*-finite, and Corollary 3.12(c) shows that A is of type $\mathfrak{C}$.

(b) Suppose on contrary that A contains a non-zero C^*-finite hereditary C^*-subalgebra B and we take any $z \in B_+$ with $\|z\| = 1$. By [21, Theorem 4.16], one has $z \oplus z \precsim z \oplus 0$ relative to $M_2(A)$, and so, $z \oplus z \precsim z \oplus 0$ relative to $M_2(\mathrm{her}(z))$ (by [21, Lemma 2.2(iii)]). Thus, [29, Proposition 4.13] implies

$$p_z \oplus p_z \;=\; p_{z\oplus z} \precsim_{\mathrm{Cu}} p_{z\oplus 0} \;=\; p_z \oplus 0$$

(see [29, §3] for the meaning of $\precsim_{\mathrm{Cu}}$). Moreover, one obviously has $p_{z\oplus 0} \precsim_{\mathrm{Cu}} p_{z\oplus z}$. Since A has stable rank-one, we conclude that $p_z \oplus p_z \sim_{\mathrm{PZ}} p_z \oplus 0$ (by [29, 6.2(1)'&(2)']) and hence $p_z \oplus p_z \sim_{\mathrm{sp}} p_z \oplus 0$. This means that $M_2(\mathrm{her}(z))$ is spatially isomorphic (and hence *-isomorphic) to its hereditary C^*-subalgebra $\mathrm{her}(z) \oplus (0)$, which is not essential in $M_2(\mathrm{her}(z))$ (because $(0) \oplus \mathrm{her}(z)$ is a non-zero hereditary C^*-subalgebra and we can apply Remark 3.2(d)). As $\mathrm{her}(z)$ is *-isomorphic to $\mathrm{her}(z) \oplus (0)$ and hence to $M_2(\mathrm{her}(z))$, we know that $\mathrm{her}(z)$ is also spatially isomorphic to an inessential hereditary C^*-subalgebra. Consequently, $\mathrm{her}(z)$ is not C^*-finite, which contradicts the fact that B is C^*-finite. □

One may regard parts (a) and (b) of the above as two extremes, because any real rank-zero C^*-algebras has plenty of projections, while a purely infinite C^*-algebra with stable rank-one is stably projectionless. Let us make the following conjecture.

Conjecture 4.12. *Every purely infinite C^*-algebra is of type $\mathfrak{C}$.*

On the other hand, by Proposition 4.11 and Corollary 4.10, we know that any separable purely infinite C^*-algebra A having real rank-zero or stable rank-one is of type III. This implication actually holds without these extra assumptions, as can be seen in the following proposition, which gives another evidence for Conjecture 4.12. Note that this proposition also implies [21, Proposition 4.4]. To show this result, let us recall the following notation from [29, p. 3476]. For any $\epsilon > 0$, let $f_\epsilon : \mathbb{R}_+ \to \mathbb{R}_+$ be the function

$$f_\epsilon(t) \;=\; \begin{cases} t/\epsilon & \text{if } t \in [0,\epsilon) \\ 1 & \text{if } t \in [\epsilon,\infty). \end{cases}$$

If $\mu \in T_s(A)$ and $a \in A_+$, we define

$$d_\mu(a) \;:=\; \sup\nolimits_{\epsilon>0} \mu(f_\epsilon(a))$$

(note that the definition in [29] is for tracial weights but we only need tracial states here).

Proposition 4.13. *Any purely infinite C^*-algebra A is of type III.*

Proof. Suppose on the contrary that $\mathscr{F}^A \neq \{0\}$. By the argument of the necessity of Corollary 4.8(a), there is $z \in A_+$ with $\|z\| = 1$ and $\mathrm{her}(z)$ being a finite C^*-algebra. By the argument of Proposition 4.11(b), one has $z \oplus z \precsim z \oplus 0$ relative to $M_2(\mathrm{her}(z))$. By [29, Remark 2.5], we see that $d_\mu(z \oplus z) \leq d_\mu(z \oplus 0)$ for each $\mu \in T_s(M_2(\mathrm{her}(z)))$. Now, if $\tau \in T_s(\mathrm{her}(z))$, then $\tau \otimes \mathrm{Tr}_2 \in T_s(M_2(\mathrm{her}(z)))$ (where Tr_2 is the canonical tracial state on M_2), and the above tells us that

$$\sup_{\epsilon>0} \tau(f_\epsilon(z)) = \sup_{\epsilon>0}(\tau\otimes\mathrm{Tr}_2)(f_\epsilon(z)\oplus f_\epsilon(z)) \leq \sup_{\epsilon>0}(\tau\otimes\mathrm{Tr}_2)(f_\epsilon(z)\oplus 0) = \sup_{\epsilon>0}\frac{\tau(f_\epsilon(z))}{2},$$

which gives $d_\tau(z) = 0$ and hence $\tau(z) = 0$. This contradicts [14, Theorem 3.4]. □

If one can show that $\mathrm{her}(a)$ is not C^*-finite, for every properly infinite positive element a in any C^*-algebra, then the above conjecture is verified. Let us recall from [21, Proposition 3.3(iv)] that $a \in A_+$ is properly infinite if and only if there are sequences $\{x_n\}_{n\in\mathbb{N}}$ and $\{y_n\}_{n\in\mathbb{N}}$ in $\mathrm{her}(a)$ such that $x_n^* x_n \to a$, $y_n y_n^* \to a$ and $x_n^* y_n \to 0$. The following remark tells us that if $a \in A_+$ satisfies a stronger condition than the above, then $\mathrm{her}(a)$ is indeed non-C^*-finite.

Remark 4.14. Let $a \in A_+$ such that there exist $x, y \in \mathrm{her}(a)$ with $x^*x = a = y^*y$ as well as $x^*y = 0$. By Example 2.9(a)&(b), we see that $\mathrm{her}(a)$ is spatially isomorphic to its hereditary C^*-subalgebra $\mathrm{her}(x^*)$. As $\mathrm{her}(x^*)\,\mathrm{her}(y^*) = (0)$, we see that $\mathrm{her}(x^*)$ is not essential in $\mathrm{her}(a)$. Thus, $\mathrm{her}(a)$ is not C^*-finite.

Example 4.15. For any AF-algebra B, the C^*-algebra $\mathcal{O}_2 \otimes B$ is purely infinite (by [21, Proposition 4.5]) and is of real rank-zero (by [12, Theorem 3.2]), which means that $\mathcal{O}_2 \otimes B$ is of type $\mathfrak{C}$ (by Proposition 4.11(a)). Note that one may replace $\mathcal{O}_2$ with any unital, simple, separable, purely infinite, nuclear C^*-algebra (which has real rank-zero because of [40, Theorem 1.2(ii)]).

4.4. The case of von Neumann algebras

In this subsection, we consider the case of von Neumann algebras. Let us start with the following lemma. Note that the necessity of part (a) of this result follows directly from Proposition 4.7, but we give an alternative proof here as this argument is also interesting (see Remark 4.17 below).

Lemma 4.16.

(a) *Let M be a von Neumann algebra. Then $p \in \mathrm{Proj}(M)$ is finite as a projection in M if and only if it is C^*-finite.*

(b) *The ideal $\mathcal{F}(M)$ in Proposition 3.5 is a dense subalgebra of the ideal $J(M)$ generated by finite projections (as defined in [19]).*

Proof. (a) Assume that p is finite. Let $\Lambda_M : M^{**} \to M$ be the canonical *-epimorphism. If $q \in \mathrm{OP}(pMp)$, then $\mathrm{her}_M(q) \subseteq \mathrm{her}_M(\Lambda_M(q))$ and $\Lambda_M(q) \leq p$, which imply that $\Lambda_M(q) = \bar{q}^p$ (notice that $\bar{q}^p \in pMp$ because of [2, Theorem II.1]).

Suppose that $r, q \in \mathrm{OP}(pMp)$ such that $r \leq q$ and $r \sim_{\mathrm{sp}} q$. Consider $w \in M^{**}$ satisfying

$$q = ww^*, \quad r = w^*w, \quad w^*\,\mathrm{her}(q)w = \mathrm{her}(r) \quad \text{and} \quad w\,\mathrm{her}(r)w^* = \mathrm{her}(q).$$

Define $v := \Lambda_M(w)$. Then $\Lambda_M(q) = vv^*$ and $\Lambda_M(r) = v^*v$. Since $\Lambda_M(r) \leq \Lambda_M(q) \leq p$, the finiteness of p tells us that $\bar{r}^p = \Lambda_M(r) = \Lambda_M(q) = \bar{q}^p$. If $\bar{r}^q \lneq q$, there is $e \in \mathrm{OP}(\mathrm{her}(q)) \setminus \{0\}$ with $re = 0$. Since $e \in \mathrm{OP}(\mathrm{her}(p))$, we obtain a contradiction that $\bar{r}^p \neq \bar{q}^p$ (as $r \leq p - e$ but $q \nleq p - e$). This shows that p is C^*-finite.

Conversely, if p is C^*-finite, then Remark 3.13 implies that p is finite.

(b) This follows from part (a) and Corollary 3.14. □

Remark 4.17. (a) Let $p \in M$ be a finite projection. If $r \in \mathrm{Proj}(pMp)$ with $r \sim_{\mathrm{sp}} p$, then Lemma 4.16(a) and Remark 3.13 tell us that $r = p$. The same is true if we relax the assumption to $r \in \mathrm{OP}(pMp)$. In fact, we first notice that the C^*-finiteness of p gives $\bar{r}^p = p$. Moreover, suppose that $w \in M^{**}$ and $v \in M$ are as in the proof of Lemma 4.16 for the case when $q = p$. Then $vv^* = p = \bar{r}^p = v^*v$. This means that v is a unitary in pMp. As $v\,\mathrm{her}(r)v^* = \Lambda_M(w\,\mathrm{her}(r)w^*) = pMp$, we have $\mathrm{her}(r) = pMp$ and hence $r = p$.

(b) If A is a C^*-algebra and $p \in \mathrm{OP}(A)$ satisfying $\bar{r}^p = \bar{q}^p$ for any $r, q \in \mathrm{OP}(\mathrm{her}(p))$ with $r \leq q$ and $r \sim_{\mathrm{sp}} q$, then by the argument of Lemma 4.16, we see that p is C^*-finite.

The following is a direct consequence of Lemma 4.16 and Corollary 3.12.

Theorem 4.18. *Let M be a von Neumann algebra.*

(a) *M is of type $\mathfrak{A}$ if and only if M is a type I von Neumann algebra.*
(b) *M is of type $\mathfrak{B}$ if and only if M is a type II von Neumann algebra.*
(c) *M is of type $\mathfrak{C}$ if and only if M is a type III von Neumann algebra.*
(d) *M is C^*-semi-finite if and only if M is a semi-finite von Neumann algebra.*

5. Factorisations

In this section, we give two factorization type results for general C^*-algebras. Let us first state the following easy lemma. Notice that if A contains a non-zero abelian hereditary C^*-subalgebra B, the closed ideal generated by B is of type $\mathfrak{A}$ (by Corollary 3.15(b) and Remark 3.7(b)), and the same is true for C^*-finite hereditary C^*-subalgebra.

Lemma 5.1. *If A is not of type $\mathfrak{C}$, then A contains a non-zero closed ideal of either type $\mathfrak{A}$ or type $\mathfrak{B}$.*

The following is our first factorization type result, which mimics the corresponding situation for von Neumann algebras.

Theorem 5.2. *Let A be a C^*-algebra.*

(a) *There is a largest type $\mathfrak{A}$ (respectively, type $\mathfrak{B}$, type $\mathfrak{C}$ and C^*-semi-finite) hereditary C^*-subalgebra $J_{\mathfrak{A}}$ (respectively, $J_{\mathfrak{B}}$, $J_{\mathfrak{C}}$ and $J_{\mathfrak{sf}}$) of A, which is also an ideal of A.*
(b) *$J_{\mathfrak{A}}$, $J_{\mathfrak{B}}$ and $J_{\mathfrak{C}}$ are mutually disjoint such that $J_{\mathfrak{A}} + J_{\mathfrak{B}} + J_{\mathfrak{C}}$ is an essential closed ideal of A. If $e_{\mathfrak{A}}, e_{\mathfrak{B}}, e_{\mathfrak{C}} \in \mathrm{OP}(A) \cap Z(A^{**})$ with $J_{\mathfrak{A}} = \mathrm{her}(e_{\mathfrak{A}})$, $J_{\mathfrak{B}} = \mathrm{her}(e_{\mathfrak{B}})$ and $J_{\mathfrak{C}} = \mathrm{her}(e_{\mathfrak{C}})$, then*
$$1 = \overline{e_{\mathfrak{A}} + e_{\mathfrak{B}}}^{1} + e_{\mathfrak{C}}.$$
(c) *$J_{\mathfrak{A}} + J_{\mathfrak{B}}$ is an essential closed ideal of $J_{\mathfrak{sf}}$. If $e_{\mathfrak{sf}} \in \mathrm{OP}(A)$ with $J_{\mathfrak{sf}} = \mathrm{her}(e_{\mathfrak{sf}})$, then*
$$e_{\mathfrak{sf}} = \overline{e_{\mathfrak{A}}}^{e_{\mathfrak{sf}}} + e_{\mathfrak{B}}.$$
(d) *The closure of $\mathcal{C}(A)$ and $\mathcal{F}(A)$ (in Proposition 3.5) are essential closed ideals of $J_{\mathfrak{A}}$ and $J_{\mathfrak{sf}}$, respectively.*

Proof. (a) We first consider the situation of type $\mathfrak{B}$ hereditary C^*-subalgebra. Let $\mathfrak{J}_{\mathfrak{B}}$ be the set of all type $\mathfrak{B}$ closed ideals of A. If $\mathfrak{J}_{\mathfrak{B}} = \{(0)\}$, then $J_{\mathfrak{B}} := (0)$ is the largest type $\mathfrak{B}$ hereditary C^*-subalgebra of A (see Remark 3.11(b)). Suppose that there exist distinct elements J_1 and J_2 in $\mathfrak{J}_{\mathfrak{B}}$. If $J_1 + J_2$ contains a non-zero abelian hereditary C^*-algebra B, then by Lemma 3.17(b), one of the two abelian hereditary

C^*-subalgebras $B \cap J_1$ and $B \cap J_2$ is non-zero, which contradicts $J_1, J_2 \in \mathfrak{J}_{\mathfrak{B}}$. On the other hand, consider a non-zero closed ideal I of $J_1 + J_2$. Again, by Lemma 3.17(b), we may assume that the closed ideal $I \cap J_1$ is non-zero. Thus, $I \cap J_1$ contains a non-zero C^*-finite hereditary C^*-subalgebra B. This shows that $J_1 + J_2 \in \mathfrak{J}_{\mathfrak{B}}$ and $\mathfrak{J}_{\mathfrak{B}}$ is a directed set.

For any ideal J of A, we consider $e_J \in \mathrm{OP}(A) \cap \mathrm{Z}(A^{**})$ with $J = \mathrm{her}(e_J)$. Set

$$J_{\mathfrak{B}} := \overline{\sum\nolimits_{J \in \mathfrak{J}_{\mathfrak{B}}} J}.$$

Then $e_{J_{\mathfrak{B}}} = w^*\text{-}\lim_{J \in \mathfrak{J}_{\mathfrak{B}}} e_J$. If there is $p \in \mathrm{OP}_{\mathfrak{C}}(A) \setminus \{0\}$ such that $\mathrm{her}(p) \subseteq J_{\mathfrak{B}}$, then

$$p = pe_{J_{\mathfrak{B}}} = pe_{J_{\mathfrak{B}}}p = w^*\text{-}\lim_{J \in \mathfrak{J}_{\mathfrak{B}}} pe_J p,$$

and one can find $J \in \mathfrak{J}_{\mathfrak{B}}$ with the abelian algebra $\mathrm{her}(p) \cap J$ being non-zero (because of Lemma 3.17(a)), which is absurd. On the other hand, suppose that I is a non-zero closed ideal of $J_{\mathfrak{B}}$. The argument above tells us that $I \cap J \neq (0)$ for some $J \in \mathfrak{J}_{\mathfrak{B}}$, and hence it contains a non-zero C^*-finite hereditary C^*-subalgebra. Consequently, $J_{\mathfrak{B}} \in \mathfrak{J}_{\mathfrak{B}}$. Finally, if $B \subseteq A$ is a hereditary C^*-subalgebra of type $\mathfrak{B}$, then, by Remark 3.11(b), one has $B \subseteq J_{\mathfrak{B}}$.

The arguments for the statements concerning $J_{\mathfrak{A}}$, $J_{\mathfrak{C}}$ and $J_{\mathfrak{sf}}$ are similar and easier.

(b) The first statement follows directly from Lemma 5.1 (any non-type $\mathfrak{C}$ ideal interests either $J_{\mathfrak{A}}$ or $J_{\mathfrak{B}}$). For the second statement, one obviously has $e_{\mathfrak{A}} + e_{\mathfrak{B}} \leq 1 - e_{\mathfrak{C}}$. Suppose that $p \in \mathrm{OP}(A)$ with $e_{\mathfrak{A}} + e_{\mathfrak{B}} \leq 1 - p$. We have $p(e_{\mathfrak{A}} + e_{\mathfrak{B}}) = 0$. If $p \not\leq e_{\mathfrak{C}}$, then $\mathrm{her}(p)$ will contain a hereditary C^*-subalgebra of either type $\mathfrak{A}$ or type $\mathfrak{B}$ (by Lemma 5.1) and Lemma 3.17(a) will give a contradiction that either $pe_{\mathfrak{A}} \neq 0$ or $pe_{\mathfrak{B}} \neq 0$. Thus, $1 - e_{\mathfrak{C}}$ is the smallest closed projection dominating $e_{\mathfrak{A}} + e_{\mathfrak{B}}$.

(c) This follows from a similar (but easier) argument as part (b).

(d) Clearly, $\mathcal{F}(A) \subseteq J_{\mathfrak{sf}}$ and $\mathcal{C}(A) \subseteq J_{\mathfrak{A}}$ (see Remark 3.11(b)). Their closures are both essential because of Proposition 3.19. □

By Proposition 3.19, there is an abelian (respectively, a C^*-finite) hereditary C^*-subalgebra that generates an essential ideal of $J_{\mathfrak{A}}$ (respectively, of $J_{\mathfrak{B}}$). Moreover, by [32, Theorem 2.3(vi)], the largest type I closed ideal A_{postlim} of A is an essential ideal of $J_{\mathfrak{A}}$.

Remark 5.3. For any closed ideal J of A, we write $J^{\perp}$ for the closed ideal $\{a \in A : aJ = (0)\}$. It is easy to see that if J_0 is an essential ideal of J, then $J_0^{\perp} = J^{\perp}$.

(a) $J_{\mathfrak{A}}^{\perp} = A_{\mathrm{postlim}}^{\perp}$ is the largest anti-liminary hereditary C^*-subalgebra of A (note that $\overline{aJ_{\mathfrak{A}}a}$ is a hereditary C^*-subalgebra of $J_{\mathfrak{A}}$ for every $a \in A_+$). Furthermore, $J_{\mathfrak{B}} + J_{\mathfrak{C}}$ is an essential ideal of $J_{\mathfrak{A}}^{\perp}$ (by Lemma 5.1).

(b) $J_{\mathfrak{sf}}^{\perp} = (J_{\mathfrak{A}} + J_{\mathfrak{B}})^{\perp} = J_{\mathfrak{C}}$.

(c) $J_{\mathfrak{A}}^{\perp} \cap J_{\mathfrak{sf}} = J_{\mathfrak{B}}$ (compare with Corollary 3.10(c)).

From now on, we denote by $J_{\mathfrak{A}}^A$, $J_{\mathfrak{B}}^A$, $J_{\mathfrak{C}}^A$ and $J_{\mathfrak{sf}}^A$, respectively, the largest type $\mathfrak{A}$, the largest type $\mathfrak{B}$, the largest type $\mathfrak{C}$ and the largest C^*-semi-finite closed ideals of a C^*-algebra A.

The following is a direct application of Theorem 4.18.

Corollary 5.4. *Let M be a von Neumann algebra. If M_I, M_{II} and M_{III} are respectively the type I summand, the type II summand and the type III summand of M, then $J_{\mathfrak{A}}^M = M_I$, $J_{\mathfrak{B}}^M = M_{II}$ and $J_{\mathfrak{C}}^M = M_{III}$.*

Our next theorem is the second factorization type result, which seems to be more interesting for C^*-algebra (cf. [14, Proposition 4.13]).

Theorem 5.5. *Let A be a C^*-algebra.*

(a) *$A/J_{\mathfrak{C}}^A$ is C^*-semi-finite and $A/(J_{\mathfrak{A}}^A)^\perp$ is of type $\mathfrak{A}$.*
(b) *If A is C^*-semi-finite, then $A/J_{\mathfrak{B}}^A$ is of type $\mathfrak{A}$.*

Proof. (a) Assume, without loss of generality, that $A/J_{\mathfrak{C}}^A \neq (0)$ and consider $Q : A \to A/J_{\mathfrak{C}}^A$ to be the canonical map. Let I be a non-zero closed ideal of $A/J_{\mathfrak{C}}^A$ and $J := Q^{-1}(I)$. Since $J \supsetneq J_{\mathfrak{C}}^A$, one knows that J contains a non-zero C^*-finite hereditary C^*-subalgebra B. Since $B \cap J_{\mathfrak{C}}^A = (0)$, the *-homomorphism Q restricts to an injection on B. Thus, $Q(B) \subseteq I$ is also a non-zero C^*-finite hereditary C^*-subalgebra, and $A/J_{\mathfrak{C}}^A$ is C^*-semi-finite (by Corollary 3.10(a)). The proof of the second statement is similar.

(b) This follows from part (a) and Remark 5.3(c). □

Remark 5.6. Let $\mathcal{S}$ be a statement concerning C^*-algebras that is stable under extensions of C^*-algebras (i.e., if I is a closed ideal of a C^*-algebra A such that $\mathcal{S}$ is true for both I and A/I, then $\mathcal{S}$ is true for A).

(a) If $\mathcal{S}$ is true for all type $\mathfrak{A}$ and all type $\mathfrak{B}$ C^*-algebras, $\mathcal{S}$ is true for all C^*-semi-finite C^*-algebras. If, in addition, $\mathcal{S}$ is true for all type $\mathfrak{C}$ C^*-algebras, it is true for all C^*-algebras.

(b) If $\mathcal{S}$ is true for all discrete C^*-algebras and all anti-liminary C^*-algebras, then $\mathcal{S}$ is true for all C^*-algebras.

The following results follows from Theorem 3.9(a).

Corollary 5.7. *If A and B are strongly Morita equivalent, then the closed ideal of B that corresponds to $J_{\mathfrak{A}}^A$ (respectively, $J_{\mathfrak{B}}^A$, $J_{\mathfrak{C}}^A$ and $J_{\mathfrak{sf}}^A$) under the strong Morita equivalence (see the paragraph preceding Theorem 3.9) is precisely $J_{\mathfrak{A}}^B$ (respectively, $J_{\mathfrak{B}}^B$, $J_{\mathfrak{C}}^B$ and $J_{\mathfrak{sf}}^B$).*

Remark 5.8. It is natural to ask if the closure $\overline{\mathcal{C}(\cdot)}$ of $\mathcal{C}(\cdot)$ (see Proposition 3.5) is also stable under strong Morita equivalence. Unfortunately, it is not the case. Suppose that A is any type I C^*-algebra. Then by [6, Theorems 1.8 and 2.2], there is a commutative C^*-algebra B that is strongly Morita equivalent to A. Notice that $\mathcal{C}(B) = B$ and $\overline{\mathcal{C}(A)}$ is of type I_0 (by [31, Proposition 6.1.7]). Thus, if $\overline{\mathcal{C}(\cdot)}$ is stable under strong Morita equivalence, then any type I C^*-algebra A will coincide with $\overline{\mathfrak{C}(A)}$ and hence is liminary (see, e.g., [31, Corollary 6.1.6]), which is absurd.

To end this section, we compare J_*^A with $J_*^{M(A)}$.

Proposition 5.9.

(a) *If $B \subseteq A$ is a hereditary C^*-subalgebra, then* $J_{\mathfrak{A}}^B = J_{\mathfrak{A}}^A \cap B$, $J_{\mathfrak{B}}^B = J_{\mathfrak{B}}^A \cap B$, $J_{\mathfrak{C}}^B = J_{\mathfrak{C}}^A \cap B$ *and* $J_{\mathfrak{sf}}^B = J_{\mathfrak{sf}}^A \cap B$.

(b) $J_{\mathfrak{A}}^{M(A)} = \{x \in M(A) : xA \subseteq J_{\mathfrak{A}}^A\}$. *Similar statements hold for* $J_{\mathfrak{B}}$, $J_{\mathfrak{C}}$ *and* $J_{\mathfrak{sf}}$.

(c) $J_{\mathfrak{B}}^{M(A)} = \{x \in M(A) : xJ_{\mathfrak{A}}^A = (0)$ *and* $xA \subseteq J_{\mathfrak{sf}}^A\}$.

(d) $J_{\mathfrak{C}}^{M(A)} = \{x \in M(A) : xJ_{\mathfrak{sf}}^A = (0)\} = \{x \in M(A) : xJ_{\mathfrak{A}}^A = (0)$ *and* $xJ_{\mathfrak{B}}^A = (0)\}$.

Proof. (a) Clearly, $J_{\mathfrak{A}}^B \subseteq B \cap J_{\mathfrak{A}}^A$. Conversely, since $B \cap J_{\mathfrak{A}}^A$ is a type $\mathfrak{A}$ closed ideal of B (by Corollary 3.15(a)), we have $B \cap J_{\mathfrak{A}}^A \subseteq J_{\mathfrak{A}}^B$. The other cases follow from similar arguments.

(b) We will only consider the case of $J_{\mathfrak{B}}$ (since the other cases follow from similar and easier arguments). Notice that $J_{\mathfrak{B}}^{M(A)} \cdot A = J_{\mathfrak{B}}^{M(A)} \cap A = J_{\mathfrak{B}}^A$ (by part (a)) and

$$J_{\mathfrak{B}}^{M(A)} \subseteq J_0 := \{x \in M(A) : xA \subseteq J_{\mathfrak{B}}^A\}.$$

Suppose that the closed ideal $J_0 \subseteq M(A)$ contains a non-zero abelian hereditary C^*-subalgebra B. The abelian hereditary C^*-subalgebra $B \cap A = B \cdot A \cdot B$ is contained in $J_{\mathfrak{B}}^A$ and so, $B \cdot A = (0)$, which contradicts the fact that A is essential in $M(A)$ (see Remark 3.2(d)). Furthermore, let I be a non-zero closed ideal of J_0. Then $I \cdot A = I \cap A \neq (0)$ and is a closed ideal of $J_{\mathfrak{B}}^A$. Thus, $I \cap A$ contains a non-zero C^*-finite hereditary C^*-subalgebra. Consequently, J_0 is of type $\mathfrak{B}$ and is a subset of $J_{\mathfrak{B}}^{M(A)}$.

(c) Obviously, $xJ_{\mathfrak{A}}^A = (0)$ if and only if $xAJ_{\mathfrak{A}}^A = (0)$. Thus, this part follows from part (b) and Remark 5.3(c).

(d) This part follows from a similar argument as part (c) as well as Remark 5.3(b). □

Acknowledgement

The authors would like to thank L. Brown, E. Effros and G. Elliott for giving some comments.

References

[1] C.A. Akemann, *The General Stone–Weierstrass problem*, J. Funct. Anal., **4** (1969), 277–294.

[2] C.A. Akemann, *Left ideal structure of C^*-algebras*, J. Funct. Anal., **6** (1970), 305–317.

[3] C.A. Akemann and S. Eilers, *Regularity of projections revisited*, J. Oper. Theory, **48** (2002), 515–534.

[4] C.A. Akemann and G.K. Pedersen, *Complications of semicontinuity in C^*-algebra theory*, Duke Math. J., **40** (1973), 785–795.

[5] C.A. Akemann, G.K. Pedersen and J. Tomiyama, *Multipliers of C^*-algebras*, J. Funct. Anal., **13** (1973), 277–301.

[6] W. Beer, *On Morita equivalence of nuclear C^*-algebras*, J. Pure Appl. Alg., **26** (1982), 249–267.

[7] M. Bekka, M. Cowling and P. de la Harpe, *Some groups whose reduced C^*-aglebra is simple*, Publ. Math. I.H.E.S., **80** (1994), 117–134.

[8] B. Blackadar, Operator algebras – theory of C^*-algebras and von Neumann algebras, in *Operator Algebras and Non-commutative Geometry III*, Encyc. Math. Sci., **122**, Springer-Verlag, Berlin (2006).

[9] L.G. Brown, *Semicontinuity and multipliers of C^*-algebras*, Canad. J. Math., **XL, 4** (1988), 865–988.

[10] L.G. Brown, *Determination of A from $M(A)$ and related matters*, C. R. Math. Rep. Acad. Sci. Canada, **10** (1988), 273–278.

[11] L.G. Brown and M.A. Rieffel, *Stable isomorphism and strong Morita equivalence of C^*-algebras*, Pacific J. Math., **71** (1977), 349–363.

[12] L.G. Brown and G.K. Pedersen, *C^*-algebras of real rank zero*, J. Funct. Anal., **99** (1991), 131–149.

[13] J. Cuntz, *K-theory for certain C^*-algebras*, Ann. of Math., **113** (1981), 181–197.

[14] J. Cuntz and G.K. Pedersen, *Equivalence and traces on C^*-algebras*, J. Funct. Anal., **33** (1979), 135–164.

[15] E.G. Effros, *Order ideals in C^*-algebras and its dual*, Duke Math. J., **30** (1963), 391–412.

[16] G.A. Elliott, *On the classification of inductive limits of sequences of semisimple finite-dimensional algebras*, J. Alg. **38** (1976), 29–44.

[17] G.A. Elliott, *On the classification of C^*-algebras of real rank zero*, J. Reine Angew. Math., **443** (1993), 179–219.

[18] G.A. Elliott and A.S. Toms, *Regularity properties in the classification program for separable amenable C^*-algebras*, Bull. Amer. Math. Soc., **45** (2008), 229–245.

[19] H. Halpern, V. Kaftal, P.W. Ng and S. Zhang, Finite sums of projections in von Neumann algebras, Trans. Amer. Math. Soc. **365** (2013), 2409–2445.

[20] R.V. Kadison and J.R. Ringrose, *Fundamentals of the theory of operator algebras Vol. II: Advanced theory*, Pure and Applied Mathematics **100**, Academic Press (1986).

[21] E. Kirchberg and M. Rørdam, *Non-simple purely infinite C^*-algebras*, Amer. J. Math., **122** (2000), 637–666.

[22] E.C. Lance, *Hilbert C^*-modules – A toolkit for operator algebraists*, Lond. Math. Soc. Lect. Note Ser. **210**, Camb. Univ. Press (1995).

[23] H. Lin, *Equivalent open projections and corresponding hereditary C^*-subalgebras*, J. Lond. Math. Soc., **41** (1990), 295–301.

[24] H. Lin and S. Zhang, *On infinite simple C^*-algebras*, J. Funct. Anal., **100** (1991), 221–231.

[25] G.J. Murphy, *C^*-algebras and operator theory*, Academic Press (1990).

[26] F.J. Murray, *The rings of operators papers*, in *The legacy of John von Neumann*, (Hempstead, NY, 1988), 57–60, Proc. Sympos. Pure Math. **50**, Amer. Math. Soc., Providence, R.I. (1990).

[27] F.J. Murray and J. von Neumann, *On rings of operators*, Ann. of Math. (2), **37** (1936), 116–229.

[28] C.K. Ng and N.C. Wong, *Comparisons of equivalence relations on open projections*, J. Oper. Theory, to appear.

[29] E. Ortega, M. Rørdam and H. Thiel, *The Cuntz semigroup and comparison of open projections*, J. Funct. Anal., **260** (2011), 3474–3493.

[30] G.K. Pedersen, *Applications of weak-*-semicontinuity in C^*-algebra theory*, Duke Math. J., **39** (1972), 431–450.

[31] G.K. Pedersen, *C^*-algebras and their automorphism groups*, Academic Press (1979).

[32] C. Peligrad and L. Zsidó, Open projections of C^*-algebras: Comparison and Regularity, *Operator Theoretical Methods, 17th Int. Conf. on Operator Theory, Timisoara (Romania), June* 23–26, 1998, Theta Found. Bucharest (2000), 285–300.

[33] R.T. Prosser, *On the ideal structure of operator algebras*, Memoirs Amer. Math. Soc., **45** (1963).

[34] M.A. Rieffel, *Morita equivalence for C^*-algebras and W^*-algebras*, J. Pure Appl. Alg., **5** (1974), 51–96.

[35] M.A. Rieffel, Morita equivalence for operator algebras, in *Operator algebras and applications, Part I (Kingston, Ont.,* 1980), Proc. Sympos. Pure Math. **38**, Amer. Math. Soc., Providence, R.I. (1982), 285–298.

[36] M. Rørdam, Classification of nuclear, simple C^*-algebras, in *Classification of nuclear C^*-algebras, Entropy in operator algebras*, Encyclopaedia Math. Sci. **126**, Springer, Berlin (2002), 1–145.

[37] M. Takesaki, *Theory of Operator algebras I*, Springer-Verlag New York (1979).

[38] A.S. Toms, *On the classification problem for nuclear C^*-algebras*, Ann. of Math. (2), **167** (2008), 1029–1044.

[39] S. Zhang, *Stable isomorphism of hereditary C^*-subalgebras and stable equivalence of open projections*, Proc. Amer. Math. Soc., **105** (1989), 677–682.

[40] S. Zhang, *Certain C^*-algebras with real rank zero and their corona and multiplier algebras part I*, Pac. J. Math. **155** (1992), 169–197.

Chi-Keung Ng
Chern Institute of Mathematics and LPMC
Nankai University
Tianjin 300071, China
e-mail: ckng@nankai.edu.cn

Ngai-Ching Wong
Department of Applied Mathematics
National Sun Yat-sen University
Kaohsiung, 80424, Taiwan
e-mail: wong@math.nsysu.edu.tw

Operator Theory:
Advances and Applications, Vol. 250, 397–452

Well-posedness via Monotonicity – an Overview

Rainer Picard, Sascha Trostorff and Marcus Waurick

Abstract. The idea of monotonicity is shown to be the central theme of the solution theories associated with problems of mathematical physics. A "grand unified" setting is surveyed covering a comprehensive class of such problems. We illustrate the applicability of this setting with a number of examples. A brief discussion of stability and homogenization issues within this framework is also included.

Mathematics Subject Classification (2010). 3502, 35D30, 35F16, 35F25, 35F61, 35M33, 35Q61, 35Q79, 35Q74, 74Q05, 74Q10, 74C10, 74B05, 78M40.

Keywords. Positive definiteness, monotonicity, material laws, coupled systems, multiphysics.

0. Introduction

In this paper we shall survey a particular class of problems, which we like to refer to as "evolutionary equations" (to distinguish it from the class of explicit first-order ordinary differential equations with operator coefficients predominantly considered under the heading of *evolution equations*). This problem class is spacious enough to include not only classical evolution equations but also partial differential algebraic systems, functional differential equations and integro-differential equations. Indeed, by thinking of elliptic systems as time-dependent, for example as constant with respect to time on the connected components of $\mathbb{R} \setminus \{0\}$, they also can be embedded into this class. The setting is – in its present state – largely limited to a Hilbert space framework. As a matter of convenience the discussion will indeed be set in a complex Hilbert space framework. For the concept of monotonicity it is, however, more appropriate to consider complex Hilbert spaces as real Hilbert spaces, which can canonically be achieved by reducing scalar multiplication to real numbers and replacing the inner product by its real part. So, a binary relation R in a complex Hilbert space H with inner product $\langle \cdot | \cdot \rangle_H$ would be called strictly monotone if

$$\mathfrak{Re} \langle x - y | u - v \rangle_H \geq \gamma \langle x - y | x - y \rangle_H$$

for all $(x, u), (y, v) \in R$ holds and γ is some positive real number. In case $\gamma = 0$ the relation R would be called monotone.

The importance of strict monotonicity, which in the linear operator case reduces to strict positive definiteness[1], is of course well known from the elliptic case. By a suitable choice of space-time norm this key to solving elliptic partial differential equations also allows us to establish well-posedness for dynamic problems in exactly the same fashion.

The crucial point for this extension is the observation that the one-dimensional derivative itself, acting as the time derivative[2] ∂_0 (on the full time line $\mathbb{R}$), can be realized as a maximal strictly positive definite operator in an appropriately exponentially weighted L^2-type Hilbert space over the real time-line $\mathbb{R}$. It is in fact this strict positive definiteness of ∂_0 which opens access to the problem class we shall describe later.

Indeed, ∂_0 simply turns out to be a *normal* operator with $\mathfrak{Re}\,\partial_0$ being just multiplication by a positive constant. Moreover, this time-derivative ∂_0 is continuously invertible and, as a normal operator, admits a straightforwardly defined functional calculus, which can canonically be extended to operator-valued functions. Indeed, since we have control over the positivity constant via the choice of the weight, the norm of ∂_0^{-1} can be made as small as wanted. This observation is the Hilbert space analogue to the technical usage of the exponentially weighted sup-norm as introduced by D. Morgenstern, [26], and allows for the convenient inclusion of a variety of perturbation terms.

Having established time-differentiation ∂_0 as a normal operator, we are led to consider evolutionary problems as operator equations in a space-time setting, rather than as an ordinary differential equation in a spatial function space. The space-time operator equation perspective implies that we are dealing with sums of unbounded operators, which, however, in our particular context is – due to the limitation of remaining in a Hilbert space setting and considering only sums, where one of the terms is a function of the normal operator ∂_0 – not so deep an issue. For more general operator sums or for a Banach space setting more sophisticated and powerful tools from the abstract theory of operator sums initiated by the influential papers by da Prato and Grisvard, [11], and Brezis and Haraux, [8], may have to be employed. In these papers operator sums $\frac{d}{dt} + A$ typically occurring in the context of explicit first-order differential equations in Banach spaces are considered as applications of the abstract theory, compare also, e.g., [21, Chapter 2, Section 7]. The obvious overlap with the framework presented in this paper would be the Hilbert space situation in the case $\mathcal{M} = 1$. We shall, however, not

[1]We use the term *strict positive definiteness* for a linear operator A in a real or complex Hilbert space X in the sense naturally induced by the classification of the corresponding quadratic form Q_A given by $u \mapsto \langle u|Au\rangle_X$ on its domain $D(A)$. So, if Q_A is non-negative (mostly called positive semi-definite), positive definite, strictly positive definite, then the operator A will be called non-negative (usually called positive), positive definite, strictly positive definite, respectively. If X is a complex Hilbert space it follows that A must be Hermitian. Note that we do *not* restrict the definition of non-negativity, positive definiteness, strict positive definiteness to Hermitian or symmetric linear operators.

[2]We follow here the time-honored convention that physicists practice by labeling the partial time derivative by index zero.

pursue exploration of how the strategies developed in this context may be expanded to include more complicated material laws, which indeed has been done extensively in the wake of these ideas, but rather stay with our limited problem class, which covers a variety of diverse problems in a highly unified setting. Naturally the results available for specialized cases are likely to be stronger and more general for this particular situation.

For introductory purposes let us consider the typical linear case of such a space-time operator equation

$$\partial_0 V + AU = f, \tag{0.1}$$

where f are given data, A is a – usually – purely spatial – prototypically *skew-selfadjoint*[3] – operator and the quantities U, V are linked by a so-called material law

$$V = \mathcal{M}U.$$

Solving such an equation would involve establishing the bounded invertibility of $\overline{\partial_0 \mathcal{M} + A}$. As a matter of "philosophy" we shall think of the – here linear – material law operator $\mathcal{M}$ as encoding the complexity of the physical material whereas A is kept simple and usually only contains spatial derivatives. If $\mathcal{M}$ commutes with ∂_0 we shall speak of an autonomous system, otherwise we say the system is non-autonomous.

Another – more peripheral – observation with regards to the classical problems of mathematical physics is that they are predominantly of first order not only with respect to the time derivative, which is assumed in the above, but frequently even in *both* the temporal and spatial derivatives. Indeed, acoustic waves, heat transport, visco-elastic and electro-magnetic waves etc. are governed by first-order systems of partial differential operators, i.e., A is a first-order differential operator in spatial derivatives, which only after some elimination of unknowns turn into the more common second-order equations, i.e., the wave equation for the pressure field, the heat equation for the temperature distribution, the visco-elastic wave equation for the displacement field and the vectorial wave equation for the electric (or magnetic) field. It is, however, only in the direct investigation of the first-order system that, as we shall see, the unifying feature of monotonicity becomes easily visible. Moreover, the first-order formulation reveals that the spatial derivative operator A is of a Hamiltonian type structure and consequently, by imposing suitable boundary conditions, turn out – in the standard cases – to lead to skew-selfadjoint A in a suitable Hilbert space H. So, from this perspective there is also undoubtedly a flavor of the concept of *symmetric hyperbolic systems* as introduced by K.O. Friedrichs, [16], and of Petrovskii well-posedness, [29], at the roots of this approach.

---

[3]Note that in our canonical reference situation A is *skew*-selfadjoint rather than selfadjoint and so we have $\mathfrak{Re}\,\langle u|Au\rangle_H = 0$ for all $u \in D(A)$ and coercitivity of A is out of the question. To make this concrete: let ∂_1 denote the weak $L^2(\mathbb{R})$-derivative. Then our paradigmatic reference example on this elementary level would be the transport operator $\partial_0 + \partial_1$ rather than the heat conduction operator $\partial_0 - \partial_1^2$.

For illustrational purposes let us consider from a purely heuristic point of view the $(1+1)$-dimensional system

$$\left(\partial_0 M_0 + M_1 + A\right)\begin{pmatrix} p \\ s \end{pmatrix} = \begin{pmatrix} f \\ 0 \end{pmatrix}, \tag{0.2}$$

where

$$M_0 := \begin{pmatrix} \eta & 0 \\ 0 & \alpha \end{pmatrix},\ M_1 := \begin{pmatrix} (1-\eta) & 0 \\ 0 & (1-\alpha) \end{pmatrix},\ \alpha,\eta \in \{0,1\},\ A := \begin{pmatrix} 0 & \partial_1 \\ \partial_1 & 0 \end{pmatrix},$$

and ∂_1 is simply the weak $L^2(\mathbb{R})$-derivative, compare Footnote 3. Assuming $\eta = 1$, $\alpha = 1$, in (0.2) clearly results in a (symmetric) hyperbolic system and eliminating the unknown s yields the wave equation in the form

$$\left(\partial_0^2 - \partial_1^2\right) p = \partial_0 f.$$

For $\eta = 1, \alpha = 0$ we obtain a differential algebraic system, which represents the parabolic case in the sense that after eliminating s we obtain the heat equation

$$\left(\partial_0 - \partial_1^2\right) p = f.$$

Finally, if both parameters vanish, we obtain a 1-dimensional elliptic system and as expected after eliminating the unknown s a 1-dimensional elliptic equation for p results:

$$\left(1 - \partial_1^2\right) p = f.$$

Allowing now α, η to be $L^\infty(\mathbb{R})$-multiplication operators with values in $\{0,1\}$, which would allow the resulting equations to jump in space between elliptic, parabolic and hyperbolic "material properties", could be a possible scenario we envision for our framework. As will become clear, the basic idea of this simple "toy" example can be carried over to general evolutionary equations. Also in this connotation there are stronger and more general results for specialized cases. A problem of this flavor of "degeneracy" has been for example discussed for a non-autonomous, degenerate integro-differential equation of parabolic/elliptic type in [22, 23].

A prominent feature distinguishing general operator equations from those describing dynamic processes is the specific role of time, which is not just another space variable, but characterizes dynamic processes via the property of causality[4]. Requiring causality for the solution operator $\left(\overline{\partial_0 \mathcal{M} + A}\right)^{-1}$ results in very specific types of material law operators $\mathcal{M}$, which are causal and compatible with causality of $\left(\overline{\partial_0 \mathcal{M} + A}\right)^{-1}$. This leads to deeper insights into the structural properties of mathematically viable models of physical phenomena.

The solution theory can be extended canonically to temporal distributions with values in a Hilbert space. In this perspective initial value problems, i.e., prescribing $V(0+)$ in (0.1), amount to allowing a source term f of the form $\delta \otimes V_0$ defined by

$$(\delta \otimes V_0)(\varphi) := \langle V_0 | \varphi(0) \rangle_H$$

[4]Note that this perspective specifically excludes the case of a periodic time interval, where "before" and "after" makes little sense.

for φ in the space $C_c(\mathbb{R}, H)$ of continuous H-valued functions with compact support. This source term encodes the classical initial condition $V(0+) = V_0$. For the constant coefficient case – say – $\mathcal{M} = 1$, it is a standard approach to establish the existence of a fundamental solution (or more generally, e.g., in the non-autonomous case, a Green functions) and to represent general solutions as convolution with the fundamental solution. This is of course nothing but a description of the continuous one-parameter semi-group approach. Indeed, such a semi-group $U = (U(t))_{t\in[0,\infty[}$ is, if extended by zero to the whole real time line, nothing but the fundamental solution

$$G = (G(t))_{t\in\mathbb{R}} \quad \text{with} \quad G(t) := \begin{cases} U(t) & \text{for } t \in [0, \infty[\,, \\ 0 & \text{for } t \in \,]-\infty, 0[\,. \end{cases}$$

In the non-autonomous case, the role of U is played by a so-called evolution family. The regularity properties of such fundamental solutions results in stronger regularity properties of the corresponding solutions. Since we allow $\mathcal{M}$ to be more general, constructing such fundamental solutions/Green functions is not always available or feasible. Indeed, we shall focus for sake of simplicity on the case that the data f do not contain such Dirac type sources, which can be achieved simply by subtracting the initial data or by including distributional objects such as $\delta \otimes V_0$ in the Hilbert space structure via extension to extrapolation spaces, which for sake of simplicity we will not burden this presentation with.

As a trade-off for our constraint, which in the simplest linear case would reduce our discussion to considering $\partial_0 + A$ as a sum of commuting normal operators, which clearly cannot support any claim of novelty, see, e.g., [54], we obtain by allowing for a large class of material law operators $\mathcal{M}$ access to a large variety of problems including such diverse topics as partial differential-algebraic systems, integro-differential equations and evolutionary equations of changing type in one unified setting.

Based on the linear theory one has of course a first access to non-linear problems by including Lipschitz continuous perturbations. A different generalization towards a non-linear theory can be done by replacing the (skew-selfadjoint) operator A by a maximal monotone relation or allowing for suitable maximal monotone material law relations (rather than material law operators). In this way the class of evolutionary problems also comprises evolutionary *inclusions*.

Having established well-posedness, qualitative properties associated with the solution theory come into focus. A first step in this direction is done for the autonomous case by the discussion of the issue of "exponential stability". One can give criteria with regards to the material law $\mathcal{M}$ ensuring exponential stability.

Another aspect in connection with the discussion of partial differential equations of mathematical physics is the problem of continuous dependence of the solution on the coefficients. A main application of results in this direction is the theory of homogenization, i.e., the study of the behavior of solutions of partial differential equations having large oscillatory coefficients. It is natural to discuss the weak operator topology for the coefficients and it turns out that the problem class under consideration is closed under limits in this topology if further suitable structural assumptions are imposed. The closedness of the problem class is a remarkable feature of the problem class, which is spacious enough to also include – hidden in the generality of the material law operator – integro-differential evolutionary problems. In this regard it is worth recalling that there are examples already for ordinary differential equations, for which the resulting limit equations are of integro-differential type, showing that differential equations are in this respect too small a problem class.

Although links to the core concepts which have entered the described approach are too numerous to be recorded here to any appropriate extent, we shall try modestly to put them in a bibliographical context. The concept of the time-derivative considered as a continuously invertible operator in a suitably weighted Hilbert space has its source in [30]. It has been employed in obtaining a solution theory for evolutionary problems in the spirit described above only more recently, compare, e.g., [33, Chapter 6]. General perspectives for well-posedness to partial differential equations via strict positive definiteness are of course at the heart of the theory of elliptic partial differential equations.

For the theory of maximal monotone operators/relations, we refer to [7, 18, 19, 27]. For non-autonomous equations, we refer to [40, 41] and – with a focus on maximal regularity – to [3]. Note that due to the generality of our approach, one cannot expect maximal regularity of the solution operator in general. In fact, maximal regularity for the solution operator just means that the operator sum is already closed with its natural domain. This is rarely the case neither in the paradigmatic examples nor in our expanded general setting.

For results regarding exponential stability for a class of hyperbolic integro-differential equations, we refer to [38] and to [13, 17, 15] for the treatment of this issue in the context of one-parameter semi-groups. A detailed introduction to the theory of homogenization can be found in [4] and in [10]. We also refer to [43, 42], where homogenization for ordinary differential equations has been discussed extensively.

The paper itself is structured as follows. We begin our presentation with a description of the underlying prerequisites, even to the extent that we review the celebrated well-posedness requirements due to Hadamard, which we found inspirational for a deeper understanding of the case of differential inclusions. A main point in this first section is to introduce the classical concept of maximal strictly monotone relations and to recall that such relations are inverse relations of Lipschitz continuous mappings (Minty's Theorem 1.1). Specializing to the linear case we recall in particular the Lax–Milgram lemma (Corollary 1.6) and as a by-product derive a variant of the classical solution theory for elliptic type equations. Moreover, we comment on a general solution theory for (non-linear) elliptic type equations in divergence form relying only on the validity of a Poincaré type estimate (Theorem 1.8). We conclude this section with an example for an elliptic type equation with possible degeneracies in the coefficients as an application of the ideas presented.

Based on the first section's general findings, Section 2 deals with the solution theory for linear evolutionary equations. After collecting some guiding examples in Subsection 2.1, we rigorously establish in Subsection 2.2 the time-derivative as a strictly monotone, normal operator in a suitably weighted Hilbert space. Based on this and with resulting structural properties, such as a functional calculus, at hand, in Subsection 2.3 (Theorem 2.5) we formulate a solution theory for autonomous, linear evolutionary equations. The subsequent examples review some of those mentioned in Subsection 2.1 in a rigorous functional analytic setting to illustrate the applicability of the solution theory. As further applications we show that Theorem 2.5 also covers integro-differential equations (Theorem 2.9) and equations containing fractional time-derivatives (Theorem 2.12). We conclude this subsection with a conceptual study of exponential stability (Definition 2.13 and Theorem 2.14) in our theoretical context.

In Subsection 2.4, starting out with a short motivating introductory part concerning homogenization issues, we discuss the closedness of the problem class under the weak operator topology for the coefficients. A first theorem in this direction is then obtained as Theorem 2.26. After presenting some examples, we continue our investigation of homogenization problems first for ordinary differential equations (Theorems 2.32 and 2.34). Then we formulate a general homogenization result (Theorem 2.37), which is afterwards exemplified by considering Maxwell's equations and in particular the so-called eddy current problem of electro-magnetic theory.

In Subsection 2.5 we extend the solution theory to include the non-autonomous case. A first step in this direction is provided by Theorem 2.42, for which the illustrative Example 2.43 is given as an application. A common generalization of the Theorems 2.5 and 2.42 is given in Theorem 2.40. This is followed by an adapted continuous dependence result Theorem 2.44, which in particular is applicable to homogenization problems. A detailed example of a mixed type problem concludes this section.

Section 3 gives an account for a non-linear extension of the theory. Similarly to the previous section, the results are considered in the autonomous case first (Subsection 3.1) and then generalized to the non-autonomous case (Subsection 3.2). Subsection 3.3 concludes this section and the paper with a discussion of an application to evolutionary problems with non-linear boundary conditions. One of the guiding conceptual ideas here is to avoid regularity assumptions on the boundary of the underlying domain. This entails replacing the classical boundary trace type data spaces, by a suitable generalized analogue of 1-harmonic functions. We exemplify our results with an impedance type problem for the wave equation and with the elastic equations with frictional boundary conditions.

Note that inner products, indeed all sesqui-linear forms, are – following the physicists habits – assumed to be conjugate-linear in the first component and linear in the second component.

1. Well-posedness and monotonicity

To begin with, let us recall the well-known Hadamard requirements for well-posedness. It is appropriate for our purposes, however, to formulate them for the case of relations rather than – as usually done – for mappings. Hadamard proposed to define what "reasonably solvable" should entail. Solving a problem involves to establish a binary relation $P \subseteq X \times Y$ between "data" in a topological space Y and corresponding "solutions" in a topological space X, which is designed to cover a chosen pool of examples to our satisfaction. Finding a solution then means, given $y \in Y$ find $x \in X$ such that $(x, y) \in P$. If we wish to supply a solution for all possible data, there are some natural requirements that the problem class P should have to ensure that this task is reasonably conceived. To exclude cases of trivial failure to describe a solution theory for P, we assume first that P is already *closed* in $X \times Y$. Then well-posedness in the spirit of Hadamard requires the following three properties.

1. ("Uniqueness" of solution) the inverse relation P^{-1} is right-unique, thus, giving rise to a mapping[5]

$$P^{-1} : P[X] \subseteq Y \to X$$

performing the association of "data" to "solutions".
2. ("Existence" for every given data) we have that

$$P[X] = Y,$$

i.e., P^{-1} is defined on the whole data space Y.
3. ("Continuous dependence" of the solution on the data) The mapping P^{-1} is continuous.

In case of P being a mapping then $[Y]P = P^{-1}[Y]$ is the *domain* $D(P)$ *of* P. For our purposes here we shall assume that $X = Y$ and that X is a complex Hilbert space.

A very particular but convenient instance of well-posedness, which nevertheless appears to dominate in applications, is the maximal monotonicity of $P - c := \{(x, y - cx) \in X \times X \mid (x, y) \in P\}$ for some $c \in]0, \infty[$. Recall that a relation $Q \subseteq X \times X$ is called *monotone* if

$$\mathfrak{Re}\, \langle x_0 - x_1 | y_0 - y_1 \rangle_X \geq 0$$

for all $(x_0, y_0), (x_1, y_1) \in Q$. Such a relation Q is called *maximal* if there exists no proper monotone extension in $X \times X$. In other words, if $(x_1, y_1) \in X \times X$ is such that $\mathfrak{Re}\, \langle x_0 - x_1 | y_0 - y_1 \rangle_X \geq 0$ for all $(x_0, y_0) \in Q$, then $(x_1, y_1) \in Q$.

[5]For subsets $M \subseteq X$, $N \subseteq Y$ the *post-set of* M *under* P and the *pre-set of* N *under* P is defined as $P[M] := \{y \in Y \mid \bigvee_{x \in M} (x, y) \in P\}$ and $[N]P := \left\{x \in X \mid \bigvee_{y \in N} (x, y) \in P\right\}$, respectively. The post-set $P[X]$ of the whole space X under P is then the domain of the mapping P^{-1}.

Theorem 1.1 (Minty, [25]). *Let $(P-c) \subseteq X \times X$ be a maximal monotone relation*[6] *for some $c \in \,]0,\infty[$*[7]*. Then the inverse relation P^{-1} defines a Lipschitz continuous mapping with domain $D(P^{-1}) = X$ and $\frac{1}{c}$ as possible Lipschitz constant.*

Proof. We first note that the monotonicity of $P - c$ implies

$$\bigwedge_{(x_0,y_0),(x_1,y_1)\in P} \mathfrak{Re}\,\langle x_0 - x_1 | y_0 - y_1\rangle_X \geq c\,\langle x_0 - x_1 | x_0 - x_1\rangle_X\,. \tag{1.1}$$

Hence, if $y_0 = y_1$ then x_0 must equal x_1, i.e., the uniqueness requirement is satisfied, making $P^{-1} : P[X] \to X$ a well-defined mapping. Moreover, $P[X]$ is closed, since from (1.1) we get

$$\bigwedge_{(x_0,y_0),(x_1,y_1)\in P} |y_0 - y_1|_X \geq c\,|x_0 - x_1|_X\,.$$

The actually difficult part of the proof is to establish that $P[X] = X$. This is the part we will omit and refer to [25] instead. To establish Lipschitz continuity of $P^{-1} : X \to X$ we observe that

$$\begin{aligned}\bigwedge_{y_0,y_1\in X} c\left|P^{-1}(y_0) - P^{-1}(y_1)\right|_X^2 &\leq \mathfrak{Re}\left\langle P^{-1}(y_0) - P^{-1}(y_1)\,|y_0 - y_1\right\rangle_X \\ &\leq \left|P^{-1}(y_0) - P^{-1}(y_1)\right|_X |y_0 - y_1|_X\,,\end{aligned}$$

holds, from which the desired continuity estimate follows. □

For many problems, the strict monotonicity is easy to obtain. The maximality, however, needs a deeper understanding of the operators involved. In the linear case, writing now A for P, there is a convenient set-up to establish maximality by noting that

$$([\{0\}]\,A^*)^{\perp} = \overline{A[X]}$$

according to the projection theorem. Here we denote by A^* the *adjoint of A*, given as the binary relation

$$A^* := \left\{(u,v) \in X \times X \,|\, \bigwedge_{(x,y)\in A} \langle y|u\rangle_X = \langle x|v\rangle_X\right\}.$$

Thus, maximality for the strictly monotone linear mapping (i.e., strictly accretive) A is characterized[8] by

$$[\{0\}]\,A^* = \{0\}\,, \tag{1.2}$$

i.e., the uniqueness for the adjoint problem. Characterization (1.2) can be established in many ways, a particularly convenient one being to require that A^* is also strictly monotone. With this we arrive at the following result.

---

[6]Note here that maximal monotone relations are automatically closed, see, e.g., [7, Proposition 2.5].

[7]In this case P would be called *maximal strictly monotone.*

[8]Recall that A has closed range.

Theorem 1.2. *Let A and A^* be closed linear strictly monotone relations in a Hilbert space X. Then for every $f \in X$ there is a unique $u \in X$ such that*

$$(u, f) \in A.$$

Indeed, the solution depends continuously on the data in the sense that we have a (Lipschitz-) continuous linear operator $A^{-1} : X \to X$ with

$$u = A^{-1} f.$$

Of course, the case that A is a closed, densely defined linear operator is a common case in applications.

Corollary 1.3. *Let A be a closed, densely defined, linear operator and A, A^* strictly accretive in a Hilbert space X. Then for every $f \in X$ there is a unique $u \in X$ such that*

$$Au = f.$$

Indeed, solutions depend continuously on the data in the sense that we have a (Lipschitz-) continuous linear operator $A^{-1} : X \to X$ with

$$u = A^{-1} f.$$

In the case that A and A^* are linear operators with $D(A) = D(A^*)$ the situation simplifies, since then strict accretivity of A implies strict accretivity of A^* due to

$$\mathfrak{Re}\,\langle x|Ax\rangle_X = \mathfrak{Re}\,\langle A^*x|x\rangle_X = \mathfrak{Re}\,\langle x|A^*x\rangle_X$$

for all $x \in D(A) = D(A^*)$.

Corollary 1.4. *Let A be a closed, densely defined, linear strictly accretive operator in a Hilbert space X with $D(A) = D(A^*)$. Then for every $f \in X$ there is a unique $u \in X$ such that*

$$Au = f.$$

Indeed, the solution depends continuously on the data in the sense that we have a continuous linear operator $A^{-1} : X \to X$ with

$$u = A^{-1} f.$$

The domain assumption of the last corollary is obviously satisfied if $A : X \to X$ is a continuous linear operator. This observation leads to the following simple consequence.

Corollary 1.5. *Let $A : X \to X$ be a strictly accretive, continuous, linear operator in the Hilbert space X. Then for every $f \in X$ there is a unique $u \in X$ such that*

$$Au = f.$$

Indeed, the solution depends continuously on the data in the sense that we have a continuous linear operator $A^{-1} : X \to X$ with

$$u = A^{-1} f.$$

Note that since continuous linear operators and continuous sesqui-linear forms are equivalent, the last corollary is nothing but the so-called Lax–Milgram theorem. Indeed, if $A : X \to X$ is in the space $L(X)$ of continuous linear operators then

$$(u, v) \mapsto \langle u|Av\rangle_X$$

is in turn a continuous sesqui-linear form on X, i.e., an element of the space $S(X)$ of continuous sesqui-linear forms on X, and conversely if $\beta\langle\,\cdot\,|\,\cdot\,\rangle \in S(X)$ then $\overline{\beta\langle\,\cdot\,|v\rangle} \in X^*$ and utilizing the unitary[9] Riesz map $R_X : X^* \to X$ we get via the Riesz representation theorem $\beta\langle u|v\rangle = \langle u|A_\beta v\rangle_X$, where $A_\beta v := R_X\overline{\beta\langle\,\cdot\,|v\rangle}$, $v \in X$, defines indeed a continuous linear operator on X. Moreover,

$$\begin{aligned} S(X) &\to L(X) \\ \beta &\mapsto A_\beta \end{aligned}$$

is not only a bijection but also an isometry. Indeed,

$$|\beta|_{S(X)} := \sup_{x,y\in B_X(0,1)} |\beta(x, y)| = \sup_{x,y\in B_X(0,1)} \left|\langle x|A_\beta y\rangle_X\right| = \|A_\beta\|_{L(X)}\,.$$

Strict accretivity for the corresponding operator A_β results in the so-called *coercitivity*[10] of the sesqui-linear form β:

$$\mathfrak{Re}\,\beta\langle u|u\rangle \geq c\langle u|u\rangle_X \tag{1.3}$$

for some $c \in\,]0, \infty[$ and all $u \in X$. Thus, as an equivalent formulation of the previous corollary we get the following.

Corollary 1.6 (Lax–Milgram theorem). *Let $\beta\langle\,\cdot\,|\,\cdot\,\rangle$ be a continuous, coercive sesqui-linear form on a Hilbert space X. Then for every $f \in X^*$ there is a unique $u \in X$ such that*

$$\beta\langle u|v\rangle = f(v)$$

for all $v \in X$.

Keeping in mind that the latter approach has been utilized extensively for elliptic type problems, it may be interesting to note that its generalization in the form of Corollary 1.3 is perfectly sufficient to solve elliptic, parabolic and hyperbolic systems in a single approach. For further illustrating the Lax–Milgram theorem in its abstract form, we discuss an example, which is related to the sesqui-linear forms method.

Example 1.7. Let H_0, H_1 be Hilbert spaces. Denote by H_{-1} the dual of H_1 and let $R_{H_1} : H_{-1} \to H_1$ be the corresponding Riesz-isomorphism. Consider a continuous linear bijection $C : H_1 \to H_0$ and a continuous linear operator $A : H_0 \to H_0$ with

$$\mathfrak{Re}\,\langle x|Ax\rangle_{H_0} \geq \alpha_0\langle x|x\rangle_{H_0} \quad (x \in H_0)$$

[9]Recall that for this we have to define the complex structure of X^* accordingly as $(\alpha f)(x) := \overline{\alpha}\, f(x)$ for every $x \in X$ and every continuous linear functional f on X.

[10]The strict positivity in (1.1) can be weakened to requiring merely $\bigwedge_{u\in X} |\beta\langle u|u\rangle| \geq c\langle u|u\rangle_X$, which yields in an analogous way a corresponding well-posedness result. This option is used in some applications.

for some $\alpha_0 \in \mathbb{R}_{>0}$. Denoting

$$C^\diamond : H_0 \to H_{-1}, y \mapsto C^\diamond y := \langle y|C\,\cdot\,\rangle_{H_0},$$

we consider

$$C^\diamond AC : H_1 \to H_{-1}.$$

Now, from

$$R_{H_1} C^\diamond AC = C^* AC,$$

we read off that $C^\diamond AC$ is an isomorphism. This may also be seen as an application of the Lax–Milgram theorem, since the equation

$$C^\diamond ACw = f,$$

for given $f \in H_{-1}$ amounts to being equivalent to the discussion of the sesqui-linear form

$$(v, w) \mapsto \beta\,\langle v|w\rangle := \langle ACv|Cw\rangle_{H_0} = (C^\diamond ACv)\,(w)$$

similar to the way it was done above.

In order to establish a solution theory for elliptic type equations, it is possible to go a step further. For stating an adapted well-posedness theorem we recall the following. Let $G : D(G) \subseteq H_1 \to H_2$ be a densely defined closed linear operator with closed range $R(G) = G[H_1]$. Then, the operator $B_G\colon D(G) \cap N(G)^\perp \subseteq N(G)^\perp \to R(G), x \mapsto Gx$, where $N(G) = [\{0\}]G$ denotes the null-space of G, is continuously invertible as it is one-to-one, onto and closed. Consequently, the modulus $|B_G|$ of B_G is continuously invertible on $N(G)^\perp$. We denote by $H_1(|B_G|)$ the domain of $|B_G|$ endowed with the norm $||B_G|\cdot|_{H_1}$, which can be shown to be equivalent to the graph norm of $|B_G|$. We denote by $H_{-1}(|B_G|)$ the dual of $H_1(|B_G|)$ with the pivot space $H_0(|B_G|) := N(G)^\perp$[11]. It is possible to show that the range of G^* is closed as well. Thus, the above reasoning also applies to G^* in the place of G.
Moreover, the operators B_G and $B_G^\diamond$, defined as in Example 1.7, are unitary transformations from $H_1(|B_G|)$ to $H_0(|B_{G^*}|)$ and from $H_0(|B_{G^*}|)$ to $H_{-1}(|B_G|)$, respectively. Moreover, note that $B_G^\diamond$ is the continuous extension of $B_{G^*}(= B_G^*)$. The abstract result asserting a solution theory for homogeneous elliptic boundary value problems reads as follows.

Theorem 1.8 ([52, Theorem 3.1.1]). *Let H_1, H_2 be Hilbert spaces and let $G : D(G) \subseteq H_1 \to H_2$ be a densely defined closed linear operator, such that $R(G) \subseteq H_2$ is closed. Let $a \subseteq R(G) \oplus R(G)$ such that $a^{-1} : R(G) \to R(G)$ is Lipschitz-continuous. Then for all $f \in H_{-1}(|B_G|)$ there exists a unique $u \in H_1(|B_G|)$ such*

[11] We use $H_1(A)$ also as a notation for the graph space of some continuously invertible operator A endowed with the norm $|A\cdot|$. Similarly, we write $H_{-1}(A)$ for the respective dual space with the pivot space $\overline{D(A)}$.

that the following inclusion holds:

$$(u,f)\in G^{\diamond}aG := \Big\{(x,z)\in H_1(|G|+\mathrm{i})\times H_{-1}(|G|+\mathrm{i}) \mid \bigvee_{y\in H_0(|G|+\mathrm{i})} (Gx,y)\in a\wedge z=G^{\diamond}y\Big\}.$$

Moreover, the solution u depends Lipschitz-continuously on the right-hand side with Lipschitz constant $|a^{-1}|_{\mathrm{Lip}}$ denoting the smallest Lipschitz constant of a^{-1}.

In other words, the relation $(B_G^{\diamond}aB_G)^{-1}\subseteq H_{-1}(|B_G|)\oplus H_1(|B_G|)$ defines a Lipschitz-continuous mapping with $\left|(B_G^{\diamond}aB_G)^{-1}\right|_{\mathrm{Lip}} = |a^{-1}|_{\mathrm{Lip}}$.

Proof. It is easy to see that $(u,f)\in G^{\diamond}aG$ for $u\in H_1(|B_G|)$ and $f\in H_{-1}(|B_G|)$ if and only if $(u,f)\in B_G^{\diamond}aB_G$. Hence, the assertion follows from $(B_G^{\diamond}aB_G)^{-1} = B_G^{-1}a^{-1}(B_G^{\diamond})^{-1}$, the unitarity of B_G and $B_G^{\diamond}$, and the fact that a^{-1} is Lipschitz-continuous on $R(G)$. □

To illustrate the latter result, we give an example.

Definition 1.9. Let $\Omega\subseteq\mathbb{R}^n$ open. We define

$$\widetilde{\operatorname{div}}_c\colon C_{\infty,c}(\Omega)^n\subseteq\bigoplus_{k=1}^{n}L^2(\Omega)\to L^2(\Omega)$$

$$\phi=(\phi_1,\ldots,\phi_n)\mapsto\sum_{k=1}^{n}\partial_k\phi_k,$$

where ∂_k denotes the derivative with respect to the k'th variable ($k\in\{1,\ldots,n\}$) and $C_{\infty,c}(\Omega)$ is the space of arbitrarily differentiable functions with compact support in Ω. Furthermore, define

$$\widetilde{\operatorname{grad}}_c\colon C_{\infty,c}(\Omega)\subseteq L^2(\Omega)\to\bigoplus_{k=1}^{n}L^2(\Omega)$$

$$\phi\mapsto(\partial_1\phi,\ldots,\partial_n\phi).$$

Integration by parts gives $\widetilde{\operatorname{div}}_c\subseteq-\left(\widetilde{\operatorname{grad}}_c\right)^*$ and consequently $\widetilde{\operatorname{grad}}_c\subseteq-\left(\widetilde{\operatorname{div}}_c\right)^*$. We set $\operatorname{div}:=-\left(\widetilde{\operatorname{grad}}_c\right)^*$, $\operatorname{grad}:=-\left(\widetilde{\operatorname{div}}_c\right)^*$, $\operatorname{div}_c:=-\operatorname{grad}^*$ and $\operatorname{grad}_c:=-\operatorname{div}^*$. In the particular case $n=1$ we set $\partial_{1,c}:=\operatorname{grad}_c=\operatorname{div}_c$ and $\partial_1:=\operatorname{grad}=\operatorname{div}$.

With the latter operators, in order to apply the solution theory above, one needs to impose certain geometric conditions on the open set Ω. Indeed, the above theorem applies to grad or grad_c in the place of G for the homogeneous Neumann and Dirichlet case, respectively (in this case $G^{\diamond}$ is then the canonical extension of $-\operatorname{div}_c$ and $-\operatorname{div}$, respectively). The only thing that has to be guaranteed is the closedness of the range of grad (grad_c, resp.). This in turn can be warranted, e.g., if Ω is bounded, connected and satisfies the segment property for the Neumann case or if Ω is bounded in one direction for the Dirichlet case. In both cases, one can prove the Poincaré inequality, which especially implies the closedness of the corresponding ranges (see, e.g., [68, Satz 7.6, p. 120] and [1, Theorem 3.8, p. 24] for the Poincaré inequality for the Dirichlet case and Rellich's theorem for the Neumann case, respectively. Note that if the domain of the gradient endowed with the graph norm is compactly embedded into the underlying space, a Poincaré type estimate can be derived by a contradiction argument).

In the remainder of this section, we discuss an elliptic-type problem in one dimension with indefinite coefficients. A similar result in two dimensions can be found in [6].

Example 1.10. Let $\Omega := \left[-\frac{1}{2}, \frac{1}{2}\right]$ and set

$$\widetilde{a}(x) := \begin{cases} \alpha & x \geq 0, \\ \beta & x < 0 \end{cases} \quad \left(x \in \left[-\frac{1}{2}, \frac{1}{2}\right]\right)$$

for some $\alpha, \beta \in \mathbb{R} \setminus \{0\}$. We denote the corresponding multiplication-operator on $L^2\left(\left[-\frac{1}{2}, \frac{1}{2}\right]\right)$ by $\widetilde{a}(\mathrm{m})$ and consider the following equation in divergence-form:

$$-\partial_1 \widetilde{a}(\mathrm{m}) \partial_{1,c} u = f \tag{1.4}$$

for some $f \in H_{-1}(|\partial_{1,c}|)$. Clearly, $R(\partial_{1,c})$ is closed and we denote the canonical embedding from $R(\partial_{1,c})$ into $L^2\left(\left[-\frac{1}{2}, \frac{1}{2}\right]\right)$ by $\iota_{R(\partial_{1,c})}$. Then $\iota_{R(\partial_{1,c})}^* : L^2\left(\left[-\frac{1}{2}, \frac{1}{2}\right]\right) \to R(\partial_{1,c})$ is the orthogonal projection onto $R(\partial_{1,c})$ (see, e.g., [35, Lemma 3.2]) and we can rewrite (1.4) as

$$-\partial_1 \iota_{R(\partial_{1,c})} \iota_{R(\partial_{1,c})}^* \widetilde{a}(\mathrm{m}) \iota_{R(\partial_{1,c})} \iota_{R(\partial_{1,c})}^* \partial_{1,c} u = f,$$

where we have used that ∂_1 vanishes on $R(\partial_{1,c})^\perp$. Hence, we are in the setting of Theorem 1.8, where $a = \iota_{R(\partial_{1,c})}^* \widetilde{a}(\mathrm{m}) \iota_{R(\partial_{1,c})} : R(\partial_{1,c}) \to R(\partial_{1,c})$. The only thing we have to show is that a^{-1} defines a Lipschitz-continuous mapping on $R(\partial_{1,c})$. As $R(\partial_{1,c})$ is closed, it follows that

$$R(\partial_{1,c}) = N(\partial_1)^\perp = \{\mathbf{1}\}^\perp,$$

where $\mathbf{1}$ denotes the constant function $\mathbf{1}(x) = 1$ for $x \in \left[-\frac{1}{2}, \frac{1}{2}\right]$. To show that a is invertible, we have to solve the problem

$$a\varphi = \psi$$

for given $\psi \in \{\mathbf{1}\}^\perp$. The latter can be written as

$$\psi = a\varphi = \widetilde{a}(\mathrm{m})\varphi - \langle \mathbf{1} | \widetilde{a}(\mathrm{m})\varphi \rangle \mathbf{1}.$$

As $\widetilde{a}(\mathrm{m})$ is continuously invertible (since $\alpha, \beta \neq 0$) we derive

$$\varphi = \widetilde{a}(\mathrm{m})^{-1}\psi + \langle \mathbf{1} | \widetilde{a}(\mathrm{m})\varphi \rangle \widetilde{a}(\mathrm{m})^{-1}\mathbf{1}.$$

Since $\varphi \in \{\mathbf{1}\}^\perp$ we obtain

$$0 = \langle \mathbf{1} | \widetilde{a}(\mathrm{m})^{-1}\psi \rangle + \langle \mathbf{1} | \widetilde{a}(\mathrm{m})\varphi \rangle \langle \mathbf{1} | \widetilde{a}(\mathrm{m})^{-1}\mathbf{1} \rangle,$$

yielding

$$\langle \mathbf{1}|\widetilde{a}(\mathrm{m})\varphi\rangle = -\frac{\langle \mathbf{1}|\widetilde{a}(\mathrm{m})^{-1}\psi\rangle}{\langle \mathbf{1}|\widetilde{a}(\mathrm{m})^{-1}\mathbf{1}\rangle},$$

provided that $\langle \mathbf{1}|\widetilde{a}(\mathrm{m})^{-1}\mathbf{1}\rangle \neq 0$. The latter holds if and only if $\alpha \neq -\beta$. Thus, assuming that $\alpha \neq -\beta$ we get

$$a^{-1}\psi = \widetilde{a}(\mathrm{m})^{-1}\psi - \frac{\langle \mathbf{1}|\widetilde{a}(\mathrm{m})^{-1}\psi\rangle}{\langle \mathbf{1}|\widetilde{a}(\mathrm{m})^{-1}\mathbf{1}\rangle}\widetilde{a}(\mathrm{m})^{-1}\mathbf{1},$$

which clearly defines a Lipschitz-continuous mapping. Summarizing, if $\alpha, \beta \neq 0$ and $\alpha \neq -\beta$, then for each $f \in H_{-1}(|\partial_{1,c}|)$ there exists a unique $u \in H_1(|\partial_{1,c}|)$ satisfying (1.4).

Remark 1.11. (a) If (1.4) is replaced by the problem with homogeneous Neumann boundary conditions, then the constraint $\alpha \neq -\beta$ can be dropped, since in this case $R(\partial_1) = N(\partial_{1,c})^\perp = L^2\left(\left[-\frac{1}{2}, \frac{1}{2}\right]\right)$, and thus, a is invertible if $\widetilde{a}(\mathrm{m})$ is invertible.

(b) Of course in view of Theorem 1.8, the coefficient a in Example 1.10 may also be induced by a relation such that its inverse relation is a (nonlinear) Lipschitz-continuous mapping in $R(\partial_{1,c})$.

2. Linear evolutionary equations and strict positivity

In this section we shall discuss equations of the form

$$(\partial_{0,\nu}\mathcal{M} + \mathcal{A})\,U = F, \tag{2.1}$$

where $\partial_{0,\nu}$ is the *time-derivative operator* to be introduced and specified below, $\mathcal{M}$ and $\mathcal{A}$ are linear operators, the former – the *material law operator* – being bounded, and the latter being possibly unbounded. The task is in finding the unknown U for a given right-hand side F. This is done by showing that both (the closure of) $(\partial_{0,\nu}\mathcal{M} + \mathcal{A})$ and $(\partial_{0,\nu}\mathcal{M} + \mathcal{A})^*$ are strictly accretive operators in a suitable Hilbert space and then using Corollary 1.3. We will comment on the specific assumptions on $\mathcal{M}$ and $\mathcal{A}$ in the subsequent sections as well as on the rigorous (Hilbert space) framework the equation (2.1) should be considered in. Before we discuss the abstract theory, we give four elementary guiding examples which shall lead us through the development of the abstract theory.

2.1. Guiding examples

Ordinary (integro-)differential equations. We shall consider the following easy form of an ordinary differential equation. For a given right-hand side $f \in C_c(\mathbb{R}\times\mathbb{R})$, i.e., f is a continuous function on $\mathbb{R}\times\mathbb{R}$ with compact support, and a coefficient $a \in L^\infty(\mathbb{R})$ we consider the problem of finding u in a suitable Hilbert space such that for (a.e.) $(t, x) \in \mathbb{R}\times\mathbb{R}$ the equation

$$u(\cdot, x)'(t) + a(x)u(t, x) = f(t, x)$$

holds. We will also have the opportunity to consider an integro-differential equation of the form

$$u(\cdot,x)'(t)+a(x)u(t,x)+\int_{-\infty}^{t} k(t-s)u(s,x)ds=f(t,x)$$

for a suitable kernel $k\colon \mathbb{R}\to\mathbb{R}$.

The heat equation. The heat ϑ in a given body $\Omega\subseteq\mathbb{R}^n$ can be described by the conservation law

$$\vartheta(\cdot,x)'(t)+\operatorname{div}q(t,x)=f(t,x)\quad ((t,x)\in\mathbb{R}\times\Omega\text{ a.e.}),$$

where f is a given heat source and q is the heat flux given by Fourier's law as

$$q(t,x)=-k(x)\operatorname{grad}\vartheta(t,x)\quad ((t,x)\in\mathbb{R}\times\Omega\text{ a.e.}),$$

where k is a certain coefficient matrix describing the specific conductivities of the underlying material varying over Ω. Here div and grad are the canonical extensions of the spatial operators div and grad defined in the previous section (Definition 1.9) to the space $L^2(\mathbb{R}\times\Omega,\mu\otimes\lambda)=L^2(\mathbb{R},\mu)\otimes L^2(\Omega,\lambda)$, where μ is a Borel measure on $\mathbb{R}$ and λ denotes the n-dimensional Lebesgue measure, i.e.,

$$\begin{aligned}\operatorname{div}q(t,x)&=(\operatorname{div}q(t,\cdot))(x)\\ \operatorname{grad}\vartheta(t,x)&=(\operatorname{grad}\vartheta(t,\cdot))(x)\quad ((t,x)\in\mathbb{R}\times\Omega\text{ a.e.}).\end{aligned}$$

In a block operator matrix form, recalling that ∂_0 denotes the derivative with respect to time, we get

$$\left(\partial_0\begin{pmatrix}1&0\\0&0\end{pmatrix}+\begin{pmatrix}0&0\\0&k^{-1}\end{pmatrix}+\begin{pmatrix}0&\operatorname{div}\\ \operatorname{grad}&0\end{pmatrix}\right)\begin{pmatrix}\vartheta\\q\end{pmatrix}=\begin{pmatrix}f\\0\end{pmatrix},$$

assuming that the coefficient k is invertible. Imposing suitable assumptions on data and coefficients and boundary conditions for the operators div and/or grad will be seen to warrant well-posedness of the resulting system. We will comment on the precise details in our discussion of abstract well-posedness results.

The elastic equations. In the theory of elasticity, the open set $\Omega\subseteq\mathbb{R}^n$, being the underlying domain, models a body in its non-deformed state (of course, in applications $n=3$). The displacement field u assigns to each space-time coordinate $(t,x)\in\mathbb{R}\times\Omega$ direction and size of the displacement at time t of the material point at position x. The displacement field u satisfies the balance of momentum equation (again writing ∂_0 for the time-derivative)

$$\partial_0^2u-\operatorname{Div}\sigma=f,$$

with f being an external forcing term, σ being the (symmetric) stress tensor and Div being the row-wise (distributional) divergence acting on suitable elements in the space $H_{\mathrm{sym}}(\Omega)$ of symmetric $n\times n$ matrices of $L^2(\Omega)$-functions as an operator from $H_{\mathrm{sym}}(\Omega)$ to $L^2(\Omega)^n$ with maximal domain[12]. Endowing $H_{\mathrm{sym}}(\Omega)$ with the

[12]The precise definition will be given later.

Frobenius inner product, we get that the negative adjoint of Div is the symmetrized gradient or strain tensor given by

$$\varepsilon(u) := \operatorname{Grad} u := \frac{1}{2}\left(\partial_i u_j + \partial_j u_i\right)_{i,j}$$

with Dirichlet boundary conditions as induced constraint on the domain. Neumann boundary conditions can be modeled similarly. The stress tensor satisfies the constitutive relation involving the elasticity tensor C in the way that

$$\sigma = C\varepsilon(u).$$

Introducing the displacement velocity $v := \partial_0 u$ as a new unknown, we write the elastic equations formally as the block operator matrix equation

$$\left(\partial_0 \begin{pmatrix} 1 & 0 \\ 0 & C^{-1} \end{pmatrix} - \begin{pmatrix} 0 & \operatorname{Div} \\ \operatorname{Grad} & 0 \end{pmatrix}\right)\begin{pmatrix} v \\ \sigma \end{pmatrix} = \begin{pmatrix} f \\ 0 \end{pmatrix},$$

where we assume that C is invertible.

Maxwell's equations. The equations for electro-magnetic theory describe evolution of the electro-magnetic field (E, H) in a 3-dimensional open set Ω. As Gauss' law can be incorporated by a suitable choice of initial data, we think of Maxwell's equations as Faraday's law of induction (the Maxwell–Faraday equation), which reads as

$$\partial_0 B + \operatorname{curl}_c E = 0,$$

where curl_c denotes the (distributional) curl operator in $L^2(\Omega)^3$ with the electric boundary condition of vanishing tangential components. The magnetic field B satisfies the constitutive equation

$$B = \mu H,$$

where μ is the magnetic permeability. Faraday's law is complemented by Ampère's law

$$\partial_0 D + J_c - \operatorname{curl} H = J_0$$

for J_0, D, J_c being the external currents, the electric displacement and the charge, respectively. The latter two quantities satisfy the two equations

$$D = \varepsilon E, \quad \text{and} \quad J_c = \sigma E.$$

The former is a constitutive equation involving the dielectricity ε and the latter is Ohm's law with conductivity σ. Plugging the constitutive relations and Ohm's law into Faraday's law of induction and Ampère's law and arranging them in a block operator matrix equation, we arrive at

$$\left(\partial_0 \begin{pmatrix} \varepsilon & 0 \\ 0 & \mu \end{pmatrix} + \begin{pmatrix} \sigma & 0 \\ 0 & 0 \end{pmatrix} + \begin{pmatrix} 0 & -\operatorname{curl} \\ \operatorname{curl}_c & 0 \end{pmatrix}\right)\begin{pmatrix} E \\ H \end{pmatrix} = \begin{pmatrix} J_0 \\ 0 \end{pmatrix}.$$

Having these examples in mind, we develop the abstract theory a bit further and discuss the time-derivative operator in the next section. After having done so, we aim at giving a unified solution theory for all of the latter examples. In fact we show that all of these equations are of the general form (2.1).

2.2. The time-derivative

When considering evolutionary equations, we need a distinguished direction of time. Anticipating this fact, we define a time-derivative $\partial_{0,\nu}$ as an operator in a weighted Hilbert space.

Beforehand, we recall some well-known facts from the (time-)derivative in the unweighted space $L^2(\mathbb{R})$. We denote the Sobolev space of $L^2(\mathbb{R})$-functions f with distributional derivative f' representable as a $L^2(\mathbb{R})$-function by $H_1(\mathbb{R})$. Then, the operator

$$\partial\colon H_1(\mathbb{R}) \subseteq L^2(\mathbb{R}) \to L^2(\mathbb{R}), f \mapsto f'$$

is skew-selfadjoint. Indeed, using that the space $C_{\infty,c}(\mathbb{R})$ is a core for ∂, we immediately verify with integration by parts that ∂ is skew-symmetric. With the help of some elementary computations it is possible to show that the range of both the operators $\partial + 1$ and $\partial - 1$ contains $C_{\infty,c}(\mathbb{R})$. The closedness of ∂ thus implies the skew-selfadjointness of ∂ (see also [20, Example 3.14]). Moreover, it is well known that ∂ admits an explicit spectral representation given by the Fourier transformation $\mathcal{F}$, being the unitary extension on $L^2(\mathbb{R})$ of the mapping given by

$$\mathcal{F}\phi(\xi) := \frac{1}{\sqrt{2\pi}} \int_{\mathbb{R}} e^{-\mathrm{i}\xi x}\phi(x)dx \quad (\xi \in \mathbb{R})$$

for $\phi \in C_{\infty,c}(\mathbb{R})$. Denoting by $\mathrm{m}\colon D(\mathrm{m}) \subseteq L^2(\mathbb{R}) \to L^2(\mathbb{R})$ the multiplication-by-argument operator given by

$$\mathrm{m}f := (\xi \mapsto \xi f(\xi))$$

for f belonging to the maximal domain $D(\mathrm{m})$ of m, we find the following unitary equivalence of differentiation and multiplication (see [2, Volume 1, p. 161-163]):

$$\partial = \mathcal{F}^*\mathrm{i}\mathrm{m}\mathcal{F},$$

which, due to the selfadjointness of m, confirms the skew-selfadjointness of ∂.

As mentioned above, in evolutionary processes there is a particular bias for the forward time direction. As $L^2(\mathbb{R})$ has no such bias, we choose a suitable weight, which serves to express this bias. For $\nu \in \mathbb{R}$ we let

$$L^2_\nu(\mathbb{R}) := \left\{ f \in L^{2,\mathrm{loc}}(\mathbb{R}) \mid \int_{\mathbb{R}} |f(t)|^2 \exp(-2\nu t)\, dt < \infty \right\},$$

the Hilbert space of (equivalence classes of) functions with $(t \mapsto \exp(-\nu t)\, f(t)) \in L^2(\mathbb{R})$ (with the obvious norm). In particular, $L^2_0(\mathbb{R}) = L^2(\mathbb{R})$. Moreover, it is easily seen that the mapping

$$e^{-\nu\mathrm{m}}\colon L^2_\nu(\mathbb{R}) \to L^2(\mathbb{R}), f \mapsto \left(t \mapsto e^{-\nu t} f(t)\right)$$

defines a unitary mapping. To carry differentiation over to the exponentially weighted L^2-spaces, we observe that for $\phi \in C_{\infty,c}(\mathbb{R})$ we have

$$\begin{aligned}\left(e^{-\nu\mathrm{m}}\right)^{-1} \partial e^{-\nu\mathrm{m}}\phi &= \left(e^{-\nu\mathrm{m}}\right)^{-1} (-\nu e^{-\nu\mathrm{m}}\phi + e^{-\nu\mathrm{m}}\phi') \\ &= -\nu\phi + \phi',\end{aligned}$$

or

$$\left(e^{-\nu\mathrm{m}}\right)^{-1}(\partial+\nu)\,e^{-\nu\mathrm{m}}\phi=\phi'.$$

Defining $\partial_{0,\nu}:=\left(e^{-\nu\mathrm{m}}\right)^{-1}(\partial+\nu)\,e^{-\nu\mathrm{m}}=\left(e^{-\nu\mathrm{m}}\right)^{-1}\partial e^{-\nu\mathrm{m}}+\nu$, we read off that $\partial_{0,\nu}$ is a realization of the (distributional) derivative operator in the weighted space $L^2_\nu(\mathbb{R})$. Moreover, we see that $\partial_{0,\nu}$ is a normal operator, i.e., $\partial_{0,\nu}$ and $\partial_{0,\nu}^*$ commute. In particular, $\mathfrak{Re}\,\partial_{0,\nu}=\frac{1}{2}\overline{\left(\partial_{0,\nu}+\partial_{0,\nu}^*\right)}=\nu$, due to the skew-selfadjointness of ∂. This also shows that $\partial_{0,\nu}$ is continuously invertible if $\nu\neq 0$. Indeed, we find the following explicit formula for the inverse: For $t\in\mathbb{R}$, $\nu\in\mathbb{R}\setminus\{0\}$, $f\in L^2_\nu(\mathbb{R})$ we have

$$\partial_{0,\nu}^{-1}f(t)=\begin{cases}\int_{-\infty}^t f(\tau)d\tau, & \nu>0,\\ -\int_t^\infty f(\tau)d\tau, & \nu<0.\end{cases}$$

For positive ν, the latter formula also shows that the values of $\partial_{0,\nu}^{-1}f$ at time t only depend on the values of f up to time t. This is the nucleus of the notion of causality, where the sign of ν switches the forward and backward time direction. Later, we will comment on this in more detail.

The spectral representation of ∂ induces a spectral representation for $\partial_{0,\nu}$. Indeed, introducing the Fourier–Laplace transformation[13] $\mathcal{L}_\nu:=\mathcal{F}e^{-\nu\mathrm{m}}:L^2_\nu(\mathbb{R})\to L^2(\mathbb{R})$ for $\nu\in\mathbb{R}$, which itself is a unitary operator as a composition of unitary operators, we get that

$$\partial_{0,\nu}=\mathcal{L}_\nu^*(\mathrm{im}+\nu)\mathcal{L}_\nu.$$

This shows that for $\nu\neq 0$ the normal operator $\partial_{0,\nu}^{-1}$ is unitarily equivalent to the multiplication operator[14] $(\mathrm{im}+\nu)^{-1}$ with spectrum $\sigma(\partial_{0,\nu}^{-1})=\partial B(\frac{1}{2\nu},\frac{1}{2\nu})$, where $\nu>0$ and $B(a,r):=\{z\in\mathbb{C}||z-a|<r\}$, $a\in\mathbb{C}$, $r>0$. We will use this fact in the next section by establishing a functional calculus for $\partial_{0,\nu}^{-1}$. Henceforth, if not otherwise stated, the parameter ν will always be a positive real number.

2.3. The autonomous case

In order to formulate the Hilbert space framework of (2.1) properly, we need to consider the space of H-valued L^2_ν-functions. Consequently, we need to invoke the (canonical) extension of $\partial_{0,\nu}$ to $L^2_\nu(\mathbb{R},H)$ for some Hilbert space H. For convenience, we re-use the notation for the respective extension since there is no risk of confusion. Moreover, we shall do so for the Fourier–Laplace transform $\mathcal{L}_\nu$ which is then understood as a unitary operator from $L^2_\nu(\mathbb{R},H)$ onto $L^2(\mathbb{R},H)$.

---

[13]For the classical Fourier–Laplace transformation, also known as two-sided Laplace transformation, this unitary character is rarely invoked. In fact, it is mostly considered as an integral expression acting on suitably integrable functions, whereas the unitary Fourier–Laplace transformation is continuously extended (thus, of course, including also some non-integrable functions).

[14]In the sense of the induced functional calculus we have

$$(\mathrm{im}+\nu)^{-1}=\frac{1}{\mathrm{im}+\nu}.$$

In this section we treat a particular example for the choice of the operators $\mathcal{M}$ and $\mathcal{A}$, namely that of autonomous operators. For this we need to introduce the operator of time translation: For $h \in \mathbb{R}$, $\nu \in \mathbb{R}$ we define the *time translation operator* τ_h on $L^2_\nu(\mathbb{R})$ by $\tau_h f := f(\cdot + h)$. Again, we identify τ_h with its extension to the H-valued case.

Definition 2.1. We call an operator $B\colon D(B) \subseteq L^2_\nu(\mathbb{R}, H) \to L^2_\nu(\mathbb{R}, H)$ *autonomous* or *time translation invariant*, if it commutes with τ_h for all $h \in \mathbb{R}$, i.e.,

$$\tau_h B \subseteq B\tau_h \quad (h \in \mathbb{R}).$$

For an evolution to take place, a physically reasonable property is *causality*. Denoting by $\chi_{_{\mathbb{R}_{\leq a}}}(\mathrm{m}_0)$ the multiplication operator on $L^2_\nu(\mathbb{R}, H)$ given by $\left(\chi_{_{\mathbb{R}_{\leq a}}}(\mathrm{m}_0)u\right)(t) := \chi_{_{\mathbb{R}_{\leq a}}}(t)u(t)$ for $u \in L^2_\nu(\mathbb{R}, H)$ and $t \in \mathbb{R}$, the definition of causality reads as follows.

Definition 2.2. We call a closed mapping $M\colon D(M) \subseteq L^2_\nu(\mathbb{R}, H) \to L^2_\nu(\mathbb{R}, H)$ *causal*, if for all $f, g \in D(M)$ and $a \in \mathbb{R}$ we have

$$\chi_{\mathbb{R}_{\leq a}}(\mathrm{m}_0)f = \chi_{\mathbb{R}_{\leq a}}(\mathrm{m}_0)g \Rightarrow \chi_{\mathbb{R}_{\leq a}}(\mathrm{m}_0)M(f) = \chi_{\mathbb{R}_{\leq a}}(\mathrm{m}_0)M(g). \tag{2.2}$$

Remark 2.3.

(a) Property (2.2) reflects the idea that the "future behavior does not influence the past", which may be taken as the meaning of *causality.*

(b) If, in addition, M in the latter definition is linear, then M is causal if and only if for all $u \in D(M)$

$$\chi_{_{\mathbb{R}_{\leq a}}}(\mathrm{m}_0)u = 0 \Rightarrow \chi_{_{\mathbb{R}_{\leq a}}}(\mathrm{m}_0)Mu = 0 \qquad (a \in \mathbb{R}).$$

(c) For continuous mappings M with full domain $L^2_\nu(\mathbb{R}, H)$, causality can also equivalently be expressed by the equation

$$\chi_{_{\mathbb{R}_{\leq a}}}(\mathrm{m}_0)M = \chi_{_{\mathbb{R}_{\leq a}}}(\mathrm{m}_0)M\chi_{\mathbb{R}_{\leq a}}(\mathrm{m}_0)$$

holding for all $a \in \mathbb{R}$. If, in addition, M is autonomous then this condition is in turn equivalent to

$$\chi_{\mathbb{R}_{\leq a}}(\mathrm{m}_0)M = \chi_{\mathbb{R}_{\leq a}}(\mathrm{m}_0)M\chi_{\mathbb{R}_{\leq a}}(\mathrm{m}_0)$$

for some $a \in \mathbb{R}$, e.g., $a = 0$.

(d) The respective concept for causality for closable mappings is a bit more involved, see [57].

In order to motivate the problem class discussed in this section a bit further we state the following well-known representation theorem:

Theorem 2.4 (see, e.g., [65, Theorem 2.3], [44, Theorem 9.1] or [45]). *Let H be a Hilbert space, $\mathcal{M}\colon L^2_\nu(\mathbb{R}, H) \to L^2_\nu(\mathbb{R}, H)$ bounded, linear, causal and autonomous. Then $\mathcal{M}$ admits a continuous extension as a continuous linear operator* *in* $L^2_{\nu'}(\mathbb{R}, H)$ *for all* $\nu' > \nu$. *Moreover there exists a unique* $\widetilde{M}\colon \{z\in\mathbb{C}|\mathfrak{Re}\, z>\nu'\}\to L(H)$ *bounded and analytic, such that for all* $u \in L^2_{\nu'}(\mathbb{R}, H)$ *we have*

$$\mathcal{M}u = \mathcal{L}^*_{\nu'}\widetilde{M}(\mathrm{im} + \nu')\mathcal{L}_{\nu'}u,$$

where $\left(\widetilde{M}(\mathrm{im} + \nu')\phi\right)(\xi) := \widetilde{M}(\mathrm{i}\xi + \nu')\phi(\xi)$ *for* $\phi \in L^2(\mathbb{R}, H)$, $\xi \in \mathbb{R}$, $\nu' > \nu$.

This theorem tells us that the class of bounded, linear, causal, autonomous operators is described by bounded and analytic functions of $\partial_{0,\nu}$ or equivalently of $\partial_{0,\nu}^{-1}$. Thus, we are led to introduce the Hardy space for some open $E \subseteq \mathbb{C}$ and Hilbert space H:

$$\mathcal{H}^\infty(E, L(H)) := \{M\colon E \to L(H) \,|\, M \text{ bounded, analytic}\}.$$

Clearly, $\mathcal{H}^\infty(E, L(H))$ (or briefly $\mathcal{H}^\infty$ if E and H are clear from the context) is a Banach space with norm

$$\mathcal{H}^\infty \ni M \mapsto \|M\|_\infty := \sup\{\|M(z)\| | z \in E\}.$$

In the particular case of $M \in \mathcal{H}^\infty(E, L(H))$ with $E = B(r, r)$ for some $r > 0$, we define for $\nu > \frac{1}{2r}$,

$$\begin{aligned} M(\partial_{0,\nu}^{-1})\colon L^2_\nu(\mathbb{R}, H) &\to L^2_\nu(\mathbb{R}, H),\\ \phi &\mapsto \mathcal{L}^*_\nu M\left(\frac{1}{\mathrm{im} + \nu}\right)\mathcal{L}_\nu\phi. \end{aligned}$$

Here $M\left(\frac{1}{\mathrm{im}+\nu}\right) \in L(L^2(\mathbb{R}, H))$ is given by

$$\left(M\left(\frac{1}{\mathrm{im} + \nu}\right)w\right)(t) := M\left(\frac{1}{\mathrm{i}t + \nu}\right)w(t)$$

for $w \in L^2(\mathbb{R}, H)$ and $t \in \mathbb{R}$.

Note that $\sup_{z\in B\left(\frac{1}{2\nu},\frac{1}{2\nu}\right)} \|M(z)\| = \|M(\partial_{0,\nu}^{-1})\|_{L(H_{\nu,0}(\mathbb{R},H))}$ according to [44, Theorem 9.1].

Our first theorem asserting a solution theory for evolutionary equations reads as follows.

Theorem 2.5 ([31, Solution Theory], [33, Chapter 6]). *Let* $\nu > 0$, $r > \frac{1}{2\nu}$ *and* $M \in \mathcal{H}^\infty(B(r, r), L(H))$, $A\colon D(A) \subseteq H \to H$. *Assume that*

$$A \text{ is skew-selfadjoint, and}$$

$$\bigvee_{c>0} \bigwedge_{z\in B(r,r)} z^{-1}M(z) - c \text{ is monotone.} \tag{2.3}$$

Then $\overline{\partial_{0,\nu}M(\partial_{0,\nu}^{-1}) + A}$ *is continuously invertible in* $L^2_\nu(\mathbb{R}, H)$. *The closure of the inverse is causal. Moreover, the solution operator is independent of the choice of* ν *in the sense that for* $\varrho > \nu$ *and* $f \in L^2_\nu(\mathbb{R}, H) \cap L^2_\varrho(\mathbb{R}, H)$ *we have that*

$$\overline{\left(\partial_{0,\nu}M(\partial_{0,\nu}^{-1}) + A\right)^{-1}}f = \overline{\left(\partial_{0,\varrho}M(\partial_{0,\varrho}^{-1}) + A\right)^{-1}}f.$$

Sketch of the proof. First, one proves that the operator $\partial_{0,\nu}M(\partial_{0,\nu}^{-1}) + A$ is closable and its closure is strictly monotone. The same holds for its adjoint, which turns out to have the same domain. Thus, the well-posedness of (2.1), where A is skew-selfadjoint and $\mathcal{M} = M(\partial_{0,\nu}^{-1})$ follows by Corollary 1.4. The causality of the solution operator can be shown by a Paley–Wiener type result (see, e.g., [39, Theorem 19.2]). □

According to the latter theorem, the solution operator associated with an evolutionary problem is independent of the particular choice of ν. Therefore, we usually will drop the index ν and write instead ∂_0 and $M(\partial_0^{-1})$.

Since the positive definiteness condition in (2.3) will occur several times, we define $\mathcal{H}^{\infty,c}$ to be the set of $M \in \mathcal{H}^\infty$ satisfying condition (2.3) with the constant $c \in \mathbb{R}_{>0}$.

Note that with $A = 0$ in Theorem 2.5, ordinary differential equations are covered. We shall further elaborate this fact in Subsection 2.4. Here, let us illustrate the versatility of this well-posedness result, by applying the result to several (partial) differential equations arising in mathematical physics.

Example 2.6 (The heat equation). Recall the definition of the operators grad_c, grad, div_c and div from Definition 1.9. The domain of grad_c coincides with the classical Sobolev space $H_{0,1}(\Omega)$, the space of $L^2(\Omega)$-functions with distributional gradients lying in $L^2(\Omega)^n$ and having vanishing trace, while the domain of grad is $H_1(\Omega)$, the Sobolev space of weakly differentiable functions in $L^2(\Omega)$. Analogously, $v \in D(\operatorname{div}_c)$ is a $L^2(\Omega)$-vector field, whose distributional divergence is in $L^2(\Omega)$ and satisfies a generalized Neumann condition $v \cdot N = 0$ on $\partial\Omega$, where N denotes the outward unit normal vector field on $\partial\Omega$.[15] Recall the conservation of energy equation, given by

$$\partial_0 \vartheta + \operatorname{div} q = f, \tag{2.4}$$

where $\vartheta \in L^2_\nu(\mathbb{R}, L^2(\Omega))$ denotes the (unknown) heat, $q \in L^2_\nu(\mathbb{R}, L^2(\Omega)^3)$ stands for the heat flux and $f \in L^2_\nu(\mathbb{R}, L^2(\Omega))$ models an external heat source. This equation is completed by a constitutive relation, called Fourier's law

$$q = -k \operatorname{grad} \vartheta, \tag{2.5}$$

where $k \in L^\infty(\Omega)^{n\times n}$ denotes the heat conductivity and is assumed to be selfadjoint-matrix-valued and strictly positive, i.e., there is some $c > 0$ such that $k(x) \geq c$ for almost every $x \in \Omega$. Plugging Fourier's law into the conservation of energy equation, we end up with the familiar form of the heat equation

$$\partial_0 \vartheta - \operatorname{div} k \operatorname{grad} \vartheta = f.$$

However, it is also possible to write the equations (2.4) and (2.5) as the system

$$\left(\partial_0 \begin{pmatrix} 1 & 0 \\ 0 & 0 \end{pmatrix} + \begin{pmatrix} 0 & 0 \\ 0 & k^{-1} \end{pmatrix} + \begin{pmatrix} 0 & \operatorname{div} \\ \operatorname{grad} & 0 \end{pmatrix}\right) \begin{pmatrix} \vartheta \\ q \end{pmatrix} = \begin{pmatrix} f \\ 0 \end{pmatrix}.$$

[15] Note that the definition of div_c still makes sense, even if the boundary of Ω is non-smooth and hence, the normal vector field N does not exist. Thus, for rough domains Ω the condition $v \in D(\operatorname{div}_c)$ is a suitable substitute for the condition $v \cdot N = 0$ on $\partial\Omega$.

Requiring suitable boundary conditions, say Dirichlet boundary conditions for the temperature density ϑ (i.e., $\vartheta \in D(\mathrm{grad}_c)$), the system becomes an evolutionary equation of the form discussed in Theorem 2.5 with

$$M(z) = \begin{pmatrix} 1 & 0 \\ 0 & 0 \end{pmatrix} + z \begin{pmatrix} 0 & 0 \\ 0 & k^{-1} \end{pmatrix}$$

and the skew-selfadjoint operator

$$A = \begin{pmatrix} 0 & \mathrm{div} \\ \mathrm{grad}_c & 0 \end{pmatrix}.$$

By our assumptions on the coefficient k, the solvability condition (2.3) can easily be verified for our material law M. Indeed, with k being bounded and strictly positive, the inverse operator k^{-1} is bounded and strictly positive as well. Now, since for $z \in B(r,r)$ for some $r > 0$ the real part of z^{-1} is bounded below by $\frac{1}{2r}$, we deduce that

$$\mathfrak{Re}\left(z^{-1}M(z)\right) = \mathfrak{Re}\left(z^{-1}\begin{pmatrix} 1 & 0 \\ 0 & 0 \end{pmatrix} + \begin{pmatrix} 0 & 0 \\ 0 & k^{-1} \end{pmatrix}\right) \geq \begin{pmatrix} \frac{1}{2r} & 0 \\ 0 & \frac{1}{|k|_\infty} \end{pmatrix}.$$

Example 2.7 (Maxwell's equations). To begin with, we formulate the functional analytic setting for the operator curl with and without the electric boundary condition. For this let $\Omega \subseteq \mathbb{R}^3$ be a non-empty open set and define

$$\widetilde{\mathrm{curl}_c} \colon C_{\infty,c}(\Omega)^3 \subseteq L^2(\Omega)^3 \to L^2(\Omega)^3$$
$$\phi \mapsto \begin{pmatrix} \partial_2\phi_3 - \partial_3\phi_2 \\ \partial_3\phi_1 - \partial_1\phi_3 \\ \partial_1\phi_2 - \partial_2\phi_1 \end{pmatrix}.$$

Analogously to the previous example, we let $\mathrm{curl} := \left(\widetilde{\mathrm{curl}_c}\right)^*$ and $\mathrm{curl}_c := \mathrm{curl}^*$. Now, let ε, μ, σ be bounded linear operators in $L^2(\Omega)^3$. We assume that both ε and μ are selfadjoint and strictly positive. As a consequence, the operator function

$$z \mapsto \begin{pmatrix} \varepsilon & 0 \\ 0 & \mu \end{pmatrix} + z \begin{pmatrix} \sigma & 0 \\ 0 & 0 \end{pmatrix}$$

belongs to $\mathcal{H}^{\infty,c}(B(r,r), L(L^2(\Omega)^6))$ for some $c \in \mathbb{R}_{>0}$, if r is chosen small enough. Thus, due to the skew-selfadjointness of the operator $\begin{pmatrix} 0 & -\mathrm{curl} \\ \mathrm{curl}_c & 0 \end{pmatrix}$ in $L^2(\Omega)^6$, Theorem 2.5 applies to the operator sum

$$\partial_0 \begin{pmatrix} \varepsilon & 0 \\ 0 & \mu \end{pmatrix} + \begin{pmatrix} \sigma & 0 \\ 0 & 0 \end{pmatrix} + \begin{pmatrix} 0 & -\mathrm{curl} \\ \mathrm{curl}_c & 0 \end{pmatrix},$$

which yields continuous invertibility of the closure of the latter operator in $L^2_\nu(\mathbb{R}, L^2(\Omega)^3)$ for sufficiently large ν. It is noteworthy that the well-posedness theorem of course also applies to the case, where $\varepsilon = 0$ and the real part of σ is strictly positive definite, i.e., to the operator of the form

$$\partial_0 \begin{pmatrix} 0 & 0 \\ 0 & \mu \end{pmatrix} + \begin{pmatrix} \sigma & 0 \\ 0 & 0 \end{pmatrix} + \begin{pmatrix} 0 & -\operatorname{curl} \\ \operatorname{curl}_c & 0 \end{pmatrix}.$$

In the literature this arises when dealing with the so-called "eddy current problem".

Example 2.8 (The equations of elasticity and visco-elasticity). We begin by introducing the differential operators involved. Let $\Omega \subseteq \mathbb{R}^3$ open. We consider the Hilbert space $H_{\mathrm{sym}}(\Omega)$ given as the space of symmetric $L^2(\Omega)^{3\times 3}$-matrices equipped with the inner product

$$\langle \Phi | \Psi \rangle_{H_{\mathrm{sym}}(\Omega)} := \int_\Omega \operatorname{trace}\left(\Phi(x)^* \Psi(x)\right) dx,$$

where $\Phi(x)^*$ denotes the adjoint matrix of $\Phi(x)$. Using this Hilbert space we define the operator Grad_c as the closure of

$$\begin{aligned} \widetilde{\operatorname{Grad}_c} : C_{\infty,c}(\Omega)^3 \subseteq L^2(\Omega)^3 &\to H_{\mathrm{sym}}(\Omega) \\ (\phi_i)_{i\in\{1,2,3\}} &\mapsto \left(\frac{\partial_i \phi_j + \partial_j \phi_i}{2}\right)_{i,j\in\{1,2,3\}} \end{aligned}$$

and likewise we define Div_c as the closure of

$$\begin{aligned} \widetilde{\operatorname{Div}_c} : C_{\infty,c}(\Omega)^{3\times 3} \cap H_{\mathrm{sym}}(\Omega) \subseteq H_{\mathrm{sym}}(\Omega) &\to L^2(\Omega)^3 \\ (\psi_{ij})_{i,j\in\{1,2,3\}} &\mapsto \left(\sum_{j=1}^3 \partial_j \psi_{ij}\right)_{i\in\{1,2,3\}} . \end{aligned}$$

Integration by parts yields that $\operatorname{Grad}_c \subseteq -(\operatorname{Div}_c)^*$ as well as $\operatorname{Div}_c \subseteq -(\operatorname{Grad}_c)^*$ and we define

$$\begin{aligned} \operatorname{Grad} &:= -(\operatorname{Div}_c)^*, \\ \operatorname{Div} &:= -(\operatorname{Grad}_c)^*. \end{aligned}$$

Similar to the case of grad and div, elements u in the domain of Grad_c satisfy an abstract Dirichlet boundary condition of the form $u = 0$ on $\partial\Omega$, while elements σ in the domain of Div_c satisfy an abstract Neumann boundary condition of the form $\sigma N = 0$ on $\partial\Omega$, where N denotes the unit outward normal vector field on $\partial\Omega$.

The equation of linear elasticity is given by

$$\partial_0^2 u - \operatorname{Div} \sigma = f, \tag{2.6}$$

where $u \in L^2_\nu(\mathbb{R}, L^2(\Omega)^3)$ denotes the displacement field of the elastic body Ω, $\sigma \in L^2_\nu(\mathbb{R}, H(\Omega))$ is the stress tensor and $f \in L^2_\nu(\mathbb{R}, L^2(\Omega)^3)$ is an external force. Equation (2.6) is completed by Hooke's law

$$\sigma = C \operatorname{Grad} u, \tag{2.7}$$

where $C \in L(H_{\text{sym}}(\Omega))$ is the so-called elasticity tensor, which is assumed to be selfadjoint and strictly positive definite (which in particular gives the bounded invertibility of C). Defining $v := \partial_0 u$ as the displacement velocity, (2.6) and (2.7) can be written as a system of the form

$$\left(\partial_0 \begin{pmatrix} 1 & 0 \\ 0 & C^{-1} \end{pmatrix} + \begin{pmatrix} 0 & -\operatorname{Div} \\ -\operatorname{Grad} & 0 \end{pmatrix}\right) \begin{pmatrix} v \\ \sigma \end{pmatrix} = \begin{pmatrix} f \\ 0 \end{pmatrix}.$$

Imposing boundary conditions on v or σ, say for simplicity Neumann boundary conditions for σ, we end up with an evolutionary equation with

$$M(z) = \begin{pmatrix} 1 & 0 \\ 0 & C^{-1} \end{pmatrix} \quad \text{and} \quad A = \begin{pmatrix} 0 & -\operatorname{Div}_c \\ -\operatorname{Grad} & 0 \end{pmatrix}.$$

Since C^{-1} is also selfadjoint and strictly positive definite, we obtain that M satisfies the solvability condition (2.3).
In order to incorporate viscous materials, i.e., materials showing some memory effects, one modifies Hooke's law (2.7) for example by

$$\sigma = C \operatorname{Grad} u + D \operatorname{Grad} \partial_0 u, \tag{2.8}$$

where $D \in L(H_{\text{sym}}(\Omega))$. This modification is known as the Kelvin–Voigt model for visco-elastic materials (see, e.g., [14, p. 163], [5, Section 1.3.3]). Using v instead of u, the latter equation reads as

$$\sigma = \left(\partial_0^{-1} C + D\right) \operatorname{Grad} v = D\left(\partial_0^{-1} D^{-1} C + 1\right) \operatorname{Grad} v,$$

where we assume that D is selfadjoint and strictly positive definite, while $C \in L(H_{\text{sym}}(\Omega))$ is arbitrary (the assumption on D can be relaxed, by requiring suitable positivity constraints on C, see, e.g., [35, Theorem 4.1]). Since $\|\partial_{0,\nu}^{-1}\| = \nu^{-1}$ we may choose $\nu_0 > 0$ large enough in order to get that $\|\partial_{0,\nu}^{-1} D^{-1} C\| < 1$ for all $\nu \geq \nu_0$. Then, by the Neumann series, we end up with

$$\left(1 + \partial_0^{-1} D^{-1} C\right)^{-1} D^{-1} \sigma = \operatorname{Grad} v.$$

Hence, our new material law operator becomes

$$\begin{aligned} M(z) &= \begin{pmatrix} 1 & 0 \\ 0 & z(1 + zD^{-1}C)^{-1} D^{-1} \end{pmatrix} \\ &= \begin{pmatrix} 1 & 0 \\ 0 & 0 \end{pmatrix} + z \begin{pmatrix} 0 & 0 \\ 0 & D^{-1} \end{pmatrix} + \sum_{k=1}^{\infty} (-1)^k z^{k+1} \begin{pmatrix} 0 & 0 \\ 0 & \left(D^{-1}C\right)^k D^{-1} \end{pmatrix}, \end{aligned}$$

which satisfies the condition (2.3) for $z \in B(r, r)$, where $r > 0$ is chosen sufficiently small.

Another way to model materials with memory is to add a convolution term on the right-hand side of Hooke's law (2.7), see, e.g., [14, Section III.7], [12]. The new constitutive relation then reads as

$$\sigma = C \operatorname{Grad} u - k * \operatorname{Grad} u = \partial_0^{-1} (C - k*) \operatorname{Grad} v,$$

where $k : \mathbb{R}_{\geq 0} \to L(H_{\text{sym}}(\Omega))$ is a strongly measurable function satisfying $\int_0^\infty \|k(t)\| e^{-\mu t}\, \mathrm{d}t < \infty$ for some $\mu \geq 0$ (note that in [12] k was assumed to be absolutely continuous). Again, choosing $\nu > 0$ large enough, we end up with a material law operator (see [51, Subsection 4.1])

$$M(z) = \begin{pmatrix} 1 & 0 \\ 0 & C^{-\frac{1}{2}}(1-\sqrt{2\pi}C^{-\frac{1}{2}}\,\widehat{k}(-\mathrm{i}z^{-1})C^{-\frac{1}{2}})^{-1}C^{-\frac{1}{2}} \end{pmatrix}, \tag{2.9}$$

where $\widehat{k}$ denotes the Fourier transform of k and where we have used that

$$\mathcal{L}_\nu\,(k*)\,\mathcal{L}_\nu^* = \sqrt{2\pi}\,\widehat{k}(\mathrm{m}-\mathrm{i}\nu)$$

for $\nu \geq \mu$ (see, e.g., [51, Lemma 3.4]). According to the solution theory presented in Theorem 2.5, we have to find suitable assumptions on the kernel k in order to obtain the positivity condition (2.3). This is done in the following theorem.

Theorem 2.9 (Integro-differential equations, [51]). *Let H be a separable Hilbert space and $k:\mathbb{R}_{\geq 0}\to L(H)$ a strongly measurable function satisfying $\int_0^\infty \|k(t)\|e^{-\mu t} < \infty$ for some $\mu \in \mathbb{R}$. If*

(a) *$k(t)$ is selfadjoint for almost every $t \in \mathbb{R}_{\geq 0}$,*
(b) *there exists $d \geq 0, \nu_0 \geq \mu$ such that*

$$t\,\mathfrak{Im}\,\widehat{k}(t-\mathrm{i}\nu_0) \leq d$$

for all $t \in \mathbb{R}$,

then there exists $r > 0$ such that the material law $M(z) := 1 + \sqrt{2\pi}\,\widehat{k}(-\mathrm{i}z^{-1})$ with $z \in B(r,r)$ satisfies (2.3). If, in addition, k satisfies

(c) *$k(t)k(s) = k(s)k(t)$ for almost every $s,t \in \mathbb{R}_{\geq 0}$,*

then there exists $r' > 0$ such that the material law $\widetilde{M}(z) := \left(1-\sqrt{2\pi}\,\widehat{k}(-\mathrm{i}z^{-1})\right)^{-1}$ with $z \in B(r',r')$ satisfies (2.3).

Remark 2.10. For real-valued kernels k, a typical assumption is that k should be non-negative and non-increasing (see, e.g., [38]). This, however implies the assumptions (a)–(c) of Theorem 2.9 for k. Moreover, kernels k of bounded variation satisfy the assumptions of the latter theorem (see [51, Remark 3.6]).

Using Theorem 2.9 we get that M given by (2.9) satisfies the solvability condition (2.3), if k satisfies the assumptions (a)–(c) of Theorem 2.9 and $k(t)$ and C commute for almost every $t \in \mathbb{R}_{\geq 0}$ (see [51, Subsection 4.1] for a detailed study).

Convolutions as discussed in the previous theorem need to be incorporated due to the fact that, for instance, the elastic behavior of a solid body depends on the stresses the body experienced in the past. One also speaks of so-called memory effects. From a similar type of nature is the modeling of material behavior with the help of fractional (time-)derivatives. In fact, in recent years, material laws for visco-elastic solids were described by using fractional derivatives (see, e.g., [5, 28]) as an ansatz to better approximate a polynomial in ∂_0^{-1} by potentially fewer terms of real powers of ∂_0^{-1}. In [61] a model for visco-elasticity with fractional derivatives has been analyzed mainly in the context of homogenization issues, which will be discussed below. For the moment, we stick to the presentation of the model and sketch the idea of the well-posedness result for this type of equation presented in [61, Theorem 2.1 and 2.2].

Example 2.11 (Visco-elasticity with fractional derivatives). In this model the Kelvin–Voigt model (2.8) is replaced by a fractional analogue of the form

$$\sigma = C \operatorname{Grad} u + D \operatorname{Grad} \partial_0^\alpha u,$$

for some $\alpha \in]0,1]$. We emphasize here that ∂_0^α has a natural meaning as a function of a normal operator. We refer to [35, Subsection 2.1] for a comparison with classical notions of fractional derivatives (see, e.g., [37] for an introduction to fractional derivatives). As in the case $\alpha = 1$ we get that ∂_0^α is boundedly invertible for each $\alpha \in]0,1]$ and $\nu > 0$ and by [35, Lemma 2.1] we can estimate the norm of its inverse by

$$\|\partial_{0,\nu}^{-\alpha}\| \leq \nu^{-\alpha}.$$

Again we assume for simplicity that D is selfadjoint and strictly positive definite, and thus we may rewrite the constitutive relation above as

$$\sigma = (C + D\partial_0^\alpha) \operatorname{Grad} u = D\partial_0^\alpha \left(\partial_0^{-\alpha} D^{-1} C + 1\right) \operatorname{Grad} u$$

yielding, for large enough $\nu > 0$,

$$\partial_0^{1-\alpha} \left(\partial_0^{-\alpha} D^{-1} C + 1\right)^{-1} D^{-1} \sigma = \operatorname{Grad} v.$$

Hence, our material law operator is given by

$$\begin{aligned} &\begin{pmatrix} 1 & 0 \\ 0 & z^\alpha \left(1 + z^\alpha D^{-1} C\right)^{-1} D^{-1} \end{pmatrix} \\ &\quad = \begin{pmatrix} 1 & 0 \\ 0 & z^\alpha D^{-1} \end{pmatrix} + \sum_{k=1}^\infty (-1)^k z^{(k+1)\alpha} \begin{pmatrix} 0 & 0 \\ 0 & \left(D^{-1} C\right)^k D^{-1} \end{pmatrix}. \end{aligned} \tag{2.10}$$

Note that the visco-elastic model under consideration is slightly different from the one treated in [61] as there is no further restriction on the parameter α. In [61], we assumed $\alpha \geq \frac{1}{2}$ and showed the positive definiteness of the sum in (2.10) with the help of a perturbation argument. This argument does not apply to the situation discussed here. However, assuming selfadjointness and non-negativity of the operators C and D well-posedness can be warranted even for $\alpha < \frac{1}{2}$.

More generally, if one considers material law operators containing fractional derivatives of the form

$$M(\partial_0^{-1}) = M_0 + \sum_{\alpha \in \Pi} \partial_0^{-\alpha} M_\alpha + \partial_0^{-1} M_1, \tag{2.11}$$

where $\Pi \subseteq\ ]0,1[$ is finite and $M_\alpha \in L(H)$ for some Hilbert space H and each $\alpha \in \{0,1\} \cup \Pi$, one imposes the following conditions on the operators M_α in order to get an estimate of the form (2.3) for the material law (2.11):

Theorem 2.12 ([35, Theorem 3.5]). *Let $(\alpha_0, \ldots, \alpha_k)$ be a monotonically increasing enumeration of Π. Assume that the operators M_0 and M_{α_j} are selfadjoint for each $j \in \{0, \ldots, k\}$. Moreover, let $P, Q, F \in L(H)$ be three orthogonal projectors satisfying*

$$P + Q + F = 1,$$

and assume that M_0 and M_{α_j} commute with P, Q and F for every $j \in \{1, \ldots, k\}$. If $PM_{\alpha_j}P \geq 0$, $QM_{\alpha_j}Q \geq 0$, $M_0 \geq 0$ and $M_0, \mathfrak{Re}\, M_1$ and M_{α_0} are strictly positive definite on the ranges of P, Q and F respectively, then the material law (2.11) *satisfies the solvability condition* (2.3).

With this theorem we end our tour through different kinds of evolutionary equations, which are all covered by the solution theory stated in Theorem 2.5.

Besides the well-posedness of evolutionary equations, it is also possible to derive a criterion for (exponential) stability in the abstract setting of Theorem 2.5. Since the systems under consideration do not have any regularizing property, we are not able to define exponential stability as it is done classically, since our solutions u do not have to be continuous. So, point-wise evaluation of u does not have any meaning. Indeed, the problem class discussed in Theorem 2.5 covers also purely algebraic systems, where definitely no (time-)regularity of the solutions is to be expected unless the given data is regular. Thus, we are led to define a weaker notion of exponential stability as follows.

Definition 2.13. Let $A\colon D(A) \subseteq H \to H$ be skew-selfadjoint[16] and let $M \in \mathcal{H}^\infty(B(r,r), L(H))$ satisfying (2.3) for some $r > 0$. Let $\nu > \frac{1}{2r}$. We call the operator $\left(\overline{\partial_0 M(\partial_0^{-1}) + A}\right)^{-1}$ *exponentially stable* with stability rate $\nu_0 > 0$ if for each $0 \leq \nu' < \nu_0$ and $f \in L^2_{-\nu'} \cap L^2_\nu(\mathbb{R}, H)$ we have

$$u := \left(\overline{\partial_0 M(\partial_0^{-1}) + A}\right)^{-1} f \in L^2_{-\nu'}(\mathbb{R}, H),$$

which in particular implies that $\int_\mathbb{R} e^{2\nu' t} \left|u(t)\right|^2 dt < \infty$.

As it turns out, this notion of exponential stability yields the exponential decay of the solutions, provided the solution u is regular enough. For instance, this can be achieved by assuming more regularity on the given right-hand side (see [49, Remark 3.2 (a)]). The result for exponential stability reads as follows.

Theorem 2.14 ([50, Theorem 3.2]). *Let $A\colon D(A) \subseteq H \to H$ be a skew-selfadjoint operator and M be a mapping satisfying the following assumptions for some $\nu_0 > 0$:*

(a) *$M\colon \mathbb{C} \setminus \overline{B\left(-\frac{1}{2\nu_0}, \frac{1}{2\nu_0}\right)} \to L(H)$ is analytic;*

[16] For sake of presentation, we assume A to be skew-selfadjoint. However, in [50, 49] A was assumed to be a linear maximal monotone operator. We will give a solution theory for this type of problem later on. One then might replace the condition of skew-selfadjointness in this definition and the subsequent theorem by the condition of being linear and maximal monotone.

(b) *for every* $0 < \nu' < \nu_0$ *there exists* $c > 0$ *such that for all* $z \in \mathbb{C}\setminus\overline{B\left(-\frac{1}{2\nu'}, \frac{1}{2\nu'}\right)}$ *we have*

$$\mathfrak{Re}\, z^{-1}M(z) \geq c.$$

Then for each $\nu > 0$ *the solution operator* $\left(\overline{\partial_0 M(\partial_0^{-1}) + A}\right)^{-1}$ *is exponentially stable with stability rate* ν_0.

Example 2.15 (A parabolic-hyperbolic system, [49, Example 4.2]). Let $\Omega \subseteq \mathbb{R}^n$ be an open subset and $\Omega_0, \Omega_1 \subseteq \Omega$ measurable, disjoint, non-empty with $\Omega = \Omega_0 \cup \Omega_1$, $c > 0$. Then the solution operator for the equation

$$\left(\partial_0 \begin{pmatrix} \chi_{\Omega_0} + \chi_{\Omega_1} & 0 \\ 0 & \chi_{\Omega_0} \end{pmatrix} + \begin{pmatrix} c & 0 \\ 0 & c \end{pmatrix} + \begin{pmatrix} 0 & \operatorname{div}_c \\ \operatorname{grad} & 0 \end{pmatrix}\right)\begin{pmatrix} v \\ q \end{pmatrix} = \begin{pmatrix} f \\ 0 \end{pmatrix}$$

for suitable f is exponentially stable with stability rate c.

Remark 2.16. As in [49, Initial Value Problems], the stability of the corresponding initial value problems can be discussed similarly.

Example 2.17 (Example 2.15 continued (see also [49, Theorem 4.4])). Let $h < 0$ and assume, in addition, that $c > 1$. Then the solution operator for the equation

$$\left(\partial_0 \begin{pmatrix} \chi_{\Omega_0} + \chi_{\Omega_1} & 0 \\ 0 & \chi_{\Omega_0} \end{pmatrix} + \tau_h + \begin{pmatrix} c & 0 \\ 0 & c \end{pmatrix} + \begin{pmatrix} 0 & \operatorname{div}_c \\ \operatorname{grad} & 0 \end{pmatrix}\right)\begin{pmatrix} v \\ q \end{pmatrix} = \begin{pmatrix} f \\ 0 \end{pmatrix}$$

is exponentially stable with stability rate $\nu_0 > 0$ such that

$$\nu_0 + e^{-\nu_0 h} = c.$$

Remark 2.18. We note that the exponential stability of integro-differential equations can be treated in the same way, see [49, Section 4.3].

2.4. The closedness of the problem class and homogenization

In this section we discuss the closedness of the problem class under consideration with respect to perturbations in the material law M. We will treat perturbations in the weak operator topology, which will also have strong connections to issues stemming from homogenization theory. For illustrational purposes we discuss the one-dimensional case of an elliptic type equation first.

Example 2.19 (see, e.g., [4, Example 1.1.3]). Let $A\colon \mathbb{R} \to \mathbb{R}$ be a bounded, uniformly strictly positive, measurable, 1-periodic function. We denote the multiplication operator on $L^2(]0,1[)$ associated with A by $A(\mathrm{m}_1)$. Denoting the one-dimensional gradient on $]0,1[$ with homogeneous Dirichlet boundary conditions by $\partial_{1,c}$ (see also Definition 1.9) and ∂_1 for its skew-adjoint, we consider the problem of finding $u_\varepsilon \in D(\partial_{1,c})$ such that for given $f \in L^2(]0,1[)$ and $\varepsilon > 0$ we have

$$-\partial_1 A\left(\frac{1}{\varepsilon}\mathrm{m}_1\right)\partial_{1,c} u_\varepsilon = f.$$

Of course, the solvability of the latter problem is clear due to Corollary 1.6. Now, we address the question whether $(u_\varepsilon)_{\varepsilon>0}$ is convergent in any particular sense and if so, whether the limit satisfies a differential equation of "similar type". Before, however, doing so, we need the following result.

Proposition 2.20 (see, e.g., [10, Theorem 2.6]). *Let $A\colon \mathbb{R}^N \to \mathbb{C}$ be bounded, measurable and $]0,1[^N$-periodic, i.e., for all $k \in \mathbb{Z}^N$ and a.e. $x \in \mathbb{R}^N$ we have $A(x+k) = A(x)$. Then*

$$A\left(\frac{\cdot}{\varepsilon}\right) \to \int_{]0,1[^N} A(x)dx \quad (\varepsilon \to 0)$$

in the weak-$*$-topology $\sigma(L^\infty(\mathbb{R}^N), L^1(\mathbb{R}^N))$ of $L^\infty(\mathbb{R}^N)$.

Example 2.21 (Example 2.19 continued). For $\varepsilon > 0$, we define $\xi_\varepsilon := A\left(\frac{1}{\varepsilon}\mathrm{m}_1\right)\partial_{1,c}u_\varepsilon$. It is easy to see that $(\xi_\varepsilon)_\varepsilon$ is bounded in $L^2(]0,1[)$ and also in $H_1(]0,1[) = D(\partial_1)$. The Arzelà–Ascoli theorem implies that $(\xi_\varepsilon)_\varepsilon$ has a convergent subsequence (again labeled with ε), which converges in $L^2(]0,1[)$. We denote $\xi := \lim_{\varepsilon\to 0}\xi_\varepsilon$. In consequence, by Proposition 2.20, we deduce that

$$\partial_{1,c}u_\varepsilon = \frac{1}{A\left(\frac{1}{\varepsilon}\mathrm{m}_1\right)}\xi_\varepsilon \rightharpoonup \left(\int_0^1 \frac{1}{A(x)}dx\right)\xi$$

weakly in $L^2(]0,1[)$ as $\varepsilon \to 0$. Hence, $(\partial_{1,c}u_\varepsilon)_{\varepsilon>0}$ weakly converges in $L^2(]0,1[)$, which, again by compact embedding, implies that $(u_\varepsilon)_\varepsilon$ converges in $L^2(]0,1[)$. Denoting the respective limit by u, we infer

$$f = -\partial_1\xi = -\partial_1\left(\int_0^1 \frac{1}{A(x)}dx\right)^{-1}\left(\int_0^1 \frac{1}{A(x)}dx\right)\xi = -\partial_1\left(\int_0^1 \frac{1}{A(x)}dx\right)^{-1}\partial_{1,c}u.$$

Now, unique solvability of the latter equation together with a subsequence argument imply convergence of $(u_\varepsilon)_\varepsilon$ without choosing subsequences.

Remark 2.22. Note that examples in dimension 2 or higher are far more complicated. In particular, the computation of the limit (if there is one) is more involved. To see this, we refer to [10, Sections 5.4 and 6.2], where the case of so-called laminated materials and general periodic materials is discussed. In the former the limit may be expressed as certain integral means, whereas in the latter so-called local problems have to be solved to determine the effective equation. Having these issues in mind, we will only give structural (i.e., compactness) results on homogenization problems of (evolutionary) partial differential equations. In consequence, the compactness properties of the differential operators as well as the ones of the coefficients play a crucial role in homogenization theory.

Regarding Proposition 2.20, the right topology for the operators under consideration is the weak operator topology. Indeed, with the examples given in the previous section in mind and modeling local oscillations as in Example 2.19, we shall consider the weak-$*$-topology of an appropriate L^∞-space. Now, if we identify any L^∞-function with the corresponding multiplication operator on L^2, we see that convergence in the weak-$*$-topology of the functions is equivalent to convergence of the associated multiplication operator in the weak operator topology of $L(L^2)$. This general perspective also enables us to treat problems with singular perturbations and problems of mixed type.

Before stating a first theorem concerning the issues mentioned, we need to introduce a topology tailored for the case of autonomous and causal material laws.

Definition 2.23 ([55, Definition 3.1]). For Hilbert spaces H_1, H_2 and an open set $E \subseteq \mathbb{C}$, we define τ_{w} to be the initial topology on $\mathcal{H}^\infty(E, L(H_1, H_2))$ induced by the mappings

$$\mathcal{H}^\infty \ni M \mapsto (z \mapsto \langle \phi, M(z)\psi\rangle) \in \mathcal{H}(E)$$

for $\phi \in H_2, \psi \in H_1$, where $\mathcal{H}(E)$ is the set of holomorphic functions endowed with the compact open topology, i.e., the topology of uniform convergence on compact sets. We write $\mathcal{H}^\infty_{\mathrm{w}} := (\mathcal{H}^\infty, \tau_{\mathrm{w}})$ for the topological space and re-use the notation $\mathcal{H}^\infty_{\mathrm{w}}$ for the underlying set.

We note the following remarkable fact.

Theorem 2.24 ([55, Theorem 3.4], [61, Theorem 4.3]). *Let H_1, H_2 be Hilbert spaces, $E \subseteq \mathbb{C}$ open. Then*

$$B_{\mathcal{H}^\infty} := \{M \in \mathcal{H}^\infty(E, L(H_1, H_2)) \mid \|M\|_\infty \leq 1\} \subseteq \mathcal{H}^\infty_{\mathrm{w}}$$

is compact. If, in addition, H_1 and H_2 are separable, then $B_{\mathcal{H}^\infty}$ is metrizable.

Sketch of the proof. For $s \in [0, \infty[$ introduce the set $B_{\mathcal{H}(E)}(s) := \{f \in \mathcal{H}(E) | \forall z \in E : |f(z)| \leq s\}$. The proof is based on the following equality

$$B_{\mathcal{H}^\infty} = \left(\prod_{\phi \in H_1, \psi \in H_2} B_{\mathcal{H}(E)} (\|\phi\| \|\psi\|) \right)$$
$$\cap \{M \colon E \to L(H_1, H_2) \mid M(z) \text{ sesquilinear } (z \in E)\},$$

which itself follows from a Dunford type theorem ensuring the holomorphy (with values in the space $L(H_1, H_2)$) of the elements on the right-hand side and the Riesz-Fréchet representation theorem for sesqui-linear forms. Now, invoking Montel's theorem, we deduce that $B_{\mathcal{H}(E)}(s)$ is compact for every $s \in [0, \infty[$. Thus, Tikhonov's theorem applies to deduce the compactness of $B_{\mathcal{H}^\infty}$. The proof for metrizability is standard. □

Recall for $r, c \in \mathbb{R}_{>0}$, and a Hilbert space H, we set

$$\mathcal{H}^{\infty,c}(B(r,r), L(H)) = \left\{ M \in \mathcal{H}^\infty(B(r,r), L(H)) \mid \bigwedge_{z \in B(r,r)} \Re z^{-1} M(z) \geq c \right\}.$$

In accordance to Definition 2.23, we will also write $\mathcal{H}^{\infty,c}_{\mathrm{w}}$ for the set $\mathcal{H}^{\infty,c}$ endowed with τ_{w}. The compactness properties from $\mathcal{H}^\infty_{\mathrm{w}}$ are carried over to $\mathcal{H}^{\infty,c}_{\mathrm{w}}$. The latter follows from the following proposition:

Proposition 2.25 ([60, Proposition 1.3]). *Let $c \in \mathbb{R}_{>0}$. Then the set $\mathcal{H}^{\infty,c}_{\mathrm{w}} \subseteq \mathcal{H}^\infty_{\mathrm{w}}$ is closed.*

We are now ready to discuss a first theorem on the continuous dependence on the coefficients for autonomous and causal material laws, which particularly covers a class of homogenization problems in the sense mentioned above. For a linear operator A in some Hilbert space H, we denote $D(A)$ endowed with the graph norm of A by D_A. If a Hilbert space H_1 is compactly embedded in H, we write $H_1 \hookrightarrow\hookrightarrow H$. A subset $M \subseteq \mathcal{H}^\infty$ is called *bounded*, if there is $\lambda > 0$ such that $M \subseteq \lambda B_{\mathcal{H}^\infty}$. The result reads as follows.

Theorem 2.26 ([60, Theorem 3.5], [61, Theorem 4.1]). *Let $\nu, c \in \mathbb{R}_{>0}$, $r > \frac{1}{2\nu}$, $(M_n)_n$ be a bounded sequence in $\mathcal{H}^{\infty,c}(B(r,r), L(H))$, $A\colon D(A) \subseteq H \to H$ skew-selfadjoint. Assume that $D_A \hookrightarrow\hookrightarrow H$. Then there exists a subsequence of $(M_n)_n$ such that $(M_{n_k})_k$ converges in $\mathcal{H}^\infty_{\mathrm{w}}$ to some $M \in \mathcal{H}^{\infty,c}$ and*

$$\left(\overline{\partial_0 M_{n_k}(\partial_0^{-1}) + A}\right)^{-1} \to \left(\overline{\partial_0 M(\partial_0^{-1}) + A}\right)^{-1}$$

in the weak operator topology.

We first apply this theorem to an elliptic type equation.

Example 2.27. Let $\Omega \subseteq \mathbb{R}^n$ be open and bounded. Let grad_c and div be the operators introduced in Definition 1.9. Let $(a_k)_{k\in\mathbb{N}}$ be a sequence of uniformly strictly positive bounded linear operators in $L^2(\Omega)^n$. For $f \in L^2(\Omega)$ consider for $k \in \mathbb{N}$ the problem of finding $u_k \in L^2(\Omega)$ such that the equation

$$u_k - \operatorname{div} a_k \operatorname{grad}_c u_k = f$$

holds. Observe that if $\iota\colon R(\operatorname{grad}_c) \to L^2(\Omega)^n$ denotes the canonical embedding, this equation is the same as

$$u_k - \operatorname{div} \iota\iota^* a_k \iota\iota^* \operatorname{grad}_c u_k = f. \tag{2.12}$$

Indeed, by Poincaré's inequality $R(\operatorname{grad}_c) \subseteq L^2(\Omega)^n$ is closed, the projection theorem ensures that $\iota\iota^*$ is the orthogonal projection on $R(\operatorname{grad}_c)$. Moreover, $N(\operatorname{div}) = R(\operatorname{grad}_c)^\perp$ yields that $\operatorname{div} = \operatorname{div}(\iota\iota^* + (1 - \iota\iota^*)) = \operatorname{div} \iota\iota^*$. Now, we realize that due to the positive definiteness of a_k so is $\iota^* a_k \iota$. Consequently, the latter operator is continuously invertible. Introducing $v_k := \iota^* a_k \iota \operatorname{grad}_c u_k$ for $k \in \mathbb{N}$, we rewrite the equation (2.12) as follows:

$$\left(\begin{pmatrix} 1 & 0 \\ 0 & (\iota^* a_k \iota)^{-1} \end{pmatrix} - \begin{pmatrix} 0 & \operatorname{div} \iota \\ \iota^* \operatorname{grad}_c & 0 \end{pmatrix}\right) \begin{pmatrix} u_k \\ v_k \end{pmatrix} = \begin{pmatrix} f \\ 0 \end{pmatrix}.$$

Now, let $\nu > 0$ and lift the above problem to the space $L^2_\nu(\mathbb{R}, L^2(\Omega) \oplus R(\operatorname{grad}_c))$ by interpreting $\begin{pmatrix} f \\ 0 \end{pmatrix}$ as $\left(t \mapsto \chi_{\mathbb{R}_{>0}}(t) \begin{pmatrix} f \\ 0 \end{pmatrix}\right) \in L^2_\nu(\mathbb{R}, L^2(\Omega) \oplus R(\operatorname{grad}_c))$. Then this equation fits into the solution theory stated in Theorem 2.5 with

$$M_k(\partial_0^{-1}) := \partial_0^{-1} \begin{pmatrix} 1 & 0 \\ 0 & (\iota^* a_k \iota)^{-1} \end{pmatrix}, \quad A := \begin{pmatrix} 0 & \operatorname{div} \iota \\ \iota^* \operatorname{grad}_c & 0 \end{pmatrix}.$$

Note that the skew-selfadjointness of A is easily obtained from $\operatorname{div}^* = -\operatorname{grad}_c$. In order to conclude the applicability of Theorem 2.26, we need the following observation.

Proposition 2.28 ([60, Lemma 4.1]). *Let H_1, H_2 be Hilbert spaces, $C\colon D(C) \subseteq H_1 \to H_2$ densely defined, closed, linear. Assume that $D_C \hookrightarrow\hookrightarrow H_1$. Then $D_{C^*} \cap N(C^*)^{\perp_{H_2}} \hookrightarrow\hookrightarrow H_2$.*

Example 2.29 (Example 2.27 continued). With the help of the theorem of Rellich–Kondrachov and Proposition 2.28, we deduce that A has compact resolvent. Thus, Theorem 2.26 is applicable.

We find a subsequence such that $\mathfrak{a} := \tau_{\mathrm{w}} - \lim_{l\to\infty} \left(\iota^* a_{k_l} \iota\right)^{-1}$ exists, where we denoted by τ_{w} the weak operator topology. Therefore, $(u_{k_l})_l$ weakly converges to some u, which itself is the solution of

$$u - \operatorname{div} \iota \mathfrak{a}^{-1} \iota^* \operatorname{grad}_c u = f.$$

In fact it is possible to show that $\iota\mathfrak{a}^{-1}\iota^*$ coincides with the usual homogenized matrix (if the possibly additional assumptions on the sequence $(a_k)_k$ permit the computation of a limit in the sense of H- or G-convergence, see, e.g., [10, Chapter 13] and the references therein).

As a next example let us consider the heat equation.

Example 2.30. Recall the heat equation introduced in Example 2.6:

$$\left(\partial_0 \begin{pmatrix} 1 & 0 \\ 0 & 0 \end{pmatrix} + \begin{pmatrix} 0 & 0 \\ 0 & k^{-1} \end{pmatrix} + \begin{pmatrix} 0 & \operatorname{div} \\ \operatorname{grad}_c & 0 \end{pmatrix}\right) \begin{pmatrix} \vartheta \\ q \end{pmatrix} = \begin{pmatrix} f \\ 0 \end{pmatrix}.$$

To warrant the compactness condition in Theorem 2.26, we again assume that the underlying domain Ω is bounded. Similarly to Example 2.27, we assume that we are given $(k_l)_l$, a bounded sequence of uniformly strictly monotone linear operators in $L(L^2(\Omega)^n)$. Consider the sequence of equations

$$\left(\partial_0 \begin{pmatrix} 1 & 0 \\ 0 & 0 \end{pmatrix} + \begin{pmatrix} 0 & 0 \\ 0 & k_l^{-1} \end{pmatrix} + \begin{pmatrix} 0 & \operatorname{div} \\ \operatorname{grad}_c & 0 \end{pmatrix}\right) \begin{pmatrix} \vartheta_l \\ q_l \end{pmatrix} = \begin{pmatrix} f \\ 0 \end{pmatrix}.$$

Now, focussing *only* on the behavior of the temperature $(\vartheta_l)_l$, we can proceed as in the previous example.

Assuming more regularity of Ω, e.g., the segment property and finitely many connected components, we can apply Theorem 2.26 also to the corresponding homogeneous Neumann problems of Examples 2.27 and 2.30. Moreover, the aforementioned theorem can also be applied to the homogenization of (visco-)elastic problems (see also Example 2.8). For this we need criteria ensuring the compactness condition $D_{\mathrm{Grad}_c} \hookrightarrow\hookrightarrow L^2(\Omega)^n$ (or $D_{\mathrm{Grad}} \hookrightarrow\hookrightarrow L^2(\Omega)^n$). The latter is warranted for a bounded Ω for the homogeneous Dirichlet case or an Ω satisfying suitable geometric requirements (see, e.g., [64]). An example of a different type of nature is that of Maxwell's equations:

Example 2.31. Recall Maxwell's equation as introduced in Example 2.7:

$$\left(\partial_0 \begin{pmatrix} \varepsilon & 0 \\ 0 & \mu \end{pmatrix} + \begin{pmatrix} \sigma & 0 \\ 0 & 0 \end{pmatrix} + \begin{pmatrix} 0 & -\operatorname{curl} \\ \operatorname{curl}_c & 0 \end{pmatrix}\right) \begin{pmatrix} E \\ H \end{pmatrix} = \begin{pmatrix} J \\ 0 \end{pmatrix}.$$

In this case, we also want to consider sequences $(\varepsilon_n)_n, (\mu_n)_n, (\sigma_n)_n$ and corresponding solutions $(E_n, H_n)_n$. In any case the nullspaces of both curl_c and curl are infinite dimensional. Thus, the projection mechanism introduced above for the heat and the elliptic equation cannot apply in the same manner. Moreover, considering the Maxwell's equations on the nullspace of $\begin{pmatrix} 0 & -\operatorname{curl} \\ \operatorname{curl}_c & 0 \end{pmatrix}$, we realize that the equation amounts to be an *ordinary* differential equation in an *infinite-dimensional* state space. For the latter we have not stated any homogenization or continuous dependence result yet. Thus, before dealing with Maxwell's equations in full generality, we focus on ordinary (integro-)differential equations next.

Theorem 2.32 ([56, Theorem 4.4]). *Let $\nu, \varepsilon \in \mathbb{R}_{>0}$, $r > \frac{1}{2\nu}$, $(M_n)_n$ in $\mathcal{H}^{\infty,c}(B(r,r), L(H)) \cap \mathcal{H}^\infty(B(0,\varepsilon), L(H))$ bounded, H separable Hilbert space. Assume that*[17]

$$M_n(0) \geq c \text{ on } R(M_n(0)) = R(M_1(0))$$

for all $n \in \mathbb{N}$. Then there exists a subsequence $(n_k)_k$ of $(n)_n$ and some $M \in \mathcal{H}^\infty$ such that

$$\left(\partial_0 M_{n_k}(\partial_0^{-1})\right)^{-1} \to \left(\partial_0 M(\partial_0^{-1})\right)^{-1}$$

in the weak operator topology.

Remark 2.33. Note that in the latter theorem, in general, the sequence $\left(M_{n_k}(\partial_0^{-1})\right)_k$ does *not* converge to $M(\partial_0^{-1})$. The reason for that is that the computation of the inverse is not continuous in the weak operator topology. So, even if one chose a further subsequence $(n_{k_l})_l$ of $(n_k)_k$ such that $\left(M_{n_{k_l}}(\partial_0^{-1})\right)_l$ converges in the weak operator topology, then, in general,

$$M_{n_{k_l}}(\partial_0^{-1}) \nrightarrow M(\partial_0^{-1})$$

in τ_{w}. Indeed, the latter can be seen by considering the periodic extensions of the mappings a^1, a^2 to all of $\mathbb{R}$ with

$$a^1(x) := \begin{cases} \frac{1}{2}, & 0 \leq x < \frac{1}{2}, \\ 1, & \frac{1}{2} \leq x < 1, \end{cases} \quad a^2(x) := \frac{3}{4}, \quad (x \in [0,1]).$$

We let $a_n := a^1(n\cdot)$ for odd $n \in \mathbb{N}$ and $a_n := a^2(n\cdot)$ if $n \in \mathbb{N}$ is even. Then, by Proposition 2.20, we conclude that $a_n \to \frac{3}{4}$, $a_{2n+1}^{-1} \to \frac{3}{2}$, and $a_{2n}^{-1} \to \frac{4}{3}$ as $n \to \infty$ in $\sigma(L^\infty, L^1)$.

In a way complementary to the latter theorem is the following. The latter theorem assumes analyticity of the M_n's at 0. But the zeroth-order term in the power series expansion of the M_n's may be non-invertible. In the next theorem,

[17]Note that $M_n \in \mathcal{H}^{\infty,c}(B(r,r), L(H)) \cap \mathcal{H}^\infty(B(0,\varepsilon), L(H))$ implies that $M_n(0)$ is selfadjoint.

the analyticity at 0 is not assumed any more. The (uniform) positive definiteness condition, however, is more restrictive.

Theorem 2.34 ([55, Theorem 5.2]). *Let $\nu, \varepsilon \in \mathbb{R}_{>0}$, $r > \frac{1}{2\nu}$, $(M_n)_n$ in $\mathcal{H}^{\infty,c}(B(r,r), L(H))$ bounded, H separable Hilbert space. Assume that*

$$\mathfrak{Re}\, M_n(z) \geq c \quad (z \in B(r,r))$$

for all $n \in \mathbb{N}$. Then there exists a subsequence $(M_{n_k})_k$ of $(M_n)_n$ and some $M \in \mathcal{H}^\infty$ such that

$$\left(\partial_0 M_{n_k}(\partial_0^{-1})\right)^{-1} \to \left(\partial_0 M(\partial_0^{-1})\right)^{-1}$$

in the weak operator topology.

Now, we turn to more concrete examples. With the methods developed, we can characterize the convergence of a particular ordinary equation. In a slightly more restrictive context these types of equations have been discussed by Tartar in 1989 (see [43, 42]) using the notion of Young measures, see also the discussion in [59, Remark 3.8].

Proposition 2.35. *Let $(a_n)_n$ in $L(H)$ be bounded, H a separable Hilbert space, $\nu > 2\sup_{n\in\mathbb{N}} \|a_n\| + 1$. Then*

$$\left((\partial_0 + a_n)^{-1}\right)_n$$

converges in the weak operator topology if and only if for all $\ell \in \mathbb{N}$

$$\left(a_n^\ell\right)_n$$

converges in the weak operator topology to some $b_\ell \in L(H)$. In the latter case $\left((\partial_0 + a_n)^{-1}\right)_n$ converges to

$$\left(\partial_0 + \partial_0 \sum_{j=1}^{\infty} \left(-\sum_{\ell=1}^{\infty} \left(-\partial_0^{-1}\right)^\ell b_\ell\right)^j\right)^{-1}$$

in the weak operator topology.

Proof. The 'if'-part is a straightforward application of a Neumann series expansion of $(\partial_0 + a_n)^{-1}$, see, e.g., [63, Theorem 2.1]. The 'only-if'-part follows from the representation

$$(\partial_0 + a_n)^{-1} = \sum_{j=0}^{\infty} \left(-\partial_0^{-1}\right)^j a_n^j \partial_0^{-1} =: M_n(\partial_0^{-1}) \quad (n \in \mathbb{N}),$$

the application of the Fourier–Laplace transform and Cauchy's integral formulas for the derivatives of holomorphic functions. For the latter argument note that $(M_n)_n$ is a bounded sequence in $\mathcal{H}^\infty(B(r,r), L(H))$ for $r > \frac{1}{2\nu}$ and, thus, contains a $\mathcal{H}^\infty_{\mathrm{w}}$-convergent subsequence, whose limit M satisfies $M(\partial_0^{-1}) = \tau_{\mathrm{w}} - \lim_{n\to\infty} (\partial_0 + a_n)^{-1}$. □

One might wonder under which circumstances the conditions in the latter theorem happen to be satisfied. We discuss the following example initially studied by Tartar.

Example 2.36 (Ordinary differential equations). Let $a \in L^\infty(\mathbb{R})$. If a is 1-periodic then $a(n\cdot)$ converges to $\int_0^1 a$ in the $\sigma(L^\infty, L^1)$-topology. Regard a as a multiplication operator $a(\mathrm{m}_1)$ on $L^2(\mathbb{R})$. Now, we have the explicit formula

$$(\partial_0 + a(n\mathrm{m}_1))^{-1} \overset{\tau_w}{\to} \left(\partial_0 + \partial_0 \sum_{j=1}^{\infty} \left(- \sum_{\ell=1}^{\infty} (-\partial_0^{-1})^\ell \int_0^1 a^\ell \right)^j \right)^{-1}.$$

We should remark here that the classical approach to this problem uses the theory of Young measures to express the limit equation. This is not needed in our approach.

With the latter example in mind, we now turn to the discussion of a general theorem also working for Maxwell's equation. As mentioned above, these equations can be reduced to the cases of Theorem 2.26 and 2.32. Consequently, the limit equations become more involved. For sake of this presentation, we do not state the explicit formulae for the limit expressions and instead refer to [56, Corollary 4.7].

Theorem 2.37 ([56, Corollary 4.7]). *Let $\nu, \varepsilon \in \mathbb{R}_{>0}$, $r > \frac{1}{2\nu}$, $(M_n)_n$ in $\mathcal{H}^{\infty,c}(B(r,r), L(H)) \cap \mathcal{H}^\infty(B(0,\varepsilon), L(H))$ bounded, $A\colon D(A) \subseteq H \to H$ skew-selfadjoint, H separable. Assume that $D_A \cap N(A)^\perp \hookrightarrow\hookrightarrow H$ and, in addition,*

$$\begin{aligned} &M_n(0) \geq c \text{ on } R(M_n(0)) = R(M_1(0)) \\ &\iota_{N(A)^\perp}^* M_n'(0) \iota_{N(A)} \left(\iota_{N(A)}^* M_n'(0) \iota_{N(A)} \right)^{-1} \\ &\quad = \iota_{N(A)^\perp}^* M_n'(0)^* \iota_{N(A)} \left(\iota_{N(A)}^* M_n'(0)^* \iota_{N(A)} \right)^{-1}, \end{aligned}$$

for all $n \in \mathbb{N}$, where $\iota_{N(A)^\perp}\colon N(M_1(0)) \cap N(A)^\perp \to H$, $\iota_{N(A)}\colon N(M_1(0)) \cap N(A) \to H$ denote the canonical embeddings. Then there exists a subsequence $(M_{n_k})_k$ of $(M_n)_n$ and $M \in \mathcal{H}^\infty(E; L(H))$ for some $0 \in E \subseteq B(0,\varepsilon)$ such that

$$\left(\overline{\partial_0 M_{n_k}(\partial_0^{-1}) + A} \right)^{-1} \to \left(\overline{\partial_0 M(\partial_0^{-1}) + A} \right)^{-1}$$

converges in the weak operator topology.

Remark 2.38. It should be noted that, similarly to the case of ordinary differential equations, in general, we do *not* have $M_{n_k}(\partial_0^{-1}) \to M(\partial_0^{-1})$ in the weak operator topology.

Before we discuss possible generalizations of the above results to the non-autonomous case, we illustrate the applicability of Theorem 2.37 to Maxwell's equations:

Example 2.39 (Example 2.31 continued). Consider

$$\left(\partial_0 \begin{pmatrix} \varepsilon_n & 0 \\ 0 & \mu_n \end{pmatrix} + \begin{pmatrix} \sigma_n & 0 \\ 0 & 0 \end{pmatrix} + \begin{pmatrix} 0 & -\operatorname{curl} \\ \operatorname{curl}_c & 0 \end{pmatrix}\right)\begin{pmatrix} E_n \\ H_n \end{pmatrix} = \begin{pmatrix} J \\ 0 \end{pmatrix}$$

for bounded sequences of bounded linear operators $(\varepsilon_n)_n, (\mu_n)_n, (\sigma_n)_n$. Assuming suitable geometric requirements on the underlying domain Ω, see, e.g., [67], we realize that the compactness condition is satisfied. Thus, we only need to guarantee the compatibility conditions: Essentially, there are two complementary cases. On the one hand, one assumes uniform strict positive definiteness of the (selfadjoint) operators $\begin{pmatrix} \varepsilon_n & 0 \\ 0 & \mu_n \end{pmatrix}$. On the other hand, we may also consider the eddy current problem, which results in $\varepsilon_n = 0$. Then, in order to apply Theorem 2.37, we have to assume selfadjointness of σ_n and the existence of some $c > 0$ such that $\sigma_n \geq c$ for all $n \in \mathbb{N}$. In this respect our homogenization theorem only works under additional assumptions on the material laws apart from (uniform) well-posedness conditions. We also remark that the limit equation is of integro-differential type, see [56, Corollary 4.7] or [66].

2.5. The non-autonomous case

The non-autonomous case is characterized by the fact that the operators $\mathcal{M}$ and $\mathcal{A}$ in (2.1) do not have to commute with the translation operators τ_h. A rather general abstract result concerning well-posedness reads as follows:

Theorem 2.40 ([62, Theorem 2.4]). *Let $\nu > 0$ and $\mathcal{M}, \mathcal{N} \in L(L^2_\nu(\mathbb{R}, H))$. Assume that there exists $M \in L(L^2_\nu(\mathbb{R}, H))$ such that*

$$\mathcal{M}\partial_{0,\nu} \subseteq \partial_{0,\nu}\mathcal{M} + M.$$

Let $\mathcal{A}\colon D(\mathcal{A}) \subseteq L^2_\nu(\mathbb{R}, H) \to L^2_\nu(\mathbb{R}, H)$ be densely defined, closed, linear and such that $\partial_{0,\nu}^{-1}\mathcal{A} \subseteq \mathcal{A}\partial_{0,\nu}^{-1}$. Assume there exists $c > 0$ such that the positivity conditions

$$\mathfrak{Re}\langle(\partial_{0,\nu}\mathcal{M} + \mathcal{N} + \mathcal{A})\,\phi|\chi_{\mathbb{R}_{\leq a}}(\mathrm{m}_0)\phi\rangle \geq c\langle\phi|\chi_{\mathbb{R}_{\leq a}}(\mathrm{m}_0)\phi\rangle$$

and

$$\mathfrak{Re}\langle((\partial_{0,\nu}\mathcal{M} + \mathcal{N})^* + \mathcal{A}^*)\,\psi|\psi\rangle \geq c\langle\psi|\psi\rangle$$

hold for all $a \in \mathbb{R}$, $\phi \in D(\partial_{0,\nu}) \cap D(\mathcal{A})$, $\psi \in D(\partial_{0,\nu}) \cap D(\mathcal{A}^*)$. Then $\mathcal{B} := \partial_{0,\nu}\mathcal{M} + \mathcal{N} + \mathcal{A}$ is continuously invertible, $\left\|\mathcal{B}^{-1}\right\| \leq \frac{1}{c}$, and the operator $\mathcal{B}^{-1}$ is causal in $L^2_\nu(\mathbb{R}, H)$.

In order to capture the main idea of this general abstract result, we consider the following special non-autonomous problem of the form

$$(\partial_{0,\nu}M_0(\mathrm{m}_0) + M_1(\mathrm{m}_0) + A)\,u = f, \tag{2.13}$$

where $\partial_{0,\nu}$ denotes the time-derivative as introduced in Subsection 2.2, and A denotes a skew-selfadjoint operator on some Hilbert space H (and its canonical extension to the space $L^2_\nu(\mathbb{R}, H)$). Moreover, $M_0, M_1 : \mathbb{R} \to L(H)$ are assumed to be strongly measurable and bounded (in symbols $M_0, M_1 \in L_s^\infty(\mathbb{R}, L(H))$) and therefore, they give rise to multiplication operators on $L_\nu^2(\mathbb{R}, H)$ by setting

$$(M_i(\mathrm{m}_0)u)(t) := M_i(t)u(t) \quad (\text{a.e. } t \in \mathbb{R})$$

for $u \in L_\nu^2(\mathbb{R}, H)$, where $\nu \geq 0$ and $i \in \{0,1\}$. Of course, the so-defined multiplication operators are bounded with

$$\|M_i(\mathrm{m}_0)\|_{L(L_\nu^2(\mathbb{R},H))} \leq |M_i|_\infty = \operatorname*{ess-sup}_{t\in\mathbb{R}} \|M_i(t)\|_{L(H)}$$

for $i \in \{0,1\}$ and $\nu \geq 0$. In order to formulate the theorem in a less cluttered way, we introduce the following hypotheses.

Hypotheses 2.41. We say that $T \in L_s^\infty(\mathbb{R}, L(H))$ satisfies the property

(a) if $T(t)$ is selfadjoint ($t \in \mathbb{R}$),
(b) if $T(t)$ is non-negative ($t \in \mathbb{R}$),
(c) if the mapping T is Lipschitz-continuous, where we denote the smallest Lipschitz-constant of T by $|T|_{\mathrm{Lip}}$, and
(d) if there exists a set $N \subseteq \mathbb{R}$ of measure zero such that for each $x \in H$ the function

$$\mathbb{R} \setminus N \ni t \mapsto T(t)x$$

is differentiable[18].

If $T \in L_s^\infty(\mathbb{R}, H)$ satisfies the hypotheses above, then for each $t \in \mathbb{R} \setminus N$ the operator

$$\begin{aligned}\dot{T}(t) : H &\to H \\ x &\mapsto (T(\cdot)x)'(t)\end{aligned}$$

becomes a selfadjoint linear operator satisfying $\|\dot{T}(t)\|_{L(H)} \leq |T|_{\mathrm{Lip}}$ for every $t \in \mathbb{R} \setminus N$ and consequently $\dot{T} \in L_s^\infty(\mathbb{R}, L(H))$. We are now able to state the well-posedness result for non-autonomous problems of the form (2.13).

Theorem 2.42 ([36, Theorem 2.13]). *Let $A\colon D(A) \subseteq H \to H$ be skew-selfadjoint and $M_0, M_1 \in L_s^\infty(\mathbb{R}, L(H))$. Furthermore, assume that M_0 satisfies the hypotheses* (a)–(d) *and that there exists a set $N_1 \subseteq \mathbb{R}$ of measure zero with $N \subseteq N_1$ such that*

$$\bigvee_{c_0>0,\nu_0>0} \ \bigwedge_{t\in\mathbb{R}\setminus N_1,\nu\geq\nu_0} \nu M_0(t) + \frac{1}{2}\dot{M}_0(t) + \mathfrak{Re}\, M_1(t) \geq c_0. \tag{2.14}$$

Then the operator $\overline{\partial_{0,\nu} M_0(\mathrm{m}_0) + M_1(\mathrm{m}_0) + A}$ is continuously invertible in $L_\nu^2(\mathbb{R}, H)$ for each $\nu \geq \nu_0$. A norm bound for the inverse is $1/c_0$. Moreover, we get that

$$(\partial_{0,\nu} M_0(\mathrm{m}_0) + M_1(\mathrm{m}_0) + A)^* = \left(\overline{M_0(\mathrm{m}_0)\,\partial_{0,\nu}^* + M_1(\mathrm{m}_0)^* - A}\right). \tag{2.15}$$

Proof. The result can easily be established, when observing that $M_0(\mathrm{m}_0)\partial_{0,\nu} \subseteq \partial_{0,\nu} M_0(\mathrm{m}_0) - \dot{M}_0(\mathrm{m}_0)$ and using Theorem 2.40. □

---

[18] If H is separable, then the strong differentiability of T on $\mathbb{R} \setminus N$ for some set N of measure zero already follows from the Lipschitz-continuity of T by Rademachers theorem.

Independently of Theorem 2.40, note that condition (2.14) is an appropriate non-autonomous analogue of the positive definiteness constraint (2.3) in the autonomous case. With the help of (2.14) one can prove that the operator $\partial_{0,\nu}M_0(\mathrm{m}_0)+M_1(\mathrm{m}_0)+A$ is strictly monotone and after establishing the equality (2.15), the same argumentation works for the adjoint. Hence, the well-posedness result may also be regarded as a consequence of Corollary 1.3.

Example 2.43. As an illustrating example for the applicability of Theorem 2.42 we consider a non-autonomous evolutionary problem, which changes its type in time and space. Let $\nu > 0$. Consider the $(1+1)$-dimensional wave equation:

$$\partial_{0,\nu}^2 u - \partial_1^2 u = f \text{ on } \mathbb{R}\times\mathbb{R}.$$

As usual we rewrite this equation as a first-order system of the form

$$\left(\partial_{0,\nu}\begin{pmatrix}1&0\\0&1\end{pmatrix}+\begin{pmatrix}0&-\partial_1\\-\partial_1&0\end{pmatrix}\right)\begin{pmatrix}u\\v\end{pmatrix}=\begin{pmatrix}\partial_{0,\nu}^{-1}f\\0\end{pmatrix}. \tag{2.16}$$

In this case we can compute the solution by Duhamel's formula in terms of the unitary group generated by the skew-selfadjoint operator

$$\begin{pmatrix}0&-\partial_1\\-\partial_1&0\end{pmatrix}.$$

Let us now, based on this, consider a slightly more complicated situation, which is, however, still autonomous:

$$\begin{gathered}\left(\partial_{0,\nu}\begin{pmatrix}\chi_{\mathbb{R}\setminus]-\varepsilon,0[}(\mathrm{m}_1)&0\\0&\chi_{\mathbb{R}\setminus]-\varepsilon,\varepsilon[}(\mathrm{m}_1)\end{pmatrix}+\begin{pmatrix}\chi_{]-\varepsilon,0[}(\mathrm{m}_1)&0\\0&\chi_{]-\varepsilon,\varepsilon[}(\mathrm{m}_1)\end{pmatrix}\right.\\ \left.+\begin{pmatrix}0&-\partial_1\\-\partial_1&0\end{pmatrix}\right)\begin{pmatrix}u\\v\end{pmatrix}=\begin{pmatrix}\partial_{0,\nu}^{-1}f\\0b\end{pmatrix},\end{gathered} \tag{2.17}$$

where $\chi_I(\mathrm{m}_1)$ denotes the spatial multiplication operator with the cut-off function χ_I, given by $(\chi_I(\mathrm{m}_1)f)(t,x)=\chi_I(x)f(t,x)$ for almost every $(t,x)\in\mathbb{R}\times\mathbb{R}$, every $f\in L_\nu^2(\mathbb{R},L^2(\mathbb{R}))$ and $I\subseteq\mathbb{R}$. Hence, (2.17) is an equation of the form (2.13) with

$$M_0(\mathrm{m}_0):=\begin{pmatrix}\chi_{\mathbb{R}\setminus]-\varepsilon,0[}(\mathrm{m}_1)&0\\0&\chi_{\mathbb{R}\setminus]-\varepsilon,\varepsilon[}(\mathrm{m}_1)\end{pmatrix}$$

and

$$M_1(\mathrm{m}_0):=\begin{pmatrix}\chi_{]-\varepsilon,0[}(\mathrm{m}_1)&0\\0&\chi_{]-\varepsilon,\varepsilon[}(\mathrm{m}_1)\end{pmatrix}$$

and both are obviously not time-dependent. Note that our solution condition (2.14) is satisfied and hence, problem (2.17) is well posed in the sense of Theorem 2.42.[19] By the dependence of the operators $M_0(\mathrm{m}_0)$ and $M_1(\mathrm{m}_0)$ on the spatial parameter, we see that (2.17) changes its type from hyperbolic to elliptic to parabolic and back

[19] Indeed, the well-posedness already follows from Theorem 2.5, since M is autonomous and satisfies (2.3).

to hyperbolic and so standard semigroup techniques are not at hand to solve the equation. Indeed, in the subregion $]-\varepsilon,0[$ the problem reads as

$$\begin{pmatrix} u \\ v \end{pmatrix} + \begin{pmatrix} 0 & -\partial_1 \\ -\partial_1 & 0 \end{pmatrix}\begin{pmatrix} u \\ v \end{pmatrix} = \begin{pmatrix} \partial_{0,\nu}^{-1} f \\ 0 \end{pmatrix},$$

which may be rewritten as an elliptic equation for u of the form

$$u - \partial_1^2 u = \partial_{0,\nu}^{-1} f.$$

For the region $]0,\varepsilon[$ we get

$$\left(\partial_{0,\nu}\begin{pmatrix} 1 & 0 \\ 0 & 0 \end{pmatrix} + \begin{pmatrix} 0 & 0 \\ 0 & 1 \end{pmatrix} + \begin{pmatrix} 0 & -\partial_1 \\ -\partial_1 & 0 \end{pmatrix}\right)\begin{pmatrix} u \\ v \end{pmatrix} = \begin{pmatrix} \partial_{0,\nu}^{-1} f \\ 0 \end{pmatrix},$$

which yields a parabolic equation for u of the form

$$\partial_{0,\nu} u - \partial_1^2 u = \partial_{0,\nu}^{-1} f.$$

In the remaining sub-domain $\mathbb{R}\setminus]-\varepsilon,\varepsilon[$ the problem is of the original form (2.16), which corresponds to a hyperbolic problem for u.
To turn this into a genuinely time-dependent problem we now make a modification to problem (2.17). We define the function

$$\varphi(t) := \begin{cases} 0 & \text{if } t \leq 0, \\ t & \text{if } 0 < t \leq 1, \\ 1 & \text{if } 1 < t \end{cases} \quad (t \in \mathbb{R})$$

and consider the material-law operator

$$M_0(\mathrm{m}_0) = \varphi(\mathrm{m}_0)\begin{pmatrix} \chi_{\mathbb{R}\setminus]-\varepsilon,0[}(\mathrm{m}_1) & 0 \\ 0 & \chi_{\mathbb{R}\setminus]-\varepsilon,\varepsilon[}(\mathrm{m}_1) \end{pmatrix},$$

which now also degenerates in time. Moreover we modify $M_1(\mathrm{m}_0)$ by adding a time-dependence of the form

$$\begin{aligned} &M_1(\mathrm{m}_0) \\ &= \begin{pmatrix} \chi_{]-\infty,0[}(\mathrm{m}_0) + \chi_{[0,\infty[}(\mathrm{m}_0)\chi_{]-\varepsilon,0[}(\mathrm{m}_1) & 0 \\ 0 & \chi_{]-\infty,0[}(\mathrm{m}_0) + \chi_{[0,\infty[}(\mathrm{m}_0)\chi_{]-\varepsilon,\varepsilon[}(\mathrm{m}_1) \end{pmatrix}. \end{aligned}$$

We show that this time-dependent material law still satisfies our solvability condition. Note that

$$\varphi'(t) = \begin{cases} 1 & \text{if } t \in]0,1[, \\ 0 & \text{otherwise} \end{cases}$$

and thus, for $t \leq 0$ we have

$$\nu M_0(t) + \frac{1}{2}\dot{M}_0(t) + \mathfrak{Re}\, M_1(t) = \begin{pmatrix} 1 & 0 \\ 0 & 1 \end{pmatrix} \geq 1.$$

For $0 < t \leq 1$ we estimate

$$\nu M_0(t) + \frac{1}{2}\dot{M}_0(t) + \mathfrak{Re}\, M_1(t)$$
$$= \left(\frac{1}{2} + \nu t\right)\begin{pmatrix} \chi_{\mathbb{R}\setminus]-\varepsilon,0[}(\mathrm{m}_1) & 0 \\ 0 & \chi_{\mathbb{R}\setminus]-\varepsilon,\varepsilon[}(\mathrm{m}_1)\end{pmatrix} + \begin{pmatrix} \chi_{]-\varepsilon,0[}(\mathrm{m}_1) & 0 \\ 0 & \chi_{]-\varepsilon,\varepsilon[}(\mathrm{m}_1)\end{pmatrix} \geq \frac{1}{2}$$

and, finally, for $t > 1$ we obtain that

$$\nu M_0(t) + \frac{1}{2}\dot{M}_0(t) + \mathfrak{Re}\, M_1(t)$$
$$= \nu\begin{pmatrix} \chi_{\mathbb{R}\setminus]-\varepsilon,0[}(\mathrm{m}_1) & 0 \\ 0 & \chi_{\mathbb{R}\setminus]-\varepsilon,\varepsilon[}(\mathrm{m}_1)\end{pmatrix} + \begin{pmatrix} \chi_{]-\varepsilon,0[}(\mathrm{m}_1) & 0 \\ 0 & \chi_{]-\varepsilon,\varepsilon[}(\mathrm{m}_1)\end{pmatrix} \geq \min\{\nu, 1\}.$$

There is also an adapted result on the closedness of the problem class for the non-autonomous situation. The case $\mathcal{A} = 0$ is thoroughly discussed in [59]. We give the corresponding result for the situation where $\mathcal{A}$ is non-zero and satisfies a certain compactness condition.

Theorem 2.44 ([58, Theorem 3.1]). *Let $\nu > 0$. Let $(\mathcal{M}_n)_n$ be a bounded sequence in $L(L^2_\nu(\mathbb{R}, H))$ such that $([\mathcal{M}_n, \partial_{0,\nu}])_n$ is bounded in $L\big(L^2_\nu(\mathbb{R}, H)\big)$. Moreover, let $\mathcal{A}\colon D(\mathcal{A}) \subseteq L^2_\nu(\mathbb{R}, H) \to L^2_\nu(\mathbb{R}, H)$ be linear and maximal monotone commuting with $\partial_{0,\nu}$ and assume that $\mathcal{M}_n$ is causal for each $n \in \mathbb{N}$. Moreover, assume the positive definiteness conditions*

$$\mathfrak{Re}\left\langle \partial_{0,\nu}\mathcal{M}_n u | \chi_{\mathbb{R}_{\leq a}}(\mathrm{m}_0)u\right\rangle \geq c\left\langle u|\chi_{\mathbb{R}_{\leq a}}(\mathrm{m}_0)u\right\rangle, \quad \mathfrak{Re}\left\langle \mathcal{A}u|\chi_{\mathbb{R}_{\leq 0}}(\mathrm{m}_0)u\right\rangle \geq 0 \tag{2.18}$$

for all $u \in D(\partial_{0,\nu}) \cap D(\mathcal{A})$, $a \in \mathbb{R}$, $n \in \mathbb{N}$ and some $c > 0$.

Assume that there exists a Hilbert space K such that $K \hookrightarrow\hookrightarrow H$ and $D_{\mathcal{A}} \hookrightarrow L^2_\nu(\mathbb{R}, K)$ and that $(\mathcal{M}_n)_n$ converges in the weak operator topology to some $\mathcal{M}$.

Then $\partial_{0,\nu}\mathcal{M} + \mathcal{A}$ is continuously invertible in $L^2_\nu(\mathbb{R}, H)$ and $\left(\overline{\partial_{0,\nu}\mathcal{M}_n + \mathcal{A}}\right)^{-1} \to \left(\overline{\partial_{0,\nu}\mathcal{M} + \mathcal{A}}\right)^{-1}$ in the weak operator topology of $L^2_\nu(\mathbb{R}, H)$ as $n \to \infty$.

As in [58], we illustrate the latter theorem by the following example, being an adapted version of Example 2.43.

Example 2.45 ([58, Section 1]). Recalling the definition of ∂_1, $\partial_{1,c}$ on $L^2([0,1])$ from Definition 1.9, we treat the following system written in block operator matrix form:

$$\left(\partial_{0,\nu}\begin{pmatrix} \chi_{[0,\frac{1}{4}]\cup[\frac{1}{2},\frac{3}{4}]}(\mathrm{m}_1) & 0 \\ 0 & \chi_{[0,\frac{1}{4}]\cup[\frac{3}{4},1]}(\mathrm{m}_1)\end{pmatrix} + \begin{pmatrix} \chi_{[\frac{1}{4},\frac{1}{2}]\cup[\frac{3}{4},1]}(\mathrm{m}_1) & 0 \\ 0 & \chi_{[\frac{1}{4},\frac{1}{2}]\cup[\frac{1}{2},\frac{3}{4}]}(\mathrm{m}_1)\end{pmatrix}\right.$$
$$\left. + \begin{pmatrix} 0 & \partial_1 \\ \partial_{1,c} & 0\end{pmatrix}\right)\begin{pmatrix} u \\ v\end{pmatrix} = \begin{pmatrix} f \\ g\end{pmatrix} \tag{2.19}$$

where f, g are thought of being given. We find that $\mathcal{M}$ is given by

$$\mathcal{M} = \begin{pmatrix} \chi_{[0,\frac{1}{4}]\cup[\frac{1}{2},\frac{3}{4}]}(\mathrm{m}_1) & 0 \\ 0 & \chi_{[0,\frac{1}{4}]\cup[\frac{3}{4},1]}(\mathrm{m}_1) \end{pmatrix} + \partial_0^{-1} \begin{pmatrix} \chi_{[\frac{1}{4},\frac{1}{2}]\cup[\frac{3}{4},1]}(\mathrm{m}_1) & 0 \\ 0 & \chi_{[\frac{1}{4},\frac{1}{2}]\cup[\frac{1}{2},\frac{3}{4}]}(\mathrm{m}_1) \end{pmatrix}.$$

We realize that

$$\mathcal{A} = \begin{pmatrix} 0 & \partial_1 \\ \partial_{1,c} & 0 \end{pmatrix}$$

is skew-selfadjoint and, thus, maximal monotone. Note that the system describes a mixed type equation. The system varies between hyperbolic, elliptic and parabolic type equations either with homogeneous Dirichlet or Neumann data. Well-posedness of the system (2.19) can be established in $L^2_\nu(\mathbb{R}, L^2(]0,1[))$.

Now, instead of (2.19), we consider the sequence of problems

$$\Bigg(\partial_{0,\nu} \begin{pmatrix} \chi_{[0,\frac{1}{4}]\cup[\frac{1}{2},\frac{3}{4}]}(n\cdot \mathrm{m}_1 \mod 1) & 0 \\ 0 & \chi_{[0,\frac{1}{4}]\cup[\frac{3}{4},1]}(n\cdot \mathrm{m}_1 \mod 1) \end{pmatrix} + \begin{pmatrix} \chi_{[\frac{1}{4},\frac{1}{2}]\cup[\frac{3}{4},1]}(n\cdot \mathrm{m}_1 \mod 1) & 0 \\ 0 & \chi_{[\frac{1}{4},\frac{1}{2}]\cup[\frac{1}{2},\frac{3}{4}]}(n\cdot \mathrm{m}_1 \mod 1) \end{pmatrix} + \begin{pmatrix} 0 & \partial_1 \\ \partial_{1,c} & 0 \end{pmatrix}\Bigg) \begin{pmatrix} u_n \\ v_n \end{pmatrix} = \begin{pmatrix} f \\ g \end{pmatrix} \tag{2.20}$$

for $n \in \mathbb{N}$, where $x \bmod 1 := x - \lfloor x \rfloor$, $x \in \mathbb{R}$. With the same arguments from above well-posedness of the latter equation is warranted in the space $L^2_\nu(\mathbb{R}, L^2(]0,1[))$. Now,

$$\begin{pmatrix} \chi_{[0,\frac{1}{4}]\cup[\frac{1}{2},\frac{3}{4}]}(n\cdot \mathrm{m}_1 \mod 1) & 0 \\ 0 & \chi_{[0,\frac{1}{4}]\cup[\frac{3}{4},1]}(n\cdot \mathrm{m}_1 \mod 1) \end{pmatrix} + \partial_{0,\nu}^{-1} \begin{pmatrix} \chi_{[\frac{1}{4},\frac{1}{2}]\cup[\frac{3}{4},1]}(n\cdot \mathrm{m}_1 \mod 1) & 0 \\ 0 & \chi_{[\frac{1}{4},\frac{1}{2}]\cup[\frac{1}{2},\frac{3}{4}]}(n\cdot \mathrm{m}_1 \mod 1) \end{pmatrix} \to \begin{pmatrix} \frac{1}{2} & 0 \\ 0 & \frac{1}{2} \end{pmatrix} + \partial_{0,\nu}^{-1} \begin{pmatrix} \frac{1}{2} & 0 \\ 0 & \frac{1}{2} \end{pmatrix}$$

in the weak operator topology due to periodicity. Theorem 2.44 asserts that the sequence $\begin{pmatrix} u_n \\ v_n \end{pmatrix}_n$ weakly converges to the solution $\begin{pmatrix} u \\ v \end{pmatrix}$ of the problem

$$\left(\partial_{0,\nu} \begin{pmatrix} \frac{1}{2} & 0 \\ 0 & \frac{1}{2} \end{pmatrix} + \begin{pmatrix} \frac{1}{2} & 0 \\ 0 & \frac{1}{2} \end{pmatrix} + \begin{pmatrix} 0 & \partial_1 \\ \partial_{1,c} & 0 \end{pmatrix}\right) \begin{pmatrix} u \\ v \end{pmatrix} = \begin{pmatrix} f \\ g \end{pmatrix}.$$

It is interesting to note that the latter system does not coincide with any of the equations discussed above.

Theorem 2.44 deals with coefficients $\mathcal{M}$ that live in space-time. Going a step further instead of treating (2.20), we let $(\kappa_n)_n$ in $W_1^1(\mathbb{R})$ be a $W_1^1(\mathbb{R})$-convergent sequence of weakly differentiable $L^1(\mathbb{R})$-functions with limit κ and support on the positive reals. Then it is easy to see that the associated convolution operators $(\kappa_n *)_n$ converge in $L(L^2(\mathbb{R}_{\geq 0}))$ to $\kappa *$. Moreover, using Young's inequality, we deduce that

$$\|\kappa_n * \|_{L(L^2_\nu(\mathbb{R}))}, \|\kappa_n' * \|_{L(L^2_\nu(\mathbb{R}))} \to 0 \quad (\nu \to \infty)$$

uniformly in n. Thus, the strict positive definiteness of

$$\partial_{0,\nu}(1+\kappa *)\left(\begin{pmatrix} \chi_{[0,\frac{1}{4}]\cup[\frac{1}{2},\frac{3}{4}]}(\mathrm{m}_1) & 0 \\ 0 & \chi_{[0,\frac{1}{4}]\cup[\frac{3}{4},1]}(\mathrm{m}_1) \end{pmatrix} + \partial_{0,\nu}^{-1}\begin{pmatrix} \chi_{[\frac{1}{4},\frac{1}{2}]\cup[\frac{3}{4},1]}(\mathrm{m}_1) & 0 \\ 0 & \chi_{[\frac{1}{4},\frac{1}{2}]\cup[\frac{1}{2},\frac{3}{4}]}(\mathrm{m}_1) \end{pmatrix}\right)$$

in the truncated form as in (2.18) in Theorem 2.44 above follows from the respective inequality for

$$\partial_{0,\nu}\begin{pmatrix} \chi_{[0,\frac{1}{4}]\cup[\frac{1}{2},\frac{3}{4}]}(\mathrm{m}_1) & 0 \\ 0 & \chi_{[0,\frac{1}{4}]\cup[\frac{3}{4},1]}(\mathrm{m}_1) \end{pmatrix} + \begin{pmatrix} \chi_{[\frac{1}{4},\frac{1}{2}]\cup[\frac{3}{4},1]}(\mathrm{m}_1) & 0 \\ 0 & \chi_{[\frac{1}{4},\frac{1}{2}]\cup[\frac{1}{2},\frac{3}{4}]}(\mathrm{m}_1) \end{pmatrix}.$$

Now, the product of a sequence converging in the weak operator topology and a sequence converging in the norm topology converges in the weak operator topology. Hence, the solutions of

$$\left(\partial_{0,\nu}(1+\kappa_n *)\left(\begin{pmatrix} \chi_{[0,\frac{1}{4}]\cup[\frac{1}{2},\frac{3}{4}]}(n\cdot \mathrm{m}_1 \mod 1) & 0 \\ 0 & \chi_{[0,\frac{1}{4}]\cup[\frac{3}{4},1]}(n\cdot \mathrm{m}_1 \mod 1) \end{pmatrix} + \partial_{0,\nu}^{-1}\begin{pmatrix} \chi_{[\frac{1}{4},\frac{1}{2}]\cup[\frac{3}{4},1]}(n\cdot \mathrm{m}_1 \mod 1) & 0 \\ 0 & \chi_{[\frac{1}{4},\frac{1}{2}]\cup[\frac{1}{2},\frac{3}{4}]}(n\cdot \mathrm{m}_1 \mod 1) \end{pmatrix}\right) + \begin{pmatrix} 0 & \partial_1 \\ \partial_{1,c} & 0 \end{pmatrix}\right)\begin{pmatrix} u_n \\ v_n \end{pmatrix} = \begin{pmatrix} f \\ g \end{pmatrix}$$

converge weakly to the solution of

$$\left(\partial_{0,\nu}(1+\kappa *)\left(\begin{pmatrix} \frac{1}{2} & 0 \\ 0 & \frac{1}{2} \end{pmatrix} + \partial_{0,\nu}^{-1}\begin{pmatrix} \frac{1}{2} & 0 \\ 0 & \frac{1}{2} \end{pmatrix}\right) + \begin{pmatrix} 0 & \partial_1 \\ \partial_{1,c} & 0 \end{pmatrix}\right)\begin{pmatrix} u \\ v \end{pmatrix} = \begin{pmatrix} f \\ g \end{pmatrix}.$$

The latter considerations dealt with time-translation invariant coefficients. We shall also treat another example, where time-translation invariance is not warranted. For this take a sequence of Lipschitz continuous functions $(N_n\colon \mathbb{R} \to \mathbb{R})_n$ with uniformly bounded Lipschitz semi-norm and such that $(N_n)_n$ converges pointwise almost everywhere to some function $N\colon \mathbb{R} \to \mathbb{R}$. Moreover, assume that there exists $c > 0$ such that $\frac{1}{c} \geq N_n \geq c$ for all $n \in \mathbb{N}$. Then, by Lebesgue's dominated convergence theorem $N_n(\mathrm{m}_0) \to N(\mathrm{m}_0)$ in the strong operator topology, where we anticipated that $N_n(\mathrm{m}_0)$ acts as a multiplication operator with respect to the temporal variable. The strict monotonicity in the above-truncated sense of

$$\partial_{0,\nu}\left(N_n(\mathrm{m}_0)\begin{pmatrix}\chi_{[0,\frac{1}{4}]\cup[\frac{1}{2},\frac{3}{4}]}(n\cdot \mathrm{m}_1 \mod 1) & 0\\ 0 & \chi_{[0,\frac{1}{4}]\cup[\frac{3}{4},1]}(n\cdot \mathrm{m}_1 \mod 1)\end{pmatrix}\right.$$
$$\left.+\,\partial_{0,\nu}^{-1}\begin{pmatrix}\chi_{[\frac{1}{4},\frac{1}{2}]\cup[\frac{3}{4},1]}(n\cdot \mathrm{m}_1 \mod 1) & 0\\ 0 & \chi_{[\frac{1}{4},\frac{1}{2}]\cup[\frac{1}{2},\frac{3}{4}]}(n\cdot \mathrm{m}_1 \mod 1)\end{pmatrix}\right)$$

is easily seen using integration by parts, see, e.g., [36, Lemma 2.6]. Our main convergence theorem now yields that the solutions of

$$\left(\partial_{0,\nu}\left(N_n(\mathrm{m}_0)\begin{pmatrix}\chi_{[0,\frac{1}{4}]\cup[\frac{1}{2},\frac{3}{4}]}(n\cdot \mathrm{m}_1 \mod 1) & 0\\ 0 & \chi_{[0,\frac{1}{4}]\cup[\frac{3}{4},1]}(n\cdot \mathrm{m}_1 \mod 1)\end{pmatrix}\right.\right.$$
$$\left.+\,\partial_{0,\nu}^{-1}\begin{pmatrix}\chi_{[\frac{1}{4},\frac{1}{2}]\cup[\frac{3}{4},1]}(n\cdot \mathrm{m}_1 \mod 1) & 0\\ 0 & \chi_{[\frac{1}{4},\frac{1}{2}]\cup[\frac{1}{2},\frac{3}{4}]}(n\cdot \mathrm{m}_1 \mod 1)\end{pmatrix}\right)$$
$$\left.+\begin{pmatrix}0 & \partial_1\\ \partial_{1,c} & 0\end{pmatrix}\right)\begin{pmatrix}u_n\\ v_n\end{pmatrix}=\begin{pmatrix}f\\ g\end{pmatrix}$$

converge weakly to the solution of

$$\left(\partial_{0,\nu}\left(N(\mathrm{m}_0)\begin{pmatrix}\frac{1}{2} & 0\\ 0 & \frac{1}{2}\end{pmatrix}+\partial_{0,\nu}^{-1}\begin{pmatrix}\frac{1}{2} & 0\\ 0 & \frac{1}{2}\end{pmatrix}\right)+\begin{pmatrix}0 & \partial_1\\ \partial_{1,c} & 0\end{pmatrix}\right)\begin{pmatrix}u\\ v\end{pmatrix}=\begin{pmatrix}f\\ g\end{pmatrix}.$$

3. Nonlinear monotone evolutionary problems

This last section is devoted to the generalization of the well-posedness results of the previous sections to a particular case of non-linear problems. Instead of considering differential equations we turn our attention to the study of differential inclusions. As in the previous section, we begin to consider the autonomous case and present the well-posedness result.

3.1. The autonomous case

Let $\nu > 0$. The problem class under consideration is given as follows

$$(u,f)\in \partial_{0,\nu}M\left(\partial_{0,\nu}^{-1}\right)+A, \tag{3.1}$$

where $M(\partial_{0,\nu}^{-1})$ is again a linear material law, arising from an analytic and bounded function $M : B_{\mathbb{C}}(r,r)\to L(H)$ for some $r>\frac{1}{2\nu}$, $f\in L_\nu^2(\mathbb{R},H)$ is a given right-hand side and $u\in L_\nu^2(\mathbb{R},H)$ is to be determined. In contrast to the above problems, $A\subseteq L_\nu^2(\mathbb{R},H)\oplus L_\nu^2(\mathbb{R},H)$ is now a maximal monotone relation, which in particular needs not to be linear. By this lack of linearity we cannot argue as in the previous section, where the maximal monotonicity of the operators were shown by proving the strict monotonicity of their adjoints (in other words, we cannot apply Corollary 1.3). Thus, the maximal monotonicity has to be shown by employing other techniques and the key tools are perturbation results for maximal monotone operators.

In the autonomous case, our hypotheses read as follows:

Hypotheses 3.1. We say that A satisfies the hypotheses (H1) and (H2) respectively, if

(H1) A is maximal monotone and *translation-invariant*, i.e., for every $h \in \mathbb{R}$ and $(u,v) \in A$ we have $(u(\cdot + h), v(\cdot + h)) \in A$.

(H2) for all $(u,v),(x,y) \in A$ the estimate $\int_{-\infty}^{0} \mathfrak{Re}\langle u(t)-x(t)|v(t)-y(t)\rangle e^{-2\nu t}\mathrm{d}t \geq 0$ holds.

Assuming the standard assumption (2.3) for the function M, the operator $\partial_{0,\nu}M\left(\partial_{0,\nu}^{-1}\right) - c$ is maximal monotone on $L^2_\nu(\mathbb{R},H)$. Thus, the well-posedness of (3.1) just relies on the maximal monotonicity of the sum of $\partial_{0,\nu}M\left(\partial_{0,\nu}^{-1}\right) - c$ and A. Since A is assumed to be maximal monotone, we can apply well-known perturbation results in the theory of maximal monotone operators to prove that $\partial_{0,\nu}M\left(\partial_{0,\nu}^{-1}\right) + A - c$ is indeed maximal monotone, which in particular yields that

$$\left(\overline{\partial_{0,\nu}M\left(\partial_{0,\nu}^{-1}\right) + A}\right)^{-1}$$

is a Lipschitz-continuous mapping on $L^2_\nu(\mathbb{R},H)$ (see Theorem 1.1). Moreover, using hypothesis (H2) we can prove the causality of the corresponding solution operator $\left(\overline{\partial_{0,\nu}M\left(\partial_{0,\nu}^{-1}\right) + A}\right)^{-1}$. The well-posedness result reads as follows:

Theorem 3.2 (Well-posedness of autonomous evolutionary inclusions, [48]). *Let H be a Hilbert space, $M : B_{\mathbb{C}}\left(\frac{1}{2\nu_0}, \frac{1}{2\nu_0}\right) \to L(H)$ a linear material law for some $\nu_0 > 0$ satisfying* (2.3). *Let $\nu > \nu_0$ and $A \subseteq L^2_\nu(\mathbb{R},H) \oplus L^2_\nu(\mathbb{R},H)$ a relation satisfying* (H1). *Then for each $f \in L^2_\nu(\mathbb{R},H)$ there exists a unique $u \in L^2_\nu(\mathbb{R},H)$ such that*

$$(u,f) \in \overline{\partial_{0,\nu}M\left(\partial_{0,\nu}^{-1}\right) + A}. \tag{3.2}$$

Moreover, $\left(\overline{\partial_{0,\nu}M\left(\partial_{0,\nu}^{-1}\right) + A}\right)^{-1}$ is Lipschitz-continuous with a Lipschitz constant less than or equal to $\frac{1}{c}$. If in addition A satisfies (H2), *then the solution operator $\left(\overline{\partial_{0,\nu}M\left(\partial_{0,\nu}^{-1}\right) + A}\right)^{-1}$ is causal.*

A typical example for a maximal monotone relation satisfying (H1) and (H2) is an extension of a maximal monotone relation $A \subseteq H \oplus H$ satisfying $(0,0) \in A$. Indeed, if $A \subseteq H \oplus H$ is maximal monotone and $(0,0) \in A$, we find that

$$A_\nu := \left\{(u,v) \in L^2_\nu(\mathbb{R},H) \oplus L^2_\nu(\mathbb{R},H) \,|\, (u(t),v(t)) \in A \text{ for a.e. } t \in \mathbb{R}\right\} \tag{3.3}$$

is maximal monotone (see, e.g., [27, p. 31]). Moreover, A_ν obviously satisfies (H1) and (H2).

Remark 3.3. It is possible to drop the assumption $(0,0) \in A$, if one considers the differential inclusion on the half-line $\mathbb{R}_{\geq 0}$ instead of $\mathbb{R}$. In this case, an analogous definition of the time derivative on the space $L^2_\nu(\mathbb{R}_{\geq 0}, H)$ can be given and the well-posedness of initial value problems of the form

$$
\begin{aligned}
(u, f) &\in (\partial_{0,\nu} M_0 + M_1 + A_\nu) \\
M_0 u(0+) &= u_0,
\end{aligned}
$$

where A_ν is given as the extension of a maximal monotone relation $A \subseteq H \oplus H$ and $M_0, M_1 \in L(H)$ satisfy a suitable monotonicity constraint, can be shown similarly (see [46]).

The general coupling mechanism as illustrated, e.g., in [31] also works for the non-linear situation. This is also illustrated in the following example.

Example 3.4 ([46, Section 5.1]). We consider the equations of thermo-plasticity in a domain $\Omega \subseteq \mathbb{R}^3$, given by

$$
M\partial_{0,\nu}^2 u - \operatorname{Div} \sigma = f, \tag{3.4}
$$

$$
\varrho \partial_{0,\nu} \vartheta - \operatorname{div} \kappa \operatorname{grad} \vartheta + \tau_0 \operatorname{trace} \operatorname{Grad} \partial_{0,\nu} u = g. \tag{3.5}
$$

The functions $u \in L^2_\nu(\mathbb{R}, L^2(\Omega)^3)$ and $\vartheta \in L^2_\nu(\mathbb{R}, L^2(\Omega))$ are the unknowns, standing for the displacement field of the medium and its temperature, respectively. $f \in L^2_\nu(\mathbb{R}, L^2(\Omega)^3)$ and $g \in L^2_\nu(\mathbb{R}, L^2(\Omega))$ are given source terms. The stress tensor $\sigma \in L^2_\nu(\mathbb{R}, H_{\mathrm{sym}}(\Omega))$ is related to the strain tensor and the temperature by the following constitutive relation, generalizing Hooke's law,

$$
\sigma = C(\operatorname{Grad} u - \varepsilon_p) - c \operatorname{trace}^* \vartheta, \tag{3.6}
$$

where $c > 0$ and $C : H_{\mathrm{sym}}(\Omega) \to H_{\mathrm{sym}}(\Omega)$ is a linear, selfadjoint and strictly positive definite operator (the elasticity tensor). The operator trace : $H_{\mathrm{sym}}(\Omega) \to L^2(\Omega)$ is the usual trace for matrices and its adjoint can be computed by $\operatorname{trace}^* f = \begin{pmatrix} f & 0 & 0 \\ 0 & f & 0 \\ 0 & 0 & f \end{pmatrix}$. The function $\varrho \in L^\infty(\Omega)$ describes the mass density and is assumed to be real-valued and uniformly strictly positive, $M, \kappa \in L^\infty(\Omega)^{3\times 3}$ are assumed to be uniformly strictly positive definite and selfadjoint and $\tau_0 > 0$ is a real numerical parameter. The additional term ε_p models the inelastic strain and is related to σ by

$$
(\sigma, \partial_{0,\nu} \varepsilon_p) \in \mathbb{I}, \tag{3.7}
$$

where $\mathbb{I} \subseteq H_{\mathrm{sym}}(\Omega) \oplus H_{\mathrm{sym}}(\Omega)$ is a maximal monotone relation satisfying

$$
\operatorname{trace} [\mathbb{I}[H_{\mathrm{sym}}(\Omega)]] = \{0\},
$$

i.e., each element in the post-set of $\mathbb{I}$ is trace-free. If $\varepsilon_p = 0$, then (3.4)–(3.6) are exactly the equations of thermo-elasticity (see [33, p. 420 ff.]). The quasi-static case was studied in [9] for a particular relation $\mathbb{I}$, depending on the temperature ϑ under the additional assumption that the material possesses the linear kinematic hardening property. We complete the system (3.4)–(3.7) by suitable boundary conditions for u and ϑ, for instance $u, \vartheta = 0$ on $\partial\Omega$. We set $v := \partial_{0,\nu} u$ and $q := \tau_0^{-1} c\kappa \operatorname{grad}_c \vartheta$.

Following [46, Subsection 5.1], the system (3.4)–(3.7) can be written as

$$\left(\begin{pmatrix} \vartheta \\ q \\ v \\ \sigma \end{pmatrix}, \begin{pmatrix} c\tau_0^{-1} f \\ 0 \\ F \\ 0 \end{pmatrix}\right) \in \partial_{0,\nu} M(\partial_{0,\nu}^{-1}) + \begin{pmatrix} 0 & -\operatorname{div} & 0 & 0 \\ -\operatorname{grad}_c & 0 & 0 & 0 \\ 0 & 0 & 0 & -\operatorname{Div} \\ 0 & 0 & -\operatorname{Grad}_c & \mathbb{I} \end{pmatrix},$$

where

$$M(\partial_{0,\nu}^{-1}) = \begin{pmatrix} c\tau_0^{-1} w + \operatorname{trace} cC^{-1} c \operatorname{trace}^* & 0 & 0 & \operatorname{trace} cC^{-1} \\ 0 & 0 & 0 & 0 \\ 0 & 0 & M & 0 \\ C^{-1} c \operatorname{trace}^* & 0 & 0 & C^{-1} \end{pmatrix} + \partial_{0,\nu}^{-1} \begin{pmatrix} 0 & 0 & 0 & 0 \\ 0 & \kappa^{-1} c^{-1} \tau_0 & 0 & 0 \\ 0 & 0 & 0 & 0 \\ 0 & 0 & 0 & 0 \end{pmatrix}.$$

Thus, we have that $M(\partial_{0,\nu}^{-1}) = M_0 + \partial_{0,\nu}^{-1} M_1$ with

$$M_0 = \begin{pmatrix} c\tau_0^{-1} w + \operatorname{trace} cC^{-1} c \operatorname{trace}^* & 0 & 0 & \operatorname{trace} cC^{-1} \\ 0 & 0 & 0 & 0 \\ 0 & 0 & M & 0 \\ C^{-1} c \operatorname{trace}^* & 0 & 0 & C^{-1} \end{pmatrix},$$

$$M_1 = \begin{pmatrix} 0 & 0 & 0 & 0 \\ 0 & \kappa^{-1} c^{-1} \tau_0 & 0 & 0 \\ 0 & 0 & 0 & 0 \\ 0 & 0 & 0 & 0 \end{pmatrix}.$$

It can easily be verified, that the material law $M(\partial_{0,\nu}^{-1})$ satisfies (2.3). Thus, we only have to check that

$$A := \begin{pmatrix} 0 & -\operatorname{div} & 0 & 0 \\ -\operatorname{grad}_c & 0 & 0 & 0 \\ 0 & 0 & 0 & -\operatorname{Div} \\ 0 & 0 & -\operatorname{Grad}_c & 0 \end{pmatrix} + \begin{pmatrix} 0 & 0 & 0 & 0 \\ 0 & 0 & 0 & 0 \\ 0 & 0 & 0 & 0 \\ 0 & 0 & 0 & \mathbb{I} \end{pmatrix}$$

is maximal monotone (note that the other assumptions on A are trivially satisfied, since A is given as in (3.3)). Since A is the sum of two maximal monotone operators its maximal monotonicity can be obtained by assuming suitable boundedness constraints on $\mathbb{I}$ and applying classical perturbation results for maximal monotone operators.[20]

[20]The easiest assumption would be the boundedness of $\mathbb{I}$, i.e., for every bounded set M the post-set $\mathbb{I}[M]$ is bounded. For more advanced perturbation results we refer to [18, p. 331 ff.].

3.2. The non-autonomous case

We are also able to treat non-autonomous differential inclusions. Consider the following problem

$$(u, f) \in (\partial_{0,\nu} M_0(\mathrm{m}_0) + M_1(\mathrm{m}_0) + A_\nu), \tag{3.8}$$

where $M_0, M_1 \in L_s^\infty(\mathbb{R}, L(H))$ and A_ν is the canonical extension of a maximal monotone relation $A \subseteq H \oplus H$ with $(0,0) \in A$ as defined in (3.3). As in Subsection 2.5 we assume that M_0 satisfies Hypotheses 2.41 (a)–(d).

Our well-posedness result reads as follows:

Theorem 3.5 (Solution theory for non-autonomous evolutionary inclusions, [53]). *Let $M_0, M_1 \in L_s^\infty(\mathbb{R}; L(H))$, where M_0 satisfies Hypotheses 2.41 (a)–(d). Moreover, we assume that $N(M_0(t)) = N(M_0(0))$ for every $t \in \mathbb{R}$ and*[21]

$$\bigvee_{c>0} \bigwedge_{t\in\mathbb{R}} \iota_{R(M_0(0))}^* M_0(t) \iota_{R(M_0(0))} \geq c \text{ and } \iota_{N(M_0(0))}^* \mathfrak{Re}\, M_1(t) \iota_{N(M_0(0))} \geq c. \tag{3.9}$$

Let $A \subseteq H \oplus H$ be a maximal monotone relation with $(0,0) \in A$. Then there exists $\nu_0 > 0$ such that for every $\nu \geq \nu_0$

$$\left(\overline{\partial_{0,\nu} M_0(\mathrm{m}_0) + M_1(\mathrm{m}_0) + A_\nu}\right)^{-1} : L_\nu^2(\mathbb{R}, H) \to L_\nu^2(\mathbb{R}, H)$$

is a Lipschitz-continuous, causal mapping. Moreover, the mapping is independent of ν in the sense that, for $\nu, \nu' \geq \nu_0$ and $f \in L_{\nu'}^2(\mathbb{R}, H) \cap L_\nu^2(\mathbb{R}, H)$ we have that

$$\left(\overline{\partial_{0,\nu'} M_0(\mathrm{m}_0) + M_1(\mathrm{m}_0) + A_{\nu'}}\right)^{-1}(f) = \left(\overline{\partial_{0,\nu} M_0(\mathrm{m}_0) + M_1(\mathrm{m}_0) + A_\nu}\right)^{-1}(f).$$

Note that in Subsection 2.5 we do not require that $N(M_0(t))$ is t-independent. However, in order to apply perturbation results, which are the key tools for proving the well-posedness of (3.8), we need to impose this additional constraint (compare [36, Theorem 2.19]).

3.3. Problems with non-linear boundary conditions

As we have seen in Subsection 3.1 the maximal monotonicity of the relation $A \subseteq L_\nu^2(\mathbb{R}, H) \oplus L_\nu^2(\mathbb{R}, H)$ plays a crucial role for the well-posedness of the corresponding evolutionary problem (3.1). Motivated by several examples from mathematical physics, we might restrict our attention to (possibly non-linear) operators $A : D(A) \subseteq L_\nu^2(\mathbb{R}, H) \to L_\nu^2(\mathbb{R}, H)$ of a certain block structure. As a motivating example, we consider the wave equation with impedance-type boundary conditions, which was originally treated in [32].

[21] We denote by $\iota_{R(M_0(0))}$ and $\iota_{N(M_0(0))}$ the canonical embeddings into H of $R(M_0(0))$ and $N(M_0(0))$, respectively.

Example 3.6. Let $\Omega \subseteq \mathbb{R}^n$ be open and consider the following boundary value problem

$$\partial_{0,\nu}^2 u - \operatorname{div}\operatorname{grad} u = f \text{ on } \Omega, \tag{3.10}$$

$$\left(\partial_{0,\nu}^2 a(\mathrm{m})u + \operatorname{grad} u\right) \cdot N = 0 \text{ on } \partial\Omega, \tag{3.11}$$

where N denotes the outward normal vector field on $\partial\Omega$ and $a \in L^\infty(\Omega)^n$ such that $\operatorname{div} a \in L^\infty(\Omega)$[22]. Formulating (3.10) as a first-order system we obtain

$$\partial_{0,\nu}\begin{pmatrix} v \\ q \end{pmatrix} + \begin{pmatrix} 0 & \operatorname{div} \\ \operatorname{grad} & 0 \end{pmatrix}\begin{pmatrix} v \\ q \end{pmatrix} = \begin{pmatrix} f \\ 0 \end{pmatrix},$$

where $v := \partial_{0,\nu}u$ and $q := -\operatorname{grad} u$. The boundary condition (3.11) then reads as

$$(\partial_{0,\nu}a(\mathrm{m})v - q) \cdot N = 0 \text{ on } \partial\Omega.$$

The latter condition can be reformulated as

$$a(\mathrm{m})v - \partial_{0,\nu}^{-1}q \in D(\operatorname{div}_c),$$

where div_c is defined as in Definition 1.9. Thus, we end up with a problem of the form

$$(\partial_{0,\nu} + A)\begin{pmatrix} v \\ q \end{pmatrix} = \begin{pmatrix} f \\ 0 \end{pmatrix},$$

where $A \subseteq \begin{pmatrix} 0 & \operatorname{div} \\ \operatorname{grad} & 0 \end{pmatrix}$ with

$$D(A) := \left\{(v,q) \in D(\operatorname{grad}) \times D(\operatorname{div}) \,|\, a(\mathrm{m})v - \partial_{0,\nu}^{-1}q \in D(\operatorname{div}_c)\right\}.$$

In order to apply the solution theory, we have to ensure that the operator A, defined in that way, is maximal monotone as an operator in $L_\nu^2(\mathbb{R}, L^2(\Omega) \oplus L^2(\Omega)^n)$.

Remark 3.7. In [32] a more abstract version of Example 3.6 was studied, where the vector field a was replaced by a suitable material law operator $a(\partial_{0,\nu}^{-1})$ as it is defined in Subsection 2.3.

Following this guiding example, we are led to consider restrictions A of block operator matrices

$$\begin{pmatrix} 0 & D \\ G & 0 \end{pmatrix},$$

where $G : D(G) \subseteq H_0 \to H_1$ and $D : D(D) \subseteq H_1 \to H_0$ are densely defined closed linear operators satisfying $D^* \subseteq -G$ and consequently $G^* \subseteq -D$. We set $D_c := -G^*$ and $G_c := -D^*$ and obtain densely defined closed linear restrictions of D and G, respectively. Regarding the example above, $G = \operatorname{grad}$ and $D = \operatorname{div}$, whereas $G_c = \operatorname{grad}_c$ and $D_c = \operatorname{div}_c$. Having this guiding example in mind, we interpret G_c and D_c as the operators with vanishing boundary conditions and G and D as the operators with maximal domains. This leads to the following definition of so-called abstract boundary data spaces.

[22]Here we mean the divergence in the distributional sense.

Definition 3.8 ([34, Subsection 5.2]). Let G_c, D_c, G and D as above. We define

$$\begin{aligned} BD(G) &:= D(G_c)^{\perp_{D(G)}} = N(1 - DG), \\ BD(D) &:= D(D_c)^{\perp_{D(D)}} = N(1 - GD), \end{aligned}$$

where $D(G_c)$ and $D(D_c)$ are interpreted as closed subspaces of the Hilbert spaces $D(G)$ and $D(D)$, respectively, equipped with their corresponding graph norms. Consequently, we have the following orthogonal decompositions

$$\begin{aligned} D(G) &= D(G_c) \oplus BD(G) \\ D(D) &= D(D_c) \oplus BD(D). \end{aligned} \tag{3.12}$$

Remark 3.9. The decomposition (3.12) could be interpreted as follows: Each element u in the domain of G can be uniquely decomposed into two elements, one with vanishing boundary values (the component lying in $D(G_c)$) and one carrying the information of the boundary value of u (the component lying in $BD(G)$). In the particular case of $G = \operatorname{grad}$ a comparison of $BD(G)$ and the classical trace space $H^{\frac{1}{2}}(\partial\Omega)$ can be found in [47, Section 4].

Let $\iota_{BD(G)} : BD(G) \to D(G)$ and $\iota_{BD(D)} : BD(D) \to D(D)$ denote the canonical embeddings. An easy computation shows that $G[BD(G)] \subseteq BD(D)$ and $D[BD(D)] \subseteq BD(G)$ and thus, we may define

$$\begin{aligned} \overset{\bullet}{G} &:= \iota_{BD(D)}^{*} G \iota_{BD(G)} : BD(G) \to BD(D) \\ \overset{\bullet}{D} &:= \iota_{BD(G)}^{*} D \iota_{BD(D)} : BD(D) \to BD(G). \end{aligned}$$

These two operators share a surprising property.

Proposition 3.10 ([34, Theorem 5.2]). *The operators $\overset{\bullet}{G}$ and $\overset{\bullet}{D}$ are unitary and*

$$\left(\overset{\bullet}{G}\right)^{*} = \overset{\bullet}{D} \quad \text{as well as} \quad \left(\overset{\bullet}{D}\right)^{*} = \overset{\bullet}{G}.$$

Coming back to our original question, when $A \subseteq \begin{pmatrix} 0 & D \\ G & 0 \end{pmatrix}$ defines a maximal monotone operator, we find the following characterization.

Theorem 3.11 ([47, Theorem 3.1]). *Let G and D be as above. A restriction $A \subseteq \begin{pmatrix} 0 & D \\ G & 0 \end{pmatrix}$ is maximal monotone, if and only if there exists a maximal monotone relation $h \subseteq BD(G) \oplus BD(G)$ such that*

$$D(A) = \left\{(u, v) \in D(G) \times D(D) \,\middle|\, \left(\iota_{BD(G)}^{*} u, \overset{\bullet}{D}\, \iota_{BD(D)}^{*} v\right) \in h\right\}.$$

Example 3.12.

(a) In Example 3.6, the operators G and D are grad and div, respectively and the relation $h \subseteq BD(\mathrm{grad}) \oplus BD(\mathrm{grad})$ is given by

$$(x, y) \in h \Leftrightarrow \partial_{0,\nu}^{-1} y = \overset{\bullet}{\mathrm{div}}\, \iota_{BD(\mathrm{div})}^{*} a(\mathrm{m}) \iota_{BD(\mathrm{grad})} x.$$

Indeed, by the definition of the operator A in Example 3.6, a pair $(v, q) \in D(\mathrm{grad}) \times D(\mathrm{div})$ belongs to $D(A)$ if and only if

$$\begin{aligned} a(\mathrm{m})v - \partial_{0,\nu}^{-1} q \in D(\mathrm{div}_c) &\Leftrightarrow \iota_{BD(\mathrm{div})}^{*} \left(a(\mathrm{m})v - \partial_{0,\nu}^{-1} q\right) = 0 \\ &\Leftrightarrow \partial_{0,\nu}^{-1} \iota_{BD(\mathrm{div})}^{*} q = \iota_{BD(\mathrm{div})}^{*} a(\mathrm{m}) \iota_{BD(\mathrm{grad})} \iota_{BD(\mathrm{grad})}^{*} v \\ &\Leftrightarrow \partial_{0,\nu}^{-1} \overset{\bullet}{\mathrm{div}}\, \iota_{BD(\mathrm{div})}^{*} q = \overset{\bullet}{\mathrm{div}}\, \iota_{BD(\mathrm{div})}^{*} a(\mathrm{m}) \iota_{BD(\mathrm{grad})} \iota_{BD(\mathrm{grad})}^{*} v \\ &\Leftrightarrow \left(\iota_{BD(\mathrm{grad})}^{*} v, \overset{\bullet}{\mathrm{div}}\, \iota_{BD(\mathrm{div})}^{*} q\right) \in h. \end{aligned}$$

Thus, if we show that h is maximal monotone, we get the maximal monotonicity of A by Theorem 3.11. For doing so, we have to assume that the vector field a satisfies a positivity condition of the form

$$\mathfrak{Re} \int_{-\infty}^{0} \left(\langle \mathrm{grad}\, u | \partial_{0,\nu} a(\mathrm{m}) u\rangle(t) + \langle u | \,\mathrm{div}\, \partial_{0,\nu} a(\mathrm{m}) u\rangle(t)\right) e^{-2\nu t}\, dt \geq 0 \tag{3.13}$$

for all $u \in D(\partial_{0,\nu}) \cap D(\mathrm{grad})$. In case of a smooth boundary, the latter can be interpreted as a constraint on the angle between the vector field a and the outward normal vector field N. Indeed, condition (3.11) implies the monotonicity of h and also of the adjoint of h (note that here, h is a linear relation). Both facts imply the maximal monotonicity of h (the proof can be found in [48, Section 4.2]).

(b) In the theory of contact problems in elasticity we find so-called frictional boundary conditions at the contact surfaces. These conditions can be modeled for instance by sub-gradients of lower semi-continuous convex functions (see, e.g., [24, Section 5]), which are the classical examples of maximal monotone relations[23].

Let $\Omega \subseteq \mathbb{R}^n$ be a bounded domain. We recall the equations of elasticity from Example 2.8

$$\left(\partial_{0,\nu} \begin{pmatrix} 1 & 0 \\ 0 & C^{-1} \end{pmatrix} + \begin{pmatrix} 0 & -\mathrm{Div} \\ -\mathrm{Grad} & 0 \end{pmatrix}\right) \begin{pmatrix} v \\ T \end{pmatrix} = \begin{pmatrix} f \\ 0 \end{pmatrix} \tag{3.14}$$

and assume that the following frictional boundary condition should hold on the boundary $\partial\Omega$ (for a treatment of boundary conditions just holding on

[23]Note that not every maximal monotone relation can be realized as a sub-gradient of a lower semi-continuous convex function. Indeed, sub-gradients are precisely the cyclic monotone relations, see [7, Théorème 2.5].

different parts of the boundary, we refer to [47]):

$$(v, -T \cdot N) \in g, \tag{3.15}$$

where N denotes the unit outward normal vector field and $g \subseteq L^2(\partial\Omega)^n \oplus L^2(\partial\Omega)^n$ is a maximal monotone relation, which, for simplicity, we assume to be bounded. We note that in case of a smooth boundary, there exists a continuous injection $\kappa : BD(\mathrm{Grad}) \to L^2(\partial\Omega)^n$ (see [47]) and we may assume that $\kappa[BD(\mathrm{Grad})] \cap [L^2(\partial\Omega)^n]g \neq \emptyset$. Then, according to [47, Proposition 2.6], the relation

$$\widetilde{g} := \kappa^* g \kappa = \{(x, \kappa^* y) \in BD(\mathrm{Grad}) \times BD(\mathrm{Grad}) \,|\, (\kappa x, y) \in g\}$$

is maximal monotone as a relation on $BD(\mathrm{Grad})$ and the boundary condition (3.15) can be written as

$$(\iota^*_{BD(\mathrm{Grad})} v, -\overset{\bullet}{\mathrm{Div}}\, \iota^*_{BD(\mathrm{Div})} T) \in \widetilde{g}.$$

Thus, by Theorem 3.11, the operator

$$A \subseteq \begin{pmatrix} 0 & -\mathrm{Div} \\ -\mathrm{Grad} & 0 \end{pmatrix}$$

$$D(A) := \left\{ (v, T) \in D(\mathrm{Grad}) \times D(\mathrm{Div}) \,\middle|\, \left(\iota^*_{BD(\mathrm{Grad})} v, -\overset{\bullet}{\mathrm{Div}}\, \iota^*_{BD(\mathrm{Div})} T \right) \in \widetilde{g} \right\}$$

is maximal monotone and hence, Theorem 3.2 is applicable and yields the well-posedness of (3.14) subject to the boundary condition (3.15).

4. Conclusion

We have illustrated that many (initial, boundary value) problems of mathematical physics fit into the class of so-called evolutionary problems. Having identified the particular role of the time-derivative, we realize that many equations (or inclusions) of mathematical physics share the same type of solution theory in an appropriate Hilbert space setting. The class of problems accessible is widespread and goes from standard initial boundary value problems as for the heat equation, the wave equation or Maxwell's equations etc. to problems of mixed type and to integro-differential-algebraic equations. We also demonstrated first steps towards a discussion of issues like exponential stability and continuous dependence on the coefficients in this framework. The methods and results presented provide a general, unified approach to numerous problems of mathematical physics.

Acknowledgement

We thank the organizers, Wolfgang Arendt, Ralph Chill and Yuri Tomilov, of the conference "Operator Semigroups meet Complex Analysis, Harmonic Analysis and Mathematical Physics" held in Herrnhut in 2013 for organizing a wonderful conference dedicated to Charles Batty's 60th birthday. We also thank Charles Batty for his manifold, inspiring contributions to mathematics and of course for thus providing an excellent reason to meet experts from all over the world in evolution equations and related subjects.

References

[1] S. Agmon. *Lectures on elliptic boundary value problems.* AMS Chelsea Publishing Series. AMS Chelsea Pub., 2010.

[2] N. Akhiezer and I. Glazman. *Theory of linear operators in Hilbert space. Transl. from the Russian and with a preface by Merlynd Nestell (Two volumes bound as one). Repr. of the* 1961 *and* 1963 *transl.* New York, NY: Dover Publications. xiv, 147, iv, 1993.

[3] W. Arendt, D. Dier, H. Laasri, E.M. Ouhabaz. *Maximal regularity for evolution equations governed by non-autonomous forms.* Adv. Differential Equations 19(11-12):1043–1066, 2014.

[4] A. Bensoussan, J. Lions, and G. Papanicolaou. *Asymptotic Analysis for Periodic Structures.* North-Holland, Amsterdam, 1978.

[5] A. Bertram. *Elasticity and plasticity of large deformations: an introduction.* Berlin: Springer, 2005.

[6] A.-S. Bonnet-Bendhia, M. Dauge, and K. Ramdani. *Analyse specrale et singularités d'un problème de transmission non coercif.* C. R. Acad. Sci., Paris, Sér. I, Math., 328(8):717–720, 1999.

[7] H. Brézis. *Opérateurs maximaux monotones et semi-groupes de contractions dans les espaces de Hilbert.* North-Holland Mathematics Studies. 5. Notas de matematica (50). Amsterdam-London: North-Holland Publishing Comp. 1973.

[8] H. Brézis and A. Haraux. *Image d'une somme d'opérateurs monotones et applications.* Isr. J. Math., 23:165–186, 1976.

[9] K. Chełmiński and R. Racke. *Mathematical analysis of thermoplasticity with linear kinematic hardening.* J. Appl. Anal., 12(1):37–57, 2006.

[10] D. Cioranescu and P. Donato. *An Introduction to Homogenization.* Oxford University Press, New York, 2010.

[11] G. da Prato and P. Grisvard. *Sommes d'opérateurs linéaires et équations différentielles opérationnelles.* J. Math. Pures Appl., IX., 54:305–387, 1975.

[12] C. Dafermos. *An abstract Volterra equation with applications to linear viscoelasticity.* J. Differ. Equations, 7:554–569, 1970.

[13] R. Datko. *Extending a theorem of A.M. Liapunov to Hilbert space.* J. Math. Anal. Appl., 32:610–616, 1970.

[14] G. Duvaut and J.L. Lions. *Inequalities in mechanics and physics.* Grundlehren der mathematischen Wissenschaften. Springer-Verlag, 1976.

[15] K. Engel and R. Nagel. *One-Parameter Semigroups for Evolution Equations.* 194. Springer-Verlag, New York, Berlin, Heidelberg, 1999.

[16] K. Friedrichs. *Symmetric hyperbolic linear differential equations.* Comm. Pure Appl. Math. 7, 345–392, 1954.

[17] L. Gearhart. *Spectral theory for contraction semigroups on Hilbert space.* Trans. Am. Math. Soc., 236:385–394, 1978.

[18] S. Hu and N.S. Papageorgiou. *Handbook of Multivalued Analysis*, volume 1: Theory. Springer, 1997.

[19] S. Hu and N.S. Papageorgiou. *Handbook of Multivalued Analysis*, volume 2: Applications, of *Mathematics and its applications*. Kluwer Academic Publishers, 2000.

[20] T. Katō. *Perturbation theory for linear operators*. Grundlehren der mathematischen Wissenschaften. Springer, 1995.

[21] J.L. Lions. Quelques méthodes de résolution des problèmes aux limites non linéaires. Etudes mathématiques. Paris: Dunod; Paris: Gauthier-Villars. XX, 1969.

[22] A.F.A. Lorenzi and H. Tanabe. Degenerate integrodifferential equations of parabolic type. In *Differential equations. Inverse and direct problems. Papers of the meeting, Cortona, Italy, June* 21–25, 2004, pages 91–109. Boca Raton, FL: CRC Press, 2006.

[23] A.F.A. Lorenzi and H. Tanabe. *Degenerate integrodifferential equations of parabolic type with Robin boundary conditions: L^2-theory*. J. Math. Soc. Japan, 61(1):133–176, 2009.

[24] S. Migórski, A. Ochal, and M. Sofonea. *Solvability of dynamic antiplane frictional contact problems for viscoelastic cylinders*. Nonlinear Analysis: Theory, Methods & Applications, 70(10):3738–3748, 2009.

[25] G. Minty. *Monotone (nonlinear) operators in a Hilbert space*. Duke Math. J., 29, 1962.

[26] D. Morgenstern. *Beträge zur nichtlinearen Funktionalanalysis*. PhD thesis, TU Berlin, 1952.

[27] G. Morosanu. *Nonlinear evolution equations and applications*. Springer, 2nd edition, 1988.

[28] B. Nolte, S. Kempfle, and I. Schäfer. *Does a real material behave fractionally? Applications of fractional differential operators to the damped structure borne sound in viscoelastic solids*. Journal of Computational Acoustics, 11(03):451–489, 2003.

[29] I.G. Petrovskii. On the diffusion of waves and lacunas for hyperbolic equations. In *Selected Works. Systems of Partial Differential Equations. Algebraic Geometry.* Nauka Moscow, 1986.

[30] R. Picard. *Hilbert space approach to some classical transforms*. John Wiley, New York, 1989.

[31] R. Picard. *A structural observation for linear material laws in classical mathematical physics*. Math. Methods Appl. Sci., 32(14):1768–1803, 2009.

[32] R. Picard. A class of evolutionary problems with an application to acoustic waves with impedance type boundary conditions. In *Spectral Theory, Mathematical System Theory, Evolution Equations, Differential and Difference Equations*, volume 221 of *Operator Theory: Advances and Applications*, pages 533–548. Springer Basel, 2012.

[33] R. Picard and D. McGhee. *Partial differential equations. A unified Hilbert space approach.* de Gruyter Expositions in Mathematics 55. Berlin: de Gruyter. xviii, 2011.

[34] R. Picard, S. Trostorff, and M. Waurick. On a comprehensive Class of linear control problems. *IMA Journal of Mathematical Control and Information*, accepted, 2014.

[35] R. Picard, S. Trostorff, and M. Waurick. On evolutionary equations with material laws containing fractional integrals. *Math. Meth. Appl. Sci.*, DOI: 10.1002/mma.3286, 2014

[36] R. Picard, S. Trostorff, M. Waurick, and M. Wehowski. *On non-autonomous evolutionary problems.* J. Evol. Equ., 13(4):751–776, 2013.

[37] I. Podlubny. *Fractional differential equations. An introduction to fractional derivatives, fractional differential equations, to methods of their solution and some of their applications.* San Diego, CA: Academic Press, 1999.

[38] J. Prüss. *Decay properties for the solutions of a partial differential equation with memory.* Arch. Math., 92(2):158–173, 2009.

[39] W. Rudin. *Real and complex analysis.* Mathematics series. McGraw-Hill, 1987.

[40] H. Sohr. *Über die Existenz von Wellenoperatoren für zeitabhängige Störungen.* Monatsh. Math., 86:63–81, 1978.

[41] H. Tanabe. *Equations of evolution. Translated from Japanese by N. Mugibayashi and H. Haneda.* Monographs and Studies in Mathematics. 6. London-San Francisco-Melbourne: Pitman. XII, 1979.

[42] L. Tartar. *Nonlocal effects induced by homogenization.* Partial Differential Equations and the Calculus of Variations, Essays in Honor of Ennio De Giorgi, 2:925–938, 1989.

[43] L. Tartar. *Memory Effects and Homogenization.* Arch. Rational Mech. Anal., 111:121–133, 1990.

[44] E.G.F. Thomas. *Vector-valued integration with applications to the operator-valued H^∞ space.* IMA Journal of Mathematical Control and Information, 14(2):109–136, 1997.

[45] J.S. Toll. *Causality and the dispersion relation: logical foundations.* Phys. Rev. (2), 104:1760–1770, 1956.

[46] S. Trostorff. *An alternative approach to well-posedness of a class of differential inclusions in Hilbert spaces.* Nonlinear Anal., Theory Methods Appl., Ser. A, Theory Methods, 75(15):5851–5865, 2012.

[47] S. Trostorff. *A characterization of boundary conditions yielding maximal monotone operators.* J. Funct. Anal., 267(8), 2787-2822, 2014.

[48] S. Trostorff. *Autonomous evolutionary inclusions with applications to problems with nonlinear boundary conditions.* Int. J. Pure Appl. Math., 85(2):303–338, 2013.

[49] S. Trostorff. *Exponential stability for linear evolutionary equations.* Asymptotic Anal., 85:179–197, 2013.

[50] S. Trostorff *A Note on Exponential Stability for Evolutionary Equations.* Proc. Appl. Math. Mech. 14:983–984, 2014.

[51] S. Trostorff *On integro-differential inclusions with operator valued kernels.* Math. Methods Appl. Sci. 38(5):834–850, 2015.

[52] S. Trostorff and M. Waurick. *A note on elliptic type boundary value problems with maximal monotone relations.* Math. Nachr, 287, 1545–1558, 2014.

[53] S. Trostorff and M. Wehowski. *Well-posedness of non-autonomous evolutionary inclusions.* Nonlinear Anal., Theory Methods Appl., Ser. A, Theory Methods, 101:47–65, 2014.

[54] B. von Szökefalvy-Nagy. Spektraldarstellung linearer Transformationen des Hilbertschen Raumes. Ergebn. Math. Grenzgeb. Bd. 5, Nr. 5, Springer, Berlin, 1942.

[55] M. Waurick. *A Hilbert Space Approach to homogenization of linear ordinary differential equations including delay and memory terms.* Math. Methods Appl. Sci., 35:1067–1077, 2012.

[56] M. Waurick. How far away is the harmonic mean from the homogenized matrix? Technical report, TU Dresden, 2012. arXiv:1204.3768, submitted.

[57] M. Waurick. *A note on causality in Banach spaces.* Indag. Math. 26(2):404–412, 2015.

[58] M. Waurick. Continuous dependence on the coefficients for a class of non-autonomous evolutionary equations. Technical report, TU Dresden, 2013. arXiv:1308.5566.

[59] M. Waurick. *G-convergence of linear differential operators.* Journal of Analysis and its Applications 33(4): 385-415, 2014.

[60] M. Waurick. *Homogenization of a class of linear partial differential equations.* Asymptotic Analysis, 82:271–294, 2013.

[61] M. Waurick. *Homogenization in fractional elasticity.* SIAM J. Math. Anal., 46(2): 1551–1576, 2014.

[62] M. Waurick. *On non-autonomous integro-differential-algebraic evolutionary problems.* Math. Methods Appl. Sci. 38(4):665–676, 2015.

[63] M. Waurick and M. Kaliske. *A note on homogenization of ordinary differential equations with delay term.* PAMM, 11:889–890, 2011.

[64] N. Weck. *Local compactness for linear elasticity in irregular domains.* Math. Methods Appl. Sci., 17(2):107–113, 1994.

[65] G. Weiss. *Representation of shift-invariant operators on L^2 by H^∞ transfer functions: An elementary proof, a generalization to L^p, and a counterexample for L^∞.* Math. Control Signals Syst., 4(2):193–203, 1991.

[66] N. Wellander. *Homogenization of the Maxwell equations. Case I: Linear theory.* Appl. Math., Praha, 46(1):29–51, 2001.

[67] K.J. Witsch. *A remark on a compactness result in electromagnetic theory.* Math. Methods Appl. Sci., 16(2):123–129, 1993.

[68] J. Wloka. *Partielle Differentialgleichungen.* B.G. Teubner Stuttgart, 1982.

Rainer Picard, Sascha Trostorff and Marcus Waurick
Institut für Analysis
Fachrichtung Mathematik
Technische Universität Dresden
Zellescher Weg 12-14
D-01062 Dresden, Germany
e-mail: `rainer.picard@tu-dresden.de`
`sascha.trostorff@tu-dresden.de`
`marcus.waurick@tu-dresden.de`

Operator Theory:
Advances and Applications, Vol. 250, 453–461

Perturbations of Exponential Dichotomies for Hyperbolic Evolution Equations

Jan Prüss

Dedicated to Professor Charles Batty on the occasion of his 60th anniversary

Abstract. We consider perturbations of linear autonomous evolution equations which preserve the properties of either exponential stability or exponential dichotomy.

Mathematics Subject Classification (2010). 34G20, 35K55, 35B35, 37D10, 35R35.

Keywords. C_0-semigroups, exponential stability, exponential dichotomy, perturbations, operator-valued Paley–Wiener lemma, cell division models.

1. Introduction

Let X be a complex Banach space and A the generator of a C_0-semigroup e^{At} in X. By

$$\omega_0(A) = \lim_{t\to\infty} t^{-1} \log |e^{At}|$$

we denote its type; then A (or e^{At}) is called *exponentially stable* if and only if $\omega_0(A) < 0$.

Recall that A (or e^{At}) is said to admit an *exponential dichotomy* if there is a bounded projection P_+ in X which commutes with e^{At}, and constants $M \geq 1$ and $\eta > 0$ such that with $P_- = I - P_+$

(i) $e^{At}P_-$ extends to a C_0-group on $R(P_-)$, and
(ii) $|e^{At}P_+|_{\mathcal{B}(X)} + |e^{-At}P_-|_{\mathcal{B}(X)} \leq Me^{-\eta t}$ for $t > 0$.

P_+ is then called the *dichotomy* for A (or e^{At}).

Note that $P_+ = I$ works in case $\omega_0(A) < 0$; this is the trivial exponential dichotomy.

Consider a bounded perturbation $B \in \mathcal{B}(X)$, the space of bounded linear operators in X. Then it is well known that $A + B$ is again the generator of a

C_0-semigroup $e^{(A+B)t}$ in X. Assuming that e^{At} is exponentially stable, we ask for the same property for $e^{(A+B)t}$, and if A admits an exponential dichotomy, we ask whether this is also true for the perturbed generator $A+B$.

By elementary perturbation theory, these questions are easily answered in the affirmative, provided the norm of B is sufficiently small, but they are nontrivial if this is not the case. There are satisfactory results in case A generates an analytic C_0-semigroup, even for $B \in \mathcal{B}(D_A(\alpha,p);X)$ where $\alpha < 1$ and $p \in [1,\infty]$; here $D_A(\alpha,p)$ denote the real interpolation spaces between X and $D(A)$. In fact, again by standard perturbation theory, it is well known that the conditions $\bar{\mathbb{C}}_+ \subset \rho(A+B)$ resp. $i\mathbb{R} \subset \rho(A+B)$ are equivalent to the corresponding properties of the unperturbed operator A. Here $\rho(A)$ denotes the resolvent set of A.

In this paper we are dealing with the hyperbolic case, which means that e^{At} is allowed to be a C_0-group. In this situation things are much more involved. This is due to the facts that the semigroup e^{At} does not smooth, and the spectral mapping theorem is in general not valid. In this case perturbation results beyond the class $B \in \mathcal{B}(X)$ which preserve the C_0-semigroup property are rare, and typically require special structure conditions for A and B.

Even more, there are simple examples which show even in a Hilbert space setting that the aforementioned spectral conditions are not sufficient to preserve neither exponential stability nor exponential dichotomies.

Example 1.1. Let $X = l_2(\mathbb{N})$, $(Ax)_n = (in-1)x_n$ and $(Bx)_n = (1-1/n)x_n$, $n \in \mathbb{N}$. Then A generates an exponentially stable C_0-semigroup, B is bounded, and $\sigma(A+B) \subset \{\lambda \in \mathbb{C} : \operatorname{Re}\lambda < 0\}$. But $(i\rho - A - B)^{-1}$ is unbounded in X for $\rho \in \mathbb{R}$, hence $A+B$ is not exponentially stable, by the Gearhart–Prüss theorem.

It is the purpose of this paper to present some new affirmative answers to the raised questions, imposing a condition on the perturbation B which is only slightly stronger than its boundedness.

2. Main results

The main result on exponential dichotomies reads as follows.

Theorem 2.1. *Let A denote the generator of a C_0-semigroup e^{At} in a Banach space X, and let $B \in \mathcal{B}(X)$. Assume*

(i) *A admits an exponential dichotomy;*
(iia) *$B \in \mathcal{B}(X; D_A(\alpha,\infty))$, for some $\alpha > 0$; or*
(iib) *B is compact;*
(iii) *$i\rho - A - B$ is invertible in X, for each $\rho \in \mathbb{R}$.*

Then $A+B$ generates a C_0-semigroup $e^{(A+B)t}$ which admits an exponential dichotomy, as well.

Observe that the spectral condition (iii) is necessary for $A+B$ to admit an exponential dichotomy.

The second main result concerns exponential stability of the perturbed semigroup.

Theorem 2.2. *Let A denote the generator of a C_0-semigroup e^{At} in a Banach space X, and let $B \in \mathcal{B}(X)$. Assume*

(i) *A is exponentially stable;*
(iia) *$B \in \mathcal{B}(X; D_A(\alpha, \infty))$, for some $\alpha > 0$; or*
(iib) *B is compact;*
(iii) *$\lambda - A - B$ is invertible in X, for each* $\operatorname{Re}\lambda \geq 0$.

Then $A + B$ generates an exponentially stable C_0-semigroup, as well.

Note that also here, (iii) is a necessary condition.

Remark 2.3. Conditions (iia) or (iib) imply the weaker condition

$$\lim_{h\to 0+} |(e^{Ah} - I)B|_{\mathcal{B}(X)} = 0. \tag{1}$$

It is this condition which is actually used in the proofs.

3. Main ideas and proofs

The proof of Theorem 2.1 relies on two results. The first one is the characterization of exponential dichotomies for C_0-semigroups in [6]. This result tells that we have an exponential dichotomy if and only if the spectrum $\sigma(e^A)$ does not intersect the unit circle $S_1 = \partial B_1(0) \subset \mathbb{C}$. The Green kernel of the exponential dichotomy is then given by

$$S(t) = \begin{cases} e^{At}P_+, & t > 0 \\ -e^{At}P_-, & t < 0. \end{cases}$$

Here $P_- = I - P_+$ and P_+ is the projection onto the stable subspace given by

$$P_+ = \frac{1}{2\pi i}\int_{S_1} (z - e^A)^{-1} dz.$$

Note that there is a constant $\eta > 0$ such that

$$|S(t)|_{\mathcal{B}(X)} \leq Me^{-\eta|t|}, \quad t \in \mathbb{R}. \tag{2}$$

Another result in the paper [6] shows that A admits an exponential dichotomy if and only if the whole-line problem

$$\dot{u}(t) = Au(t) + f(t), \quad t \in \mathbb{R}, \tag{3}$$

for each $f \in BUC(\mathbb{R}; X)$ admits a unique mild solution $u \in BUC(\mathbb{R}; X)$. The solution operator is then given by $u = S * f$. Actually, in [6] this result has been stated for $BC(\mathbb{R}; X)$ instead of $BUC(\mathbb{R}; X)$; however the -if-part of the proof uses only continuous periodic functions, which are in BUC.

Recall that $u \in BUC(\mathbb{R}; X)$ is a *strong solution* of (3) with $f \in BUC(\mathbb{R}; X)$ if $u \in BUC^1(\mathbb{R}; X) \cap BUC(\mathbb{R}; D(A))$ satisfies (3) pointwise on $\mathbb{R}$. $u \in BUC(\mathbb{R}; X)$ is a *mild solution* of (3) if there are *strong solutions* $u_k \to u$ in $BUC(\mathbb{R}; X)$ with the corresponding right-hand sides $f_k := \dot{u}_k - Au_k \to f$ in $BUC(\mathbb{R}; X)$. Similarly, strong and mild solutions are defined in $L_p(\mathbb{R};X)$, $1 \le p \le \infty$, replacing the symbol BUC by L_p.

The above BUC-result easily extends to $L_\infty(\mathbb{R};X)$, i.e., we may replace the symbol BUC by L_∞. In fact, if we have an exponential dichotomy, then $S * f \in L_\infty(\mathbb{R};X)$ for any $f \in L_\infty(\mathbb{R};X)$. On the other hand, if the map $f \mapsto u =: Gf$ is bounded in $L_\infty(\mathbb{R};X)$, and $f \in BUC(\mathbb{R};X)$ then by the Friedrichs mollifier ρ_ε we obtain $\rho_\varepsilon * f \to f$ in $BUC(\mathbb{R};X)$, hence by translation invariance

$$\rho_\varepsilon * u = \rho_\varepsilon * Gf = G(\rho_\varepsilon * f) \to Gf = u$$

in $L_\infty(\mathbb{R};X)$ as $\varepsilon \to 0$. As $\rho_\varepsilon * u \in BUC(\mathbb{R};X)$, we see that $BUC(\mathbb{R};X)$ is also preserved by the solution map G.

Here we need another characterization of the existence of an exponential dichotomy which probably was not known before.

Theorem 3.1. *Let A be the generator of a C_0-semigroup in the Banach space X. Then the following are equivalent.*

(a) *A admits an exponential dichotomy in X;*
(b) *For each $f \in L_1(\mathbb{R};X)$, (3) admits a unique mild solution $u \in L_1(\mathbb{R};X)$.*

The proof of Theorem 3.1 is given in the next section.

The second result needed for the proofs of Theorems 2.1 and 2.2 is a recent variant of the Paley–Wiener lemma [2]. For this purpose we introduce the *operator-valued Wiener algebra*

$$W_X := \mathcal{B}(X, L_1(\mathbb{R};X)), \quad \|K\| := \sup_{|x|\le 1} |Kx|_{L_1(\mathbb{R};X)}.$$

This is of course a Banach space, but also a (non-commutative) Banach algebra with multiplication the convolution defined by

$$[T * S](x)(t) := \int_{\mathbb{R}} T\big(S(x)(\tau)\big)(t-\tau)d\tau, \quad x \in X,\ t \in \mathbb{R}.$$

We introduce the closed subalgebra W_X^0 by means of the following conditions:

(W1) $\|\tau_\sigma K - K\| \to 0$ as $\sigma \to 0+$;

(W2) $\sup_{|x|\le 1} \int_{|t|\ge R} |Kx(t)|dt \to 0$ as $R \to \infty$.

Here $\{\tau_\sigma\}_{\sigma\in\mathbb{R}}$ denotes the group of translations.

Unfortunately, W_X^0 is not an ideal in W_X, (W1) is ok, but (W2) destroys the ideal property. It is easy to see that $L_1(\mathbb{R};\mathcal{B}(X))$ as a set is a subalgebra of W_X^0. In fact, it is not difficult to prove that W_X^0 is the closure of $L_1(\mathbb{R};\mathcal{B}(X))$ in W_X. We observe that the Fourier transform $\tilde{K}$ of $K \in W_X$ is well defined and belongs to $\mathcal{B}(X;C_0(\mathbb{R};X))$. If $K \in W_X^0$ we even have $\tilde{K} \in C_0(\mathbb{R};\mathcal{B}(X))$, as can be shown by standard arguments.

The Paley–Wiener lemma in the space W_X^0 can be proved in the same way as Theorem 0.6 in [7] for the Banach algebra $L_1(\mathbb{R};\mathcal{B}(X))$. It reads as follows.

Theorem 3.2. *Suppose $K \in W_X^0$ and assume that the Paley–Wiener condition*

(PW) $I - \tilde{K}(\rho)$ *is invertible for each* $\rho \in \mathbb{R}$

is satisfied. Then the convolution equation

$$R = K + K * R$$

admits a unique solution $R \in W_X^0$.

Actually, in the result presented in [2] only $R \in W_X$ is claimed, but following the proof of the operator-valued Paley–Wiener lemma in [7] even yields $R \in W_X^0$.

To prove Theorem 2.1 we rewrite the equation

$$\dot{u} = Au + Bu + f$$

as the convolution equation

$$u = S * f + S * Bu,$$

where S denotes the Green kernel of the dichotomy of A. Consider $K := SB$ and the resolvent equation

$$R = K + K * R = K + R * K$$

in W_X^0. Then u is given by

$$u = S * f + R * S * f$$

as soon as $R \in W_X^0$ exists. So the Green kernel G for the perturbed problem is given by $G = S + R * S$. To find R by Theorem 3.2, we have to verify that $K = SB$ belongs to W_X^0, and that the Paley–Wiener condition is satisfied.

(W1) It is not difficult to prove the estimate

$$\int_{\mathbb{R}} |K(t+h)x - K(t)x|dt \leq 2M|B|h|x| + \frac{2M}{\eta}|(e^{Ah} - I)Bx|, \quad x \in X,\ h > 0.$$

Condition (1) then easily implies (W1).

(W2) This is a direct consequence of the exponential bound (2).

(PW) If $(I - \tilde{K}(\rho))x = 0$, then $x \in D(A)$, hence applying $(i\rho - A)$ this yields $(i\rho - A - B)x = 0$, hence $x = 0$ as this operator is injective by (iii). On the other hand, one checks easily that $x = y + (i\rho - A - B)^{-1}By$ is a solution of $(I - \tilde{K}(\rho))x = y$, hence (iii) implies also surjectivity.

As a consequence of Theorem 3.2, the problem

$$\dot{u} = Au + Bu + f$$

admits a unique mild solution $u \in L_1(\mathbb{R}; X)$ whenever $f \in L_1(\mathbb{R}; X)$ is given. Then by Theorem 3.1 the perturbed operator $A + B$ admits an exponential dichotomy, proving Theorem 2.1

Theorem 2.2 is deduced from Theorem 2.1 in the usual way, employing Liouville's theorem; see, e.g., the proof of Theorem 0.7 in [7].

4. Proof of Theorem 3.1

The implication $(a) \Rightarrow (b)$ is obvious. So we have to prove its converse. For this purpose let $G : L_1(\mathbb{R}; X) \to L_1(\mathbb{R}; X)$ denote the solution operator for (3), which maps a given inhomogeneity $f \in L_1(\mathbb{R}; X)$ to the unique mild solution $u \in L_1(\mathbb{R}; X)$. This operator is closed hence bounded by the closed graph theorem. Then by duality, the operator G^* is bounded in $L_1(\mathbb{R}; X)^*$ as well.

Recall that the dual space of $L_1(\mathbb{R}; X)$ can be represented by the space

$$\mathrm{Lip}_0(\mathbb{R}; X) = \left\{ w^* : \mathbb{R} \to X^* : |w^*|_{\mathrm{Lip}} := \sup_{t \neq \bar{t}} \frac{|w^*(t) - w^*(\bar{t})|}{|t - \bar{t}|} < \infty, w^*(0) = 0 \right\},$$

via the duality

$$\langle u | w^* \rangle := \int_{\mathbb{R}} (u(t) | dw^*(t)), \quad u \in L_1(\mathbb{R}; X),\ w^* \in \mathrm{Lip}(\mathbb{R}; X^*).$$

This well-known representation can be found, e.g., in [7], p. 169f. Observe that via the identification $w^*(t) = \int_0^t v^*(s)ds$ the space $BUC(\mathbb{R}; X^*)$ is a closed subspace of $\mathrm{Lip}_0(\mathbb{R}; X^*)$, and the duality becomes the usual one,

$$\langle u | w^* \rangle = \int_{\mathbb{R}} (u(t) | v^*(t)) dt, \quad u \in L_1(\mathbb{R}; X),\ v^* \in BUC(\mathbb{R}; X^*).$$

By translation invariance of (3), the solution operator G commutes with the translation group $\{\tau_h\}_{h \in \mathbb{R}}$, and as $\tau_h^* = \tau_{-h}$, G^* commutes with the translation group as well. This implies that G^* maps $BUC(\mathbb{R}; X^*)$ into itself, hence also $G^* BUC^1(\mathbb{R}; X^*) \subset BUC^1(\mathbb{R}; X^*)$ holds.

Next, for $f^* \in BUC^1(\mathbb{R}; X^*)$, we show that $u^* = G^* f^*$ is the unique strong solution of the equation dual to (3), i.e.,

$$-\dot{u}^*(t) = A^* u^*(t) + f^*(t), \quad t \in \mathbb{R}.$$

In fact, let $\phi \in \mathcal{D}(\mathbb{R})$, a test function, $x \in D(A)$ and set $u(t) = \phi(t)x$, $f = \dot{u} - Au$. Then we obtain

$$\begin{aligned} \langle u | f^* \rangle &= \langle Gf | f^* \rangle = \langle f | G^* f^* \rangle = \langle f | u^* \rangle \\ &= \langle \dot{u} - Au | u^* \rangle = -\langle u | \dot{u}^* \rangle - \langle Au | u^* \rangle, \end{aligned}$$

which yields

$$\int_{\mathbb{R}} \phi(t)(Ax | u^*(t)) dt = -\int_{\mathbb{R}} \phi(t)(x | \dot{u}^*(t) + f^*(t)) dt, \quad x \in D(A),\ \phi \in \mathcal{D}(\mathbb{R}).$$

As $\mathcal{D}(\mathbb{R})$ is dense in $L_1(\mathbb{R})$ we may conclude that

$$(Ax | u^*(t)) = -(x | \dot{u}^*(t) + f^*(t)), \quad t \in \mathbb{R},\ x \in D(A),$$

which implies further $u^*(t) \in D(A^*)$, and $-\dot{u}^*(t) = A^* u^*(t) + f^*(t)$, for all $t \in \mathbb{R}$.

To prove uniqueness, suppose that $u^* \in BUC(\mathbb{R}; X^*)$ is a mild solution of the homogeneous dual equation. Then there are $u_k^* \in BUC^1(\mathbb{R}; X^*) \cap BUC(\mathbb{R}; D(A^*))$

such that $u_k^* \to u^*$ and $f_k^* := -(\dot{u}_k^* + A^* u_k^*) \to 0$ in $BUC(\mathbb{R}; X^*)$. Fix a function $f \in W_1^1(\mathbb{R}; X)$ and let $u = Gf$. Then

$$\langle f | u_k^* \rangle = \langle \dot{u} - Au | u_k^* \rangle = -\langle u | \dot{u}_k^* + A^* u_k^* \rangle = \langle u | f_k^* \rangle \to 0, \quad k \to \infty,$$

hence $\langle f | u^* \rangle = 0$ for all $f \in W_1^1(\mathbb{R}; X)$, and so $u^* = 0$ since $W_1^1(\mathbb{R}; X)$ is dense in $L_1(\mathbb{R}; X)$.

With $X^\odot := \overline{D(A^*)}$, for the solution $u^* = G^* f^*$ with $f^* \in BUC^1(\mathbb{R}; X^*)$, it follows that $u^* \in BUC(\mathbb{R}; X^\odot)$, and even $u^* \in BUC^1(\mathbb{R}; X^\odot)$ as $X^\odot$ is a closed subspace of X^*. In particular, if $f^* \in BUC^1(\mathbb{R}; X^\odot)$ then $u^*(t) \in D(A^\odot)$ by the definition of $A^\odot$ as the part of A in $X^\odot$. Therefore the function $u^\odot(t) = u^*(-t)$ is a strong solution of

$$\dot{u}^\odot = A^\odot u^\odot + f^\odot \tag{4}$$

on $\mathbb{R}$, where $f^\odot(t) = f^*(-t)$. As $BUC^1(\mathbb{R}; X^\odot)$ is dense in $BUC(\mathbb{R}; X^\odot)$ this shows that the dual problem (4) admits a unique solution $u^\odot \in BUC(\mathbb{R}; X^\odot)$, for any given $f^\odot \in BUC(\mathbb{R}; X^\odot)$

By the aforementioned characterization of exponential dichotomies in the paper [6], this implies that the spectrum $\sigma((e^A)^\odot)$ does not intersect the unit circle S_1, hence the same is true for the spectrum $\sigma(e^A) = \sigma((e^A)^*) = \sigma((e^A)^\odot)$; for the last equality we refer to [5], Proposition IV.2.18. Using again the result in [6] this implies that A has an exponential dichotomy.

5. An illustrative example

Let $u(t, x)$ denote the size distribution of a population of cells at time t. Here the size or mass x of the cells is assumed in between $a > 0$ and $b > 2a$. The following model for the evolution of the size distribution is popular in mathematical biology.

$$\begin{aligned} \partial_t u(t,x) + \partial_x (q(x) u(t,x)) = &-(\beta(x) + \mu(x)) u(t,x) \\ &+ 2 \int_a^b \kappa(x,y) \beta(y) u(t,y) dy, \quad t > 0,\ a < x < b, \\ u(t,a) = 0, &\quad t > 0, \\ u(0,x) = u_0(x), &\quad a < x < b. \end{aligned} \tag{5}$$

Here $q \in C^1[a,b]$ denotes the intrinsic growth rate of the cells, we assume $q(x) > 0$ for $a \le x < b$. The function $\mu \in L_\infty(a,b)$, $\mu \ge 0$, means the natural death rate, $\beta \in L_\infty(a,b)$, $\beta \ge 0$, the rate of cell division, and finally $\kappa \in L_\infty((a,b)^2)$, $\kappa \ge 0$, the distribution of daughter cells after a cell division. The boundary condition means that there are no cells with minimal size. Depending on the type of cell division, κ is subject to several further conditions which are not important here, we refer to [3], [4] or [8] for further background and discussion.

We want to apply our main results to this model problem. To this end we choose as a state space $X = L_1(a,b)$, which is the natural choice as $|u|_1$ measures the total mass of cells if u is nonnegative. Then we define the operator A by means of

$$Au(x) := -\partial_x(q(x)u(x)) - (\beta(x) + \mu(x))u(x), \quad x \in (a,b), \tag{6}$$
$$u \in D(A) := \{u \in W_1^1(a,b) : u(a) = 0\}.$$

It is easy to see that A is dissipative and that the range condition $R(I - A) = X$ is valid. Therefore by the Lumer–Phillips theorem, cf. [1], or [5], A generates a C_0-semigroup of contractions in X. If we impose the condition

$$\mu(x) + \beta(x) \geq \mu_0 > 0, \quad x \in (a,b), \tag{7}$$

then the semigroup satisfies $|e^{At}| \leq e^{-\mu_0 t}$, in particular it is exponentially stable, and therefore A has a (trivial) exponential dichotomy. Note that the semigroup e^{At} is not analytic, it is even not continuous in operator-norm, the problem is hyperbolic. In fact, e^{At} is a damped translation semigroup which can be explicitly computed by means of the method of characteristics. But on the other hand, due to boundedness of the interval (a,b), the generator A has compact resolvent, and so its spectrum only consists of eigenvalues of finite multiplicity. The operator B is defined according to

$$Bu(x) = 2\int_a^b \kappa(x,y)\beta(y)u(y)dy, \quad x \in (a,b)\ u \in X. \tag{8}$$

This way (5) is reformulated as the abstract Cauchy problem

$$\dot{u} = Au + Bu,\ t > 0, \quad u(0) = u_0.$$

Obviously B is bounded in X, hence $A+B$ is also the generator of a C_0-semigroup, and the resolvent of $A + B$ is also compact. Further, B is also compact, provided κ is subject to the mild regularity condition

$$\lim_{h\to 0+} \sup_{a<y<b} \Big[\int_a^{a+h} \kappa(x,y)dx + \int_a^{b-h} |\kappa(x+h,y) - \kappa(x,y)|dx + \int_{b-h}^b \kappa(x,y)dx\Big] = 0. \tag{9}$$

This follows from Kolmogorov's compactness criterion. Therefore Theorems 2.1 and 2.2 apply. Let us see what news comes out for (5).

As the semigroup generated by A is positive and B is positive, $e^{(A+B)t}$ is positive as well; cf. [1], Section 5.3. It is known that a positive C_0-semigroup on an L_p-space has the spectrum determined growth property which means that its growth bound equals its spectral bound; cf. [1], Section 5.3. This means in this application that the largest real eigenvalue λ_0 of $A + B$ equals the growth bound $\omega_0(A+B)$. In other words, if $\lambda_0 < 0$ then the problem (5) is exponentially stable, and if $\lambda_0 > 0$ it is unstable. So Theorem 2.2 for this problem gives no new information. We remark in passing that there has been quite an interest in the number $\mathcal{R}_0 = e^{\lambda_0}$ which is called *net reproduction rate* of the cell population. Here we do not want to discuss this number any further, and refer to the specific literature, e.g., [4].

On the other hand, if $A + B$ has no imaginary eigenvalues, then Theorem 2.1 proves that (5) allows for an exponential dichotomy, which is a new result. It shows, for example, that if we add an immigration term $f \in BUC(\mathbb{R}; L_1(a,b))$ on the right-hand side of the first equation in (5), then the problem (5) considered on the whole time horizon $t \in \mathbb{R}$ (without initial condition) has a unique solution $u \in BUC(\mathbb{R}; L_1(a,b))$. The same assertion applies to subspaces of $BUC(\mathbb{R}; X)$ which are translation-invariant, like $C_0(\mathbb{R}; X)$ or $AP(\mathbb{R}; X)$, etc.

Acknowledgment

I am indebted to Christoph Schwerdt for bringing reference [2] to my attention. I would also like to thank the anonymous referee for his careful reading and his comments, which led to several improvements in the presentation of the paper.

References

[1] W. Arendt, Ch. Batty, M. Hieber, F. Neubrander, *Vector-Valued Laplace Transforms and Cauchy problems*, Monographs in Mathematics **96**, Birkhäuser Verlag, Basel 2001

[2] M. Beceanu, M. Goldberg, *Schrödinger dispersive estimates for a scaling-critical class of potentials.* Comm. Math. Phys. **314**, 471–482 (2012)

[3] G.I. Bell, E.C. Anderson, *Cell growth and division. I. A mathematical model with applications to cell volume distributions in mammalian suspension cultures.* Biophys. J. **7**, 329–351 (1967).

[4] O. Diekmann, H.J.A.M. Heijmans, H.R. Thieme, *On the stability of the cell size distribution.* J. Math. Biology **19**, 227–248 (1984).

[5] K.-J. Engel, R. Nagel, *One-Parameter Semigroups foor Linear Evolution Equations*, Graduate Texts in Mathematics **194**, Springer Verlag, Berlin 2000.

[6] J. Prüss, *On the spectrum of C_0-semigroups.* Trans. Amer. Math. Soc. **284** , 847–857 (1984)

[7] J. Prüss, *Evolutionary Integral Equations and Applications*, Monographs in Mathematics **87**, Birkhäuser Verlag, Basel 1993.

[8] J. Prüss, R. Schnaubelt, R. Zacher, *Mathematische Modelle in der Biologie*, Mathematik Kompakt, Birkhäuser Verlag, Basel, 2008.

Jan Prüss
Martin-Luther-Universität Halle-Wittenberg
Institut für Mathematik
Theodor-Lieser-Strasse 5
D-06120 Halle, Germany
e-mail: `jan.pruess@mathematik.uni-halle.de`

Operator Theory:
Advances and Applications, Vol. 250, 463–481

Gaussian and non-Gaussian Behaviour of Diffusion Processes

Derek W. Robinson

Abstract. We survey some recent results on Gaussian and non-Gaussian behaviour for the solutions of second-order diffusion equations on $\mathbf{R}^d$. Our emphasis is on non-Gaussian aspects of the diffusion corresponding to degenerate operators. In particular we describe

- the equivalence of strong ellipticity and Gaussian upper and lower bounds,
- the deduction of non-ergodic behaviour from integrated Gaussian upper bounds, and
- the relationships between volume doubling, the Poincaré inequality and Gaussian estimates.

To place these results into context we also summarize some well-established structural properties of diffusion phenomena.

Mathematics Subject Classification (2010). 47B25, 47D07, 35J70.

Keywords. Gaussian bounds, diffusion equations, strong ellipticity, Poincareé inequality, Grushin operators.

1. Introduction

Our aim is to describe and discuss some recent results on Gaussian and non-Gaussian behaviour for the solutions of second-order diffusion equations on $\mathbf{R}^d$. In particular we describe the relationships between strong ellipticity and Gaussian upper and lower bounds, the deduction of non-ergodic behaviour from integrated Gaussian upper bounds and the relationships between volume doubling, the Poincaré inequality and Gaussian estimates. In addition, to place these results into context, we also summarize some well-established structural properties of diffusion phenomena. Although much of the activity in this area in the last fifty years has concentrated on the Gaussian behaviour of non-degenerate diffusion our emphasis is on the non-Gaussian behaviour of degenerate diffusion. Non-Gaussian behaviour of non-degenerate diffusion on manifolds with ends has been considered by various authors [CF91] [BCF96] [Dav97] [GSC09] but some additional features arise for degenerate diffusions on $\mathbf{R}^d$.

There have been two principal areas of development since Aronson's original derivation [Aro67] of Gaussian upper and lower bounds. (A more complete picture with references to the literature is given by the books [Dav89] [SC02] [Ouh05] [Gri09].) The first area is that considered by Aronson, the analysis of diffusion phenomena described by strongly elliptic second-order operators in divergence-form on the Euclidean space $\mathbf{R}^d$. The strong ellipticity assumption ensures that the rate of diffusion is both strictly positive and uniformly bounded. These conditions are sufficient to guarantee Gaussian behaviour. If, however, the strong ellipticity assumption is relaxed then non-Gaussian phenomena can occur. Local degeneracies or global growth can both lead to variation away from the canonical Gaussian model.

The second area of activity has been analysis of the diffusion described by the Laplace–Beltrami operator acting on a Riemannian manifold. Then it follows by arguments of Grigor'yan and Saloff-Coste [Gri92] [SC92a] that the diffusion is Gaussian if and only if the manifold satisfies two specific regularity properties, one geometric and one analytic. In fact non-Gaussian behaviour can occur on quite simple manifolds such as a catenoid.

Following Aronson's original paper [Aro67] on Gaussian bounds we consider pure second-order operators in divergence-form on $\mathbf{R}^d$. In particular we consider quadratic forms

$$h(\varphi) = \sum_{i,j=1}^{d} (\partial_i\varphi, c_{ij}\, \partial_j\varphi) \tag{1}$$

with domain $D(h) = C_c^\infty(\mathbf{R}^d)$ where the coefficients c_{ij} are real $L_{\infty,\mathrm{loc}}$-functions, $c_{ij} = c_{ji}$ and $C = (c_{ij}) \geq 0$, in the sense of matrix ordering, almost everywhere. If the form h is closable then its closure $\overline{h}$ is a local Dirichlet form [BH91] [MR92] [FOT94]. The corresponding positive self-adjoint operator H on $L_2(\mathbf{R}^d)$ generates a submarkovian semigroup S on the spaces $L_p(\mathbf{R}^d)$ and the action of S is determined by a positive, i.e., non-negative, distributional kernel K. Explicitly, $(S_t\varphi)(x) = \int_{\mathbf{R}^d} dy\, K_t(x\,;y)\varphi(y)$.

The coefficient matrix C is defined to be strongly elliptic if there exist $\lambda, \mu > 0$ such that

$$\lambda I \geq C(x) \geq \mu I > 0 \tag{2}$$

for almost all $x \in \mathbf{R}^d$. It follows immediately from (2) that h is indeed closable and the domain of $\overline{h}$ is given by $D(\overline{h}) = W^{1,2}(\mathbf{R}^d)$. Then Aronson's arguments establish that there are $a, a', b, b' > 0$ such that

$$a'\, G_{b';t}(x-y) \leq K_t(x\,;y) \leq a\, G_{b;t}(x-y) \tag{3}$$

for almost all $x, y \in \mathbf{R}^d$ and all $t > 0$ where $G_{b;t}(x) = t^{-d/2}\, e^{-b\,|x|^2 t^{-1}}$. It is remarkable that these Gaussian bounds encapsulate a great deal of information concerning the semigroup kernel and the diffusion process. For example, Fabes and Stroock [FS86] demonstrated that the Gaussian bounds were sufficient to derive the Nash–De Giorgi [DeG57] [Nas58] results on the Hölder continuity of K. The bounds also suffice to establish that the semigroup S is conservative, i.e., its extension to $L_\infty(\mathbf{R}^d)$ satisfies $S_t\mathbb{1} = \mathbb{1}$ for all $t > 0$.

Our first topic of discussion is a converse of Aronson's result, the deduction of strong ellipticity of C from Gaussian bounds on K and a mild growth hypothesis on the coefficients.

2. Gaussian bounds: strong ellipticity

The simplest characterization of strong ellipticity by Gaussian bounds involves the semigroup conservation property, a property which places an implicit restriction on the growth of the c_{ij} coefficients at infinity.

Theorem 2.1. *Assume h is the quadratic form with $L_{\infty,\mathrm{loc}}$-coefficients defined by* (1)*. Then the following conditions are equivalent:*

I. *the matrix of coefficients C is strongly elliptic,*
II. *the form h is closable, the semigroup S is conservative and the semigroup kernel K satisfies the Gaussian bounds* (3).

The implication I⇒II is the classic result originating with Aronson. The converse implication is proved in [Rob13]. The theorem extends an earlier result of [ERZ06] in which it was assumed that the coefficients c_{ij} were uniformly bounded.

It might appear surprising that the implication II⇒I only appears to require local boundedness of the coefficients. But the conservation condition restricts the possible growth at infinity. To quantify the allowed growth let $\|C(x)\|$ denote the matrix norm, define $\nu(s) = \operatorname{ess\,sup}_{|x|\le s} \|C(x)\|$. Then introduce the positive increasing function ρ on $[0,\infty\rangle$ by

$$\rho(s) = \int_0^s dt\, (1+\nu(t))^{-1/2}$$

and the corresponding balls $B_\rho(r)$ by

$$B_\rho(r) = \{x \in \mathbf{R}^d : \rho(|x|) < r\} \;.$$

The Tikhonov growth condition is the requirement that there exist $a, b \ge 0$ such that

$$|B_\rho(r)| \le a\, e^{b\, r^2} \tag{4}$$

for all $r \ge 1$ where $|B_\rho(r)|$ is the volume, i.e., Lebesgue measure, of the ball. This is an implicit condition on the growth of the coefficients c_{ij}. It automatically implies that $\lim_{s\to\infty} \rho(s) = \infty$ because if the latter condition is false then ρ is bounded and $|B_\rho(r)| = \infty$ for all large r. One readily checks that (4) is satisfied if $\|C(x)\| \le c\,(1+|x|)^2 \log(2+|x|)$ and this is essentially the maximal growth allowed by the condition.

Theorem 2.2. *Let h be the quadratic form with $L_{\infty,\mathrm{loc}}$-coefficients defined by* (1). *Assume h is closable. If the Tikhonov growth condition* (4) *is satisfied then the submarkovian semigroup S associated with $\overline{h}$ is conservative.*

Combination of these two theorems gives an explicit characterization of strong ellipticity.

Corollary 2.3. *Assume h is the quadratic form with $L_{\infty,\mathrm{loc}}$-coefficients defined by* (1). *Further assume the Tikhonov growth condition* (4) *is satisfied. Then the following conditions are equivalent:*

I. *the matrix of coefficients C is strongly elliptic,*

II. *the form h is closable and the semigroup kernel K satisfies the Gaussian bounds* (3).

Remark 2.4. In the foregoing statements it is assumed that h is closable. This is not absolutely necessary. If h is not closable then similar statements are valid with the closure replaced by the relaxation. The relaxation h_0 of h is defined as the largest closed form which is dominated by h, i.e., the largest quadratic form in the set of closed forms k with $D(h) \subseteq D(k)$ and $k(\varphi) = h(\varphi)$ for all $\varphi \in D(h)$. The relaxation is automatically a Dirichlet form but it is an extension of h if and only if h is closable and then $h_0 = \overline{h}$. (For further details on the relaxation see [ET76] [Dal93] [Mos94] [Jos98] [Bra02]. The earlier version of Theorem 2.1 in [ERZ06] is phrased in terms of the relaxation.)

Corollary 2.3 clearly establishes that Gaussian kernel bounds of the type (3) are only possible for strongly elliptic diffusion processes. They are not valid if the diffusion coefficients have local degeneracies or grow at infinity. This shortcoming is emphasized by observing that the bounds imply that the action of S_t is ergodic on each of the spaces $L_p(\mathbf{R}^d)$, i.e., there are no non-trivial S_t-invariant subspaces. Nevertheless the Gaussian bound technique can be adapted to the description of some classes of degenerate processes by modifying the definition of the Gaussian function to match it with the inherent geometry of the process. Alternatively a different L_2-Gaussian technique can be used to derive information about the possible breakdown of ergodicity. We next discuss some aspects of these two approaches.

3. Integrated Gaussian bounds

In 1982 Cheeger, Gromov and Taylor [CGT82] introduced an alternative type of Gaussian upper bound to the pointwise bounds of (3). The new bound gives a direct estimate on the semigroup S acting on $L_2(\mathbf{R}^d)$ which takes the form

$$|(\varphi_A, S_t\varphi_B)| \leq e^{-d(A;B)^2(4t)^{-1}} \|\varphi_A\|_2 \, \|\varphi_B\|_2 \tag{5}$$

for all $\varphi_A \in L_2(A)$, $\varphi_B \in L_2(B)$ and $t > 0$ where A and B are two measurable subsets of $\mathbf{R}^d$ and $d(A\,;B)$ is a measure of the distance between the subsets A and B. These integrated bounds can be valid for a variety of choices of the distance but there is an optimal choice which we next describe.

The first and most commonly used distance is defined in terms of the coefficient matrix $C = (c_{ij})$ of the form h by

$$d_C(x\,;y) = \sup\{\psi(x) - \psi(y) : \psi \in W^{1,\infty}_{\rm loc}(\mathbf{R}^d)\,,\, \Gamma(\psi) \le 1\} \tag{6}$$

where the *carré du champ* Γ is given by $\Gamma(\varphi) = \sum_{i,j=1}^d c_{ij}\,(\partial_j\varphi)\,(\partial_i\varphi)$ for all $\varphi \in W^{1,\infty}_{\rm loc}(\mathbf{R}^d)$. The function $x, y \in \mathbf{R}^d \mapsto d_C(x\,;y)$ is often referred to as the control distance corresponding to H or to C. The associated set-theoretic distance is defined by

$$d_C(A\,;B) = \operatorname*{ess\,inf}_{x\in A, y\in B} d_C(x\,;y)\;. \tag{7}$$

If the inverse C^{-1} of the coefficient matrix defines a Riemannian metric then the corresponding Riemannian distance and the control distance coincide. But in general $d_C(\,\cdot\,;\,\cdot\,)$ is not strictly a distance. If C is merely elliptic, i.e., if it only satisfies the mild positivity assumption $C \ge 0$, the function can take the value ∞. If, however, $C \ge \mu I > 0$ then $d_C(x\,;y) \le \mu^{-1/2}|x-y|$. Moreover, if C is strongly elliptic, i.e., if $\lambda I \ge C \ge \mu I$ with $\lambda \ge \mu > 0$, then d_C is equivalent to the Euclidean distance since $\lambda^{-1/2}|x-y| \le d_C(x\,;y) \le \mu^{-1/2}|x-y|$. Therefore in the strongly elliptic case the bounds (5) are an integrated form of the pointwise Gaussian upper bound of (3). It is of interest that in the integrated form there are no arbitrary constants.

The derivation in [CGT82] of bounds of the form (5) was based on the property of finite speed of propagation of the associated wave equation. But Davies [Dav92] subsequently observed that these bounds could be derived by a method of Gaffney [Gaf59]. Hence the bounds are often referred to as Davies–Gaffney bounds. All these authors considered the Laplace–Beltrami operator on a Riemannian manifold and the bounds were expressed in terms of the Riemannian distance. Similar bounds can, however, be derived for general submarkovian semigroups as we next discuss. Then, however, there is some freedom of choice of the distance involved. We continue to adopt the assumptions of the introduction and, for simplicity, assume the form h given by (1) is closable.

The efficacity of the bounds (5) clearly depends on the choice of the distance $d(A\,;B)$, the larger the distance the better the bounds. But there is an optimal distance function associated with the closure $\overline{h}$. This function which we next describe can take values in $[0,\infty]$. Consequently the integrated bounds (5) can give information about invariant subspaces and non-ergodic behaviour of the semigroup S.

The closure $\overline{h}$ of h is a Dirichlet form and $D(\overline{h}) \cap L_\infty(\mathbf{R}^d)$ is an algebra. Therefore, for each $\varphi, \psi \in D(\overline{h}) \cap L_\infty(\mathbf{R}^d)$, one can define the truncated form Γ_φ by

$$\Gamma_\varphi(\psi) = \overline{h}(\psi\,\varphi, \psi) - 2^{-1}\,\overline{h}(\psi^2, \varphi)\;.$$

If $\varphi \ge 0$ it follows that $\psi \mapsto \Gamma_\varphi(\psi) \in \mathbf{R}$ is a form with domain $D(\Gamma_\varphi) = D(\overline{h}) \cap L_\infty(\mathbf{R}^d)$ and $0 \le \Gamma_\varphi(\psi) \le \|\varphi\|_\infty\,\overline{h}(\psi)$ for all $\psi \in D(\Gamma_\varphi)$ (see [BH91] or [FOT94]).

The definition is motivated by the observation that

$$\Gamma_\varphi(\psi) = \int_{\mathbf{R}^d} dx\, \varphi(x)\, \Gamma(\psi)(x)$$

for all $\varphi, \psi \in D(\overline{h}) \cap L_\infty(\mathbf{R}^d)$.

Next define $D(\overline{h})_{\mathrm{loc}}$ as the space of all measurable functions such that for every compact subset K of X there is a $\hat{\psi} \in D(\overline{h})$ with $\psi|_K = \hat{\psi}|_K$. Then for $\varphi \in D(\overline{h}) \cap L_{\infty,c}(\mathbf{R}^d)$, where $L_{\infty,c}(\mathbf{R}^d)$ denotes the L_∞-functions with compact support, one can define $\widehat{\Gamma}_\varphi$ by $D(\widehat{\Gamma}_\varphi) = D(\overline{h})_{\mathrm{loc}} \cap L_\infty(\mathbf{R}^d)$ and

$$\widehat{\Gamma}_\varphi(\psi) = \Gamma_\varphi(\hat{\psi})$$

for all $\psi \in D(\widehat{\Gamma}_\varphi)$ where $\hat{\psi} \in D(\Gamma_\varphi)$ is such that $\psi|_{\mathrm{supp}\,\varphi} = \hat{\psi}|_{\mathrm{supp}\,\varphi}$. Now set

$$|||\widehat{\Gamma}(\psi)||| = \sup\{|\widehat{\Gamma}_\varphi(\psi)| : \varphi \in D(\overline{h}) \cap L_{\infty,c}(\mathbf{R}^d),\ \|\varphi\|_1 \le 1\} \in [0, \infty]$$

for all $\psi \in D(\widehat{\Gamma}_\varphi)$. Finally, for all $\psi \in L_\infty(\mathbf{R}^d)$ and measurable sets $A, B \subset \mathbf{R}^d$ introduce

$$d_\psi(A\,;B) = \operatorname*{ess\,inf}_{x \in A} \psi(x) - \operatorname*{ess\,sup}_{y \in B} \psi(y) \in \langle -\infty, \infty] \,.$$

Then following [AH05] and [ERSZ06] the optimal distance function for the integrated Gaussian bounds is defined by

$$d(A\,;B) = \sup\{d_\psi(A\,;B) : \psi \in D_0(h)\} \tag{8}$$

where

$$D_0(h) = \{\psi \in D(\overline{h})_{\mathrm{loc}} \cap L_\infty(\mathbf{R}^d) : |||\widehat{\Gamma}(\psi)||| \le 1\}\,. \tag{9}$$

The function $A, B \mapsto d(A\,;B)$ has all the properties appropriate for a set-theoretic distance except one can have $d(A\,;B) = \infty$.

Theorem 3.1. *Let S be the submarkovian semigroup associated with the closure $\overline{h}$ of h. Further let $d(\,\cdot\,;\,\cdot\,)$ be defined by* (8) *and* (9)*. If A and B are measurable subsets of $\mathbf{R}^d$ then*

$$|(\varphi_A, S_t\varphi_B)| \le e^{-d(A;B)^2(4t)^{-1}}\, \|\varphi_A\|_2\, \|\varphi_B\|_2$$

for all $\varphi_A \in L_2(A)$, $\varphi_B \in L_2(B)$ and all $t > 0$ with the convention $e^{-\infty} = 0$.

The theorem is a special case of more general statements for submarkovian semigroups established by [AH05] Theorem 4.1 and [ERSZ06] Theorem 1.2 (see also [HR03] Theorem 2.8). One can also establish that the distance defined by (8) and (9) is the largest function for which the integrated Gaussian bounds are valid. This is achieved with the aid of a function which corresponds to the distance to a given measurable set.

Theorem 3.2. *Let A be a measurable set with $|A| > 0$. Then there exists a unique measurable function $d_A \in [0, \infty]$ such that*

I. *$d_A \wedge R \in D_0(h)$ for any $R \ge 0$,*

II. *$d_A = 0$ almost everywhere on A,*

III. *d_A is the largest function satisfying the previous two conditions.*
Moreover, if B is another measurable set then

$$d(A\,;B) = \operatorname*{ess\,sup}_{x\in B} d_A(x)$$

where $d(A\,;B)$ is given by (8) *and* (9).

It follows from this result that $d(A\,;B) \geq d_C(A\,;B)$ where $d_C(\,\cdot\,;\,\cdot\,)$ is the control distance defined by (6) and (7). The bounds in Theorem 3.1 were initially proved with respect to the control distance and the proof of the stronger result is similar but more delicate.

One can exploit the bounds of Theorem 3.1 to obtain criteria for the failure of ergodicity of the semigroup S. There are several equivalent definitions of ergodicity of the semigroup. We define S to be ergodic if there are no non-trivial measurable subsets A of $\mathbf{R}^d$ such that the subspaces $L_2(A)$ are left invariant by S_t for one $t > 0$ or, equivalently, for all $t > 0$. Since the semigroup is positive this is equivalent to the condition that $(\varphi, S_t\psi) > 0$ for each pair of non-zero, non-negative $\varphi, \psi \in L_2(\mathbf{R}^d)$ and for one, or for all, $t > 0$. This in turn is equivalent to the distributional kernel K_t of S_t being strictly positive for one, or for all, $t > 0$.

Alternatively, ergodicity of S is equivalent to irreducibility of the family of operators $S \cup L_\infty$. Here $S \cup L_\infty$ indicates the family of operators S_t, $t > 0$, together with the operators of multiplication by $L_\infty(\mathbf{R}^d)$-functions. Moreover, a family of operators acting on $L_2(\mathbf{R}^d)$ is defined to be irreducible if there is no non-trivial closed subspace of $L_2(\mathbf{R}^d)$ which is left invariant by the action of the family. Alternatively, the family is irreducible if and only if there are no non-trivial bounded operators which commute with each member of the family.

Finally if S is a submarkovian semigroup whose generator is determined by the local Dirichlet form k then the subset A is invariant if and only if the characteristic function $\mathbb{1}_A$ is a multiplier for the domain of k, i.e., if and only if $\mathbb{1}_A D(k) \subset D(k)$.

The next theorem gives alternative criteria for invariance of a set under the semigroup S in terms of the distance $d(\,\cdot\,;\,\cdot\,)$ or the subset $D_0(h)$.

Theorem 3.3. *Adopt the assumptions of Theorem* 3.1.

The following conditions are equivalent for each measurable subset A:

I. (I′.) $S_tL_2(A) \subseteq L_2(A)$ *for all* $t > 0$ (*for one* $t > 0$),
II. (II′.) $d(A\,;A^{\mathrm{c}}) = \infty$ $(d(A\,;A^{\mathrm{c}}) > 0)$,
III. (III′.) $\mathbb{1}_A \in D_0(h)$ *and* $|||\widehat{\Gamma}(\mathbb{1}_A)||| = 0$ $(\mathbb{1}_A \in D_0(h))$.

Variations of this statement are given by [AH05] Proposition 5.1, which is an extension of Lemma 2.16 in [HR03], or Theorem 1.3 of [ERSZ06]. In particular the equivalence of the first four conditions is given by the latter theorem. But the equivalence of these four conditions with the last two is straightforward. Suppose $\mathbb{1}_A \in D_0(h)$. Then

$$d(A\,;A^{\mathrm{c}}) \geq d_{\mathbb{1}_A}(A\,;A^{\mathrm{c}}) = \operatorname*{ess\,inf}_{x\in A} \mathbb{1}_A(x) - \operatorname*{ess\,sup}_{y\in A^{\mathrm{c}}} \mathbb{1}_A(y) = 1$$

so III⇒II′. Conversely suppose $S_tL_2(A) \subseteq L_2(A)$ for all $t > 0$. Then it follows that $\mathbb{1}_A\varphi, \mathbb{1}_A\psi \in D(\overline{h}) \cap L_\infty(\mathbf{R}^d)$ for each pair $\varphi, \psi \in D(\overline{h}) \cap L_\infty(\mathbf{R}^d)$. Since the form $\overline{h}$ is local it follows immediately that $\Gamma_\varphi(\mathbb{1}_A\psi) = \Gamma_{\mathbb{1}_A\varphi}(\psi)$. Hence if $\psi \in D(\overline{h})_{\mathrm{loc}} \cap L_\infty(\mathbf{R}^d)$ and $\varphi \in D(\overline{h}) \cap L_{\infty,c}(\mathbf{R}^d)$ then $\widehat{\Gamma}_\varphi(\mathbb{1}_A\psi) = \widehat{\Gamma}_{\mathbb{1}_A\varphi}(\psi)$. But $\mathbb{1} \in D(\overline{h})_{\mathrm{loc}} \cap L_\infty(\mathbf{R}^d)$ and $\widehat{\Gamma}_\varphi(\mathbb{1})$ for all $\varphi \in D(\overline{h}) \cap L_{\infty,c}(\mathbf{R}^d)$. Therefore $\widehat{\Gamma}_\varphi(\mathbb{1}_A) = \widehat{\Gamma}_{\mathbb{1}_A\varphi}(\mathbb{1}) = 0$. Hence I⇒III.

Ergodicity of S is clearly a prerequisite for the existence of Gaussian lower bounds on the corresponding kernel because the lower bounds imply strict positivity of the kernel.

Corollary 3.4. *The following conditions are equivalent:*

I. *S is ergodic,*

II. *the set-theoretic distance $d(\,\cdot\,;\,\cdot\,)$ defined by* (8) *and* (9) *is finite valued.*

Proof. I⇒II If Condition II is false then, by Theorem 3.1, there are measurable subsets A, B and $\varphi_A \in L_2(A)$, $\varphi_B \in L_2(B)$ such that $(\varphi_A, S_t\varphi_B) = 0$. Thus Condition I is false.

II⇒I If Condition I is false then there is a measurable subset A such that $S_tL_2(A) \subseteq L_2(A)$ for some $t > 0$. Then $d(A; A^{\mathrm{c}}) = \infty$ by Theorem 3.3 and so Condition II is false. □

The following one-dimensional example illustrates that local degeneracies can lead to the existence of non-trivial invariant subspaces, i.e., the breakdown of ergodicity.

Example 3.5. Let $d = 1$. Then there is a single non-negative coefficient c. Consider the specific case $c(s) = |s|^{2\delta_+} \wedge 1$ if $s \geq 0$ and $c(s) = |s|^{2\delta_-} \wedge 1$ if $s \leq 0$ where $\delta_\pm \in \langle 0, 1\rangle$. In particular $c(0) = 0$. Since c is bounded and continuous it follows that the corresponding form $h(\varphi) = \int_{-\infty}^{\infty} c\,(\varphi')^2$ is closable.

Now suppose $\delta_- \in [1/2, 1\rangle$ and define $\chi_n\colon \mathbf{R} \to [0, 1]$ by

$$\chi_n(x) = \begin{cases} 0 & \text{if } x \leq -1\,, \\ \eta_n^{-1}\eta(x) & \text{if } x \in \langle -1, -n^{-1}\rangle\,, \\ 1 & \text{if } x \geq -n^{-1}\,. \end{cases} \tag{10}$$

where $\eta(x) = \int_{-1}^{x} c^{-1}$ and $\eta_n = \eta(-n^{-1})$. It follows that χ_n is absolutely continuous, increasing and $\lim_{n\to\infty} \chi_n = \mathbb{1}_{[0,\infty\rangle}$ pointwise. But $\chi_n = 1$ on $[0, \infty\rangle$ so one readily computes that $\overline{h}(\chi_n\varphi - \chi_m\varphi) \to 0$ as $n, m \to \infty$ for all $\varphi \in D(h) \cap L_\infty(\mathbf{R})$. Thus $\mathbb{1}_{[0,\infty\rangle}\varphi \in D(\overline{h})$ and $L_2(0, \infty)$ is invariant under the corresponding semigroup S. Hence $L_2(-\infty, 0)$ is also S-invariant. A similar conclusion is valid if $\delta_+ \in [1/2, 1\rangle$. Therefore one deduces that the subspaces $L_2(-\infty, 0)$ and $L_2(0, \infty)$ are both S-invariant whenever $\delta_+ \vee \delta_- \in [1/2, 1\rangle$. Thus the latter condition implies that S is not ergodic. It can also be verified that if the condition fails then S is ergodic.

Since the half-lines $A = \langle -\infty, 0]$ and $B = [0, \infty\rangle$ are left invariant by the semigroup one must have $d(A\,;B) = \infty$. Nevertheless the 'Riemannian distance' is given by $d_c(x\,;y) = |\int_y^x c^{-1/2}|$ and since $\delta_\pm \in \langle 0, 1\rangle$ one has $d_c(x\,;y) < \infty$ for all $x, y \in \mathbf{R}$. Hence $d_c(A\,;B) < \infty$ and consequently $d_c(A\,;B) < d(A\,;B)$. The critical feature in this example is the condition $\delta_+ \vee \delta_- \in [1/2, 1\rangle$ which ensures that c^{-1} is not integrable at the origin.

In higher dimensions it is also possible to have 'approximate' breakdown of ergodicity. Such phenomena occur if there are large but finitely extended surfaces of degeneracy which present obstacles to diffusion but which can be circumvented. In such cases one can expect that $d_C(A\,;B) < d(A\,;B) < \infty$. In the next section we give an example in which global degeneracy gives rise to approximately invariant subspaces.

4. Gaussian bounds: Riemannian geometry

In this section we discuss a different form of pointwise Gaussian bound formulated in terms of the control distance $d_C(\,\cdot\,;\,\cdot\,)$ defined by (6). In particular the Gaussian functions are defined in terms of the geometry corresponding to the control distance. The essential idea is to identify $d_C(\,\cdot\,;\,\cdot\,)$ as the distance related to the 'metric' C^{-1}. There are two problems with this approach. The first problem is that $d_C(\,\cdot\,;\,\cdot\,)$ is not necessarily a *bona fide* distance since it may happen that $d_C(x\,;y) = \infty$. Secondly, C^{-1} does not necessarily define a metric. This difficulty can occur either because C has strong local degeneracies or because it grows too rapidly at infinity. Consequently the discussion requires more detailed assumptions on the coefficients than were hitherto necessary.

Throughout the section we assume the coefficients are locally bounded and that $C > 0$ almost everywhere. Then C^{-1} is almost everywhere defined but not necessarily bounded nor bounded away from zero. Note that even the assumption $C > 0$ is quite restrictive since it rules out simple examples such as the Heisenberg sublaplacian $-\partial_1^2 - (\partial_2 + x_1\,\partial_3)^2$ or the sublaplacian $-\partial_1^2 - (c_1\partial_2 + s_1\,\partial_3)^2$ of the Euclidean motions group on $L_2(\mathbf{R}^3)$. Here $c_1 = \cos x_1, s_1 = \sin x_1$. Both these operators have the property that the lowest eigenvalue of the coefficient matrix is identically zero. We assume, however, that d_C has the following two basic properties:

$$\left.\begin{array}{ll} 1. & d_C(x\,;y) < \infty \text{ for all } x, y \in \mathbf{R}^d, \\ 2. & d_C \text{ is continuous and induces the Euclidean topology.} \end{array}\right\} \tag{11}$$

In the special case that C^{-1} is everywhere invertible it follows that $d_C(\,\cdot\,;\,\cdot\,)$ coincides with the corresponding Riemannian distance. In particular $d_I(x\,;y) = |x-y|$, the Euclidean distance.

Our aim is to describe a characterization by Grigor'yan [Gri92] and Saloff-Coste [SC92a, SC92b, SC95] of Gaussian bounds formulated in terms of the control distance by two general conditions, volume doubling and the Poincaré inequality. We begin by introducing these conditions.

The ball $B_C(x\,;r)$ with centre x and radius r corresponding to $d_C(\,\cdot\,;\,\cdot\,)$ is defined by

$$B_C(x\,;r) = \{y : d_C(x\,;y) < r\}\,. \tag{12}$$

The volume doubling property is then defined by the condition

$$|B_C(x\,;2\,r)| \le a\,|B_C(x\,;r)| \tag{13}$$

for some $a \ge 1$, for all $x \in \mathbf{R}^d$ and $r > 0$ where $|B_C|$ denotes the volume (Lebesgue measure) of the ball B_C. Condition (13) is in fact equivalent to the seemingly stronger condition that there is a $D > 0$ such that

$$|B_C(x\,;s)| \le a\,(s/r)^D |B_C(x\,;r)|$$

for all $r \le s$. The latter condition shows that the volume can grow at most polynomially and the effective dimension is given by D. (One can in fact choose $D = \log a/\log 2$ although this is not always optimal.)

The volume doubling property has the inherent drawback that it limits the applicability of subsequent results since it places a constraint on the uniformity of growth. This is illustrated by the following example.

Example 4.1. Let $d = 1$. Then the matrix C is replaced by a single real non-negative function c. Assume c is strictly positive almost everywhere. It follows from (6) that $d_c(x\,;y) = |\int_y^x ds\,c(s)^{-1/2}|$ for all $x, y \in \mathbf{R}$. Now consider the specific case $c(s) = |s|^{2\delta_-}$ if $s < 0$ and $c(s) = |s|^{2\delta_+}$ if $s \ge 0$ where $\delta_\pm \in [0, 1\rangle$. Then $d_c(x\,;0) = (1-\delta_-)^{-1}|x|^{1-\delta_-}$ if $x < 0$ and $d_c(x\,;0) = (1-\delta_+)^{-1}|x|^{1-\delta_+}$ if $x \ge 0$.

Now fix $r > 0$ and set $x = -((1-\delta_-)r)^{1/(1-\delta_-)}$. Thus $d_c(x\,;0) = r$ and $B_c(x\,;r) = \langle\gamma, 0\rangle$ with $\gamma = -((1-\delta_-)2r)^{1/(1-\delta_-)}$. Hence $|B_c(x\,;r)| \sim r^{1/(1-\delta_-)}$ as $r \to \infty$. But $B_c(x\,;2r) = \langle\gamma_1, \gamma_2\rangle$ with $\gamma_1 = -((1-\delta_-)3r)^{1/(1-\delta_-)}$ and $\gamma_2 = ((1-\delta_+)r)^{1/(1-\delta_+)}$. Hence $|B_c(x\,;2r)| = \gamma_1 + \gamma_2 \sim r^{1/(1-\delta_-\vee\delta_+)}$ as $r \to \infty$. Thus the doubling property fails for large r if $\delta_+ > \delta_-$. Alternatively by considering balls centred on the right half-line one finds that doubling fails if $\delta_- > \delta_+$. Therefore the doubling property holds if and only if $\delta_- = \delta_+$.

Note that a similar conclusion follows with $c(s) = (1+|s|)^{2\delta_-}$ if $s < 0$ and $c(s) = (1+|s|)^{2\delta_+}$ if $s \ge 0$. But in this latter case $c \ge 1$.

The second property of importance for the characterization of Gaussian bounds is the Poincaré inequality. This requires that there is a $b > 0$ and a $\kappa \in \langle 0, 1]$ such that

$$\int_{B_C(x;r)} \Gamma(\varphi) \ge b\,r^{-2} \inf_{M\in\mathbf{R}} \int_{B_C(x;\kappa r)} (\varphi - M)^2 \tag{14}$$

for all $x \in \mathbf{R}^d$, $r > 0$ and $\varphi \in D(h)$. Note that the infimum is attained with $M = \langle\varphi\rangle_{B_C} = |B_C(x;\kappa r)|^{-1} \int_{B_C(x;\kappa r)} \varphi$. In the classic formulation of the Poincaré inequality for the Laplacian one has $\kappa = 1$. It was, however, established by Jerison [Jer86] (see also [Lu94]) that under the foregoing assumptions, and in particular the volume doubling property, the validity of the inequality is independent of the value of $\kappa \in \langle 0, 1]$. The seemingly weaker formulation of the Poincaré inequality is convenient since it allows one to establish an important stability property.

If f and g are two functions with values in a real ordered space then the equivalence relation $f \sim g$ is defined to mean there are a, $b > 0$ such that $a\,f \leq g \leq b\,f$. In particular two strictly positive matrices C and C_0 are equivalent, $C \sim C_0$, if there exist $\lambda \geq \mu > 0$ such that $\lambda\, C_0 \geq C \geq \mu\, C_0$. It then follows that $d_C \sim d_{C_0}$. Explicitly one has

$$\mu^{-1/2} d_{C_0}(x\,;y) \geq d_C(x\,;y) \geq \lambda^{-1/2} d_{C_0}(x\,;y) \tag{15}$$

for all $x, y \in \mathbf{R}^d$. Therefore d_C and d_{C_0} satisfy (11) simultaneously, i.e., the conditions (11) are stable under the equivalence relation $C \sim C_0$. It also follows from (15) that

$$B_{C_0}(x\,;\mu^{1/2}r) \subseteq B_C(x\,;r) \subseteq B_{C_0}(x\,;\lambda^{1/2}r) \tag{16}$$

for all $x \in \mathbf{R}^d$ and $r > 0$. Hence the balls B_C satisfy the volume doubling if and only if the balls B_{C_0} satisfy the property. Thus volume doubling is also stable under the equivalence relation. Finally if Γ_C and Γ_{C_0} denote the *carré du champ* corresponding to C and C_0, respectively, then $C \sim C_0$ implies $\Gamma_C \sim \Gamma_{C_0}$. Hence it follows straightforwardly with the aid of (15) that the Poincaré inequality (14) is valid for Γ_C if and only if it is valid for Γ_{C_0}. Thus the Poincaré inequality is also stable under the equivalence relation.

The Gaussian function corresponding to the C^{-1} metric is now defined by

$$G_{b;t}(x\,;y) = \Big(B(x\,;t^{1/2})\,B(y\,;t^{1/2})\Big)^{-1/2} e^{-b\,d_C(x;y)^2/t} \tag{17}$$

for all $x, y \in \mathbf{R}^d$ and $b, t > 0$.

Theorem 4.2 (Grigor'yan, Saloff-Coste). *Assume the form h defined by* (1) *is closable. Let K be the distributional kernel of the submarkovian semigroup S associated with the closure $\overline{h}$ of h. The following conditions are equivalent:*

I. *the volume doubling property* (13) *and the Poincaré inequality* (14) *are both satisfied,*
II. *there are $a, a', b, b' > 0$ such that*

$$a'\,G_{b';t}(x\,;y) \leq K_t(x\,;y) \leq a\,G_{b;t}(x\,;y) \tag{18}$$

for all $x, y \in \mathbf{R}^d$ and $t > 0$ where $G_{b;t}$ is given by (17).

The theorem is a special case of a result obtained independently by Grigor'yan [Gri92] and Saloff-Coste [SC92a, SC92b, SC95]. The general result is for operators on manifolds and it has subsequently been extended to Dirichlet spaces [Stu95, Stu96]. The key observation is that the combination of volume doubling and the Poincaré inequality is equivalent to the parabolic Harnack inequality introduced by Moser [Mos64] in his derivation of the Nash–De Giorgi regularity theorem [Nas58] [DeG57] for strongly elliptic second-order operators with measurable coefficients. Theorem 4.2 gives a useful and insightful characterization of Gaussian upper and lower bounds. It not only gives a criterion for the validity of the bounds but it can also be used to characterize situations for which the Gaussian bounds fail and to understand the obstructions which lead to the failure.

Example 4.1 shows that volume doubling can fail if the asymptotic growth is inhomogeneous. The following examples show that the Poincaré inequality can also fail either because the global growth of the coefficients is too rapid or because their local degeneracy is too strong.

Example 4.3. Let $d = 1$. Then there is a single coefficient c which is strictly positive almost everywhere and again $d_c(x\,;y) = |\int_y^x ds\, c(s)^{-1/2}|$ for all $x, y \in \mathbf{R}$. Now consider the case $c(s) = (1+|s|)^{2\delta}$ with $\delta \in \langle 1/2, 1\rangle$. Then $d_c(x\,;0) = (1-\delta)^{-1}(1+|x|)^{1-\delta}$. Hence $|B_c(x\,;r)| \sim r^{1/(1-\delta)}$ as $r \to \infty$.

Let $\chi \in C^1(\mathbf{R})$ be an odd increasing function with $\chi(x) = 1$ if $x \geq 1$. Then $\Gamma(\chi)$ is a positive bounded function with support in the interval $[-1, 1]$. Hence

$$\int_{B_c(0;r)} dx\, \Gamma(\chi)(x) \leq \int_{-1}^{1} dx\, \Gamma(\chi)(x)$$

for all $r > 0$. On the other hand

$$r^{-2}\int_{B_c(0;r)} dx\, (\chi(x) - \langle\chi\rangle)^2 = r^{-2}\int_{B_c(0;r)} dx\, (\chi(x_1))^2 \sim |B_c(0\,;r)|\, r^{-2}$$
$$\sim r^{(2\delta-1)/(1-\delta)}$$

as $r \to \infty$. Since $(2\delta-1)/(1-\delta) > 0$ the Poincaré inequality must again fail. In fact the same conclusion follows for $\delta = 1/2$ by a similar argument.

In this example the Poincaré inequality fails because of the rapid growth of the coefficient at infinity or the concomitant rapid volume growth. The next example shows that the inequality can also fail because of local degeneracy.

Example 4.4. Again consider the case $d = 1$ but with $c(s) = |s|^{2\delta}$ for $s \in [-1, 1]$ where $\delta \in [1/2, 1\rangle$. The value of c for $|s| > 1$ is irrelevant since our aim is to argue that the Poincaré inequality fails on the ball $B_c(0\,;r)$ where r is chosen such that $B_c(0\,;r) = \langle -1, 1\rangle$. (Since $d_c(x\,;0) = (1-\delta)^{-1}|x|^{1-\delta}$ one has $r = (1-\delta)^{-1}$.)

For each $n = 1, 2, \ldots$ define χ_n as an odd increasing function on $[-1, 1]$ with

$$\chi_n(x) = \begin{cases} 0 & \text{if } 0 \leq x \leq n^{-1} \\ 1 - \eta_n^{-1}\,\eta(x) & \text{if } n^{-1} \leq x \leq 1 \end{cases}$$

where $\eta(x) = \int_x^1 ds\, |s|^{-2\delta}$ and $\eta_n = \eta(n^{-1})$. It follows that $\lim_{n\to\infty} \chi_n(x) = 1$ if $x > 0$ and $\lim_{n\to\infty} \chi_n(x) = -1$ if $x < 0$, e.g., if $\delta = 1/2$ then $\eta_n^{-1}\eta(x) \sim (\log n)^{-1}\log|x| \to 0$. Therefore

$$\lim_{n\to\infty} \int_{-\kappa}^{\kappa} dx\; (\chi_n(x) - \langle\chi_n\rangle)^2 = \lim_{n\to\infty} \int_{-\kappa}^{\kappa} dx\, \chi_n(x)^2 = 2\,\kappa$$

for all $\kappa \in \langle 0, 1]$. But

$$\lim_{n\to\infty} \int_{-1}^{1} dx\, \Gamma(\chi_n)(x)^2 = \lim_{n\to\infty} \int_{-1}^{1} dx\, |x|^{2\delta}\, (\chi_n'(x))^2 = 2 \lim_{n\to\infty} \eta_n = 0\,.$$

One concludes that the Poincaré inequality must fail for χ_n on the ball $B_c(0\,;r)$ if n is sufficiently large.

Although the three examples we have considered are all one-dimensional they do illustrate general phenomena which persist in higher dimensions. We consider more complicated and more interesting examples in Section 5.

Remark 4.5. Theorem 4.2 assumes that h is closable. Again this is not absolutely necessary. A similar result is valid with the closure replaced by the relaxation h_0. In particular if $K^{(0)}$ is the distributional kernel of the submarkovian semigroup $S^{(0)}$ associated with h_0 then $K^{(0)}$ satisfies the Gaussian bounds of Theorem 4.2 if and only if the volume doubling property and the Poincaré inequality are valid.

5. Models of behaviour

Theorem 4.2 provides conceptual insight into the Gaussian character of diffusion and, in principle, provides a method to verify Gaussian bounds. Unfortunately there is a large divide between practise and principle. In order to establish the validity or invalidity of volume doubling or the Poincaré inequality it is first necessary to estimate the control distance and the growth properties of the corresponding balls. The examples of Sections 4 and 3 indicate that this is not difficult in one-dimension. There are, however, many more difficulties in higher dimensions because of the possibility of more complicated phenomena. Nevertheless there is a fairly realistic family of operators which describe the diffusion associated with flows around a surface which can be analyzed in detail. We conclude with a brief description of this family.

Assume $d = n + m$ and $C \sim C_\delta$ where C_δ is a block diagonal matrix, $C_\delta(x_1, x_2) = c_1(x_1)\, I_n + c_2(x_1)\, I_m$, on $\mathbf{R}^d = \mathbf{R}^n \times \mathbf{R}^m$ with c_1, c_2 positive functions and

$$c_i(x_1) \sim |x_1|^{(2\delta_i, 2\delta_i')} \,. \tag{19}$$

The indices $\delta_1, \delta_2, \delta_1', \delta_2'$ are all non-negative and $\delta_1, \delta_1' < 1$ but there is no upper bound on δ_2 and δ_2'. Here $a^{(\alpha,\alpha')} = a^\alpha$ if $a \in [0,1]$ and $a^{(\alpha,\alpha')} = a^{\alpha'}$ if $a \geq 1$. It follows that $\Gamma \sim \Gamma_\delta$ where

$$\Gamma_\delta(\varphi) = c_1\, |\nabla_{x_1}\varphi|^2 + c_2\, |\nabla_{x_2}\varphi|^2 \tag{20}$$

for all $\varphi \in W^{1,2}_{\mathrm{loc}}(\mathbf{R}^d)$. Note that we do not assume any regularity of the coefficients $C = (c_{ij})$ but since C is only defined up to equivalence with C_δ there is a freedom of choice of the coefficients c_1, c_2. In particular they may be chosen to be continuous. Then the corresponding quadratic form h_δ is closable by standard reasoning (see, for example, [MR92] Section II.2b). Since $h_\delta \sim h$, the form with coefficients C, it follows that h is also closable. Throughout the remainder of the section H and H_δ denote the operators associated with the closures of the forms h and h_δ. Note that $H \sim H_\delta$ in the sense of ordering of positive self-adjoint operators.

One can exploit the equivalence $C \sim C_\delta$ and the special form of the C_δ-coefficients to characterize the corresponding control distances up to equivalence.

Proposition 5.1. *The control distances d_C and d_{C_δ} corresponding to the coefficients C and C_δ, respectively, are equivalent to the 'distance' D_δ given by*

$$D_\delta((x_1,x_2);(y_1,y_2)) =$$
$$\frac{|x_1-y_1|}{(|x_1|+|y_1|)^{(\delta_1,\delta_1')}} + \frac{|x_2-y_2|}{(|x_1|+|y_1|)^{(\delta_2,\delta_2')}+(|x_2|+|y_2|)^{(\gamma,\gamma')}} \tag{21}$$

where $x_1,y_1 \in \mathbf{R}^n$ and $x_2,y_2 \in \mathbf{R}^m$ and $\gamma = \delta_2(1+\delta_2-\delta_1)^{-1}$ and $\gamma' = \delta_2'(1+\delta_2'-\delta_1')^{-1}$.

In fact D_δ is a quasi-distance. It has all the metric properties of a distance except the triangle inequality. But it does satisfy the weaker version $D_\delta(x+x';y) \le a\,(D_\delta(x;y)+D_\delta(x';y))$ etc. with $a>1$. Nevertheless the balls defined by D_δ are equivalent to those defined by d_C or to those defined by d_{C_δ} in the sense of inclusions analogous to (16).

Although the function D_δ looks quite complicated its structure can be understood by examining the simplest case $n=1$. Then the operator H_δ describes the diffusion corresponding to a 'flow' $(c_1^{1/2}\partial_{x_1}, c_2^{1/2}\nabla_{x_2})$. The component $c_1^{1/2}\partial_{x_1}$ describes the flow normal to the $(d-1)$-dimensional hypersurface $x_1=0$ and the component $c_2^{1/2}\nabla_{x_2}$ describes the tangential flow. The first term on the right of (21) is equivalent to the distance measured in the normal direction, $|\int_{x_1}^{y_1} c_1^{-1/2}|$. In particular it is independent of the tangential flow, i.e., independent of δ_2,δ_2'. The second term is a measure of the distance in the tangential direction. This contribution depends on both the tangential and the normal flows. If $\delta_2>0$ the tangential component of the flow is zero on the hypersurface $x_1=0$. Thus a geodesic from $(0,x_2)$ to $(0,y_2)$ must leave the hypersurface, under the impetus of the normal flow, at one endpoint and return at the other. This explains the dependence of the second term on the normal flow and indicates why the geometry is relatively complicated.

The identification of the control distance given by Proposition 5.1 allows one to verify the doubling property and to estimate the corresponding local and global dimensions.

Corollary 5.2. *The balls B_C corresponding to the distance d_C satisfy the doubling property*

$$|B_C(x;s)| \le a\,(s/r)^{(D,D')}|B_C(x;r)|$$

for all $s>r$ where

$$D=(n+m(1+\delta_2-\delta_1))\,(1-\delta_1)^{-1} \quad \text{and} \quad D'=(n+m(1+\delta_2'-\delta_1'))\,(1-\delta_1')^{-1}\,.$$

Detailed proofs of both these results are given in [RS08]. It is evident that D and D' correspond to local and global dimensions. This is borne out by the estimates of Proposition 3.1 of [RS08] which establishes that the semigroup S generated by H, the operator with coefficients C, satisfies the L_1 to L_∞ bounds $\|S_t\|_{1\to\infty} \le a\,t^{(-D/2,-D'/2)}$ for all $t>0$. Thus the corresponding kernel K is bounded on $\mathbf{R}^d \times \mathbf{R}^d$ and $\|K_t\|_\infty \le a\,t^{(-D/2,-D'/2)}$ for all $t > 0$. The semigroup and kernel S^δ and K^δ corresponding to H_δ satisfy similar bounds.

The volume doubling property established by Corollary 5.2 is the first key property in the criterion for Gaussian behaviour given by Theorem 4.2. It should, however, be emphasized that this property can fail if one has asymmetry of growth. This was illustrated in one dimension by Example 4.1 and similar behaviour can occur in the multi-dimensional case.

Example 5.3. Assume $n = 1$. Further assume $c_1(x_1) \sim |x_1|^{2\delta_\pm}$ as $x_1 \to \pm\infty$ with $\delta_\pm \in [0,1\rangle$ and $c_2(x_1) \sim |x_1|^{2\delta_2'}$ as $|x_1| \to \infty$ with $\delta_2' \ge 0$. Now for each $r > 0$ choose $x_1 < 0$ such that $d_C((0,0)\,;(x_1,0)) = r$. Then the ball $B_{C_\delta}((x_1,0)\,;r)$ is a subset of the left half-space, $\Omega_- = \{y = (y_1,y_2) : y_1 < 0\}$. Next choose $y_1 > z_1 > 0$ such that $d_C((0,0)\,;(y_1,0)) = 2\,r$ and $d_C((0,0)\,;(z_1,0)) = r = d_C((z_1,0)\,;(y_1,0))$. Then the ball $B_{C_\delta}((z_1,0)\,;r)$ is a subset of the right half-space, $\Omega_+ = \{y = (y_1,y_2) : y_1 > 0\}$. Moreover, $B_{C_\delta}((z_1,0)\,;r) \subseteq B_{C_\delta}((x_1,0)\,;3\,r)$. Therefore if volume doubling is valid there is an $a > 0$ such that $|B_{C_\delta}((x_1,0)\,;3\,r)| \le a\,|B_{C_\delta}((x_1,0)\,;r)|$ for all $r > 0$. This then implies that $|B_{C_\delta}((z_1,0)\,;r)| \le a\,|B_{C_\delta}((x_1,0)\,;r)|$ for all $r > 0$. But the growth in volume of the left-hand and right-hand balls, as $r \to \infty$, is dependent on δ_+' and δ_-', respectively. It follows from a variation of the proof of Proposition 5.1 in [RS08] that $|B_{C_\delta}((z_1,0)\,;r)| \sim r^{D_+'}$ and $|B_{C_\delta}((x_1,0)\,;r)| \sim r^{D_-'}$ as $r \to \infty$ where $D_\pm' = m + (1 + m\,\delta_2')(1 - \delta_\pm')^{-1}$. Therefore this is a contradiction if $\delta_+' > \delta_-'$. Alternatively arguing with $x_1 > 0$ one finds a contradiction if $\delta_-' > \delta_+'$. Hence for volume doubling the condition $\delta_+' = \delta_-'$ is necessary.

It is also possible to analyze the Poincaré inequality for the operator H or, equivalently, for H_δ. The situation is straightforward if $n \ge 2$ but there are two distinct interesting effects that can arise if $n = 1$. The situation is summarized by the following result from [RS13].

Theorem 5.4. **I.** *If $n \ge 2$, or if $n = 1$ and $\delta_1 \vee \delta_1' \in [0,1/2\rangle$, then the Poincaré inequality* (14) *is valid.*

II. *If $n = 1$ and $\delta_1 \vee \delta_1' \in [1/2,1\rangle$ then the Poincaré inequality* (14) *fails.*

III. *If $n = 1$ and $\delta_1 \in [1/2,1\rangle$ then the Poincaré inequality is valid on the half-spaces $\Omega_\pm$.*

The last statement means that if $n = 1$ and $\delta_1 \in [1/2,1\rangle$ then there exist $b > 0$ and $\kappa \in \langle 0,1]$ such that

$$\int_{B_\pm(x;r)} dy\,\Gamma(\varphi)(y) \ge b\,r^{-2} \int_{B_\pm(x;\kappa r)} dy\,\bigl(\varphi(y) - \langle\varphi\rangle_{B_\pm}\bigr)^2 \qquad (22)$$

for all $x \in \Omega_\pm$, $r > 0$ and $\varphi \in D(h)$ where $B_\pm(x\,;r) = B_C(x\,;r) \cap \Omega_\pm$.

The theorem establishes that the Poincaré inequality is always valid if $n \ge 2$ but if $n = 1$ then there are three distinct cases to consider. The implications for Gaussian behaviour of the corresponding diffusion is not evident. We next summarize the conclusions one can draw from the foregoing results.

Corollary 5.5. *Assume $n \geq 2$. Then the distributional kernel K of the semigroup S generated by H satisfies the Gaussian bounds* (18).

Proof. It follows from Corollary 5.2 that the balls B_C satisfy the volume doubling property (13) and the Poincaré inequality (14) is valid by the first statement of Theorem 5.4. Therefore the corollary is an immediate consequence of Theorem 4.2. □

Finally consider the case $n = 1$. As explained above this corresponds to the diffusion around the $(d-1)$-dimensional hypersurface $x_1 = 0$. The properties of the diffusion are then dictated by the local and global degeneracies of the normal flow, i.e., to the values of δ_1 and δ_1'. At the risk of pedantry we consider the three cases separately.

Corollary 5.6. *Assume $n = 1$ and $\delta_1, \delta_1' \in [0, 1/2\rangle$ Then the distributional kernel K of the semigroup S generated by H satisfies the Gaussian bounds* (18).

This follows by the same reasoning as for $n = 2$. The Gaussian behaviour of the diffusion persists if the normal flow does not slow too suddenly at the hypersurface, i.e., if $\delta_1 \in [0, 1/2\rangle$, and if it also does not accelerate too swiftly at infinity, i.e., if $\delta_1' \in [0, 1/2\rangle$. The situation changes dramatically if the normal flow is strongly degenerate.

Corollary 5.7. *Assume $n = 1$ and $\delta_1 \in [1/2, 1\rangle$. Then the half-spaces $\Omega_\pm$ are invariant under the semigroup S. Moreover, the distributional kernels $K^{(\pm)}$ of the restrictions of S to $L_2(\Omega_\pm)$ satisfy the Gaussian bounds*

$$a'\, G^{(\pm)}_{b';t}(x\,;y) \leq K^{(\pm)}_t(x\,;y) \leq a\, G^{(\pm)}_{b;t}(x\,;y)$$

for all $x, y \in \Omega_\pm$ and $t > 0$ where

$$G^{(\pm)}_{b;t}(x\,;y) = \big(B_\pm(x\,;t^{1/2})\, B_\pm(y\,;t^{1/2})\big)^{-1/2}\, e^{-b\, d_C(x;y)^2/t}\ .$$

The failure of ergodicity of the semigroup follows from the assumption $\delta_1 \in [1/2, 1\rangle$ by a modification of the argument given in Example 3.5 for the one-dimensional case. The Gaussian bounds then follow from a modification of the arguments of Grigor'yan and Saloff-Coste which establish Theorem 4.2 or from Sturm's extension of this theorem to Dirichlet spaces. The argument is based on volume doubling and the Poincaré inequality (22) on the half-spaces given by Statement III of Theorem 5.4. Thus the Gaussian characteristic of the diffusion survive on the ergodic components $\Omega_\pm$.

The remaining case with $n = 1$ not covered by the last two corollaries is given by $\delta_1 \in [0, 1/2\rangle$ and $\delta_1' \in [1/2, 1\rangle$. It follows from the second statement of Theorem 5.4 that the Poincaré inequality is no longer valid. Nevertheless one can establish a local version of the inequality (see Theorem 5.1 of [RS13]). In particular for each $R > 0$ there is a $b > 0$ and $\kappa \in \langle 0, 1]$ such that the Poincaré inequality (14) is valid for all balls $B(x\,;r)$ with $x \in \mathbf{R}^{1+m}$ and $r \in \langle 0, R]$. Now the value of b depends on R and tends to zero as $R \to \infty$. More precisely $b \sim R^{-\alpha(\delta_1' - \delta_1)}$ for some $\alpha > 0$ as $R \to \infty$. This weaker version of the Poincaré inequality, combined with the volume doubling property, is sufficient to establish lower kernel bounds

$$K_t(x\,;y) \geq a\,|B_C(x\,;t^{1/2})|^{-1}$$

valid for all x, y, t with $d_C(x\,;y) \leq t^{1/2} \leq R$. These bounds imply that K_t is strictly positive for all $t > 0$ and consequently the semigroup S is ergodic. In addition the local version of the Poincaré inequality suffices to deduce local Hölder continuity of the semigroup kernel K by Moser's arguments and then Gaussian upper bounds

$$K_t(x\,;y) \leq a\,G_{b;t}(x\,;y)$$

follow for all $x, y \in \mathbf{R}^{1+m}$ and $t > 0$ by Corollary 6.6 of [RS08]. There are, however, no matching lower bounds as this would imply a contradiction with the second statement of Theorem 5.4. The detailed behaviour of the diffusion is not completely understood but the general intuition is that an approximate failure of ergodicity occurs. For example, if the one-dimensional diffusion process determined by $-d_x\,(1 \vee |x|)\,d_x$ begins at the right (left) of the origin then with large probability it diffuses to infinity on the right (left). Therefore the two half-lines, $x \geq 0$ and $x \leq 0$ are approximately invariant. This behaviour is analogous to diffusion on manifolds with ends (see, for example, [GSC09]) which leads to more complicated lower bounds.

Acknowledgement

I am indebted to Adam Sikora for his many contributions to our various collaborations on the topics of these notes.

References

[AH05] Ariyoshi, T., and Hino, M., *Small-time asymptotic estimates in local Dirichlet spaces.* Elec. J. Prob. **10** (2005), 1236–1259.

[Aro67] Aronson, D.G., *Bounds for the fundamental solution of a parabolic equation.* Bull. Amer. Math. Soc. **73** (1967), 890–896.

[BCF96] Benjamini, I., Chavel, I., and Feldman, E.A., *Heat kernel lower bounds on Riemannian manifolds using the old ideas of Nash.* Proc. London Math. Soc. **72** (1996), 215–240.

[BH91] Bouleau, N., and Hirsch, F., *Dirichlet forms and analysis on Wiener space,* vol. 14 of de Gruyter Studies in Mathematics. Walter de Gruyter & Co., Berlin, 1991.

[Bra02] Braides, A., *Γ-convergence for beginners,* vol. 22 of Oxford Lecture Series in Mathematics and its Applications. Oxford University Press, Oxford, 2002.

[CF91] Chavel, I., and Feldman, E.A., *Isoperimetric constants, the geometry of ends, and large time heat diffusion in Riemannian manifolds.* Proc. London Math. Soc. **62** (1991), 427–448.

[CGT82] Cheeger, J., Gromov, M., and Taylor, M., *Finite propagation speed, kernel estimates for functions of the Laplace operator, and the geometry of complete Riemannian manifolds.* J. Differential Geom. **17** (1982), 15–53.

[Dal93] Dal Maso, G., *An introduction to Γ-convergence*, vol. 8 of Progress in Nonlinear Differential Equations and their Applications. Birkhäuser Boston Inc., Boston, MA, 1993.

[Dav89] Davies, E.B., *Heat kernels and spectral theory.* Cambridge Tracts in Mathematics 92. Cambridge University Press, Cambridge etc., 1989.

[Dav92] ——, *Heat kernel bounds, conservation of probability and the Feller property.* J. Anal. Math. **58** (1992), 99–119. Festschrift on the occasion of the 70th birthday of Shmuel Agmon.

[Dav97] ——, *Non-Gaussian aspects of heat kernel behaviour.* J. London Math. Soc. **55** (1997), 105–125.

[DeG57] De Giorgi, E., *Sulla differenziabilità e l'analiticità delle estremali degli integrali multipli regolari.* Mem. Accad. Sci. Torino cl. Sci. Fis. Mat. Nat. **3** (1957), 25–43.

[ERSZ06] Elst, A.F.M. ter, Robinson, D.W., Sikora, A., and Zhu, Y., Dirichlet forms and degenerate elliptic operators. In Koelink, E., Neerven, J. van, Pagter, B. de, and Sweers, G., eds., *Partial Differential Equations and Functional Analysis*, vol. 168 of Operator Theory: Advances and Applications. Birkhäuser, 2006, 73–95. Philippe Clement Festschrift.

[ERZ06] Elst, A.F.M. ter, Robinson, D. W., and Zhu, Y., *Positivity and ellipticity.* Proc. Amer. Math. Soc. **134** (2006), 707–714.

[ET76] Ekeland, I., and Temam, R., *Convex analysis and variational problems.* North-Holland Publishing Co., Amsterdam, 1976.

[FOT94] Fukushima, M., Oshima, Y., and Takeda, M., *Dirichlet forms and symmetric Markov processes*, vol. 19 of de Gruyter Studies in Mathematics. Walter de Gruyter & Co., Berlin, 1994.

[FS86] Fabes, E.B., and Stroock, D.W., *A new proof of Moser's parabolic Harnack inequality using the old ideas of Nash.* Arch. Rat. Mech. and Anal. **96** (1986), 327–338.

[Gaf59] Gaffney, M.P., *The conservation property of the heat equation on Riemannian manifolds.* Comm. Pure Appl. Math. **12** (1959), 1–11.

[Gri92] Grigor'yan, A., *The heat equation on noncompact Riemannian manifolds.* Math. USSR-Sb. **72**, No. 1 (1992), 47–77. Mat. Sb. **182**, No. 1, (1991), 55–87.

[Gri09] ——, *Heat kernel and analysis on manifolds.* Studies in Advanced Mathematics 47. American Mathematical Society, International Press, 2009.

[GSC09] Grigor'yan, A., and Saloff-Coste, L., *Heat kernel on manifolds with ends.* Ann. Inst. Fourier, Grenoble **59** (2009), 1917–1997.

[HR03] Hino, M., and Ramírez, J.A., *Small-time Gaussian behavior of symmetric diffusion semigroups.* Ann. Prob. **31** (2003), 254–1295.

[Jer86] Jerison, D., *The Poincaré inequality for vector fields satisfying Hörmander's condition.* Duke Math. J. **53** (1986), 503–523.

[Jos98] Jost, J., Nonlinear Dirichlet forms. In *New directions in Dirichlet forms*, vol. 8 of AMS/IP Stud. Adv. Math., 1–47. Amer. Math. Soc., Providence, RI, 1998.

[Lu94] Lu, G., *The sharp Poincaré inequality for free vector fields: an endpoint result.* Revista Matemática Iberoamericana **10** (1994), 453–466.

[Mos64] MOSER, J., *A Harnack inequality for parabolic differential equations.* Commun. Pure Appl. Math. **17** (1964), 101–134. Correction to: "A Harnack inequality for parabolic differential equations", Comm. Pure Appl. Math. **20** (1967), 231–236.

[Mos94] MOSCO, U., *Composite media and asymptotic Dirichlet forms.* J. Funct. Anal. **123** (1994), 368–421.

[MR92] MA, Z.M., and RÖCKNER, M., *Introduction to the theory of (non-symmetric) Dirichlet Forms.* Universitext. Springer-Verlag, Berlin etc., 1992.

[Nas58] NASH, J., *Continuity of solutions of parabolic and elliptic equations.* Amer. J. Math. **80** (1958), 931–954.

[Ouh05] OUHABAZ, E.M., *Analysis of heat equations on domains*, vol. 31 of London Mathematical Society Monographs Series. Princeton University Press, Princeton, NJ, 2005.

[Rob13] ROBINSON, D.W., *Gaussian bounds, strong ellipticity and uniqueness criteria.* Bull. London Math. Soc. (2014), in press.

[RS08] ROBINSON, D.W., and SIKORA, A., *Analysis of degenerate elliptic operators of Grušin type.* Math. Z. **260** (2008), 475–508.

[RS13] ——, *The limitations of the Poincaré inequality.* J. Evol. Equ. (2014), in press.

[SC92a] SALOFF-COSTE, L., *A note on Poincaré, Sobolev, and Harnack inequalities.* Internat. Math. Res. Notices **1992**, No. 2 (1992), 27–38.

[SC92b] ——, *Uniformly elliptic operators on Riemannian manifolds.* J. Diff. Geom. **36** (1992), 417–450.

[SC95] ——, *Parabolic Harnack inequality for divergence-form second-order differential operators.* Potential Anal. **4** (1995), 429–467.

[SC02] ——, *Aspects of Sobolev-type inequalities.* London Math. Soc. Lect. Note Series 289. Cambridge University Press, Cambridge, 2002.

[Stu95] STURM, K.-T., *Analysis on local Dirichlet spaces. II. Upper Gaussian estimates for the fundamental solutions of parabolic equations.* Osaka J. Math. **32** (1995), 275–312.

[Stu96] ——, *Analysis on local Dirichlet spaces. III. The parabolic Harnack inequality.* J. Math. Pures Appl. **75** (1996), 273–297.

Derek W. Robinson
Centre for Mathematics and its Applications
Mathematical Sciences Institute
Australian National University
Canberra, ACT 0200, Australia
e-mail: Derek.Robinson@anu.edu.au

Operator Theory:
Advances and Applications, Vol. 250, 483–489

Functional Calculus for C_0-semigroups Using Infinite-dimensional Systems Theory

Felix L. Schwenninger and Hans Zwart

Dedicated to Charles Batty on the occasion of his sixtieth birthday

Abstract. In this short note we use ideas from systems theory to define a functional calculus for infinitesimal generators of strongly continuous semigroups on a Hilbert space. Among others, we show how this leads to new proofs of (known) results in functional calculus.

Mathematics Subject Classification (2010). (Primary 47A60; Secondary 93C25.

Keywords. Functional calculus; H^∞-calculus; C_0-semigroups; infinite-dimensional systems theory; admissibility.

1. Introduction

Let A be a linear operator on the linear space X. In essence, a functional calculus provides for every (scalar) function f in the algebra $\mathcal{A}$ a linear operator $f(A)$ from (a subspace of) X to X such that

- $f \mapsto f(A)$ is linear;
- $f(s) \equiv 1$ is mapped on the identity I;
- If $f(s) = (s-r)^{-1}$, then $f(A) = (A - rI)^{-1}$;
- For $f = f_1 \cdot f_2$ we have $f(A) = f_1(A)f_2(A)$.

As the domains of the operators $f(A)$ might differ, the above properties have to be seen formally, and, in general, need to be made rigorous. It is well known that self-adjoint (or unitary operators) on a Hilbert space have a functional calculus with $\mathcal{A}$ being the set of continuous functions from $\mathbb{R}$ (or the torus $\mathbb{T}$ respectively) to $\mathbb{C}$, (von Neumann [10]). This theory has been further extended to different operators and algebras, see, e.g., [7], [3], and [2]. For an excellent overview, in particular on the $\mathcal{H}^\infty$-calculus, we refer to the book by Markus Haase, [5].

The first named author has been supported by the Netherlands Organisation for Scientific Research (NWO), grant no. 613.001.004.

For the algebra of bounded analytic functions on the left half-plane and A the infinitesimal generator of a strongly continuous semigroup, we show how to build a functional calculus using infinite-dimensional systems theory.

2. Functional calculus for $\mathcal{H}_\infty^-$

We choose our class of functions to be $\mathcal{H}_\infty^-$, i.e., the algebra of bounded analytic functions on the left half-plane. For A we choose the generator of an exponentially stable strongly continuous semigroup on the Hilbert space X. This semigroup will be denoted by $\left(e^{At}\right)_{t\geq 0}$. We refer to [4] for a detailed overview on C_0-semigroups. In the following all semigroups are assumed to be strongly continuous. To explain our choice/set-up we start with the following observation.

Let h be an integrable function from $\mathbb{R}$ to $\mathbb{C}$ which is zero on $(0,\infty)$ and let $t \mapsto \mathbb{1}(t)$ denote the indicator function of $[0,\infty)$, i.e., $\mathbb{1}(t) = 1$ for $t \geq 0$ and $\mathbb{1}(t) = 0$ for $t < 0$. Then for $t > 0$,

$$\begin{aligned}\left(h * e^{A\cdot} x_0\, \mathbb{1}(\cdot)\right)(t) &= \int_{-\infty}^{\infty} h(\tau) e^{A(t-\tau)} x_0\, \mathbb{1}(t-\tau) d\tau \\ &= \left[\int_{-\infty}^{t} h(\tau) e^{-A\tau} d\tau\right] e^{At} x_0 \\ &= \left[\int_{-\infty}^{0} h(\tau) e^{-A\tau} d\tau\right] e^{At} x_0.\end{aligned}$$

Hence the convolution of h with the semigroup gives an operator times the semigroup. We denote this operator by $g(A)$, with g the Laplace transform of h.

Now we want to extend the mapping $g \mapsto g(A)$. Therefore we need the Hardy space $H^2(X) = H^2(\mathbb{C}_+; X)$, i.e., the set of X-valued functions, analytic on the right half-plane which are uniformly square integrable along every line parallel to the imaginary axis. By the (vector-valued) Paley–Wiener Theorem, this space is isomorphic to $L^2((0,\infty); X)$ under the Laplace transform, see [1, Theorem 1.8.3].

Definition 2.1. Let X be a Hilbert space. For $g \in \mathcal{H}_\infty^-$ and $f \in L^2((0,\infty); X)$ we define the *Toeplitz operator*

$$M_g(f) = \mathfrak{L}^{-1}\left[\Pi(g(\mathfrak{L}(f))\right], \tag{1}$$

where $\mathfrak{L}$ and $\mathfrak{L}^{-1}$ denotes the Laplace transform and its inverse, respectively, and Π is the projection from $L^2(i\mathbb{R}, X)$ onto $H^2(X)$.

Remark 2.2. If we take $f(t) = e^{At}x_0$, $t \geq 0$, and "$g = \mathfrak{L}(h)$", then this extends the previous convolution.

The following norm estimate is easy to see.

Lemma 2.3. *Under the conditions of Definition* 2.1, *we have that M_g is a bounded linear operator from $L^2((0,\infty); X)$ to itself with norm satisfying*

$$\|M_g\| \leq \|g\|_\infty. \tag{2}$$

To show that Definition 2.1 leads to a functional calculus, we need the following concept from infinite-dimensional systems theory, see, e.g., [16].

Definition 2.4. Let Y be a Hilbert space, and C a linear operator bounded from $D(A)$, the domain of A, to Y. C is an *admissible* output operator if the mapping $x_0 \mapsto Ce^{A\cdot}x_0$ can be extended to a bounded mapping from X to $L^2([0,\infty);Y)$.

Since in this paper only admissible output operators appear, we shall sometimes omit "output". In [17] the following was proved.

Theorem 2.5. *Let A be the generator of an exponentially stable semigroup on the Hilbert space X. For every $g \in \mathcal{H}_\infty^-$ there exists a linear mapping $g(A) : D(A) \to X$ such that*

$$\left(\left(M_g(e^{A\cdot}x_0)\right)(t) = g(A)e^{At}x_0, \qquad x_0 \in D(A).$$

Furthermore,

- *$g(A)$ is an admissible operator;*
- *$g(A)e^{At}$ extends to a bounded operator for $t > 0$;*
- *$g(A)$ commutes with the semigroup;*
- *$g(A)$ can be extended to a closed operator $g_\Lambda(A)$ such that $g \mapsto g_\Lambda(A)$ has the properties of an (unbounded) functional calculus;*
- *This (unbounded) calculus extends the Hille–Phillips calculus.*

Hence in general the functional calculus constructed in this way will contain unbounded operators. However, they may not be "too unbounded", as the product with any admissible operator is again admissible.

Theorem 2.6 (Lemma 2.1 in [17]). *Let A be the generator of an exponentially stable semigroup on the Hilbert space X and let C be an admissible operator, then*

$$\left(M_g(Ce^{A\cdot}x_0)\right)(t) = Cg(A)e^{At}x_0, \qquad x_0 \in D(A^2).$$

Moreover, $Cg(A)$ extends to an admissible output operator.

3. Analytic semigroups

From Theorem 2.5 we know that $g(A)e^{At}$ is a bounded operator for $t > 0$. In this section we show that for analytic semigroups the norm of $g(A)e^{At}$ behaves like $|\log(t)|$ for t close to zero. Let A generate an exponentially stable, analytic semigroup on the Hilbert space X. Then there exists a $M, \omega > 0$ such that, see [11, Theorem 2.6.13],

$$\|(-A)^{\frac{1}{2}}e^{At}\| \leq M\frac{1}{\sqrt{t}}e^{-\omega t}, \qquad t > 0. \tag{3}$$

Using this inequality, we prove the following estimate.

Theorem 3.1. *Let A generate an exponentially stable, analytic semigroup on the Hilbert space X. There exists $m, \varepsilon_0 > 0$ such that for every $g \in \mathcal{H}_\infty^-$, $\varepsilon \in (0, \varepsilon_0)$*

$$\|g(A)e^{A\varepsilon}\| \leq m\|g\|_\infty |\log(\varepsilon)|. \tag{4}$$

If we assume that $(-A^*)^{\frac{1}{2}}$ or $(-A)^{\frac{1}{2}}$ is admissible, then

$$\|g(A)e^{A\varepsilon}\| \leq m\|g\|_\infty \sqrt{|\log(\varepsilon)|} \quad \text{for} \quad \varepsilon \in (0, \varepsilon_0). \tag{5}$$

If both $(-A^*)^{\frac{1}{2}}$ and $(-A)^{\frac{1}{2}}$ are admissible, then $g(A)$ is bounded.

Proof. For $y \in D(A^*)$, $x \in D(A^2)$ we have

$$\begin{aligned}
\frac{1}{2}\langle y, g(A)e^{A2\varepsilon}x\rangle &= \int_0^\infty \langle y, (-A)e^{A2t}g(A)e^{A2\varepsilon}x\rangle dt \\
&= \int_0^\infty \langle (-A^*)^{\frac{1}{2}}e^{A^*\varepsilon}e^{A^*t}y, g(A)(-A)^{\frac{1}{2}}e^{A\varepsilon}e^{At}x\rangle dt,
\end{aligned}$$

where we used that $g(A)$ commutes with the semigroup. Using the Cauchy–Schwarz inequality, we find

$$\begin{aligned}
\frac{1}{2}|\langle y, g(A)e^{A2\varepsilon}x\rangle| &\leq \|(-A^*)^{\frac{1}{2}}e^{A^*\varepsilon}e^{A^*\cdot}y\|_{L^2}\|g(A)(-A)^{\frac{1}{2}}e^{A\varepsilon}e^{A\cdot}x\|_{L^2} \qquad (6)\\
&= \|(-A^*)^{\frac{1}{2}}e^{A^*\varepsilon}e^{A^*\cdot}y\|_{L^2} \cdot \|M_g\left((-A)^{\frac{1}{2}}e^{A\varepsilon}e^{A\cdot}x\right)\|_{L^2} \\
&\leq \|(-A^*)^{\frac{1}{2}}e^{A^*\varepsilon}e^{A^*\cdot}y\|_{L^2} \cdot \|g\|_\infty \cdot \|(-A)^{\frac{1}{2}}e^{A\varepsilon}e^{A\cdot}x\|_{L^2},
\end{aligned}$$

where we used Lemma 2.3. Hence it remains to estimate the two L^2-norms. Since X is a Hilbert space $\left(e^{A^*t}\right)_{t\geq 0}$ is an analytic semigroup as well. Hence both L^2-norms behave similarly. We do the estimate for e^{At}. For $\omega\varepsilon < 1/4$,

$$\begin{aligned}
\|(-A)^{\frac{1}{2}}e^{A\varepsilon}e^{A\cdot}x\|_{L^2}^2 &= \int_0^\infty \|(-A)^{\frac{1}{2}}e^{A\varepsilon}e^{At}x\|^2 dt \\
&= \int_\varepsilon^\infty \|(-A)^{\frac{1}{2}}e^{At}x\|^2 dt \\
&\leq M^2 \int_\varepsilon^\infty \frac{e^{-2\omega t}}{t}\|x\|^2 dt \\
&= M^2\|x\|^2 \int_1^\infty \frac{e^{-2\varepsilon\omega t}}{t} dt \\
&\leq M^2\|x\|^2 m_1 |\log(\varepsilon\omega)|,
\end{aligned}$$

where we used (3) and m_1 is an absolute constant.

Combining the estimates and using the fact that ω is fixed, we find that there exists a constant $m_3 > 0$ such that for all $x \in D(A^2)$ and $y \in D(A^*)$ there holds

$$|\langle y, g(A)e^{A2\varepsilon}x\rangle| \leq m_3|\log(\varepsilon)|\|g\|_\infty\|x\|\|y\|.$$

Since $D(A^2)$ and $D(A^*)$ are dense in X, we have proved the estimate (4).

We continue with the proof of inequality (5). If $(-A^*)^{\frac{1}{2}}$ is admissible, then (6) implies that

$$\begin{aligned}\frac{1}{2}|\langle y, g(A)e^{A2\varepsilon}x\rangle| &\le \|(-A^*)^{\frac{1}{2}}e^{A^*\varepsilon}e^{A^*\cdot}y\|_{L^2}\|g(A)(-A)^{\frac{1}{2}}e^{A\varepsilon}e^{A\cdot}x\|_{L^2}\\ &\le m_2\|y\|\cdot\|M_g\left((-A)^{\frac{1}{2}}e^{A\varepsilon}e^{A\cdot}x\right)\|_{L^2}.\end{aligned}$$

The estimate follows as shown previously. Let us now assume that $(-A)^{\frac{1}{2}}$ is admissible. Then by Theorem 2.6 there holds

$$\begin{aligned}\|g(A)(-A)^{\frac{1}{2}}e^{A\varepsilon}e^{A\cdot}x\|_{L^2} &\le \|g(A)(-A)^{\frac{1}{2}}e^{A\cdot}x\|_{L^2}\\ &= \|M_g\left((-A)^{\frac{1}{2}}e^{A\cdot}x\right)\|_{L^2}\\ &\le \|g\|_\infty\|(-A)^{\frac{1}{2}}e^{A\cdot}x\|_{L^2}\\ &\le \|g\|_\infty m\|x\|,\end{aligned}$$

where we have used Lemma 2.3 and the admissibility of $(-A)^{\frac{1}{2}}$. Now the proof of (5) follows similarly as in the first part.

If $(-A)^{\frac{1}{2}}$ and $(-A^*)^{\frac{1}{2}}$ are both admissible, then we see from the above that the epsilon disappears from the estimate, and since the semigroup is strongly continuous, $g(A)$ extends to a bounded operator. □

In [13], it is shown that for any $\delta \in (0,1)$ there exists an analytic, exponentially stable semigroup on a Hilbert space, and $g \in \mathcal{H}_\infty^-$ such that $(-A)^{\frac{1}{2}}$ is admissible and $\|g(A)e^{A\varepsilon}\| \sim (\sqrt{|\log(\varepsilon)|})^{1-\delta}$. Similarly, the sharpness of (4) is shown.

In the next section we relate the above theorem to results in the literature.

4. Closing remarks

A natural question is whether the calculus above coincides with other definitions of the $\mathcal{H}_\infty^-$-calculus. As the construction extends the Hille–Phillips calculus, the answer is "yes", see [14].

In [15], Vitse showed a similar estimate as in (4) for analytic semigroups on general Banach spaces by using the Hille–Phillips calculus. The setting there is slightly different since bounded analytic semigroups and functions $g \in \mathcal{H}_\infty^-$ with bounded Fourier spectrum are considered. In [13], the first named author improves Vitse's result with a more direct technique. In the course of that work, the approach to Theorem 3.1 via the calculus construction used here was obtained. Moreover, the techniques here and in Vitse's work [15] require that the functions f are bounded, analytic on a half-plane. In [13], it is shown that the corresponding result is even true for functions f that are only bounded, analytic on sectors which are larger than the sectorality sector of the generator A.

Furthermore, Haase and Rozendaal proved that (4) holds for general (exponentially stable) semigroups on Hilbert spaces, see [6]. Their key tool is a *transference principle.* More general, they show that on general Banach spaces one has to consider the analytic multiplier algebra $\mathcal{AM}_2(X)$, as the function space to obtain a corresponding result. Note that $\mathcal{AM}_2(X)$ is continuously embedded in $\mathcal{H}_\infty^-$ with equality if X is a Hilbert space.

The difference in the transference principle and the approach followed here is that in the transference principle, estimates are first proved for "nice" functions and than extended to the whole space $\mathcal{H}_\infty^-$. Whereas we prove the result first for "nice" elements in X, and then extend the operators $g(A)$.

The fact that the calculus is bounded for analytic semigroups when both $(-A)^{\frac{1}{2}}$ and $(-A^*)^{\frac{1}{2}}$ are admissible, can already be found in [8]. However, as the admissibility of $(-A)^{\frac{1}{2}}$ is equivalent to A satisfying *square function estimates*, the result is much older and goes back to McIntosh, [9].

The construction of the $\mathcal{H}_\infty^-$-calculus followed here can be adapted to general Banach spaces, see [12, 14].

References

[1] W. Arendt, C.J.K. Batty, M. Hieber, and F. Neubrander, *Vector-valued Laplace transforms and Cauchy problems*, volume 96 of *Monographs in Mathematics.* Second Edition. Birkhäuser Verlag, Basel, 2011.

[2] D. Albrecht, X. Duong and A. McIntosch, Operator theory and harmonic analysis, appeared in: Workshop on Analysis and Geometry, 1995, Part III, *Proceedings of the Centre for Mathematics and its Applications*, ANU, Canberra, 34 (1996) 77–136.

[3] N. Dunford and J.T. Schwartz, *Linear Operators, Part III: Spectral Operators*, Wiley, 1971.

[4] K.-J. Engel and R. Nagel, *One-parameter Semigroups for Linear Evolution Equations*, Graduate Texts in Mathematics, vol. 194, Springer-Verlag, New York, 2000.

[5] M. Haase, *The Functional Calculus for Sectorial Operators*, Operator Theory, Advances and Applications, vol. 169, Birkhäuser, Basel, 2006.

[6] M. Haase and J. Rozendaal, *Functional calculus for semigroup generators via transference*, Journal of Functional Analysis, 265 (2013) 3345–3368.

[7] E. Hille and R.S. Phillips, *Functional Analysis and Semi-Groups*, AMS, 1957.

[8] C. Le Merdy, *The Weiss conjecture for bounded analytic semigroups*, J. London Math. Soc., 67 (2003), 715–738.

[9] A. McIntosh, Operators which have an H_∞ functional calculus, *Miniconference on Operator Theory and Partial Differential Equations, Proceedings of the Centre for Mathematical Analysis, Australian National University*, 14 (1986) 220–231.

[10] J. von Neumann, *Mathematische Grundlagen der Quantummechanik*, zweite Auflage, Springer Verlag, reprint 1996.

[11] A. Pazy, *Semigroups of linear operators and applications to partial differential equations*, volume 44 of *Applied Mathematical Sciences.* Springer-Verlag, New York, 1983.

[12] F.L. Schwenninger and H. Zwart, *Weakly admissible $\mathcal{H}_\infty^-$-calculus on reflexive Banach spaces*, Indag. Math. (N.S.), 23(4) (2012), 796–815.

[13] F.L. Schwenninger, *On measuring unboundedness of the H^∞-calculus for generators of analytic semigroups*, submitted 2015.

[14] F.L. Schwenninger, *On Functional Calculus Estimates*, PhD thesis, University of Twente, Enschede, September 2015.

[15] P. Vitse, *A Besov class functional calculus for bounded holomorphic semigroups*, Journal of Functional Analysis, 228 (2005), 245–269.

[16] G. Weiss, *Admissible observation operators for linear semigroups*, Israel Journal of Mathematics, 65-1 (1989) 17–43.

[17] H. Zwart, *Toeplitz operators and $\mathcal{H}_\infty$ calculus*, Journal of Functional Analysis, 263 (2012) 167–182.

Felix L. Schwenninger and Hans Zwart
Department of Applied Mathematics
P.O. Box 217
NL-7500 AE Enschede, The Netherlands
e-mail: f.l.schwenninger@utwente.nl
h.j.zwart@utwente.nl

Operator Theory:
Advances and Applications, Vol. 250, 491–496

On Self-adjoint Extensions of Symmetric Operators

Jürgen Voigt

Abstract. For a closed symmetric operator in a Hilbert space and a real regular point of this operator we obtain two 'natural' self-adjoint extensions, in terms of the von Neumann method. One of these extensions is used in order to describe the Friedrichs extension of a positive symmetric operator in the context of the von Neumann theory. The theory is illustrated by an example.

Mathematics Subject Classification (2010). Primary 47B25.

Keywords. Symmetric, self-adjoint, von Neumann extension, Friedrichs extension.

Introduction

The theory of constructing self-adjoint extensions of symmetric operators is well established. If H is a closed symmetric operator in a complex Hilbert space, then the task is transferred to finding unitary operators between $N(\mathrm{i}-H^*)$ and $N(-\mathrm{i}-H^*)$. Then the method going back to von Neumann yields the description of self-adjoint extensions of H. On the other hand, if H is bounded below, then there is a distinguished extension, the Friedrichs extension. It consists in associating a closed form with H and then obtaining the extension by a representation theorem.

It is the main objective of this note to present a description of the Friedrichs extension in terms of the von Neumann theory.

In Section 1 we show that for each regular point $a \in \mathbb{R}$ of a closed self-adjoint operator H there exist 'natural' associated unitary operators V_a and U_a between $N(\mathrm{i} - H^*)$ and $N(-\mathrm{i} - H^*)$; they are obtained from the orthogonal projections onto $N(\pm\mathrm{i} - H^*)$, restricted to $N(a - H^*)$.

In Section 2 we recall important known facts which lead to the result that, for a positive symmetric operator H, and for $a \to -\infty$, the unitary operators U_a converge to the unitary operator corresponding to the Friedrichs extension of H.

In Section 3 we illustrate the theory by an example.

The author is indebted to K. Schmüdgen for pointing out the reference [2].

1. Kilpi's 'Hilfssatz 2'

We assume that H is a closed symmetric operator in a complex Hilbert space $\mathcal{H}$ and that $a \in \mathbb{R}$ is a regular point for H, i.e., there exists a constant $c > 0$ such that $\|(a - H)f\| \geqslant c\|f\|$ for all $f \in D(H)$.

Following the notation of Weidmann [6; 8.2] we define

$$N_\pm := N(\pm\mathrm{i} - H^*) = R(\mp\mathrm{i} - H)^\perp.$$

The following is partly a version and also an extension of [3; Hilfssatz 2].

Theorem 1.1. *Let $P_\pm \in L(\mathcal{H})$ be the orthogonal projections onto $N_\pm$, and define $\check{P}_\pm := P_\pm|_{N(a-H^*)}$.*

Then the mappings

$$\check{P}_\pm \colon N(a - H^*) \to N_\pm$$

are bijective, $\|\check{P}_\pm^{-1}\| \leqslant |a \pm \mathrm{i}| + \frac{a^2+1}{c}$, and the mapping

$$V_a := \check{P}_- \check{P}_+^{-1} \colon N_+ \to N_-$$

is unitary.

Proof. We recall that

$$D(H^*) = D(H) \oplus N(\mathrm{i} - H^*) \oplus N(-\mathrm{i} - H^*) = D(H) \oplus N_+ \oplus N_-$$

is an orthogonal direct sum, with respect to the scalar product

$$(\varphi \,|\, \psi)_{H^*} := (\varphi \,|\, \psi) + (H^*\varphi \,|\, H^*\psi) \tag{1.1}$$

on $D(H^*)$; cf. [4; Section X.1].

Let $\varphi \in N(a - H^*)$ ($\subseteq D(H^*)$). Then there exist $f \in D(H)$, $\varphi_\pm \in N_\pm$ such that

$$\varphi = \varphi_+ + \varphi_- + f. \tag{1.2}$$

Applying $\mathrm{i} + H^*$ and $\mathrm{i} - H^*$ one obtains

$$\varphi = \frac{2}{1 - \mathrm{i}a}\varphi_+ + \frac{1}{\mathrm{i} + a}(\mathrm{i} + H)f \in N_+ \oplus R(-\mathrm{i} - H) \tag{1.3}$$

and

$$\varphi = \frac{2}{1 + \mathrm{i}a}\varphi_- + \frac{1}{\mathrm{i} - a}(\mathrm{i} - H)f \in N_- \oplus R(\mathrm{i} - H), \tag{1.4}$$

respectively. This implies that $P_\pm\varphi = \frac{2}{1\mp\mathrm{i}a}\varphi_\pm$, and from

$$\|\varphi\|^2 = \left\|\frac{2}{1 \mp \mathrm{i}a}\varphi_\pm\right\|^2 + \left|\frac{1}{\mathrm{i} \pm a}\right|^2 \|(\mathrm{i} \pm H)f\|^2$$

and $\|(\mathrm{i} + H)f\|^2 = \|(\mathrm{i} - H)f\|^2$ we obtain that $\|P_+\varphi\| = \|P_-\varphi\|$. (So far, the computations are essentially as in [3; proof of Hilfssatz 2].)

Next we show that $\check{P}_\pm$ are injective and that $\check{P}_\pm^{-1}$ are continuous. Let φ, $\varphi_\pm$, f be as above. Subtracting (1.4) from (1.3) one obtains

$$P_+\varphi - P_-\varphi = \frac{2\mathrm{i}}{a^2 + 1}(H - a)f,$$

and this implies the estimate

$$\|f\| \leqslant c^{-1}\|(H-a)f\| = \frac{a^2+1}{2c}\|P_+\varphi - P_-\varphi\| \leqslant \frac{a^2+1}{c}\|P_\pm\varphi\|.$$

Inserting this estimate into (1.2) one finally obtains

$$\|\varphi\| \leqslant \Big(|a \pm \mathrm{i}| + \frac{a^2+1}{c}\Big)\|P_\pm\varphi\|.$$

This shows the injectivity of $\check{P}_\pm$ as well as the asserted norm estimate.

It remains to show that $R(\check{P}_\pm) = N_\pm$. By what is shown above it is sufficient to show that $R(\check{P}_\pm)$ is dense in $N_\pm$. We will only show this for $\check{P}_+$, the proof for $\check{P}_-$ being analogous.

So, let $\psi \in N_+ \cap R(\check{P}_+)^\perp$. Then $(\eta\,|\,\psi) = 0$ for all $\eta \in N(a - H^*)$, and this shows that $\psi \in N(a-H^*)^\perp = R(a-H)$. (This was not used so far: $R(a-H)$ is closed because H is closed and a is regular for H.)

As a consequence, there exists $\varphi \in D(H)$ such that $(a-H)\varphi = \psi$. It then follows that

$$\begin{aligned} 0 &= ((\mathrm{i}-H^*)\psi\,|\,\varphi) = ((a-H)\varphi\,|\,(-\mathrm{i}-H)\varphi) \\ &= (\mathrm{i}-a)((a-H)\varphi\,|\,\varphi) + \|(a-H)\varphi\|^2. \end{aligned}$$

This implies that $((a-H)\varphi\,|\,\varphi) = 0$ (because $((a-H)\varphi\,|\,\varphi) \in \mathbb{R}$), and then that $\psi = (a-H)\varphi = 0$. □

Remark 1.2. We recall that by the von Neumann method of extending a symmetric operator to self-adjoint operators, the unitary operator V_a from Theorem 1.1 gives rise to the self-adjoint extension $\tilde{H}_a$ of H defined by

$$D(\tilde{H}_a) := D(H) + \{g + V_a g;\, g \in N_+\},$$

and $\tilde{H}_a$ the restriction of H^* to $D(H_a)$ (see [6; Theorem 8.12], for instance).

The operator V_a defined in Theorem 1.1 arises in a natural way. Another natural unitary operator, which will turn out to be more important for the problem described in the Introduction, is obtained as follows. Let $P_\pm^{(*)}$ be the orthogonal projections from $D(H^*)$ onto $N_\pm$ with respect to the scalar product $(\cdot\,|\,\cdot)_{H^*}$ defined in (1.1), and denote by $\check{P}_\pm^{(*)}$ their restrictions to $N(a-H^*)$. Then the proof of Theorem 1.1 shows that

$$\check{P}_\pm^{(*)} = \frac{1 \mp \mathrm{i}a}{2}\check{P}_\pm.$$

Hence, $\check{P}_\pm^{(*)} \colon N(a-H^*) \to N_\pm$ are bijective, and

$$U_a := \check{P}_-^{(*)}\big(\check{P}_+^{(*)}\big)^{-1} = \frac{1+\mathrm{i}a}{1-\mathrm{i}a}V_a$$

is a unitary operator from N_+ to N_-.

The self-adjoint extension H_a of H, corresponding to U_a, has the domain

$$D(H_a) = D(H) + \{g + U_a g;\, g \in N_+\} = D(H) + N(a-H^*), \tag{1.5}$$

where the last equality is a consequence of the decomposition (1.2) and the bijectivity of the operators $\check{P}_\pm^{(*)}$. This kind of extension of a symmetric operator was also constructed in [1; VIII.107, proof of Satz 3′]. Note that a is an eigenvalue of H_a, with eigenspace $N(a - H^*)$. A novel aspect in our treatment is the description of this extension in terms of the von Neumann theory.

2. The Friedrichs extension and the von Neumann extension theory

We start by recalling two important results concerning the extension of positive symmetric operators. As before, let $\mathcal{H}$ be a complex Hilbert space, and let now $H \geqslant 0$ be a closed (densely defined) symmetric operator.

The first result we recall is that there exist a largest positive self-adjoint extension H_{F}, the Friedrichs extension, and a smallest positive self-adjoint extension H_{N}, the Krein–von Neumann extension, of H. We refer to [2], [5; Theorems 10.17 and 13.12] for these facts.

In order to describe the second result we introduce the notation

$$H_a := (H - a)_{\mathrm{N}} + a,$$

for $a < 0$. (Note that then $H_a \geqslant a$.) The hypothesis that $H \geqslant 0$ implies that a is a regular point for H, and in fact, the operator H_a introduced above is the same as H_a defined in Remark 1.2; we refer to [2; Section 4], [5; (14.67)] for this circumstance. What we want to recall is that the net $(H_a)_{a<0}$ converges to H_{F} in the strong resolvent sense, as $a \to -\infty$. This was shown by Ando and Nishio in [2; Theorem 3]. (Convergence in the strong resolvent sense means that $(z - H_a)^{-1} \to (z - H_{\mathrm{F}})^{-1}$ $(a \to -\infty)$ in the strong operator topology, for some/all $z \in \mathbb{C} \setminus \mathbb{R}$.)

Remarks 2.1. (a) The second result recalled above is quite remarkable. Indeed, the operators H_a have the eigenvalue $a < 0$, with a tending to $-\infty$, whereas the limiting operator H_{F} is positive.

(b) We note that for a net $(H_\iota)_{\iota\in I}$ of self-adjoint operators and a self-adjoint operator H one has convergence of $(H_\iota)_{\iota\in I}$ to H in the strong resolvent sense if and only if the Cayley transforms $(\mathrm{i} - H_\iota)(-\mathrm{i} - H_\iota)^{-1} = 2\mathrm{i}(-\mathrm{i} - H_\iota)^{-1} + I$ of H_ι converge strongly to the Cayley transform $(\mathrm{i} - H)(-\mathrm{i} - H)^{-1} = 2\mathrm{i}(-\mathrm{i} - H)^{-1} + I$ of H.

Theorem 2.2. *Let $H \geqslant 0$ be a closed symmetric operator in a complex Hilbert space $\mathcal{H}$. For $a < 0$ let $U_a \colon N_+ \to N_-$ be the unitary operator defined in Remark* 1.2.

Then $U_{-\infty} := \text{s-lim}_{a\to-\infty} U_a$ exists and defines a unitary operator from N_+ to N_-, and the Friedrichs extension of H is the self-adjoint extension of H corresponding to $U_{-\infty}$ in the von Neumann extension theory.

Proof. This follows from [2; Theorem 3], recalled above as 'the second result', and Remark 2.1(b). □

3. An example

The example presented in this section should serve as an illustration for the convergence proved in Theorem 2.2.

In $\mathcal{H} := L_2(0,\infty)$ we define

$$H := -\partial^2, \quad D(H) := H_0^2(0,\infty)$$

(∂ denoting differentiation). Then H is symmetric and closed, $H \geqslant 0$. Further

$$H^* = -\partial^2, \quad D(H^*) = H^2(0,\infty).$$

For a discussion of different aspects concerning boundary conditions for this example we refer to [5; Examples 14.9 and 14.15]. We recall that the Friedrichs extension H_F of H is given by

$$H_F = -\partial^2, \quad D(H_F) = \{f \in H^2(0,\infty);\ f(0) = 0\}.$$

In this example, the deficiency index of H is 1; therefore the spaces $N_\pm$, $N(a-H^*)$, for $a<0$, are one-dimensional.

Looking for elements $\psi \in N_\pm$, i.e., solving $-\psi'' = \pm\mathrm{i}\psi$, we find that the functions $\psi_\pm$, given by

$$\psi_\pm(x) := \mathrm{e}^{-\frac{x}{\sqrt{2}}(1\mp\mathrm{i})},$$

span the spaces $N_\pm$, respectively. An elementary computation shows that $\|\psi_\pm\|^2 = \frac{1}{\sqrt{2}}$, and therefore the projections onto $N_\pm$ are given by

$$P_\pm = \sqrt{2}(\,\cdot \mid \psi_\pm)\psi_\pm.$$

For $b>0$, the function φ_b, given by

$$\varphi_b(x) := \mathrm{e}^{-bx},$$

spans the space $N(a-H^*)$, for $a=-b^2$. We compute

$$P_\pm\varphi_b = \sqrt{2}\int_0^\infty \mathrm{e}^{-bx}\mathrm{e}^{-\frac{1}{\sqrt{2}}(1\pm\mathrm{i})x}\,dx\,\psi_\pm = \frac{2}{\sqrt{2}b+1\pm\mathrm{i}}\psi_\pm.$$

We denote by V_a the unitary operator defined in Remark 1.2, corresponding to $a=-b^2$. The above computation shows that V_a maps $P_+\varphi_b = \frac{2}{\sqrt{2}b+1+\mathrm{i}}\psi_+$ to $P_-\varphi_b = \frac{2}{\sqrt{2}b+1-\mathrm{i}}\psi_-$,

$$V_a\psi_+ = \frac{\sqrt{2}b+1+i}{\sqrt{2}b+1-i}\psi_-,$$

which implies that $V_{-\infty}\psi_+ := \lim_{a\to-\infty} V_a\psi_+ = \psi_-$.

The unitary map U_{F} belonging to the Friedrichs extension is given by $U_{\mathrm{F}}\psi_+ = -\psi_-$. (Note that then the function $\psi_+ + U_{\mathrm{F}}\psi_+ = \psi_+ - \psi_-$ is 0 at the left boundary, and therefore belongs to the domain of H_{F}.)

Recalling that $U_a = \frac{1+\mathrm{i}a}{1-\mathrm{i}a}V_a$, we obtain

$$U_{-\infty}\psi_+ := \lim_{a\to-\infty} U_a\psi_+ = \lim_{a\to-\infty}\frac{1+\mathrm{i}a}{1-\mathrm{i}a}V_a\psi_+ = -\psi_- = U_{\mathrm{F}}\psi_+,$$

$U_{-\infty} = U_{\mathrm{F}}$.

References

[1] N.I. Achieser and I.M. Glasmann: *Theorie der linearen Operatoren im Hilbert-Raum.* 5. Aufl., Akademie-Verlag, Berlin, 1968.

[2] T. Ando and K. Nishio: *Positive selfadjoint extensions of positive symmetric operators.* Tôhoku Math. J. **22**, 65–75 (1970).

[3] Y. Kilpi: *Über selbstadjungierte Fortsetzungen symmetrischer Transformationen im hilbertschen Raum.* Ann. Acad. Sci. Fennicae, Ser. A.1, Mathematica; 264 (1959).

[4] M. Reed and B. Simon: *Methods of Modern Mathematical Physics II: Fourier Analysis, Self-Adjointness.* Academic Press, New York, 1975.

[5] K. Schmüdgen: *Unbounded Self-adjoint Operators in Hilbert Space.* Springer, Dordrecht, 2012.

[6] J. Weidmann: *Linear Operators in Hilbert Spaces.* Springer-Verlag, New York, 1980.

Jürgen Voigt
Technische Universität Dresden
Fachrichtung Mathematik
D-01062 Dresden, Germany
e-mail: `juergen.voigt@tu-dresden.de`

Printed by Printforce, the Netherlands